Campylobacter Ecology and Evolution

Edited by

Samuel K. Sheppard

Associate Editor: Guillaume Méric

Medical Microbiology and Infectious Diseases
Swansea University
Swansea
UK

Caister Academic Press

Caister Academic Press
Norfolk, UK

www.caister.com

British Library Cataloguing-in-Publication Data
A catalogue record for this book is available from the British Library

ISBN: 978-1-908230-36-2 (hardback)
ISBN: 978-1-908230-98-0 (ebook)

Cover design adapted from photo courtesy of Michael E. Konkel and Jason M. Neal-McKinney (School of Molecular Biosciences, Washington State University, Pullman, WA, USA)

Printed and bound in Great Britain

Contents

Contributors vii

Foreword xiii

Preface xv

Part I Mechanisms of Evolution

1 A Candidate Hopeful Monster in the Genus *Campylobacter* 3
Samuel K. Sheppard and Daniel Falush

2 Learning About Recombination in *Campylobacter* 9
Paul Fearnhead, Patrick J. Biggs and Nigel French

3 Within-host Evolution of *Campylobacter jejuni* 23
John P. Jerome and Linda S. Mansfield

4 Concerted Evolution in *Campylobacter jejuni* and *Campylobacter coli* 33
Richard J. Meinersmann

5 Genome and Transcriptome Evolution in the Genus *Campylobacter* 43
My Thanh Le and Arnoud H.M. van Vliet

6 Repetitive DNA: A Major Source of Genetic Diversity in *Campylobacter* Populations? 55
Jack Aidley and Christopher D. Bayliss

Part II Adaptations to Host Colonization

7 *Campylobacter* Strategies for Colonizing the Host Gastrointestinal Tract 75
Annika Flint, James Butcher and Alain Stintzi

8 The Intricate Relationship Between *Campylobacter* and Mucus 89
Claire Shortt, Gina Duggan and Billy Bourke

9 *Campylobacter* Association with the Human Host 99
Abiyad Baig and Georgina Manning

10	Siderophore-mediated Iron Acquisition for *Campylobacter* Infection Ximin Zeng and Jun Lin	111
11	Genetic Mechanisms Involved in *Campylobacter jejuni* Survival Under Oxidative Stress Conditions Ozan Gundogdu, Brendan W. Wren and Nick Dorrell	125
12	Oxidative Stress Survival During *Campylobacter* Transmission and Infection Rebecca Handley, Mark Reuter and Arnoud H.M. van Vliet	143
13	The Role of the Flagellum in *Campylobacter jejuni* Colonization and Disease Jason L. O'Loughlin and Michael E. Konkel	159
14	Does Biofilm Formation Aid Colonization and Infection in *Campylobacter*? Ben Pascoe, Hilary Lappin-Scott, Samuel K. Sheppard and Hiroshi Asakura	177
15	Alternative Infection Models for *Campylobacter jejuni* Nicola J. Senior, Sok Kiang Lau, Roberto M. La Ragione and Olivia L. Champion	189
Part III	Molecular Epidemiology	
16	Exploring the Population Dynamics of Multiple *Campylobacter* Strains in Poultry Frances M. Colles and Martin C.J. Maiden	213
17	Evolution of *Campylobacter* Species in New Zealand Nigel French, Shoukai Yu, Patrick Biggs, Barbara Holland, Paul Fearnhead, Barbara Binney, Andrew Fox, Dai Grove-White, Jessica W. Leigh, William Miller, Petra Muellner and Philip Carter	221
18	How Molecular Typing has Changed our Understanding on Sources and Transmission Routes of Campylobacteriosis in Finland Caroline P.A. de Haan, Rauni I. Kivistö, Hilpi Rautelin and Marja-Liisa Hänninen	241
19	Extensive Spatial and Temporal Clustering of *Campylobacter* Infections Norval J.C. Strachan and Ken Forbes	253
20	Ecology and Host Associations of *Campylobacter* in Wild Birds Jonas Waldenström and Petra Griekspoor	265
Part IV	Ecology in Poultry	
21	Epidemiology and Control of *Campylobacter* in Modern Broiler Production Ana B. Vidal, Rob H. Davies, John D. Rodgers, Anne Ridley and Felicity Clifton-Hadley	287
22	Conditional Commensalism of *Campylobacter* in Chickens Lisa K. Williams, Emma K. Trantham and Tristan A. Cogan	313

23 The Chicken Reservoir and Specific Alleles at Conserved Contingency Loci Can Promote Host Colonization and Disease by *Campylobacter jejuni* 321
John E. Linz

24 The Long Road Ahead: Unravelling the Immune Response to *Campylobacter* in the Chicken 347
Paul Wigley and Suzanne Humphrey

Index 357

Colour Plate A1

Current books of interest

Contributors

Jack Aidley
Department of Genetics
University of Leicester
Leicester
UK

jba9@le.ac.uk

Hiroshi Asakura
Division of Biomedical Food Research
National Institute of Health Sciences
Tokyo
Japan

hasakura@nihs.go.jp

Abiyad Baig
Department of Veterinary Medicine
University of Cambridge
Cambridge
UK

ab2168@cam.ac.uk

Christopher D. Bayliss
Department of Genetics
University of Leicester
Leicester
UK

cdb12@le.ac.uk

Patrick J. Biggs
Infectious Disease Research Centre
Institute of Veterinary, Animal, and Biomedical Sciences
Massey University
Palmerston North
New Zealand

p.biggs@massey.ac.nz

Barbara Binney
Infectious Disease Research Centre
Institute of Veterinary, Animal, and Biomedical Sciences
Massey University
Palmerston North
New Zealand

b.m.binney@massey.ac.nz

Billy Bourke
School of Medicine and Medical Science and School of Biological Science
University College
Dublin
Ireland

billy.bourke@ucd.ie

James Butcher
Ottawa Institute of Systems Biology
Department of Biochemistry, Microbiology and Immunology
Faculty of Medicine
University of Ottawa
Ottawa, Ontario
Canada

jbutc076@uottawa.ca

Philip Carter
ESR Keneperu Science Centre
Porirua
New Zealand

philip.carter@esr.cri.nz

Olivia L. Champion
University of Exeter
Biosciences
College of Life and Environmental Science
Exeter
UK

o.l.champion@exeter.ac.uk

Felicity Clifton-Hadley
Department of Bacteriology and Food Safety
Animal Health and Veterinary Laboratories Agency (AHVLA)
Surrey
UK

felicity.clifton-hadley@ahvla.gsi.gov.uk

Tristan A. Cogan
Bristol Veterinary school
University of Bristol
Langford
Bristol
UK

tristan.cogan@bristol.ac.uk

Frances M. Colles
Department of Zoology
University of Oxford
Oxford
UK

frances.colles@zoo.ox.ac.uk

Rob H. Davies
Department of Bacteriology and Food Safety
Animal Health and Veterinary Laboratories Agency (AHVLA)
Surrey
UK

rob.davies@ahvla.gsi.gov.uk

Nick Dorrell
Faculty of Infectious and Tropical Diseases
London School of Hygiene and Tropical Medicine
Keppel Street
London
UK

nick.dorrell@lshtm.ac.uk

Gina Duggan
School of Medicine and Medical Science
University College Dublin
Belfield
Ireland

gina.duggan@ucdconnect.ie

Daniel Falush
Max Planck Institute for Evolutionary Anthropology
Leipzig
Germany

daniel_falush@eva.mpg.de

Paul Fearnhead
Department of Mathematics and Statistics
Lancaster University
Lancaster
UK

p.fearnhead@lancaster.ac.uk

Annika Flint
Ottawa Institute of Systems Biology
Department of Biochemistry, Microbiology and Immunology
Faculty of Medicine
University of Ottawa
Ottawa, Ontario
Canada

aflin053@uottowa.ca

Andrew Fox
Health Protection Agency Microbiology Services
Food Water and Environmental Microbiology Laboratory
Royal Preston Hospital
Preston
UK

andrew.fox@hpa.org.uk

Nigel French
Infectious Disease Research Centre
Institute of Veterinary, Animal, and Biomedical Sciences
Massey University
Palmerston North
New Zealand

n.p.french@massey.ac.nz

Petra Griekspoor
Centre for Ecology and Evolution in Microbial Model Systems (EEMiS)
Linnaeus University
Sweden

petra.griekspoor@lnu.se

Dai Grove-White
School of Veterinary Science
University of Liverpool
Liverpool
UK

daigw@liv.ac.uk

Ozan Gundogdu
Faculty of Infectious and Tropical Diseases
London School of Hygiene and Tropical Medicine
London
UK

ozan.gundogdu@lshtm.ac.uk

Caroline P.A. de Haan
Department of Food Hygiene and Environmental Health
University of Helsinki
Helsinki
Finland

astrid.dehaan@helsinki.fi

Rebecca Handley
School of Chemistry
University of East Anglia
Norwich
UK

rebecca.handley@ifr.ac.uk

Marja-Liisa Hänninen
Department of Food Hygiene and Environmental Health
University of Helsinki
Helsinki
Finland

marja-liisa.hanninen@helsinki.fi

Barbara Holland
School of Mathematics and Physics
University of Tasmania
Hobart
Australia

barbara.holland@utas.edu.au

Suzanne Humphrey
Zoonotic Infections in Pigs and Poultry Group
Department of Infection Biology
Institute for Infection and Global Health and School of Veterinary Science
University of Liverpool
Neston
UK

s.humphrey@liverpool.ac.uk

John P. Jerome
BEACON Center for the Study of Evolution in Action
Comparative Enteric Diseases Laboratory
Department of Microbiology and Molecular Genetics
Michigan State University
East Lansing, MI
USA

jeromejo@msu.edu

Rauni I. Kivistö
Department of Food Hygiene and Environmental Health
University of Helsinki
Helsinki
Finland

rauni.kivisto@helsinki.fi

Michael E. Konkel
School of Molecular Biosciences
Life Sciences Building
College of Veterinary Medicine
Washington State University
Pullman, WA
USA

konkel@vetmed.wsu.edu

Roberto M. La Ragione
University of Surrey
Department of Microbial and Cellular Sciences
Guildford
UK

r.laragione@surrey.ac.uk

Hilary Lappin-Scott
College of Medicine
Swansea University
Swansea
UK

h.m.lappin-scott@swansea.co.uk

Sok Kiang Lau
University of Exeter
Biosciences
College of Life and Environmental Science
Exeter
UK

skl205@exeter.ac.uk

My Thanh Le
Gut Health and Food Safety Programme
Institute of Food Research
Norwich
UK

thanh.le@ifr.ac.uk

Jessica W. Leigh
Department of Mathematics and Statistics
University of Otago
Dunedin
New Zealand

jessica.w.leigh@gmail.com

Jun Lin
Department of Animal Science
The University of Tennessee
Knoxville, TN
USA

jlin6@utk.edu

John E. Linz
Department of Food Science and Human Nutrition
Michigan State University
East Lansing, MI
USA

jlinz@msu.edu

Martin C.J. Maiden
Department of Zoology
University of Oxford
Oxford
UK

martin.maiden@zoo.ox.ac.uk

Georgina Manning
School of Science and Technology
Nottingham Trent University
Nottingham
UK

georgina.manning@ntu.ac.uk

Linda S. Mansfield
BEACON Center for the Study of Evolution in Action
Comparative Enteric Diseases Laboratory
Department of Microbiology and Molecular Genetics
Michigan State University
East Lansing, MI
USA

mansfie4@cvm.msu.edu

Richard J. Meinersmann
USDA Agricultural Research Service
Athens, GA
USA

rick.meinersmann@ars.usda.gov

William Miller
Produce Safety and Microbiology Research Unit
Western Regional Research Center
Agricultural Research Service
US Department of Agriculture
Albany, CA
USA

william.miller@ars.usda.gov

Petra Muellner
Epi-interactive
Miramar
Wellington
New Zealand

petra.epi-interactive.com

Jason L. O'Loughlin
School of Molecular Biosciences
Life Sciences Building
College of Veterinary Medicine
Washington State University
Pullman, WA
USA

oloughlin@vetmed.wsu.edu

Ben Pascoe
Medical Microbiology and Infectious Disease
College of Medicine
Swansea University
Institute of Life Science
Swansea
UK

b.pascoe@swansea.ac.uk

Hilpi Rautelin
Department of Bacteriology and Immunology
Haartman Institute
University of Helsinki
Helsinki
Finland

hilpi.rautelin@helsinkifi

Mark Reuter
Gut Health and Food Safety Programme
Institute of Food Research
Norfolk
UK

mark.reuter@ifr.ac.uk

Anne Ridley
Department of Bacteriology and Food Safety
Animal Health and Veterinary Laboratories Agency (AHVLA)
Surrey
UK

anne.ridley@ahvla.gis.gov.uk

John D. Rodgers
Department of Bacteriology and Food Safety
Animal Health and Veterinary Laboratories Agency (AHVLA)
Surrey
UK

john.rodgers@ahvla.gov.uk

Nicola J. Senior
University of Exeter
Biosciences
College of Life and Environmental Science
Exeter
UK

n.j.senior@exeter.ac.uk

Samuel K. Sheppard
Medical Microbiology and Infectious Diseases
College of Medicine
Swansea University
Swansea
UK

s.k.sheppard@swansea.ac.uk

Claire Shortt
School of Medicine and Medical Science and School of Biological Science
University College
Dublin
Ireland

claire.shortt@ucdconnect.ie

Alain Stintzi
Ottawa Institute of Systems Biology
Department of Biochemistry, Microbiology and Immunology
Faculty of Medicine
University of Ottawa
Ottawa, Ontario
Canada

astintzi@uottawa.ca

Emma K. Trantham
Bristol Veterinary school
University of Bristol
Langford
Bristol
UK

emma.trantham@bristol.ac.uk

Ana B. Vidal
Department of Bacteriology and Food Safety
Animal Health and Veterinary Laboratories Agency (AHVLA)
Surrey
UK

ana.vidal@ahvla.gsi.gov.uk

Arnoud H.M. van Vliet
Gut Health and Food Safety Programme
Institute of Food Research
Norwich
UK

arnoud.vanvliet@ifr.ac.uk

Jonas Waldenström
Centre for Ecology and Evolution in Microbial Model Systems (EEMiS)
Linnaeus University
Sweden

jonas.waldenstrom@lnu.se

Paul Wigley
Zoonotic Infections in Pigs and Poultry Group
Department of Infection Biology
Institute for Infection and Global Health and School of Veterinary Science
University of Liverpool
Neston
UK

paul.wigley@live.ac.uk

Lisa K. Williams
Bristol Veterinary school
University of Bristol
Langford
Bristol
UK

lisa.williams@bristol.ac.uk

Brendan W. Wren
Faculty of Infectious and Tropical Diseases
London School of Hygiene and Tropical Medicine
Keppel Street
London
UK

brendan.wren@lshtm.ac.uk

Shoukai Yu
Infectious Disease Research Centre
Institute of Veterinary, Animal, and Biomedical Sciences
Massey University
Palmerston North
New Zealand

s.yu1@massey.ac.nz

Ximin Zeng
Department of Animal Science
The University of Tennessee
Knoxville, TN
USA

xzeng3@utk.edu

Foreword

Towards an integrated understanding of *Campylobacter* biology

My first exposure to *Campylobacter* was over 30 years ago at the University of Reading, through my undergraduate tutor, Bob Park, who was one of the pioneers of research into these fascinating, yet enigmatic, organisms (Morris and Park, 1971). Despite causing a wide range of human and animal pathologies, *Campylobacter* was a relative latecomer to microbiology with its importance as a human pathogen fully recognized only in the late 1970s (Skirrow, 1977). Since that time, progress in research on these organisms has been hindered by a variety of features of their biology, which were also responsible for their late rise to prominence. They are fastidious organisms, a property that makes their isolation difficult (Bolton and Robertson, 1982; Humphrey, 1989), yet they are ubiquitous, being found in a wide variety of animal and environmental sources. They cause a large burden of human disease (Blaser *et al.*, 1983), including an association with neuropathy (Nachamkin *et al.*, 1998), yet most of this is sporadic, making investigation by classical epidemiological approaches difficult. Finally, they are highly diverse both genetically and phenotypically which has further added to difficulties and confusion in their study (Wassenaar *et al.*, 1998).

It was the problem of their diversity which resulted in my reacquaintance with the campylobacters nearly 20 years later, when Eric Bolton and Andrew Fox asked if I could apply multilocus sequence typing (MLST), which we had recently developed for the meningococcus (Maiden *et al.*, 1998), to the problem of *Campylobacter* diversity. Following the completion of the first *Campylobacter jejuni* genome sequence (Parkhill *et al.*, 2000) it was the application of molecular techniques, such as MLST (Dingle *et al.*, 2001), which provided a major stimulus to the research on the epidemiology and evolution of *Campylobacter*. Sequence-based approaches were necessary to provide an accurate and portable description of the diversity of all members of the genus, and these data have also enabled investigation of the population structure and evolution of these highly recombining organisms (Wilson *et al.*, 2009).

The current volume appears at another crucial juncture in the development of our understanding of *Campylobacter*. Following what can be regarded as periods of recognition of the problem, consolidation of understanding, and exploitation of microbiological and molecular methods, genomic approaches provide the prospect of unifying the various areas of research in an evolutionary and ecological framework. The various sections of the book comprehensively describe the basic and applied biology of the pathogenic *Campylobacter*, ranging from mechanisms of evolution through the processes of host colonization and within host adaptation, and epidemiology to considerations of their broader biochemical and ecological properties. Many of the authors have contributed to the development of the science of *Campylobacter* over the years but the authorship also includes individuals who have recently begun their researches in this area and who will doubtless play a part in the exciting developments to come.

In my view the major contribution of genomics will be to stimulate the development of an organismal approach to bacteriology in general and the study of *Campylopbacter* in particular. In this approach, concepts ideas and techniques from different areas of research will be effectively combined. In this respect the very diversity of *Campylobacter* populations, so long a hindrance to understanding, can be seen as offering many research opportunities. We now know, for example, that certain *Campylobacter* genotypes are associated with particular hosts, as seen with the association of different wild bird species with particular genotypes (Griekspoor *et al.*, 2013). An intriguing exception to this is the existence of 'multihost' genotypes which are found in chicken and ruminant sources, which perhaps represent organisms that have evolved since the advent of farming or intensive farming methods (Sheppard *et al.*, 2010). It is even possible that the introgression of *C. jejuni* genes into *C. coli* clade one is an evolutionary consequence of expansion into a novel agricultural niche (Sheppard *et al.*, 2011).

Whatever the ultimate causes of the extent and structuring of genetic and phenotypic diversity of these organisms, their existence provides the prospect of genomic investigations that identify the genomic determinants for particular phenotypic properties using gene association studies. The diversity of the candidate genes identified can be explored in populations in terms of sequence variation and distribution and their biological and immunological roles probed in functional investigations. As the fortieth anniversary of Skirrow's seminal paper approaches we can look forward to an ever more detailed appreciation of the biology of these important pathogens and, in addition to a comprehensive view of the achievements of the past, this volume provides a tantalizing glimpse of the future.

Martin C.J. Maiden
Oxford

References

Blaser, M.J., Taylor, D.N., and Feldman, R.A. (1983). Epidemiology of Campylobacter jejuni infections. Epidemiol. Rev. *5*, 157–176.

Bolton, F.J., and Robertson, L. (1982). A selective medium for isolating *Campylobacter jejuni/coli*. J. Clin. Pathol. *35*, 462–467.

Dingle, K.E., Colles, F.M., Wareing, D.R.A., Ure, R., Fox, A.J., Bolton, F.J., Bootsma, H.J., Willems, R.J.L., Urwin, R., and Maiden, M.C.J. (2001). Multilocus sequence typing system for *Campylobacter jejuni*. J. Clin. Microbiol. *39*, 14–23.

Griekspoor, P., Colles, F.M., McCarthy, N.D., Hansbro, P.M., Ashhurst-Smith, C., Olsen, B., Hasselquist, D., Maiden, M.C.J., and Waldenstrom, J. (2013). Marked host specificity and lack of phylogeographic population structure of *Campylobacter jejuni* in wild birds. Mol. Ecol. (in press).

Humphrey, T.J. (1989). An appraisal of the efficacy of pre-enrichment for the isolation of *Campylobacter jejuni* from water and food. J. Appl. Bacteriol. *66*, 119–126.

Maiden, M.C.J., Bygraves, J.A., Feil, E., Morelli, G., Russell, J.E., Urwin, R., Zhang, Q., Zhou, J., Zurth, K., Caugant, D.A., *et al.* (1998). Multilocus sequence typing: a portable approach to the identification of clones within populations of pathogenic microorganisms. Proc. Natl. Acad. Sci. U.S.A. *95*, 3140–3145.

Morris, J.A., and Park, R.W. (1971). Metabolism of *Campylobacter* spp. (*Vibrio* spp.) in connexion with pathogenicity and the use of neuraminidase to suppress pregnancy. Nature *232*, 132–133.

Nachamkin, I., Allos, B.M., and Ho, T. (1998). *Campylobacter* species and Guillain-Barré syndrome. Clin. Microbiol. Rev. *11*, 555–567.

Parkhill, J., Wren, B.W., Mungall, K., Ketley, J.M., Churcher, C., Basham, D., Chillingworth, T., Davies, R.M., Feltwell, T., Holroyd, S., *et al.* (2000). The genome sequence of the food-borne pathogen *Campylobacter jejuni* reveals hypervariable sequences. Nature *403*, 665–668.

Sheppard, S.K., Dallas, J.F., Wilson, D.J., Strachan, N.J., McCarthy, N.D., Colles, F.M., Rotariu, O., Ogden, I.D., Forbes, K.J., and Maiden, M.C.J. (2010). Evolution of an agriculture-associated disease causing *Campylobacter coli* clade: evidence from national surveillance data in Scotland. PLoS ONE e15708.

Sheppard, S.K., McCarthy, N.D., Jolley, K.A., and Maiden, M.C.J. (2011). Introgression in the genus *Campylobacter*: generation and spread of mosaic alleles. Microbiology *157*, 1066–1074.

Skirrow, M.B. (1977). Campylobacter enteritis: a 'new' disease. BMJ 2, 9–11.

Wassenaar, T.M., Geilhausen, B., and Newell, D.G. (1998). Evidence for genome instability in *Campylobacter jejuni* isolated from poultry. Appl. Environ. Microbiol. *64*, 1816–1821.

Wilson, D.J., Gabriel, E., Leatherbarrow, A.J., Cheesbrough, J., Gee, S., Bolton, E., Fox, A., Hart, C.A., Diggle, P.J., and Fearnhead, P. (2009). Rapid evolution and the importance of recombination to the gastroenteric pathogen *Campylobacter jejuni*. Mol. Biol. Evol. *26*, 385–397.

Preface

Of all the bacterial causes of food poisoning, *Campylobacter* remains one of the most pervasive and difficult to eradicate. In many industrialized countries it is by far the most common cause of bacterial gastroenteritis, responsible for many more cases than *Listeria, Escherichia coli* and *Salmonella* combined. Not only does this impact on public health, but it also imposes a significant economic burden on agencies tasked with monitoring and treating infection. Because of this, control of *Campylobacter* is currently a priority and enhanced understanding of the biology of this organism is the key to effective interventions.

The success of this organism as a human pathogen depends on many factors. *Campylobacter* is common in all kinds of host reservoirs and has frequently been isolated from wild bird and mammal species as well as their domesticated counterparts. Unlike some food-borne pathogens, *Campylobacter* is poorly adapted to survival outside the host animal gut, showing, for example, a particular intolerance for atmospheric oxygen levels. However, in spite of this, it is regularly isolated from food, the farm environment and environmental waters – presumably associated with faecal contamination. This large-scale host and habitat variation is matched by finer scale phenotypic variation between strains in terms of physiological and metabolic versatility, all of which portrays an organism with complex ecology.

Related to the high level of phenotype and niche variation is the considerable capacity for *Campylobacter* to rapidly adapt. For example, some strains are able to proliferate in avian and mammalian hosts which are clearly very different in many factors associated with their gut morphology, physiology and immune response. Their ecological complexity and capacity to adapt to new niches may appear surprising for an organism which has a relatively small genome (1.6 Mb), presumably limiting available function. To overcome this, *Campylobacter* is highly evolvable with various mechanisms for augmenting or varying function including horizontal acquisition of DNA via natural transformation, phase variation, plasmid transfer, and other genetic and epigenetic factors.

Excellent books are available that describe aspects of *Campylobacter* biology. In particular the book entitled '*Campylobacter*' edited by Irving Nachamkin, Christine Szymanski, and Martin Blaser (2006) from ASM is essential reading for those with an interest in this organism. However, in recent years there have been numerous advances in our understanding of this organism, not least those associated with the application of modern genetics and genomics techniques. What we hope to achieve with this book is a synthesis of current research in the field that will help to train current and future generations of scientists. This book has been subdivided into four parts. In the first part we have given emphasis to the evolutionary and genetic processes that shape variation in *Campylobacter* populations. In Parts 2 and 3 we focus on the ecology of *Campylobacter*, describing detailed interactions with the host (Chapter 2) and population-scale patterns of genetic variation (Chapter 3). In Chapter 4, particular attention is given to domestic chickens – an important host for *Campylobacter*. An audience for this book could be found among both students and professionals who work on this important zoonotic organism.

Samuel K. Sheppard

Part I

Mechanisms of Evolution

A Candidate Hopeful Monster in the Genus *Campylobacter*

1

Samuel K. Sheppard and Daniel Falush

Abstract

Analysis of genetic variation in *Campylobacter* – through identification of mosaic alleles, mixed species multilocus sequence types and genome-wide introgression – has shown evidence for progressive hybridization of the species *C. jejuni* and *C. coli*. The resultant hybrid *C. coli* lineages have proliferated in agricultural animals and have emerged as the major cause of *C. coli* infection in humans. It is entirely unexpected that bacteria with such high sequence divergence, as far from each other at the nucleotide level as marmoset and human, could exchange almost a quarter of their genome and this finding challenges common assumptions within evolutionary biology and influential theories about how gene networks evolve. Here we discuss the possibility that agricultural *C. coli* originated as a hopeful monster and consider this natural setting as a model system within which the fitness effects of introgression can be investigated in unprecedented detail.

Introduction

In evolutionary biology, saltation is a process by which there is a sudden change in an organism, compared to the previous generation, which can result in single-step speciation. Because of its contrast with gradualism assumed in standard evolutionary theory, the importance of saltational evolution via hybridization has been doubted by many evolutionary biologists, for example Ernst Mayr:

> The occurrence of genetic monstrosities by mutation ... is well-substantiated, but they are such evident freaks that these monsters can be designated only as 'hopeless.' They are so utterly unbalanced that 'they would not have the slightest chance of escaping elimination through selection.' Giving a thrush the wings of a falcon does not make it a better flyer. Indeed, having all the equipment of a thrush, it would probably hardly be able to fly at all ... To believe that such a drastic mutation would produce a viable new type, capable of occupying a new adaptive zone, 'is equivalent to believing in miracles.'
>
> (E. Mayr, 1970)

In fact, there are many examples of such hopeful monsters. The case we will make in this chapter is that agricultural *C. coli* may be among the most interesting and the most amenable to experimentation.

The phenomenon of hybridization, broadly defined as the movement of DNA across species boundaries, has been documented in natural populations across all domains of life. When developing taxonomic theory, Linnaeus himself described how new plant genera could arise from the hybridization of existing species (Linnaeus, 1736). When hybridization occurs it can lead to new phenotypes such drought resistance in plants (Rieseberg *et al.*, 2003), larger size in animals (Berger, 1970, 1973).

The patterns of interspecific hybridization are of interest to biologists because they provide insight into the evolutionary processes that generate different phenotypes and the nature of barriers to gene flow that maintain species. For example, in many plants, sperm competition ensures that species form hybrids only when sperm from their own species are not available. Therefore, if one species is rare, hybridization is almost unidirectional with the maternal parent almost always coming from the rarer species (Arnold *et al.*, 1993; Carney *et al.*, 1994; Rieseberg, 1995; Rieseberg and Carney, 1998; Wirtz, 1999). However, there are also cases where the direction of hybridization is opposite to that predicted by numerical dominance, for example within the mangrove genera *Sonneratia*, *Bruguiera* and *Rhizophora*, and the alpine genus *Ligularia* (Zhou *et al.*, 2008), providing a first example of the complexity of patterns of gene flow in nature.

Hybridization by horizontal gene transfer

Because of the nature of meiotic reproduction in eukaryotes, hybrids are typically thought to first arise as F1 intermediates resulting from interspecies sexual reproduction of two parent individuals although there are exceptions to this simple mechanism, such as polyploidy speciation (Coyne and Orr, 2004). In contrast to meiosis-associated recombination, horizontal gene transfer (HGT) in bacteria is unidirectional and is not necessarily associated with cell division. It occurs by three processes: (i) bacteriophage-mediated transduction of donor DNA into a recipient cell; (ii) conjugation of donor DNA, typically in the form of a circular plasmid, directly from cell-to-cell; (iii) transformation by direct uptake of DNA from the environment. These mechanisms can introduce a spectrum of genomic changes, from replacement of short stretches of DNA with homologous sequence, to large conjugation events accounting for more than 10% of the genome (Brochet *et al.*, 2008). HGT can introduce genes, plasmids and gene islands that were previously absent bestowing novel metabolic capabilities that allow expansion into new niches (Lawrence, 1999; Ochman *et al.*, 2000).

Adaptation by introgression?

Recent work analysing mosaic alleles (Sheppard *et al.*, 2011), mixed species multilocus sequence types (Sheppard *et al.*, 2008) and genome-wide introgression (Sheppard *et al.*, 2013) has shown evidence for progressive hybridization of the species *C. jejuni* and *C. coli*. One of the key questions when considering introgressive hybridization is to what extent gene flow is adaptive.

Whole genome studies have shown evidence that some regions are selected and others are not. A genomics survey of the hybrid zone of natural populations of the mouse species *Mus musculus* and *Mus domesticus* showed evidence for positive and negative selection of introgressed DNA (Payseur *et al.*, 2004; Teeter *et al.*, 2008). Another study based upon North American salamander species provides strong evidence for the rapid fixation of beneficial adaptive traits compared to total introgressed DNA. The barred tiger salamander *Amystoma mavortium* was introduced to the habitat of the native California Tiger salamander *A. californiense* in the 1940s forming a hybrid zone between these species. By comparing SNP frequencies in introgressed genomes with those from animals outside of the hybrid zone it was possible to detect deviations from expected allele frequencies over a relatively short time period. Such deviations are potentially indicative of selection for introgressed DNA. The frequency of introgressed alleles varied but 3 *A. mavortium* alleles (out of 64) where present at significantly higher frequencies in all *A californiense* populations. This evidence of rapid allele fixation is consistent with selection driven introgression.

Perhaps, the most interesting consequence of hybridization is when it creates new phenotypes that allow the colonization of new environments. By looking at examples of introgression in an ecological context we can try and understand if they are adaptive. In *Yersinia pestis*, the causative agent of plague, acquisition of plasmids on successive occasions allowed a single clone to expand into a rat and then rat flea hosts (Achtman *et al.*, 2004). In common with other highly specialized pathogenic bacteria such as *Bacillus anthracis* (Van Ert *et al.*, 2007) and *Mycobacterium tuberculosis* (Liu *et al.*, 2006), the aetiological agents of anthrax and TB respectively, there is little evidence of on-going introgression in *Y. pestis*. However, the generalist ancestors of all of these pathogens show evidence of higher rates of introgression consistent with the idea that introgression may have played a role in niche colonization.

The most well known example of an extreme environment being colonized by hybrid lineages is among sunflower species of the genus *Helianthus*. The species *H. anomalus*, *H. deserticola* and *H. paradoxus* are found in the most extreme habitats including the deserts in Arizona, Utah and Nevada and brackish salt marshes in Texas (*H. paradoxus*). These species are ancient hybrids of the species *H. annuus* and *H. petiolaris* and therefore in this example hybridization has facilitated ecological divergence and colonization of the new niches (Rieseberg *et al.*, 2003).

Introgressive hybridization can have important implications. For this reason, considerable effort has gone into explaining the factors that promote genetic exchange across large genetic distances. The importance of the ecological niche in maintaining species has been widely studied among Galapagos finches (Grant and Grant, 1992). In this case, diversification is based largely on differences in body size and beak shape, features that are maintained by natural selection for resource use, specifically the size of the seeds they consume. Hybrids between the species *Geospiza fortis* and *G. fuliginosa* are typically rare but during the El Niño event in 1982–1983, the environmental conditions changed dramatically as did the size distribution of seeds. This change to the forces maintaining species separation resulted in a period of hybrid proliferation with species intermediates in some cases showing greater fitness than parental species (Grant and Grant, 1993).

As well as climatic factors, manmade environments can promote the emergence of hybrid genotypes. For example, acid mine drainage sites such as the Richmond mine, California, USA. In this extreme environment exchange of large blocks of gene variants has occurred between coexisting bacterial and archaeal members of the naturally occurring acidophilic biofilm (Lo *et al.*, 2007). This study suggests that introgression of hundreds of kb of DNA has been crucial for adaptation to niches within the acidic and metal-rich environment.

There is evidence of introgressive hybridization from *C. jejuni* to *C. coli* that are approximately 10% divergent at the nucleotide level. Unrecombined *C. coli* are common inhabitants of the gut of wild birds such as ducks and geese and display three phylogenetic clades (Sheppard *et al.*, 2010; Colles *et al.*, 2011). However, the great majority of *C. coli* found in agricultural hosts and human disease are hybrids between clade 1 lineages and *C. jejuni* (Sheppard *et al.*, 2008, 2011, 2013). Sequenced agricultural and clinical *C. coli* isolates belong almost entirely to two clonal complexes. The ST-828 complex is the most common and has an average of approximately 10% introgression from *C. jejuni*, with variation between isolates in which regions are introgressed. The ST-1150 complex has an average of 20% introgression, also with variation between isolates. This pattern implies that interspecies recombination is an on-going process but does not preclude large recombination events in the common ancestors of each.

Is introgression in *Campylobacter* adaptive?

It is interesting to speculate about the effect of introgression in *Campylobacter* with reference to the examples above. Highly introgressed *C. coli* do not appear to form a hybrid zone as in the mouse example (Payseur *et al.*, 2004; Teeter *et al.*, 2008). Furthermore the pattern in *Campylobacter* is different to that in *Yersinia*, which involves acquisition of plasmid DNA that augments function (Achtman *et al.*, 2004) rather than disrupting species boundaries to gene flow. The example of hybridization in Galapagos finches shows some similarity to the situation in *Campylobacter*, as changes in ecology resulted in the disruption of barriers to gene flow (Grant and Grant, 1993). However, in this example it was the loss of niche separation rather than the creation of a new niche that was important. Perhaps the best comparison is with hybrid sunflower species of the genus *Helianthus* where hybridization is associated with colonization of a new niche. In evolutionary terms, agricultural hosts are a new niche for *Campylobacter*, with very unusual factors of habitation including diet, generation time, and stocking density compared to the natural pre-agricultural hosts.

Novel or extreme environments can provide an opportunity for large and rapid adaptive transitions. To enter a new discrete niche may require several simultaneous phenotypic changes involving alleles at multiple loci. Mutation may be too slow to achieve this and under these circumstances introgressive hybridization may provide a solution potentially introducing large numbers of substitutions in few generations. A further advantage is that when introgression introduces DNA that spans an entire locus, it is more likely that the genes within the recombinant fragment will be functional as they have been subject to selection in the genetic background of the donor genome.

Adaptive introgression in *Helianthus* was shown experimentally to involve the movement of three small genomic regions of *H. debilis* into *H. annuus*. This was sufficient to recreate the drought resistant phenotype of the subspecies *H. annuus texanus* (Kim and Rieseberg, 1999) – a known hybrid of these species (Heiser, 1954). Significantly, the introduction of *H. debilis* DNA onto an *H. annuus* genetic background conferred morphological traits associated with increased resistance to herbivore predation (Whitney *et al.*, 2006).

Colonization of the agricultural niche by *C. coli* has been shown to be associated with genome wide introgression but clues about the possible adaptive effects of recombination can be inferred from genes that are only present in *C. jejuni* and introgressed *C. coli* (Sheppard *et al.*, 2013). These included, several of the genes putatively involved L-fucose metabolism. Metabolism of this common animal tissue-associated sugar has been shown to confer a competitive advantage to certain strains *in vitro* (Stahl *et al.*, 2011).

Agriculture clearly represented a new colonization opportunity for *Campylobacter* and there are factors associated with this niche that are unusual compared to the ancestral niche. As in *Helianthus*, rapid genome shuffling could have facilitated colonization through high levels of introgression across relatively large genetic distances. However, unlike in *Helianthus*, defining the core features of the hybrid phenotype remains a challenge. It is difficult to see the world through the eyes of a bacterium. Therefore attempts to define which of the distinctive features of agriculture are most important to the colonization success of bacteria are currently no more than guesses, especially since there has been very little study of any sort of non-agricultural *Campylobacter* to date.

Future work on *Campylobacter* should establish the phenotypic differences between agricultural and non-agricultural strains for both *C. jejuni* and *C. coli*. It should characterize the on-going pattern of introgression; is it random within the genome or is there a progressive pattern being independently realized in different sublineages within the successful clonal complexes. And as far as possible, the likely pattern of initial introgression should be established and ideally recreated in laboratory conditions. If these modest aims are realized, then *Campylobacter* would become a model system for ecological genomics equivalent to *Helianthus*.

Acknowledgement

SKS is a Wellcome Trust Fellow.

References

Achtman, M., Morelli, G., Zhu, P., Wirth, T., Diehl, I., Kusecek, B., Volger, A.J., Wagner, D.M., Allender, C.J., Easterday, W.R., *et al.* (2004). Microevolution and history of the plague bacillus, *Yersinia pestis*. Proc. Natl. Acad. Sci. U.S.A. *101*, 17837–17842.

Arnold, M.L., Hamrick, J.L., and Bennett, B.D. (1993). Interspecific pollen competition and reproductive isolation in Iris. J. Hered. *84*, 13–16.

Berger, L. (1970). Some characteristics of the crossess within Rana esculenta complex in postlarval development. Ann. Zool. *27*, 374–416.

Berger, L. (1973). Systematics and hybridization in European green frogs of *Rana esculenta* complex. J. Herpetol. *7*, 1–10.

Brochet, M., Rusniok, C., Couve, E., Dramsi, S., Poyart, C., Trieu-Cuot, P., Kunst, F., and Glaser, P. (2008). Shaping a bacterial genome by large chromosomal replacements, the evolutionary history of Streptococcus agalactiae. PNAS *105*, 15961–15966.

Carney, S.E., Cruzan, M.B., and Arnold, M.L. (1994). Reproductive interactions between hybridizing irises: Analyses of pollen-tube growth and fertilization success. Am. J. Botany *81*, 1169–1175.

Colles, F.M., Ali, J.S., Sheppard, S.K., McCarthy, N.D., and Maiden, M.C.J. (2011). *Campylobacter* populations in wild and domesticated Mallard ducks (*Anas platyrhynchos*). Environ. Microbiol. *3*, 574–580.

Coyne, J.A., and Orr, H.A. (2004). Speciation, 1 edition. Sinauer Associates Inc., Sunderland, MA.

Grant, P.R., and Grant, B.R. (1992). Hybridization of Bird Species. Science *256*, 193–197.

Grant, B.R., and Grant, P.R. (1993). Evolution of Darwin's finches caused by a rare climatic event. Proc. R. Soc. London B *251*, 111–117.

Heiser, C.B. (1954). Variation and subspeciation in the common sunflower, *Helianthus annuus*. Am. Midland Natural. J. *51*, 287–305.

Kim, S.C., and Rieseberg, L.H. (1999). Genetic architecture of species differences in annual sunflowers: implications for adaptive trait introgression. Genetics *153*, 965–977.

Lawrence, J.G. (1999). Gene transfer, speciation, and the evolution of bacterial genomes. Curr. Opin. Microbiol. *2*, 519–523.

Linnaeus, C. (1736). Fundamenta Botanica, quae Majorum Operum Prodromi instar Theoriam Scientiae Botanices by breves Aphorismos tradunt.

Liu, X., Gutacker, M.M., Musser, J.M., and Fu, Y.X. (2006). Evidence for recombination in *Mycobacterium tuberculosis*. J. Bacteriol. *188*, 8169–8177.

Lo, I., Denef, V.J., VerBerkmoes, N.C., Shah, M.B., Goltsman, D., DiBartolo, G., Tyson, G.W., Allen, E.E., Ram, R.J., Detter, C., *et al.* (2007). Strain-resolved community proteomics reveals recombining genomes of acidophilic bacteria. Nature *446*, 537–41.

Mayr, E. (1970). Multiplication of species. In: Populations, Species and Evolution, Mayr, E. ed. (Boston, MA, Harvard University Press), p. 253.

Ochman, H., Lawrence, J.G., and Groisman, E.A. (2000). Lateral gene transfer and the nature of bacterial innovation. Nature *405*, 299–304.

Payseur, B.A., Krenz, J.G., and Nachman, M.W. (2004). Differential patterns of introgression across the X chromosome in a hybrid zone between two species of house mice. Evolution *58*, 2064–2078.

Rieseberg, L.H. (1995). The role of hybridization in evoltuion: old wine in new skins. Am J. Botany *82*, 944–953.

Rieseberg, L.H., and Carney, S.C. (1998). Plant Hybridization. N. Phytologist *140*, 598–624.

Rieseberg, L.H., Raymond, O., Rosenthal, D.M., Lai, Z., Livingston, K., Nakazato, T., Durphy, J.L., Schwarzbach, A.E., Donovan, L.A., and Lexer, C. (2003). Major Ecological Transitions in Wild Sunflowers Facilitated by Hybridization. Science *301*, 1211.

Sheppard, S.K., McCarthy, N.D., Falush, D., and Maiden, M.C. (2008). Convergence of *Campylobacter* species: implications for bacterial evolution. Science *320*, 237–239.

Sheppard, S.K., Dallas, J.F., Wilson, D.J., Strachan, N.J.C., McCarthy, N.D., Jolley, K.A., Colles, F.M., Rotariu, O., Ogden, I.D., Forbes, K.J., *et al.* (2010). Evolution of an agriculture-associated disease causing *Campylobacter coli* clade: evidence from national surveillance data in Scotland. PLoS ONE, e15708.

Sheppard, S.K., McCarthy, N.D., Jolley, K.A., and Maiden, M.C.J. (2011). Introgression in the genus *Campylobacter*: generation and spread of mosaic alleles. Microbiology *157*, 1066–1074.

Sheppard, S.K., Didelot, X., Jolley, K.A., Darling, A.E., Pascoe, B., Meric, G., Kelly, D.J., Cody, A., Colles, F.M., Strachan, N.J.C., *et al.* (2013). Progressive genome-wide introgression in agricultural *Campylobacter coli*. Mol. Ecol. *22*, 1051–1064.

Teeter, K.C., Payseur, B.A., Harris, L.W., Bakewell, M.A., Thibodeau, L.M., O'Brien, J.E., Krenz, J.G., Sans-Fuentes, M.A., Nachman, M.W., and Tucker, P.K. (2008). Genome-wide patterns of gene flow across a house mouse hybrid zone. Genome Res. *18*, 67–76.

Van Ert, M.N., Easterday, W.R., Huynh, L.Y., Okinaka, R.T., Hugh-Jones, M.E., Ravel, J., Zanecki, S.R., Pearson, T., Simonson, T.S., U'Ren, J.M., *et al.* (2007). Global genetic population structure of *Bacillus anthracis*. PLoS ONE *2*, e461.

Whitney, K.D., Randell, R.A., and Rieseberg, L.H. (2006). Adaptive introgression of herbivore resistance traits in the weedy sunflower *Helianthus annuus*. Am. Nat. *167*.

Wirtz, P. (1999). Mother species-father species: unidirectional hybridization in animals with female choice. Anim. Behav. *58*, 1–12.

Zhou, R., Gong, X., Boufford, D., Wu, C.I., and Shi, S. (2008). Testing a hypothesis of unidirectional hybridization in plants: observations on Sonneratia, Bruguiera and Ligularia. BMC Evol. Biol. *8*, 149.

Learning About Recombination in *Campylobacter*

2

Paul Fearnhead, Patrick J. Biggs and Nigel French

Abstract

How important a role does recombination play in the evolution of *Campylobacter*? In this chapter we overview some of the methods people have used to answer this question. The focus is on the importance of recombination relative to mutation in terms of generating genetic diversity within *C. jejuni*. We see that recombination events appear to occur at a similar rate to mutation, but can potentially have a much greater effect in terms of the changes it makes on the DNA sequence. Looking forward, the increasing availability of full-genome data has the potential to transform our detailed knowledge about the recombination process, and its role in the evolution, of *Campylobacter*.

Introduction

Homologous recombination is a process through which new DNA can be incorporated within an organism's genome. Within bacteria, there are three mechanisms through which this can occur: conjugation, the uptake of DNA from other bacteria; transformation, the uptake of naked DNA from the remains of bacteria that exists in the living environment; and transduction, where DNA is implanted by bacteriophage (Low and Porter, 1978). Although different, each of these mechanisms results in the introduction of a new DNA sequence within a region of the genome, and thus is potentially an important factor in the evolution of a given bacteria. How important it is varies considerably across bacterial species: with estimates of the relative effect of recombination to mutation varying by over three orders of magnitude in Vos and Didelot (2009).

How important is recombination to the evolution of *Campylobacter*? This chapter will look at how researchers have tried to answer this question, and to learn about recombination in *Campylobacter* from DNA sequences of isolates randomly sampled in different animal and environmental sources. The focus will be primarily on recombination in *Campylobacter jejuni*, and on estimating the relative rate of recombination to mutation.

There are two alternative definitions of the relative rate of recombination to mutation. The first is to just consider the ratio of the rates at which either event occurs within a region of the genome of interest. Alternatively, we can look at the ratio of the probabilities that a single nucleotide is changed by each event. This involves weighting recombination events by the average number of nucleotide changes that they introduce, and thus is affected also by the size of region that is imported at a recombination event, and the diversity between the original and imported sequence. Both definitions have been used, and each tells us something slightly different about the role that recombination plays in the evolution of *C. jejuni*. The first of these is often called the ρ/θ ratio, and the latter r/m. For notational convenience we will sometimes use $\lambda=\rho/\theta$ and $\gamma=r/m$ in the following.

In this chapter we will consider two sorts of population genetic data that have been used to address these questions. The first is multilocus sequence type (MLST) data (Maiden *et al.*, 1998), which describes the DNA sequence of an isolate within approximately 500 bp fragments of seven housekeeping genes. These genes were chosen because they each have an important function and are part of the core genome of all *C. jejuni* isolates. These data are often summarized in terms of an allelic profile, which gives the allele number of an isolate at each gene locus. Having the allelic profile of a set of isolates just contains information as to which isolates share the same DNA sequence (allele) for each of the seven genes, but does not contain information about the differences in DNA sequence for isolates which have different alleles at a given gene. However, the actual DNA sequence is known for each allele at each locus, and it is this latter information that is often used in analyses that try and learn about recombination in *C. jejuni*. Until very recently, MLST data were the primary type of population genetic data collected for *C. jejuni*, and it has been collected for tens of thousands of isolates, sampled from a range of geographical and source locations. The pubMLST database, http://pubmlst.org/campylobacter/, currently contains information from over 20,000 isolates, which between them have over 6000 distinct allelic profiles.

One problem with MLST data is that they describes genetic variation in less than 1% of the genome. This both limits the resolution of such data, and also suffers from concerns that the patterns observed in the housekeeping genes may not be representative of patterns across the genome. Thus, in the past few years, full-genome data have started to be collected for sizeable numbers of *C. jejuni* isolates (e.g. Timms *et al.*, 2010; Lef′ebure *et al.*, 2010; Biggs *et al.*, 2011). Only recently have these data been used to learn about recombination (Biggs *et al.*, 2011).

The methods for estimating recombination that we discuss in this chapter can be classified into two groups. In the next section we describe methods based on analysing data sampled at random from the *C. jejuni* population. To date these methods have been used to analyse MLST data from *C. jejuni*. In the following section we then describe methods based on trying to identify mutation and recombination events in the history of closely related pairs of isolates. We look at analyses of both MLST and full-genome data. The chapter then concludes by looking at the different estimates of recombination that these methods have led to, trying both to explain any inconsistencies in results and to discuss what we can learn about recombination in *C. jejuni* from them.

Estimating recombination from population data

If we are going to use population genetic data (data collected from a sample of isolates from some population) to infer recombination, we need to know, and be able to model, how recombination affects the data. One way of doing this is to consider the unobserved genealogical relationship of our sample. At a single site there will be a tree that describes this genealogy, with the data we observe at that site being the effect of mutations within that genealogy. If there were no recombination then all sites would have the same genealogy. The effect of recombination is to allow the genealogy to change as you move around the genome. Thus sites close together are more likely to share the same, or very similar genealogies, than sites further apart. Furthermore, the smaller the amount of recombination, the more similar the genealogies at different sites will be. The more similar the genealogy of a pair of sites, the greater dependence we would expect in the patterns of DNA at those sites.

For bacteria, where each recombination event affects only a small region of the genome, you can define a single tree for a sample of isolates. This tree is obtained by 'ignoring' recombination events: that is following the ancestry of regions unaffected by the recombination at each event. This tree has been called the clonal frame of the sample (Didelot and Falush, 2007). The genealogies of sites that are far apart, so that they will be not affected by the same recombination events, will still be correlated with one another due to the shared clonal frame. See Fig. 2.1 for an example of the genealogies within one region, and the clonal frame for a sample of isolates.

As recombination affects the dependence in nucleotide patterns at different sites across the genome, it is possible to use this dependence to learn about recombination. Simple versions of this include calculating a measure of dependence, such as one of the measures of linkage disequilibrium (LD) for all pairs of sites, and then to plot these values as a function of the distance between

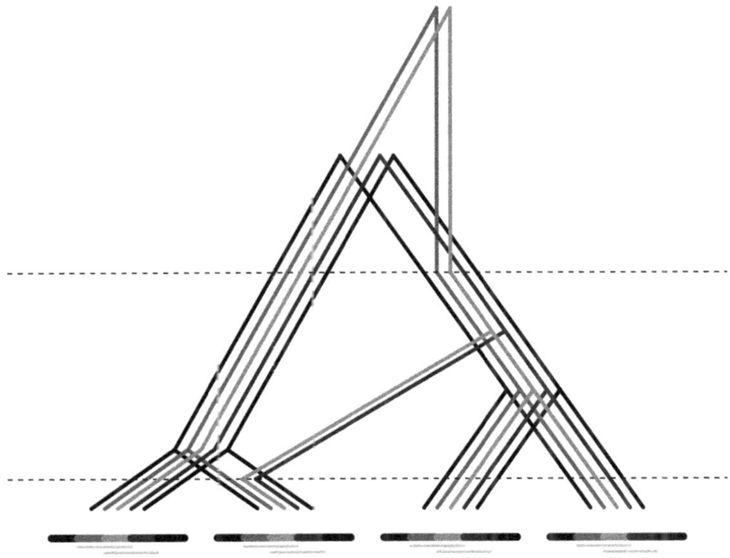

Figure 2.1 Example of the genealogical trees for four isolates in one gene. The effect of recombination is that different sites in the gene have different genealogical trees. Here we have two recombination events that have affected the genealogical relationship of the isolates (the position back in time of these are denoted by the dashed lines, the regions they affect by the light blue lines under the gene fragments). The black tree is the clonal frame of the sample: the genealogy at regions unaffected by recombination. The other trees represent the genealogies of regions affected by the either one or both recombination events. A colour version of this figure is available in the plate section at the back of the book.

the sites. Whilst there are numerous measures of LD, all have the property that larger values imply stronger dependence in the nucleotide patterns observed at the two sites, which in turn suggest greater dependence in the genealogies at the two sites. Thus for *C. jejuni* we would expect the average value of LD to initially decay as the distance increases. As this distance increases beyond the size of region affected by a single recombination event, the average value of LD should then become roughly constant.

However more accurate approaches are available through using stochastic models that describe the probability of observing a specific pattern of DNA at a set of sites. These models are often based on the coalescent (Nordborg, 2001; Stephens, 2001), or approximations to it. One issue with such approaches is that the models often make assumptions and thus we have to be careful about how robust estimates are to violations of these assumptions. We will now describe two such model-based approaches to estimating the rate of recombination in *C. jejuni*, both of which are based on models that assume there is no structure to the population, that the population has remained constant in time, and that there is no selection.

Using pairwise LD

Our first approach analyses the patterns we observe at each pair of sites. For each isolate we can define a haplotype as being the nucleotide at each of these two sites. Then coalescent models are used to obtain a family of distributions for the haplotypes of a sample of isolates as a function of parameters that govern the rate of recombination and mutation, and the distribution of the length of regions affected by a single recombination event. If we denote the haplotype data at two sites i, j by $\mathbf{X}_{i,j}$, the distance between the sites by $d_{i,j}$, and the set of parameters in the model by φ, then our model will give a series of probabilities $\Pr(\mathbf{X}_{i,j} = \mathbf{x}_{i,j} | \varphi, d_{i,j})$, for all possible haplotypes $\mathbf{x}_{i,j}$. We can then estimate the parameters φ by maximizing the sum of the log of these probabilities for the observed data for all pairs of sites (Hudson, 2001):

$$\sum_{i,j} \log \Pr\left(X_{i,j} = x_{i,j} \middle| \varphi, d_{i,j} \right) \qquad (2.1)$$

Calculating eqn. 2.1 is computationally efficient, for moderate numbers of isolates (up to of the order of 100), through the use of a look-up table (Hudson, 2001). Such an estimator has been shown empirically to be one of the most accurate ways of estimating the rate of recombination (Smith and Fearnhead, 2005), and has been successfully applied to learn about fine-scale variation in recombination rates within humans (McVean *et al.*, 2004; Myers *et al.*, 2005). The function (eqn. 2.1) that is maximized is often called a composite log-likelihood. It is based on summing the log-likelihood of different subsets of the data, and is similar to the idea of the log-likelihood function, except that it ignores the dependence between the data for different pairs of sites. There is substantial theory underpinning the use of composite likelihoods (Varin *et al.*, 2011; Cox and Reid, 2004) and they have been used substantially in genetic applications (Larribe and Fearnhead, 2011). The key idea behind the rationale for maximizing eqn. 2.1 is that, if data were generated from a model for which $\Pr(\mathbf{X}_{i,j} = \mathbf{x}_{i,j} \mid \varphi, d_{i,j})$ gave the correct probabilities of data at a pair of sites $d_{i,j}$ apart for some parameter values φ_0 (i.e. our underlying model for the data is correct, with φ_0 being the true parameter values), then the expected value of eqn. 2.1 will have its maximum at φ_0. If we have data from many sites, we can hope that eqn. 2.1 will be roughly proportional to this expected value, and thus when we maximize we will get estimates of φ that are close to the value φ_0.

Note that while it is easy to use this approach to estimate φ, for population genetic applications it is often difficult, if not impossible, to get appropriate measures of uncertainty in our estimates. Most studies that use this, or similar, composite likelihoods have to resort to some form of simulation procedure to assess uncertainty. How accurate the resulting estimate of the uncertainty is in practice is not clear.

Fearnhead *et al.* (2005) used this approach to estimate recombination rates in *C. jejuni* for isolates sampled from farm and wildlife sources in a 100 km^2 area in the UK (French *et al.*, 2005). They separately analysed isolates from different animal and wildlife sources. Averaging the estimates from each source gives an overall estimate of ρ/θ of 0.74 (with estimates for the different sources ranging from 0.3 to 1.2). Whilst they do not estimate r/m directly, we can obtain an estimate from the results in Fearnhead *et al.* (2005). The estimated average tract length of a recombination event is 430 bp, with a mutation rate of 12.6 per kb. This suggests one recombination event will introduce 5.4 nucleotide differences on average. Thus we get an estimate of r/m of 4. The original paper assesses uncertainty in estimates of ρ and of the tract length, but it is not possible to translate these into formal confidence intervals for either ρ/θ or r/m. However they suggest the estimates are accurate to within a factor of 2 for ρ/θ and of 4 for r/m.

As mentioned above, the model used in this analysis makes assumptions, such as a randomly mating population, that are not valid for *C. jejuni*. The effect of the violation of such assumptions on the estimates of the recombination rate was investigated by Smith and Fearnhead (2005). They considered a range of scenarios for population structure, or population growth. Their simulations suggest that significant population structure, or a recent bottleneck, could to lead to underestimating the rate of recombination by close to a factor of 2.

Clonal frame

An alternative model for the data was suggested by Didelot and Falush (2007), based directly on the idea of the clonal frame. They model the joint probability of the DNA sequences observed at each gene for the sample of isolates. Their model consists of a probability distribution for the clonal frame of the sample, together with a conditional distribution for the DNA sequences given the clonal frame. For the latter they make the simplification of not considering the ancestry of recombinant regions, instead they model such regions as introducing a random number of changes to the DNA sequence. See Fig. 2.2 for an example. This simplification greatly reduces the computational burden of fitting the model to data. Simulation results in Didelot and Falush (2007) suggest that the resulting method can be accurate in terms of estimating r/m, at least for data simulated under the assumptions of a constant sized, randomly mating population.

This method has been applied to data from a range of bacteria by Vos and Didelot (2009).

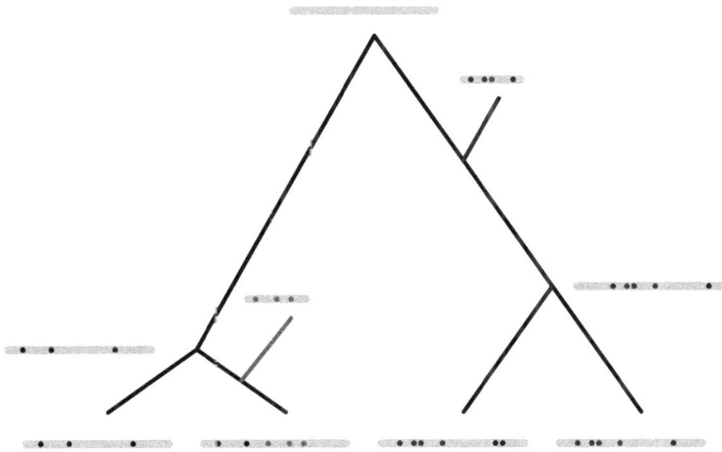

Figure 2.2 Simplified clonal frame model for genealogies of Fig. 2.1. Origin of recombinant regions is ignored, and the two sequences imported at recombination events are just modelled as introducing a number of new nucleotide differences. In this schematic the grey bars represent the ancestral sequence, black dots indicate nucleotide changes introduced by mutation on the clonal frame, and red and blue dots those introduced by the two respective recombination events. A colour version of this figure is available in the plate section at the back of the book.

Their results for *C. jejuni* suggest $r/m = 2.2$ (95% credible interval of 1.7–2.8), suggesting just over twice as many nucleotides are changed by recombination as by mutation. This is close to the median estimate obtained for the roughly 50 bacteria studied. This result is noticeably smaller than that obtained using pairwise LD by Fearnhead et al. (2005) of $r/m = 4$, though within the level of uncertainty that there was in that estimate.

As well as giving estimates of r/m, the method of Didelot and Falush (2007) can be used to estimate the clonal frame for a dataset. The clonal frame for a set of *C. jejuni* and *C. coli* was estimated by Sheppard *et al.* (2008) and then used as part of an analysis which suggests that whilst there must have been an historic barrier to gene-flow that caused the creation of these two species of *Campylobacter*, that barrier has recently broken down and now recombination has led to noticeable gene flow between *C. jejuni* and *C. coli*.

Estimating recombination from closely related isolates

An alternative approach to making inference about recombination is to compare the DNA sequences from closely related isolates. For such a pair there will have been fewer mutation and recombination events since they shared a common ancestor, which potentially makes it easier to infer which changes in the DNA sequence are due to mutation and which are due to recombination. This is particularly the case when the effect of recombination is very different from that of mutation, which is when recombination events introduce a comparatively large number of changes in nucleotides. For bacteria where it is common that recombination events only change one or two nucleotides, looking at patterns from closely related isolates is unlikely to be informative. Fortunately, the size of recombination events and the levels of genetic diversity in *C. jejuni* populations means that recombination events are likely to introduce many changes in DNA sequence, and hence this is a very good approach for learning about the importance and effect of recombination.

When performing a statistical analysis of data from closely related isolates we need to be careful that the way we have selected the isolates does not lead to biases in our estimates. This is most easily done by using different and unrelated data or information to choose the pairs of isolates to analyse. Alternatively, we need to account for the method of choosing isolates when performing the statistical analysis. To see why this is important, assume we choose pairs of isolates to analyse based on the number of nucleotide differences between them across the seven MLST housekeeping genes. If, say, we only chose pairs that differed at a single nucleotide, then this would be biasing our data towards pairs, which differ only due to a single mutation, as most pairs that differ due to recombination would differ at more than one nucleotide. If we do not account for this selection bias in our statistical analysis we would over-estimate the rate of mutation relative to that of recombination.

MLST data

Firstly we will describe approaches that use MLST data. The general idea is to define closeness, and select pairs of isolates, based on the allelic profiles; but then to estimate the relative rate of recombination to mutation based on the DNA sequences at the gene loci or locus at which the pair differs. If the pair differs at a large number of nucleotides that is indicative of there having been a recombination event; if they differ at just one, or may be a small number of nucleotides then that is more likely to be solely due to mutations. This has the property that the information being used to select the closely related isolates is different from that being used to infer whether there has been mutation or recombination.

This approach was suggested by Feil *et al.* (1999, 2000). Their approach was to cluster isolates into clonal complexes. Each clonal complex consists of a consensus allelic profile, together with all further allelic profiles that differ at just one locus from this consensus. These are called single locus variants (SLVs) of the consensus. The idea is that the consensus is an estimate of the common ancestor strain of the clonal complex, and all SLVs will be closely related to this common ancestor. They then look at the number of nucleotide differences between each SLV and the consensus at the locus at which it differs. Based on the number of differences, they classify each SLV as being caused by mutation or recombination. A general rule is that if there are three or more nucleotide differences then the SLV is caused by recombination, whereas if there is just one or two it could be caused by either recombination or mutation. For the latter, the authors look whether the nucleotide differences are observed in isolates in other clonal complexes. This is motivated by the idea that mutations will predominantly produce a novel nucleotide at that site, whereas recombination is more likely to produce a nucleotide that is observed in other (distantly related) isolates. Feil *et al.* (2000), who looked at estimating recombination in *Streptococcus pneumoniae,* use the rule that an SLV with a single, novel, nucleotide difference is caused by mutation. All other SLVs are due to recombination. The estimate of r/m is just the ratio of the sum of nucleotide differences in the sets of SLVs that have been classified as due to recombination as compared to those due to mutation.

We can also estimate the relative rate of recombination to mutation: but this will be in terms of the rate at which recombination and mutation affect a locus. If we denote these rates by $\tilde{\rho}$ and $\tilde{\theta}$ respectively, then we can get an estimate of $\tilde{\lambda}=\tilde{\rho}/\tilde{\theta}$ by the number of SLVs assumed to be created by recombination as opposed to those due to mutation. If we have that a locus is of length l bps, and the mean recombination tract length is kl bps then whilst $\tilde{\theta}=l\theta$, as l bases can mutate, we have $\rho\tilde{\rho}=(k+1)l\rho$, as the locus is affected not just by recombination events that start within the locus, but also those that start outside the locus but whose tract overlaps with the locus. Thus, we have $\tilde{\lambda}=(k+1)\lambda$. So we are estimating a quantity that depends not only on the relative rate of recombination to mutation, but also on the relative size of the tract length to the size of locus.

More recently, an alternative approach to inference was implemented by Yu *et al.* (2012). They avoid clustering by clonal complex, and considered analysing all pairs of SLVs. For all SLV pairs that differ at a given gene, the number of nucleotide differences at that gene are recorded. For a single SLV pair we can model the probability distribution of the number of differences as a function of the relative rate of recombination to mutation. Such a model can be derived using coalescent theory to give a distribution on the number of mutation and recombination events in the history of the SLV pair, and then a model for the distribution of the number of nucleotide differences introduced by a recombination event. A simple proxy for the latter can be obtained by looking at the distribution of nucleotide differences between a randomly chosen pair of isolates in the population.

Together these give a family of distributions for N, the number of nucleotide differences for an SLV pair, parameterized by the relative rate of recombination to mutation affecting the locus, $\tilde{\lambda} = \tilde{\rho}/\tilde{\theta}$, which we denote $\Pr(N = n \mid \tilde{\lambda})$. Given data $n_1, \ldots, n_m$ of the observed numbers of nucleotide differences from m SLV pairs, we can then estimate $\tilde{\lambda}$ by maximizing

$$\sum_{i=1}^{m} \log \Pr\left(N = n_i \middle| \tilde{\lambda}\right) \quad (2.2)$$

Like eqn. 2.1, this is a composite-likelihood. As mentioned previously, whilst there is theory

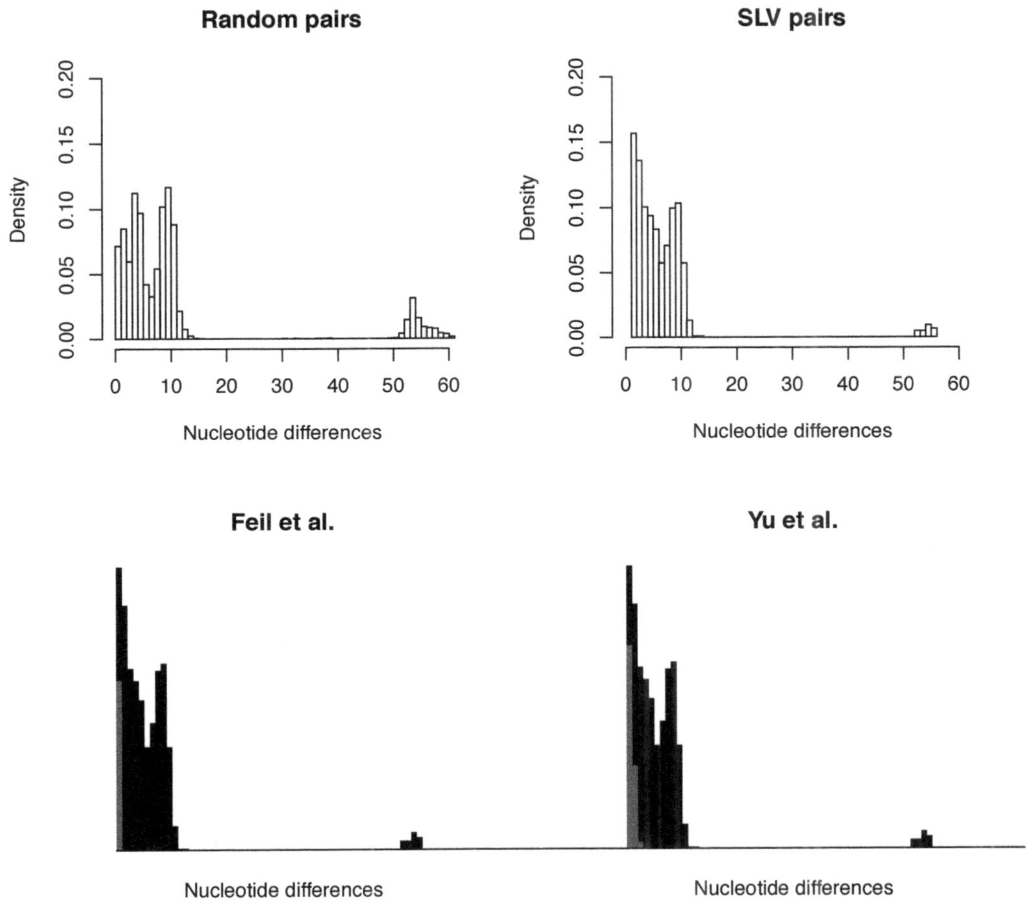

Figure 2.3 Nucleotide differences at one of the MLST alleles for a random sample of pairs of *C. jejuni* isolates (top left), and for SLV pairs (top right). The method of Feil *et al.* (2000), which was used to analyse SLVs from *C. jejuni*, is to assign a proportion of the SLVs with one nucleotide difference to be due to mutation (bottom left plot; red), and all others to be due to recombination (bottom left plot; black). The method of Yu *et al.* (2012) uses a simple model that assumes the data from SLV pairs is a mixture of pairs where the changes are due solely to mutation, and those that are due to recombination. Simplistically this models the data for SLV pairs to be a mixture of a geometric distribution for the number of differences (mutation only case) and the nucleotide differences observed from random pairs (recombination case). It then calculates the expected proportion of SLV that are due to mutation only (bottom right plot; red) or recombination (bottom right plot; black) for each value of the number of nucleotide difference. Note the two methods have different ways of collecting the SLV pairs to analyse (see text for details). A colour version of this figure is available in the plate section at the back of the book.

underpinning the use of such an estimator, it can be difficult to accurately measure uncertainty in the estimates that are obtained. In practice people resort to simulation methods to evaluate this.

We would expect the accuracy of both approaches will depend on the number of nucleotide differences introduced by recombination events. Fig. 2.3 gives an example of the data for *C. jejuni*, for one of the MLST genes (similar patterns are observed for other genes). Firstly note that the pattern for SLV pairs is similar to that for randomly chosen pairs of isolates, except for a larger proportion of SLV pairs with just a few nucleotide differences. This is precisely the pattern that would be expected under the model of Yu *et al.* (2012). Furthermore, well over half the SLV pairs have four or more nucleotide differences, and thus there is high probability that they have been the result of recombination. As a result we would expect methods for estimating recombination based on SLVs to be reasonably accurate and also to be robust to modelling assumptions.

The two methods described above differ in two main regards. The first is which SLV pairs they use. The second is how they decide which of the SLV pairs with few nucleotide differences are due to recombination. For the method of Feil *et al.* (2000), all SLVs are assumed to be due to recombination except for a subset of those with just a single difference (see bottom left plot of Fig. 2.3). For that of Yu *et al.* (2012), the statistical model makes it possible to calculate an expected proportion of SLV pairs that are due to recombination for each value of the number of nucleotide differences observed. In practice, this proportion is close to 1 for all SLVs with more than 3 nucleotide differences, but there is a noticeable proportion for both three and two nucleotide differences, as well as the majority of pairs with a single difference, that are due to mutation (see bottom right plot of Fig. 2.3). This simple schematic suggests that the method of Feil et al. (2000) will tend to give larger estimates for the rate of recombination than that of Yu *et al.* (2012).

The method of Feil *et al.* (2000) was used to estimate both $\tilde{\rho}/\tilde{\theta}$ and r/m in *C. jejuni* by Schouls *et al.* (2003). Their results suggest $\tilde{\rho}/\tilde{\theta}=8$ and $r/m=47$, so that recombination events are eight times more likely than mutation events to affect an MLST locus, but a single nucleotide is roughly 50 times more likely to be changed by these recombination events. By comparison, the results of Yu *et al.* (2012) give estimates of $\tilde{\rho}/\tilde{\theta}=7$ and $r/m\approx 50$. These results are remarkably consistent, and suggest a much bigger role for recombination than the results based on analysing population data. The results in Yu *et al.* (2012) also suggest there is evidence for the value of $\tilde{\rho}/\tilde{\theta}$ to differ across the MSLT loci, with the estimated value for the *glyA* gene being significantly smaller than the estimates for other genes. Also Yu *et al.* (2012) observed a different pattern within *C. coli*, with the estimated value of $\tilde{\rho}/\tilde{\theta}$ being close to 1.

Full-genome data

With the advent of full-genome data it is now possible to use comparisons of closely related isolates to learn about recombination across the whole genome. One instance of this is given by Biggs *et al.* (2011), where they compared full genome data from two isolates that were identical in terms of their MLST allelic profiles. In total, sequence data from 1502 genes was obtained for both isolates. The isolates analysed had identical sequences at nearly 95% of these genes. The remaining 83 genes had 672 point differences.

The approach in Biggs *et al.* (2011) was based around inferring which nucleotide differences between the two isolates where due to recombination and which where due to mutation and to inferring specific regions of the genome where recombination had occurred. This was performed using a variant of the PACL approach of Li and Stephens (2003), which models the sequence of an isolate within a gene as a mosaic of the sequences of the other isolate and of a number of reference genomes, together with mutations (see Fig. 2.4 for an example). The idea of the model, is that as the pair of isolates are very similar they will be closely related on the clonal frame of the sample of the two isolates and the reference genomes. Thus at most regions in the genome the two isolates will have a recent common ancestor. However within recombinant regions the two isolates will either be more distantly related, or one of the isolates will be most closely related to one of the reference genomes. Thus, at each site the model has an unobserved state which classifies

REF1 ACTAGCCAAAGGC

REF2 ATGAGGCCGACCC

ISO2 ACGTGGCCTATGC

ISO1 ACGTGGCCGACTC

Figure 2.4 Example of the model in Biggs *et al.* (2011), for a pair of isolates and two reference genomes. The model gives a distribution for the DNA sequence one isolate (ISO1) conditional on the sequence of a closely related isolate (ISO2), and two more distantly related reference genomes (REF1 and REF2). The DNA sequence of ISO1 is a mosaic of the DNA sequences of ISO2, REF1 and REF2 with mutation. In this example the nucleotide at the first eight sites are copied from ISO2, and for the remaining five sites from REF2. Additionally there is a mutation at site 12. The last five sites would be classified as a recombinant region. A colour version of this figure is available in the plate section at the back of the book.

the site as being most closely related to the other isolate, with few nucleotide differences, being most closely related to the other isolate but with a substantially higher frequency of nucleotide differences, or most closely related to one of the reference strains. All except the first of these states correspond to the site being in a recombinant region. Analysis of data involves calculating the joint probability distribution of each site being in each state, given the observed sequences. This then enables us to calculate the probability of any site or region being affected by recombination, and also the probability of whether each of the nucleotide differences between the two isolates is due to recombination or mutation.

While this approach does not directly estimate ρ/θ and r/m, it gives information about both quantities. We can obtain an estimate for r/m by looking at the ratio of the expected number of nucleotides changed by recombination as opposed to mutation. Estimating ρ/θ is harder, as the method infers recombinant regions rather than recombination events. Biggs *et al.* (2011) used the output of the analysis to get a lower bound on the number of recombination events, which could in turn be used to get an estimated lower bound for ρ/θ.

Biggs *et al.* (2011) estimated that just 22 of the 672 nucleotide differences were in non-recombinant regions. The remaining 650 nucleotide differences were within recombinant regions created by at least 19 distinct recombination events. As recombination was relatively rare, we can make the simplifying assumption that these 22 changes were the only changes due to mutation, an assumption that would tend to overestimate the role of recombination. These results then suggest a crude estimate of $r/m \approx 30$ and that recombination rates are similar to mutation rates. They found evidence for large recombinant regions, including two regions spanning 11 and seven genes respectively. These could either be the effect of large recombination events, or be due to multiple recombination events, and hence evidence of genomic regions with higher rates of recombination. Finally, they observed that within one of these regions, recombination had introduced two completely new genes.

Further full-genome analyses

We repeated the analysis of Biggs *et al.* (2011), but with full-genome data from four different *C. jejuni* isolates. These were three isolates found in poultry P104, P544 and P722, and one from cattle, S150. They were chosen from a set of isolates sampled from New Zealand, based on their MLST profile. Three isolates (P104, P544 and S150) had identical MLST profiles (ST45), and the other (P722) had an MLST profile (ST25) that differed at a single gene, glyA. The three poultry isolates were sampled from the same poultry supplier, in July 2005 (P104), January 2008 (P544) and November 2008 (P722); S150 was sampled in May 2007.

We considered four pairwise comparisons of the full-genome data for these isolates. Each comparison involved over 1300 genes for which we have sequences from both isolates and at least one reference strain. Over 99% of genes had the same sequences for P104 and P544, and thus we omitted comparisons between P544 and the other two isolates, as results would be almost identical to those when comparing P104 to P722 and S150 respectively. A summary of the data is given in Table 2.1. A histogram of the percentage of nucleotides that differ for each gene and pairwise comparison is shown in Fig. 2.5. The most striking

Table 2.1 Summary of data from pairwise comparisons

				Differences in		
Isolate pair	Same source?	No. of genes	No. of bp (Mb)	MLST loci	Genes	Nucleotides (%)
P544–P104	Yes	1353	1.19	0	8	0.0025
P722–P104	Yes	1322	1.17	1	418	0.43
S150–P104	No	1308	1.15	0	458	0.62
S150–P722	No	1304	1.15	1	563	0.74

For each pair of isolates we give whether they were both sampled from the same source; the number of genes and bp used in the comparison; and the number of MLST loci, all genes, and bp at which the pairs differ. For the last two calculations we ignore any differences due to insertions or deletions.

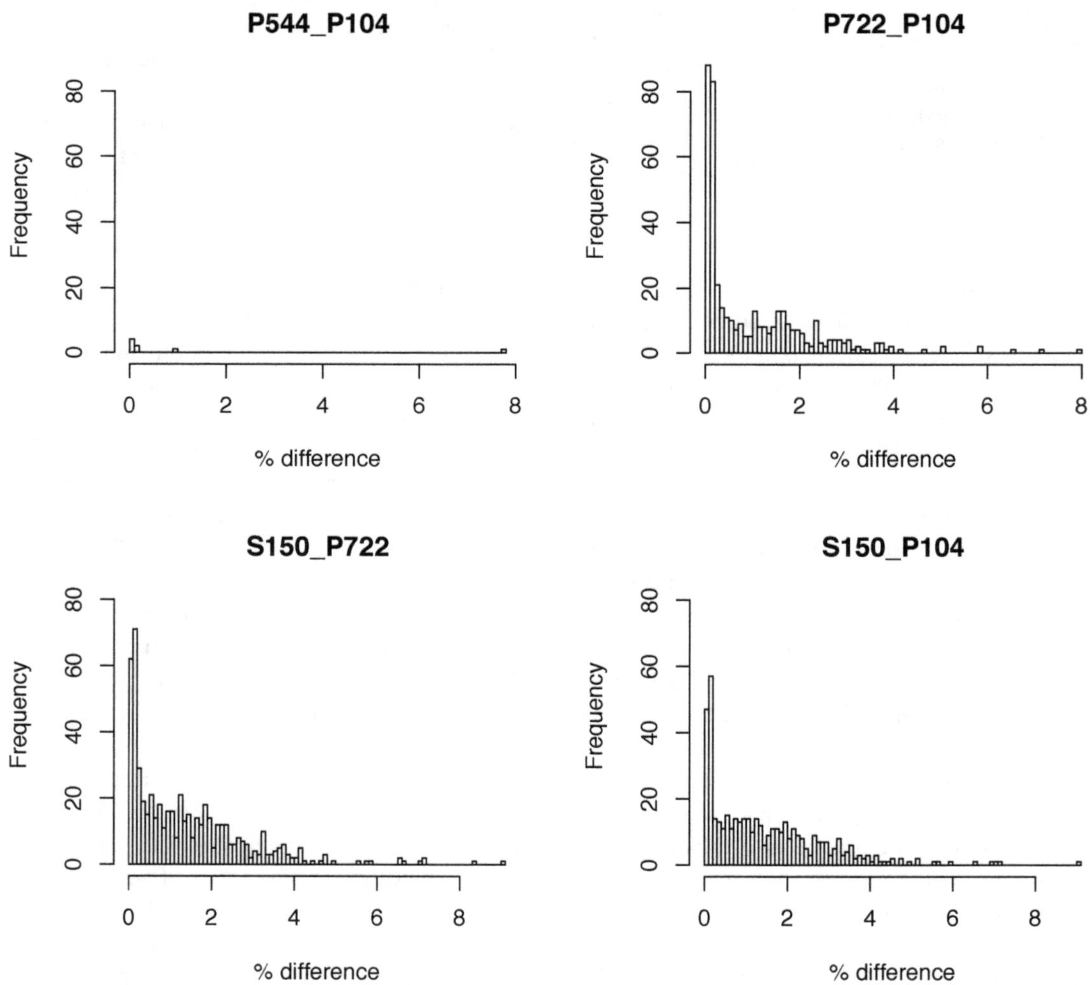

Figure 2.5 Per cent difference in nucleotide sequences for genes in pairwise comparison. Each figure is for a difference comparison of pairs of isolates. Per cent difference is only plotted for genes where the pair of isolates were not identical. For ease of presentation we ignored genes with more than 10% difference in sequence. There were three such genes for the S150-P104 comparison, and four for both the comparisons involving P722.

feature is that for three of the four comparisons we have a number of genes with substantial differences between two relatively closely related isolates.

We repeated the analysis of Biggs *et al.* (2011) for each of the pairwise comparisons. This gives us an estimate of the proportion of each gene that has been affected by recombination with the ancestry of the pair of isolates, see Fig. 2.6. The amount of recombination we observe between P104, P722 and S150 makes it difficult to get quantitative estimates of the relative rate of recombination to mutation. For each pairwise comparison we can estimate the number of nucleotide differences that lie in non-recombinant regions, but it is not possible to quantify how many of the nucleotide

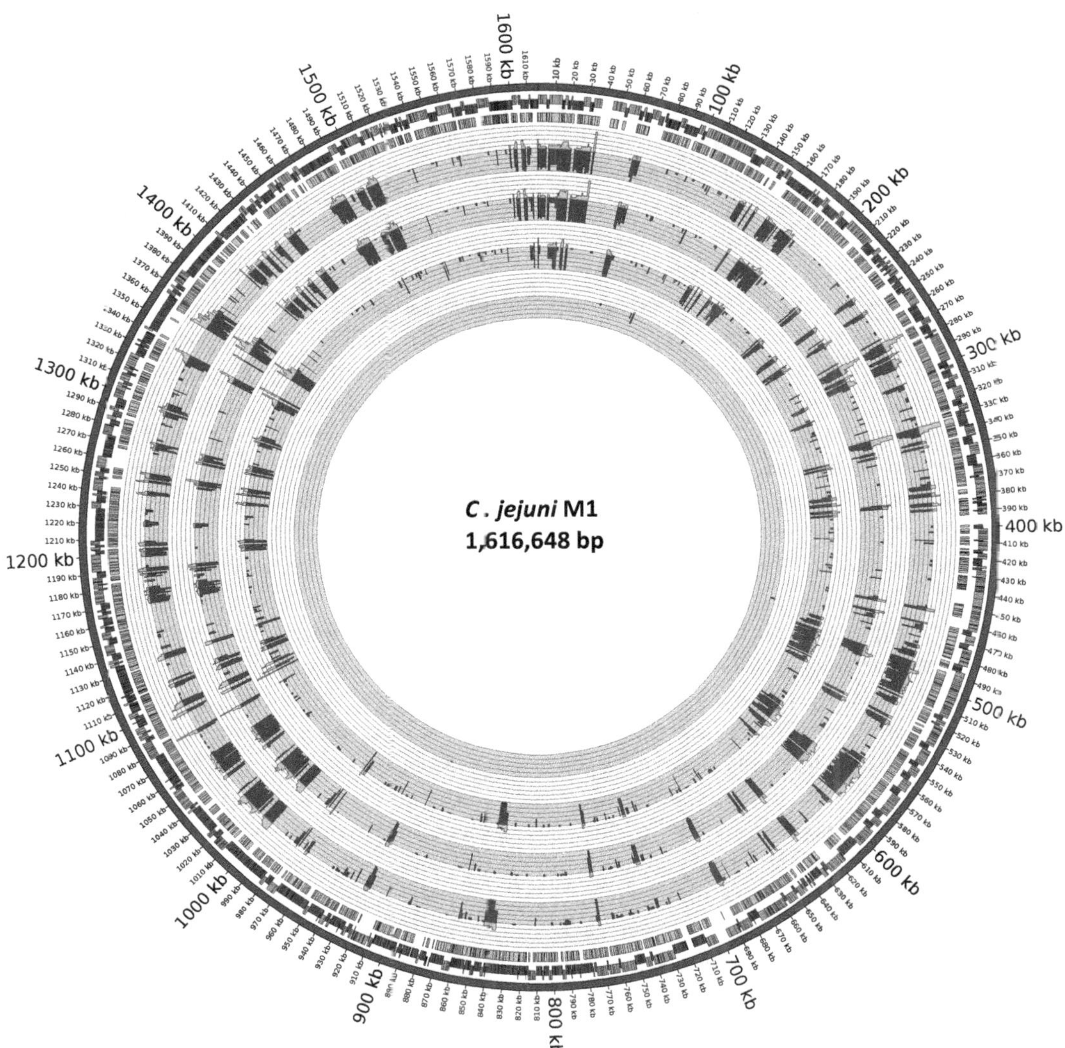

Figure 2.6 Genome pairs mapped against the M1 *C. jejuni* genome. The outer three rings are: M1 ideogram (red); coding genes (orange: forward; blue: reverse); genes under analysis (green). For the remainder of the plot, results for the 4 pairs of *C. jejuni* isolates, from outside to inside: P722b and S150a; P104a and S150a; P722b and P104a; P544b and P104a. The yellow histogram represents the percentage nucleotide difference between the orthologous genes (with a maximum of 10%), and the purple represents the estimated proportion of the gene that is in a recombinant region. A colour version of this figure is available in the plate section at the back of the book.

differences in recombinant regions are due to mutation as opposed to recombination.

The most striking aspect of the plot is the way that genes that have undergone recombination appear to cluster. Many such clusters can be explained by recombination in the ancestry of just one of the isolates. For example, the region from 1400 kb to 1450 kb, which could be explained by recombination just in S150 producing evidence of recombination in the comparisons between S150 and P104 and S150 and P722. The two most natural explanations for such regions would be that they are caused by a single recombination event, or that these are regions with higher recombination rates and are due to a larger number of smaller recombination events. The former of these appear more plausible, as if the regions have higher recombination rates we would not expect all the recombination to have occurred solely in the ancestry of one of the isolates. However, this then suggests that recombination can affect large regions (of the order of tens of kilobases), which is much greater than many previous estimates.

There are also clusters which show evidence of recombinations between three pairs of isolates, and which would need recombination to have affected the ancestry in at least two of the three isolates. One explanation for the latter regions, for example 1330 kb to 1380 kb, or for numerous genes between 1090 kb and 1130 kb, would be that these are regions with a higher mutation rate. However, this is not supported with the location of mutations between P104 and P544, which do not appear to occur more often within such regions (only one of the 8 genes for which there are nucleotide differences between P104 and P544 are in a region with strong evidence of recombination between each of the pairs involving P104, P722 and S150). The patterns we observe appear to be more consistent with either two large recombination events, or maybe regions with higher recombination rates and hence multiple recombination events. The patterns in region 1090 kb to 1130 kb, which has a combination of genes with strong evidence of recombination interspersed with genes with little evidence, perhaps supports the latter explanation.

Table 2.2 Point estimates of *r/m* from five published analyses of *C. jejuni* discussed in this chapter

Reference	Section	*r/m*
Fearnhead *et al.* (2005)	Using pairwise LD	4
Vos and Didelot (2009)	Clonal frame	2.2
Schouls *et al.* (2003)	MLST data	47
Yu *et al.* (2012)	MLST data	50
Biggs *et al.* (2011)	Full-genome data	30

Discussion

A summary of the different estimates of r/m from the five published papers we have described in this chapter is given in Table 2.2. One of the most striking aspects of these estimates is their degree of inconsistency. Why do we get such different estimates?

There are a number of explanations for the differences in these estimates. Firstly there is natural statistical variability of the estimates. It can be difficult to quantify the degree of uncertainty of estimates of parameters linked to recombination. In particular, the estimates of Schouls *et al.* (2003) included no quantification of uncertainty, and were based on a small dataset. Furthermore, the results of Biggs *et al.* (2011) is based on a small number of actual inferred recombination events, and the estimate of r/m may quite variable due to the potential effect of a small number of large recombination events (which introduce a large number of nucleotide differences). Even in cases where we can quantify uncertainty in estimates, this will be based upon many simplifying assumptions in our model for the data, and the full extent of model errors on estimates is not known.

Secondly, each of the methods uses different types of data and hence may be estimating different features of the recombination process. For example the analysis of Biggs *et al.* (2011) is looking at the full genome, whereas the other analyses just consider data at the MLST loci. The analysis of SLVs is basing inference on differences between closely related isolates, and thus will be affected only by recent mutation and recombination events. Estimates based on population data will be affected by all mutation and recombination events in the ancestry of the sample. If recombination rates have increased recently then we will get higher estimates of recombination rates from using SLVs. Similarly, if there is weak purifying selection, then this may tend to purge the population of recombination events which introduce many changes to the sequence. Such selection will have had more chance to act, and hence have greater effect, on older recombination events; again suggesting higher estimates of recombination rates when you consider data from SLVs.

The full-genome analysis in the previous section, whilst not giving direct estimates of the rate of recombination to mutation, does still highlight the important role that recombination is playing in the evolution of *C. jejuni*. We observe instances where recombination has introduced large changes to the sequences of certain genes. In particular there is evidence for large recombination events (of the order of tens of kilobases), and also for regions of the genome with increased recombination rates. However, it is impossible to draw out strong conclusions from such a limited and more qualitative study as presented here.

Whilst the work reviewed in this chapter gives only a crude picture of the role of recombination in the evolution of *C. jejuni*, the availability of many full-genome sequences suggests that we can start to learn in much more detail about recombination in the coming years. We should be able to see how recombination rates vary around the genome, how the tract length of recombination events vary, what affect does the host-type have on recombination, and maybe start to link whether, and how, the genome sequence itself affects the recombination process. To really address these sorts of question will require the further development of appropriate statistical methods (e.g. Didelot *et al.*, 2010) to analyse the full-genome data that are and will be generated.

References

Biggs, P.J., Fearnhead, P., Hotter, G., Mohan, V., Collins-Emerson, J., Kwan, E., Besser, T.E., Cookson, A., Carter, P.E., and French, N.P. (2011). Whole-Genome Comparison of Two *Campylobacter jejuni* Isolates of the Same Sequence Type Reveals Multiple Loci of Different Ancestral Lineage. PLoS ONE *6*(11), e27121.

Cox, D.R., and Reid, N. (2004). A note on pseudolikelihood constructed from marginal densities. Biometrika *91*(3), 729–737.

Didelot, X., and Falush, D. (2007). Inference of bacterial microevolution using multilocus sequence data. Genetics *175*, 1251–1266.

Didelot, X., Lawson, D., Darling, A., and Falush, D. (2010). Inference of homologous recombination in bacteria using whole-genome sequences. Genetics *186*(4), 1435–1449.

Fearnhead, P., Smith, N.G.C., Barrigas, M., Fox, A., and French, N. (2005). Analysis of recombination in *Campylobacter jejuni* from MLST population data. J. Mol. Evol. *61*, 333–340.

Feil, E.J., Maiden, M.C.J., Achtman, M., and Spratt, B.G. (1999). The relative contributions of recombination and mutation to the divergence of clones of *Neisseria meningitidis*. Mol. Biol. Evol. *16*, 1496–1502.

Feil, E.J., Smith, J.M., Enright, M.C., and Spratt, B.G. (2000). Estimating recombinational parameters in *Streptococcus pneumoniae* from multilocus sequence typing data. Genetics *154*, 1439–1450.

French, N.P., Barrigas, M., Brown, P., Ribiero, P., Williams, N.J., Leatherbarrow, H., Birtles, R., Bolton, E., Fearnhead, P., and Fox, A. (2005). Spatial epidemiology and natural population structure of *Campylobacter jejuni* colonising a farmland ecosystem. Environ. Microbiol. *7*, 1116–1126.

Hudson, R.R. (2001). Two-locus sampling distributions and their application. *Genetics 159*, 1805–1817.

Larribe, F., and Fearnhead, P. (2011). On composite likelihood in statistical genetics. Statistica Sinica *21*, 43–69.

Lef'ebure, T., Pavinski Bitar, P.D., Suzuki, H., and Stanhope, M.J. (2010). Evolutionary dynamics of complete campylobacter pan-genomes and the bacterial species concept. Genome Biol. Evol. 2, 646–655.

Li, N., and Stephens, M. (2003). Modelling LD, and identifying recombination hotspots from SNP data. Genetics *165*, 2213–2233.

Low, K.B., and Porter, D.D. (1978). Modes of gene transfer and recombination in bacteria. Annu. Rev. Genet. *12*, 249–287.

McVean, G.A.T., Myers, S.R., Hunt, S., Deloukas, P., Bentley, D.R., and Donnelly, P. (2004). The fine-scale structure of recombination rate variation in the human genome. Science *304*, 581–584.

Maiden, M.C.J., Bygraves, J.A., Feil, E., Morelli, G., Russell, J.E., Urwin, R., Zhang, Q., Zhou, J., Zurth, K., Caugant, D.A., *et al.* (1998). Multilocus sequence typing: A portable approach to the identification of clones within populations of pathogenic microorganisms. Proc. Natl Acad. Sci. U.S.A. *95*, 3140–3145.

Myers, S., Bottolo, L., Freeman, C., McVean, G.A.T., and Donnelly, P. (2005). A fine-scale map of recombination rates and hotspots across the human genome. Science *310*, 321–324.

Nordborg, M. (2001). Coalescent theory. In: Handbook of Statistical Genetics, Balding, D.J., Bishop, M., and Cannings, C., eds. (John Wiley and Sons, Chichester), pp. 179–212.

Schouls, L.M., Reulen, S., Duim, B., Wagenaar, J.A., Willems, R.J.L., Dingle, K.E., Colles, F.M., and Van

Embden, J.D.A. (2003). Comparative genotyping of *Campylobacter jejuni* by amplified fragment length polymorphism, multilocus sequence typing, and short repeat sequencing: Strain diversity, host range, and recombination. J. Clin. Microbiol. *41*, 15–26.

Sheppard, S.K., McCarthy, N.D., Falush, D., and Maiden, M.C.J. (2008). Convergence of *Campylobacter* species: Implications for bacterial evolution. Science *320*(5873), 237–239.

Smith, N.G.C., and Fearnhead, P. (2005). A comparison of three estimators of the population-scaled recombination rate: accuracy and robustness. Genetics *171*, 2051–62.

Stephens, M. (2001). Inference under the coalescent. In Handbook of Statistical Genetics, Balding, D.J., Bishop, M., and Cannings, C., eds. (John Wiley and Sons, Chichester), pp. 213–238.

Timms, A., Cambray-Young, J., Scott, A., Petty, N., Connerton, P., Clarke, L., Seeger, K., Quail, M., Cummings, N., Maskell, D., *et al.* (2010). Evidence for a lineage of virulent bacteriophages that target *Campylobacter*. BMC Genomics *11*(1), 214.

Varin, C., Reid, N., and Firth, D. (2011). An overview of composite likelihood methods. Statistica Sinica *21*, 5–42.

Vos, M., and Didelot, X. (2009). A comparison of homologous recombination rates in bacteria and archaea. ISME J. *3*, 199–208.

Yu, S., Fearnhead, P., Holland, B., Biggs, P., Maiden, M., and French, N. (2012). Estimating the relative roles of recombination and point mutation in the generation of single locus variants in *Campylobacter jejuni* and *Campylobacter coli*. J. Mol. Evol. *74*.

Within-host Evolution of *Campylobacter jejuni*

3

John P. Jerome and Linda S. Mansfield

Abstract

Pathogens can adapt to the host environment through changes in gene expression or by selection for mutations. For bacterial pathogens within-host gene expression has been studied extensively, but adaptation through mutation during infection has only been lightly researched. However, by comparing ancestral and host-passaged populations, researchers have shown that *C. jejuni* is able to rapidly adapt during infection by heritable changes. Host passage of *C. jejuni* often leads to increased flagellar motility – an important phenotype that enhances *C. jejuni*–host interactions. In addition, host passage results in *C. jejuni* with increased rates of colonization, and with higher virulence, during subsequent host infections. Recently, homopolymeric DNA tract mutations, and large-scale genomic rearrangements, have been implicated as potential mechanisms for these phenotypic changes. In the future, advances in sequencing technologies and predictive models of homopolymeric DNA mutability will help to elucidate the mechanisms underlying the ability of *C. jejuni* to rapidly adapt by genetic change during host infection.

Introduction

Campylobacter jejuni can colonize a broad range of animal hosts. Based on multilocus sequence typing some researchers have argued that natural *C. jejuni* populations have host-specific genotypic adaptations indicative of a long-term association with specific host species (Falush, 2009; Manning *et al.*, 2003; McCarthy *et al.*, 2007). However, based on extensive phenotypic and genetic assays of isolates within a single, large multilocus sequence type, others have suggested *C. jejuni* may have an opportunistic, or generalist lifestyle (Gripp *et al.*, 2011). Irrespective of the long-term evolutionary history of *C. jejuni*, direct comparisons of ancestral and host-adapted *C. jejuni* populations suggest an ability to adapt during host infection through heritable changes.

Heritable adaptation is a stochastic process since it is dependent on the generation of random mutations for adaptive change to occur by natural selection (Moxon *et al.*, 1994). This is distinct from deterministic adaptation that is prescriptive through gene regulation in response to environmental cues (Moxon *et al.*, 2006). Describing the prescriptive responses of pathogenic bacteria by classical gene regulation has been a main focus of infectious disease microbiology (DiRita and Mekalanos, 1989; Miller *et al.*, 1989), and for *C. jejuni* pathogenesis (Gilbreath *et al.*, 2011). However, recent work with *C. jejuni* has also highlighted the ability to adapt to novel environments by mutational mechanisms. This review will focus on the phenotypic and genotypic changes that have been observed in host-passaged *C. jejuni*.

Several studies have documented increased fitness of specific *C. jejuni* isolates after an adaptive period in the host. These studies have demonstrated fitness increases for the homologous host, and in some cases, for divergent hosts with different physiologies. For example, in a chicken colonization model, Ringoir and Korolik (2003) showed that *C. jejuni* which had been re-isolated after a single chicken passage was able to stably colonize subsequent chickens with a 1000-fold lower dose than the ancestral inoculum. In addition, Jones *et al.* (2004) reported that *C. jejuni* adapted from low to high-level colonization during a 5-week colonization period within the chicken host. Furthermore, when the C57 BL/6 IL-10-deficient mouse model of campylobacteriosis was developed (Mansfield *et al.*, 2007), it was noted that a single passage of *C. jejuni* resulted in a variant that was better able to colonize and elicit an inflammatory response compared to the ancestral, unpassaged *C. jejuni* (Bell *et al.*, 2009). Kim *et al.* (2012) then showed that passage through chickens led to an increased ability to colonize mice, suggesting that in some cases the increased fitness after host passage is not due to narrow host-specific adaptations. The hosts used in this study – a bird and a mammal – had major differences in host anatomy and physiology including the nature of hind gut fermentation. All of these phenotypic observations suggested that *C. jejuni* was able to adapt through heritable changes during host infection, but data concerning the genetic basis of this phenomenon was lacking.

The first genetic evidence that *C. jejuni* could evolve during host infection came from a bacterial genotyping technique used to investigate *C. jejuni* epidemiology: pulse-field gel electrophoresis (PFGE). It was shown that in some instances the PFGE banding pattern was altered during an initially clonal *C. jejuni* infection of chickens (de Boer *et al.*, 2002; Wassenaar *et al.*, 1998). These results suggested that large-scale genome rearrangements had occurred during host passage. Although these first PFGE studies of host-passaged *C. jejuni* variants did not link the changes in PFGE profile to increased fitness after passage, they did demonstrate that the *C. jejuni* genome was not static during host infection.

Interrogation of genetic adaptations across the entire genome has been challenging. Only with the recent advent of new sequencing technologies has the ability to sufficiently probe the entire genome of ancestral and host-passaged *C. jejuni* variants become feasible. Well-designed sequencing studies of ancestral and host-passaged isolates are beginning to shed light on the genetic mechanisms that underlie the ability of *C. jejuni* to rapidly adapt by mutations during infection. These studies and others have so far implicated homopolymeric DNA – the simplest form of simple sequence DNA repeats (SSRs) – and large-scale genome rearrangements, as possible causes for the rapid mutation-based adaptation of *C. jejuni* during infection. The specific phenotypic changes that have been observed after within-host evolution of *C. jejuni*, and their potential genetic basis are discussed below.

Phenotypic changes of *Campylobacter jejuni* after host passage

Motility and a greater ability for host colonization

Researchers have noted changes in *C. jejuni* phenotypes after passage through chickens (Jones *et al.*, 2004; Sang *et al.*, 1989), mice (Bell *et al.*, 2009), rabbits (Caldwell *et al.*, 1985), and humans (Black *et al.*, 1988). In most of these studies it was found that the motility of *C. jejuni* increased during host passage. Most often higher rates of motility (higher frequency of highly motile cells) in the *C. jejuni* population recovered post infection are observed. In previous *C. jejuni* pathogenesis studies there was direct evidence that non-flagellated, non-motile mutants are colonization deficient in chickens, piglets, mice and humans (Black *et al.*, 1988; Diker *et al.*, 1992; Hendrixson, 2006; Josenhans and Suerbaum, 2002; Malik-Kale *et al.*, 2007). Thus, it is not surprising that flagellar motility is selected during host passage because of its key role in maintaining place in the host and in enhancing bacterial host interactions.

The genetic basis of the motility increase is unknown, but hypermutable homopolymeric DNA within some motility accessory factor (*maf*) genes may play a role (Karlyshev *et al.*, 2002a). In addition, homopolymeric DNA tracts within a two-component system necessary for flagellar biosynthesis, FlgSR, have been shown to result in motility phase variation and the ability to reach higher colonization loads during chicken colonization (Hendrixson, 2006, 2008). It is likely that in some instances standing variation in motility, as has been observed during soft agar spreading assays (Karlyshev *et al.*, 2002b), is present in the *C. jejuni* population during inoculation and is then selected during passage. For example, in our lab some of the stock cultures of *Campylobacter jejuni* NCTC11168, originally obtained from the American Type Culture Collection (ATCC 700819), are composed of both motile and non-motile cells (Jerome and Mansfield, unpublished).

Given the necessity of motility for host colonization it is not surprising that motile or hypermotile variants are selected by host passage. Although not tested, it is possible that selection for motility also explains the observation that chicken passage generates a *C. jejuni* variant with approximately 1000-fold lower infectious dose to yield sustained colonization (Ringoir and Korolik, 2003). That is, if only a fraction of *C. jejuni* cells are motile before passage then a higher dose will be needed for colonization compared to the host-passaged *C. jejuni* population that is mostly, or completely, made up of motile cells. Much of the work described here, suggests that the increase in motility that occurs during host passage may be the main reason that many researchers also observe an increase in colonization ability.

From our work using the C57BL/6 IL-10-deficient mouse model, nearly all recovered cells from the re-isolated *C. jejuni* population after a single 35 day *in vivo* passage are motile (J.P. Jerome, unpublished). However, when those re-isolated motile populations are cultured in broth media, non-motile cells become detectable after just three 24 hour passages. This suggests that although motility is necessary and selected for in the mouse model, the opposite is true during growth in liquid media. From whole genome sequencing we found that the loss in motility during broth experimental evolution is attributable to multiple types of mutations, including many frameshifting indel mutations in homopolymeric adenine and thymine tracts, and including some mutations that have been described previously (Hendrixson, 2006, 2008).

Altered host–pathogen interactions

Serial passage through mice leads to an increased virulence phenotype of *C. jejuni* in a C57BL/6 IL-10-deficient mouse model of campylobacteriosis (Bell *et al.*, 2009). Passaged *C. jejuni* strains included NCTC11168 (ATCC 700819), which was isolated in the 1970s, sequenced in 2000, and is widely used in *Campylobacter* research (Parkhill *et al.*, 2000), along with four recent clinical isolates of *C. jejuni* from patients with gastroenteritis: D0835, D2586, D2600 and NW. After three rounds of passage there was an increase in faecal *C. jejuni* populations (NW and 11168), and an increase in the number of mice harbouring *C. jejuni* in the jejunum (D0835 and 11168). Survivorship of infected mice was decreased by host passage, and there was an increase in the fraction of mice exhibiting gross pathological changes, and histopathological changes in the gut (Bell *et al.*, 2009). Collectively these results suggested that some passaged *C. jejuni* strains had increased fitness (based on colonization levels) and increased virulence (based on host damage) as a result of serial passage, which is consistent with numerous other pathogen serial passage experiments (Ebert, 1998). Surprisingly, the genetic basis of adaptation in this case was solely attributable to changes in homopolymeric DNA tracts (Jerome *et al.*, 2011). The role of these tracts for within-host evolution of *C. jejuni* is further discussed below.

Aside from an increased virulence phenotype, it has been documented that antigenic variation in lipooligosaccharide (LOS) modifications may occur in *C. jejuni* during infection of humans. Prendergast *et al.* (2004) noted that *C. jejuni* strain 81-176 inoculated into human volunteers initially lacked ganglioside mimics in the LOS that are linked to the development of the post-infection neuropathy, Guillain–Barré syndrome (GBS). However, *C. jejuni* that had been recovered from volunteers had undergone antigenic variation to express multiple ganglioside mimics. It was previously shown that strain 81-176 was capable of producing various potential ganglioside mimics since some genes responsible for LOS modification were variably expressed due to the presence of mutable homopolymeric DNA within their open reading frames (Guerry *et al.*, 2002). Though the hypothesis was not directly tested or considered in the study by Prendergast *et al.*, it seems likely that the variation in LOS antigenicity that occurred during human infection was driven by mutations in these LOS biosynthesis genes containing homopolymeric tracts of DNA. Future work should focus on understanding why certain ganglioside mimicking LOS structures appear to have been selected by passage through humans, as this may have implications for the pathogenesis of *C. jejuni*-induced GBS.

Genotypic changes of *Campylobacter jejuni* during host passage

Few studies have described in detail the genetic changes that occur during *C. jejuni* host passage. However, with the relative ease and low cost of new sequencing technologies, using host passage as a forward genetic screen will likely become more common. This should lead to more analysis of the genetic basis of adaptation by *C. jejuni* during host passage that is contributing to the phenotypic changes described above. As should be expected considering the phenotypic changes noted, the *C. jejuni* genome is not static during infection. Evidence for genotypic change during host passage comes from pulse field gel electrophoresis (PFGE) experiments (de Boer *et al.*, 2002; Scott *et al.*, 2007; Wassenaar *et al.*, 1998), chicken co-infection of two antibiotic resistance marked strains (de Boer *et al.*, 2002), targeted sequencing of chicken passaged isolates (Bayliss *et al.*, 2012; Kim *et al.*, 2012; Wilson *et al.*, 2010), and whole genome resequencing of mouse passaged *C. jejuni* (Jerome *et al.*, 2011). The challenge when considering such forward genetic screens in *C. jejuni* will be the follow-up documentation of contributing genes when multiple genes act in concert, and some but not all genes with heritable changes are expected to be necessary and sufficient to produce the phenotype. It is likely that standard complementation following gene knockouts will not suffice to provide these proofs.

Intra- and intergenomic rearrangements

Multiple studies have shown that *C. jejuni* is capable of intragenomic recombination during chicken colonization (de Boer *et al.*, 2002; Nuijten *et al.*, 2000; Scott *et al.*, 2007; Wassenaar *et al.*, 1998), and in some cases the functional consequences of DNA rearrangements have been reported. Ridley *et al.* showed that chicken passaged *C. jejuni* isolates with different PFGE profiles had differential fitness during *in vivo* competition experiments (Ridley *et al.*, 2008). This suggested an association of the altered PFGE profiles to colonization phenotype. Furthermore, Scott *et al.* showed that a rearrangement of the *C. jejuni* genome occurred during chicken colonization between prophage DNA sequences. These genomic changes, involving approximately one-third of the genome, were associated with the avoidance of phage predation and the generation of infectious bacteriophage (Scott *et al.*, 2007). Interestingly, although *C. jejuni* with the rearranged genome was more resistant to phage attack, it had a decreased ability to colonize chickens, suggesting a trade-off between phage resistance and the ability to colonize. In addition, Nuijten *et al.* documented another example of functional DNA rearrangements that occurred during chicken colonization (Nuijten *et al.*, 2000). This group showed that deletions in *flaA*, which is necessary for flagellar motility, could be repaired through rearrangements with the *flaB* duplication located immediately downstream during chicken colonization. The importance of flagellar motility for *C. jejuni* colonization is undisputed, so there is a great selective pressure to regain motility during chicken colonization. The restoration of a functional *flaA* likely represents one possible reason for DNA mutation repair in *C. jejuni* occurring via intragenomic recombination, and is another mechanism by which *C. jejuni* has been shown to adapt during infection. Taken together these studies suggest that the *C. jejuni* genome is constantly changing via mutation and intragenomic recombination during chicken colonization to generate genomic diversity for rapid genetic adaptation. Alternatively, intragenomic recombination may play a role in the stability of functional DNA sequences in the genome, since *C. jejuni* lacks some DNA mutation repair genes present in some other bacteria (Gaasbeek *et al.*, 2009). More work is needed to determine the extent of intragenomic recombination that occurs during host colonization, and the mechanisms by which these changes contribute to within-host adaptation and/or DNA repair of the *C. jejuni* genome.

Other studies have further examined the potential for *C. jejuni* to generate genetic diversity through intergenomic recombination during chicken colonization. It has been known for some time that *C. jejuni* is naturally transformable (Wang and Taylor, 1990), but the importance of this phenotype during host infection had not been assessed. De Boer *et al.* created two *C. jejuni* strains that were distinguishable by different antibiotic resistance genes, but otherwise isogenic (de Boer *et al.*, 2002). When these strains were used to co-colonize chickens, strains were recovered after colonization that contained both antibiotic resistance genes. Based on this finding the authors concluded that *C. jejuni* inter-strain recombination occurred *in vivo*, which should be considered another potential source of genomic diversity. As in other bacteria, natural transformation in *C. jejuni* occurs preferentially between similar genotypes. That is, natural transformation is most efficient between very similar (or identical) DNA sequences, and not very efficient when incorporating DNA from other species into *C. jejuni* (Wang and Taylor, 1990). In the work by de Boer *et al.*(2002), the only difference between the genomes of the experimental *C. jejuni* strains was in a single antibiotic resistance gene, and the rate of recombination between more distant *C. jejuni* strains, or other *Campylobacter* species, during co-colonization was not tested. These studies clearly show that *C. jejuni* is capable of both intra- and inter- genomic recombination during infection. These are two potential genetic mechanisms by which the *C. jejuni* genome may adapt through heritable changes during infection.

Hypermutable contingency loci and other mutable homopolymeric DNA

Mutability of *C. jejuni* homopolymeric DNA was first discovered when Parkhill and colleagues sequenced the genome of strain NCTC11168 (Parkhill *et al.*, 2000). This genome sequence was published under a title that highlighted the presence of hypervariable tracts of homopolymeric DNA (Parkhill *et al.*, 2000). The authors defined these tracts as variable since traditional Sanger genome sequencing could not resolve their length to one specific value – indicative of a high rate of insertion and deletion (indel) mutations accumulating during routine laboratory culture. Most of the mutations described by Parkhill *et al.* occurred in 8–12 base tracts of homopolymeric cytosine/guanine DNA (polyC/G tracts), but one variable homopolymeric adenine/thymine (polyA/T) tract was also reported. Subsequent analyses by whole genome sequencing (WGS) (Fouts *et al.*, 2005; Jerome *et al.*, 2011) and direct length analysis of PCR products (Bayliss *et al.*, 2012; Kim *et al.*, 2012; Wassenaar *et al.*, 2002), have confirmed that polyC/G tracts are unstable due to indel mutations in *C. jejuni*. In addition, multiple researchers have since described instances of variation in *C. jejuni* polyA/T tracts (Hendrixson, 2006, 2008; Hyytiainen and Hanninen, 2012; Jerome *et al.*, 2011; Wiesner *et al.*, 2003).

Loci containing mutable homopolymeric tracts in *C. jejuni* may be considered contingency loci because of their increased rate of mutation relative to the average rate across the genome (Moxon *et al.*, 1994). In *C. jejuni*, homopolymeric DNA tracts are most often within the open-reading frame of a contingency locus and dictate gene expression (ON versus OFF phases) through frameshifting indel mutations. Since homopolymeric DNA in *C. jejuni* is highly mutable, it may be a source for rapid genetic adaptation. Indeed, evidence is accumulating that shows homopolymeric DNA mutations are one of the main drivers of *C. jejuni* genetic adaptation during infection (Jerome *et al.*, 2011; Bayliss *et al.*, 2012; Kim *et al.*, 2012).

PolyC/G tract length distributions of contingency loci may change as a result of host passage (Wilson *et al.*, 2010; Jerome *et al.*, 2011; Bayliss *et al.*, 2012; Kim *et al.*, 2012). Genome sequencing of mouse passaged *C. jejuni*, and targeted polyC/G tract analyses of chicken passaged *C. jejuni* have shown changes in the frequency of ON and OFF phases in contingency loci (Jerome *et al.*, 2011; Kim *et al.*, 2012; Wilson *et al.*, 2010). In a mouse model of campylobacteriosis the contingency loci changes that occurred during infection were associated with increased virulence; passage of the same *C. jejuni* strain in the model resulted in increased inflammation of the gastrointestinal (GI) tract (Bell *et al.*, 2009). Many of the observed changes occurred in a locus involved in flagellar glycosylation, but the molecular implications of these changes are still unknown. In another study, changes in *C. jejuni* polyC/G tracts occurring during chicken passage were associated with the successful colonization of mice relative to the unpassaged control that colonized poorly (Kim *et al.*, 2012). This suggests that the microevolution of contingency loci in the avian intestine, a natural reservoir of *C. jejuni*, may promote subsequent infection of different hosts such as mice, or potentially, humans. However, it is not clear whether all the changes to the genome of the chicken passaged *C. jejuni* were discovered since the genome was not sequenced.

The observed changes in polyC/G length distributions during animal infection suggest that specific phase variants have differential fitness *in vivo*. However, alternative mechanisms leading to these changes could also include genetic drift due to polyC/G mutation rates and patterns, bottlenecks of *C. jejuni* cells during initial colonization or by active immune system clearance, epistasis in a genome with the potential to encode millions of genotypes ($2^{\text{no. of contingency loci}}$), and/or genetic hitchhiking. In chickens and mice, the changes in contingency loci that occur during gut passage are not explained by mathematical models of polyC/G genetic drift (Bayliss *et al.*, 2012; J.P. Jerome, unpublished). Also, in a mouse serial passage experiment, hitchhiking did not seem to drive the observed contingency loci changes, since the differences in polyC/G tract distributions between ancestral and mouse-adapted *C. jejuni* were the only genetic changes detectable by whole genome sequencing, expression microarray, and pulse-field gel electrophoresis (Jerome *et al.*, 2011). In addition, when resequencing of *C. jejuni* populations isolated from multiple independent infected mice, many polyC/G tract mutations changed in parallel (J.P. Jerome, unpublished), providing further evidence that these mutations are under selection. However, some combination of these effects may contribute to the variability in contingency locus changes sometimes observed between replicate experimental animals (Bayliss *et al.*, 2012; Kim *et al.*, 2012; J.P. Jerome, unpublished). The combined use of next-generation sequencing technologies, improved mathematical models that incorporate bottleneck effects, and targeted analysis of contingency loci dynamics during controlled evolution experiments, should be able to better elucidate the evolutionary forces driving the observed polyC/G tract changes during host infection.

Aside from polyC/G tract containing contingency loci, researchers have observed polyA/T tract changes during host passage. Hendrixson showed that a non-motile, non-flagellated *C. jejuni* isolate restored motility during chick passage, which resulted in increased bacterial loads in the caeca (Hendrixson, 2006). In this study the genetic basis for the switch from the non-motile to motile phenotype was found to be due to polyA/T tract phase variation in *flgR* – one part of a two-component system necessary for flagellar biosynthesis. Indeed, multiple researchers have noted polyA/T tract mutations in the genome of different isolates of the same *C. jejuni* strain (Hyytiainen and Hanninen, 2012; Jerome *et al.*, 2011; Wiesner *et al.*, 2003). Our laboratory has also seen numerous instances of polyA/T tract mutations that accumulate during *C. jejuni* experimental evolution in broth media (J.P. Jerome, unpublished). Thus, polyA/T tract mutations are another mechanism by which *C. jejuni* may adapt during host passage.

Conclusions

Multiple researchers have shown that when *C. jejuni* infects a host, heritable phenotypic changes may occur. The phenomenon most often observed is increases in flagellar motility after host passage. This phenotype appears to be selected *in vivo* and quickly lost during *in vitro* culture (J.P. Jerome, unpublished). In addition to increased motility, passaging *C. jejuni* has been shown to result in a greater potential to colonize subsequent hosts, in altered antigenic properties, and in some cases in an increase in virulence. Thus far, studies to determine the genetic bases of these phenotypic changes have implicated homopolymeric DNA tract changes and genomic rearrangements.

Future studies

Future efforts should be made to understand the importance of rapid, mutation-based adaptation for *C. jejuni* pathogenesis. Although it has been shown that phenotypic and genotypic changes occur during host passage, in most cases the effects of change(s) in a specific gene on *C. jejuni* fitness and virulence have not been shown. Furthermore, the introductory studies of *C. jejuni* evolvability have revealed that multiple genes may be involved, and that each strain will have numerous different genes with homopolymeric tracts that can undergo rapid change. Thus, because of the mutability of homopolymeric DNA in a variety of open reading frames and rapid genomic rearrangement, researchers should consider a single *C. jejuni* strain to encode multiple genotypes/phenotypes. That is, a single isolate may quickly diversify into a population of genotypes and phenotypes during growth in the laboratory or within a host. Understanding how this diversity responds to selection and other evolutionary forces will be of utmost importance for future studies of *C. jejuni* pathogenesis and evolution.

The mutability of the *C. jejuni* genome also has implications for assessing the importance of genetic factors in *C. jejuni* through traditional molecular cloning techniques to satisfy Falkow's molecular Koch's postulates (Falkow, 1988). Researchers need always be aware of the high probability that a defined mutant strain will not be otherwise isogenic to the wild-type strain from which it was derived. This issue was made evident by Gaynor *et al.* in 2004 by an extensive investigation of the virulence differences in the original clinical isolate of strain NCTC11168 (11168-O) and the genome sequenced variant (11168-GS) (Gaynor *et al.*, 2004). Subsequently, other works have shown differences in genotype and phenotype of the NCTC11168 strain being used in distinct laboratories (Cooper *et al.*, 2013; Jerome *et al.*, 2011). The instability of the *C. jejuni* genome makes mutant complementation important for reliably assessing the necessity of *C. jejuni* genes for a phenotype of interest, but at the same time, this instability will also make complementation difficult. Essentially this leads to a catch-22 situation that may make robust *C. jejuni* gene knockout studies difficult.

Recent work has implicated mutations in homopolymeric DNA tracts of *C. jejuni* contingency loci to confer the ability to rapidly adapt through heritable changes during infection. Still, much about the role of contingency loci in *C. jejuni* biology is unknown. Although we have presented compelling evidence that the ON/OFF frequencies of multiple contingency loci change during mouse passage (Jerome *et al.*, 2011), and others have described similar changes during chicken passage (Bayliss *et al.*, 2012; Kim *et al.*, 2012), we still do not know why hypermutability in these specific loci has arisen during the course of *C. jejuni* evolutionary history. Based on the genomic locations, and *in silico* predicted functions of most contingency loci in *C. jejuni*, surface carbohydrate structures are most affected by contingency loci mutability.

Did mutability via contingency loci evolve as a mechanism for host immune system evasion? *C. jejuni* only replicates naturally within a host. To remain within a host to replicate, *C. jejuni* must evade clearance by the immune system. Therefore, the ability to present variable sets of antigens on the cell surface via contingency loci mutations might be advantageous. If the immune system recognizes specific *C. jejuni* surface structures that are phase-variable, then subpopulations of cells with mutations that turn the relevant genes OFF may go unrecognized, and proliferate within the host.

It may be possible to screen for contingency loci that are important for immune system evasion in mice by studying the populations of *C. jejuni* cells present during the time course of infection in wild-type and immune system deficient (e.g. recombination activating gene(s) (RAG)-deficient) mice. For example, if a specific contingency locus was responsible for expression of an antigen that is recognized by the immune system, then it would be expected that *C. jejuni* cells with that gene in the OFF phase would be selected during infection and increase in frequency during the wild-type host's immune response. However, the same pattern would not occur in mice deficient in immune system activity. Contingency genes that appear to be important could be analysed further in these mouse models by targeted mutations that disrupt mutability of the gene in question.

High rates and non-random patterns of homopolymeric DNA mutations alone may lead to the changes in contingency loci ON/OFF expression patterns observed during experimental host adaptation. Also, it is possible that these changes might be interpreted as being the result of selective pressures in the host, when they are actually just an effect of genetic drift. To understand the importance of drift for contingency loci evolution, Bayliss *et al.* developed a mathematical model of contingency loci mutability (Bayliss *et al.*, 2012). Their model predicts the outcome of effects due to the mutation rates and patterns of homopolymeric DNA in *C. jejuni* contingency loci in a dividing population. Using this model they showed that changes in *C. jejuni* contingency loci that occur during chicken passage are not explained by the model of genetic drift, suggesting that selection is indeed acting on these loci (Bayliss *et al.*, 2012). Furthermore, this model does not explain the patterns observed to occur during C57BL/6 IL-10-deficient mouse passage (J.P. Jerome, unpublished). Using mathematical modelling to understand the genetic drift of homopolymeric DNA mutations is an important step towards elucidating the importance of contingency loci mutations for *C. jejuni* within-host evolution.

Future modelling efforts should be directed at incorporating more evolutionary forces that may influence the microevolution of homopolymeric DNA to fully understand the importance of contingency loci for *C. jejuni* host adaptation. For example, a bottleneck effect during initial colonization whereby only a small number of *C. jejuni* cells successfully establish colonization may result in significant contingency loci ON/OFF expression pattern changes in re-isolated populations. A bottleneck effect would cause random contingency loci mutations to increase and decrease in frequency, such that the same changes would not be predicted to occur in multiple experimentally infected animals. In our recent work we have shown that parallel changes in contingency loci occur in multiple mice. As an example, there is a significant increase in frequency of the ON phase for contingency locus *Cj1295* in all mice analysed (J.P. Jerome, unpublished). It is likely that mathematical models that incorporate effects from evolutionary forces such as a colonization bottleneck, along with analyses of multiple, independent experimental animals, will be important for future research in *C. jejuni* experimental host-adaptation studies.

It is very difficult to rule out the possibility that some of the associations made between contingency loci mutations and *C. jejuni* phenotypic changes after passage are due to genetic hitchhiking. That is, it is possible that many of the observed contingency locus changes are hitchhiking on other beneficial mutations. In our work we could not detect mutations in mouse-adapted populations that were outside of contingency loci (Jerome *et al.*, 2011), but our analyses are only as good as the current state of technology. As more advanced sequencing techniques are realized, currently undetectable mutations might be found.

Throughout years of research, anecdotal evidence had accumulated to suggest the *C. jejuni* genome was highly mutable. Initially researchers noted phenotypic changes in *C. jejuni* re-isolated after host passage, which suggested heritable adaptations to the host environment were occurring during infection. Thus far, genome rearrangements and homopolymeric DNA tract mutations have been associated with these phenotypic changes. Technological advances in genome sequencing and evolutionary modelling are expected to continue to provide researchers with new tools as they seek to understand the evolvability of the *C. jejuni* genome.

Acknowledgements

The work by Jerome and Mansfield in this chapter was supported by a U19AI090872 Enterics Research Investigational Network, Cooperative Research Center (ERIN CRC) grant to Linda S. Mansfield, Principal Investigator at Michigan State University from the National Institute of Allergy and Infectious Disease at the National Institutes of Health. Furthermore, this material is based in part upon work supported by the National Science Foundation under Cooperative Agreement No. DBI-0939454. Any opinions, findings, and conclusions or recommendations expressed in this material are those of the authors and do not necessarily reflect the views of the National Science Foundation. Dr Jerome's stipend was paid in part by matching grants to the ERIN CRC from the College of Veterinary Medicine and AgBioResearch at Michigan State University.

References

Bayliss, C.D., Bidmos, F.A., Anjum, A., Manchev, V.T., Richards, R.L., Grossier, J.P., Wooldridge, K.G., Ketley, J.M., Barrow, P.A., Jones, M.A., *et al.* (2012). Phase variable genes of *Campylobacter jejuni* exhibit high mutation rates and specific mutational patterns but mutability is not the major determinant of population structure during host colonization. Nucleic Acids Res. *40*, 5876–5889.

Bell, J.A., St Charles, J.L., Murphy, A.J., Rathinam, V.A., Plovanich-Jones, A.E., Stanley, E.L., Wolf, J.E., Gettings, J.R., Whittam, T.S., and Mansfield, L.S. (2009). Multiple factors interact to produce responses resembling spectrum of human disease in *Campylobacter jejuni* infected C57BL/6 IL-10–/– mice. BMC Microbiol. *9*, 57.

Black, R.E., Levine, M.M., Clements, M.L., Hughes, T.P., and Blaser, M.J. (1988). Experimental *Campylobacter jejuni* infection in humans. J. Infect. Dis. *157*, 472–479.

de Boer, P., Wagenaar, J.A., Achterberg, R.P., van Putten, J.P.M., Schouls, L.M., and Duim, B. (2002). Generation of *Campylobacter jejuni* genetic diversity in vivo. Mol. Microbiol. *44*, 351–359.

Caldwell, M.B., Guerry, P., Lee, E.C., Burans, J.P., and Walker, R.I. (1985). Reversible expression of flagella in *Campylobacter jejuni*. Infect. Immun. *50*, 941–943.

Cooper, K.K., Cooper, M.A., Zuccolo, A., and Joens, L.A. (2013). Re-sequencing of a virulent strain of *Campylobacter jejuni* NCTC11168 reveals potential virulence factors. Res. Microbiol. *164*, 6–11.

Diker, K.S., Hascelik, G., and Diker, S. (1992). Colonization of infant mice with flagellar variants of *Campylobacter jejuni*. Acta Microbiol. Hung. *39*, 133–136.

DiRita, V.J., and Mekalanos, J.J. (1989). Genetic regulation of bacterial virulence. Annu. Rev. Genet. *23*, 455–482.

Ebert, D. (1998). Experimental evolution of parasites. Science *282*, 1432–1435.

Falkow, S. (1988). Molecular Koch's postulates applied to microbial pathogenicity. Rev. Infect. Dis. *10* (Suppl. 2), S274–276.

Falush, D. (2009). Toward the use of genomics to study microevolutionary change in bacteria. PLoS Genet. *5*, e1000627.

Fouts, D.E., Mongodin, E.F., Mandrell, R.E., Miller, W.G., Rasko, D.A., Ravel, J., Brinkac, L.M., DeBoy, R.T., Parker, C.T., Daugherty, S.C., *et al.* (2005). Major structural differences and novel potential virulence mechanisms from the genomes of multiple *Campylobacter* species. PLoS Biol. *3*, e15.

Gaasbeek, E.J., van der Wal, F.J., van Putten, J.P., de Boer, P., van der Graaf-van Bloois, L., de Boer, A.G., Vermaning, B.J., and Wagenaar, J.A. (2009). Functional characterization of excision repair and RecA-dependent recombinational DNA repair in *Campylobacter jejuni*. J. Bacteriol. *191*, 3785–3793.

Gaynor, E.C., Cawthraw, S., Manning, G., MacKichan, J.K., Falkow, S., and Newell, D.G. (2004). The genome-sequenced variant of *Campylobacter jejuni* NCTC 11168 and the original clonal clinical isolate differ markedly in colonization, gene expression, and virulence-associated phenotypes. J. Bacteriol. *186*, 503–517.

Gilbreath, J.J., Cody, W.L., Merrell, D.S., and Hendrixson, D.R. (2011). Change is good: variations in common biological mechanisms in the epsilonproteobacterial genera *Campylobacter* and *Helicobacter*. Microbiol. Mol. Biol. Rev. *75*, 84–132.

Gripp, E., Hlahla, D., Didelot, X., Kops, F., Maurischat, S., Tedin, K., Alter, T., Ellerbroek, L., Schreiber, K., Schomburg, D., *et al.* (2011). Closely related *Campylobacter jejuni* strains from different sources reveal a generalist rather than a specialist lifestyle. BMC Genom. *12*, 584.

Guerry, P., Szymanski, C.M., Prendergast, M.M., Hickey, T.E., Ewing, C.P., Pattarini, D.L., and Moran, A.P. (2002). Phase variation of *Campylobacter jejuni* 81-176 lipooligosaccharide affects ganglioside mimicry and invasiveness *in vitro*. Infect. Immun. *70*, 787–793.

Hendrixson, D.R. (2006). A phase-variable mechanism controlling the *Campylobacter jejuni* FlgR response regulator influences commensalism. Mol. Microbiol. *61*, 1646–1659.

Hendrixson, D.R. (2008). Restoration of flagellar biosynthesis by varied mutational events in *Campylobacter jejuni*. Mol. Microbiol. *70*, 519–536.

Hyytiainen, H., and Hanninen, M.L. (2012). Quality control strain *Campylobacter jejuni* ATCC 33560 contains a frameshift mutation in the CmeR regulator. Antimicrob. Agents Chemother. *56*, 1148.

Jerome, J.P., Bell, J.A., Plovanich-Jones, A.E., Barrick, J.E., Brown, C.T., and Mansfield, L.S. (2011). Standing genetic variation in contingency loci drives the rapid adaptation of *Campylobacter jejuni* to a novel host. PLoS One *6*, e16399.

Jones, M.A., Marston, K.L., Woodall, C.A., Maskell, D.J., Linton, D., Karlyshev, A.V., Dorrell, N., Wren, B.W., and Barrow, P.A. (2004). Adaptation of *Campylobacter jejuni* NCTC11168 to high-level colonization of the avian gastrointestinal tract. Infect. Immun. *72*, 3769–3776.

Josenhans, C., and Suerbaum, S. (2002). The role of motility as a virulence factor in bacteria. Int. J. Med. Microbiol. *291*, 605–614.

Karlyshev, A.V., Linton, D., Gregson, N.A., and Wren, B.W. (2002a). A novel paralogous gene family involved in phase-variable flagella-mediated motility in *Campylobacter jejuni*. Microbiology *148*, 473–480.

Karlyshev, A.V., Linton, D., Gregson, N.A., and Wren, B.W. (2002b). A novel paralogous gene family involved in phase-variable flagella-mediated motility in *Campylobacter jejuni*. Microbiol.-Sgm. *148*, 473–480.

Kim, J.S., Artymovich, K.A., Hall, D.F., Smith, E.J., Fulton, R., Bell, J., Dybas, L., Mansfield, L.S., Tempelman, R., Wilson, D., *et al.* (2012). Passage of *Campylobacter jejuni* through the chicken reservoir or mice promotes phase variation in contingency genes Cj0045 and Cj0170 that strongly associates with colonization and disease in a mouse model. Microbiology *158*, 1304–1316.

McCarthy, N.D., Colles, F.M., Dingle, K.E., Bagnall, M.C., Manning, G., Maiden, M.C., and Falush, D. (2007). Host-associated genetic import in *Campylobacter jejuni*. Emerg. Infect. Dis. *13*, 267–272.

Malik-Kale, P., Raphael, B.H., Parker, C.T., Joens, L.A., Klena, J.D., Quinones, B., Keech, A.M., and Konkel, M.E. (2007). Characterization of genetically matched isolates of *Campylobacter jejuni* reveals that mutations in genes involved in flagellar biosynthesis alter the organism's virulence potential. Appl. Environ. Microbiol. *73*, 3123–3136.

Manning, G., Dowson, C.G., Bagnall, M.C., Ahmed, I.H., West, M., and Newell, D.G. (2003). Multilocus sequence typing for comparison of veterinary and human isolates of *Campylobacter jejuni*. Appl. Environ. Microbiol. *69*, 6370–6379.

Mansfield, L.S., Bell, J.A., Wilson, D.L., Murphy, A.J., Elsheikha, H.M., Rathinam, V.A., Fierro, B.R., Linz, J.E., and Young, V.B. (2007). C57BL/6 and congenic interleukin-10-deficient mice can serve as models of *Campylobacter jejuni* colonization and enteritis. Infect. Immun. *75*, 1099–1115.

Miller, J.F., Mekalanos, J.J., and Falkow, S. (1989). Coordinate regulation and sensory transduction in the control of bacterial virulence. Science *243*, 916–922.

Moxon, E.R., Rainey, P.B., Nowak, M.A., and Lenski, R.E. (1994). Adaptive evolution of highly mutable loci in pathogenic bacteria. Curr. Biol. *4*, 24–33.

Moxon, R., Bayliss, C., and Hood, D. (2006). Bacterial contingency loci: the role of simple sequence DNA repeats in bacterial adaptation. Annu. Rev. Genet. *40*, 307–333.

Nuijten, P.J., van den Berg, A.J., Formentini, I., van der Zeijst, B.A., and Jacobs, A.A. (2000). DNA rearrangements in the flagellin locus of an flaA mutant of *Campylobacter jejuni* during colonization of chicken ceca. Infect. Immun. *68*, 7137–7140.

Parkhill, J., Wren, B.W., Mungall, K., Ketley, J.M., Churcher, C., Basham, D., Chillingworth, T., Davies, R.M., Feltwell, T., Holroyd, S., *et al.* (2000). The genome sequence of the food-borne pathogen *Campylobacter jejuni* reveals hypervariable sequences. Nature *403*, 665–668.

Prendergast, M.M., Tribble, D.R., Baqar, S., Scott, D.A., Ferris, J.A., Walker, R.I., and Moran, A.P. (2004). *In vivo* phase variation and serologic response to lipooligosaccharide of *Campylobacter jejuni* in experimental human infection. Infect. Immun. *72*, 916–922.

Ridley, A.M., Toszeghy, M.J., Cawthraw, S.A., Wassenaar, T.M., and Newell, D.G. (2008). Genetic instability is associated with changes in the colonization potential of *Campylobacter jejuni* in the avian intestine. J. Appl. Microbiol. *105*, 95–104.

Ringoir, D.D., and Korolik, V. (2003). Colonisation phenotype and colonisation potential differences in *Campylobacter jejuni* strains in chickens before and after passage *in vivo*. Vet. Microbiol. *92*, 225–235.

Sang, F., Shane, S., Yogasundram, K., Hagstad, H., and Kearney, M. (1989). Enhancement of *Campylobacter jejuni* virulence by serial passage in chicks. Avian Dis. *33*, 425–430.

Scott, A.E., Timms, A.R., Connerton, P.L., Loc Carrillo, C., Adzfa Radzum, K., and Connerton, I.F. (2007). Genome dynamics of *Campylobacter jejuni* in response to bacteriophage predation. PLoS Pathog. *3*, e119.

Wang, Y., and Taylor, D.E. (1990). Natural transformation in *Campylobacter* species. J. Bacteriol. *172*, 949–955.

Wassenaar, T.M., Geilhausen, B., and Newell, D.G. (1998). Evidence of genomic instability in *Campylobacter jejuni* isolated from poultry. Appl. Environ. Microbiol. *64*, 1816–1821.

Wassenaar, T.M., Wagenaar, J.A., Rigter, A., Fearnley, C., Newell, D.G., and Duim, B. (2002). Homonucleotide stretches in chromosomal DNA of *Campylobacter jejuni* display high frequency polymorphism as detected by direct PCR analysis. FEMS Microbiol. Lett. *212*, 77–85.

Wiesner, R.S., Hendrixson, D.R., and DiRita, V.J. (2003). Natural transformation of *Campylobacter jejuni* requires components of a type II secretion system. J. Bacteriol. *185*, 5408–5418.

Wilson, D.L., Rathinam, V.A., Qi, W., Wick, L.M., Landgraf, J., Bell, J.A., Plovanich-Jones, A., Parrish, J., Finley, R.L., Mansfield, L.S., *et al.* (2010). Genetic diversity in *Campylobacter jejuni* is associated with differential colonization of broiler chickens and C57BL/6J IL10-deficient mice. Microbiology *156*, 2046–2057.

Concerted Evolution in *Campylobacter jejuni* and *Campylobacter coli*

4

Richard J. Meinersmann

Abstract

Concerted evolution is the phenomenon in which multiple copies of genes maintain sequence similarity in a single individual while the genes continue to diverge between individuals. Concerted evolution has been described in *Campylobacter jejuni* and *C. coli* for the pair of flagellin genes, which are notable for their diversity, the 23S ribosomal RNA genes, which are notable for the degree that they are conserved, and for the portion of the ribosomal operon between the 16S and 23S genes. The evidence for concerted evolution is presented and the biological implications are discussed in this review. An interesting case is the concerted evolution of the 23S rRNA genes that is needed to rapidly become resistant to macrolides. Understanding which genes are affected by concerted evolution will allow proper assessment of probable phylogenetic distortions that will be seen with such genes.

Concerted evolution

Concerted evolution is the result of the mechanism(s) for genes that are present in multiple copies on a given genome to maintain sequence similarity. Concerted evolution is generally regarded as a phenomenon that applies to repeats that are longer than about 500 bp with no real restriction on where the repeats are located on the genome relative to each other (Santoyo and Romero, 2005). The mechanisms for the creation of gene duplication in bacteria are manifold and complex, but are not the focus of this review. For this, the reader is referred to Treangen *et al.* (2009). It is important to note that irrespective of the genesis of multiple gene copies, gene duplication often represents the origin of new genes with new functions or functional subdivisions of genes with multiple functions. This can happen because selective pressure is relaxed on the members of a gene duplication, allowing at least one copy of the duplication to accumulate mutations without affecting the fitness of the organism. Having multiple copies of a gene presents specific problems to the organism, to be considered here only in light of how the gene evolution is affected. If a gene is duplicated without increasing the selective advantage of the carrying organism, i.e. there is no advantage to increased gene dosage, the duplicate is usually rapidly lost (Santoyo and Romero, 2005; Treangen *et al.*, 2009). However, it turns out that every duplicated gene has at least a period post-duplication of concerted evolution and defining when the duplicated genes escape from concerted evolution remains a research problem (Walsh, 1987).

The defining model for concerted evolution is the maintenance of the ribosomal RNA sequences among multiple copies. The growth rate of bacteria is tightly dependent on the ribosomal content (Goelzer and Fromion, 2011). A ribosome is a complex structure consisting of many different proteins (e.g. 53 in *Eschericha coli*; Wittmann, 1982a; Korobeinikova *et al.*, 2012) and three separately encoded pieces of RNA (23S, 16S, and 5S) (Wittmann, 1982b). The rate limiting step on the production of ribosomes is the transcription of the stable RNA components since the protein components are amplified during translation. Thus there is benefit to having multiple copies of stable RNA genes to amplify their rate of production.

Various species of eubacteria have been noted to have between one to 15 copies of the ribosomal RNA genes (Klappenbach *et al.*, 2001). *Campylobacter* spp. have three copies (Friis *et al.*, 2010; Karlyshev *et al.*, 1998; Kim *et al.*, 1993; Nuijten *et al.*, 1990a).

In the case of ribosomes, the selective pressure for homogeneity is apparent since there are a multitude of parameters that are optimized for protein production. The normal expectation is that all genes on a genome evolve independently and, if they did, that would result in duplicated genes diverging from each other to an extent that reflects the age of the duplication. Concerted evolution is when the multiple copies of a gene do not diverge. There is editing of the gene copies based on the sequence of the other copies. In concerted evolution, this occurs by gene conversion. Santoyo and Romero (2005) provide a review of several plausible mechanisms for gene conversion, all of which involve the formation of Holliday junctions without crossing over. In fact, several genes involved in recombination and Holliday junction resolution have been documented as needed for gene conversion in *Escherichia coli* (Santoyo and Romero, 2005). These include *recA*, *recG*, *ruvA*, *ruvB ruvC*, and *mutS*; all of which are found in the genome sequence for *C. jejuni* NCTC11168 (Parkhill *et al.*, 2000; Gundogdu *et al.*, 2007). However, in some experimental systems in *E. coli*, genes for RecBCD, RecF, and/or RecE pathways were also needed for gene conversion (Kobayashi and Takahashi, 1988. Yamamoto *et al.*, 1992), all of which are absent from the *Campylobacter* genome. In *Salmonella enterica* Typhimurium, the *tufA* and *tufB* genes have been extensively studied for the concerted evolution they display. Interestingly, mutations of the *mutSLH* genes increased the rate of gene conversions between these genes (Abdulkarim and Hughes, 1996). The *mutS* gene is the only one of these that is found in the *Campylobacter* genome.

An alternative hypothesis to explain maintenance of sequence among paralogous genes, is the birth and death model (Nei and Rooney, 2005). In this instance, copies of genes are repeatedly gained and lost and the copies maintain similarity because of relatively recent gene duplications. However, the number of copies at any given moment is random and thus a signature of birth and death evolution is frequent differences in copy number. Many examples of birth-and-death evolution have been found in eukaryotes, but the only example documented to date in prokaryotes is among the internalin genes of *Listeria* spp. (Rooney and Ward, 2008).

On the other hand, it has been suggested that most paralogous genes in bacteria arise from lateral gene transfers from outside donors, i.e. without a duplication event in the recipient cell (Ttreangen and Rocha, 2011; Grassi *et al.*, 2012). Then the likelihood that the newly introduced gene will homogenize by gene conversion with the original copy depends largely on the amount of similarity that exists between the two copies. The more similar the imported gene, the more likely it will participate in concerted evolution. Otherwise the imported gene is free to evolve independently and become specialized, pseudogenized, or lost (Treangen *et al.*, 2009). Also, the likelihood of which copy will be the surviving sequence will depend on selective pressures (Innan, 2009). Pressure for diversification will favour the newly introduced sequence.

Other classes of duplicated sequences that do not undergo concerted evolution are short sequence repeats and variable number tandem repeats. The paradigm for their creation and homogenization is slip-strand replication, in which there is mispairing of strands during DNA replication that results in creation and losses of repeats that are close to each other.

Concerted evolution is sometimes confused with the evolution of genes for proteins that work in concert, i.e. are members of an essential pathway or subunits of an essential structure. This is a situation in which selective pressure pushes towards conservation of function but synonymous changes will continue independently in the separate copies. In concerted evolution, synonymous bases are maintained as well as the non-synonymous bases.

No matter the means by which duplicated genes coordinate their evolution, the phylogenetic signal that can be derived from them will be distorted. Subtle inferences such as differentiation are often not possible. The new wave of analyses for most studies in genetics has been stimulated by the immense amount of new data coming from high-throughput sequencing. Unfortunately, repeat sequences are challenging for sequence assemblers to accurately put together and are often missing or truncated in draft level sequences (Treangen and Salzberg, 2011). Thus, the phylogenies of these genes are infrequently studied.

Duplicated genes in *Campylobacter*

In the Integrated Microbial Genomes (IMG) database (Markowitz *et al.*, 2012) (accessed 20 December 2012) there are nine finished *C. jejuni* genome sequences. The IMG tools estimate that approximately 34% of the genes in these genomes are members of paralogue families. Potentially, any of these genes could be involved in concerted evolution, but the documented cases of concerted evolution in *Campylobacter* are limited to two gene families: the 23S ribosomal genes and the *flaA*/*flaB* gene family. It is likely that the intergenic sequence fragment between the 16S and 23S ribosomal genes also engages in concerted evolution, but some major aberrations were noted for that fragment. These examples will be discussed below.

Ribosomal genes

Evidence for the presence of three copies of ribosomal operons in *C. jejuni* was documented in early studies using restriction fragment mapping with Southern blot hybridization analyses (Nuijten *et al.*, 1990a; Taylor *et al.*, 1992; Kim *et al.*, 1993; Karlyshev *et al.*, 1998), although there was confusion whether all three loci had the 16S, 23S and 5S ribosomal sequences in the canonical tandem order. The arrangement of the ribosomal operons has been confirmed by total genome sequencing and in every case the operon included the 16S ribosomal gene sequence followed by $tRNA_{Ala}$, then $tRNA_{Ile}$, then a spacer, followed by the 23S ribosomal gene sequence, and ending with the 5S ribosomal gene sequence (finished *C. jejuni* genome sequences have been published by Friis *et al.*, 2010; Zhang *et al.*, 2010; Pearson *et al.*, 2007; Parkhill *et al.*, 2000; Fouts *et al.*, 2005; Cooper et al., 2011).

An interesting feature of the 23S ribosomal gene sequence that was independently discovered by Konkel *et al.* (1994) and Trust *et al.* (1994) was that some isolates of *C. jejuni* and *C. coli* had transcribed intervening sequences (IVS) in helix 45 of the 23S ribosomal Domain II. The spacer is apparently cleaved from the ribosomal gene product during ribosome construction without re-ligation of the new ends of the product. This yields a 23S gene product, the ribosomal RNA, that is fragmented and shows up in a denaturing gel as two smaller bands. Thus there are two methods for detecting whether a given isolate of *Campylobacter* has an IVS, by denaturing gel electrophoresis of ribosome extracts (Konkel *et al.*, 1994) or by polymerase chain reaction (PCR; Trust *et al.*, 1994; Meinersmann *et al.*, 2009). It was immediately noted from a small pool of *C. jejuni* isolates that some isolates had the IVS and others did not (Konkel *et al.*, 1994). In a slightly larger pool of isolates, Trust *et al.* (1994) found that 8 out of 11 (72%) *C. coli* had the IVS and 14 out of 21 (67%) *C. jejuni* had the IVS. Meinersmann *et al.* (2009) found the IVS in 40 out of 43 (93%) *C. coli* and 34 out of 82 (42%) *C. jejuni*. Conversely, Tazumi *et al.* (2009) found the IVS in 5 out of 11 (45%) *C. coli* and 30 out of 56 (54%) of the tested *C. jejuni*. The variations in the proportions of isolates with the IVS probably reflect differences in the populations that the isolates were drawn from. For instance, Meinersmann *et al.* (2009) found that *C. jejuni* isolates from cattle were more likely to have an IVS than were *C. jejuni* isolates from chickens. The remarkable point of these observations is that the presence of the IVS was not congruent with the species, suggesting that the fragment is exchanged in lateral gene transfers. Furthermore, sequence analyses showed no fixed differences in the IVS when comparing *C. coli* with *C. jejuni* (Meinersmann *et al.*, 2009). Meinersmann *et al.* (2009) performed an additional analysis of the copy number of the IVS in each isolate, and this is where the evidence for concerted evolution is found. They found that 7 out of 82 (8%) *C. jejuni* and 12 out of 43 (28%) *C. coli* carried 23S ribosomal genes both with and without the IVS. This indicates two possible models that both require gene conversion to homogenize the copies. Either the strains had no copies with the IVS, had an IVS introduced by lateral gene transfer and then by gene conversion introduced the IVS to other copies, or the strains had all three copies with an IVS, lost one which was then homogenized to the other copies by gene conversion. It was felt that the strains with mixes of copies with and without an IVS may be deficient in gene conversion capacity. This was supported by the fact that those isolates were also much more likely to have sequence differences between copies in the parts of the 23S ribosomal gene outside of the IVS as well.

A more direct illustration of gene conversion of the 23S ribosomal gene leading to concerted evolution, that also had implications on the rate of gene conversion, was provided by Ladely *et al.* (2008). Chickens were artificially colonized with *C. coli* or *C. jejuni* and then fed tylosin (a macrolide antibiotic) in order to select for macrolide resistant isolates. To assay for the single base changes at positions 2058 or 2059 of 23S ribosomal RNA that are known to create high level resistance (Jensen and Aarestrup, 2001; Vester and Douthwaite, 2001), this segment of the ribosomal gene was sequenced for each operon of isolates recovered from the treated birds. For an experiment done with *C. coli*, 17 isolates were analysed; one of which had the resistance mutation at two loci and another one that had the resistance mutation at all three loci. For an experiment done with *C. jejuni*, 36 isolates were analysed; one of which had the resistance mutation at two loci and eight had the mutation at all three loci. The mutation rate for ribosomal genes is not known, but based on other genes in *Campylobacter* (Wilson *et al.*, 2009; Yu *et al.*, 2012) a high end estimate of 10^{-9} per generation is probably a reasonable high-end estimate. If the mutations for each locus are independent, at that rate the estimate for two specific mutations would be 10^{-18} per generation and for three specific mutations would be 10^{-27} per generation. Based on the number *Campylobacter* typically recovered from a colonized chicken (Berrang *et al.*, 2000), a reasonable estimate is that there were about 10^{11} generations of *Campylobacter* in the course of the experiment. That means that gene conversion happened at least 10^{7} to 10^{16} times faster than random mutation could demonstrate. Note that this was under strong selective pressure that has unknown effect on the rate of gene conversion.

The segment of genome sequence between the 16S and 23S rRNA genes, variously known as the 'internal spacer region' (Christensen *et al.*, 1999) or the 'intergenic sequence region' (Meinersmann *et al.*, 2010) (ISR in both cases), has been investigated in ways that give clues that it also has concerted evolution playing a part in its lineage. Both labs used PCR to analyse for heterogeneity of product size in the ISR of select isolates. Christensen *et al.* (1999) tested 39 isolates and found three isolates with ISR heterogeneity. Meinersmann *et al.* (2010) used primers based on the genome sequence of *C. coli* RM2228 (Fouts *et al.* 2005) that demonstrated the strain had two copies of the ISR that were typical of the ISR from *C. coli* and the third was typical of the ISR sequence found in *C. jejuni*. 134 independent isolates of *C. jejuni* and 133 independent isolates of *C. coli* were tested and the only one with the species-specific heterogeneity was *C. coli* RM2228. Thus, fragments of DNA can be transferred within the ISR portion of the ribosomal operon, but with prevalence well under 1%. This contrasts with Sheppard *et al.* (2008) who showed cross-species sharing of alleles for housekeeping genes, all of which are single copy, ranging from 0.8% to 18.6%. Since lateral gene transfer may share the same mechanics as gene conversion, it is reasonable to conclude that these differences may be due to the differences in availability of the donor fragments, i.e. every duplicated gene represents a potential in-house source of donor fragments that could outnumber exogenous fragments.

Flagellin genes

The remaining example of documented concerted evolution in *Campylobacter* is of the pair of flagellin genes, *flaA* and *flaB* (Meinersmann *et al.*, 2000). *flaA*, a gene usually about 1731 nucleotides, was noted to be duplicated in *C. coli* as early as 1989 (Logan *et al.*, 1989). We know of no naturally occurring *C. coli* or *C. jejuni* that has been documented to have more or less than two copies; thus, a birth and death model of evolution is unlikely. The so-called *flaC* flagellin annotated in the genome sequence for *C. jejuni* NCTC11168 (Parkhill *et al.*, 2000) is not in the same gene family as *flaA* and *flaB* by our analyses (data not published) contrary to the conclusion presented by Song *et al.* (2004).

As the number of *Campylobacter* flagellin sequences increased (Nuijten *et al.*, 1990b; Fischer and Nachamkin, 1991; Khawaja *et al.*, 1992; Guerry *et al.*, 1990; Alm *et al.*, 1993a; Meinersmann *et al.*, 1997; Harrington *et al.*, 1997) it became clear that the diversity and dissimilarity of the flagellin sequences was high, exceeding expectations for the accumulation of point mutations and thus is probably remodelled by recombination (Wassenaar *et al.*, 1995). It was also found that the sequence could be divided into five domains or blocks (Meinersmann *et al.*, 1997). Variation between strains was concentrated in two segments of the flagellin, the so-called short variable region, corresponding to Block II, and the long variable region, denoted Block IV. The amino acid sequence of Block IV, from residue 290 to residue 480, had only 71 of its residues conserved in the aligned *flaA* or the aligned *flaB* amino acid sequences. These correspond to the portions of Salmonella flagellin sequence that in the protein product are surface exposed and presumably are available for antibody attack (Yamashita *et al.*, 1995). Apparently the flagellin protein has little functional dependence on primary amino acid sequence in that region. It was hypothesized that two copies of *fla* serves to make a donor for re-assortment and recombination of the DNA as a mechanism to create new antigenic variants for immune avoidance (Alm *et al.*, 1993b). However, a combined phylogenetic analysis of both *flaA* and *flaB* gene sequences showed this could not be the case (Meinersmann and Hiett, 2000). The *flaA* and *flaB* sequences from the same isolates clustered together in dendrogram. This is a signal of concerted evolution (Li, 1997). Remarkably, analyses of the polymorphism patterns in the flagellin genes within and between strains of *Campylobacter* showed that the two parts of the protein that were most variable between strains were the parts that were most conserved between the *flaA* and *flaB* within a strain (Meinersmann and Hiett, 2000). From the synonymous to non-synonymous mutation ratios (K_S/K_A) there appears to be the greatest functional constraint on sequence in Block V and the lowest functional constraint in Block II and Block IV. Block IV had almost no differences between the *flaA* and *flaB* within individual strains.

The extremely low number of intrastrain *flaA* to *flaB* synonymous differences in the region of Block IV DNA sequence, bases 832 and 1515, between the *flaA* and *flaB* of individual strains implies editing or gene conversion with selection at the DNA level (Endo *et al.*, 1996). The higher level of intrastrain *flaA* to *flaB* synonymous differences between bases 244 and 735 suggests selection for conservation at the protein or functional level without gene conversion (Endo *et al.*, 1996).

Thus inter-strain variability is greatest within the segment that is conserved intra-strain, perhaps indicating that the generation of diversity in that region is not dependent on recombination between the duplicate copies. More likely, an external source contributed some portion of the variable region and then the two copies were homogenized by gene conversion. Analyses of the SVR have been done for many strains of *C. jejuni* in conjunction with multi-locus sequence analyses (Dingle *et al.*, 2008). These studies have shown that the SVR sequence is commonly imported into lineages by horizontal gene transfer. It is unknown if the imported sequence is simultaneously placed in the two flagellin genes or if there is an ordered two step introduction. It cannot be said which copy of the flagellin genes is more likely to originally accept the foreign sequence, but it probably makes little difference. The concerted evolution occurring in Blocks II and IV must happen by gene conversion events that occur almost as fast as the mutation rate since the Block IV sequences of *flaA* and *flaB* are almost identical within a strain.

The expression of *flaA* is regulated by a sigma-28 promoter whereas *flaB* is regulated by a sigma-54 promoter (Guerry *et al.*, 1990; Nuijten *et al.*, 1995; Wassenaar *et al.*, 1994), each of which is stimulated by environmental factors (Allen and Griffiths, 2001; Alm *et al.*, 1993c). Furthermore, it has been found that motility is dependent on the expression of a complete flaA and motility is at best only partially maintained by an intact *flaB* (Guerry *et al.*, 1991; Wassenaar *et al.*, 1991, 1993). Doig *et al.* (1996) have demonstrated that *Campylobacter* flagellins are glycosylated. Based on the maps of *flaA* and *flaB* for 14 strains of

C. jejuni and one strain of *C. coli* (Meinersmann and Hiett, 2000), six of the 12 amino acids that are strictly conserved in each copy of *flaA* but differ between *flaA* to *flaB* represent differences in amino acids that can be O-linked (serine or threonine). Five of these modifiable amino acids are in the carboxy-terminal region. It would seem that the different copies of the flagellin gene offer the greatest opportunity for antigenic phase variation by changing the glycosylation sites. However, the amino acids in flagellin that have been identified as sites for glycosylation (Ewing *et al.*, 2009. Zampronio *et al.*, 2011) are either not conserved from strain to strain or are unchanged from *flaA* to *flaB*.

Unresolved questions

Tibayrenc and Ayala (2012) have asked, 'recombination: what is it good for?' They recognize that the main role may be for DNA repair and introduction of laterally transferred genes may be a by-product of that mechanism. However, they did not recognize the need for recombination for concerted evolution. It is clear that there is strong selective pressure to maintain similar sequences of at least the multiple copies of the ribosomal operons, which is done by concerted evolution. We hypothesize that recombination is the by-product of the mechanism used for that sequence homogenization.

Since so little is known about the mechanisms that result in concerted evolution, speculative answers are all we have for many questions derived from observations of apparent concerted evolution. It appears that concerted evolution is well regulated at least in some ways because the margins of sequences that are involved are very precise. This is seen in the case of the *flaA* and *flaB* genes discussed above, with the two segments of the gene that are consistently homogenized while the distinguishing segments of *flaA* and *flaB* are untouched. A necessary hallmark of concerted evolution is that the gene conversion events are non-reciprocal, which is the most congruent explanation of the natural evolution of *flaA* and *flaB*. However, in the direct demonstrations of intra-genomic recombination between *flaA* and *flaB* (Alm *et al.*, 1993b; Nuijten *et al.*, 2000) there was reciprocal crossover resulting in rearrangements of the flagellin genes. It is plausible that the antimicrobial resistance genes that were used in those studies for transfer signals disrupted the regulation of the margins of gene conversion.

Without regulation it is hypothesized that the edges of duplicated regions would be imprecise in participation of recombination, thus leading to ragged edges of conserved sequence (Santoyo and Romero, 2005). If this was the case with the *fla* genes the differences between *flaA* and *flaB* would be eroded. The margins of the ribosomal genes also do not appear to be eroded although we have noted that in a small set of strains sections of the inter-genic sequence have been observed to depart from sequence homogenization. However, as noted above, those may be aberrations due to a mutational loss of the concerted evolution ability in specific lineages. But it is unknown how large the fragment can be that is involved in a single gene conversion event.

Another question is whether the same mechanism using the same enzymes are responsible for concerted evolution of different gene families, which seems likely (Santoyo and Romero, 2005). If this is the case, does the mechanism have ways to distinguish those duplicated genes that are appropriate for concerted evolution and those that need to be left alone? This is related to the question above on defining the margins of concertedly evolved segments of sequence. Inappropriate gene conversions may lead to the generation of new clones that are at a selective disadvantage and soon lost. But if distinction of appropriate genes is not controlled, this leaves the possibility that any duplication will undergo concerted evolution while they exist. Thus a mechanism that arose to serve one purpose would become applied to all duplicate genes. The most parsimonious speculation is that the mechanism for concerted evolution arose for the purpose that has the strongest selective pressure, and the strongest example that we can identify today is the maintenance of ribosomal sequences.

But if the mechanism is the same for ribosomal concerted evolution and *flaA*/*flaB* concerted evolution in *C. jejuni*, why are ribosomal sequences so well conserved across the species while the *fla* sequences vary so much from strain to strain? Given that recombination is stochastically related to the similarity of the imported sequence, another possibility that deserves attention is that multiple DNA-sequence copies within a genome can increase the odds for successful lateral gene transfer of similar sequences from other lineages, e.g. two copies would double the likelihood of successful recombination. The recipient has twice as many sites to stash an incoming sequence by homologous recombination. A necessary assumption is that if a DNA fragment is not rapidly recombined, it will be digested and lost, i.e. successful recombination has to occur before degradation and having twice as many landing sites will increase that chance.

Recombination needs donors. As an organism grows and multiplies, most of its neighbours are going to be siblings. The DNA they can donate will be little or no different than what the recipient already has and thus recombination will not be recognized even if it happens. Most recombination is homologous, not illegitimate. How similar is similar enough for recombination probably depends on the gene with regards to selective pressure. Flagellin probably has more relaxed selection than housekeeping genes, and thus the arrival of sequence from a more distant cousin could still make it, perhaps with twice the chance of recombination compared to an equally relaxed (with respect to selective pressure) single-copy gene. If there is pressure for diversification, recipient survival will be favoured. There are plenty of reasons for speculating on advantage of diversifying flagellin. This would explain why the *fla* genes change so rapidly, whereas the strong selection for preserved sequences of the ribosomal genes would protect them from such changes.

References

Abdulkarim, F., and Hughes, D. (1996). Homologous recombination between the *tuf* genes of *Salmonella typhimurium*. J. Mol. Biol. *260*, 506–522.

Allen, K.J., and Griffiths, M.W. (2001). Effect of environmental and chemotactic stimuli on the activity of the *Campylobacter jejuni flaA* sigma(28) promoter. FEMS Microbiol. Lett. *205*, 43–48.

Alm, R.A., Guerry, P., and Trust, T.J. (1993a). Distribution and polymorphism of the flagellin genes from isolates of *Campylobacter coli* and *Campylobacter jejuni*. J. Bacteriol. *175*, 3051–3057.

Alm, R.A., Guerry, P., and Trust, T.J. (1993b). Significance of duplicated flagellin genes in *Campylobacter*. J. Mol. Biol. *230*, 359–363.

Alm, R.A., Guerry, P., and Trust, T.J. (1993c). The *Campylobacter* sigma 54 *flaB* flagellin promoter is subject to environmental regulation. J. Bacteriol. *175*, 4448–4455.

Berrang, M.E., Buhr, R.J., and Cason, J.A. (2000). *Campylobacter* recovery from external and internal organs of commercial broiler carcass prior to scalding. Poult. Sci. *79*, 286–290.

Christensen, H., Jorgensen, K., and Olsen, J.E. (1999). Differentiation of *Campylobacter coli* and *C. jejuni* by length and DNA sequence of the 16S-23S rRNA internal spacer region. Microbiology *145*, 99–105.

Cooper, K.K., Cooper, M.A., Zuccolo, A., Law, B., and Joens, L.A. (2011). Complete genome sequence of *Campylobacter jejuni* strain S3. J. Bacteriol. *193*, 1491–1492.

Dingle, K.E., McCarthy, N.D., Cody, A.J., Peto, T.E., and Maiden, M.C. (2008). Extended sequence typing of *Campylobacter* spp., United Kingdom. Emerg. Infect. Dis. *14*, 1620–1622.

Doig, P., Kinsella, N., Guerry, P., and Trust, T.J. (1996). Characterization of a post-translational modification of *Campylobacter flagellin*: Identification of a sero-specific glycosyl moiety. Mol. Microbiol. *19*, 379–387.

Endo, T., Ikeo, K., and Gojobori, T. (1996). Large-scale search for genes on which positive selection may operate. Mol. Biol. Evol. *13*, 685–690.

Ewing, C.P., Andreishcheva, E., and Guerry, P. (2009). Functional characterization of flagellin glycosylation in *Campylobacter jejuni* 81-176. J. Bacteriol. *191*, 7086–7093.

Fischer, S.H., and Nachamkin, I. (1991). Common and variable domains of the flagellin gene, *flaA*, in *Campylobacter jejuni*. Mol. Microbiol. *5*, 1151–1158.

Fouts, D.E., Mongodin, E.F., Mandrell, R.E., Miller, W.G., Rasko, D.A., Ravel, J., Brinkac, L.M., DeBoy, R.T., Parker, C.T., Daugherty, S.C., *et al.* (2005). Major structural differences and novel potential virulence mechanisms from the genomes of multiple *Campylobacter* species. PLoS Biol. 3, e15.

Friis, C., Wassenaar, T.M., Javed, M.A., Snipen, L., Lagesen, K., Hallin, P.F., Newell, D.G., Toszeghy, M., Ridley, A., Manning, G., and Ussery, D.W. (2010). Genomic Characterization of *Campylobacter jejuni* Strain M1. PLoS ONE. *5*, e12253.

Goelzer, A., and Fromion, V. (2011). Bacterial growth rate reflects a bottleneck in resource allocation. Biochim. Biophys. Acta *1810*, 978–988.

Grassi, L., Caselle, M., Lercher, M.J., and Lagomarsino, M.C. (2012). Horizontal gene transfers as metagenomic gene duplications. Mol. Biosyst. *8*, 790–795.

Guerry, P., Logan, S.M., Thornton, S., and Trust, T.J. (1990). Genomic organization and expression of *Campylobacter flagellin* genes. J. Bacteriol. *172*, 1853–1860.

Guerry, P., Alm, R.A., Power, M.E., Logan, S M., and Trust, T.J. (1991). Role of two flagellin genes in *Campylobacter* motility. J. Bacteriol. 173, 4757–4764.

Gundogdu, O., Bentley, S.D., Holden, M.T., Parkhill, J., Dorrell, N., and Wren, B.W. (2007). Re-annotation and re-analysis of the *Campylobacter jejuni* NCTC11168 genome sequence. BMC Genomics *8*, 162.

Harrington, C.S., Thomson-Carter, F.M., and Carter, P.E. (1997). Evidence for recombination in the flagellin locus of *Campylobacter jejuni*: implications for the flagellin gene typing scheme. J. Clin. Microbiol. *35*, 2386–2392.

Innan, H. (2009). Population genetic models of duplicated genes. Genetica *137*, 19–37.

Jensen, L.B., and Aarestrup, F.M. (2001). Macrolide resistance in *Campylobacter coli* of animal origin in Denmark. Antimicrob. Agents Chemother. *45*, 371–372.

Karlyshev, A.V., Henderson, J., Ketley, J.M., and Wren, B.W. (1998). An improved physical and genetic map of *Campylobacter jejuni* NCTC 11168 (UA580). Microbiology *144*, 503–508.

Khawaja, R., Neote, K., Bingham, H.L., Penner, J.L., and Chan, V.L. (1992). Cloning and sequence analysis of the flagellin gene of *Campylobacter jejuni* TGH9011. Curr. Microbiol. *24*, 213–221.

Kim, N.W., Lombardi, R., Bingham, H., Hani, E., Louie, H., Ng, D., and Chan, V.L. (1993). Fine mapping of the three rRNA operons on the updated genomic map of *Campylobacter jejuni* TGH9011 (ATCC 43431). J. Bacteriol. *175*, 7468–7470.

Klappenbach, J.A., Saxman, P.R., Cole, J.R., and Schmidt, T.M. (2001). rrndb: the Ribosomal RNA Operon Copy Number Database. Nucleic Acids Res. *29*, 181–184.

Kobayashi, I., and Takahashi, N. (1988). Double-stranded gap repair of DNA by gene conversion in *Escherichia coli*. Genetics *119*, 751–757.

Konkel, M.E., Marconi, R.T., Mead, D.J., and Cieplak, W., Jr. (1994). Identification and characterization of an intervening sequence within the 23S ribosomal RNA genes of *Campylobacter jejuni*. Mol. Microbiol. *14*, 235–241.

Korobeinikova, A.V., Garber, M.B., and Gongadze, G.M. (2012). Ribosomal proteins: structure, function, and evolution. Biochemistry (Mosc) *77*, 562–574.

Ladely, S.R., Meinersmann, R.J., Englen, M.D., Fedorka-Cray, P.J., and Harrison, M.A. (2008). 23S rRNA Gene Mutations Contributing to Macrolide Resistance in *Campylobacter jejuni* and *Campylobacter coli*. Foodborne Pathog. Dis. *6*, 91–98.

Li, W.H. (1997). Molecular Evolution. Sunderland, MA, Sinauer Associates, Inc.

Logan, S.M., Trust, T.J., and Guerry, P. (1989). Evidence for posttranslational modification and gene duplication of *Campylobacter flagellin*. J. Bacteriol. *171*, 3031–3038.

Markowitz, V.M., Chen, I.M., Palaniappan, K., Chu, K., Szeto, E., Grechkin, Y., Ratner, A., Jacob, B., Huang, J., Williams, P., *et al.* (2012). IMG: the Integrated Microbial Genomes database and comparative analysis system. Nucleic Acids Res. *40*(Database issue), D115–122. doi: 10.1093/nar/gkr1044

Meinersmann, R.J., and Hiett, K.L. (2000). Concerted evolution of duplicate *fla* genes in *Campylobacter*. Microbiology *146*, 2283–2290.

Meinersmann, R.J., Helsel, L.O., Fields, P.I., and Hiett, K.L. (1997). Discrimination of *Campylobacter jejuni* isolates by *fla* gene sequencing. J. Clin. Microbiol. *35*, 2810–2814.

Meinersmann, R.J., Phillips, R.W., and Ladely, S.R. (2009). Inter- and intra-genomic heterogeneity of the intervening sequence in the 23S ribosomal RNA gene of *Campylobacter jejuni* and *Campylobacter coli*. Syst. Appl. Microbiol. *32*, 91–100.

Meinersmann, R.J., Ladely, S.R., and Lindsey, R.L. (2010). Ribosomal operon intergenic sequence region (ISR) heterogeneity in *Campylobacter coli* and *Campylobacter jejuni*. Lett. Appl. Microbiol. *51*, 539–545.

Nei, M., and Rooney, A.P. (2005). Concerted and birth-and-death evolution of multigene families. Annu. Rev. Genet. *39*, 121–152.

Nuijten, P.J., Bartels, C., Bleumink-Pluym, N.M., Gaastra, W., and van der Zeijst, B.A. (1990a). Size and physical map of the *Campylobacter jejuni* chromosome. Nucleic Acids Res. *18*, 6211–6214.

Nuijten, P., van Asten, F.J., Gaastra, W., and van der Zeijst, B.A. (1990b). Structural and functional analysis of two *Campylobacter jejuni* flagellin genes. J. Biol. Chem. *265*, 17798–17804.

Nuijten, P.J., Marquez-Magana, L., and van der Zeijst, B.A. (1995). Analysis of flagellin gene expression in flagellar phase variants of *Campylobacter jejuni* 81116. Antonie Van Leeuwenhoek *67*, 377–383.

Nuijten, P.J., van den Berg, A.J., Formentini, I., van der Zeijst, B.A., and Jacobs, A.A. (2000). DNA rearrangements in the flagellin locus of an *flaA* mutant of *Campylobacter jejuni* during colonization of chicken ceca. Infect. Immun. *68*, 7137–7140.

Parkhill, J., Wren, B.W., Mungall, K., Ketley, J.M., Churcher, C., Basham, D., Chillingworth, T., Davies, R.M., Feltwell, T., Holroyd, S., *et al.* (2000). The genome sequence of the food-borne pathogen *Campylobacter jejuni* reveals hypervariable sequences. Nature *403*, 665–668.

Pearson, B.M., Gaskin, D.J., Segers, R.P., Wells, J.M., Nuijten, P.J., and van Vliet, A.H. (2007). The complete genome sequence of *Campylobacter jejuni* strain 81116 (NCTC11828). J. Bacteriol. *189*, 8402–8403.

Rooney, A.P., and Ward, T.J. (2008). Birth-and-death evolution of the internalin multigene family in *Listeria*. Gene *427*, 124–128.

Santoyo, G., and Romero, D. (2005). Gene conversion and concerted evolution in bacterial genomes. FEMS Microbiol. Rev. *29*, 169–183.

Sheppard, S.K., McCarthy, N.D., Falush, D., and Maiden, M.C. (2008). Convergence of *Campylobacter* species: implications for bacterial evolution. Science *320*, 237–239.

Song, Y.C., Jin, S., Louie, H., Ng, D., Lau, R., Zhang, Y., Weerasekera, R., Al Rashid, S., Ward, L.A., Der, S.D., and Chan, V.L. (2004). *FlaC*, a protein of *Campylobacter jejuni* TGH9011 (ATCC43431) secreted through the flagellar apparatus, binds epithelial cells and influences cell invasion. Mol. Microbiol. *53*, 541–553.

Taylor, D.E., Eaton, M., Yan, W., and Chang, N. (1992). Genome maps of *Campylobacter jejuni* and *Campylobacter coli*. J. Bacteriol. *174*, 2332–2337.

Tazumi, A., Kakinuma, Y., Misawa, N., Moore, J.E., Millar, B.C., and Matsuda, M. (2009). Identification and characterization of intervening sequences within 23S rRNA genes from more than 200 *Campylobacter* isolates from seven species including atypical *campylobacters*. BMC Microbiol. *9*, 256.

Tibayrenc, M., and Ayala, F.J. (2012). Reproductive clonality of pathogens: A perspective on pathogenic viruses, bacteria, fungi, and parasitic protozoa. Proc. Natl. Acad. Sci. U.S.A. *109*, E3305–13.

Treangen, T.J., and Rocha, E.P. (2011). Horizontal transfer, not duplication, drives the expansion of protein families in prokaryotes. PLoS Genet. *7*, e1001284.

Treangen, T.J., and Salzberg, S.L. (2011). Repetitive DNA and next-generation sequencing: computational challenges and solutions. Nat. Rev. Genet. *13*, 36–46.

Treangen, T.J., Abraham, A.L., Touchon, M., and Rocha, E.P. (2009). Genesis, effects and fates of repeats in prokaryotic genomes. FEMS Microbiol. Rev. *33*, 539–371.

Trust, T.J., Logan, S.M., Gustafson, C.E., Romaniuk, P.J., Kim, N.W., Chan, V.L., Ragan, M.A., Guerry, P., and Gutell, R.R. (1994). Phylogenetic and molecular characterization of a 23S rRNA gene positions the genus *Campylobacter* in the epsilon subdivision of the *Proteobacteria* and shows that the presence of transcribed spacers is common in *Campylobacter* spp. J. Bacteriol. *176*, 4597–4609.

Vester, B., and Douthwaite, S. (2001). Macrolide resistance conferred by base substitutions in 23S rRNA. Antimicrob. Agents Chemother. *45*, 1–12.

Walsh, J.B. (1987). Sequence-dependent gene conversion: can duplicated genes diverge fast enough to escape conversion? Genetics *117*, 543–57.

Wassenaar, T.M., Bleumink-Pluym, N.M., and van der Zeijst, B.A. (1991). Inactivation of *Campylobacter jejuni* flagellin genes by homologous recombination demonstrates that *flaA* but not *flaB* is required for invasion. EMBO J. *10*, 2055–2061.

Wassenaar, T.M., van der Zeijst, B.A., Ayling, R., and Newell, D.G. (1993). Colonization of chicks by motility mutants of *Campylobacter jejuni* demonstrates the importance of flagellin A expression. J. Gen. Microbiol. *139*, 1171–1175.

Wassenaar, T.M., Bleumink-Pluym, N.M., Newell, D.G., Nuijten, P.J., and van der Zeijst, B.A. (1994). Differential flagellin expression in a *flaA flaB*+ mutant of *Campylobacter jejuni*. Infect. Immun. *62*, 3901–3906.

Wassenaar, T.M., Fry, B.N., and van der Zeijst, B.A. (1995). Variation of the flagellin gene locus of *Campylobacter jejuni* by recombination and horizontal gene transfer. Microbiology *141*, 95–101.

Wilson, D.J., Gabriel, E., Leatherbarrow, A.J., Cheesbrough, J., Gee, S., Bolton, E., Fox, A., Hart, C.A., Diggle, P.J., and Fearnhead, P. (2009). Rapid evolution and the importance of recombination to the gastroenteric pathogen *Campylobacter jejuni*. Mol. Biol. Evol. *26*, 385–397.

Wittmann, H.G. (1982a). Structure and evolution of ribosomes. Proc. R. Soc. Lond. B Biol. Sci. *216*(1203), 117–135.

Wittmann, H.G. (1982b). Components of bacterial ribosomes. Annu. Rev. Biochem. *51*, 155–183.

Yamamoto, K., Kusano, K., Takahashi, N.K., Yoshikura, H., and Kobayashi, I. (1992). Gene conversion in the *Escherichia coli RecF* pathway: a successive half crossing-over model. Mol. Gen. Genet. *234*, 1–13.

Yamashita, I., Vonderviszt, F., Mimori, Y., Suzuki, H., Oosawa, K., and Namba, K. (1995). Radial mass analysis of the flagellar filament of *Salmonella*: implications for the subunit folding. J. Mol. Biol. *253*, 547–558.

Yu, S., Fearnhead, P., Holland, B.R., Biggs, P., Maiden, M., and French, N. (2012). Estimating the relative roles of recombination and point mutation in the generation of single locus variants in *Campylobacter jejuni* and *Campylobacter coli*. J. Mol. Evol. *74*, 273–280.

Zampronio, C.G., Blackwell, G., Penn, C.W., and Cooper, H.J. (2011). Novel glycosylation sites localized in *Campylobacter jejuni* flagellin FlaA by liquid chromatography electron capture dissociation tandem mass spectrometry. J. Proteome. Res. *10*, 1238–1245.

Zhang, M., He, L., Li, Q., Sun, H., Gu, Y., You, Y., Meng, F., and Zhang, J. (2010). Genomic Characterization of the Guillain-Barre Syndrome-Associated *Campylobacter jejuni* ICDCCJ07001 Isolate. PLoS ONE *5*(11), e15060.

Genome and Transcriptome Evolution in the Genus *Campylobacter*

5

My Thanh Le and Arnoud H.M. van Vliet

Abstract

The genus *Campylobacter* belongs to the Epsilon subdivision of the Proteobacteria, and comprises a large number of species, which includes the important human pathogen *Campylobacter jejuni*. The recent progress in high-throughput sequencing technologies are leading to a rapid increase in information on *C. jejuni* and other *Campylobacter* species, and these data are beginning to feed into studies on transmission, infection and evolution of this highly variable group of bacteria. The genomes of *Campylobacter* species display significant levels of variation at the level of genes, promoters, operons, and gene order. This may be explained by the relative paucity of DNA repair mechanisms and horizontal gene transfer via natural transformation, phase variation, plasmid transfer and infection with bacteriophages. In this chapter we will review the current state of genomics and transcriptomics of *Campylobacter* species, and highlight opportunities where further progress may be made in our understanding of the evolution of *Campylobacter*.

Introduction

The genus *Campylobacter* belongs to the family *Campylobacteraceae*, together with the genera *Arcobacter* and *Sulfurospirillum*, and is placed in the order Campylobacterales in the Epsilon subdivision of the Proteobacteria. The Epsilonproteobacteria colonize a wide variety of habitats, ranging from deep-sea hydrothermal vents to the gastrointestinal tract of mammals and birds (Campbell *et al.*, 2006). Within the Campylobacterales, most attention has been on the species pathogenic to humans and animals, such as *Helicobacter pylori* and *Campylobacter jejuni*, but the advances in 'omics' technologies (e.g. genome sequencing, transcriptomics) have now opened up the possibilities to study other members of the Campylobacterales, and gain insight into the evolution of these fascinating bacterial species.

Members of the genus *Campylobacter* are commonly found in the gastrointestinal tract of mammals and birds, and can be commensal or pathogenic in nature, which often depends on the interplay between the bacterial behaviour and the immune status of the host (Young *et al.*, 2007; Janssen *et al.*, 2008). The best studied members of the genus *Campylobacter* are the thermophilic species *Campylobacter jejuni* and *Campylobacter coli*, which are mostly seen as commensal organisms (and sometimes opportunistic pathogens) in poultry and other avian and animal hosts, but are also zoonotic human pathogens that are the major cause of bacterial gastroenteritis in many industrialized countries (Janssen *et al.*, 2008). The gastroenteritis can be very severe but is usually self-limiting, while sequelae of *Campylobacter* infection include the development of neurodegenerative diseases like Guillain-Barré syndrome and Miller-Fisher Syndrome (Hughes, 2004). Other well-known pathogenic members of the genus *Campylobacter* are *Campylobacter upsaliensis* (Bourke *et al.*, 1998; Couturier *et al.*, 2012), which is found in dogs but can cause disease in humans, and *Campylobacter fetus*, which may cause abortion and venereal disease in cattle and sheep and is also considered to be an opportunistic human pathogen (Umehara *et al.*, 2009; Gorkiewicz *et al.*, 2010; Kienesberger *et al.*, 2011).

However, the genus *Campylobacter* also contains non-pathogenic species like *Campylobacter hominis*, which has been isolated from the intestinal tract of both healthy and diseased individuals (Lawson *et al.*, 2001). In addition, there are species for which associations with disease have not been unambiguously established, such as *Campylobacter gracilis*, which has been isolated from periodontal tissues and may be associated with endodontic infections (Vandamme *et al.*, 1995; Siqueira and Rocas, 2003). The members of the genus *Campylobacter* are commonly subdivided into the thermophilic species (like *C. jejuni*, which has an optimal growth temperature of 42°C), and the non-thermophilic species (like *C. fetus* and *C. hominis*). This subdivision is clearly visible in a phylogenetic tree based on 16S rDNA sequences of the *Campylobacter* species for which a genome sequence is available (Fig. 5.1).

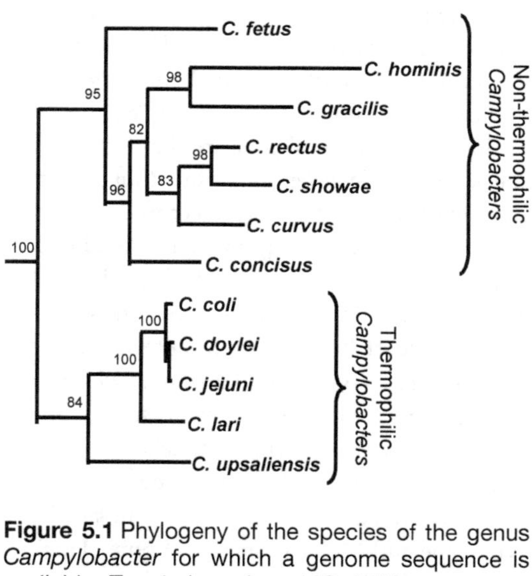

Figure 5.1 Phylogeny of the species of the genus *Campylobacter* for which a genome sequence is available. Tree is based on 16S rDNA sequences, which were aligned using ClustalX2. The tree was generated using the neighbour-joining method of the Phylip software suite v3.69 (Felsenstein, 1989), using *H. pylori* as an out group. The two subgroups within the genus *Campylobacter* (thermophilic and non-thermophilic) are indicated on the right. The percentages of replicate trees in which the associated taxa clustered together in the bootstrap test (1000 replicates) are shown next to the branches. *C. doylei*=*C. jejuni* subsp. *doylei*.

Characteristics of *Campylobacter* genome sequences

The first complete genome sequence of a member of the genus *Campylobacter* was published in 2000 with the release of the *C. jejuni* NCTC 11168 genome (Parkhill *et al.*, 2000; Gundogdu *et al.*, 2007), and since then the genome sequences of major reference strains used in studies of *C. jejuni* (81-176, 81116 and RM1221) (Fouts *et al.*, 2005; Hofreuter *et al.*, 2006; Pearson *et al.*, 2007) have been released, as well as many other genomes of *C. jejuni* (Table 5.1). The *C. jejuni* genome is relatively small (1.6–1.8 Mbp) and has an overall percentage G + C of ~ 32%. The genomes of the *C. jejuni* reference strains show good conservation of overall sequences, as displayed in Fig. 5.2A, but also clear differences, which will be further discussed below. While several *C. jejuni* genomes have been available for some years now, until recently only a single *C. coli* genome sequence was available (RM2228) (Fouts *et al.*, 2005), as well as the genome sequence of a *C. jejuni* subspecies (*doylei*) (Parker *et al.*, 2007). While these genomes do share large regions with *C. jejuni* isolate NCTC 11168, there are also clear differences including inversion of genomic regions (Fig. 5.2B), as well as more specific differences in gene content which require further investigation. Finally, although phylogenetically closely related, the genomes of *C. lari* and *C. upsaliensis* differ significantly, with multiple rearrangements and regions that lack homology when compared with *C. jejuni* (Fig. 5.2C).

Plasmids and insertion sequences

A subset of *C. jejuni* strains contains plasmids and integrated mobile elements similar to insertion sequences. Strains NCTC 11168 and 81116 do not contain plasmids, whereas strain 81-176 can contain two plasmids, named pVir and pTet (Bacon *et al.*, 2000), and although the pVir plasmid name suggests a role in virulence due to the presence of a potential type IV secretion system (Bacon *et al.*, 2000), its contribution to *C. jejuni* virulence is unclear (Louwen *et al.*, 2006). The pTet plasmid confers tetracycline resistance to *C. jejuni* (Batchelor *et al.*, 2004). Another plasmid found in *C. jejuni* and *C. coli* is pCC31, which is > 90% identical to pTet, but also contains genes

Table 5.1 Overview of available genome sequences for *Campylobacter* species in the Genbank/EMBL/DDBJ databases[a]

Campylobacter species	No. of genomes (complete)	Genome size (Mbp)	% G+C
C. jejuni (subsp. *jejuni*)	87 (13)	1.7	31
C. jejuni (subsp. *doylei*)	1 (1)	1.8	31
C. coli	64 (0)	1.7	32
C. lari	1 (1)	1.6	30
C. upsaliensis	2 (0)	1.7	35
C. fetus (subsp. *fetus*)	1 (1)	1.9	34
C. fetus (subsp. *venerealis*)	2 (0)	2.0	34
C. concisus [b]	3 (1)	1.9	42
C. curvus [c]	2 (1)	2.0	45
C. rectus	1 (0)	2.5	46
C. showae	3 (1)	2.1	47
C. gracilis	1 (0)	2.3	48
C. hominis	1 (1)	1.7	33

[a]Genome sequences available via Genbank (http://www.ncbi.nlm.nih.gov/pubmed) and WGC Shotgun Assembly (http://0-www.ncbi.nlm.nih.gov.ilsprod.lib.neu.edu/Traces/wgs/) webpages, 14 April 2013.
[b]Includes *Campylobacter* sp. 10_1_50.
[c]Includes *Campylobacter* sp. FOBRC14.

encoding for another putative type IV secretion system for conjugative transfer (Batchelor *et al.*, 2004; Friis *et al.*, 2007). *C. jejuni* strain RM1221 contains four insertion elements named CJIE1-CJIE4 (Fouts *et al.*, 2005; Parker *et al.*, 2006), which may represent prophages, and three of these insertion elements contain genes for the production of extracellular DNases, which make these strains non-competent for natural transformation (Gaasbeek *et al.*, 2009, 2010). CJIE1 is present in up to half of the *C. jejuni* isolates, and the presence of CJIE1 has been proposed as a marker for virulence in *C. jejuni* isolates (Clark *et al.*, 2012), although this requires further study.

Plasticity zones

After the release of the *C. jejuni* NCTC 11168 genome sequence, microarrays were used for comparative genomic hybridization (CGH) to study the gene content of *C. jejuni* strains, based on presence or absence of sequences homologous to the *C. jejuni* NCTC 11168 genes. Several regions of the genome were described as highly variable (e.g. plasticity regions (Pearson *et al.*, 2003), hypervariable regions (Taboada *et al.*, 2004)), and some of the genes present in plasticity/hypervariable regions have been associated with clinical outcome or environmental source (Pearson *et al.*, 2003; Taboada *et al.*, 2004; Champion *et al.*, 2005; Fouts *et al.*, 2005; Rodin *et al.*, 2008; Stabler *et al.*, 2013). The most pronounced differences have been observed in regions containing genes involved in LOS and capsule biosynthesis, as well as flagellar modification (Pearson *et al.*, 2003; Champion *et al.*, 2005; Rodin *et al.*, 2008; Richards *et al.*, 2013), where gene content has a clear effect on the glycosylation or other sugar-based modification of outer structures of *C. jejuni* and *C. coli*. The composition of the LOS loci has been used for typing purposes, especially to predict whether strains may be involved in GBS syndrome (van Belkum *et al.*, 2001; Godschalk *et al.*, 2004), and LOS sialylation is associated with increased virulence (Louwen *et al.*, 2008, 2012; Habib *et al.*, 2009). However, the gene differences also have implications for metabolism, such as the *Cj0480c-0490c* region, which was recently shown to encode the capability to utilize fucose, and as such conveys an advantage to strains containing this locus (Muraoka and Zhang, 2011; Stahl *et al.*, 2011; Sheppard *et al.*, 2013). Other differences between strains are the presence or absence

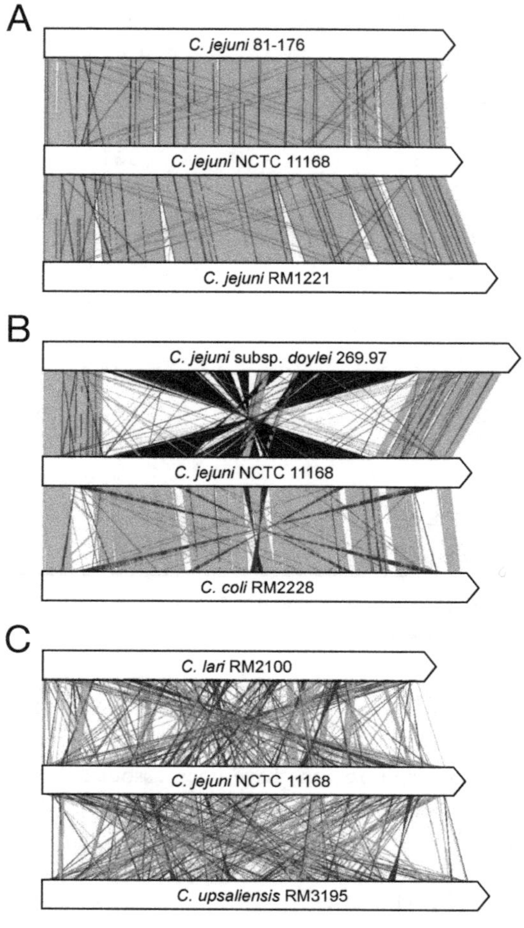

Figure 5.2 Visualization of comparisons of *C. jejuni* NCTC 11168 at the DNA level with genomes of (A) *C. jejuni* strains 81-176 and RM1221; (B) *C. jejuni* subsp. *doylei* 269.97 and *C. coli* RM2228 and (C) *C. lari* RM2100 and *C. upsaliensis* RM3195. Homologous sequences are identified by light grey (same orientation) and dark grey (opposite orientation) and visualized using the Artemis Comparison Tool (Carver *et al.*, 2005).

of the CRISPR-Cas system, which can convey phage defence, but may also modulate virulence of *C. jejuni* (Louwen *et al.*, 2013).

Other *Campylobacter* species and the *Campylobacter* core genome

While there are now multiple genome sequences available for *C. jejuni* and *C. coli*, this is unfortunately not the case for most of the other *Campylobacter* species (Table 5.1). This makes it difficult to make any definite statement about genome content of these species, as it is not known how representative the available genomes are for the specific species. However, it is likely that these species display similar genomic variation both in sequence and gene content as *C. jejuni* and *C. coli*. As G + C content and nucleotide sequences have significantly diverged within these *Campylobacter* species (from 30% in *C. lari* up to 48% in *C. gracilis*), we have used the predicted proteome to compare gene content of the available *Campylobacter* genome sequences (Fig. 5.3). The differences in gene content observed are reflected in the branching of a *Campylobacter* 16S rDNA-based phylogenetic tree (Fig. 5.1), and show that the non-thermophilic *Campylobacter* species show a much wider divergence than the thermophilic *Campylobacter* species. Three-way comparisons of the predicted proteins of different *Campylobacter* shows that depending on the species used and their relative phylogenetic diversity, about half of the predicted proteome is present in the compared *Campylobacter* species (Fig. 5.3A). Using these homology tables, we have predicted the core genome of *Campylobacter* species based on comparison of five species for which complete genome sequences are available (*C. jejuni*, *C. lari*, *C. fetus*, *C. curvus* and *C. hominis*). A core genome of ~650 genes is predicted (Fig 5.3B), with the largest number of core genes involved in translation (111/651) and membrane biogenesis (54/651) (Fig. 5.3B), while genes which are annotated as 'function unknown' and those lacking a COG annotation are underrepresented in the core genome. It should be noted that due to the lack of flagellar motility in *C. hominis*, motility-associated genes are absent from the core genome. Finally, visualization of the position of the core genes on the respective genomes demonstrates that these core genes are mostly scattered over the genome, and are not clustered (Fig. 5.3C). This is consistent with the differences observed between the genomes of *C. jejuni* and non-jejuni *Campylobacter* species (Fig. 5.2C). Each of the visualized core genomes in Fig. 5.3B has gaps where no core genes are present, which may be areas of interest for further study, similar to the plasticity regions in *C. jejuni* genomes (Pearson *et al.*, 2003; Champion *et al.*, 2005; Parker *et al.*, 2006; Clark *et al.*, 2012; Sheppard *et al.*, 2013; Stabler *et al.*, 2013).

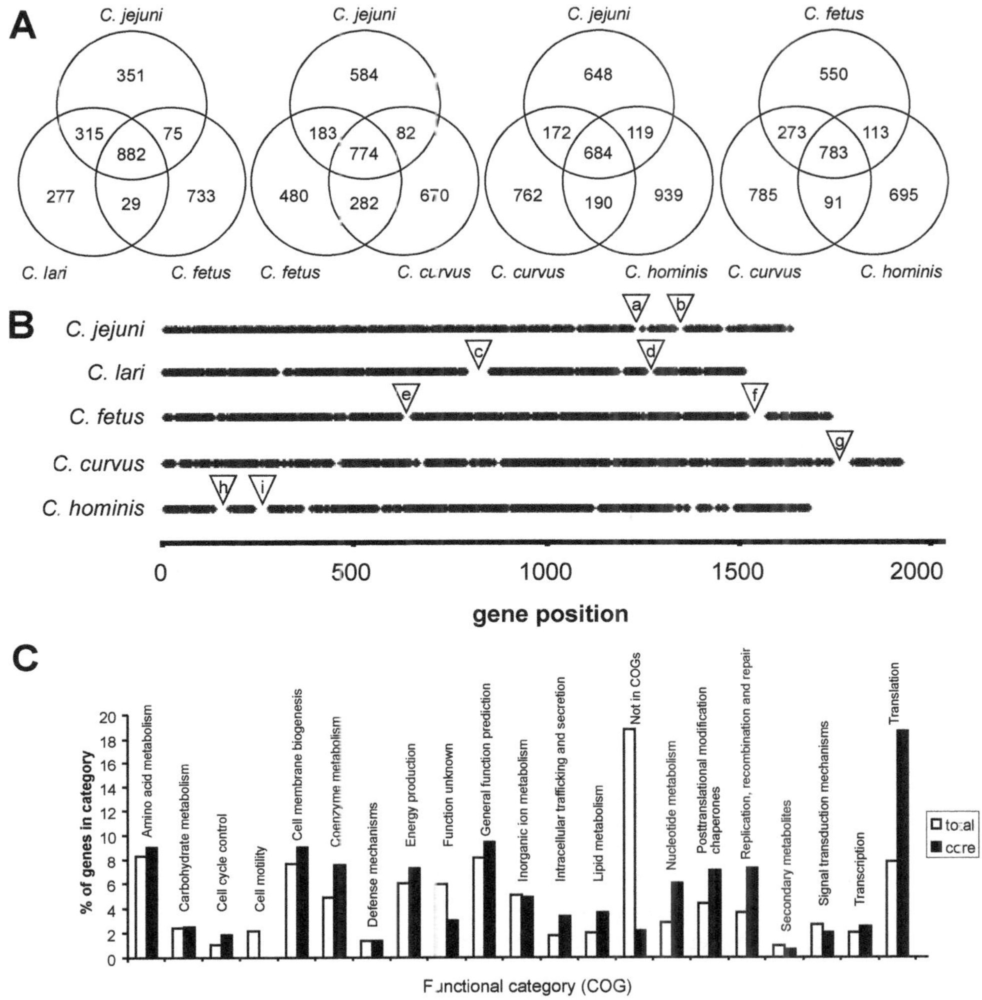

Figure 5.3 Identification of the *Campylobacter* core genome. (A) Venn diagrams showing the number of shared genes between three *Campylobacter* species each, based on pairwise comparisons output of the GeneOrder 4.0 software suite (Mahadevan and Seto, 2010). (B) Visualization of the position of 652 core genes (defined as orthologues present in the complete genome sequences of *C. jejuni* NCTC 11168, *C. lari* RM2100, *C. fetus* subsp. *fetus* 82–40, *C. curvus* 525.92 and *C. hominis* ATCC BAA-381), showing that core genes are spread over the whole genome in each species and not specifically localized. Inverted triangles highlight larger areas of the genome not containing any core genes, and represent flagellar modification regions (a, d and f), capsular genes (b), prophage (c, i), potential secretion systems (h) and regions of unknown function (e.g.). (C) Comparison of the relative number of *C. jejuni* total genes (white bars) and core genes (black bars), classified per functional category (COG), displaying the overrepresentation of core genes in several metabolic categories (e.g. 'Translation'), and underrepresentation in the categories 'Function unknown' and 'Not in COGs'.

Transcriptome evolution in the genus *Campylobacter*

Although differences at the level of the genome sequence may be relatively small between *C. jejuni* and *C. coli* strains, this does not necessarily translate into an absence of phenotypic differences. Due to the relative rarity of repair mechanisms and the opportunity for genetic transfer by natural transformation, there may be significant differences in the expression profiles from different strains. One such example has been described for *C. jejuni* strain 81-176, where a single point mutation in the asparaginase gene (Hofreuter *et al.*, 2008) has resulted in the protein being located in the periplasm rather than the cytoplasm, which is thought to give this strain more metabolic flexibility. Also, point mutations may result in modification of promoter motifs or regulatory sequences, as described for *H. pylori* Fur (Ernst *et al.*, 2005; Carpenter *et al.*, 2009), and thus alter regulatory or expression patterns. In this section we will briefly describe transcription mechanisms involved in *Campylobacter* species.

Sigma factors

Transcription in prokaryotes requires the concerted action of the RNA polymerase apo-enzyme and a sigma factor, the latter being responsible for promoter recognition (Browning and Busby, 2004). Only three sigma factors have been identified in *Campylobacter* species: the vegetative sigma factor σ^{70}, and the alternative sigma-factors σ^{28} and σ^{54} (Wosten, 1998). Both σ^{28} and σ^{54} appear to be involved largely in flagella biosynthesis and regulation, and are absent from *C. gracilis* and *C. hominis*. Previous studies on σ^{28} and σ^{54} dependent genes have resulted in the definition of consensus sequences (σ^{28} –10: CGATwt and σ^{54} –24 and –12: GGaa-N_6-TTGCTT) (Carrillo *et al.*, 2004; Gilbreath *et al.*, 2011), whereas a bioinformatic prediction of σ^{70} promoters in *C. jejuni* and *H. pylori* suggested a poorly conserved promoter sequence, with only the –10 sequence conserved (gnTanaAT) (Petersen *et al.*, 2003). Consensus sequences for all three promoters have now been experimentally confirmed in *H. pylori* and *C. jejuni* (Sharma *et al.*, 2010; Porcelli *et al.*, 2013), and do not differ significantly. This suggests that differences in expression profiles between species, at the transcriptional or at the post-transcriptional level, may be important.

Post-transcriptional regulators of gene expression

Next to global regulatory systems such as the Fur and PerR proteins, which will be discussed elsewhere, a new and exciting development in transcriptomics has been in the field of post-transcriptional regulation. This can be mediated either by regulatory proteins or by regulatory RNAs, and we will briefly discuss the latter here.

Non-protein-coding RNAs (ncRNAs or small RNAs) are established post-transcriptional regulators of gene expression, functioning directly on nucleic acids or indirectly on proteins (Storz *et al.*, 2011). In prokaryotes, riboregulation is required in many biological processes, including regulation of virulence factors, and is important in all kingdoms of life. Broad classes of ncRNA exist in prokaryotes and can be grouped based on their mechanism of function. Perhaps the simplest regulatory RNAs are riboswitches and *C. jejuni* contains a thiamine pyrophosphate riboswitch, which is an extended 5′ untranslated region that regulates mRNA *in cis* (Gundogdu *et al.*, 2007). This was predicted upstream of the gene encoding the thiamine biosynthesis protein ThiC (Cj0453), but has not been evaluated experimentally in *C. jejuni*.

As mentioned previously, some *C. jejuni* genomes contain clustered regularly interspaced short palindrome repeats (CRISPR). These are repeated sequences of DNA that are transcribed together in a single transcript and are subsequently cleaved into short repeats. CRISPR loci act as an immune mechanism against viruses guiding CRISPR associated proteins (cas) to recognize and cleave invading foreign DNA (Makarova *et al.*, 2011). The CRISPR-cas system may also be required for virulence, as inactivation of the *csn1* (*cas9*, *cj1523*) marker gene reduces invasion of human intestinal cell lines (Louwen *et al.*, 2013). There is significant interest in the Cas9 proteins of bacteria, both for genome editing and gene regulation studies (Chylinski *et al.*, 2013; Jiang *et al.*, 2013; Qi *et al.*, 2013), and hence the *C. jejuni* Cas9 protein deserves further study for its role in gene regulation.

Discrete ncRNAs can bind proteins and nucleic acids. Those that bind proteins serve as molecular mimics, which compete for regulatory sites of proteins and inhibit their activity (Marzi and Romby, 2012). The CrsA family of protein post-transcriptional regulators are regulated by ncRNAs CsrB/CrsC (Suzuki *et al.*, 2002), but these and the two-component system that regulates their expression have not been identified in *C. jejuni* (Fields and Thompson, 2008).

Nucleic acid binding ncRNAs can be located *in trans* or *in cis* (antisense) in relation to their targets. These are commonly located in intergenic regions and encode RNAs typically ~50 to 250 nt in length (Papenfort and Vogel, 2010). They silence or promote gene expression via complementary base pairing with the mRNA of their target gene (Papenfort and Vogel, 2010). The ncRNA binding regions are surprisingly short and perhaps explain the involvement of a RNA stabilizing protein chaperone, Hfq, which is required for ncRNAs to function in many prokaryotes (Vogel and Luisi, 2011). However, *C. jejuni* lacks Hfq and so either has novel RNA stabilizing chaperones or ncRNAs are able to act independently of protein chaperones. Putative *C. jejuni hfq* genes have been mutated, but did not result in any pleiotropic effects, suggesting that they were not major ncRNA protein chaperones. Identification of such a protein chaperone would aid in the study of *C. jejuni* ncRNAs. RNA-seq using Illumina sequencing confirmed the presence of the ThiC riboswitch, signal recognition particle RNA, which is involved in the co-translational targeting of proteins to membranes (Chaudhuri *et al.*, 2011). RNA-seq also revealed transcripts for the RNA component of RNaseP and for 10S RNA, which rescues stalled ribosomes and tags proteins for degradation (Chaudhuri *et al.*, 2011). In addition, the Rfam database was used to screen all large intergenic regions and has identified five putative ncRNAs (Chaudhuri *et al.*, 2011). Over 60 riboregulators have been discovered in *H. pylori* 26695 (Sharma *et al.*, 2010) and have been shown to regulate virulence genes (Pernitzsch and Sharma, 2012). Therefore, regulatory RNAs clearly have relevant roles in bacterial survival and host interactions of Epsilon-proteobacteria.

In *C. jejuni*, ncRNAs are conserved between different *Campylobacter* species, but are not found outside this genus. This seems to be the case for *H. pylori* also (Pernitzsch and Sharma, 2012), and this suggests that Epsilon-proteobacteria have developed their ncRNA repertoires independently of other prokaryotes. Future ncRNA studies will give great insight into how a relatively small genome can hold large amounts of complex information and how the organism is highly adaptable.

Genome synteny and expression patterns

Next to the location of promoter sequences, transcriptional patterns also depend on genome organization, as operon structures and concerted regulation depend on genes being located in close vicinity. One interesting aspect of *Campylobacter* genomes is the apparent lack of strong genome structure conservation (e.g. gene order) when different species are compared, which may be due to significant levels of horizontal gene transfer, recombination and reorganization, and lack of DNA repair mechanisms. We have compared genome synteny using the GeneOrder 4.0 program (Mahadevan and Seto, 2010), and have quantified the level of genome synteny by calculating the distance between two genes on both the tester and comparator genomes (Fig. 5.4). This comparison highlights the lack of genome synteny within the genus *Campylobacter*. For example, although *C. curvus* and *C. jejuni* have more than half of their gene complement in common (Figs. 5.3A and 5.4A), only half of these genes are found neighbouring, or even in close vicinity to, similar genes in both species. This suggests high levels of genome rearrangement during species divergence. In contrast, *C. jejuni* has high levels of genome synteny, with >90% of genes neighbouring the same gene as in *C. jejuni* NCTC 11168 (Fig. 5.4B). Synteny is focused in certain areas such as the genes encoding the ribosomal proteins, which suggests a strong evolutionary pressure on the co-localization of these genes. Future studies will have to focus on identifying regulatory structures in *Campylobacter* species, and should include the identification of operon structures, for instance using RNA-seq technologies (van Vliet, 2010; Chaudhuri *et al.*, 2011).

Conclusions

The rapid development in genomic and transcriptomic technologies in the last few years now allows for an insight into *Campylobacter* genome evolution. One of the striking findings is the lack of conservation of genome structure, which is both visible in the virtual absence of functional gene clustering in the genome, and the large differences in gene order. This in turn will affect

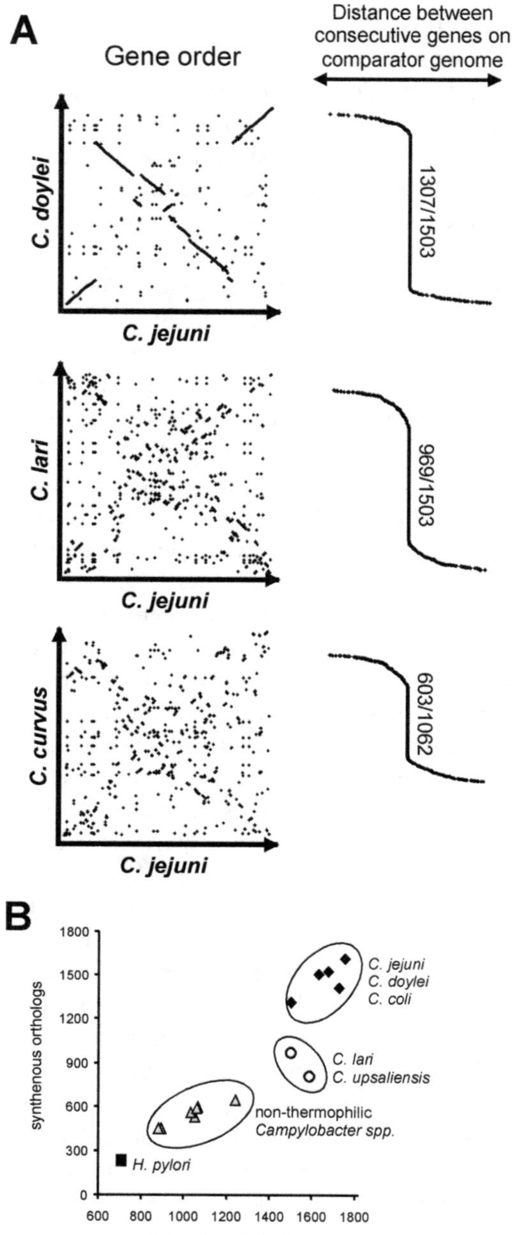

◁ **Figure 5.4** Lack of genome synteny in the genus *Campylobacter*. (A) Pairwise genome synteny of translated annotated features (protein sequences) was determined using GeneOrder 4.0 (Mahadevan and Seto, 2010) and visualized as scatter plots (left side). For quantification, the distance between two genes on the tester genome (*C. jejuni* NCTC 11168) and the corresponding distance between the orthologues on the comparator genome (*C. doylei*, *C. lari* and *C. curvus*) was plotted as line graph (right side), with the plateau representing conservation of synteny. The numbers above the platform indicate number of genes in close vicinity on both genomes and the total number of features present on both genomes. Please note that the latter number includes genes encoding proteins with multiple orthologues (e.g. methyl-accepting chemotaxis proteins). (B) Graph showing the quantification of synteny for comparing *C. jejuni* NCTC 11168 with other *Campylobacter* species. The different subdivisions are highlighted next to the respective data points (filled diamonds: *C. jejuni* and *C. coli*; open circles: other thermophilic *Campylobacter* species, grey triangles: non-thermophilic *Campylobacter* species, filled square: *H. pylori* as out group).

operon structure and transcriptome evolution. However, there are still gaps in our understanding of some of the basic aspects of the biology and virulence of *Campylobacter* species, especially for the non-jejuni species. Targets of future *Campylobacter* research should include further elucidation of its association with different hosts and disease, and should aid the identification of novel targets for prevention and intervention strategies. The big strides in sequencing of *Campylobacter* genomes will have to be matched by the development of other high-throughput phenotypic and molecular approaches to test the hypotheses generated from genomics approaches, and this will be a major challenge in the coming years.

Acknowledgements

The authors wish to thanks members of the IFR Campylobacter research group for helpful discussions. We gratefully acknowledge the support of the Biotechnology and Biological Sciences Research Council (BBSRC) via the Institute Strategic Programme (IFR/08/3 and BB/J004529/1) and a DTG studentship (BB/F016816/1) to M.T.L. We apologize for not being able to cite many publications due to space limitations.

References

Bacon, D.J., Alm, R.A., Burr, D.H., Hu, L., Kopecko, D.J., Ewing, C.P., Trust, T.J., and Guerry, P. (2000). Involvement of a plasmid in virulence of *Campylobacter jejuni* 81-176. Infect. Immun. *68*, 4384–4390.

Batchelor, R.A., Pearson, B.M., Friis, L.M., Guerry, P., and Wells, J.M. (2004). Nucleotide sequences and comparison of two large conjugative plasmids from different *Campylobacter* species. Microbiology *150*, 3507–3517.

van Belkum, A., van den Braak, N., Godschalk, P., Ang, W., Jacobs, B., Gilbert, M., Wakarchuk, W., Verbrugh, H., and Endtz, H. (2001). A *Campylobacter jejuni* gene associated with immune-mediated neuropathy. Nat. Med. *7*, 752–753.

Bourke, B., Chan, V.L., and Sherman, P. (1998). *Campylobacter upsaliensis*: waiting in the wings. Clin. Microbiol. Rev. *11*, 440–449.

Browning, D.F., and Busby, S.J. (2004). The regulation of bacterial transcription initiation. Nat. Rev. Microbiol. *2*, 57–65.

Campbell, B.J., Engel, A.S., Porter, M.L., and Takai, K. (2006). The versatile epsilon-proteobacteria: key players in sulfidic habitats. Nat. Rev. Microbiol. *4*, 458–468.

Carpenter, B.M., Gancz, H., Gonzalez-Nieves, R.P., West, A.L., Whitmire, J.M., Michel, S.L., and Merrell, D.S. (2009). A single nucleotide change affects fur-dependent regulation of sodB in *H. pylori*. PLoS One *4*, e5369.

Carrillo, C.D., Taboada, E., Nash, J.H.E., Lanthier, P., Kelly, J., Lau, P.C., Verhulp, R., Mykytczuk, O., Sy, J., Findlay, W.A., *et al.* (2004). Genome-wide Expression Analyses of *Campylobacter jejuni* NCTC11168 Reveals Coordinate Regulation of Motility and Virulence by flhA. J. Biol. Chem. *279*, 20327–20338.

Carver, T.J., Rutherford, K.M., Berriman, M., Rajandream, M.A., Barrell, B.G., and Parkhill, J. (2005). ACT: the Artemis Comparison Tool. Bioinformatics *21*, 3422–3423.

Champion, O.L., Gaunt, M.W., Gundogdu, O., Elmi, A., Witney, A.A., Hinds, J., Dorrell, N., and Wren, B.W. (2005). Comparative phylogenomics of the food-borne pathogen *Campylobacter jejuni* reveals genetic markers predictive of infection source. Proc. Natl. Acad. Sci. U.S.A. *102*, 16043–16048.

Chaudhuri, R.R., Yu, L., Kanji, A., Perkins, T.T., Gardner, P.P., Choudhary, J., Maskell, D.J., and Grant, A.J. (2011). Quantitative RNA-seq analysis of the *Campylobacter jejuni* transcriptome. Microbiology *157*, 2922–2932.

Chylinski, K., Le Rhun, A., and Charpentier, E. (2013). The tracrRNA and Cas9 families of type II CRISPR-Cas immunity systems. RNA Biol. *10*.

Clark, C.G., Grant, C.C., Pollari, F., Marshall, B., Moses, J., Tracz, D.M., and Gilmour, M.W. (2012). Effects of the *Campylobacter jejuni* CJIE1 prophage homologs on adherence and invasion in culture, patient symptoms, and source of infection. BMC Microbiol. *12*, 269.

Couturier, B.A., Hale, D.C., and Couturier, M.R. (2012). Association of *Campylobacter upsaliensis* with persistent bloody diarrhea. J. Clin. Microbiol. *50*, 3792–3794.

Ernst, F.D., Homuth, G., Stoof, J., Mäder, U., Waidner, B., Kuipers, E.J., Kist, M., Kusters, J.G., Bereswill, S., and van Vliet, A.H.M. (2005). Iron-responsive regulation of the *Helicobacter pylori* iron-cofactored superoxide dismutase SodB is mediated by Fur. J. Bacteriol. *187*, 3687–3692.

Felsenstein, J. (1989). PHYLIP – Phylogeny Inference Package (Version 3.2). Cladistics *5*, 164–166.

Fields, J.A., and Thompson, S.A. (2008). *Campylobacter jejuni* CsrA mediates oxidative stress responses, biofilm formation, and host cell invasion. J. Bacteriol. *190*, 3411–3416.

Fouts, D.E., Mongodin, E.F., Mandrell, R.E., Miller, W.G., Rasko, D.A., Ravel, J., Brinkac, L.M., DeBoy, R.T., Parker, C.T., Daugherty, S.C., *et al.* (2005). Major structural differences and novel potential virulence mechanisms from the genomes of multiple *campylobacter* species. PLoS Biol. *3*, e15.

Friis, L.M., Pin, C., Taylor, D.E., Pearson, B.M., and Wells, J.M. (2007). A role for the tet(O) plasmid in maintaining *Campylobacter* plasticity. Plasmid *57*, 18–28.

Gaasbeek, E.J., Wagenaar, J.A., Guilhabert, M.R., Wosten, M.M., van Putten, J.P., van der Graaf-van Bloois, L., Parker, C.T., and van der Wal, F.J. (2009). A DNase encoded by integrated element CJIE1 inhibits natural transformation of *Campylobacter jejuni*. J. Bacteriol. *191*, 2296–2306.

Gaasbeek, E.J., Wagenaar, J.A., Guilhabert, M.R., van Putten, J.P., Parker, C.T., and van der Wal, F.J. (2010). Nucleases encoded by the integrated elements CJIE2 and CJIE4 inhibit natural transformation of *Campylobacter jejuni*. J. Bacteriol. *192*, 936–941.

Gilbreath, J.J., Cody, W.L., Merrell, D.S., and Hendrixson, D.R. (2011). Change is good: variations in common biological mechanisms in the epsilonproteobacterial genera *Campylobacter* and *Helicobacter*. Microbiol. Mol. Biol. Rev. *75*, 84–132.

Godschalk, P.C., Heikema, A.P., Gilbert, M., Komagamine, T., Ang, C.W., Glerum, J., Brochu, D., Li, J., Yuki, N., Jacobs, B.C., *et al.* (2004). The crucial role of *Campylobacter jejuni* genes in anti-ganglioside antibody induction in Guillain-Barre syndrome. J. Clin. Invest. *114*, 1659–1665.

Gorkiewicz, G., Kienesberger, S., Schober, C., Scheicher, S.R., Gully, C., Zechner, R., and Zechner, E.L. (2010). A genomic island defines subspecies-specific virulence features of the host-adapted pathogen *Campylobacter fetus* subsp. *venerealis*. J. Bacteriol. *192*, 502–517.

Gundogdu, O., Bentley, S.D., Holden, M.T., Parkhill, J., Dorrell, N., and Wren, B.W. (2007). Re-annotation and re-analysis of the *Campylobacter jejuni* NCTC11168 genome sequence. BMC Genomics *8*, 162.

Habib, I., Louwen, R., Uyttendaele, M., Houf, K., Vandenberg, O., Nieuwenhuis, E.E., Miller, W.G., van Belkum, A., and De Zutter, L. (2009). Correlation between genotypic diversity, lipooligosaccharide gene locus class variation, and caco-2 cell invasion potential of *Campylobacter jejuni* isolates from chicken meat and humans: contribution to virulotyping. Appl. Environ. Microbiol. *75*, 4277–4288.

Hofreuter, D., Tsai, J., Watson, R.O., Novik, V., Altman, B., Benitez, M., Clark, C., Perbost, C., Jarvie, T., Du, L., and Galan, J.E. (2006). Unique features of a highly pathogenic *Campylobacter jejuni* strain. Infect. Immun. *74*, 4694–4707.

Hofreuter, D., Novik, V., and Galan, J.E. (2008). Metabolic diversity in *Campylobacter jejuni* enhances specific tissue colonization. Cell Host Microbe *4*, 425–433.

Hughes, R. (2004). *Campylobacter jejuni* in Guillain–Barré syndrome. Lancet Neurol. 3, 644.

Janssen, R., Krogfelt, K.A., Cawthraw, S.A., van Pelt, W., Wagenaar, J.A., and Owen, R.J. (2008). Host-pathogen interactions in *Campylobacter* infections: the host perspective. Clin. Microbiol. Rev. *21*, 505–518.

Jiang, W., Bikard, D., Cox, D., Zhang, F., and Marraffini, L.A. (2013). RNA-guided editing of bacterial genomes using CRISPR-Cas systems. Nat. Biotechnol. *31*, 233–239.

Kienesberger, S., Gorkiewicz, G., Wolinski, H., and Zechner, E.L. (2011). New molecular microbiology approaches in the study of *Campylobacter fetus*. Microb. Biotechnol. *4*, 8–19.

Lawson, A.J., On, S.L., Logan, J.M., and Stanley, J. (2001). *Campylobacter hominis* sp. *nov.*, from the human gastrointestinal tract. Int. J. Syst. Evol. Microbiol. *51*, 651–660.

Louwen, R.P., van Belkum, A., Wagenaar, J.A., Doorduyn, Y., Achterberg, R., and Endtz, H.P. (2006). Lack of association between the presence of the pVir plasmid and bloody diarrhea in *Campylobacter jejuni* enteritis. J. Clin. Microbiol. *44*, 1867–1868.

Louwen, R., Heikema, A., van Belkum, A., Ott, A., Gilbert, M., Ang, W., Endtz, H.P., Bergman, M.P., and Nieuwenhuis, E.E. (2008). The sialylated lipooligosaccharide outer core in *Campylobacter jejuni* is an important determinant for epithelial cell invasion. Infect. Immun. *76*, 4431–4438.

Louwen, R., Nieuwenhuis, E.E., van Marrewijk, L., Horst-Kreft, D., de Ruiter, L., Heikema, A.P., van Wamel, W.J., Wagenaar, J.A., Endtz, H.P., Samsom, J., *et al.* (2012). *Campylobacter jejuni* translocation across intestinal epithelial cells is facilitated by ganglioside-like lipooligosaccharide structures. Infect. Immun. *80*, 3307–3318.

Louwen, R., Horst-Kreft, D., de Boer, A.G., van der Graaf, L., de Knegt, G., Hamersma, M., Heikema, A.P., Timms, A.R., Jacobs, B.C., Wagenaar, J.A., *et al.* (2013). A novel link between *Campylobacter jejuni* bacteriophage defence, virulence and Guillain-Barre syndrome. Eur. J. Clin. Microbiol. Infect. Dis. *32*, 207–226.

Mahadevan, P., and Seto, D. (2010). Rapid pair-wise synteny analysis of large bacterial genomes using web-based GeneOrder4.0. BMC Res. Notes 3, 41.

Makarova, K.S., Haft, D.H., Barrangou, R., Brouns, S.J., Charpentier, E., Horvath, P., Moineau, S., Mojica, F.J., Wolf, Y.I., Yakunin, A.F., *et al.* (2011). Evolution and classification of the CRISPR-Cas systems. Nat. Rev. Microbiol. *9*, 467–477.

Marzi, S., and Romby, P. (2012). RNA mimicry, a decoy for regulatory proteins. Mol. Microbiol. *83*, 1–6.

Muraoka, W.T., and Zhang, Q. (2011). Phenotypic and genotypic evidence for L-fucose utilization by *Campylobacter jejuni*. J. Bacteriol. *193*, 1065–1075.

Papenfort, K., and Vogel, J. (2010). Regulatory RNA in bacterial pathogens. Cell Host Microbe *8*, 116–127.

Parker, C.T., Quinones, B., Miller, W.G., Horn, S.T., and Mandrell, R.E. (2006). Comparative genomic analysis of *Campylobacter jejuni* strains reveals diversity due to genomic elements similar to those present in *C. jejuni* strain RM1221. J. Clin. Microbiol. *44*, 4125–4135.

Parker, C.T., Miller, W.G., Horn, S.T., and Lastovica, A.J. (2007). Common genomic features of *Campylobacter jejuni* subsp. *doylei* strains distinguish them from *C. jejuni* subsp. *jejuni*. BMC Microbiol. *7*, 50.

Parkhill, J., Wren, B.W., Mungall, K., Ketley, J.M., Churcher, C., Basham, D., Chillingworth, T., Davies, R.M., Feltwell, T., Holroyd, S., *et al.* (2000). The genome sequence of the food-borne pathogen *Campylobacter jejuni* reveals hypervariable sequences. Nature *403*, 665–668.

Pearson, B.M., Pin, C., Wright, J., I'Anson, K., Humphrey, T., and Wells, J.M. (2003). Comparative genome analysis of *Campylobacter jejuni* using whole genome DNA microarrays. FEBS Lett. *554*, 224–230.

Pearson, B.M., Gaskin, D.J., Segers, R.P., Wells, J.M., Nuijten, P.J., and van Vliet, A.H.M. (2007). The complete genome sequence of *Campylobacter jejuni* strain 81116 (NCTC11828). J. Bacteriol. *189*, 8402–8403.

Pernitzsch, S.R., and Sharma, C.M. (2012). Transcriptome complexity and riboregulation in the human pathogen *Helicobacter pylori*. Front. Cell Infect. Microbiol. 2, 14.

Petersen, L., Larsen, T.S., Ussery, D.W., On, S.L., and Krogh, A. (2003). RpoD promoters in *Campylobacter jejuni* exhibit a strong periodic signal instead of a –35 box. J. Mol. Biol. *326*, 1361–1372.

Porcelli, I., Reuter, M., Pearson, B.M., Wilhelm, T., and van Vliet, A.H.M. (2013). Parallel evolution of genome structure and transcriptional landscape in the Epsilonproteobacteria. BMC Genomics (in press).

Qi, L.S., Larson, M.H., Gilbert, L.A., Doudna, J.A., Weissman, J.S., Arkin, A.P., and Lim, W.A. (2013). Repurposing CRISPR as an RNA-guided platform for sequence-specific control of gene expression. Cell *152*, 1173–1183.

Richards, V.P., Lefebure, T., Pavinski Bitar, P.D., and Stanhope, M.J. (2013). Comparative characterization of the virulence gene clusters (lipooligosaccharide [LOS] and capsular polysaccharide [CPS]) for *Campylobacter coli*, *Campylobacter jejuni* subsp. *jejuni* and related *Campylobacter* species. Infect. Genet. Evol. *14*, 200–213.

Rodin, S., Andersson, A.F., Wirta, V., Eriksson, L., Ljungstrom, M., Bjorkholm, B., Lindmark, H., and Engstrand, L. (2008). Performance of a 70-mer oligonucleotide microarray for genotyping of *Campylobacter jejuni*. BMC Microbiol. *8*, 73.

Sharma, C.M., Hoffmann, S., Darfeuille, F., Reignier, J., Findeiß, S., Sittka, A., Chabas, S., Reiche, K., Hackermüller, J., Reinhardt, R., Stadler, J.F., and Vogel, J. (2010). The primary transcriptome of the major human pathogen *Helicobacter pylori*. Nature *464*, 250–255.

Sheppard, S.K., Didelot, X., Jolley, K.A., Darling, A.E., Pascoe, B., Meric, G., Kelly, D.J., Cody, A., Colles, F.M., Strachan, N.J., *et al.* (2013). Progressive genome-wide introgression in agricultural *Campylobacter coli*. Mol. Ecol. *22*, 1051–1064.

Siqueira, J.F., Jr., and Rocas, I.N. (2003). *Campylobacter gracilis* and *Campylobacter rectus* in primary endodontic infections. Int. Endod. J. *36*, 174–180.

Stabler, R.A., Larsson, J.T., Al-Jaberi, S., Nielsen, E.M., Kay, E., Tam, C.C., Higgins, C.D., Rodrigues, L.C., Richardson, J.F., O'Brien, S.J., and Wren, B.W. (2013). Characterization of water and wildlife strains as a subgroup of *Campylobacter jejuni* using DNA microarrays. Environ. Microbiol. *15*, 2371–2183.

Stahl, M., Friis, L.M., Nothaft, H., Liu, X., Li, J., Szymanski, C.M., and Stintzi, A. (2011). L-fucose utilization provides *Campylobacter jejuni* with a competitive advantage. Proc. Natl. Acad. Sci. U.S.A. *108*, 7194–7199.

Storz, G., Vogel, J., and Wassarman, K.M. (2011). Regulation by small RNAs in bacteria: expanding frontiers. Mol. Cell *43*, 880–891.

Suzuki, K., Wang, X., Weilbacher, T., Pernestig, A.K., Melefors, O., Georgellis, D., Babitzke, P., and Romeo, T. (2002). Regulatory circuitry of the CsrA/CsrB and BarA/UvrY systems of *Escherichia coli*. J. Bacteriol. *184*, 5130–5140.

Taboada, E.N., Acedillo, R.R., Carrillo, C.D., Findlay, W.A., Medeiros, D.T., Mykytczuk, O.L., Roberts, M.J., Valencia, C.A., Farber, J.M., and Nash, J.H. (2004). Large-scale comparative genomics meta-analysis of *Campylobacter jejuni* isolates reveals low level of genome plasticity. J. Clin. Microbiol. *42*, 4566–4576.

Umehara, Y., Kudo, M., and Kawasaki, M. (2009). *Campylobacter fetus* meningitis in a patient with Crohn's disease. Inflamm. Bowel Dis. *15*, 645–646.

van Vliet, A.H.M. (2010). Next generation sequencing of microbial transcriptomes: challenges and opportunities. FEMS Microbiol. Lett. *302*, 1–7.

Vandamme, P., Daneshvar, M.I., Dewhirst, F.E., Paster, B.J., Kersters, K., Goossens, H., and Moss, C.W. (1995). Chemotaxonomic analyses of *Bacteroides gracilis* and *Bacteroides ureolyticus* and reclassification of *B. gracilis* as *Campylobacter gracilis* comb. nov. Int. J. Syst. Bacteriol. *45*, 145–152.

Vogel, J., and Luisi, B.F. (2011). Hfq and its constellation of RNA. Nat. Rev. Microbiol. *9*, 578–589.

Wosten, M.M. (1998). Eubacterial sigma-factors. FEMS Microbiol. Rev. *22*, 127–150.

Young, K.T., Davis, L.M., and Dirita, V.J. (2007). *Campylobacter jejuni*: molecular biology and pathogenesis. Nat. Rev. Microbiol. *5*, 665–679.

Repetitive DNA: A Major Source of Genetic Diversity in *Campylobacter* Populations?

6

Jack Aidley and Christopher D. Bayliss

Abstract

A marked feature of the genomes of *Campylobacter jejuni* and other *Campylobacter* species is the presence of long polyG tracts within the reading frames of genes. Instability in these repetitive DNA sequences leads to frequent, reversible switches in expression of the associated genes termed phase variation. High rates of phase variation have been detected for *C. jejuni* genes using reporter constructs and antibodies and these high switching rates enable rapid access to hundreds of different phenotypes. Observations of *C. jejuni* populations from *in vitro* and *in vivo* populations indicate the presence of many of these phase-variable phenotypes. We speculate on the forces driving the distributions of these phase-variable genotypes and on the potential impact of phase variation on the commensal and disease-causing attributes of *C. jejuni*.

Introduction

Mechanisms of phase variation in bacterial commensals and pathogens

Programmed alterations in gene expression are utilized by bacteria for adaptation to fluctuations in their external environments and to a range of selective pressures. These programmed alterations frequently involve transcriptional, post-transcriptional or post-translational regulation of gene expression and are not accompanied by alterations in the genetic material. In general these responses are rapid and require a signal input for initiation of a regulatory alteration in gene expression. Mutational processes – *de novo* mutation and recombination – can also facilitate adaptation to environmental fluctuation but are accompanied by alterations in the genetic material, require a population of cells and occur over long time scales. Localized hypermutation refers to a group of mutational or epigenetic mechanisms that decrease the adaptation time normally associated with mutational processes (Moxon *et al.*, 1994). In this case, adaptation is mediated by selection of the 'fittest' variant in the new environment and does not require integration of a signal. The loci subject to these heightened mutation rates are referred to as 'contingency loci' differentiating them from the 'house-keeping' genes with mutation rates close to the average across the genome (Moxon *et al.*, 1994). This name also highlights the stochastic nature of the appearance of mutants or variants and the potential for variants to arise before the imposition of the selective pressure to which they mediate adaptation.

Localized hypermutation has been extensively reviewed elsewhere (Bart *et al.*, 1999; Deitsch *et al.*, 2009; Moxon *et al.*, 2007; van der Woude and Baumler, 2004). There are five major mechanisms: high level intragenomic recombination (e.g. neisserial *pilE*/*pilS* genes); gene amplification (e.g. capsule locus of Hib); site-specific recombination (e.g. Fim locus in *Escherichia coli*); Dam-mediated epigenetic regulation (e.g. p-pili in uropathogenic *E. coli*); and simple sequence repeats (SSR). The latter three processes have the distinction of being able to generate reversible, high frequency ($>1\times10^{-5}$ variants/generation) changes in gene expression referred to as phase variation (PV) (Andrewes, 1922; Moxon *et al.*, 1994) (Fig. 6.1).

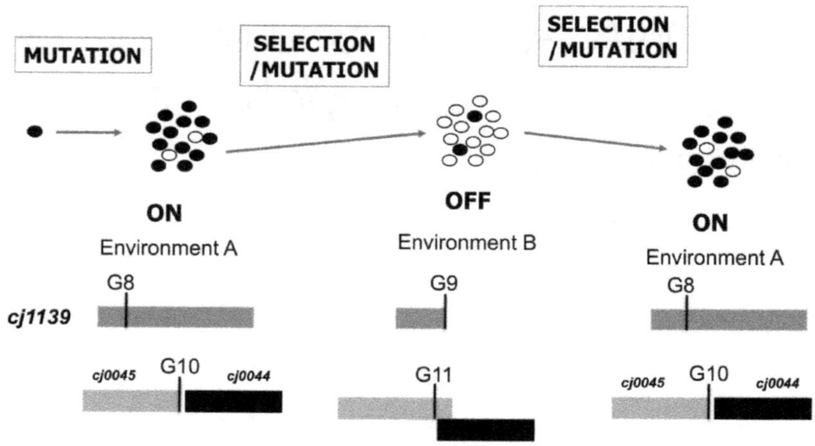

Figure 6.1 PolyG tract mediated phase variation in *C. jejuni*. A single cell in the ON state for a phase-variable gene (black oval) replicates to form a population. In the absence of selection, the hypermutable polyG tract of the phase-variable generates a high frequency of OFF (white oval) phase variants. The ON variants are adapted for survival in environment A. An external change to environment B exerts a selective pressure on this population and gives the OFF variants a fitness advantage leading to OFF variants dominating the population. A reversal to the original environment results in resetting of the population due to the presence of ON variants generated by phase variation. Phase variation is usually mediated by alterations in polyG tracts present within the reading frame (as shown for *cj1139*), which alter the reading the reading frame, truncating the gene product and switching OFF gene function. Gene *cj0045* has a tract at the end of the gene, which may either alter the function of the gene product or influence expression of *cj0044* (see text).

PV is a major feature of *Campylobacter* species and is mediated by SSRs (Fouts *et al.*, 2005; Parkhill *et al.*, 2000). Multiple genes encoding modifiers of surface determinants are potentially subject to PV in *Campylobacter*. In this chapter, we will review the mechanism of switching, the functions of phase-variable genes, and the impact of selective and population-based factors on the proportions of phase variants in *Campylobacter jejuni* populations.

Repetitive DNA in bacterial species

Repetitive DNA sequences are extremely common in eukaryotes, with one recent estimate suggesting they make up over two thirds of the human genome (de Koning *et al.*, 2012). In contrast, repetitive DNA is comparatively rarer in bacterial species. This is because bacterial repeat sequences are generally restricted to intergenic regions, which constitute a small subfraction of bacterial genomes. The types of repetitive DNA found in bacterial species also differ with repeat arrays containing 10s (instead of 100–1000s) of repeat units and wide variations between species in the numbers of insertion sequences or other mobile elements (Treangen *et al.*, 2009).

Some forms of repetitive DNA in bacterial species appear to be molecular parasites with no known function [for example, RS elements in neisserial species (Schoen *et al.*, 2009) or STAR repeats in staphylococci (Purves *et al.*, 2012)], whilst others have important adaptive functions. One example is CRISPR repeats that have been characterized as a form of adaptive immune system protecting bacteria against phage infection (Marraffini and Sontheimer, 2010).

This review will concentrate on just one class of repetitive DNA: simple sequence repeats (SSRs). Simple sequence repeats consist of multiple copies of a short, identical sequence or repeat unit, typically consisting of 1–9 nucleotides (equivalent repeats in eukaryotes are usually referred to as microsatellites). SSR tracts consist of arrays of directly arranged repeats usually consisting of less than 50 units (Moxon *et al.*, 2007). SSRs should not be confused with variable number tandem repeats (VNTRs), which consist of non-identical repeats also arranged as arrays of direct repeats.

VNTRs have much lower mutation rates and have not been associated with PV but are widely utilized for epidemiological purposes (Feschenko *et al.*, 2003; Schouls *et al.*, 2009).

SSRs and simple sequence contingency loci

SSRs are highly susceptible to mutation resulting in variations in the number of repeats. Mutations are thought to occur during DNA replication as a result of slipped-strand mispairing (Levinson and Gutman, 1987) and all the available evidence points to the existence of this mechanism (Bayliss, 2009). Mispairing occurs during duplication of DNA when a loop forms in either the new or template strand, meaning that extra or fewer repeat units are included in the new DNA strand. If the SSR is located within the reading frame of a gene and consists of non-triplet repeat units then the change in repeat number will result in a frameshift mutation. SSR may also be located within the promoter region where repeat number changes alter the relative positioning of key elements outside of their functional bounds and hence increase/decrease transcription. Both of these processes can effectively switch the relative gene ON or OFF and hence cause PV (Fig. 6.1) (Moxon *et al.*, 2007). In a few cases the repeat tracts are positioned at the end of the gene leading to a potential for alterations in gene function or changes in regulation of downstream genes (Ritz *et al.*, 2001; Snyder *et al.*, 2010).

SSR-containing loci are a subset of the contingency loci referred to as simple sequence contingency loci (SSCL) (Bayliss *et al.*, 2001). Bacterial species vary in the number of SSCL and the predominant repeat type present in these loci. Amongst the species with multiple SSCL (i.e. > 10), *Haemophilus influenzae* contains tetranucleotide, *Neisseria meningitidis* has mono-, tetra- and pentanucleotide and *Helicobacter pylori* has mono- and dinucleotide repeats (Bayliss *et al.*, 2001; Hood *et al.*, 1996; Saunders *et al.*, 1998, 2000). *Campylobacter* species only have mononucleotide repeats, suggesting a separate evolutionary pathway has led to evolution of SSCL in this genus (Parkhill *et al.*, 2000).

Types, prevalence and mutability of SSRs in *Campylobacter* genomes

Campylobacter genomes contain multiple mononucleotide SSRs

Campylobacter genomes are approximately 1.7 Mbp in size and have a low GC content of around 30% (Fouts *et al.*, 2005). Outside of repetitive rRNA clusters, simple sequence repeats (SSRs) form the major class of repetitive DNA in *Campylobacter* genomes. The majority of these SSRs take the form of a single nucleotide repeated 6–12 times (i.e. a mononucleotide tract). Bioinformatic analysis has demonstrated an over-representation of these SSRs within the amino- and carboxy-termini of protein coding sequences of *C. jejuni* genomes, implying that their distribution is not simply down to random sequence variation, and a higher mutability for polyG or polyC tracts as compared to polyA and polyT tracts (Lin and Kussell, 2012). The higher mutability of polyG tracts may be connected with DNA structure as these tracts are less tightly twisted than equivalent A or T sequences and general mixed sequences of DNA, resulting in lower stability and a higher propensity for displacement of nucleotides (Dornberger *et al.*, 1999).

A particular feature of *Campylobacter* genomes is the over-representation of long tracts (eight or more nucleotides) of guanine–cytosine pairs with the guanine in the sense direction for transcription (hence polyG tracts). In *C. jejuni* genomes, the median lengths of the polyG tracts are 9 and 10 repeats with very few 7G tracts, suggesting a strong selection or molecular driver for the longer tract lengths (Bayliss *et al.*, 2012). PolyG tracts of eight or more units are therefore regarded as the major mechanism of PV in *C. jejuni*. It is notable that other *Campylobacter* species are characterized by the presence of much longer polyG tracts than detected in *C. jejuni* (Fouts *et al.*, 2005). Long tracts of A or T nucleotides are also present within these genomes but are present as a continuum of shorter tracts. This indicates that the longer polyAT tracts are present due to chance, associated with the AT-rich nature of these genomes, and are not mediators of PV.

The advent of whole genome sequencing has allowed the identification of all polyG SSRs in the *C. jejuni* genomes. The first *C. jejuni* strain to be sequenced was NCTC11168 (Parkhill *et al.*, 2000) followed subsequently by completely closed genome sequences for other key research strains (Fouts *et al.*, 2005; Hofreuter *et al.*, 2006; Pearson *et al.*, 2007). Multiple other genome sequences have been generated and while sequence coverage is 90% or more break points between contigs are often associated with polyG tracts. Comparisons of the number of polyG/polyC tracts with eight or more repeats detected considerable variability: e.g. NCTC11168 has 29 identified SSRs, whilst 81116 has 17 and the non-closed M1 genome sequence has 11. Analysis of the conservation of repeat-containing loci in four genomes indicated that only 5 of the 55 loci are present and phase-variable in all strains whilst some have a distributive presence in genomes and other loci are conserved but lack repeats. This variability is an exemplar of genes that provide an adaptive advantage in a restricted set of selective environments and is typical of contingency loci (Moxon *et al.*, 1994).

The majority of the long polyG SSRs are present within the N-terminal coding regions of *Campylobacter* genes (Fig. 6.1). These tracts code for glycine and are presumed to exert neutral effects on protein function. An exception to this rule is *cj0045* in the NCTC11168 genome where the tract is present at the end of the reading frame. In this case changes in tract length add 0 to 15 amino acids to the end of the protein, which may alter protein function. However, one of the termination codons is adjacent to the initiation codon of *cj0044* whilst the others are located beyond this codon. It is possible that changes in tract length alter the efficiency of translation of this downstream gene as the ribosomes inefficiently scan back to the *cj0044* initiation codon from the downstream termination codons. Another exception is *cj0685* This gene has a polyC tract coding for proline at its centre. Changes in this tract may influence protein function as prolines can disrupt alpha helices and other types of protein structures. Alternatively, the mutability may be influenced by differential effects of transcription on polyC and polyG tracts. Finally, a small number of tracts are present in intergenic regions and pseudogenes. The intergenic tracts may influence gene expression although there is no evidence to support such a role as the tracts are not obviously located within promoters. These tracts provide useful controls as alterations are unlikely to be driven by selection and can be utilized for assessing the impact of mutability and population parameters on repeat changes.

DNA repair mechanisms and phase variation in *Campylobacter*

C. jejuni appears to lack the mismatch repair mechanism of similar species, as they possess no homologues of the *mutS* and *mutL* genes usually considered central to this kind of repair mechanism (Ambur *et al.*, 2009). The lack of such a mechanism increases mutation rates and has a major destabilizing effect on mono-, di-, and trinucleotide SSR. Moreover, whereas as in species with mismatch repair there is an overall bias towards contraction in homonucleotide sequences of eight base pairs or less, *C. jejuni* shows an equal prevalence of insertions and deletions (Kumar and Nagarajaram, 2012). However, point mutation rates are comparable to species with mismatch repair (Hanninen and Hannula, 2007) suggesting the presence of mechanisms capable of correcting mismatches. Whilst the impact of some DNA repair pathways on basal mutation rates has been examined in *C. jejuni* (Gaasbeek *et al.*, 2009), it is not known how these pathways influence switching rates of the phase-variable genes.

Rates of variation in SSRs of *C. jejuni*

Longer SSR tracts have higher mutation rates than shorter tracts. This can be intuitively understood as longer tracts have more locations at which strand slippage can occur and thus more opportunities for mutation. Empirical data on PV rates has been obtained for three *C. jejuni* genes (Bayliss *et al.*, 2012) by preparing serial dilutions of an initial colony grown on an agar plate and then testing the resultant colonies for phenotypic differences from the initial colony. This testing utilized either immunoblotting (*capA*) or *lacZ* reporter constructs. Rates of variation were then calculated from these raw frequencies using the method described in (Drake, 1991). With mutation rates of between 4×10^{-3} and 4×10^{-4}, these results highlight the high level of variation in *C. jejuni* poly-G tracts. These PV rates are 10- to 100-fold higher than the rates detected for equivalent tracts in *N. meningitidis* (Martin *et al.*, 2004; Richardson *et al.*, 2002). This difference was postulated as being due to the divergent absence and presence of mismatch repair in these species and extrapolated from the observation of a 100- to 1000-fold increase in mononucleotide repeat-mediated PV rates of *N. meningitidis mutS* mutants. These rates are, however, equivalent to the switching rates of the tetranucleotide repeats in *H. influenzae* phase-variable genes suggesting commonality in the selection for particular switching rates (De Bolle *et al.*, 2000).

As well as investigating variation in the naturally occurring tract lengths, a modified version of the *cj1139-lacZ* reporter was created by site directed mutagenesis with a G11 rather than G8 tract, allowing the effect of altering tract length to be studied in isolation (Bayliss *et al.*, 2012). This modification produced a tenfold increase in mutability, emphasizing the importance of tract length in controlling of PV rates in *C. jejuni*. These experimental measurements of PV rates have been critical for providing the basis for modelling of the impact of PV on how the proportions of phase variants change when *C. jejuni* populations are subject to a range of selective pressures.

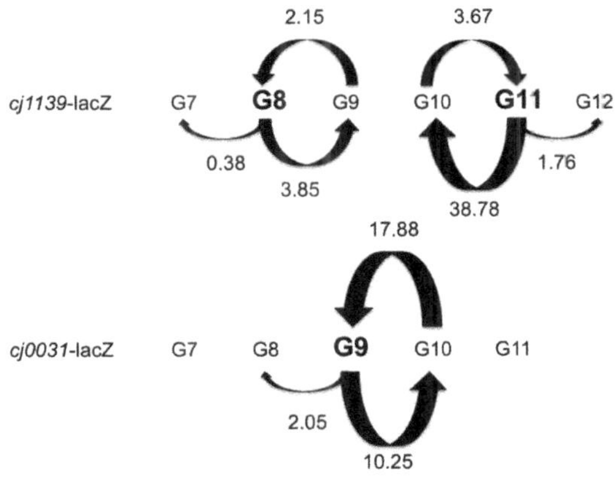

Figure 6.2 Switching rates of *C. jejuni* phase-variable genes. This figure summarizes the results of Bayliss *et al.* (2012) showing experimentally derived mutation rates (mutations/division) for insertions and deletions in the repeat tracts of two genes. Rates were determined using *lacZ* reporter constructs and a combination of phenotypic observations of blue-white phase variation combined with sequence/GeneScan analysis of repeat tracts.

Mutational patterns of phase-variable *C. jejuni* genes

The relative proportions of insertions and deletions in the SSRs of phase-variable genes are critical for understanding their evolution and for modelling the behaviour of these loci. Alterations in phase variants were examined during PV experiments by a combined sequencing and GeneScan approach (Bayliss *et al.*, 2012). The latter technique involves accurate determination of the sizes for PCR products generated using fluorescently labelled primers spanning the repeat tracts. All the examined variants exhibited alterations by a single nucleotide showing that mutations in these tracts occur by small step-wise changes. A bias for insertions over deletions was detected for a G8 tract with inversion towards a higher propensity for deletions in G10 and G11 tracts. These analyses led to predictions for specific types of mutational events (Fig. 6.2). Based on these observations, it was suggested that molecular drivers rather than selection are responsible for the preponderance of G9 and G10 tracts in *C. jejuni* genomes (Bayliss *et al.*, 2012).

Variation in SSRs has functionally important phenotypic effects

Phase-variable genes are not randomly located throughout the *Campylobacter* genome but are particularly associated with genes involved in modifying the lipooligosaccharide layer, the capsular polysaccharides and the flagella (Parkhill *et al.*, 2000; Wren *et al.*, 2001) (Table 6.1). These modifications influence motility, auto-agglutination, virulence, immune-invasion and phage susceptibility among other things. An important consideration particular to phase-variable loci is the fitness advantage associated with both the ON and OFF states (Fig. 6.1). It is assumed that a selective pressure for both states is required for evolution of PV in a particular locus. The

Table 6.1 Distribution of phase-variable genes among functional groups of *C. jejuni* genes

Strain (LOS class) (total no. of phase-variable genes)	LOS locus	Flagella locus	Capsule locus	Outer membrane protein	Other[2]
11168 (C) (29)[1]	2	9	6	1 (*capA*)	1 (RM), 1 (*clpX*), 1 (cipA), 3 (intergenic) 3 (unknown), 2 (pseudogene)
1221 (F) (24)[1]	2	7	2	0	4 (unknown), 1 (*clpX*), 1 (cipA), 5 (intergenic), 2 (pseudo)
81-176 (?) (19)	1	5	4	0	1 (cipA), 3 (unknown), 2 (intergenic) 3 (pseudogene
81116 (E) (17)	2	4	3	0	1 (cipA), 2 (unknown), 5 (intergenic)

[1]LOS, flagella and capsule loci of strain NCTC11168 are located *between cj1134-cj1146c*, *cj1293–1343* and *cj1414–1449*, respectively; the relative positions in strain 1221 are *CJE1277–CJE1282*, *CJE1485–CJE1532* and *CJE1601–CJE1622*. [2]RM, restriction-modification system.

following section will discuss the phase-variable genes in this context.

Phase variable flagellar genes of *C. jejuni*

Reversible variation in the flagellar genes of *C. jejuni* was identified many years before the advent of whole genome sequencing by Caldwell *et al.* (1985). These researchers observed reversible switching between flagellate and aflagellate strains. The motile, flagellate strains were able to successfully infect mice while the aflagellate strains were not. A possible explanation for the persistence of the aflagellate variant is that the flagella appears to be an important immune system target (Blaser *et al.*, 1986; Pavlovskis *et al.*, 1991) and thus aflagellate variants may have an advantage in immune system evasion once infection has occurred. The mechanism of switching was unclear until subsequent work demonstrated that the *flgR* and *flgS* gene products form a two-component system controlling expression of σ^{28}- and σ^{54}-dependent flagellar genes. Mutations in these genes were responsible for the reversible expression of *flaA* and hence of motility in *C. jejuni* strains (Hendrixson, 2006, 2008). However, no reproducible reversible changes in SSRs of *flgR* and *flgS* were detected and the switching rates were 10–100-fold lower than measured for the phase-variable polyG tracts. These genes cannot, therefore, be classed as 'true' phase-variable loci and fall into a grey area, which consists of many loci with intermediately high mutation rates but lacking a specific hypermutable sequence.

While the *flgR/flgS* variation results in the total absence of flagella, many more subtle variations in the flagella structure and motion are also under phase-variable control with the flagella locus containing the highest concentration of polyG tract-containing genes (Table 6.1). *C. jejuni* was the bacterial species in which post-translation modification of flagellin was first demonstrated (Logan *et al.*, 1989) and subsequent work has demonstrated involvement of phase-variable genes in the pattern of modification. *C. jejuni* flagella consist mostly of repeated units of the FlaA protein, these units are decorated on the outside by a variety of sugars and sugar derivatives. In strain 81-176 there are 19 O-linked pseudaminic derived sugars added to serine/threonine residues of the flagellin protein (Thibault *et al.*, 2001), but this number varies in other strains and some strains include legionaminic acid and pseudamnic acid derivatives (Logan *et al.*, 2009).

The *maf* (motility accessory factor) family of genes is a group of paralogous genes involved in flagella biosynthesis and function. In NCTC11168, there are seven of these genes, of which three are phase variable. These genes were identified by Karlyshev *et al.* (2002) who determined that *maf5* was required for flagellar expression and motility. Work by van Alphen *et al.* (2008) into the function of the *maf4* gene, showed that switching of this gene altered the glycosylation pattern of the flagella and significantly reduced auto-agglutination when the modification was present. Thus the bacteria may be able to switch between aggregation and free-swimming phenotypes, which may alternately favour adhesion to surfaces and dispersal.

Several additional genes are predicted to contribute to glycosylation of flagella. Bioinformatics analyses has identified five phase-variable genes in the NCTC11168 genome known as the '617 family' which have no known homologues in other species but are thought to alter flagella glycosylation. The existence of multiple phase-variable glycosyltransferases for flagella suggests the presence of a large repertoire of glycosylation patterns for this important surface molecule.

This variety in the phase-variable modifications of the flagella suggests roles for PV in changing the functions of the flagella but also for altering the antigenicity of the flagella such that switching mediates immune evasion. However the major selective force for PV of these modifications is likely to be phage resistance as many phages are known or predicted to bind the flagella of *Campylobacters*.

Phase variation of LOS and autoimmune syndromes

The outer membrane of *Campylobacter* contains lipooligosaccharides (LOS). The lack of an O-antigen is compensated by high variability in the polysaccharides and other modifications of the surface-exposed component of LOS. These LOS modifier genes are located in one region of the genome, which exhibits significant plasticity. To date, 19 different LOS regions (A-S) have been recognized amongst *C. jejuni* strains (Gilbert *et al.*, 2002; Parker *et al.*, 2008). Each of these regions contains one or two phase-variable genes with only the K type lacking any evidence of polyG tracts.

As with the phase-variable flagella genes, there is only limited experimental data on the specific functions and phenotypic consequences of these modifications. The role of *C. jejuni* LOS in triggering the Guillain-Barré and Miller-Fisher autoimmune syndromes (see Nachamkin *et al.*, 1998) has however been a major area of interest in *Campylobacter* research. Here too, phase-variable genes have been implicated. Current research suggests that the causal factor triggering autoimmunity is mimicry of the ganglioside surface molecules found on human ganglion cells by the LOS of *C. jejuni*, in particular GM_1, GM_2 and GM_3. Work by Linton *et al* (Linton *et al.*, 2000) showed that elaboration of a GM_1 modification of the LOS is mediated by a gene known as *wlaN* (or *cj1139c*). A functional gene contained a poly-G tract of eight bases while loss of expression, an OFF phenotype, was associated with a G9 tract. The OFF variants elaborated a GM_2 LOS epitope. Such phase-variable genes affecting ganglioside mimicry are not limited to NCTC11168 but have also been identified in 81-176. Work by Guerry *et al.* (2002) demonstrated that variation in tract length within the *cgtA* gene altered the LOS from mimicking the GM_2 class of gangliosides to the GM_3 class. This shift from GM_2 to GM_3 mimicry produced an *in vitro* increase in infectivity when tested in INT407 cells derived from embryonic intestinal tissue. PV of these genes may therefore have functional consequences facilitating colonization of host cells in some states. It should be noted that whilst these genes and their PV will impact on *Campylobacter*-associated diseases, it is not clear if this is an evolved process as humans may or may not be one of the hosts routinely colonized during the natural life cycle of *Campylobacter*. If humans are a native host, then molecular mimicry may have evolved in order to avoid immune detection and switching may facilitate escape of the antibodies elicited in a small number of hosts. In this respect it is relevant that Guillain–Barré only occurs in a small number of people who are subject to *Campylobacter*-induced gastroenteritis. Some of these gangliosides are also present in chickens (Nakamura *et al.*, 1983) and hence molecular mimicry may have evolved in part to facilitate survival in birds.

PV of capsular modifications and phage resistance

A polysaccharide capsule is present on *C. jejuni* and other *Campylobacters*. As with flagella and LOS, there is a major locus of capsule-associated genes and this locus exhibits significant plasticity and contains multiple phase-variable genes (Table 6.1). The phase-variable genes control both expression of the capsule and addition of specific modifications (Guerry *et al.*, 2012). Thus, a polyG tract is often present in *dmhA*, whose product is proposed to facilitate conversion of heptose to deoxyheptose during synthesis of capsule in some *C. jejuni* serotypes. Loss of capsule is associated with a modest decrease in invasion and heightened susceptibility to serum (and presumably to complement). Capsular ON phase variants may therefore facilitate evasion of innate immune responses and invasion of host tissues.

Capsular polysaccharides of *C. jejuni* frequently include an unusual post-translation modification: the addition of an O-methyl phosphoramidate (MeO*P*N) group. The MeO*P*N group has a chemical structure, $CH_3OP(O)(NH_2)(OR)$, which includes a phosphorus–nitrogen bond that is extremely rare. In fact, the discovery of this bond in *C. jejuni* capsular polysaccharide modifications was the first observed in bacteria (Szymanski *et al.*, 2003). Synthesis of MeOPN is mediated by a set of conserved genes, *cj1415–cj1418*. These MeO*P*N groups can be added to various locations on the repeat unit of the capsular polysaccharide, with two principal locations identified on the repeat unit of NCTC11168: either attached to the Gal*f*NAc sugar (*N*-acetylgalactosamine in the furanose configuration) or the heptose sugar (McNally *et al.*, 2007; Sorensen *et al.*, 2012).

Work by Sørensen *et al.* (2011) identified the MeO*P*N- Gal*f*NAc moiety as being recognized by the bacteriophage F336, and linked to host sensitivity to this phage. F336 is a member of the *Myoviridae* family of DNA viruses that have distinctive icosahedral 'head' sections housing the DNA and a contractile hexagonally symmetrical 'tail' that terminates in a number of long fibres. In other members of this family of phage the host invasion process begins with the attachment of the tail fibres to surface receptors on the bacterium followed by contraction of the tail and injection of the genetic material from the head section. Sørensen *et al.* (2011) were able to demonstrate that the MeO*P*N- Gal*f*NAc modification was acting as a receptor for F336.

Intriguingly, the genes responsible for additions of MeOPN to the capsule are phase-variable. Thus, *cj1421* is responsible for adding this MeOPN group to the Gal*f*NAc sugar. When the tract length in this gene is 9G, the gene is active whereas 8G and 10G variants are non-functional and the group is not added. The presence of the phage should produce a selective pressure for the OFF state of the gene, and *in vivo* experiments confirmed that the prevalence of the OFF state rose after experiments in which the bacterium was passaged in chicks inoculated with the phage (Sorensen *et al.*, 2012). However, the interaction between F336 and the phase-variable genes of *C. jejuni* is more complex than this simplified picture suggests. In fact, three separate genes can interact to provide resistance to F336. The addition of MeO*P*N to heptose is also controlled by a phase-variable gene (*cj1422*), and is also involved in phage resistance; if both *cj1421* and *cj1422* are ON, the bacterium is resistant to F336. It is thought that the modification to the second sugar is blocking access of the phage to the site occupied by the other MeOPN moiety. Thus *cj1422* switching ON provides resistance to F336 but only when *cj1421* is switched ON. Finally Sørensen *et al.* were able to identify a third protective modification – the addition of 6-O-methyl groups – also shown to be controlled by a phase-variable gene, *cj1426* (Sternberg *et al.*, 2012). This still leaves open the involvement of *cj1420*, which is also phase-variable and is thought to encode yet another MeOPN transferase (Guerry *et al.*, 2012). These findings indicate how selection for a single phenotypic trait (resistance to bacteriophage F336) can result in different phase-variable genotypic profiles.

Phase variation of MeOPN impacts on multiple functional roles

The MeO*P*N modifications involved in phage resistance have also been shown to have insecticidal properties. Champion *et al.* (2010) investigated insect colonization by *C. jejuni* using *Galleria mellonella* (wax moth) larvae. This study did not investigate the effect of the phase-variable genes controlling specific MeO*P*N attachments but rather switching of the whole MeO*P*N process using mutants with defective *cj1416*. Cj1416 is a putative sugar nucleotidyltransferase that is required for the addition of MeO*P*N groups to the CPS (McNally *et al.*, 2007) so the defective mutant lacks MeO*P*N groups. Whereas complemented 11168H/*cj1416 cj1416*$^+$ strains killed the larvae with a very high efficiency, the 11168H/*cj1416* strain showed larval survival statistically similar to that without any bacterial inoculum. Another study has shown that loss of the MeOPN modification elicits an increase in production of pro-inflammatory cytokines by murine dendritic cells (Rose *et al.*, 2012). These findings suggest that the MeOPN variation involved in phage resistance has a wider importance and PV may permit alternation between a phage resistant (OFF) and an immune evasion phenotype (ON). The likelihood of dose-dependent effects associated with the immuno-regulatory phenotype may contribute to the variable patterns of MeOPN addition and PV.

CapA is a phase-variable outer membrane protein

It is perhaps not surprising, given the number of phase-variable genes involved in the modification of surface molecules, that examples of phase-variable outer membrane proteins have also been discovered in *Campylobacter*. Ashgar *et al.* (2007) reported the discovery of PV in the *capA* gene. Putatively recognized *in silico* as an autotransporter protein, experiments with *capA*-deficient mutants identified it as having a role in cell adhesion *in vitro* with human-derived epithelial cells and also demonstrated an affect on host colonization *in vivo* with chickens. However, phase variants in both the ON and OFF states can be readily isolated from chickens infected with wild-type strains (Bayliss *et al.*, 2012). This may indicate switching between a tissue-adhesive and a dispersal or immune evasion phenotype. Redundancy in the adhesins in *C. jejuni* may also contribute to the prevalence of OFF phase variants.

The role of phase variation in restriction–modification systems

Since their discovery, restriction–modification (RM) systems are thought to have evolved primarily as a means to protect against bacteriophage infections (Bertani and Weigle, 1953; Labrie *et al.*, 2010; Meselson *et al.*, 1972). However, the two-stage nature of the process suffers from an inherent flaw: any phage that successfully pass the restriction barrier will go on to multiply in cells carrying the methyltransferase required to modify the restriction site of the RM system and thereby become insensitive to the restriction enzyme. This means that if sufficient phage are present, one or more cells will eventually become infected and then produce modified phage that are able to attack the other cells in the colony. In fact, if the restriction–modification system is entirely uniform the entire phage population will become modified. PV in the RM system allows a defence against this saturation of resistant phage at the population level, since it ensures the presence of cells without the modification system, which will generate susceptible phage (Bayliss *et al.*, 2006). The *cj0031* gene in *C. jejuni* strain 11168 encodes a putative type IIS RM system and is subject to PV owing to a polyG tract present in the reading frame. Whilst PV of this gene has been confirmed (Bayliss *et al.*, 2012), its function has not yet been determined, it may be that it mediates phage resistance as observed for a phase-variable type I RM system in *H. influenzae* (Zaleski *et al.*, 2005) or as a 'phasevarion' regulator of gene expression described for other RM systems (Srikhanta *et al.*, 2010). The observation of selection for ON expression in two different *in vivo* experiments suggests a role for host colonization associated with expression of this gene and hence a possible role as a phasevarion (Bayliss *et al.*, 2012; Jerome *et al.*, 2011).

Combinatorial consequences of multiple phase-variable genes

Even a small number of phase-variable genes can produce a huge number of potential genotypes. If we consider the simplified case in which each phase-variable gene is considered to exist in only one of two states – ON or OFF – then *n* genes produces 2^n possible genotypes. This means just ten genes are sufficient to produce over a thousand genotypes, and the 28 phase-variable tracts of NCTC11168 could produce over a quarter of a billion possible genotypes. The potential for rapid phenotypic variation produced by the combinatoric effect of multiple PV genes within a population and high switching rates is clearly vast. However, in reality the degree of phenotypic variation is likely smaller; firstly some phase-variable tracts lie in non-coding regions and, secondly, some combinations of phase-variable expression states are either phenotypically indistinguishable or the effect of one gene overrides another. Even so, the phenotypic variation producible from just a few contingency loci is remarkable.

The high prevalence of polyG tracts within reading frames facilitates analysis of the combinatorial states of *C. jejuni* isolates. The tracts can be analysed by a multiplex PCR and GeneScan approach. The ON and OFF states can then be readily determined from the repeat tracts and these states can be turned into a binary code with 0 and 1 representing OFF and ON states, respectively. A multiplex code for multiple genes can then be explored to determine whether particular combinations of expression states are present following exposure to a specific selective pressure. The exploration of this phenomenon is at an early stage but has already detected differences between input and output populations of *C. jejuni* during infections of chickens and bird-to-bird variations in levels of certain genotypes (Bayliss *et al.*, 2012).

Influence of mutational, selective and population forces on the diversity of phase-variable genotypes

One of the primary experimental approaches for investigating the functional importance of phase-variable genes is the passage experiment. In this approach a known *input* population is passaged in a specific environment or exposed to a known selective pressure, examples include: *in vitro* growth with passage from one petri dish to another; subjection to selection with a specific bactericidal antibody; exposure to and growth in serum; or *in vivo* by inoculating a host, young chicks or mice in the case of *C. jejuni*, and then collecting samples from excretions (caecal samples), body fluids (blood) or tissues. The final sample is the *output* population. Comparisons are performed to detect differences between the input and output populations in the proportions of each phase-variable genotype. It is then tempting to conclude that any difference in the genotypic makeup of the output population can be ascribed to selective processes however there are other competing explanations. Firstly, any potential bias in the sampling method and/or effects of sample size needs to be considered. Statistical tests of the results are paramount in these investigations. The other processes may be inherent to a particular system and include mutational drift, genetic drift or bottlenecks, hitch-hiking and finally selection (Figs. 6.3 and 6.4). The following sections elaborate on the potential effects.

Mutational drift and the steady state

Mutations in phase-variable genes happen at high rates and are reversible with similar rates for both directions of switching. This permits the occurrence of mutational drift, which we outline below in a theoretical context (Bayliss *et al.*, 2012). Note that time is measured in generations as mutations occur during DNA replication and mutation rates are expressed as the number of mutational events per generation.

In the absence of selection it follows that, regardless of the initial state of a population, the proportions with particular patterns of genes in the ON and OFF states will converge over time towards proportions determined by the mutation rate of the tracts at different lengths. To see why, consider a colony that begins with a single cell and thus all cells in that colony share the same ancestral state in terms of tract lengths. Initially, within the colony the switching of phase-variable genes can only be from the state in the initial cell to the opposite state (e.g. either ON-to-OFF or OFF-to-ON). Let's further simplify by considering a single gene that is initially in the ON state and is in a constant environment with no selection so that both ON and OFF variants replicate at the same rates. Expression of this gene (i.e. the ON state) will result from the polyG tract having a particular length, which can be arbitrarily taken as 9. The probability of a mutation increasing this to 10, or decreasing this to 8, is fixed by the mutation rate such that the chance of a mutation occurring depends only on time and how many cells with a tract length of 9 are present in the population. Similar parameters apply to the stochasticity of cells with a tract length of 8 increasing in length by one, and those with a tract length of 10 decreasing in length by one. Obviously, there are originally no cells in the OFF state, so the rate at which cells are returning to the ON state is initially zero. However, as OFF variants accumulate in the population, the numbers of cells undergoing

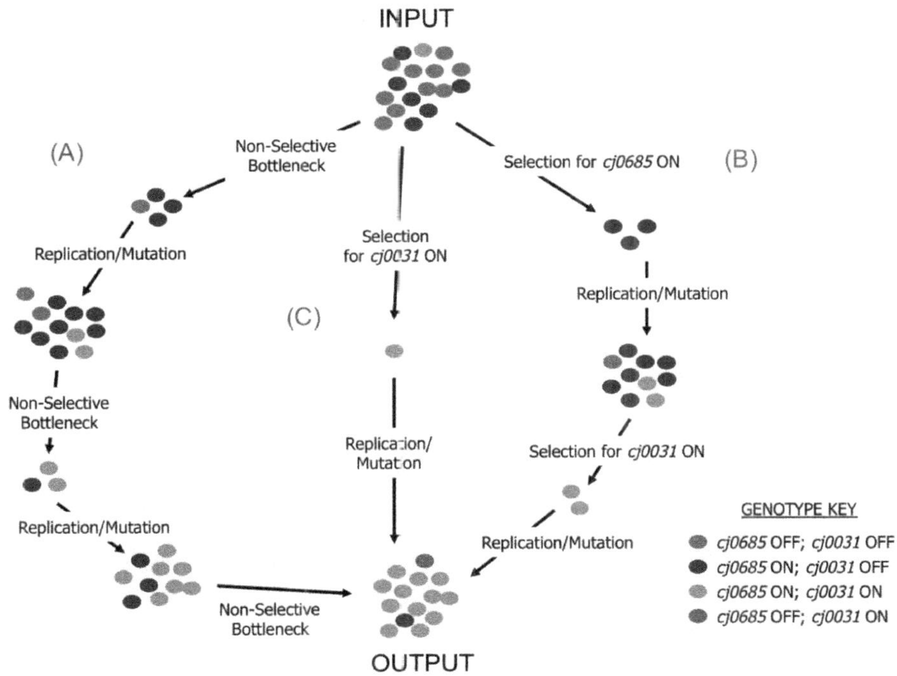

Figure 6.3 Potential pathways for altering the proportions of phase variants within a population. A bacteria containing two phase-variable genes is shown switching from one distribution of expression states to another (the *INPUT* and *OUTPUT* populations). Three pathways to change in the population are shown and a fourth, mutational drift, is outlined in Fig. 6.4. (A) Non-selective bottlenecks. The population is reduced to a small number of cells randomly selected from the initial population. A series of bottlenecks leads to a gradual change in the dominant phase variants. (B) Selection. Sequential changes in the environment select first for the ON state of one gene and then for the ON state of the second gene. This leads to a major change in population structure. (C) Hitch-hiking. Selection is imposed for the ON state of *cj0031*, which exists as a rare variant in the initial population. This variant is in the ON state for the second gene (*cj0685*), which 'hitch-hikes' to prominence in the population due to linkage with the *cj0031* ON state. Note that this population has an identical phenotype to population (B). A colour version of this figure is available in the plate section at the back of the book.

conversion from OFF to ON will increase whilst the opposing conversion from ON to OFF will decrease. This cannot continue indefinitely and eventually the population reaches a point where numbers of the conversion events are *equal*. From this time on, the mutation rates will tend to restore the proportion of each genotype to the steady state.

This can be described more concisely and precisely in a few mathematical equations, under the simplifying assumption of constant population size, k and constant rates of switching. If we take n_0, n_1 to be the current size of the population in the OFF and ON state (respectively), the initial population sizes to be k_0 and k_1 and the conversion rate to be c_0 and c_1 (where c_0 is the rate at which OFF switches to ON). Then the rates of change in the size of each population are given by eqns. 6.1 and 6.2.

$$\frac{dn_0}{dt} = n_1c_1 - n_0c_0 \tag{6.1}$$

$$\frac{dn_1}{dt} = n_0c_0 - n_1c_1 \tag{6.2}$$

It also follows that the population can only be in a stable state when the rate of change in the ON and OFF populations are 0, i.e. $dn_0/dt = dn_1/dt = 0$ and the only condition under which this can occur is when these rates are equal to each other. To solve for n_0 and n_1, we need only substitute into eqn. 6.1 and eqn. 6.2, to give eqn. 6.3 and eqn. 6.4.

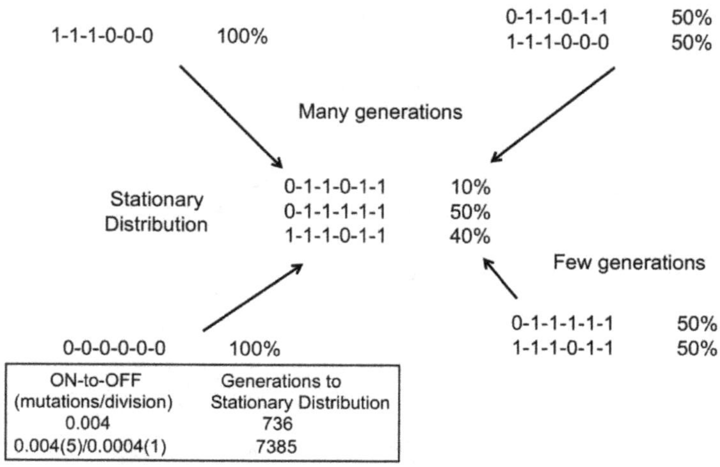

Figure 6.4 Mutational drift due to high switching rates. The high switching rates of *C. jejuni* genes will lead to the population adopting a stationary distribution. The nature of the stationary distribution is determined by the relative ON-to-OFF and OFF-to-ON phase variation rates of each phase-variable gene. The stationary distribution will only be reached in the absence of changes in the selective pressures and bottlenecks and with a relatively large population. The diagram indicates cells containing six genes in either an ON (1) or OFF (2) state. The number of divisions required to reach the stationary distribution will influenced by the initial state of the population as shown. The inset indicates the number of generation required to reach the stationary distribution for six phase-variable genes with either all genes switching at the same high rate (0.004) or one gene having a ten-fold lower rate for ON-to-OFF switching.

$$n_0 = k\frac{c_1}{c_0 + c_1} \quad (6.3)$$

$$n_1 = k\frac{c_0}{c_0 + c_1} \quad (6.4)$$

Thus, the final proportion of the population that is in each state is exactly equal to the proportion of switching that is in the direction of that state, and is independent of the starting size of the ON and OFF populations. Since the switching rates of the phase-variable genes are independent of each other, this analysis can be extended in a natural fashion to multiple genes. In reality, the switching between states is a stochastic process rather than occurring at a metronomic rate. Stochastic processes are also amenable to mathematical modelling, and simpler to solve when dealing with multiple phase-variable genes. These models produce the same overall result in terms of steady state population structure (Bayliss *et al.*, 2012).

A key aspect is that the rate of approach to the steady state is dependent on the starting mixture of the population and the switching rate of each gene for the ON-to-OFF and OFF-to-ON directions (Fig. 6.3). Thus, a gene with a particularly low PV rate will act as a limiter on the ability of a population to reach the steady state. The high PV rates of *C. jejuni* genes means that the steady state may be accessible for many populations of this bacteria and hence there is a mutational drift to a particular combination of phase variant states. The nature of the steady state in the absence of selection can be readily predicted from observed mutation rates and is amenable to modelling.

The steady state can be influenced by selection and if selection for or against a particular phase variant state of one gene was constantly applied then a different steady state would be achieved. Population size will also impact on whether and how quickly the steady state can be achieved. Finally, variable selection or bottlenecks will prevent populations from reaching the steady state.

Non-selective bottlenecks may radically alter genotype frequency

A non-selective bottleneck occurs when a population is reduced in size by random processes such as occur during transmission involving excretion of caecal or faecal material containing a gastrointestinal organism. The amount of organisms present in an excretion particle will depend on its size. Small particles with a restricted population size may mediate transmission to a new host and hence create a non-selective bottleneck. This population bottleneck is likely to significantly reduce the genetic diversity of the resulting population simply because some genotypes will be randomly excluded, and others randomly over-represented. This is equivalent to migrations known to exert strong levels of genetic drift in human and animal populations. The magnitude of these effects increases as the size of bottleneck shrinks. Since transmission events *in vivo* will involve population bottlenecks, possibly with the infection begun by a single cell (Margolis and Levin, 2007), random sampling may have a powerful effect on the makeup of the output population.

The magnitude of population effects in *C. jejuni* populations is difficult to estimate (Conlan *et al.*, 2007). Transmission between chickens, a major host, requires passage of the inoculating population through the gizzard and stomach. These environments may result in non-specific killing of a significant proportion of the inoculum. A similar effect may result from transmission through excreted caecal material resulting in a significant reduction in the population during environmental persistence and survival of desiccation. A high inoculum dose is known to be required for colonization of chickens suggesting the existence of a significant bottleneck during the early stages of bird-to-bird transmission. The high PV rates of *C. jejuni* serve to counteract these bottlenecks as small populations will still contain a significant number and variety of variants whilst variants will be rapidly regenerated during replication of small starting populations.

Hitch-hiking as a mechanism for altering population structure

The phenomenon of hitch-hiking was brought to the fore by considerations of the impact that a mutator phenotype has on adaptation by bacterial populations (Taddei *et al.*, 1997). Inactivation of mismatch repair genes in *E. coli* causes a significant increase in the genome-wide mutation rate and elevates the rate of generation of both adaptive and deleterious mutations. The linkage of a mutator allele with an adaptive mutation increases the prevalence of the mutator allele in a population even though the mutator confers no direct selective benefit. In a similar manner, one phase-variable change may hitch-hike to high proportions in a population if it happens to be present in a genome where there is selection for another beneficial phase variant state. The occurrence of hitch-hiking may occur most frequently in situations where there is selection for rare phase variants (or indeed other types of mutations) among an otherwise diverse population. This phenomenon may complicate interpretations of the impact of phase variants in particular selective environments.

Selection and phase-variable genes

The previous sections have elaborated some of our current knowledge on the potential functions of the phase-variable genes. As noted, there is a requirement to determine the selective pressures exerted on both expression states of phase-variable genes. Incorporation of selection into models of PV requires predictions of the strength of selection. Experimental determination of this variable may be difficult but possible for phenotypes such as phage sensitivity. It may also be possible to estimate selection strength by using modelling to fit observed changes in the proportions of phase variants to theoretical populations. A key component of future studies in this area will focus on identifying the particular selective pressures associated with each phase-variable gene.

Modelling phase-variable genes and selection for mutational rates

Determination of the PV rates for *C. jejuni* genes enabled derivation of a theoretical model of the interplay between switching rates and the proportions of phase variants present in a population given a particular inoculum population. Comparison of theoretical populations with observed outputs from *C. jejuni* passaged populations has shown that mutational drift alone is not responsible for changes in the phase-variable genes.

Rather selection or non-selective bottlenecks were invoked (Bayliss *et al.*, 2012).

However, a deeper understanding of the population genetics of phase-variable genes requires more detailed modelling. Of particular interest is determining under what circumstances PV is beneficial and how different environmental patterns favour different rates of switching. Palmer *et al.* (2013) developed a detailed stochastic model based on experimental data for a phase-variable *H. influenzae* gene. This model allowed investigation of the strength of selection and the periodicity of switching between environments favourable to the ON and OFF phenotypes. This model showed that selection would favour phase-variable switching over a broad range of conditions, and that longer periods between switching favoured lower mutation rates and vice versa (Bayliss and Palmer, 2012). Critically long periods of weak selection for one state combined with short periods of strong selection for the alternate state could still result in evolution of a phase-variable gene. This situation may be particularly pertinent to *C. jejuni* wherein sudden but infrequent exposure to bacteriophages may mediate strong selection for one state of a phase-variable gene.

Although not directly comparable to SSR-based phase variation, evolutionary experiments into 'bet hedging' strategies in bacteria support the overall picture from modelling the advantage of higher mutational rates under changing environmental conditions. Here bet hedging refers to stochastic variation in phenotypic state, not due to genotypic changes. Beaumont *et al.* (2009) carried out an experiment in which lines of *Pseudomonas fluorescens* – which does not exhibit bet hedging – were subjected to repeated bottlenecks based on an 'exclusion rule' where only cells of a randomly chosen phenotype were taken on to the next generation. In two of twelve experimental replicates, switching mutants evolved which exhibited stochastic phenotypic switching. Theoretical modelling of this experiment by Libby and Rainey (2011) demonstrated that switching mutants could outcompete non-switching mutants where changeable environmental pressures exerted powerful selection for variable phenotypes even when the switching mutant exhibited a substantially lower growth rate.

Experimental issues connected with phase variation

The existence of high PV rates has the potential to confound some experiments with *C. jejuni*. Analysis of mutants may be subject to misinterpretation as differences between the parental strain and mutant may be due to alternate PV states in each of them. This will arise as the bottleneck effects associated with production of a mutant can rapidly give rise to changes in the phase-variable genes. Whilst, these effects can be ameliorated by analysis of multiple independent mutants and use of revertants, there is frequently the potential for subtle differences to occur which are independent of the mutation being tested.

The potential for such variation was raised by the failure of signature tagged mutagenesis (STM). This experimental approach was successfully utilized for detecting host colonization genes in *Salmonella* (Hensel *et al.*, 1995). The approach involves inoculation of a host with a pool of transposon-insertion mutations in different genes and analysis of the output population to determine whether any mutants are absent and hence required for host colonization. A random recovery of mutants was observed when this technique was applied to *C. jejuni* (Grant *et al.*, 2005). This indicated that changes independent of the transposon-mutation were responsible for the variation between mutants in colonization of chickens. A subsequent approach using a wild-type *C. jejuni* strain, containing different tags (WITS), detected similar variation between supposedly isogenic variants (Coward *et al.*, 2008). Although not proven, it seems likely that PV may have had a significant effect in these experiments and may similarly influence other experiments on host colonization, adhesion to and invasion of eukaryotic cells, and other tests of *C. jejuni* virulence factors.

Summary

Campylobacter jejuni has a number of genes containing homopolymeric G or C tracts, these tracts are subject to high rates of mutation by insertion or deletion of a single base. This change in length can alter the expression of the repeat-containing gene, typically either enabling or disabling the expression of a functional gene product. Most of these genes are related to the expression of surface molecules, and the switch in expression has phenotypic effects. These phenotypic effects include changes in auto-agglutination related to host infectivity, shifts in phage susceptibility, changes in immune system relevant molecules (and, in particular, molecular mimicry of important gangliosides linked to Guillain-Barré syndrome) and alterations in mobility mediated by flagellar modifications. Thus, this class of repetitive DNA provides a significant source of genotypic and phenotypic variety in populations of *Campylobacter*.

Much remains to be learnt about the functional implications of the changes produced by phase variation and their wider implications in terms of population genetics and host infectivity and disease. However, a combination of experimental and theoretical approaches has revealed significant details concerning the switching rates and mutational patterns associated with SSRs in phase-variable genes and the impact of mutability in these loci on the genotypic and phenotypic variability of *C. jejuni* populations.

Future perspectives

Our understanding of many aspects of PV is in its infancy and individual studies of the functions of each phase-variable gene are a cornerstone of this research field. Global studies of phase-variable genes are also likely to generate rapid advances. These studies may reveal the loci involved in host colonization, serum resistance, evasion of adaptive immune responses and phage resistance. A global study of the binding sites of multiple phages may be particularly revealing in determining the adaptive advantages associated with phase-variable genes and how they have evolved.

The explosion of genome sequencing will also contribute to this field by identifying the repertoire of phase-variable genes in *Campylobacter* species and further highlighting the types of molecules subject to this type of stochastic variation. Molecular studies of the mutational mechanism are also required to further characterize the determinants of switching rates and whether there are strain-to-strain and species-to-species differences in mutability.

Finally, studies of phase-variable genes in *Campylobacter* have an advantage over other bacterial systems. These genes can be studied in the context of their contributions to colonization of a native host and one in which transmission can be readily achieved. Studies in this species may, therefore, facilitate investigation of how adaptive immune responses impact on phase-variable genes, a hypothesis widely touted as a major driver of evolution of such loci. Equally, competition experiments between strains with differing switching rates should be possible and may demonstrate the importance or otherwise of the actual switching rate for adaptation. These lines of research may finally begin to uncover some of the secrets of this fascinating biological phenomenon.

References

van Alphen, L.B., Wuhrer, M., Bleumink-Pluym, N.M., Hensbergen, P.J., Deelder, A.M., and van Putten, J.P. (2008). A functional *Campylobacter jejuni maf4* gene results in novel glycoforms on flagellin and altered autoagglutination behaviour. Microbiology *154*, 3385–3397.

Ambur, O.H., Davidsen, T., Frye, S.A., Balasingham, S.V., Lagesen, K., Rognes, T., and Tonjum, T. (2009). Genome dynamics in major bacterial pathogens. FEMS Microbiol. Rev. 33, 453–470.

Andrewes, F.W. (1922). Studies in group agglutination I. J. Path. Bacteriol. 25, 505.

Ashgar, S.S., Oldfield, N.J., Wooldridge, K.G., Jones, M.A., Irving, G.J., Turner, D.P., and Ala'Aldeen, D.A. (2007). CapA, an autotransporter protein of *Campylobacter jejuni*, mediates association with human epithelial cells and colonization of the chicken gut. J. Bacteriol. *189*, 1856–1865.

Bart, A., Dankert, J., and van der Ende, A. (1999). Antigenic variation of the class I outer membrane protein in hyperendemic *Neisseria meningitidis* strains in the Netherlands. Infect. Immun. 67, 3842–3846.

Bayliss, C.D. (2009). Determinants of phase variation rate and the fitness implications of differing rates for bacterial pathogens and commensals. FEMS Microbiol. Rev. 33, 504–520.

Bayliss, C.D., and Palmer, M.E. (2012). Evolution of simple sequence repeat-mediated phase variation in bacterial genomes. Ann. N.Y. Acad. Sci. *1267*, 39–44.

Bayliss, C.D., Field, D., and Moxon, E.R. (2001). The simple sequence contingency loci of *Haemophilus influenzae* and *Neisseria meningitidis*. J. Clin. Invest *107*, 657–662.

Bayliss, C.D., Callaghan, M.J., and Moxon, E.R. (2006). High allelic diversity in the methyltransferase gene of a phase-variable type III restriction-modification system has implications for the fitness of *Haemophilus influenzae*. Nucleic Acids Res. 34, 4046–4059.

Bayliss, C.D., Bidmos, F.A., Anjum, A., Manchev, V.T., Richards, R.L., Grossier, J., Wooldridge, K.G., Ketley, J.M., Barrow, P.A., Jones, M.A., and Tretyakov, M.V. (2012). Phase variable genes of *Campylobacter jejuni* exhibit high mutation rates and specific mutational patterns but mutability is not the major determinant of population structure during host colonization. Nucleic Acids Res. *40*, 5876–5889.

Beaumont, H.J., Gallie, J., Kost, C., Ferguson, G.C., and Rainey, P.B. (2009). Experimental evolution of bet hedging. Nature *462*, 90–93.

Bertani, G., and Weigle, J.J. (1953). Host controlled variation in bacterial viruses. J. Bacteriol. *65*, 113–121.

Blaser, M.J., Hopkins, J.A., Perez-Perez, G.I., Cody, H.J., and Newell, D.G. (1986). Antigenicity of *Campylobacter jejuni* flagella. Infect. Immun. *53*, 47–52.

Caldwell, M.B., Guerry, P., Lee, E.C., Burans, J.P., and Walker, R.I. (1985). Reversible expression of flagella in *Campylobacter jejuni*. Infect. Immun. *50*, 941–943.

Champion, O.L., Karlyshev, A.V., Senior, N.J., Woodward, M., La Ragione, R., Howard, S.L., Wren, B.W., and Titball, R.W. (2010). Insect infection model for *Campylobacter jejuni* reveals that O-methyl phosphoramidate has insecticidal activity. J. Infect. Dis. *201*, 776–782.

Conlan, A.J., Coward, C., Grant, A.J., Maskell, D.J., and Gog, J.R. (2007). *Campylobacter jejuni* colonization and transmission in broiler chickens: a modelling perspective. J. R. Soc. Interface *4*, 819–829.

Coward, C., van Diemen, P.M., Conlan, A.J., Gog, J.R., Stevens, M.P., Jones, M.A., and Maskell, D.J. (2008). Competing isogenic *Campylobacter* strains exhibit variable population structures in vivo. Appl. Environ. Microbiol. *74*, 3857–3867.

De Bolle, X., Bayliss, C.D., Field, D., van de Ven, T., Saunders, N.J., Hood, D.W., and Moxon, E.R. (2000). The length of a tetranucleotide repeat tract in *Haemophilus influenzae* determines the phase variation rate of a gene with homology to type III DNA methyltransferases. Mol. Microbiol. 35, 211–222.

Deitsch, K.W., Lukehart, S.A., and Stringer, J.R. (2009). Common strategies for antigenic variation by bacterial, fungal and protozoan pathogens. Nat. Rev. Microbiol. *7*, 493–503.

Dornberger, U., Leijon, M., and Fritzsche, H. (1999). High base pair opening rates in tracts of GC base pairs. J. Biol. Chem. *274*, 6957–6962.

Drake, J.W. (1991). A constant rate of spontaneous mutation in DNA-based microbes. Proc. Natl. Acad. Sci. U.S.A. *88*, 7160–7164.

Feschenko, V.V., Rajman, L.A., and Lovett, S.T. (2003). Stabilization of perfect and imperfect tandem repeats by single-strand DNA exonucleases. Proc. Natl. Acad. Sci. U.S.A. *100*, 1134–1139.

Fouts, D.E., Mongodin, E.F., Mandrell, R.E., Miller, W.G., Rasko, D.A., Ravel, J., Brinkac, L.M., DeBoy, R.T., Parker, C.T., Daugherty, S.C., *et al.* (2005). Major structural differences and novel potential virulence mechanisms from the genomes of multiple *Campylobacter* species. PLoS Biol. 3, e15.

Gaasbeek, E.J., van der Wal, F.J., van Putten, J.P., de Boer, P., van der Graaf-van Bloois, L., de Boer, A.G., Vermaning, B.J., and Wagenaar, J.A. (2009). Functional characterization of excision repair and RecA-dependent recombinational DNA repair in *Campylobacter jejuni*. J. Bacteriol. *191*, 3785–3793.

Gilbert, M., Karwaski, M.F., Bernatchez, S., Young, N.M., Taboada, E., Michniewicz, J., Cunningham, A.M., and Wakarchuk, W.W. (2002). The genetic bases for the variation in the lipo-oligosaccharide of the mucosal pathogen, *Campylobacter jejuni*. Biosynthesis of sialylated ganglioside mimics in the core oligosaccharide. J. Biol. Chem. *277*, 327–337.

Grant, A.J., Coward, C., Jones, M.A., Woodall, C.A., Barrow, P.A., and Maskell, D.J. (2005). Signature-tagged transposon mutagenesis studies demonstrate the dynamic nature of cecal colonization of 2-week-old chickens by *Campylobacter jejuni*. Appl. Environ. Microbiol. *71*, 8031–8041.

Guerry, P., Szymanski, C.M., Prendergast, M.M., Hickey, T.E., Ewing, C.P., Pattarini, D.L., and Moran, A.P. (2002). Phase variation of *Campylobacter jejuni* 81-176 lipooligosaccharide affects ganglioside mimicry and invasiveness in vitro. Infect. Immun. *70*, 787–793.

Guerry, P., Poly, F., Riddle, M., Maue, A.C., Chen, Y.H., and Monteiro, M.A. (2012). *Campylobacter* polysaccharide capsules: virulence and vaccines. Front. Cell Infect. Microbiol. 2, 7.

Hanninen, M.L., and Hannula, M. (2007). Spontaneous mutation frequency and emergence of ciprofloxacin resistance in *Campylobacter jejuni* and *Campylobacter* coli. J. Antimicrob. Chemother. *60*, 1251–1257.

Hendrixson, D.R. (2006). A phase-variable mechanism controlling the *Campylobacter jejuni* FlgR response regulator influences commensalism. Mol. Microbiol. *61*, 1646–1659.

Hendrixson, D.R. (2008). Restoration of flagellar biosynthesis by varied mutational events in *Campylobacter jejuni*. Mol. Microbiol. *70*, 519–536.

Hensel, M., Shea, J.E., Gleeson, C., Jones, M.D., Dalton, E., and Holden, D.W. (1995). Simultaneous identification of bacterial virulence genes by negative selection. Science *269*, 400–403.

Hofreuter, D., Tsai, J., Watson, R.O., Novik, V., Altman, B., Benitez, M., Clark, C., Perbost, C., Jarvie, T., Du, L., and Galan, J.E. (2006). Unique features of a highly pathogenic *Campylobacter jejuni* strain. Infect. Immun. *74*, 4694–4707.

Hood, D.W., Deadman, M.E., Jennings, M.P., Bisercic, M., Fleischmann, R.D., Venter, J.C., and Moxon, E.R. (1996). DNA repeats identify novel virulence genes in *Haemophilus influenzae*. Proc. Natl. Acad. Sci. U.S.A. *93*, 11121–11125.

Jerome, J.P., Bell, J.A., Plovanich-Jones, A.E., Barrick, J.E., Brown, C.T., and Mansfield, L.S. (2011). Standing genetic variation in contingency loci drives the rapid adaptation of *Campylobacter jejuni* to a novel host. PLoS One *6*, e16399.

Karlyshev, A.V., Linton, D., Gregson, N.A., and Wren, B.W. (2002). A novel paralogous gene family involved in phase-variable flagella-mediated motility in *Campylobacter jejuni*. Microbiology *148*, 473–480.

de Koning, A.P., Gu, W., Castoe, T.A., Batzer, M.A., and Pollock, D.D. (2012). Repetitive elements may comprise over two-thirds of the human genome. PLoS Genet. 7, e1002384.

Kumar, P., and Nagarajaram, H.A. (2012). A Study on mutational dynamics of simple sequence repeats in relation to mismatch repair system in prokaryotic genomes. J. Mol. Evol. *74*, 127–139.

Labrie, S.J., Samson, J.E., and Moineau, S. (2010). Bacteriophage resistance mechanisms. Nat. Rev. Microbiol. *8*, 317–327.

Levinson, G., and Gutman, G. (1987). Slipped-strand mispairing: a major mechanism for DNA sequence evolution. Mol. Biol. Evol. *4*, 203–221.

Libby, E., and Rainey, P.B. (2011). Exclusion rules, bottlenecks and the evolution of stochastic phenotype switching. Proc. Biol. Sci. *278*, 3574–3583.

Lin, W.H., and Kussell, E. (2012). Evolutionary pressures on simple sequence repeats in prokaryotic coding regions. Nucleic Acids Res. *40*, 2399–2413.

Linton, D., Gilbert, M., Hitchen, P.G., Dell, A., Morris, H.R., Wakarchuk, W.W., Gregson, N.A., and Wren, B.W. (2000). Phase variation of a beta-1,3 galactosyltransferase involved in generation of the ganglioside GM1-like lipo-oligosaccharide of *Campylobacter jejuni*. Mol. Microbiol. *37*, 501–514.

Logan, S.M., Hui, J.P., Vinogradov, E., Aubry, A.J., Melanson, J.E., Kelly, J.F., Nothaft, H., and Soo, E.C. (2009). Identification of novel carbohydrate modifications on *Campylobacter jejuni* 11168 flagellin using metabolomics-based approaches. FEBS. J. *276*, 1014–1023.

Logan, S.M., Trust, T.J., and Guerry, P. (1989). Evidence for posttranslational modification and gene duplication of *Campylobacter* flagellin. J. Bacteriol. *171*, 3031–3038.

McNally, D.J., Lamoureux, M.P., Karlyshev, A.V., Fiori, L.M., Li, J., Thacker, G., Coleman, R.A., Khieu, N.H., Wren, B.W., Brisson, J.R., *et al.* (2007). Commonality and biosynthesis of the O-methyl phosphoramidate capsule modification in *Campylobacter jejuni*. J. Biol. Chem. *282*, 28566–28576.

Margolis, E., and Levin, B.R. (2007). Within-host evolution for the invasiveness of commensal bacteria: an experimental study of bacteremias resulting from *Haemophilus influenzae* nasal carriage. J. Infect. Dis. *196*, 1068–1075.

Marraffini, L.A., and Sontheimer, E.J. (2010). CRISPR interference: RNA-directed adaptive immunity in bacteria and archaea. Nat. Rev. Genet. *11*, 181–190.

Martin, P., Sun, L., Hood, D.W., and Moxon, E.R. (2004). Involvement of genes of genome maintenance in the regulation of phase variation frequencies in *Neisseria meningitidis*. Microbiology *150*, 3001–3012.

Meselson, M., Yuan, R., and Heywood, J. (1972). Restriction and modification of DNA. Annu. Rev. Biochem. *41*, 447–466.

Moxon, E.R., Rainey, P.B., Nowak, M.A., and Lenski, R. (1994). Adaptive evolution of highly mutable loci in pathogenic bacteria. Curr. Biol. *4*, 24–33.

Moxon, E.R., Bayliss, C.D., and Hood, D.W. (2007). Bacterial contingency loci: the role of simple sequence DNA repeats in bacterial adaptation. Ann. Rev. Genet. *40*, 307–333.

Nachamkin, I., Allos, B.M., and Ho, T. (1998). *Campylobacter* species and Guillain-Barre syndrome. Clin. Microbiol. Rev. *11*, 555–567.

Nakamura, K., Ariga, T., Yahagi, T., Miyatake, T., Suzuki, A., and Yamakawa, T. (1983). Interspecies comparison of muscle gangliosides by two-dimensional thin-layer chromatography. J. Biochem. *94*, 1359–1365.

Palmer, M.E., Lipsitch, M., Moxon, E.R., and Bayliss, C.D. (2013). Broad conditions favor the evolution of phase-variable loci. M. Bio. *4*, e00430–12.

Parker, C.T., Gilbert, M., Yuki, N., Endtz, H.P., and Mandrell, R.E. (2008). Characterization of lipooligosaccharide-biosynthetic loci of *Campylobacter jejuni* reveals new lipooligosaccharide classes: evidence of mosaic organizations. J. Bacteriol. *190*, 5681–5689.

Parkhill, J., Wren, B.W., Mungall, K., Ketley, J.M., Churcher, C., Basham, D., Chillingworth, T., Davies, R.M., Feltwell, T., Holroyd, S., *et al.* (2000). The genome sequence of the food-borne pathogen *Campylobacter jejuni* reveals hypervariable sequences. Nature *403*, 665–668.

Pavlovskis, O.R., Rollins, D.M., Haberberger, R.L., Jr., Green, A.E., Habash, L., Strocko, S., and Walker, R.I. (1991). Significance of flagella in colonization resistance of rabbits immunized with *Campylobacter* spp. Infect. Immun. *59*, 2259–2264.

Pearson, B.M., Gaskin, D.J., Segers, R.P., Wells, J.M., Nuijten, P.J., and van Vliet, A.H. (2007). The complete genome sequence of *Campylobacter jejuni* strain 81116 (NCTC11828). J. Bacteriol. *189*, 8402–8403.

Purves, J., Blades, M., Arafat, Y., Malik, S., Bayliss, C.D., and Morrissey, J.A. (2012). Variation in the genomic locations and sequence conservation of STAR elements among staphylococcal species provides insight into DNA repeat evolution. BMC Genomics *13*, 515.

Richardson, A.R., Yu, Z., Popovic, T., and Stojiljkovic, I. (2002). Mutator clones of *Neisseria meningitidis* in epidemic serogroup A disease. Proc. Natl. Acad. Sci. U.S.A. *99*, 6103–6107.

Ritz, D., Lim, J., Reynolds, C.M., Poole, L.B., and Beckwith, J. (2001). Conversion of a peroxiredoxin into a disulfide reductase by a triplet repeat expansion. Science *294*, 158–160.

Rose, A., Kay, E., Wren, B.W., and Dallman, M.J. (2012). The *Campylobacter jejuni* NCTC11168 capsule prevents excessive cytokine production by dendritic cells. Med. Microbiol. Immunol. *201*, 137–144.

Saunders, N.J., Peden, J., Hood, D., and Moxon, E.R. (1998). Simple sequence repeats in the *Helicobacter pylori* genome. Mol. Microbiol. *27*, 1091–1098.

Saunders, N.J., Jeffries, A.C., Peden, J.F., Hood, D.W., Tettelin, H., Rappouli, R., and Moxon, E.R. (2000). Repeat-associated phase-variable genes in the complete genome sequence of *Neisseria meningitidis* strain MC58. Mol. Microbiol. *37*, 207–215.

Schoen, C., Tettelin, H., Parkhill, J., and Frosch, M. (2009) Genome flexibility in *Neisseria meningitidis*. Vaccine *27(Suppl 2)*, B103–111.

Schouls, L.M., Spalburg, E.C., van Luit, M., Huijsdens, X.W., Pluister, G.N., van Santen-Verheuvel, M.G., van der Heide, H.G., Grundmann, H., Heck, M.E., and de

Neeling, A.J. (2009). Multiple-locus variable number tandem repeat analysis of *Staphylococcus aureus*: comparison with pulsed-field gel electrophoresis and spa-typing. PLoS One 4, e5082.

Snyder, L.A., Loman, N.J., Linton, J.D., Langdon, R.R., Weinstock, G.M., Wren, B.W., and Pallen, M.J. (2010). Simple sequence repeats in *Helicobacter canadensis* and their role in phase-variable expression and C-terminal sequence switching. BMC Genomics *11*, 67.

Sorensen, M.C., van Alphen, L.B., Harboe, A., Li, J., Christensen, B.B., Szymanski, C.M., and Brondsted, L. (2011). The F336 bacteriophage recognizes the capsular phosphoramidate modification of *Campylobacter jejuni* NCTC11168. J. Bacteriol. *193*, 6742–6749.

Sorensen, M.C., van Alphen, L.B., Fodor, C., Crowley, S.M., Christensen, B.B., Szymanski, C.M., and Brondsted, L. (2012). Phase variable expression of capsular polysaccharide modifications allows *Campylobacter jejuni* to avoid bacteriophage infection in chickens. Front. Cell Infect. Microbiol. *2*, 11.

Srikhanta, Y.N., Fox, K.L., and Jennings, M.P. (2010). The phasevarion: phase variation of type III DNA methyltransferases controls coordinated switching in multiple genes. Nat. Rev. Microbiol. *8*, 196–206.

Sternberg, M.J., Tamaddoni-Nezhad, A., Lesk, V.I., Kay, E., Hitchen, P.G., Cootes, A., van Alphen, L.B., Lamoureux, M.P., Jarrell, H.C., Rawlings, C.J., *et al.* (2012). Gene function hypotheses for the *Campylobacter jejuni* glycome generated by a logic-based approach. J. Mol. Biol. *425*, 186–197.

Szymanski, C.M., Michael, F.S., Jarrell, H.C., Li, J., Gilbert, M., Larocque, S., Vinogradov, E., and Brisson, J.R. (2003). Detection of conserved N-linked glycans and phase-variable lipooligosaccharides and capsules from *Campylobacter* cells by mass spectrometry and high resolution magic angle spinning NMR spectroscopy. J. Biol. Chem. *278*, 24509–24520.

Taddei, F., Radman, M., Maynard-Smith, J., Toupance, B., Gouyon, P.H., and Godelle, B. (1997). Role of mutator alleles in adaptive evolution. Nature *387*, 700–702.

Thibault, P., Logan, S.M., Kelly, J.F., Brisson, J.R., Ewing, C.P., Trust, T.J., and Guerry, P. (2001). Identification of the carbohydrate moieties and glycosylation motifs in *Campylobacter jejuni* flagellin. J. Biol. Chem. *276*, 34862–34870.

Treangen, T.J., Abraham, A.L., Touchon, M., and Rocha, E.P. (2009). Genesis, effects and fates of repeats in prokaryotic genomes. FEMS Microbiol. Rev. *33*, 539–571.

van der Woude, M.W., and Baumler, A.J. (2004). Phase and antigenic variation in bacteria. Clin. Microbiol. Rev. *17*, 581–611.

Wren, B.W., Linton, D., Dorrell, N., and Karlyshev, A.V. (2001). Post genome analysis of *Campylobacter jejuni*. Symp. Ser. Soc. Appl. Microbiol. 36S-44S.

Zaleski, P., Wojciechowski, M., and Piekarowicz, A. (2005). The role of Dam methylation in phase variation of *Haemophilus influenzae* genes involved in defence against phage infection. Microbiology *151*, 3361–3369.

Part II

Adaptations to Host Colonization

Campylobacter Strategies for Colonizing the Host Gastrointestinal Tract

7

Annika Flint, James Butcher and Alain Stintzi

Abstract

Colonizing the gastrointestinal tract is fraught with multiple challenges for incoming bacteria. These microbes must not only evade the host immune systems and the harsh environmental conditions of the gastrointestinal tract, but also out-compete the native microbiota. *Campylobacter jejuni* is an efficient colonizer of both avian and mammalian hosts. In both these environments, *C. jejuni* is able to scavenge and metabolize the nutrients found within the mucus layer and out-compete commensal microbes. To accomplish this feat, *C. jejuni* preferentially metabolizes specific constituents within the mucus layer of the intestine as growth substrates, including key amino acids and fucose. This chapter provides an overview of the pathways and mechanisms of *C. jejuni* important for adaptation, survival and growth within the gastrointestinal tract.

Introduction

Campylobacter colonization of the gastrointestinal tract involves a multitude of cellular mechanisms to facilitate growth and circumvent the host's innate and adaptive immune systems. Campylobacteriosis commonly occurs by ingestion of contaminated food such as poultry or from raw milk or water (Young *et al.*, 2007). Following ingestion, *Campylobacter* must transit through the low pH of the stomach as well survive bile exposure in the duodenum before colonizing the host's gastrointestinal tract. Typically, the colon and caecum intestinal segments display the highest level of colonization by *Campylobacter*. In humans, once colonization of the gastrointestinal tract is established, *Campylobacter* can adhere to and invade the intestinal epithelial cells leading to the onset of pathogenesis and gastroenteritis (Humphrey *et al.*, 2007).

The host gastrointestinal tract is equipped with numerous innate and adaptive defence mechanisms to protect against invasion by microbial pathogens. Structurally, the intestine contains an epithelium that secretes a protective mucus layer. The mucus layer is further subdivided into the outer loosely adherent and inner firmly adherent mucus layers. A layer of glycocalyx is also present on the surface of the epithelial cells in the lumen and consists of membrane-bound mucins and glycoproteins. In a healthy host, the inner mucus layer is devoid of microbes with the commensal bacteria colonizing the outer mucus layer (Kim and Ho, 2010). An estimated 160–300 different bacterial species are found within the human gut (Qin *et al.*, 2010) with up to as many as 10^{14} bacterial cells (Ballal *et al.*, 2011). The mucus layer of the lumen is a crucial barrier protecting the underlying epithelium against invasion from bacterial pathogens.

The epithelium layer consists of several cell types including goblet cells, Paneth cells, absorptive enterocytes and enteroendocrine cells. Mucus is produced and secreted from goblet cells into the luminal side of the intestine. Goblet cells also secrete additional mucus components such as TFF3, RELMβ, and Fcgbp. These components are involved in wound repair (Taupin and Podolsky, 2003), up-regulation of MUC2 and induction of goblet cell hyperplasia (Krimi *et al.*, 2008), and cross-linking of MUC2 mucin networks (Johansson *et al.*, 2009; Kobayashi *et al.*, 2002) respectively. Other host-derived defences such as lysozymes and antimicrobial peptides are produced by Paneth cells (Lievin-Le Moal and Servin, 2006) and secretory IgA by enterocytes (Kim and Ho, 2010). Together these molecules form important structural defence mechanisms against invading pathogens.

The major constituents of the mucus secreted by the epithelium are mucins, specifically MUC2 (Kim and Ho, 2010). Mucins form a gel-like matrix and are composed of large macromolecular structures of glycoproteins that have highly negatively charged surfaces and high hydration capacity (Dharmani *et al.*, 2009; Kim and Ho, 2010). Physically, mucins form a semipermeable barrier allowing for the passage of nutrients, hormones, gases, and small molecules to the underlying epithelium. Larger molecules or microorganisms are excluded from penetrating the mucus layer and thus the mucus protects the epithelium of the gastrointestinal tract from bacterial adhesion and invasion (Kim and Ho, 2010).

The intestinal mucus also serves as a nutrient source and binding site for host commensal bacteria (Kim and Ho, 2010; Martens *et al.*, 2009; Salyers *et al.*, 1977). Apo-MUC2, the major mucin constituent of the intestine, is comprised of two distinct mucin domains. The first mucin domain consists of a 347 irregular amino acid repeat located closer to the N-terminus of the protein and contains high amounts of serine, threonine, and proline. The second, larger domain is located nearer the C-terminus and consists of 23 amino acid long tandem repeats that are primarily composed of threonine and proline residues. Furthermore, the serine and threonine residues are heavily glycosylated with O-linked glycans (Herrmann *et al.*, 1999). Bacteria produce carbohydrate-degrading enzymes that can digest these oligosaccharides to utilize them as an energy source (Hansson and Johansson, 2010). Conversely, the high level of glycosylation of the mucin domains helps to protect the mucins from degradation by proteolytic enzymes (Moncada *et al.*, 2003). The oligosaccharides also function as epitopes, facilitating the binding of microbes, viruses, as well as other biological molecules, such as cytokines, to mucin (Kim and Ho, 2010).

The mucus layer of the gastrointestinal tract is constantly turned over due to the peristaltic movements of the intestine, which cause shearing of the mucus layer from the passage of matter. Microbiota and digestive enzymes further degrade the intestinal mucins. Thus, goblet cells continually produce and secrete mucus to replenish the mucus lining of the intestine. Penetration of the mucus layer by bacterial pathogens results in the up-regulation of mucus secretion in an effort by the host to remove the invading bacteria (Kim and Ho, 2010). Bacterial penetration of the protective mucus barrier is accomplished by several different strategies such as those displayed by *Escherichia coli*, *Vibrio cholerae*, *Salmonella typhi*, *Shigella flexneri* and *Campylobacter jejuni*. To migrate through the intestinal mucus barrier, *E. coli* and *V. cholera* actively produce and secrete proteases and carbohydrate degrading enzymes into the surrounding environment to break down mucins in order to gain passage to the epithelium (Harrington *et al.*, 2009; Silva *et al.*, 2003). *S. typhi* and *S. flexneri* use a different strategy to infect the host by instead invading microfold cells (Kohbata *et al.*, 1986; Wassef *et al.*, 1989). Microfold cells are located in the intestinal mucosal epithelium but do not have a mucus layer which the bacteria must bypass (Kim and Ho, 2010). Alternatively, *Campylobacter* species display high motility in viscous milieu, such as mucus (Lee *et al.*, 1986), compared to other pathogens including *E. coli* and *Samonella enteritidis* (Ferrero and Lee, 1988). It has been suggested that the slender, curved-rod morphology of *Campylobacter* enables it to penetrate the viscous mucus layer (Ferrero and Lee, 1988).

The host uses microbe molecular patterns, pathogen-associated molecular patterns, and pattern recognition receptors (PRRs), consisting of toll-like receptors and NOD-like receptors, to differentiate between the commensal bacteria of the gut and invading pathogens (Fukata and Abreu, 2009; Moncada *et al.*, 2003). Recognition of invading pathogens by PPRs is speculated to elicit an inflammatory response by inducing a common signalling pathway leading to activation of the proinflammatory mediator NF-κB and production of cytokine and co-stimulatory molecules (Kim and Ho, 2010). Although *Campylobacter* pathogenesis is still not completely understood, several important molecular mechanisms have been characterized that contribute to *Campylobacter* virulence and survival of host mediated defence responses. These include a flagellar type III secretion system (Konkel *et al.*, 2004; Song *et al.*, 2004), production of cytolethal distending toxin (CDT), which is involved in apoptosis of host cells (Lara-Tejero and Galan, 2000) and induction of inflammatory responses (Fox *et al.*, 2004; Hickey *et al.*, 2000), and bacterial capsule and lipooligosaccharides, which are involved in immune evasion (Keo *et al.*, 2011). *Campylobacter* surface lipooligosaccharides have recently been shown to play a role in resistance to components of the host innate immune defence such as polymyxin B, α-defensins, cathelicidins, and bactericidal/permeability-increasing protein (Keo *et al.*, 2011). Furthermore the capsule is involved in resistance to human serum (Keo *et al.*, 2011). Iron acquisition and metabolism are also essential virulence factors (Palyada *et al.*, 2004). Indeed, deletion of several iron acquisition proteins including FeoB, CfrA/CfrB/CeuB and CtuA result in decreased colonization in chick colonization assays (Miller *et al.*, 2008; Naikare *et al.*, 2006; Palyada *et al.*, 2004; Xu *et al.*, 2010). Furthermore, FeoB and CtuA were outcompeted by wild-type *C. jejuni* in a pathogenic piglet model and a rabbit ileal loop model respectively (Naikare *et al.*, 2006; Stintzi *et al.*, 2005). Clearly *C. jejuni* must be able to scavenge iron sources for successful colonization and pathogenesis.

This chapter focuses on providing an overview of the mechanisms and pathways that facilitate *Campylobacter* survival, growth, and colonization within the host. *C. jejuni* has specific nutritional requirements that allow the bacteria to grow within the mucus layer of the intestine. Furthermore, the composition of mucus appears to influence whether *C. jejuni* establishes a commensal or pathogenic interaction with its host.

Carbon sources utilized by *Campylobacter* to support growth *in vivo*

Unlike most other enteric pathogenic bacteria, campylobacters are fastidious organisms that have specific nutrient requirements. *C. jejuni* primarily uses amino acids as sole carbon sources, although they can also grow on short chain fatty acids (lactate/acetate) and citric acid cycle intermediates (i.e. succinate, fumarate, etc.) (Stahl *et al.*, 2012). *C. jejuni* NCTC11168 shows preferential metabolism of amino acids with serine, aspartate, glutamate, and proline the preferred substrates when grown in rich media (Guccione *et al.*, 2008; Line *et al.*, 2010). Furthermore, studies using phenotypic microarrays have largely corroborated these results demonstrating that serine, aspartate, and glutamate are utilized by *C. jejuni* as sole carbon sources before switching to asparagine or proline (Line *et al.*, 2010). Interestingly, serine, proline, glutamate, and aspartate are abundant in chick faecal matter (Parsons, 1984), coinciding with *C. jejuni*'s preferential use of these amino acids for energy generation. Campylobacters have been historically believed to be unable to utilize carbohydrates (i.e. glucose, lactose) as sole carbon sources due to the lack of a functional glycolytic pathway (Parkhill *et al.*, 2000; Velayudhan and Kelly, 2002). However recent work has shown that *C. jejuni* can metabolize the carbohydrate L-fucose and use this sugar as a carbon source (Muraoka and Zhang, 2011; Stahl *et al.*, 2011). In the following subsections we discuss the metabolism and colonization role of the major nutrient sources used by *C. jejuni* (serine, glutamate/aspartate, proline, asparagine, glutamine) as well as the recently discovered nutrient L-fucose (Fig. 7.1).

Serine

Serine is one of *C. jejuni*'s preferred amino acids as determined by growth assays in minimal essential media (Guccione *et al.*, 2008; Velayudhan *et al.*, 2004) and from phenotypic microarrays (Line *et al.*, 2010). *C. jejuni* uses SdaC to transport L-serine into the cell. The *sdaC* gene is encoded as part of the *sdaC–sdaA* operon in *C. jejuni* NCTC11168 and likely functions as an H^+ symporter based on homology to *E. coli* SdaC. Characterization of a Δ*sdaC* mutant in *C. jejuni* suggests that SdaC encodes for a high-capacity, low-affinity serine-specific permease protein (Velayudhan *et al.*, 2004). Results of transport kinetic assays using the Δ*sdaC* mutant also revealed that serine is transported into *C. jejuni* through a

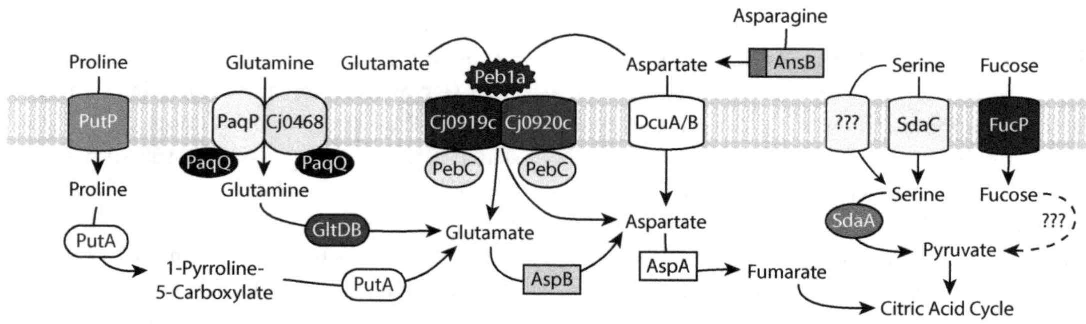

Figure 7.1 Amino acid and fucose metabolism in *C. jejuni*. Characterized (glutamine, glutamate, aspartate, asparagine, serine, and fucose) and predicted (proline) uptake complexes from the inner membrane are shown. Upon transport of amino acids into the cytoplasm, proline, glutamine, and glutamate are metabolized into aspartate. Asparagine is metabolized into aspartate in the periplasm by AnsB. Aspartate is converted by AspA into fumarate and feeds into the citric acid cycle for energy production. Serine is metabolized into pyruvate via SdaA for use in the citric acid cycle. Fucose is transported into the cell by FucP and converted into pyruvate via an uncharacterized pathway.

second, high-affinity transport protein. This putative transport protein might act as a dual serine/threonine transporter as speculated from results of amino acid competition experiments; however, the identity of this putative transporter remains unknown (Velayudhan *et al.*, 2004).

The second gene of the *sdaC-sdaA* operon, *sdaA*, encodes for an L-serine specific dehydratase, which functions to convert serine into pyruvate and ammonia. SdaA has been shown to have an essential role for *C. jejuni* growth when using serine as a sole carbon source as a Δ*sdaA* mutant was impaired in growth relative to the wild-type strain. The SdaA enzyme is likely a [4Fe–4S] containing protein, which is similar to serine dehydratases found in some strict anaerobic bacterial species (Velayudhan *et al.*, 2004). This is in contrast to the active site found in other characterized serine dehydratases, which contain pyridoxal-5′-phosphate. The *C. jejuni* SdaA enzyme is inactivated by air and thus, given the importance of SdaA for serine metabolism and growth, likely contributes to *C. jejuni*'s sensitivity towards oxygen (Velayudhan *et al.*, 2004). The importance of serine metabolism is further demonstrated using chick colonization experiments. The Δ*sdaA* mutant was completely attenuated in colonization of the caeca relative to wild-type *C. jejuni*, demonstrating the importance of SdaA during *in vivo* colonization (Velayudhan *et al.*, 2004).

Glutamate and aspartate

Glutamate and aspartate are transported into *C. jejuni* by a series of proteins consisting of Cj0919c, Cj0920c, Cj0921c (Peb1a), and Cj0922c (PebC), which together constitute the Peb1a ABC transporter uptake system (Fig. 7.1). Peb1a is a periplasmic protein, with homology to periplasmic binding protein components of bacterial ABC transporters. The Peb1a protein is able to bind both glutamate and aspartate with high affinity, consistent with a role as a dual amino acid transporter (del Rocio Leon-Kempis *et al.*, 2006). Furthermore, Peb1a has been previously classified as an adhesin protein that is important for adhesion to epithelial cells and intestinal colonization in mice (Pei *et al.*, 1998). However, immunoblotting of cell fractions did not reveal that Peb1a was associated with the outer *C. jejuni* membrane (del Rocio Leon-Kempis *et al.*, 2006). Peb1a contains a two-component signal peptide and it is therefore speculated that Peb1a might be transported across the outer membrane to the cell surface and/or into the surrounding environment using a mechanism similar to the one observed for Egl export in *Pseudomonas solanacearum* (Huang and Schell, 1992; Pei and Blaser, 1993). How Peb1a is able to act as an adhesin and contribute to *C. jejuni* colonization remains unclear, but given the role of Peb1a as a transporter of aspartate and glutamate, the colonization defect observed in mice may also be a consequence of the inability of *C. jejuni* to utilize these amino acids present within the intestine for growth (del Rocio Leon-Kempis *et al.*, 2006; Pei and Blaser, 1993).

The role of Peb1a in *C. jejuni* glutamate and aspartate metabolism was further confirmed by L-^{14}C-aspartate and L-^{14}C-glutamate uptake assays and growth assays using a Δ*peb1a* mutant. The assays revealed that Peb1a is responsible for the entire uptake of glutamate into the cell and that Peb1a is required for growth when glutamate is the sole carbon source. In contrast, although significantly attenuated, the Δ*peb1a* mutant was still able to take up aspartate into the cell but was not able to grow when aspartate was supplemented in the media (del Rocio Leon-Kempis *et al.*, 2006). *C. jejuni* is still able to transport aspartate into the cell by a second characterized transport system consisting of two partially redundant proteins, DcuA and DcuB. The *dcuA* and *dcuB* genes encode for anaerobic C4-decarboxylate transport proteins. Mutants into these proteins were observed to have reduced L-aspartate uptake into *C. jejuni* under oxygen limited conditions (Guccione *et al.*, 2008). Additionally, both the Δ*dcuA* and Δ*dcuB* mutants were still able to grow on aspartate to levels almost comparable to the wild-type strain, although a Δ*dcuA*Δ*dcuB* double mutant resulted in almost negligible growth (Guccione *et al.*, 2008). It is speculated that the inability of *C. jejuni* to excrete succinate in the double mutant, which would be essential during fumarate reduction, accounts for the observed phenotype. Therefore, the Peb1a ABC transporter is likely responsible for *C. jejuni*'s ability to grow and uptake L-aspartate in the single Δ*dcuA* and Δ*dcuB* mutants (Guccione *et al.*, 2008).

Upon glutamate uptake into *C. jejuni*, AspB is involved in conversion of glutamate and oxaloacetate into aspartate and 2-oxoglutarate (α-ketoglutarate) (Guccione *et al.*, 2008). AspB encodes an amino-transferase enzyme and deletion of *aspB* results in the inability of *C. jejuni* to grow using glutamate as a carbon source (Guccione *et al.*, 2008). The aspartate generated from cytosolic AspB can then be metabolized by the AspA aspartase enzyme. AspA deaminates aspartate into fumarate, which can be used as a citric acid cycle intermediate during respiration (Fig. 7.1). AspA plays an important central role in amino acid metabolism as a deletion mutant into *aspA* was unable to grow in the presence of glutamate, aspartate, proline or glutamine (Guccione *et al.*, 2008). The exception is serine, which as mentioned above, can be converted into pyruvate to feed into the citric acid cycle. Upon uptake of glutamate, proline or glutamine, *C. jejuni* can metabolize these amino acids into aspartate for deamination into fumarate for energy respiration. The importance of AspA in colonization of the host is further demonstrated in chick colonization assays, where mutation of *aspA* resulted in attenuated colonization of chicks (Guccione *et al.*, 2008). Given the high abundance of glutamate and aspartate in chick excreta (Parsons, 1984), it is probable that aspartate and glutamate metabolism plays an important role in *C. jejuni* colonization of chicks and thus, defects in the metabolism of these amino acids contributes to the colonization defects observed in the Δ*aspA* mutant (Guccione *et al.*, 2008).

Proline

Proline metabolism in *C. jejuni* is facilitated by *putP* and *putA* (Fig. 7.1). The *putP* gene encodes a Na$^+$/L-proline symporter and likely functions in the binding and uptake of proline into the cytoplasm (Hofreuter *et al.*, 2012). Once in the cytoplasm, proline can be metabolized by the PutA bifunctional proline dehydrogenase/delta-1-pyrroline-5-carboxylate dehydrogenase. PutA is involved in catalysing two sequential reactions that first converts proline into 1-pyrroline-5-carboxylate and then oxidizes the 1-pyrroline-5-carboxylate to glutamate (Tanner, 2008). The glutamate produced can be further metabolized as described above. *C. jejuni* NCTC11168 and 81-176 display increased growth in minimal media supplemented with proline demonstrating *C. jejuni*'s ability to use proline as a growth substrate (Guccione *et al.*, 2008; Hofreuter *et al.*, 2012). Furthermore, deletion of *putP* and *putA* in *C. jejuni* 81-176 resulted in the inability of the mutants to utilize proline as a growth substrate (Hofreuter *et al.*, 2012). The importance of proline metabolism in *C. jejuni* colonization was demonstrated using a mouse model, which found significant attenuation in intestinal colonization for the Δ*putP* mutant (Hofreuter *et al.*,

2012). Given the high abundance of proline in chick excreta (Parsons *et al.*, 1982), these genes will also likely play an important role during chick caecal colonization.

Asparagine

Asparagine utilization has also been observed for *C. jejuni*. Interestingly, *C. jejuni* 81-176 is able to grow using asparagine as the sole carbon source; however, growth was not observed for NCTC11168 (Hofreuter *et al.*, 2008). Nevertheless, phenotypic microarray analysis suggested that *C. jejuni* NCT1168 might still be able to metabolize this amino acid (Line *et al.*, 2010). The apparent discrepancy may be explained by the fact that the phenotypic microarrays used in the study by Line *et al.* only show respiration by the cells for the substrate tested (Line *et al.*, 2010). It has been previously demonstrated that although cells are able to respire, it does not necessarily follow that cells are able to grow (Bochner *et al.*, 2001). Although a transport protein for asparagine has not been identified in *C. jejuni*, most strains contain *ansB*, which encodes an asparginase enzyme to convert asparagine into aspartate. Furthermore, the *C. jejuni* 81-176 AnsB contains a sec-dependent secretion signal to allow for secretion of AnsB into the periplasmic space (Fig. 7.1). Deletion of *ansB* in *C. jejuni* 81-176 or insertion of a copy of *ansB* lacking the secretion signal inhibited the ability of 81-176 to grow using asparagine as a carbon source. The presence of AnsB in the periplasm enables 81-176 to metabolize asparagine into aspartate. Furthermore, the Peb1a transporter is then required to transport aspartate into the cytoplasm as a Δ*peb1a* mutant showed impaired growth in the presence of asparagine (Hofreuter *et al.*, 2008). Although affected in *in vitro* growth studies using asparagine as the sole carbon source, Δ*ansB* was not affected in its ability to colonize the intestine of mice relative to the wild-type. Interestingly, the Δ*ansB* mutant was affected in colonization of the liver. This result suggests a role for asparagine metabolism for growth within the liver of mice in *C. jejuni* 81-176 and thus AnsB may contribute to *C. jejuni* tissue tropism (Hofreuter *et al.*, 2008).

Glutamine

The remaining amino acid preferentially utilized by *C. jejuni* is glutamine. However, studies involving NCTC11168 report conflicting data as to the ability of the strain to grow using glutamine as a carbon source in defined minimal media (Guccione *et al.*, 2008; Hofreuter *et al.*, 2008). Furthermore, phenotypic microarray analysis using glutamine as a carbon source did not detect any respiration for NCTC11168 (Line *et al.*, 2010). Despite this, *C. jejuni* NCTC11168 contains a glutamine transport system consisting of *cj0467* (*paqP*), *cj0468*, and *cj0469* (*paqQ*). The *paqP* and *cj0468* genes encode amino-acid ABC transporter integral membrane proteins, whereas *paqQ* encodes an amino-acid ABC transporter ATP-binding protein (Fig. 7.1). Characterization of *paqP* and *paqQ* mutants has been carried out in *C. jejuni* 81-176, a strain that readily utilizes glutamine (Hofreuter *et al.*, 2008). Assays using ^{14}C-L-glutamine revealed that deletion of *paqP* or *paqQ* resulted in impaired glutamine uptake into the cell. The Δ*paqP* and Δ*paqQ* mutants also displayed varying levels of decreased uptake of glutamate, cysteine and aspartate. This suggests that the *paqP*/*paqQ* uptake system does not just specifically transport glutamine but also other amino acids, albeit to a lesser extent (Lin *et al.*, 2009). Although the Δ*paqP* and Δ*paqQ* mutants showed decreased uptake in glutamine, mice colonization assays revealed no *in vivo* defects for these mutants. This result suggests that although *paqP*/*paqQ* are involved in glutamine uptake, glutamine transport does not appear to play a crucial role in *C. jejuni* colonization of the mouse intestine and indicates that other nutrients are preferentially metabolized for growth (Lin *et al.*, 2009). Glutamine metabolism in *C. jejuni* NCTC11168 is likely accomplished by glutamate synthase (*gltBD*). GltDB converts L-glutamine and 2-oxoglutarate into two molecules of L-glutamate, which can then be metabolized as described above. Although absent in *C. jejuni* NCTC11168, some *C. jejuni* strains, such as 81-176, encode γ-glutamyltranspeptidase (Ggt). This periplasmic protein hydrolyses glutamine (as well as glutathione) to glutamate. A Δ*ggt* mutant strain was unable to grow in minimal essential media when supplemented with glutamine confirming the glutamine metabolism role of *ggt* in 81-176 (Hofreuter *et al.*, 2008). The importance of *ggt* in both mice and chick colonization experiments has been previously demonstrated using the Δ*ggt* mutants for *C. jejuni* 81-176 and *C. jejuni* 81116 strains respectively (Barnes *et al.*, 2007; Hofreuter *et al.*, 2008). Colonization of MyD88$^{-/-}$-deficient mice was attenuated for the Δ*ggt* mutant in *C. jejuni* 81-176 (Hofreuter *et al.*, 2008), whereas Ggt was found to be required for long term colonization of chicks for *C. jejuni* 81116 (Barnes *et al.*, 2007). Together these results highlight the importance of glutamine utilization during host colonization for specific *C. jejuni* strains.

L-Fucose

Despite *C. jejuni's* chemotaxic response towards fucose (Hugdahl *et al.*, 1988) as well as its ability to bind fucosylated glycan structures (Day *et al.*, 2009), *C. jejuni* has been considered to be a completely asaccharolytic organism. In order to obtain carbon sources required for growth, *C. jejuni* has been shown to primarily utilize serine, proline, aspartate and glutamate as well as short-chain fatty acids and citric acid cycle intermediates (Stahl *et al.*, 2012). Recently however, work by Muraoka *et al.* and Stahl *et al.* have demonstrated that some strains of *C. jejuni* are able to metabolize L-fucose. Fucose is highly abundant on mucin glycoproteins within the intestine and is important for supporting growth and providing binding sites for commensal microbes. Indeed, up to 14% of the oligosaccharides present on mucin intestinal glycoproteins are composed of fucose (Allen and Griffiths, 2001). Bacteria can use this fucose as a nutrient source by secreting fucosidases to remove fucose from the mucin glycoproteins as is observed for *Bacteroides thetaiotaomicron* (Hooper *et al.*, 1999). Furthermore *B. thetaiotaomicron,* a commensal microbe found in high abundance within the distal small intestine and colon, has been shown to induce the production of fucosylated glycan structures *in vivo* (Hooper and Gordon, 2001). An indication that *C. jejuni* might be able to metabolize fucose was suggested by transcriptomic studies with *C. jejuni* NCTC11168 grown in the presence of L-fucose or mucus extracts (Stahl *et al.*, 2011), and phenotype arrays to identify carbohydrates metabolized by *C. jejuni* (Muraoka and Zhang, 2011).

Transcriptome analysis of *C. jejuni* NCTC11186 supplemented with L-fucose in MEMα revealed a significant up-regulation of the genes *cj0481-cj0490,* which are co-transcribed (Stahl *et al.*, 2011). These genes encode for *cj0481 (dapA),* putative dihydrodipicolinate synthase, *cj0482 (uxaA),* putative altronate hydrolase N-terminus, *cj0483 (uxaA),* putative altronate hydrolase C-terminus, *cj0484,* putative MFS (major facilitator superfamily) transport protein, *cj0485,* putative oxidoreductase, *cj0486,* putative sugar transporter, *cj0487,* putative amidohydrolase, *cj0488,* conserved hypothetical protein, and *cj0489 (ald),* putative aldehyde dehydrogenase (Parkhill *et al.*, 2000). Given that L-fucose is highly abundant on mucin oligosaccharides, microarray analysis was also performed with *C. jejuni* NCTC11168 grown in presence of porcine gastric mucin to determine genes and pathways important for growth under these conditions. Interestingly, a significant up-regulation for *cj0481–cj0490* was also observed for growth in presence of porcine gastric mucin. Of the genes that were up-regulated in this operon, it was noted that Cj0486 shares 37% homology with *E. coli* FucP, which encodes for a fucose permease. The fucose permease gene in *E. coli* facilitates uptake of L-fucose into the cell via proton-dependent active transport (Dang *et al.*, 2010).

To confirm the role of Cj0486 as a fucose permease, the *C. jejuni* NCTC *cj0486* gene was expressed in an *E. coli* Δ*fucP* background. Expression of *cj0486* restored the ability of the *E. coli* Δ*fucP* + *cj0486* mutant to metabolize fucose. This was demonstrated by growth on MacConkey agar supplemented with fucose with the concomitant appearance of a red colour indicating fucose fermentation (Stahl *et al.*, 2011). Additionally, fucose uptake assays quantifying ^{3}H-L-fucose transport into *C. jejuni* Δ*cj0481*, Δ*cj0483*, Δ*cj0486*, and Δ*cj0487* mutant strains showed abrogation of ^{3}H-L-fucose uptake in the Δ*cj0486* mutant but not the Δ*cj0481*, Δ*cj0483*, or Δ*cj0487* mutants. In the presence of L-fucose, initial uptake of ^{3}H-L-fucose was induced in the wild-type NCTC11168 strain but not in the mutant strains. These findings support the role of *cj0486* as a fucose permease in *C. jejuni* (Fig. 7.1) but also indicate that *cj0481*, *cj0483*, and *cj0487* are not essential for successful uptake of L-fucose as mutants in these genes transported L-fucose at levels comparable to the wild-type strain. Interestingly however, although Δ*cj0481*, Δ*cj0483*, and Δ*cj0487* showed no decrease in L-fucose uptake relative to the wild-type strain, these mutants were not inducible for uptake compared to the wild-type when L-fucose was added to the growth media (Stahl *et al.*, 2011).

The ability of various different *C. jejuni* strains to utilize fucose as a growth substrate was demonstrated using L-fucose growth promotion assays. When L-fucose was supplemented to the growth media, *C. jejuni* NCTC11168 and RM1221 displayed a significant increase in growth (Stahl *et al.*, 2011). As anticipated, no discernible growth increase was observed for 81-176 or 81116, which both lack the *cj0481-cj0489* operon (Muraoka and Zhang, 2011; Stahl *et al.*, 2011). Although *cj0481* and *cj0487* are not involved in fucose uptake, the importance of *cj0481*, *cj0486* and *cj0487* in fucose metabolism was shown using these growth promotion assays. The *cj0481*, *cj0486* and *cj0487* mutant strains were unable to utilize L-fucose as a carbon source (Stahl *et al.*, 2011). However, unlike *cj0486*, which is clearly involved in fucose uptake (and is therefore unable to grow using L-fucose as a growth substrate), the function of *cj0481* and *cj0487* during L-fucose metabolism is still unclear. Furthermore, the cellular pathway that *C. jejuni* utilizes to metabolize L-fucose has yet to be characterized. Unlike *E. coli* and *Bacteroides* species, which use ATP-dependent and GTP-dependent fucose activation pathways (Baldoma and Aguilar, 1988; Coyne *et al.*, 2005), *C. jejuni* is missing several of the key enzymes essential for these pathways. Specifically, *E. coli* utilizes FucK, which encodes an L-fuculose kinase involved in ATP-dependent phosphorylation of L-fuculose (Baldoma and Aguilar, 1988). *C. jejuni* NCTC11168 was unable to produce detectable levels of phosphorylated L-fucose intermediates suggesting *C. jejuni* uses an alternative L-fucose metabolism pathway. A second known pathway for L-fucose metabolism has been characterized in *Bacteroides*. *Bacteroides* utilizes a bifunctional L-fucokinase/GDP-fucose pyrophosphorylase (FKP) to catalyse two important enzymatic steps during L-fucose metabolism. Firstly, FKP utilizes ATP to phosphorylate L-fucose and then uses one molecule of GTP to produce GDP-L-fucose (Coyne *et al.*, 2005). Similar to the result with ^{3}H-L-fucose ATP dependent phosphorylation, *C. jejuni* was unable to produce GTP-activated ^{3}H-L-fucose intermediates. Together these results suggest that *C. jejuni* NCTC11168 metabolizes fucose using an alternative pathway to those present *E. coli* and *Bacteroides* species (Stahl *et al.*, 2011). It is thought that *C. jejuni* might metabolize fucose using an alternative pathway that has been characterized in *Xanthomonas campestris* due to some amino acid identity between the enzymes in these two bacteria. This pathway functions in an ATP and GTP independent manner to metabolize L-fucose into L-lactate and pyruvate for subsequent cellular use (Yew *et al.*, 2006). Further experimentation is required to determine if *C. jejuni* does indeed follow the same pathway as *X. campestris* or if it possesses a novel pathway for L-fucose metabolism.

The advantage of fucose metabolism during host colonization was highlighted in the competitive chick colonization and piglet pathogenesis models of *C. jejuni*. Competition of wild-type *C. jejuni* NCTC11168 against the Δ*cj0486* mutant in piglets revealed significant out-competition of the mutant strain compared to the wild-type. The ability to metabolize fucose provided a distinct advantage during colonization of the piglet intestine. Competitive assays using chicks, however, did not result in a significant decrease in fitness for the Δ*cj0486* mutant strain indicating that fucose utilization provides no significant advantage when assayed in a commensal host. Interestingly, when chicks were supplemented with L-fucose in their diet, the wild-type out-competed the Δ*cj0486* mutant during the early stages of colonization. However, the competitive advantage was lost at later time points in colonization. These results demonstrate the benefit of fucose uptake and metabolism during the early stages of chick colonization (Stahl *et al.*, 2011).

The reason for the competitive advantage of fucose utilization in the pathogenic model and not in the chick commensal model is unclear. Mucus extracts from the piglet and chick caecum have similar levels of fucosylation when analysed by mass spectrometry (Stahl *et al.*, 2011). Differences in mucin sulfation, folding, charge, glycosylation or differences in host commensal microbiota between piglet and chick mucus may influence the ability of *C. jejuni* to utilize or gain access to fucose oligosaccharides present within chick mucus (Alemka *et al.*, 2012; Byrne *et al.*, 2007; Stahl *et al.*, 2011). It is also unclear if *C. jejuni* is able to cleave fucose from the mucin glycoproteins in the intestine given that *C. jejuni* does not appear to encode for a fucosidase (Muraoka and Zhang, 2011). Overall, fucose metabolism provides specific *C. jejuni* strains with an *in vivo* colonization advantage and that fucose is an important colonization factor in the intestine of the piglet pathogenesis model.

Mucins and the mucous layer

During colonization and infection of the host, *Campylobacter* must transverse the mucus layer of the intestine to be able to adhere to and invade the intestinal epithelial cells to cause onset of gastroenteritis (Humphrey *et al.*, 2007). Once reaching the intestinal mucosal epithelium, *Campylobacter* would be in close proximity to the MUC1 mucin, a membrane protein that is expressed at the intestinal epithelial cell surface of the glycocalyx (Kim and Ho, 2010). When MUC1 makes contact with pathogens such as *C. jejuni*, the surface glycoprotein is released from the cell into the mucus layer for subsequent expulsion from the host (Day *et al.*, 2012). Thus, MUC1 may function as a decoy receptor for bacterial ligands to protect the host against harmful pathogens. The effectiveness of MUC1 in preventing *C. jejuni* pathology has been previously demonstrated using MUC1-deficient mice, which show increased susceptibility to *C. jejuni* invasion (McAuley *et al.*, 2007). The MUC1 mucin is highly sialylated and mannosylated (Day *et al.*, 2012). Glycan binding arrays with *C. jejuni* have demonstrated binding of *C. jejuni* to multiple glycans including mannose, sialic acid, galactose, fucose, glucosamine, and glycosaminoglycans. Interestingly, the growth conditions under which *C. jejuni* was cultured for the glycan arrays influenced the carbohydrate structures to which *C. jejuni* specifically bound. *C. jejuni* exposed to environmental conditions (atmospheric levels of oxygen at room temperature) were able to bind sialic acid and mannose glycans. Carbohydrate binding was reduced when *C. jejuni* were grown under conditions mimicking avian hosts (42°C) and mammalian hosts (37°C) (Day *et al.*, 2009). Given the high level of sialylation and mannosylation of MUC1, this result suggests that *C. jejuni* may interact with MUC1 initially after the transition from the external environment to the host during colonization. However, the direct binding to sialic acid and mannose structures is reduced once *C. jejuni* adapts to the conditions encountered within the host (Day *et al.*, 2009). Prolonged interaction with MUC1 would be detrimental to *C. jejuni* pathogenesis due to the role of MUC1 as a decoy receptor in host defence against pathogens (Day *et al.*, 2012).

C. jejuni is also able to directly bind galactose and fucose glycans when cultured under microaerophilic conditions at 37°C or 42°C. However, binding of these carbohydrates is attenuated when *C. jejuni* is maintained under environmental stress conditions (Day *et al.*, 2009). MUC2, the major mucin present within the intestine of the host, is highly fucosylated, contains abundant terminal galactose structures and has less sialylation than MUC1. Furthermore, *C. jejuni* preferentially colonizes the crypts of the intestinal epithelium where MUC2 is highly abundant (Day *et al.*, 2012). Thus, it is hypothesized that once *C. jejuni* has adapted to the host environment it has reduced binding to MUC1 and subsequently prefers to bind to fucose or galactose structures on MUC2 for long term colonization (Day *et al.*, 2012).

Mucus composition is an important determinant in *Campylobacter* commensalism and virulence

The predominant route of *Campylobacter* infection in humans is through consumption of undercooked poultry (Skirrow, 1991). *C. jejuni* is able to colonize chickens without the onset of the clinical symptoms of gastroenteritis seen in humans. Indeed, *C. jejuni* is capable of colonizing the deep crypts of the caecum in excess of 10^{10} CFU per gram of caecal content (Young *et al.*,

2007). However, unlike in human intestinal epithelium cells, it has been previously reported that *C. jejuni* does not adhere to or invade chick intestinal cells (Young *et al.*, 2007). It was therefore not understood how *C. jejuni* could maintain such high colonization levels without being cleared from the intestine given that the bacteria were not thought to be capable of invading the underlying epithelium. Recent work involving the characterization of *C. jejuni* reproduction and growth within chick mucus has revealed a novel mechanism, which allows *C. jejuni* to persist within the chick intestine despite rapid turnover of the mucus layer (Van Deun *et al.*, 2008) (Fig. 7.2). Contrary to what had been previously published, experiments by Byrne *et al.* and Van Deun *et al.* have demonstrated that *C. jejuni* is able to adhere and invade primary chicken caecal epithelial cells (Byrne *et al.*, 2007; Van Deun *et al.*, 2008) but cannot survive within the cells for an extended duration; specifically, *C. jejuni* was not detected intracellularly 4 h post infection of the primary chicken caecal epithelial cells (Van Deun *et al.*, 2008). Interestingly, the majority of the bacteria were found to escape from the cells in as little time as 5 min post invasion. Moreover, invasion of the primary chicken caecal epithelium did not result in apoptosis, necrosis or elicit an inflammatory response of the cells (Fig. 7.2). Additionally, it was found that *C. jejuni* is able to replicate successfully utilizing the nutrients found within chick mucus. Thus, the high level of *C. jejuni* colonization within chickens is proposed to be the result of rapid replication within the mucus layer and continual invasion and escape from the epithelial cells of the intestine to prevent clearance of *C. jejuni* from the intestine by mucus turnover (Van Deun *et al.*, 2008).

During infection, *C. jejuni* is able to elicit an inflammatory response in humans, whereas no such response has been observed in avian hosts. Interestingly, the species origin of mucus has been shown to play an important role in *C. jejuni* invasion of primary human intestinal cells. Invasion of human intestinal epithelial cells in the presence of chicken small or large intestinal mucus resulted in decreased cell invasion of *C. jejuni* by approximately 98% and 54% respectively (Byrne *et al.*, 2007). Conversely, the presence of crude human mucus extracts increased *C. jejuni* invasion of primary human intestinal cells by 35% (although this increase was not statistically significant). Whether *C. jejuni* invasion is increased in primary chicken

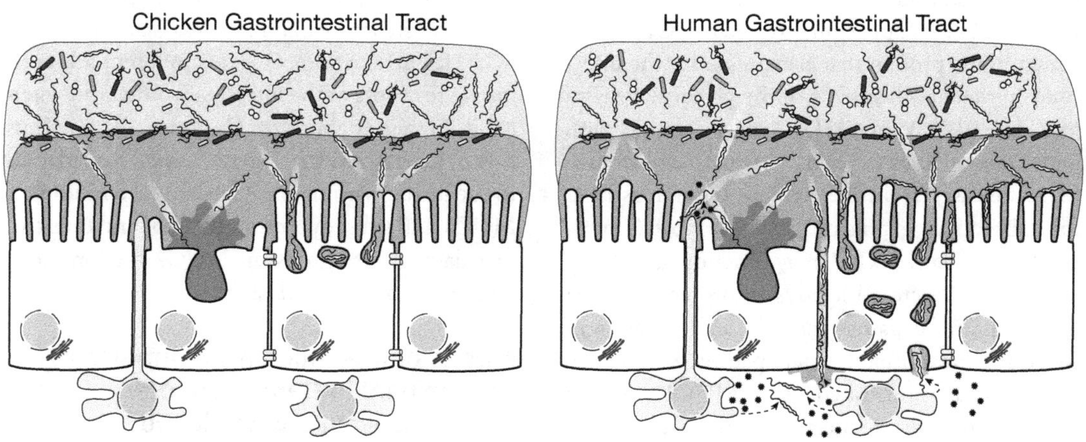

Figure 7.2 Proposed model of *C. jejuni* commensalism and pathogenesis in the chicken and human gastrointestinal tract. Left: *C. jejuni* colonization in the chicken gastrointestinal tract. *C. jejuni* replicates within the outer mucus layer and briefly invades the underlying epithelium before evading the cells and returning to the mucus to replicate. No immune response is elicited. Right: *C. jejuni* colonization of the human gastrointestinal tract. *C. jejuni* penetrates the inner mucus layer, adheres to and invades the underlying epithelium. Adherence and invasion is enhanced in the presence of human mucus. *C. jejuni* is able to replicate within the cells and translocate to the subepithelial compartment. Invasion of epithelial cells by *C. jejuni* elicits an immune response and recruitment of macrophages, neutrophils, and dendritic cells and production of pro and anti-inflammatory cytokines.

caecal epithelial cells in the presence of human intestinal mucus is currently unknown. If an increase in cell invasion were observed, it would provide additional evidence for the importance of species-derived intestinal mucus and its influence in promoting commensalism in chickens. Overall, these results suggest that differences between host-derived mucus extracts play an important role in *C. jejuni* cell invasion (Byrne *et al.*, 2007).

The variation in *C. jejuni* invasiveness in the presence of chick or human derived mucus was hypothesized by Alemka *et al.* to be due to differences in glycosylation of the mucins present within the mucus. *C. jejuni* invasion of HCT-8 cells was no longer attenuated when *C. jejuni* was incubated with mucins that had been treated with sodium metaperiodate, which oxidizes glycan structures on the mucins, prior to invasion (Alemka *et al.*, 2010). Further characterization of the mucin structures within chicken mucus is required to determine the specific differences between human and chicken mucus that account for different virulence phenotypes observed. It has been previously reported that chicken mucin oligosaccharides are more heavily sulfated than their human counterparts (Alemka *et al.*, 2012; Stahl *et al.*, 2011). Additionally, there are differences in glycosylation, folding, and charge of the mucins, as well different constituents found within the mucus between the two species that requires elucidation (Byrne *et al.*, 2007). It is plausible that the differences in composition between chicken and human mucus are an important factor in the pathogenicity of *Campylobacter*.

Conclusions and future trends

Much progress has been made towards the understanding of *Campylobacter* colonization of the host gastrointestinal tract. Investigation into *C. jejuni* nutrient metabolism has highlighted the importance of serine, glutamate, aspartate, proline, and glutamine metabolism for cell growth and *in vivo* colonization. Furthermore, the recent discovery of fucose utilization has overturned a firmly held notion that *C. jejuni* is an asaccharolytic organism. Interestingly, fucose utilization does not provide a competitive advantage in a chick commensal animal model compared to a piglet pathogenic model despite similar levels of fucosylation of mucins found within the caecum. This suggests that *C. jejuni* may preferentially metabolize alternative carbon sources while colonizing the chick caecum. The ability of *C. jejuni* to secrete enzymes to degrade mucin components for growth utilization is currently unknown. However, given the absence of an annotated fucosidase, other commensal intestinal microbes may play a role secreting enzymes to cleave fucose from mucin structures for which *C. jejuni* could utilize advantageously. The fucosylated O-glycan mucin structures from chicken have been shown to be more sulfated compared to mucin from pigs. Sulfation of mucin may decrease the accessibility to fucose (Stahl *et al.*, 2011). Thus, fucose may not be a readily usable energy source for *C. jejuni* in the chick caeca.

Clearly the importance of mucus composition plays a significant role in *C. jejuni* pathogenesis as highlighted by cell invasion assays performed in the presence of chick or human intestinal mucus. The reduction in human intestinal epithelial cell invasion in the presence of chick mucus suggests that mucus composition is an important determinant for *C. jejuni* invasion. Although *C. jejuni* invasiveness is reduced in the presence of chick mucus, *C. jejuni* is still able to invade and escape from these chicken intestinal cells. However, despite invading these cells, an inflammatory response is not elicited unlike in humans. Thus it is unclear as to why *C. jejuni* does not elicit an immune response in chicks and yet still invades and can rapidly escape from the intestinal cells. Much still remains to be deciphered to understand the fundamental differences that lead to *C. jejuni* commensalism or pathogenicity.

Acknowledgements

Research in Dr Stintzi's laboratory is supported by the Canadian Institute of Health Research (CIHR).

References

Alemka, A., Whelan, S., Gough, R., Clyne, M., Gallagher, M.E., Carrington, S.D., and Bourke, B. (2010). Purified chicken intestinal mucin attenuates *Campylobacter jejuni* pathogenicity *in vitro*. J. Med. Microbiol. 59, 898–903.

Alemka, A., Corcionivoschi, N., and Bourke, B. (2012). Defense and Adaptation: The Complex Inter-Relationship between *Campylobacter jejuni* and Mucus. Front. Cell. Infect. Microbiol. *2*, 15.

Allen, K.J., and Griffiths, M.W. (2001). Effect of environmental and chemotactic stimuli on the activity of the *Campylobacter jejuni* flaA sigma(28) promoter. FEMS Microbiol. Lett. *205*, 43–48.

Baldoma, L., and Aguilar, J. (1988). Metabolism of L-fucose and L-rhamnose in *Escherichia coli*: aerobic-anaerobic regulation of L-lactaldehyde dissimilation. J. Bacteriol. *170*, 416–421.

Ballal, S.A., Gallini, C.A., Segata, N., Huttenhower, C., and Garrett, W.S. (2011). Host and gut microbiota symbiotic factors: lessons from inflammatory bowel disease and successful symbionts. Cell. Microbiol. *13*, 508–517.

Barnes, I.H., Bagnall, M.C., Browning, D.D., Thompson, S.A., Manning, G., and Newell, D.G. (2007). Gamma-glutamyl transpeptidase has a role in the persistent colonization of the avian gut by *Campylobacter jejuni*. Microb. Pathog. *43*, 198–207.

Bochner, B.R., Gadzinski, P., and Panomitros, E. (2001). Phenotype microarrays for high-throughput phenotypic testing and assay of gene function. Genome Res. *11*, 1246–1255.

Byrne, C.M., Clyne, M., and Bourke, B. (2007). *Campylobacter jejuni* adhere to and invade chicken intestinal epithelial cells *in vitro*. Microbiology *153*, 561–569.

Coyne, M.J., Reinap, B., Lee, M.M., and Comstock, L.E. (2005). Human symbionts use a host-like pathway for surface fucosylation. Science *307*, 1778–1781.

Dang, S., Sun, L., Huang, Y., Lu, F., Liu, Y., Gong, H., Wang, J., and Yan, N. (2010). Structure of a fucose transporter in an outward-open conformation. Nature *467*, 734–738.

Day, C.J., Tiralongo, J., Hartnell, R.D., Logue, C.A., Wilson, J.C., von Itzstein, M., and Korolik, V. (2009). Differential carbohydrate recognition by *Campylobacter jejuni* strain 11168: influences of temperature and growth conditions. PLoS One *4*, e4927.

Day, C.J., Semchenko, E.A., and Korolik, V. (2012). Glycoconjugates Play a Key Role in *Campylobacter jejuni* Infection: Interactions between Host and Pathogen. Front. Cell. Infect. Microbiol. *2*, 9.

Dharmani, P., Srivastava, V., Kissoon-Singh, V., and Chadee, K. (2009). Role of intestinal mucins in innate host defense mechanisms against pathogens. J. Innate Immun. *1*, 123–135.

Ferrero, R.L., and Lee, A. (1988). Motility of *Campylobacter jejuni* in a viscous environment: comparison with conventional rod-shaped bacteria. J. Gen. Microbiol. *134*, 53–59.

Fox, J.G., Rogers, A.B., Whary, M.T., Ge, Z., Taylor, N.S., Xu, S., Horwitz, B.H., and Erdman, S.E. (2004). Gastroenteritis in NF-kappaB-deficient mice is produced with wild-type *Camplyobacter jejuni* but not with *C. jejuni* lacking cytolethal distending toxin despite persistent colonization with both strains. Infect. Immun. 72, 1116–1125.

Fukata, M., and Abreu, M.T. (2009). Pathogen recognition receptors, cancer and inflammation in the gut. Curr. Opin. Pharmacol. *9*, 680–687.

Guccione, E., Leon-Kempis Mdel, R., Pearson, B.M., Hitchin, E., Mulholland, F., van Diemen, P.M., Stevens, M.P., and Kelly, D.J. (2008). Amino acid-dependent growth of *Campylobacter jejuni*: key roles for aspartase (AspA) under microaerobic and oxygen-limited conditions and identification of AspB (Cj0762), essential for growth on glutamate. Mol. Microbiol. *69*, 77–93.

Hansson, G.C., and Johansson, M.E. (2010). The inner of the two Muc2 mucin-dependent mucus layers in colon is devoid of bacteria. Gut Microbes *1*, 51–54.

Harrington, S.M., Sheikh, J., Henderson, I.R., Ruiz-Perez, F., Cohen, P.S., and Nataro, J.P. (2009). The Pic protease of enteroaggregative *Escherichia coli* promotes intestinal colonization and growth in the presence of mucin. Infect. Immun. *77*, 2465–2473.

Herrmann, A., Davies, J.R., Lindell, G., Martensson, S., Packer, N.H., Swallow, D.M., and Carlstedt, I. (1999). Studies on the 'insoluble' glycoprotein complex from human colon. Identification of reduction-insensitive MUC2 oligomers and C-terminal cleavage. J. Biol. Chem. *274*, 15828–15836.

Hickey, T.E., McVeigh, A.L., Scott, D.A., Michielutti, R.E., Bixby, A., Carroll, S.A., Bourgeois, A.L., and Guerry, P. (2000). *Campylobacter jejuni* cytolethal distending toxin mediates release of interleukin-8 from intestinal epithelial cells. Infect. Immun. *68*, 6535–6541.

Hofreuter, D., Novik, V., and Galan, J.E. (2008). Metabolic diversity in *Campylobacter jejuni* enhances specific tissue colonization. Cell. Host Microbe *4*, 425–433.

Hofreuter, D., Mohr, J., Wensel, O., Rademacher, S., Schreiber, K., Schomburg, D., Gao, B., and Galan, J.E. (2012). Contribution of amino acid catabolism to the tissue specific persistence of *Campylobacter jejuni* in a murine colonization model. PLoS One *7*, e50699.

Hooper, L.V., and Gordon, J.I. (2001). Glycans as legislators of host–microbial interactions: spanning the spectrum from symbiosis to pathogenicity. Glycobiology *11*, 1R-10R.

Hooper, L.V., Xu, J., Falk, P.G., Midtvedt, T., and Gordon, J.I. (1999). A molecular sensor that allows a gut commensal to control its nutrient foundation in a competitive ecosystem. Proc. Natl. Acad. Sci. U.S.A. *96*, 9833–9838.

Huang, J.Z., and Schell, M.A. (1992). Role of the two-component leader sequence and mature amino acid sequences in extracellular export of endoglucanase EGL from *Pseudomonas solanacearum*. J. Bacteriol. *174*, 1314–1323.

Hugdahl, M.B., Beery, J.T., and Doyle, M.P. (1988). Chemotactic behavior of *Campylobacter jejuni*. Infect. Immun. *56*, 1560–1566.

Humphrey, T., O'Brien, S., and Madsen, M. (2007). Campylobacters as zoonotic pathogens: a food production perspective. Int. J. Food. Microbiol. *117*, 237–257.

Johansson, M.E., Thomsson, K.A., and Hansson, G.C. (2009). Proteomic analyses of the two mucus layers of the colon barrier reveal that their main component, the

Muc2 mucin, is strongly bound to the Fcgbp protein. J. Proteome Res. *8*, 3549–3557.

Keo, T., Collins, J., Kunwar, P., Blaser, M.J., and Iovine, N.M. (2011). *Campylobacter* capsule and lipooligosaccharide confer resistance to serum and cationic antimicrobials. Virulence *2*, 30–40.

Kim, Y.S., and Ho, S.B. (2010). Intestinal goblet cells and mucins in health and disease: recent insights and progress. Curr. Gastroenterol. Rep. *12*, 319–330.

Kobayashi, K., Ogata, H., Morikawa, M., Iijima, S., Harada, N., Yoshida, T., Brown, W.R., Inoue, N., Hamada, Y., Ishii, H., *et al.* (2002). Distribution and partial characterisation of IgG Fc binding protein in various mucin producing cells and body fluids. Gut *51*, 169–176.

Kohbata, S., Yokoyama, H., and Yabuuchi, E. (1986). Cytopathogenic effect of *Salmonella typhi* GIFU 10007 on M cells of murine ileal Peyer's patches in ligated ileal loops: an ultrastructural study. Microbiol. Immunol. *30*, 1225–1237.

Konkel, M.E., Klena, J.D., Rivera-Amill, V., Monteville, M.R., Biswas, D., Raphael, B., and Mickelson, J. (2004). Secretion of virulence proteins from *Campylobacter jejuni* is dependent on a functional flagellar export apparatus. J. Bacteriol. *186*, 3296–3303.

Krimi, R.B., Kotelevets, L., Dubuquoy, L., Plaisancie, P., Walker, F., Lehy, T., Desreumaux, P., Van Seuningen, I., Chastre, E., Forgue-Lafitte, M.E., *et al.* (2008). Resistin-like molecule beta regulates intestinal mucous secretion and curtails TNBS-induced colitis in mice. Inflamm. Bowel Dis. *14*, 931–941.

Lara-Tejero, M., and Galan, J.E. (2000). A bacterial toxin that controls cell cycle progression as a deoxyribonuclease I-like protein. Science *290*, 354–357.

Lee, A., O'Rourke, J.L., Barrington, P.J., and Trust, T.J. (1986). Mucus colonization as a determinant of pathogenicity in intestinal infection by *Campylobacter jejuni*: a mouse cecal model. Infect. Immun. *51*, 536–546.

Lievin-Le Moal, V., and Servin, A.L. (2006). The front line of enteric host defense against unwelcome intrusion of harmful microorganisms: mucins, antimicrobial peptides, and microbiota. Clin. Microbiol. Rev. *19*, 315–337.

Lin, A.E., Krastel, K., Hobb, R.I., Thompson, S.A., Cvitkovitch, D.G., and Gaynor, E.C. (2009). Atypical roles for *Campylobacter jejuni* amino acid ATP binding cassette transporter components PaqP and PaqQ in bacterial stress tolerance and pathogen–host cell dynamics. Infect. Immun. *77*, 4912–4924.

Line, J.E., Hiett, K.L., Guard-Bouldin, J., and Seal, B.S. (2010). Differential carbon source utilization by *Campylobacter jejuni* 11168 in response to growth temperature variation. J. Microbiol. Methods *80*, 198–202.

McAuley, J.L., Linden, S.K., Png, C.W., King, R.M., Pennington, H.L., Gendler, S.J., Florin, T.H., Hill, G.R., Korolik, V., and McGuckin, M.A. (2007). MUC1 cell surface mucin is a critical element of the mucosal barrier to infection. J. Clin. Invest. *117*, 2313–2324.

Martens, E.C., Roth, R., Heuser, J.E., and Gordon, J.I. (2009). Coordinate regulation of glycan degradation and polysaccharide capsule biosynthesis by a prominent human gut symbiont. J. Biol. Chem. *284*, 18445–18457.

Miller, C.E., Rock, J.D., Ridley, K.A., Williams, P.H., and Ketley, J.M. (2008). Utilization of lactoferrin-bound and transferrin-bound iron by *Campylobacter jejuni*. J. Bacteriol. *190*, 1900–1911.

Moncada, D.M., Kammanadiminti, S.J., and Chadee, K. (2003). Mucin and Toll-like receptors in host defense against intestinal parasites. Trends Parasitol. *19*, 305–311.

Muraoka, W.T., and Zhang, Q. (2011). Phenotypic and genotypic evidence for L-fucose utilization by *Campylobacter jejuni*. J. Bacteriol. *193*, 1065–1075.

Naikare, H., Palyada, K., Panciera, R., Marlow, D., and Stintzi, A. (2006). Major role for FeoB in *Campylobacter jejuni* ferrous iron acquisition, gut colonization, and intracellular survival. Infect Immun. *74*, 5433–5444.

Palyada, K., Threadgill, D., and Stintzi, A. (2004). Iron acquisition and regulation in *Campylobacter jejuni*. J. Bacteriol. *186*, 4714–4729.

Parkhill, J., Wren, B.W., Mungall, K., Ketley, J.M., Churcher, C., Basham, D., Chillingworth, T., Davies, R.M., Feltwell, T., Holroyd, S., *et al.* (2000). The genome sequence of the food-borne pathogen *Campylobacter jejuni* reveals hypervariable sequences. Nature *403*, 665–668.

Parsons, C.M. (1984). Influence of caecectomy and source of dietary fibre or starch on excretion of endogenous amino acids by laying hens. Br. J. Nutr. *51*, 541–548.

Parsons, C.M., Potter, L.M., and Brown, R.D. (1982). Effects of dietary protein and intestinal microflora on excretion of amino acids in poultry. Poultry Sci. *61*, 939–946.

Pei, Z., and Blaser, M.J. (1993). PEB1, the major cell-binding factor of *Campylobacter jejuni*, is a homolog of the binding component in Gram-negative nutrient transport systems. J. Biol. Chem. *268*, 18717–18725.

Pei, Z., Burucoa, C., Grignon, B., Baqar, S., Huang, X.Z., Kopecko, D.J., Bourgeois, A.L., Fauchere, J.L., and Blaser, M.J. (1998). Mutation in the peb1A locus of *Campylobacter jejuni* reduces interactions with epithelial cells and intestinal colonization of mice. Infect. Immun. *66*, 938–943.

Qin, J., Li, R., Raes, J., Arumugam, M., Burgdorf, K.S., Manichanh, C., Nielsen, T., Pons, N., Levenez, F., Yamada, T., *et al.* (2010). A human gut microbial gene catalogue established by metagenomic sequencing. Nature *464*, 59–65.

del Rocio Leon-Kempis, M., Guccione, E., Mulholland, F., Williamson, M.P., and Kelly, D.J. (2006). The *Campylobacter jejuni* PEB1a adhesin is an aspartate/glutamate-binding protein of an ABC transporter essential for microaerobic growth on dicarboxylic amino acids. Mol. Microbiol. *60*, 1262–1275.

Salyers, A.A., West, S.E., Vercellotti, J.R., and Wilkins, T.D. (1977). Fermentation of mucins and plant polysaccharides by anaerobic bacteria from the human colon. Appl. Environ. Microbiol. *34*, 529–533.

Silva, A.J., Pham, K., and Benitez, J.A. (2003). Haemagglutinin/protease expression and mucin gel penetration in El Tor biotype *Vibrio cholerae*. Microbiology *149*, 1883–1891.

Skirrow, M.B. (1991). Epidemiology of *Campylobacter* enteritis. Int. J. Food Microbiol. *12*, 9–16.

Song, Y.C., Jin, S., Louie, H., Ng, D., Lau, R., Zhang, Y., Weerasekera, R., Al Rashid, S., Ward, L.A., Der, S.D., *et al.* (2004). FlaC, a protein of *Campylobacter jejuni* TGH9011 (ATCC43431) secreted through the flagellar apparatus, binds epithelial cells and influences cell invasion. Mol. Microbiol. *53*, 541–553.

Stahl, M., Friis, L.M., Nothaft, H., Liu, X., Li, J., Szymanski, C.M., and Stintzi, A. (2011). L-fucose utilization provides *Campylobacter jejuni* with a competitive advantage. Proc. Natl. Acad. Sci. U.S.A. *108*, 7194–7199.

Stahl, M., Butcher, J., and Stintzi, A. (2012). Nutrient Acquisition and Metabolism by *Campylobacter jejuni*. Front. Cell. Infect. Microbiol. *2*, 5.

Stintzi, A., Marlow, D., Palyada, K., Naikare, H., Panciera, R., Whitworth, L., and Clarke, C. (2005). Use of genome-wide expression profiling and mutagenesis to study the intestinal lifestyle of *Campylobacter jejuni*. Infect. Immun. *73*, 1797–1810.

Tanner, J.J. (2008). Structural biology of proline catabolism. Amino Acids *35*, 719–730.

Taupin, D., and Podolsky, D.K. (2003). Trefoil factors: initiators of mucosal healing. Nat. Rev. Mol. Cell. Biol. *4*, 721–732.

Van Deun, K., Pasmans, F., Ducatelle, R., Flahou, B., Vissenberg, K., Martel, A., Van den Broeck, W., Van Immerseel, F., and Haesebrouck, F. (2008). Colonization strategy of *Campylobacter jejuni* results in persistent infection of the chicken gut. Vet. Microbiol. *130*, 285–297.

Velayudhan, J., and Kelly, D.J. (2002). Analysis of gluconeogenic and anaplerotic enzymes in *Campylobacter jejuni*: an essential role for phosphoenolpyruvate carboxykinase. Microbiology *148*, 685–694.

Velayudhan, J., Jones, M.A., Barrow, P.A., and Kelly, D.J. (2004). L-serine catabolism via an oxygen-labile L-serine dehydratase is essential for colonization of the avian gut by *Campylobacter jejuni*. Infect. Immun. *72*, 260–268.

Wassef, J.S., Keren, D.F., and Mailloux, J.L. (1989). Role of M cells in initial antigen uptake and in ulcer formation in the rabbit intestinal loop model of shigellosis. Infect. Immun. *57*, 858–863.

Xu, F., Zeng, X., Haigh, R.D., Ketley, J.M., and Lin, J. (2010). Identification and characterization of a new ferric enterobactin receptor, CfrB, in *Campylobacter*. J. Bacteriol. *192*, 4425–4435.

Yew, W.S., Fedorov, A.A., Fedorov, E.V., Rakus, J.F., Pierce, R.W., Almo, S.C., and Gerlt, J.A. (2006). Evolution of enzymatic activities in the enolase superfamily: L-fuconate dehydratase from *Xanthomonas campestris*. Biochemistry *45*, 14582–14597.

Young, K.T., Davis, L.M., and Dirita, V.J. (2007). *Campylobacter jejuni*: molecular biology and pathogenesis. Nat. Rev. Microbiol. *5*, 665–679.

The Intricate Relationship Between *Campylobacter* and Mucus

8

Claire Shortt, Gina Duggan and Billy Bourke

Abstract

For any successful mucosal pathogen a key early step in the infection sequence is to establish effective colonization of the mucus layer. Campylobacters have evolved to live as commensals in the mucus of avian species. In humans these organisms can breach the mucus defence systems and interact directly with intestinal epithelial cells. Recent insights into the role of mucus in protecting the intestine and the influence of mucus on the behaviour and adaption of *C. jejuni*, has led to a much better understanding of the ecology of *C. jejuni* in the supramucosal niche.

The stucture and function of intestinal mucus

Colonization of mucus by *Campylobacter* is associated with a number of human pathologies including enteritis, irritable bowel syndrome, inflammatory bowel disease and the post-infectious Guillan-Barré Syndrome. The mucus membrane of the intestinal lumen consists of a loose outer layer which hosts an array of microbes, and a firmly adherent, sterile inner layer (Johansson *et al.*, 2008). The mucus produced is a viscous colloid comprised of a biochemically complex mixture of immunoglobulins, α-defensins, cathlecidins, lysozymes, angiogenin 4, lectins (e.g. REGIIIγ and REGIIIα), secretory phospholipase A2, lipopolysaccharide-binding protein, collectins, and histatins (Lopez-Vidriero and Reid, 1978; Stappenbeck, 2009; Porter *et al.*, 2002). These biologically active molecules coupled with a high degree of mucus turnover promote bacterial clearance. The mucus layer ranges in depth from 150 μm in the jejunum where nutrient absorption occurs, to a thickness of 800–900 μm in the distal colon. It is this initial barrier that enteric pathogens such as *Campylobacter* must colonize in order to establish an infection.

Mucins are high molecular weight, heavily *O*-glycosylated proteins that are the main components of mucus. These highly branched, transmembrane glycoproteins are packaged into intracellular storage vesicles in a pH-dependent manner (Perez-Vilar *et al.*, 2005) and expressed at the apical surface of specialized secretory goblet cells in the columnar epithelium. They may be divided into three categories: cell-surface mucins (MUC1, MUC4, MUC3A/B, MUC12, MUC13, MUC15, MUC16, MUC17 and MUC20), secreted mucins (MUC7) and gel-forming mucins (MUC2, MUC5AC, MUC5B, MUC6, and MUC19). Their α-linked core polypeptide backbone consists of multiple serine/threonine repeats known as variable number tandem repeat (VNTR) regions that are highly polymorphic in nature. During biosynthesis, the nascent polypeptides become *N*-glycosylated in the endoplasmic reticulum and are chaperoned to the Golgi apparatus where they undergo extensive addition of *O*-glycans such as *N*-acetylgalactosamine, *N*-acetylglucosamine, fucose, galactose, and sialic acid moieties.

MUC2 is the major gel forming mucin secreted by goblet cells of the small and large intestine and has a high rate of turnover (Johansson, 2012; van Klinken *et al.*, 1999). The *MUC2* gene is located on chromosome 11p15 (Pigny *et al.*, 1996). Both the N- and C-termini of gel-forming mucins such as MUC2 contain cysteine residues that form intermolecular and intramolecular disulfide bonds, thereby causing homo-oligomers to form between the domains known as von Willebrand factor (vWF) or D-domains (Moncada *et al.*, 2003; Linden *et al.*, 2008). While there is strong interspecies sequence conservation among these vWF/D domains the VNTR regions differ markedly in the number of serine/threonine repeats.

Cell-surface mucins such as MUC1 do not directly influence the rheological properties of the mucus layer but do have important molecular interactions with microbes. For example, the extracellular domain of cell-surface mucins has been suggested to act as a releasable decoy ligand for bacterial adhesins (Linden *et al.*, 2009) preventing adherence to epithelial cell components. In addition, the cytoplasmic domains of these mucins are thought to play a role in signal transduction (Jonckheere and Van Seuningen, 2010).

The ecology of *C. jejuni* in intestinal mucus

The mucus layer is one of the first anatomical features intestinal pathogens encounter before invading the epithelium. It has a dual functioning role in that it harbours a diverse microbial population in a dynamic ecosystem that provides a source of nutrients to the host and it also acts as a protective barrier to prevent invasion of the underlying epithelial layer by pathogens (Johansson and Hansson, 2011). Therefore, the mucus layer is a key contributor to intestinal homeostasis.

As the presence of a dense inner mucus layer excludes resident microbiota, penetration into and colonization of this region is required for disease causing bacteria to invade and establish infection in the host. Many pathogenic species of bacteria, such as *Vibrio* spp., *Salmonella*, *Shigella*, *Helicobacter pylori* and *C. jejuni* and *C. coli*, have adapted to effectively colonize the human mucosal layer. In order to successfully colonize mucus many of these pathogens secrete mucin degrading enzymes such as mucinases, sialidases, glycosulfatases, N-acetyl neuraminate lyases and sialate *O*-acetylesterases (Corfield *et al.*, 1992, 1993; Haider *et al.*, 1993; Homer *et al.*, 1994). In addition, a number have structural adaptations to the mucosal niche. For example, *C. jejuni's* corkscrew shape and bi-polar flagella is thought to contribute to movement within this polymeric gel matrix.

As part of the host innate defence system, natural anti-microbial compounds and mucin glycoproteins present in mucus adhere to bacteria preventing direct cell to cell contact. As yet the details of which anti microbial compounds in mucus may target *C. jejuni* are unclear but defensins and other products of the innate defence system undoubtedly have a role in host protection (Zilbauer *et al.*, 2005; Wehkamp *et al.*, 2007).

However, it has become increasingly apparent that the relationship of *C. jejuni* with mucus is complex and extends beyond a simple struggle between host defences and pathogenic adaptation. We have been interested in the contribution of mucus and mucins to the contrasting outcomes following *C. jejuni* in different host species. For instance, we have shown that *C. jejuni* adhesion and internalization is enhanced in the presence of mucus secreting human intestinal cells (Alemka *et al.*, 2010a). Further to this, an increase in internalization of the bacterium has also been noted in the presence of crude human mucus (Byrne *et al.*, 2007). In contrast, chicken mucus and mucin have profound inhibitory affects on *C. jejuni* pathogenicity (Alemka *et al.*, 2010b) (Fig. 8.1). While many of the details are not well understood, we consider below that which is currently known of how *C. jejuni* is influenced by the properties of mucus.

Motility and chemotaxis

Motility is vital for both the pathogenic and commensal lifestyle of many enteric bacteria. Selective pressures have compelled *Campylobacter* species to evolve specific mechanisms to enable them to disseminate through highly viscous environments. Increased levels of motility are achieved by utilizing its prominent helical shape and bipolar flagella. These combined adaptations provide *C. jejuni* with the propulsive force and rotation required to travel through the mucosal layer, enabling them to persist and colonize the intestinal milieu (Jaganatthan and Penn, 2005). The high viscosity of mucus is no barrier to *Campylobacter*. In fact, *C. jejuni* has a higher velocity in solutions with increased viscosity (Ferrero and Lee, 1988;

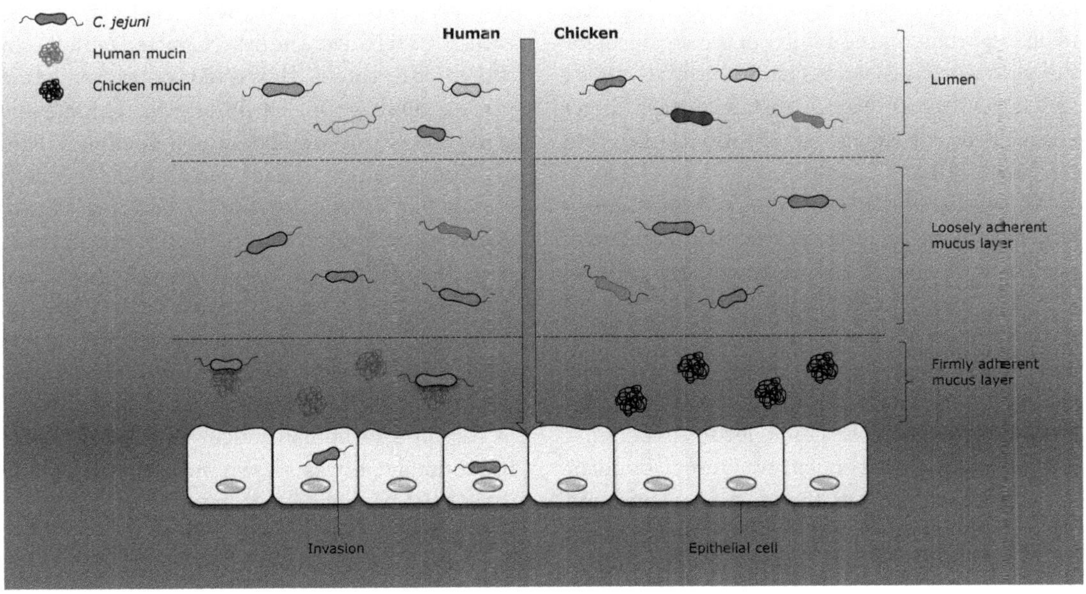

Figure 8.1 A hypothetical model of *C. jejuni* infection in the human versus chicken gut. The spiral shape and bipolar flagella of *Campylobacter* make it ideally suited to life in the mucosal layer. Penetration and colonization of the dense, inner mucus layer in humans results in a pathogenic effect on intestinal epithelium (Left panel), whereas failure to transit the dense mucus layer in the chicken intestine (Right panel) results in a commensal lifestyle. This may arise from a species-specific mucin interaction, as illustrated. A colour version of this figure is available in the plate section at the back of the book.

Szymanski *et al.*, 1995). This increase in motility also correlates with increased infectivity *in vitro*, as incubation of *C. jejuni* in growth media of increased viscosity results in increased adherence and invasion of Caco-2 cell lines (Szymanski *et al.*, 1995).

The flagellum is made up of two subunits encoded by two almost identical flagellin genes, *flaA* and *flaB*. These genes are located in close proximity to one another but have separate promoters (Wassenaar *et al.*, 1994). Transcriptional studies were carried out to determine the effect of MUC2 on a number of pathogenicity genes (Tu *et al.*, 2008). MUC2 is the most abundant gel-forming mucin and is only secreted in the human intestine (Tu *et al.*, 2008). With motility being key to the colonization of the human host, the effect of the secreted human mucin, MUC2 on the major and minor flagella filament was determined. Interestingly, MUC2 up-regulated the expression of *flaA* but had no significant effect on the expression of *flaB* (Tu *et al.*, 2008). It is reasonable to hypothesize that certain human mucins also stimulate *C. jejuni* motility during infection *in vivo*.

Chemotaxis is closely interlinked with flagellar motility. Bacterial motility is subject to sensory control enabling the bacterium to swim towards environments favourable for growth and survival, and away from environments that are harmful to the organism. This swimming bias is central to the pathogenic lifestyle of *C. jejuni*. The gastro-intestinal tract is the natural habitat for *C. jejuni*. Therefore, it is not surprising that chemotactic motility is key to its success as a colonizing pathogen. Campylobacters swim actively up and down the mucus layer using their bipolar flagella displaying a darting-like motility. This efficient mode of locomotion through a viscous environment, in combination with chemotaxis towards mucin probably allows the bacterium to remain localized close to the epithelium (Szymanski *et al.*, 1995). The importance of chemotaxis in colonization of the intestine was initially demonstrated by showing that chemotactic mutants are unable to colonize suckling mice (Takata *et al.*, 1992). The chemotactic effect of L-fucose on *C. jejuni* appears to be important for the affinity of the organism for the gastrointestinal tract (Hugdahl *et al.*, 1988) and may work at least in part through a direct effect on flagellar gene transcription (Allen and Griffiths, 2001).

Mucus- based chemotaxis behaviour in *Campylobacter* appears to be strain dependent. Burrough *et al.* (2012), compared the growth and chemotaxis of *C. jejuni* IA3902, a clone that is a common cause of ovine abortion, and the reference strain NCTC 11168 in the presence of subplacental and placental factors from pregnant and non pregnant guinea pigs. Using lectin staining, it was determined that mucin was most abundant in subplacental lacunae, the junctional zone, and visceral yolk sac placenta. L-Fucose-containing surface glycans were mainly limited to the visceral yolk sac placenta. Mucin and L-fucose were both growth promoting, while L-fucose was also found to be a chemoattractant for both strains. Interestingly, while an earlier study using different *C. jejuni* isolates (Hugdahl *et al.*, 1988) showed that gastric porcine mucin was chemotactic for the organism, Shain *et al.* could not demonstrate this effect using IA3902 and NCTC 11168 (Burrough *et al.*, 2012)

Further evidence for the strain-dependent effect of chemotaxis in *C. jejuni* was provided in a study of *C. jejuni* chemoreceptor mutants (Vegge *et al.*, 2009). Five chemoreceptor mutants, *tlp1*, *tlp2*, *tlp3*, *docB* and *docC* were studied in order to establish which proteins are vital for sensing environmental signals. Interestingly, it was found that strains from different origins have a different taxis response. Differences between chicken and clinical isolates were observed, as the clinical isolate NCTC 11168 was found to be responsive to a range of compounds such as formate and chicken mucus, but not L-fucose. Given that L-fucose is known as an important colonization factor for *C. jejuni*, it was hypothesized that variation in taxis response between isolates could lead to varying colonization ability (Vegge *et al.*, 2009).

Bacterial binding and internalization

Adherence to intestinal epithelial cells is necessary for *C. jejuni* to remain localized in the intestine and to become internalized. There are numerous types of lectin-like structures that can attach to glycosylated receptors on the host cell (Szymanski *et al.*, 2002). To date, the best characterized adhesin is the CadF protein. This 37 kDa outer membrane protein binds epithelial fibronectin (Monteville *et al.*, 2003) and stimulates the induction of Rho GTPases resulting in Rac1 and Cdc42 dependent membrane ruffling leading to bacterial uptake (Eucker and Konkel, 2012; Krause-Gruszczynska *et al.*, 2007). JlpA is a 42-kDa lipoprotein that can be N-linked glycosylated at two residues (Jin *et al.*, 2001). In 2001, a study revealed that JlpA mutants exhibit a decreased ability to bind to HEp-2 cells (Jin *et al.*, 2001). Furthermore *jlpA* is up-regulated in response to human mucin (Tu *et al.*, 2008). Interestingly it was shown in the same study that *cadF* was down-regulated in the presence of human mucin. These results point to the possibility that the CadF protein may not be necessary to penetrate the mucus barrier, while JlpA may have a pivotal role in maintaining *C. jejuni* in this hostile mucosal environment (Tu *et al.*, 2008). However, whether this hypothesis is correct awaits the study of these virulence factors in models of infection *in vivo*.

The majority of conventional intestinal cell lines are devoid of a mucus layer. Recently, a cell line that harbours an adherent mucus layer has been derived from HT29 cells. HT29-MTX-E12 (E12) cells are characterized by formation of tight junctions and a confluent monolayer when polarized on transwell inserts (Behrens *et al.*, 2001). This *in vitro* model provides a unique opportunity for the analysis of the colonization and interaction of *C. jejuni* within the mucosal layer. Bacterial cell binding and internalization were enhanced in E12 cells harbouring mucus compared to parental cells without the adherent mucus layer (Alemka *et al.*, 2010a). On the other hand, the addition of chicken intestinal mucus markedly inhibited *C. jejuni* pathogenicity. This observation may explain, at least in part, why *C. jejuni* is a commensal in the chicken intestine *in vivo*.

Reproduction and proliferation in mucus

It has been shown that *C. jejuni* is able to replicate in PBS medium supplemented with mucus but not when inoculated with chicken caecal content. Mucus appears to contain all the necessary components for growth of *C. jejuni* (Van Deun *et al.*, 2008). More recently, we have shown that *C. jejuni* can colonize and reproduce in the mucus layer of E12. Growth media supplemented using mucus derived from E12 cells was able to enhance growth and reproduction of *C. jejuni* (Alemka *et al.*, 2010a). These studies indicate that mucus provides an optimal niche for growth and survival, allowing *C. jejuni* to flourish and (in the case of avian) persist within the host.

Conventionally, *C. jejuni* was believed not to be able to utilize exogenous carbohydrates. However, two important studies (Muraoka and Zhang, 2011; Stahl *et al.*, 2011) have provided evidence that *C. jejuni* can utilize L-fucose (Muraoka and Zhang, 2011; Stahl *et al.*, 2011) as a carbohydrate source. This phenotype has been shown to be strain dependent and is linked to a genomic island found only in certain *Campylobacter* isolates. Stahl *et al.* (2011) have also demonstrated this strain specific utilization of L-fucose and how the transport of L-fucose provides a competitive advantage when colonizing the piglet model of human disease. It is possible that in the mucosal niche these adapted strains can access L-fucose substrate to their advantage.

Toxins and mucosal defence

Certain cell surface mucins contain conserved phosphorylation motifs, which are important for various cellular functions. There is evidence for the role of MUC1 in mucosal defence as there is a progressive increase in the expression of MUC1 following infection of mice with *C. jejuni* (McAuley *et al.*, 2007). *C. jejuni* was found in the systemic organs of MUC1$^{-/-}$ mice but not in MUC1$^{-/+}$ mice. MUC1 also provides protection against *C. jejuni* cytolethal distending toxin (CDT) *in vitro* (McAuley *et al.*, 2007). CDT is a genotoxin with nuclease activity that blocks the cell cycle, resulting in increased apoptosis *in vitro* (Lara-Tejero and Galan, 2000). *C. jejuni* also produces a vacuolating cytotoxin and the gene encoding this vacuolating cytotoxin is significantly up-regulated in the presence of MUC2 alongside the CDT toxin. This implies that these human mucins may act as signals to the bacteria to elaborate pathogenicity factors during the infection process (Tu *et al.*, 2008).

Mucus as a source of glycans

It is now understood that the human-bacterial glycome plays a pivotal role in host cell interactions. Glycosylation is a co-translational, post-translational and site-specific modification that involves the addition of glycans or sugars to organic molecules such as proteins and lipids. Glycosylation plays a range of functional roles in host–pathogen interactions, ranging from microbial adhesion and molecular mimicry to innate immunity in the host. Mucins are heavily glycosylated protein components of mucus that act a reservoir for microbes, providing a source of nutrients while also serving as a potential binding site for bacterial adhesins (Ilver *et al.*, 1998). The contribution of intestinal glycans to the complex interdependent relationships of resident microbiota is beginning to be understood (Marcobal *et al.*, 2011).

As noted above, recent evidence has suggested that certain *Campylobacter* strains possess a genomic island encoding a putative fucose permease that allows for the uptake and utilization of fucose as a carbohydrate source, providing it with a competitive advantage (Stahl *et al.*, 2011). This discovery is all the more remarkable for the fact that conventional dogma asserted that *C. jejuni* could not use carbohydrates as a carbon source. The relationship of *Campylobacter* with intestinal glycans generally, and fucose in particular, appears to be complex. For example, fucose seems to act as a chemoattractant for the organisms (Hugdahl *et al.*, 1988) whilst fucosylated structures present in breast milk have been identified as interaction partners for *C. jejuni* (Morrow *et al.*, 2005). In the context of human infection it is particularly intriguing that human intestinal mucin displays a decreasing fucose gradient from ileum to rectum with low levels present in the large intestine where *C. jejuni* preferentially colonizes (Robbe *et al.*, 2003). In the same study Robbe *et al.* also reported an inverse relationship of sialylated residues from ileum to colon. Together these findings suggest that binding of *C. jejuni* to fucosylated structures on mucins may play a key role in establishing disease.

Recent findings suggest that the glycosylation of mucins in the intestinal tract is region-specific (Ismail *et al.*, 2011). In the human sigmoid colon the *O*-glycan structure of MUC2 was characterized and found typically to be sialylated, harbouring a relatively uniform composition of glycans among healthy individuals (Larsson *et al.*, 2009) (Gill *et al.*, 2006). In the small bowel, however, *O*-glycosylation demonstrates a high degree of variability. A recent study using mice deficient in core 1 derived *O*-glycans exhibited a breach of the inner mucus layer and they developed spontaneous colitis (Kawashima, 2012). Why the repertoire of glycans in the distal colon is relatively constant is not well understood, but may be due to host driven selection for species specific microflora. From these data it is clear that both mucin expression and glycosylation are dynamic processes that change in response to the host microbiome. The importance of the role of these features in disease outcome following bacterial infection has yet to be established.

Campylobacter jejuni commensalism in chickens

Campylobacteriosis is a foodborne illness the primary mode of transmission of which is faecal–oral via contaminated poultry, milk or water. Following ingestion, *C. jejuni* travels to the caecum where it establishes infection within 24 h, causing symptoms such as fever, headaches, myalgia, diarrhoea, dysentery and abdominal pain. Avian species serve as the natural host for *C. jejuni* and the organism is capable of colonizing birds in high numbers (up to 10^{10} cfu/g). Despite its pathogenicity in humans, *Campylobacter* behaves as a commensal in chickens. As broiler chickens are the main reservoir for transmission of infection to humans, the relationship between colonization of the intestinal tract of chickens versus that of humans is worthy of investigation.

C. jejuni avidly colonizes the deep caecal mucosal crypts of chickens but does not adhere to or invade avian intestinal epithelial cells (Young *et al.*, 2007; Ziprin *et al.*, 1999). Despite the apparent lack of pathological effects in the avian intestine, an initial low level, pro-inflammatory response is elicited and the strength of this response may influence the intestinal bacterial load among different birds (Connell *et al.*, 2012). Nevertheless, neutrophil recruitment often does not occur following colonization and downstream IgA and T-cell recruitment fail to clear the organism. This indicates that *C. jejuni* is somehow capable of evading its hosts' immune system (Louwen *et al.*, 2008; Guerry *et al.*, 2000). However the mechanism by which this is accomplished remains unknown.

Campylobacters possess a range of surface accessible carbohydrates for example; *N*- and *O*- linked glycans, flagella, lipooligosaccharide (LOS) and capsular polysaccharide (CPS). Interestingly, a mutation in the capsular polysaccharide transporter gene *kpsM* of *C. jejuni* abolishes colonization in chickens (Jones *et al.*, 2004) while disruption of the *kpsE* gene (which results in complete loss of capsule) fails to impede colonization in chickens (Bachtiar *et al.*, 2007). As with most prokaryotes, Campylobacters use a two-component regulatory system in response to environmental stimuli. To date, five TCRs have been established as central in chick colonization; FlgRS (Wosten *et al.*, 2004) the orphan response regulator (CbrR) (Stephens *et al.*, 2006), the reduced ability to colonize (RacRS system) (Bras *et al.*, 1999), diminished capacity to colonize (DccRS) (MacKichan *et al.*, 2004) and *Campylobacter* planktonic growth regulation (CprRS) (Svensson *et al.*, 2009). A *ΔcprS* mutant has been reported to have profound dose-dependent effect on chick colonization.

There is an extremely diverse repertoire of carbohydrates present on mucin molecules. These offer an array of potential binding sites for bacteria. It has been posited that differences in human and chicken mucus composition may be influential in establishing disease states. For example, chicken mucus is highly sulfated in comparison to humans (Alemka *et al.*, 2010b). Incubation with chicken intestinal mucins has been shown to reduce the adhesion and invasiveness of some *C. jejuni* strains in human intestinal epithelial cells *in vitro*. The level of inhibition is dependent on the source of the mucus, with mucus isolated from the duodenum and ileum having a more pronounced effect compared to mucus isolated from the caecum (Byrne *et al.*, 2007). In contrast to this, human mucin actively promotes adhesion and invasion of *C. jejuni* (Byrne *et al.*, 2007). The exact mechanism by which chicken mucus alters *C. jejuni* virulence *in vivo* is not fully understood.

However, the effect of mucus is mirrored by purified mucins from the chicken intestine and this attenuation in pathogenicity appears to be mediated by glycan structures on the mucin molecules (Alemka *et al.*, 2010b). In general, it is possible that *Campylobacter* utilizes mucins as docking sites. Depending on the nature of the interaction the organisms are either 'primed' for penetration deeper into the mucosal layer in order to elaborate disease (human mucus/mucins) or 'trapped' by glycosylated decoy structures on mucin and prevented from penetrating deeper ensuring commensalism (chicken mucin) (Fig. 8.1).

Conclusions and future directions

The overlay of mucus on the intestinal epithelium plays a vital role in host defence. Numerous enteric pathogens have developed strategies to penetrate the mucus layer and come into close contact with the epithelial surface. *C. jejuni* can persist and remain colonized for long periods of time in the mucus layer of avian species. It is likely that the bacterial adaptations that facilitate chicken colonization (spiral morphology, flagella, chemotaxis) are the same ones that allow infection and disease manifestations in human hosts. It appears that glycan structures on mucins have a fundamental role in determining the outcome of disease following colonization by *Campylobacter* in different species. Future studies aimed at identifying the specific components of mucin, which alter the pathogenesis of *Campylobacter*, could be further utilized in development of intervention strategies against this unique organism.

References

Alemka, A., Clyne, M., Shanahan, F., Tompkins, T., Corcionivoschi, N., and Bourke, B. (2010a). Probiotic colonization of the adherent mucus layer of HT29MTXE12 cells attenuates *Campylobacter jejuni* virulence properties. Infect. Immun. *78*, 2812–2822.

Alemka, A., Whelan, S., Gough, R., Clyne, M., Gallagher, M.E., Carrington, S.D., and Bourke, B. (2010b). Purified chicken intestinal mucin attenuates *Campylobacter jejuni* pathogenicity in vitro. J. Med. Microbiol. *59*, 898–903.

Allen, K.J., and Griffiths, M.W. (2001). Effect of environmental and chemotactic stimuli on the activity of the *Campylobacter jejuni* flaA sigma(28) promoter. FEMS Microbiol. Lett. *205*, 43–48.

Bachtiar, B.M., Coloe, P.J., and Fry, B.N. (2007). Knockout mutagenesis of the kpsE gene of *Campylobacter jejuni* 81116 and its involvement in bacterium–host interactions. FEMS Immunol. Med. Microbiol. *49*, 149–154.

Bras, A.M., Chatterjee, S., Wren, B.W., Newell, D.G., and Ketley, J.M. (1999). A novel *Campylobacter jejuni* two-component regulatory system important for temperature-dependent growth and colonization. J. Bacteriol. *181*, 3298–3302.

Burrough, E.R., Wu, Z., Shahin, O., Zhang, Q., and Yaeger, M.J. (2012). Spatial distribution of putative growth factors in the guinea pig placenta and the effects of these factors, plasma, and bile on the growth of the chemotaxis of *Campylobacter jejuni*. Vet. Pathol. *49*, 470–481.

Byrne, C.M., Clyne, M., and Bourke, B. (2007). *Campylobacter jejuni* adhere to and invade chicken intestinal epithelial cells *in vitro*. Microbiology *153*, 561–569.

Connell, S., Meade, K.G., Allan, B., Lloyd, A.T., Kenny, E., Cormican, P., Morris, D.W., Bradley, D.G., and O'Farrelly, C. (2012). Avian resistance to *Campylobacter jejuni* colonization is associated with an intestinal immunogene expression signature identified by mRNA sequencing. PLoS ONE *7*, e40409.

Corfield, A.P., Wagner, S.A., Clamp, J.R., Kriaris, M.S., and Hoskins, L.C. (1992). Mucin degradation in the human colon: production of sialidase, sialate O-acetylesterase, N-acetylneuraminate lyase, arylesterase, and glycosulfatase activities by strains of fecal bacteria. Infect. Immun. *60*, 3971–3978.

Corfield, A.P., Wagner, S.A., O'Donnell, L.J., Durdey, P., Mountford, R.A., and Clamp, J.R. (1993). The roles of enteric bacterial sialidase, sialate O-acetyl esterase and glycosulfatase in the degradation of human colonic mucin. Glycoconjugate J. *10*, 72–81.

Eucker, T.P., and Konkel, M.E. (2012). The cooperative action of bacterial fibronectin-binding proteins and secreted proteins promote maximal *Campylobacter jejuni* invasion of host cells by stimulating membrane ruffling. Cell. Microbiol. *14*, 226–238.

Ferrero, R.L., and Lee, A. (1988). Motility of *Campylobacter jejuni* in a viscous environment: comparison with conventional rod-shaped bacteria. J. Gen. Microbiol. *134*, 53–59.

Gill, S.R., Pop, M., Deboy, R.T., Eckburg, P.B., Turnbaugh, P.J., Samuel, B.S., Gordon, J.I., Relman, D.A., Fraser-Liggett, C.M., and Nelson, K.E. (2006). Metagenomic analysis of the human distal gut microbiome. Science *312*, 1355–1359.

Guerry, P., Ewing, C.P., Hickey, T.E., Prendergast, M.M., and Moran, A.P. (2000). Sialylation of lipooligosaccharide cores affects immunogenicity and serum resistance of *Campylobacter jejuni*. Infect. Immun. *68*, 6656–6662.

Haider, K., Hossain, A., Wanke, C., Qadri, F., Ali, S., and Nahar, S. (1993). Production of mucinase and neuraminidase and binding of Shigella to intestinal mucin. J. Diarrhoeal Dis. Res. *11*, 88–92.

Homer, K.A., Whiley, R.A., and Beighton, D (1994). Production of specific glycosidase activities by

Streptococcus intermedius strain UNS35 grown in the presence of mucin. J. Med. Microbiol. *41*, 184–190.

Hugdahl, M.B., Beery, J.T., and Doyle, M.P. (1988). Chemotactic behavior of *Campylobacter jejuni*. Infect. Immun. *56*, 1560–1566.

Ilver, D., Arnqvist, A., Ogren, J., Frick, I.M., Kersulyte, D., Incecik, E.T., Berg, D.E., Covacci, A., Engstrand, L., and Boren, T. (1998). *Helicobacter pylori* adhesin binding fucosylated histo-blood group antigens revealed by retagging. Science *279*, 373–377.

Ismail, M.N., Stone, E.L., Panico, M., Lee, S.H., Luu, Y., Ramirez, K., Ho, S.B., Fukuda, M., Marth, J.D., Haslam, S.M., *et al.* (2011). High-sensitivity O-glycomic analysis of mice deficient in core 2 {beta}1,6-N-acetylglucosaminyltransferases. Glycobiology *21*, 82–98.

Jaganatthan, A., and Penn, C. (2005). Motility. In *Campylobacter:* Molecular And Cellular Biology Ketley, J.M., Konkel, M.E., eds. (Norfolk, UK, Horizon Bioscience), pp. 341–348.

Jin, S., Joe, A., Lynett, J., Hani, E.K., Sherman, P., and Chan, V.L. (2001). JlpA, a novel surface-exposed lipoprotein specific to *Campylobacter jejuni*, mediates adherence to host epithelial cells. Mol. Microbiol. 39, 1225–1236.

Johansson, M.E. (2012). Fast renewal of the distal colonic mucus layers by the surface goblet cells as measured by *in vivo* labeling of mucin glycoproteins. PLoS ONE *7*, e41009.

Johansson, M.E., and Hansson, G.C. (2011). Microbiology. Keeping bacteria at a distance. Science *334*, 182–183.

Johansson, M.E., Phillipson, M., Petersson, J., Velcich, A., Holm, L., and Hansson, G.C. (2008). The inner of the two Muc2 mucin-dependent mucus layers in colon is devoid of bacteria. Proc. Natl. Acad. Sci. U.S.A. *105*, 15064–15069.

Jonckheere, N., and Van Seuningen, I. (2010). The membrane-bound mucins: From cell signalling to transcriptional regulation and expression in epithelial cancers. Biochimie 92, 1–11.

Jones, M.A., Marston, K.L., Woodall, C.A., Maskell, D.J., Linton, D., Karlyshev, A.V., Dorrell, N., Wren, B.W., and Barrow, P.A. (2004). Adaptation of *Campylobacter jejuni* NCTC11168 to high-level colonization of the avian gastrointestinal tract. Infect. Immun. *72*, 3769–3776.

Kawashima, H. (2012). Roles of the gel-forming MUC2 mucin and its O-glycosylation in the protection against colitis and colorectal cancer. Biol. Pharma. Bull. *35*, 1637–1641.

van Klinken, B.J., Einerhand, A.W., Duits, L.A., Makkink, M.K., Tytgat, K.M., Renes, I.B., Verburg, M., Buller, H.A., and Dekker, J. (1999). Gastrointestinal expression and partial cDNA cloning of murine Muc2. Am. J. Physiol. *276*, G115–124.

Krause-Gruszczynska, M., Rohde, M., Hartig, R., Genth, H., Schmidt, G., Keo, T., Konig, W., Miller, W.G., Konkel, M.E., and Backert, S. (2007). Role of the small Rho GTPases Rac1 and Cdc42 in host cell invasion of *Campylobacter jejuni*. Cell. Microbiol. 9, 2431–2444.

Lara-Tejero, M., and Galan, J.E. (2000). A bacterial toxin that controls cell cycle progression as a deoxyribonuclease I-like protein. Science *290*, 354–357.

Larsson, J.M., Karlsson, H., Sjovall, H., and Hansson, G.C. (2009). A complex, but uniform O-glycosylation of the human MUC2 mucin from colonic biopsies analyzed by nanoLC/MSn. Glycobiology *19*, 756–766.

Linden, S.K., Sutton, P., Karlsson, N.G., Korolik, V., and McGuckin, M.A. (2008). Mucins in the mucosal barrier to infection. Mucosal Immun. *1*, 183–197.

Linden, S.K., Sheng, Y.H., Every, A.L., Miles, K.M., Skoog, E.C., Florin, T.H., Sutton, P., and McGuckin, M.A. (2009). MUC1 limits *Helicobacter pylori* infection both by steric hindrance and by acting as a releasable decoy. PLoS Pathog. *5*, e1000617.

Lopez-Vidriero, M.T., and Reid, L. (1978). Chemical markers of mucous and serum glycoproteins and their relation to viscosity in mucoid and purulent sputum from various hypersecretory diseases. The American review of respiratory disease *117*, 465–477.

Louwen, R., Heikema, A., van Belkum, A., Ott, A., Gilbert, M., Ang, W., Endtz, H.P., Bergman, M.P., and Nieuwenhuis, E.E. (2008). The sialylated lipooligosaccharide outer core in *Campylobacter jejuni* is an important determinant for epithelial cell invasion. Infect. Immun. *76*, 4431–4438.

McAuley, J.L., Linden, S.K., Png, C.W., King, R.M., Pennington, H.L., Gendler, S.J., Florin, T.H., Hill, G.R., Korolik, V., and McGuckin, M.A. (2007). MUC1 cell surface mucin is a critical element of the mucosal barrier to infection. J. Clin. Invest. *117*, 2313–2324.

MacKichan, J.K., Gaynor, E.C., Chang, C., Cawthraw, S., Newell, D.G., Miller, J.F., and Falkow, S. (2004). The *Campylobacter jejuni* dccRS two-component system is required for optimal in vivo colonization but is dispensable for in vitro growth. Mol. Microbiol. *54*, 1269–1286.

Marcobal, A., Barboza, M., Sonnenburg, E.D., Pudlo, N., Martens, E.C., Desai, P., Lebrilla, C.B., Weimer, B.C., Mills, D.A., German, J.B., *et al.* (2011). Bacteroides in the infant gut consume milk oligosaccharides via mucus-utilization pathways. Cell Host Microbe *10*, 507–514.

Moncada, D.M., Kammanadiminti, S.J., and Chadee, K. (2003). Mucin and Toll-like receptors in host defense against intestinal parasites. Trends Parasitol. *19*, 305–311.

Monteville, M.R., Yoon, J.E., and Konkel, M.E. (2003). Maximal adherence and invasion of INT 407 cells by *Campylobacter jejuni* requires the CadF outer-membrane protein and microfilament reorganization. Microbiology *149*, 153–165.

Morrow, A.L., Ruiz-Palacios, G.M., Jiang, X., and Newburg, D.S. (2005). Human-milk glycans that inhibit pathogen binding protect breast-feeding infants against infectious diarrhea. J. Nutr. *135*, 1304–1307.

Muraoka, W.T., and Zhang, Q. (2011). Phenotypic and genotypic evidence for L-fucose utilization by *Campylobacter jejuni*. J. Bacteriol. *193*, 1065–1075.

Perez-Vilar, J., Olsen, J.C., Chua, M., and Boucher, R.C. (2005). pH-dependent intraluminal organization of mucin granules in live human mucous/goblet cells. J. Biol. Chem. *280*, 16868–16881.

Pigny, P., Guyonnet-Duperat, V., Hill, A.S., Pratt, W.S., Galiegue-Zouitina, S., d'Hooge, M.C., Laine, A., Van-Seuningen, I., Degand, P., Gum, J.R., *et al.* (1996). Human mucin genes assigned to 11p15.5: identification and organization of a cluster of genes. Genomics *38*, 340–352.

Porter, E.M., Bevins, C.L., Ghosh, D., and Ganz, T. (2002). The multifaceted Paneth cell. CMLS *59*, 156–170.

Robbe, C., Capon, C., Maes, E., Rousset, M., Zweibaum, A., Zanetta, J.P., and Michalski, J.C. (2003). Evidence of regio-specific glycosylation in human intestinal mucins: presence of an acidic gradient along the intestinal tract. J. Biol. Chem. *278*, 46337–46348.

Stahl, M., Friis, L.M., Nothaft, H., Liu, X., Li, J., Szymanski, C.M., and Stintzi, A. (2011). L-fucose utilization provides *Campylobacter jejuni* with a competitive advantage. Proc. Natl. Acad. Sci. U.S.A. *108*, 7194–7199.

Stanley P., and Cummings, R.D. (2009). Structures Common to Different Glycans. In Essentials of Glycobiology, Varki, A., Cummings, R.D., Esko, J.D., Freeze, H.H., Stanley, P., Bertozzi, C.R., Hart, G.W., Etzler, M.E., and Crocker, P.R., eds. 2nd edition. Cold Spring Harbor (NY): Cold Spring Harbor Laboratory Press. Available from: http://www.ncbi.nlm.nih.gov/books/NBK1892/

Stappenbeck, T.S. (2009). Paneth cell development, differentiation, and function: new molecular cues. Gastroenterology *137*, 30–33.

Stephens, B.B., Loar, S.N., and Alexandre, G. (2006). Role of CheB and CheR in the complex chemotactic and aerotactic pathway of *Azospirillum brasilense*. J. Bacteriol. *188*, 4759–4768.

Svensson, S.L., Davis, L.M., MacKichan, J.K., Allan, B.J., Pajaniappan, M., Thompson, S.A., and Gaynor, E.C. (2009). The CprS sensor kinase of the zoonotic pathogen *Campylobacter jejuni* influences biofilm formation and is required for optimal chick colonization. Mol. Microbiol. *71*, 253–272.

Szymanski, C.M., King, M., Haardt, M., and Armstrong, G.D. (1995). *Campylobacter jejuni* motility and invasion of Caco-2 cells. Infect. Immun. *63*, 4295–4300.

Szymanski, C.M., Burr, D.H., and Guerry, P. (2002). *Campylobacter* protein glycosylation affects host cell interactions. Infect. Immun. *70*, 2242–2244.

Takata, T., Fujimoto, S., and Amako, K. (1992). Isolation of nonchemotactic mutants of *Campylobacter jejuni* and their colonization of the mouse intestinal tract. Infect. Immun. *60*, 3596–3600.

Tu, Q.V., McGuckin, M.A., and Mendz, G.L. (2008). *Campylobacter jejuni* response to human mucin MUC2: modulation of colonization and pathogenicity determinants. J. Med. Microbiol. *57*, 795–802.

Vegge, C.S., Brondsted, L., Li, Y.P., Bang, D.D., and Ingmer, H. (2009). Energy taxis drives *Campylobacter jejuni* toward the most favorable conditions for growth. Appl. Environ. Microbiol. *75*, 5308–5314.

Wassenaar, T.M., Bleumink-Pluym, N.M., Newell, D.G., Nuijten, P.J., and van der Zeijst, B.A. (1994). Differential flagellin expression in a flaA flaB+ mutant of *Campylobacter jejuni*. Infect. Immun. *62*, 3901–3906.

Wehkamp, J., Schmid, M., and Stange, E.F. (2007). Defensins and other antimicrobial peptides in inflammatory bowel disease. Curr. Op. Gastroenterol. *23*, 370–378.

Wosten, M.M., Wagenaar, J.A., and van Putten, J.P. (2004). The FlgS/FlgR two-component signal transduction system regulates the fla regulon in *Campylobacter jejuni*. J. Biol. Chem. *279*, 16214–16222.

Young, K.T., Davis, L.M., and Dirita, V.J. (2007). *Campylobacter jejuni*: molecular biology and pathogenesis. Nature Rev. Microbiol. *5*, 665–679.

Zilbauer, M., Dorrell, N., Boughan, P.K., Harris, A., Wren, B.W., Klein, N.J., and Bajaj-Elliott, M. (2005). Intestinal innate immunity to *Campylobacter jejuni* results in induction of bactericidal human beta-defensins 2 and 3. Infect. Immun. *73*, 7281–7289.

Ziprin, R.L., Young, C.R., Stanker, L.H., Hume, M.E., and Konkel, M.E. (1999). The absence of cecal colonization of chicks by a mutant of *Campylobacter jejuni* not expressing bacterial fibronectin-binding protein. Avian Dis. *43*, 586–589.

Campylobacter Association with the Human Host

9

Abiyad Baig and Georgina Manning

Abstract

Campylobacter jejuni is a foodborne pathogen and the major cause of gastroenteritis in humans worldwide. Adhesion to, and invasion of, host epithelial cells have been investigated as important mechanisms in the infection process. Many genes have been identified that are proposed to be involved in the interaction with host cells and many mechanisms suggested. *C. jejuni* seems to be an opportunistic pathogen in humans so an important question is: are there any real 'virulence factors' or are there several different mechanisms utilized by different strains of this organism to access the human niche? Furthermore, many studies have focussed on a single strain, however it is known that invasion potential varies among *C. jejuni* strains and human clinical isolates are generally more invasive compared to non-clinical isolates. There may be much to be learned from studying phenotypically related strains. This chapter will review the current understanding of the mechanisms of host cell interaction and will explore whether the use of a phenotypically related group of hyper-invasive strains can shed any light as to how *C. jejuni* has evolved to become an invasive pathogen.

Introduction

C. jejuni is a commensal bacterium in the chicken gut but causes disease when inside the human intestinal tract. To establish infection in humans *C. jejuni*, like all other organisms that colonize the human gastrointestinal tract (GIT), has to overcome several host defence mechanisms. *C. jejuni* has developed a number of strategies to survive these hostile host gut conditions and colonize and initiate infection. These bacteria have to pass through the acidic stomach environment, and when inside the intestinal tract they have to withstand high bile salt concentrations (Dasti *et al.*, 2010) and the majority of *C. jejuni* strains have developed resistance to the potent effects of bile salts in the intestine (Van Deun *et al.*, 2007). The mucin rich mucosal lining acts as an additional barrier protecting the underlying epithelium against foreign intrusion (Hugdahl *et al.*, 1988) and the helical appearance of the *C. jejuni* cell (Frirdich *et al.*, 2012) as well as the up-regulation of flagella when in contact with the intestinal mucosa (Tu *et al.*, 2008) offer an advantage to the organism in this environment.

However, not all of the organisms that colonize the human GIT lead to infection yet these organisms must also overcome similar defences to get to the site of colonization. These mechanisms should perhaps be redefined as niche rather than virulence mechanisms as they are important in the ability of any organism to get into and become established within this particular niche (Hill, 2012).

So what are the real 'virulence factors' of *C. jejuni*? It is thought that *C. jejuni* attaches, albeit transiently, to the epithelial cells before disrupting the gut lining through invasion or translocation which brings about an inflammatory response resulting in the symptoms of diarrhoea. Evidence that *C. jejuni* is invasive comes from the fact that blood and mucous are present in the stool in some patients and early *in vitro* studies (Newell *et al.*, 1985).

The ability of *C. jejuni* to adhere to and invade epithelial cells is generally strongly correlated with the severity of clinical symptoms presented (Fauchere *et al.*, 1986). Several studies have used *in vitro* cultured cell lines of human and non-human origin to investigate *C. jejuni* interaction with host cells (Konkel *et al.*, 2001; Ketley, 1997; Prasad *et al.*, 1996; Konkel and Jones, 1989). In addition, *in vivo* animal infection models including primates have been used to study the role of invasion and adhesion in *C. jejuni* pathogenesis (Yao *et al.*, 1997; Babakhani and Joens, 1993; Russell *et al.*, 1993; Senior *et al.*, 2011). All these studies have reported adhesion and invasion as important factors contributing to *C. jejuni* pathogenesis and it is apparent that *C. jejuni*'s ability to adhere to and invade epithelial cells very strongly depends on strain type and the infection model used, for example type of intestinal cell line used (Prasad *et al.*, 1996; Konkel and Jones, 1989; Newell *et al.*, 1985).

Adhesion, invasion and translocation

C. jejuni interaction with host cells has been studied as a critical process in the mediation of disease in human hosts (Hu and Kopecko, 2008) but the mechanisms that *C. jejuni* uses for adhesion and invasion are not fully understood. It was suggested that *C. jejuni* adheres to epithelial cells before internalization, however Everest *et al.* (1992) showed that *C. jejuni* and *C. coli* strains were able to adhere to epithelial cells without subsequent invasion. Other studies have also showed that adhesion to host cells was not necessary as *C. jejuni* was able to successfully remain in the intestine after colonizing the mucus suggesting that *C. jejuni* might invade the epithelial cells directly (Lee *et al.*, 1986).

Different mechanisms by which *C. jejuni* invades the host cells have been proposed but it is not fully understood. Like other invasive pathogens including *Listeria*, *Salmonella* and *Shigella*, *C. jejuni* may enter the host cells in a microfilament dependent manner (DeMelo *et al.*, 1989; Konkel and Joens, 1989; Konkel *et al.*, 1992b). Other studies found that the invasion was not affected in the presence of microfilament inhibitors (Oelschlaeger *et al.*, 1993), suggesting that invasion was via a microtubule dependent pathway (Oelschlaeger *et al.*, 1993). During *C. jejuni* strain 81-176 invasion into INT-407 cells, actin filament depolymerization and polymerization of microtubule bundles was observed in response to bacterial cells bound to the cell surface. This resulted in the formation of finger like protrusions by the cell surface, which engulfed the bacteria into membrane-bound endosomes that were then transported along the microtubules to the perinuclear region of the cell (Hu *et al.*, 1999). A novel host signal transduction pathway has also been proposed for *C. jejuni* invasion into the differentiated enterocyte like cells (Wooldridge *et al.*, 1996). Wooldridge *et al.* (1996) showed that *Campylobacter* stimulates a receptor on the host cell membrane that interacts with molecules associated with caveolae. This interaction results in a signal transduction event that causes ruffling across the membrane and subsequent actin filament-dependent endocytosis of *C. jejuni*.

Campylobacter strains are also isolated from cases of bacteraemia which suggests the organism has the ability to cross the epithelium to access extraintestinal locations. *C. jejuni* are able to translocate through the otherwise impermeable epithelial cell layer by using three reported systems. *Campylobacter* strains have been observed to translocate by a paracellular pathway or paracytosis in which bacteria move through the epithelial cells without invading them via tight junctions between them (Everest *et al.*, 1992; Konkel *et al.*, 1992a; Grant *et al.*, 1993). In the second system, *C. jejuni* are observed to use specialized intestinal epithelial cells called M-cells to translocate through the epithelial cells layer into the underlying lymphoid tissue (Walker *et al.*, 1988). Finally, the third translocation mechanism called transcellular translocation or transcytosis involves bacterial invasion into the epithelial cells from where they travel through the cytoplasm appearing at the other side (Hu *et al.*, 2008; Brás and Ketley, 1999). Within 60 min of post-infection, *C. jejuni* 81-176 was internalized into endosomes, which then travelled for 8–10 h from the apical surface to the basolateral surface where they were released by exocytosis via endosome fusion to the basolateral surface (Hu *et al.*, 2008). The transcellular movement of *C. jejuni* 81-176 (Hu *et al.*, 2008), and more recently of *C. fetus* 11686 (Baker and Graham, 2010), continued without compromising the integrity of the polarized Caco-2 cell monolayer. In contrast, a study that infected polarized human colonic HCA-7 cells with *C. jejuni* (Beltinger *et al.*, 2008) observed complete loss of colonic cell integrity and disruption of cellular tight junctions within 6 h of infection. A modification of this transcellular transport known as subvasion has recently been reported in *C. jejuni* where bacteria interact with the basal layer of infected cells before finally invading the epithelial cells (van Alpen *et al.*, 2008; Pogacar *et al.*, 2010).

Bacterial factors influencing adhesion and invasion of *C. jejuni*

Campylobacter does not possess a Type III secretion system (TTSS) and other structures including fimbriae or pili as in other Gram-negative bacteria like *Salmonella* and *E. coli*. Adherence and invasion of intestinal epithelial cells is mediated by several bacterial structures. Some of these factors directly influence adhesion and invasion of *C. jejuni* to the host surface receptors while others play an indirect role in host cell interaction. The role of these host cell interaction factors in *C. jejuni* pathogenesis has only started to be understood (Hu and Kopecko, 2008; Dasti *et al.*, 2010). Table 9.1 provides a summary of many of the genes reported to influence adhesion and invasion, some of which are discussed below.

Factors associated with adhesion

The role of adhesion in *Campylobacter* pathogenesis is not well defined yet several studies have reported that *Campylobacter* does contain adhesins or binding factors that are required for adhesion to the host epithelium. CadF (*Campylobacter* adhesion to fibronectin) is an outer membrane protein present in all *C. jejuni* and *C. coli* strains tested so far and establishes an interaction with the host extracellular fibronectin receptors (Konkel *et al.*, 1997). The *cadF* gene sequence in *C. jejuni* has a 39 bp insertion compared to that in *C. coli* and it is thought that this difference results in *C. jejuni* being more adherent to INT-407 epithelial cells compared to *C. coli* strains. Furthermore, an isogenic *cadF* mutant strain showed a significantly impaired host cell interaction ability, compared to the wild-type, confirming its role in *C. jejuni* pathogenesis (Krause-Gruszczynska *et al.*, 2007). A recent study reported that CadF along with FlaA, mediate invasion of host epithelial cells via an epidermal growth factor (EGF) pathway. This process involves activation of cytoskeleton related proteins – mainly, P13 kinase, c-Src and focal adhesion kinase (FAK) (Eucker and Konkel, 2012). Two surface exposed lipoprotein adhesins, CapA and JlpA, have also been shown to mediate adhesion of *C. jejuni* to human epithelial cell lines as well as affect *in vivo* colonization of the chicken gut (Asghar *et al.*, 2007; Jin *et al.*, 2001; Pei and Blaser, 1993).

Campylobacter contains a number of different periplasmic proteins PEB 1–4 that serve as adhesins (Pei *et al.*, 1991). PEB 1 is a 28-kDa protein and is conserved in *C. jejuni* and *C. coli* strains (Pei and Blaser, 1993; Pei *et al.*, 1991). PEB 1 has immunogenic and immunoprotective properties (Du *et al.*, 2008) and is required for *C. jejuni* binding to HeLa cells (Pei *et al.*, 1998). It is functionally homologous to periplasmic binding proteins from other Gram-negative bacteria with a role in glutamate/aspartate transport (Leon-Kempis Mdel *et al.*, 2006; Pei and Blaser, 1993). PEB 1 was detected in culture supernatants by cell fractionation and immunoblotting techniques confirming that PEB 1 in *C. jejuni* is a periplasmic associated protein and not an inner or outer membrane protein (Leon-Kempis Mdel *et al.*, 2006). PEB 3 has antigenic properties similar to PEB 1 and shares homology with class 1 pili in *Neisseria meningitidis* and the heat-labile enterotoxin B subunit in *E. coli*, and is required for interaction with intestinal epithelial cells (Pei *et al.*, 1991). PEB 4 is another periplasmic chaperone in *C. jejuni* and is as antigenic in nature as PEB 1. It is involved in adhesion to INT-407 cells, biofilm formation and mouse colonization (Asakura *et al.*, 2007) as well as invasion of cultured human epithelial cells (Kervela *et al.*, 1993). Recently, another periplasmic-binding protein encoding gene, *cj1289*, has been identified in *C. jejuni*. This protein is structurally similar to PEB 4 in *C. jejuni* and may have a role in virulence (Kale *et al.*, 2011). HtrA is a periplasmic protein in *C. jejuni* with dual functionality: a chaperone and a protease and its chaperone activity is required

Table 9.1 Bacterial factors reported to influence adhesion or invasion by *C. jejuni*

Bacterial factor	Function	Phenotype: Adhesion	Phenotype: Invasion	Reference
Flagella				
FlaA	Flagellin A		✓	Wassenaar *et al*. (1991), Jain *et al*. (2008), Eucker and Konkel (2012)
FlaC	Flagellar export apparatus		✓	Song *et al*. (2004)
cj0977	Flagellum		✓	Goon *et al*. (2006)
Sigma factors				
σ28 (*fliS*)	Transcription of flagellar genes		✓	Carrillo *et al*. (2004), Fernando *et al*. (2007)
σ54 *(rpoN)*			✓	Fernando *et al*. (2007)
Chemotaxis				
luxS	Autoinducer-2 (AI-2) producing gene (methyl recycling pathway)	✓		Quinones *et al*. (2009)
CetB	Chemoreceptor		✓	Golden and Acheson (2002)
cj0952c	Formic acid chemoreceptor		✓	Tareen *et al*. (2011)
CheR, CheB	Chemoreceptor	✓	✓	Kanungpean *et al*. (2011)
LOS				
Cst-II	Sialyltranferase		✓	Louwen *et al*. (2008)
WlaRG	Aminotransferase		✓	Holden *et al*. (2012)
WlaTB, WlaTC	Glycosyltranferase		✓	
cj1136	Putative galactosyltransferase		✓	Javed *et al*. (2012)
Periplasmic adhesions				
CadF	*Campylobacter* adhesion to fibronectin		✓	Krause-Gruszczynska *et al*. (2007), Eucker and Konkel (2012)
CapA, JlpA	Lipoprotein adhesins		✓	Asghar *et al*. (2007)
PEB1	Periplasmic adhesins	✓		Pei *et al*. (1998)
PEB 3		✓		Pei *et al*. (1991)
PEB4	Periplasmic chaperone	✓	✓	Asakura *et al*. (2007), Kervela *et al*. (1993)
HtrA	Periplasmic protein	✓		Baek *et al*. (2011)
Toxin				
CdtA,CdtC	Cytolethal distending toxin (CDT) subunits	✓		Lee *et al*. (2003)
Capsule				
KpsE	Capsular polysaccharide ABC transporter	✓	✓	Bachtiar *et al*. (2007)
JJD26997_1801	Capsular polysaccharide biosynthesis		✓	Javed *et al*. (2010)
O-linked glycosylation system				
PseAc	Pseudaminic acid	✓	✓	Ewing *et al*. (2009)
N-linked glycosylation system				
PglH	General glycosylation pathway protein	✓	✓	Karlyshev *et al*. (2004)

		Phenotype		
Bacterial factor	Function	Adhesion	Invasion	Reference
PglB	Oligosaccharide transferase	✓	✓	Szymanski *et al.* (2002)
PglE	Putative aminotransferase	✓	✓	
cj1121c	Aminotransferase	✓		Vijayakumar *et al.* (2006)
Other				
FedA	Flagellar coexpressed determinant		✓	Barrero-Tobon and Hendrixson (2012)
CiaC	*Campylobacter* invasion antigen (Cia)	✓		Christensen *et al.* (2009)
CiaI			✓	Barrero-Tobon and Hendrixson (2012)
cj0034c, *cj0404*, *cj0525c*	Antigen encoding genes		✓	Nielsen *et al.* (2012)
cj0004c	Oxidoreductase	✓	✓	Tareen *et al.* (2011)
cj0005c	Sulfite metabolism	✓	✓	
ppk2	Polyphosphate kinase 2 in inorganic polyphosphate (poly P) synthesis	✓	✓	Gangaiah *et al.* (2010)
FdhTU	Formate metabolism		✓	Pryjma *et al.* (2012)
feoB	Ferrous iron uptake		✓	Raphael and Joens (2003)

for interaction with host epithelial cells and binding to macrophages (Baek *et al.*, 2011).

Delivery of effectors via the flagella apparatus

In the absence of a TTSS, a functional *C. jejuni* flagella apparatus is required for delivery of effector proteins into host epithelial cells. The flagellin protein, FlaC, which is structurally homologous to FlaA and FlaB, has been shown to have a role in invasion by supporting the function of the flagella export apparatus (Song *et al.*, 2004). Two *Campylobacter* invasion antigen (Cia) effector proteins have been characterized, CiaB with limited homology to SipB of *Salmonella* species and CiaC with no known similarity to any other proteins (Christensen *et al.*, 2009). An insertion mutation into gene *cj1242*, which was annotated as CiaC in *C. jejuni* clinical strain F38011, showed defects in motility, adherence and internalization of INT-407 cells (Christensen *et al.*, 2009). In contrast, mutation in *ciaB* had no effect on the invasion potential of *C. jejuni* 81-176 in T84 cells (Novik *et al.*, 2010) suggesting that the role of CiaB in invasion is strain and cell line dependent. Another Cia protein called CiaI (*cj1450*) has been recently identified in *C. jejuni*. CiaI contains an amino terminal type III secretion sequence and is transported via the flagella type III secretion system into host epithelial cells. A *C. jejuni ciaI* mutant was attenuated for survival in cultured INT-407 and HeLa epithelial cells that suggest CiaI might have a role in intracellular survival (Buelow *et al.*, 2011). Not all Cia proteins are yet known nor is their exact mechanism in *C. jejuni* host interaction fully understood. However, a *C. jejuni* strain lacking the CiaB-mediated secretion ability inhibited secretion of other Cia proteins into the epithelial cells (Ko and Park, 2000).

Flagella-related factors

Motility is known to influence invasion of *C. jejuni* and mutants in flagella genes have been shown to have impaired invasion (Wassenaar *et al.*, 1991). Therefore, when considering flagella-related factors that influence invasion, consideration must be given to whether or not motility is actually being affected. In bacteria, the main role of σ28 regulated genes is in the functioning of flagella (Carrillo *et al.*, 2004; Goon *et al.*, 2006; Poly *et al.*, 2007). However, *C. jejuni* expresses some σ28 genes that are not important in motility but have been involved in invasion and damage to epithelial cells (Goon *et al.*, 2006; Poly *et al.*, 2007). The σ28 dependent gene *fspA* is expressed as two alleles *fspA1* and *fspA2* in some *C. jejuni* strains. FspA1 and FspA2 proteins are not directly involved in *C. jejuni* invasion but FspA2 has been implicated in apoptosis of INT-407 cells (Poly *et al.*, 2007). Another flagella dependent gene with a role in invasion of intestinal epithelial cells is *cj0977* (Goon *et al.*, 2006). At first the Cj0977 protein was shown to be secreted in the cytoplasm, however this secretion was independent of the flagella export apparatus and had no effect on motility (Goon *et al.*, 2006). A later contradictory study showed a *cj0977* mutant was deficient in motility and was subsequently unable to invade epithelial cells (Novik *et al.*, 2010). Inactivation of *fliS* (σ28) and *rpoN* (σ54) resulted in reduced motility and defects in invasion of HeLa cells (Fernando *et al.*, 2007). The flagella co-expressed determinant FedA, together with CiaI, have been reported recently to be involved in the invasion of human epithelial cells (Barrero-Tobon and Hendrixson, 2012). Chemotaxis and the glycosylation systems (O and N-linked) also have an important role in influencing the adhesion and invasion phenotypes in *C. jejuni*.

Surface structures and their role in host cell interaction

Other surface polysaccharides including LOS and capsule have been implicated in adherence and invasion of epithelial cells and have been shown to have a role in an *in vitro* ferret model of infection (Karlyshev *et al.*, 2000, Bacon *et al.*, 2001, Kanpies *et al.*, 2004, Bachtiar *et al.*, 2007, Louwen *et al.*, 2008). An early study by McSweegan and Walker (1986) investigated the role of *C. jejuni* LOS in host cell interaction and internalization. This study showed that the purified extract of LOS promoted adhesion to epithelial cells and this interaction was eliminated with the oxidation of LOS (McSweegan and Walker, 1986). Bacon *et al.* (2001) demonstrated the role of *C. jejuni* 81-176 LOS in invasion and serum resistance using INT-407 cell lines in an *in vitro* model. Kanipes *et al.* (2004) reported that a mutant in the *C. jejuni* 81-176 LOS locus lost its ability to invade human epithelial cell lines and others have reported that sialylated LOS exposed on the outer surface of *C. jejuni* does influence invasion (Louwen *et al.*, 2008, Habib *et al.*, 2009). Louwen *et al.* (2008) showed that a mutation in the *cst-II* gene in GBS associated *C. jejuni* strains, encoding for a sialyltranferase, made these strains defective in the invasion of intestinal epithelial cells. To further investigate the role of LOS in *C. jejuni* pathogenesis, Marsden *et al.* (2009) showed that a *C. jejuni* mutant lacking a functional LOS locus was unable to invade Caco-2 cells. Mutations in one aminotransferase encoding gene *wlaRG* and two glycosyltranferase encoding genes *wlaTB* and *wlaTC* in the LOS core in *C. jejuni* 81116 deprived the mutants of their ability to invade human epithelial cell lines *in vitro*. Moreover, mutants in these three genes showed reduced immunogenicity and invasion into chick embryonic fibroblasts (Holden *et al.*, 2012). Recently, Javed *et al.* (2012) showed that a transposon mutant in *cj1136*, which encodes for a putative galactosyltransferase, was impaired in its ability to invade cultured INT-407 and Caco-2 cells. All the above studies strongly support the role of LOS in *C. jejuni* interaction to the human cells.

Other bacterial factors recently identified that influence adhesion and invasion phenotypes include a *C. jejuni* 11168 antigen encoding gene, *cj0034c*, with a role in invasion of INT-407 epithelial cell lines. Furthermore, mutants in *cj0034c*, *cj0404*, *cj0525c* were reduced in invasion of spleen and liver in a mouse model of infection (Nielsen *et al.*, 2012). Oxidoreductase encoding genes *cj0004c* and *cj0005c* – involved in sulfite metabolism in the cytoplasm (Tareen *et al.*, 2011), and a polyphosphate kinase 2 – encoding gene (*ppk2*) required for inorganic polyphosphate (poly P) synthesis as an alternate energy source in stress conditions (Gangaiah *et al.*, 2010), are important *C. jejuni* metabolic pathways that influence *C. jejuni* attachment and internalization in *in vitro* tissue culture assays. A formate metabolism related operon (*fdh*TU), expressed as two alleles: *fdhT* and *fdhU*, has recently been identified in *C. jejuni* to have a role in adherence and invasion in an *in vitro* colony based assay. However, the *fdhTU* mutant was shown to be as invasive as the wild-type strain using fluorescence microscopy but high numbers of *C. jejuni* were recovered after invasion. Thus a role for FdhTU in bacterial internalization was concluded (Pryjma *et al.*, 2012).

As discussed above, there have been numerous studies that have identified a range of different bacterial factors with direct or indirect roles in adhesion to and invasion of host cells yet still no single consensus mechanism has been elucidated suggesting that in this opportunistic pathogen this is a multifactorial, strain-dependent process. Can anything be learned from studying larger groups of strains or groups of phenotypically related strains to try to understand how *C. jejuni* has become to be an invasive pathogen in humans?

Do all campylobacters invade to the same extent?

Research has suggested that human clinical isolates of *C. jejuni* are generally more adherent and invasive than non-clinical isolates. Prasad *et al.* (1996) demonstrated that human isolates are more pathogenic than chicken isolates for their invasion, adhesion and cytotoxin production phenotypes using HeLa and HEp-2 cells as *in vitro* models. Similar results were reported by Konkel and Jones (1989) for 11 clinical and non-clinical strains of *C. jejuni* tested for their adherence and invasion to cultured HEp-2 cells. Newell *et al.* (1985) compared the adherence to and invasion of HeLa cells by environmental *C. jejuni* isolates and those obtained from patients suffering from diarrhoea using electron microscopy and immunofluorescence. The authors concluded that clinical isolates were more invasive than the non-clinical strains. A similar difference in the invasion potential between human and animal isolates was also reported by Biswas and colleagues (2000). In contrast, other research (van Deun *et al.*, 2007) did not find any difference in the invasion potential among the set of human and poultry isolates investigated. A more recent study classified the invasion potential of *C. jejuni* strains isolated from clinical, poultry and environmental sources as hyper, high and low using an *in vitro* invasion assay. It was reported here that the hyper-invasive group of strains consisted of a greater proportion of clinical isolates compared to strains from other sources (Fearnley *et al.*, 2008). This group of hyper-invasive *C. jejuni* strains has been further investigated at the whole genome level to determine if their genomic content differs from that of the low invasive *C. jejuni* strains that might help explain why certain strains are hyper-invasive.

Genomic studies of a hyper-invasive group of strains to investigate host–cell interaction

Comparative genomic hybridization (CGH) and pooled suppressive subtractive hybridization (PSSH) comparing six hyper-invasive *C. jejuni* strains against a selection of low invasive *C. jejuni* strains did not identify any genomic regions specific only to the hyper-invasive *C. jejuni* strains. Detailed mining of the CGH data identified nine genes that were present in, and six genes that were highly divergent from, all six hyper-invasive *C. jejuni* strains (Tables 9.2 and 9.3 respectively). These included genes involved in cellular and other metabolic functions; some known virulence associated genes were also identified. This presence or absence of genes was not exclusive to the hyper-invasive strains though as it was also observed in one out of the four low-invasive strains studied. PSSH identified 11 genes that were present in the hyper-invasive *C. jejuni* strains however none of these 11 genes were present in all six hyper-invasive strains and only four genes were found to be only present in one hyper-invasive strain (out of a total of six hyper- and four low-invasive strains) following PCR confirmation. The genes detected by PSSH had a range of functions with inter- and intra-species homology and similarity with other bacteria. The genomic data suggested that hyper-invasive *C. jejuni* form a genetically diverse group of strains that seem to share just a phenotypic trait which appears to be a multifactorial process and is a function of different cellular networks rather than a single gene or functional pathway (A. Baig, unpublished).

Interestingly whole genome sequencing of two hyper-invasive *C. jejuni* strains identified a mosaic pattern within the CPS locus containing genes with similarity to the CPS genes in other pathogenic *Campylobacter* species (*C. jejuni* subspecies *doylei* and *C. lari*) (A. Baig, unpublished). This mosaicism in the capsular region has only recently

Table 9.2 Genes identified as present in all hyper-invasive and one low-invasive strain(s) following CGH analysis of six hyper-invasive and four low-invasive *C. jejuni* strains (Adapted from Abiyad Baig, PHD thesis, 2012)

Gene name/number	Role/function (strain)
Post-translational modification, protein turnover, chaperones*	
aat (*cj1109*)	Putative leucyl/phenylalanyl-tRNA-protein (*C. jejuni* NCTC11168)
cj0271	Bacterioferritin comigratory protein homologue (*C. jejuni* NCTC11168)
Translation	
hisS (*cj0765c*)	Histidyl-tRNA synthestase (*C. jejuni* NCTC11168)
Amino acid transport and metabolism	
proC (*cj1076*)	Putative pyrroline-5-carboxylate reductase (*C. jejuni* NCTC11168)
metF (*cj1202*)	5,10-Methylenetetrahydrofolate reductase (*C. jejuni* NCTC11168)
Transport and metabolism	
modA (*cj0303c*)	Putative molybdate-binding lipoprotein (*C. jejuni* NCTC11168)
Hypothetical	
cj0738	Hypothetical protein (*C. jejuni* NCTC11168)
cj0564	Putative integral membrane protein (*C. jejuni* NCTC11168)
CCOA0033	Hypothetical protein (*C. coli* RM2228)

*COG functional grouping is given.

Table 9.3 Genes identified as highly divergent/absent from all hyper-invasive and one low-invasive strain(s) following CGH analysis (Adapted from Abiyad Baig, PHD thesis, 2012)

Gene name/number	Role/Function (strain)
Posttranslational modification, protein turnover, chaperones*	
cj0701	Putative protease (*C. jejuni* NCTC11168)
Translation	
tgt (*cj1010*)	Queuine tRNA-ribosyltransferase (*C. jejuni* NCTC11168)
Probable cell wall/membrane biogenesis	
cj0266c	Putative intergral membrane protein (*C. jejuni* NCTC11168)
Replication, recombination and repair	
cje0731	Type III restriction/modification enzyme (*C. jejuni* RM1221)
Hypothetical	
cje1112	Hypothetical protein (*C. jejuni* RM1221)
cje1128	Hypothetical protein (*C. jejuni* RM1221)

*COG functional grouping is given.

been reported in one other *Campylobacter* strain (Karlyshev *et al.*, 2013). It is evident that the CPS locus is a hotspot of recombination and may be a marker for the hyper-invasive phenotype observed in these *C. jejuni* strains and this phenomenon is currently still being investigated.

Conclusion

Interaction with human epithelial cells is an important phenomenon in the pathogenesis of *C. jejuni*. Several bacterial factors (Table 9.1) have been reported to influence adhesion to and invasion of a range of epithelial cells *in vitro*. Genes associated with essential bacterial cellular functions, metabolic pathways, transport systems and cell surface structures all contribute to the variation in invasion potential of *C. jejuni* strains. It seems that *C. jejuni* adheres to the epithelial cell surface using a number of defined adhesins. The bacterial cell surface, including the capsule, is important in the interaction with the host cell and it seems that the flagella may also have a role in the delivery of effectors to bring about invasion. A number of metabolism-related genes have also been reported. However, depending on how virulence is defined, these may be required for the survival of the organism and ability of the organism to colonize the niche of the human epithelium. The exact mechanism by which these pathways interact to facilitate invasion still needs to be understood and it appears unlikely that a common pathway used by all strains will be easily identifiable, even in phenotypically related strains of this human pathogen.

References

van Alphen, L.B., Bleumink-Pluym, N.M., Rochat, K.D., van Balkom, B.W., Wosten, M.M., and van Putten, J.P. (2008). Active migration into the subcellular space precedes *Campylobacter jejuni* invasion of epithelial cells. Cell. Microbiol. *10*, 53–66.

Asakura, H., Yamasaki, M., Yamamoto, S., and Igimi, S. (2007). Deletion of peb4 gene impairs cell adhesion and biofilm formation in *Campylobacter jejuni*. FEMS Microbiol. Lett. 275, 278–285.

Ashgar, S.S., Oldfield, N.J., Wooldridge, K.G., Jones, M.A., Irving, G.J., Turner, D.P., and Ala'Aldeen, D.A. (2007). CapA, an autotransporter protein of *Campylobacter jejuni*, mediates association with human epithelial cells and colonization of the chicken gut. J. Bacteriol. *189*, 1856–1865.

Babakhani, F.K., and Joens, L.A. (1993). Primary swine intestinal cells as a model for studying *Campylobacter jejuni* invasiveness. Infect. Immun. *61*, 2723–2726.

Bachtiar, B.M., Coloe, P.J., and Fry, B.N. (2007). Knockout mutagenesis of the kpsE gene of *Campylobacter jejuni* 81116 and its involvement in bacterium–host interactions. FEMS Immunol. Med. Microbiol. *49*, 149–154.

Bacon, D.J., Szymanski, C.M., Burr, D.H., Silver, R.P., Alm, R.A., and Guerry, P. (2001). A phase-variable capsule is involved in virulence of *Campylobacter jejuni* 81-176. Mol. Microbiol. *40*, 769–777.

Baek, K.T., Vegge, C.S., and Brondsted, L. (2011). HtrA chaperone activity contributes to host cell binding in *Campylobacter jejuni*. Gut Pathog. 3,13.

Baker, N.T., and Graham, L.L. (2010). *Campylobacter fetus* translocation across Caco-2 cell monolayers. Microb. Pathog. *49*, 260–272.

Barrero-Tobon, A.M., and Hendrixson, D.R. (2012). Identification and analysis of flagellar coexpressed determinants (Feds) of *Campylobacter jejuni* involved in colonization. Mol. Microbiol. *84*, 352–369.

Beltinger, J., del Buono, J., Skelly, M.M., Thornley, J., Spiller, R.C., Stack, W.A., and Hawkey, C.J. (2008). Disruption of colonic barrier function and induction of mediator release by strains of *Campylobacter jejuni* that invade epithelial cells. World J. Gastroenterol. *14*, 7345–7352.

Biswas, D., Itoh, K., and Sasakawa, C. (2000). Uptake pathways of clinical and healthy animal isolates of *Campylobacter jejuni* into INT-407 cells. FEMS Immunol. Med. Microbiol. *29*, 203–211.

Bras, A.M., and Ketley, J.M. (1999). Transcellular translocation of *Campylobacter jejuni* across human polarised epithelial monolayers. FEMS Microbiol. Lett. *179*, 209–215.

Buelow, D.R., Christensen, J.E., Neal-McKinney, J.M., and Konkel, M.E. (2011). *Campylobacter jejuni* survival within human epithelial cells is enhanced by the secreted protein CiaI. Mol. Microbiol. *80*, 1296–1312.

Carrillo, C.D., Taboada, E., Nash, J.H., Lanthier, P., Kelly, J., Lau, P.C., Verhulp, R., Mykytczuk, O., Sy, J., Findlay, W.A., *et al.* (2004). Genome-wide expression analyses of *Campylobacter jejuni* NCTC11168 reveals coordinate regulation of motility and virulence by flhA. J. Biol. Chem. *279*, 20327–20338.

Christensen, J.E., Pacheco, S.A., and Konkel, M.E. (2009). Identification of a *Campylobacter jejuni*-secreted protein required for maximal invasion of host cells. Mol. Microbiol. *73*, 650–662.

Dasti, J.I., Tareen, A.M., Lugert, R., Zautner, A.E., and Gross, U. (2010). *Campylobacter jejuni*: a brief overview on pathogenicity-associated factors and disease-mediating mechanisms. Int. J. Med. Microbiol. 3, 205–211.

De Melo, M.A., Gabbiani, G., and Pechere, J.C. (1989). Cellular events and intracellular survival of *Campylobacter jejuni* during infection of HEp-2 cells. Infect. Immun. *57*, 2214–2222.

Du, L.F., Li, Z.J., Tang, X.Y., Huang, J.Q., and Sun, W.B. (2008). Immunogenicity and immunoprotection of recombinant PEB1 in *Campylobacter-jejuni*-infected mice. World J. Gastroenterol. *14*, 6244–6248.

Eucker, T.P., and Konkel, M.E. (2012). The cooperative action of bacterial fibronectin-binding proteins and secreted proteins promote maximal *Campylobacter jejuni* invasion of host cells by stimulating membrane ruffling. Cell. Microbiol. *14*, 226–238.

Everest, P.H., Goossens, H., Butzler, J.P., Lloyd, D., Knutton, S., Ketley, J.M., and Williams, P.H. (1992). Differentiated Caco-2 cells as a model for enteric invasion by *Campylobacter jejuni* and *C. coli*. J. Med. Microbiol. *37*, 319–325.

Ewing, C.P., Reishcheva, E., and Guerry, P. (2009). Functional characterization of flagellin glycosylation in *Campylobacter jejuni* 81-176. J. Bacteriol. *191*, 7086–7093.

Fauchere, J.L., Rosenau, A., Veron, M., Moyen, E.N., Richard, S., and Pfister, A. (1986). Association with HeLa cells of *Campylobacter jejuni* and *Campylobacter* coli isolated from human feces. Infect. Immun. *54*, 283–287.

Fearnley, C., Manning, G., Bagnall, M., Javed, M.A., Wassenaar, T.M., and Newell, D.G. (2008). Identification of hyperinvasive *Campylobacter jejuni* strains isolated from poultry and human clinical sources. J. Med. Microbiol. *57*, 570–580.

Fernando, U., Biswas, D., Allan, B., Willson, P., and Potter, A.A. (2007). Influence of *Campylobacter jejuni* fliA, rpoN and flgK genes on colonization of the chicken gut. Int. J. Food Microbiol. *118*, 194–200.

Frirdich, E., Biboy, J., Adams, C., Lee, J., Ellermeier, J., Gielda, L.D., Dirita, V.J., Girardin, S.E., Vollmer, W., and Gaynor, E.C. (2012). Peptidoglycan-Modifying Enzyme Pgp1 Is Required for Helical Cell Shape and Pathogenicity Traits in *Campylobacter jejuni*. PLoS Pathog. *8*, e1002602.

Gangaiah, D., Liu, Z., Arcos, J., Kassem, I.I., Sanad, Y., Torrelles, J.B., and Rajashekara, G. (2010). Polyphosphate kinase 2: a novel determinant of stress responses and pathogenesis in *Campylobacter jejuni*. PLoS One 5, e12142.

Golden, N.J., and Acheson, D.W. (2002). Identification of motility and autoagglutination *Campylobacter jejuni* mutants by rand Om transposon mutagenesis. Infect. Immun. *70*, 1761–1771.

Goon, S., Ewing, C.P., Lorenzo, M., Pattarini, D., Majam, G., and Guerry, P. (2006). A sigma28-regulated nonflagella gene contributes to virulence of *Campylobacter jejuni* 81-176. Infect. Immun. *74*, 769–772.

Grant, C.C., Konkel, M.E., Cieplak, W., Jr., and Tompkins, L.S. (1993). Role of flagella in adherence, internalization, and translocation of *Campylobacter jejuni* in nonpolarized and polarized epithelial cell cultures. Infect. Immun. *61*, 1764–1771.

Habib, I., Louwen, R., Uyttendaele, M., Houf, K., Vand Enberg, O., Nieuwenhuis, E.E., Miller, W.G., van Belkum, A., and De Zutter, L. (2009). Correlation between genotypic diversity, lipooligosaccharide gene locus class variation, and caco-2 cell invasion potential of *Campylobacter jejuni* isolates from chicken meat and humans: contribution to virulotyping. Appl. Environ. Microbiol. 75, 4277–4288.

Hill, C. (2012). Virulence or niche factors: what's in a name? J. Bacteriol. *194*(21), 5725–5727.

Holden, K.M., Gilbert, M., Coloe, P.J., Li, J., and Fry, B.N. (2012). The role of WlaRG, WlaTB and WlaTC in lipooligosaccharide synthesis by *Campylobacter jejuni* strain 81116. Microb. Pathog. *52*, 344–352.

Hu, L., and Kopecko, D.J. (1999). *Campylobacter jejuni* 81-176 associates with microtubules and dynein during invasion of human intestinal cells. Infect. Immun. *67*, 4171–4182.

Hu, L., and Kopecko, D.J. (2008). Cell biology of human host cell entry by *Campylobacter jejuni*. In *Campylobacter*, Nachamkin, I., Szymanski, C.M., and Blaser, M.J., eds. (ASM Press, Washington, DC), pp. 297–313.

Hu, L., Tall, B.D., Curtis, S.K., and Kopecko, D.J. (2008). Enhanced microscopic definition of *Campylobacter jejuni* 81-176 adherence to, invasion of, translocation across, and exocytosis from polarized human intestinal Caco-2 cells. Infect. Immun. *76*, 5294–5304.

Hugdahl, M.B., Beery, J.T., and Doyle, M.P. (1988). Chemotactic behavior of *Campylobacter jejuni*. Infect. Immun. *56*, 1560–1566.

Jain, D., Prasad, K.N., Sinha, S., and Husain, N. (2008). Differences in virulence attributes between cytolethal distending toxin positive and negative *Campylobacter jejuni* strains. J. Med. Microbiol. *57*, 267–272.

Javed, M.A., Grant, A.J., Bagnall, M.C., Maskell, D.J., Newell, D.G., and Manning, G. (2010). Transposon mutagenesis in a hyper-invasive clinical isolate of *Campylobacter jejuni* reveals a number of genes with potential roles in invasion. Microbiol. *156*, 1134–1143.

Javed, M.A., Cawthraw, S.A., Baig, A., Li, J., McNally, A., Oldfield, N.J., Newell, D.G., and Manning, G. (2012). Cj1136 is required for lipooligosaccharide biosynthesis, hyperinvasion, and chick colonization by *Campylobacter jejuni*. Infect. Immun. *80*, 2361–2370.

Jin, S., Joe, A., Lynett, J., Hani, E.K., Sherman, P., and Chan, V.L. (2001). JlpA, a novel surface-exposed lipoprotein specific to *Campylobacter jejuni*, mediates adherence to host epithelial cells. Mol. Microbiol. *39*, 1225–1236.

Kale, A., Phansopa, C., Suwannachart, C., Craven, C.J., Rafferty, J.B., and Kelly, D.J. (2011). The virulence factor PEB4 (Cj0596) and the periplasmic protein Cj1289 are two structurally related SurA-like chaperones in the human pathogen *Campylobacter jejuni*. J. Biol. Chem. *286*, 21254–21265.

Kanipes, M.I., Holder, L.C., Corcoran, A.T., Moran, A.P., and Guerry, P. (2004). A deep-rough mutant of *Campylobacter jejuni* 81-176 is noninvasive for intestinal epithelial cells. Infect. Immun. *72*, 2452–2455.

Kanungpean, D., Kakuda, T., and Takai, S. (2011). Participation of CheR and CheB in the chemosensory response of *Campylobacter jejuni*. Microbiol. *157*, 1279–1289.

Karlyshev, A.V., Linton, D., Gregson, N.A., Lastovica, A.J., and Wren, B.W. (2000). Genetic and biochemical evidence of a *Campylobacter jejuni* capsular polysaccharide that accounts for Penner serotype specificity. Mol. Microbiol. 35, 529–541.

Karlyshev, A.V., Everest, P., Linton, D., Cawthraw, S., Newell, D.G., and Wren, B.W. (2004). The *Campylobacter jejuni* general glycosylation system is important for attachment to human epithelial cells and in the colonization of chicks. Microbiol. *150*, 1957–1964.

Karlyshev, A.V., Quail, M.A., Parkhill, J., and Wren, B.W. (2013). Unusual features in organisation of capsular polysaccharide-related genes of *C. jejuni* strain X. Gene *522*, 37–45.

Kervella, M., Pages, J.M., Pei, Z., Grollier, G., Blaser, M.J., and Fauchere, J.L. (1993). Isolation and characterization of two *Campylobacter* glycine-extracted proteins that bind to HeLa cell membranes. Infect. Immun. *61*, 3440–3448.

Ketley, J.M. (1997). Pathogenesis of enteric infection by *Campylobacter*. Microbiology *143*, 5–21.

Ko, M., and Park, C. (2000). H-NS-Dependent regulation of flagellar synthesis is mediated by a LysR family protein. J. Bacteriol. *182*:4670–4672.

Konkel, M.E., and Joens, L.A. (1989). Adhesion to and invasion of HEp-2 cells by *Campylobacter* spp. Infect. Immun. *57*, 2984–2990.

Konkel, M.E., Babakhani, F., and Joens, L.A. (1990). Invasion-related antigens of *Campylobacter jejuni*. J. Infect. Dis. *162*, 888–895.

Konkel, M.E., Mead, D.J., Hayes, S.F., and Cieplak, W., Jr. (1992a). Translocation of *Campylobacter jejuni* across human polarized epithelial cell monolayer cultures. J. Infect. Dis. *166*, 308–315.

Konkel, M.E., Hayes, S.F., Joens, L.A., and Cieplak, W., Jr. (1992b). Characteristics of the internalization and intracellular survival of *Campylobacter jejuni* in human epithelial cell cultures. Microb. Pathog. *13*, 357–370.

Konkel, M.E., Garvis, S.G., Tipton, S.L., And Erson, D.E., Jr., and Cieplak, W., Jr. (1997). Identification and molecular cloning of a gene encoding a fibronectin-binding protein (CadF) from *Campylobacter jejuni*. Mol. Microbiol. *24*, 953–963.

Konkel, M.E., Monteville, M.R., Rivera-Amill, V., and Joens, L.A. (2001). The pathogenesis of *Campylobacter jejuni*-mediated enteritis. Curr. Issues Intest. Microbiol. *2*, 55–71.

Krause-Gruszczynska, M., van Alphen, L.B., Oyarzabal, O.A., Alter, T., Hanel, I., Schliephake, A., Konig, W., van Putten, J.P., Konkel, M.E., and Backert, S. (2007). Expression patterns and role of the CadF protein in *Campylobacter jejuni* and *Campylobacter coli*. FEMS Microbiol. Lett. *274*, 9–16.

Lee, A., O'Rourke, J.L., Barrington, P.J., and Trust, T.J. (1986). Mucus colonization as a determinant of pathogenicity in intestinal infection by *Campylobacter jejuni*: a mouse cecal model. Infect Immun. *51*, 536–546.

Lee, R.B., Hassane, D.C., Cottle, D.L., and Pickett, C.L. (2003). Interactions of *Campylobacter jejuni* cytolethal distending toxin subunits CdtA and CdtC with HeLa cells. Infect. Immun. *71*, 4883–4890.

Leon-Kempis Mdel, R., Guccione, E., Mulholland, F., Williamson, M.P., and Kelly, D.J. (2006). The *Campylobacter jejuni* PEB1a adhesin is an aspartate/glutamate-binding protein of an ABC transporter essential for microaerobic growth on dicarboxylic amino acids. Mol. Microbiol. *60*, 1262–1275.

Louwen, R., Heikema, A., van Belkum, A., Ott, A., Gilbert, M., Ang, W., Endtz, H.P., Bergman, M.P., and Nieuwenhuis, E.E. (2008). The sialylated lipooligosaccharide outer core in *Campylobacter jejuni* is an important determinant for epithelial cell invasion. Infect. Immun. *76*, 4431–4438.

McSweegan, E., and Walker, R.I. (1986). Identification and characterization of two *Campylobacter jejuni* adhesins for cellular and mucous substrates. Infect. Immun. *53*, 141–148.

Marsden, G.L., Li, J., Everest, P.H., Lawson, A.J., and Ketley, J.M. (2009). Creation of a large deletion mutant of *Campylobacter jejuni* reveals that the lipooligosaccharide gene cluster is not required for viability. J. Bacteriol. *191*, 2392–2399.

Newell, D.G., McBride, H., Saunders, F., Dehele, Y., and Pearson, A.D. (1985). The virulence of clinical and environmental isolates of *Campylobacter jejuni*. J. Hyg. (Lond) *94*, 45–54.

Nielsen, L.N., Luijkx, T.A., Vegge, C.S., Johnsen, C.K., Nuijten, P., Wren, B.W., Ingmer, H., and Krogfelt, K.A. (2012). Identification of immunogenic and virulence-associated *Campylobacter jejuni* proteins. Clin. Vacc. Immunol. *19*, 113–119.

Novik, V., Hofreuter, D., and Galan, J.E. (2010). Identification of *Campylobacter jejuni* genes involved in its interaction with epithelial cells. Infect. Immun. *78*, 3540–3553.

Oelschlaeger, T.A., Guerry, P., and Kopecko, D.J. (1993). Unusual microtubule-dependent endocytosis mechanisms triggered by *Campylobacter jejuni* and *Citrobacter freundii*. Proc. Natl. Acad. Sci. U.S.A. *90*, 6884–6888.

Pei, Z., and Blaser, M.J. (1993). PEB1, the major cell-binding factor of *Campylobacter jejuni*, is a homolog of the binding component in Gram-negative nutrient transport systems. J. Biol. Chem. *268*, 18717–18725.

Pei, Z.H., Ellison, R.T., 3rd., and Blaser, M.J. (1991). Identification, purification, and characterization of major antigenic proteins of *Campylobacter jejuni*. J. Biol. Chem. *266*, 16363–16369.

Pei, Z., Burucoa, C., Grignon, B., Baqar, S., Huang, X.Z., Kopecko, D.J., Bourgeois, A.L., Fauchere, J.L., and Blaser, M.J. (1998). Mutation in the peb1A locus of *Campylobacter jejuni* reduces interactions with epithelial cells and intestinal colonization of mice. Infect. Immun. *66*, 938–943.

Pogacar, M.S., Klancnik, A., Mozina, S.S., and Cencic, A. (2010). Attachment, invasion, and translocation of *Campylobacter jejuni* in pig small-intestinal epithelial cells. Food. Pathog. Dis. *7*, 589–595.

Poly, F., Ewing, C., Goon, S., Hickey, T.E., Rockabrand, D., Majam, G., Lee, L., Phan, J., Savarino, N.J., and Guerry, P. (2007). Heterogeneity of a *Campylobacter jejuni* protein that is secreted through the flagellar filament. Infect. Immun. *75*, 3859–3867.

Prasad, K.N., Dhole, T.N., and Ayyagari, A. (1996). Adherence, invasion and cytotoxin assay of *Campylobacter jejuni* in HeLa and HEp-2 cells. J. Diarrhoeal Dis. Res. *14*, 255–259.

Pryjma, M., Apel, D., Huynh, S., Parker, C.T., and Gaynor, E.C. (2012). FdhTU-modulated formate dehydrogenase expression and electron donor availability enhance recovery of *Campylobacter jejuni* following host cell infection. J. Bacteriol. *194*, 3803–3813.

Quinones, B., Miller, W.G., Bates, A.H., and Mand Rell, R.E. (2009). Autoinducer-2 production in *Campylobacter jejuni* contributes to chicken colonization. Appl. Environ. Microbiol. *75*, 281–285.

Raphael, B.H., and Joens, L.A. (2003). FeoB is not required for ferrous iron uptake in *Campylobacter jejuni*. Can. J. Microbiol. *49*, 727–731.

Russell, R.G., O'Donnoghue, M., Blake, D.C., Jr., Zulty, J., and DeTolla, L.J. (1993). Early colonic damage and invasion of *Campylobacter jejuni* in experimentally challenged infant Macaca mulatta. J. Infect. Dis. *168*, 210–215.

Senior, N.J., Bagnall, M.C., Champion, O.L., Reynolds, S.E., La Ragione, R.M., Woodward, M.J., Salguero, F.J., and Titball, R.W. (2011). Galleria mellonella as an infection model for *Campylobacter jejuni* virulence. J. Med. Microbiol. *60*, 661–669.

Song, Y.C., Jin, S., Louie, H., Ng, D., Lau, R., Zhang, Y., Weerasekera, R., Al Rashid, S., Ward, L.A., Der, S.D., and Chan, V.L. (2004). FlaC, a protein of *Campylobacter jejuni* TGH9011 (ATCC43431) secreted through the flagellar apparatus, binds epithelial cells and influences cell invasion. Mol. Microbiol. *53*, 541–553.

Szymanski, C.M., Burr, D.H., and Guerry, P. (2002). *Campylobacter* protein glycosylation affects host cell interactions. Infect. Immun. *70*, 2242–2244.

Tareen, A.M., Dasti, J.I., Zautner, A.E., Gross, U., and Lugert, R. (2011). Sulfite: cytochrome c oxidoreductase deficiency in *Campylobacter jejuni* reduces motility, host cell adherence and invasion. Microbiology *157*, 1776–1785.

Tu, Q.V., McGuckin, M.A., and Mendz, G.L. (2008). *Campylobacter jejuni* response to human mucin MUC2: modulation of colonization and pathogenicity determinants. J. Med. Microbiol. *57*, 795–802.

Van Deun, K., Haesebrouck, F., Heyndrickx, M., Favoreel, H., Dewulf, J., Ceelen, L., Dumez, L., Messens, W., Leleu, S., Van Immerseel, F., *et al.* (2007). Virulence properties of *Campylobacter jejuni* isolates of poultry and human origin. J. Med. Microbiol. *56*, 1284–1289.

Vijayakumar, S., Merkx-Jacques, A., Ratnayake, D.B., Gryski, I., Obhi, R.K., Houle, S., Dozois, C.M., and Creuzenet, C. (2006). Cj1121c, a novel UDP-4-keto-6-deoxy-GlcNAc C-4 aminotransferase essential for protein glycosylation and virulence in *Campylobacter jejuni*. J. Biol. Chem. *281*, 27733–27743.

Walker, R.I., Schmauder-Chock, E.A., Parker, J.L., and Burr, D. (1988). Selective association and transport of *Campylobacter jejuni* through M cells of rabbit Peyer's patches. Can. J. Microbiol. *34*, 1142–1147.

Wassenaar, T.M., Bleumink-Pluym, N.M.C., and van der Zeijst, B.A. (1991). Inactivation of *Campylobacter jejuni* flagellin genes by homologous recombination demonstrate that flaA but not flaB is required for invasion. EMBO *10*, 2055–2061.

Wooldridge, K.G., Williams, P.H., and Ketley, J.M. (1996). Host signal transduction and endocytosis of *Campylobacter jejuni*. Microb. Pathog. *21*, 299–305.

Yao, R., Burr, D.H., and Guerry, P. (1997). CheY-mediated modulation of *Campylobacter jejuni* virulence. Mol. Microbiol. 23, 1021–1031.

Siderophore-mediated Iron Acquisition for *Campylobacter* Infection

10

Ximin Zeng and Jun Lin

Abstract

Iron acquisition is critical for bacterial pathophysiology. The high-affinity iron acquisition mediated by siderophores is the most efficient and common iron scavenging mechanism in Gram-negative bacteria. While *Campylobacter* does not produce any siderophores, siderophore piracy is evident in various *Campylobacter* species. To date, enterobactin, a triscatecholate siderophore with the highest affinity for ferric iron, is the only known physiologically relevant siderophore utilized by *Campylobacter* for *in vivo* colonization. Recent studies have revealed novel features of ferric enterobactin acquisition systems in *Campylobacter*, provided new insights into the evolution of ferric enterobactin acquisition in different *Campylobacter* species, and suggested that *Campylobacter* is an ideal model organism to examine key issues of enterobactin-mediated iron scavenging. Salmochelin, the glycosylated enterobactin, is likely another significant iron source for *Campylobacter* during infection. This chapter also discusses the ecology of potential siderophores and corresponding producers as an iron source in the intestine and discusses the role of siderophore-mediated iron scavenging in the pathogenesis of *Campylobacter* and other enteric pathogens.

Introduction

Iron plays an essential role in the metabolism of many diverse biological organisms. All Gram-negative bacteria have an absolute requirement for iron to survive. However, to avoid participation of iron in generating toxic oxygen-derived free radicals, through the Haber–Weiss–Fenton chemistry (Koppenol, 2001), higher organisms have evolved mechanisms for lowering the levels of free iron to well below those required for the growth of Gram-negative bacteria. For example, most iron exists intracellularly in eukaryotic cells as ferritin or as haem compounds, while the rest of the small amount of extracellular iron that exists in host is bound by host iron-binding glycoproteins (e.g. lactoferrin and transferrin in mucosal secretions). Iron from these sources is normally not available to invading Gram-negative bacteria. In particular, the iron binding proteins in the intestine sequester free ferric iron at concentrations as low as 10^{-24} M and make it unavailable to most bacteria which demand at least 10^{-7} M for normal growth (Braun *et al.*, 1998).

To obtain sufficient iron for survival and multiplication, Gram-negative bacteria have evolved complex and aggressive genetic systems for iron uptake. Several comprehensive reviews on iron-uptake mechanisms of Gram-negative bacteria have been published (Andrews *et al.*, 2003; Koster, 2005; Miethke and Marahiel, 2007; Wandersman and Delepelaire, 2004; Wooldridge and van Vliet, 2005; Wooldridge and Williams, 1993). Four primary strategies have been identified in Gram-negative bacteria for iron acquisition in the iron-restricted environment: (i) high-affinity iron uptake mediated by iron chelators termed siderophores, such as enterobactin and ferrichrome; (ii) utilization of naturally occurring organic acids as low-affinity iron carriers, such as citrate; (iii) directly use host iron complexes (such as lactoferrin) without utilization of siderophores; and (iv) utilization of ferrous iron via the inner membrane

FeoB-type transport proteins under anaerobic and reducing environment. Among these strategies, the high-affinity iron acquisition mediated by siderophore is the most efficient and common iron scavenging mechanism in Gram-negative bacteria.

Campylobacter species, including *C. jejuni* and *C. coli*, are the most common bacterial causes of human gastroenteritis in the United States and other industrialized countries (Olson, 2008). Human *Campylobacter* infections are epidemiologically linked to ingestion of contaminated poultry meat (Friedman *et al.*, 2000; Tauxe, 2002) and the infective dose can be as low as 5–800 organisms (Black *et al.*, 1988; Mentzing, 1981). *Campylobacter* establishes persistent, benign infections in chickens (up to 10^{10} CFU per gram of faeces) but causes significant inflammation and enteritis in humans (Young *et al.*, 2007). As an enteric Gram-negative pathogen, the thermophilic *Campylobacter* species enters the host intestine via oral ingestion and colonizes in the intestine, primarily the distal ileum and colon in humans and other mammals, and the caecum in poultry. Once inside the intestine, *Campylobacter* encounters multiple levels of stresses, such as nutrient starvation (e.g. iron limitation), the action of innate antimicrobials (e.g. bile salts and antimicrobial peptides), antagonism of gut microbiota, and attack of host immunity. To successfully establish *in vivo* colonization, *Campylobacter* has evolved effective mechanisms to counteract these harsh intestinal conditions. In terms of iron acquisition, several iron-uptake systems have been identified in *Campylobacter*, which include the utilization of the ferric iron bound to siderophores, glycoproteins and haem, and ferrous iron (reviewed in Miller *et al.*, 2009; Stintzi *et al.*, 2008; Wooldridge and van Vliet, 2005). In this chapter, we will focus on siderophore-mediated iron acquisition because of its high prevalence and efficiency in *Campylobacter*. Notably, *Campylobacter* does not produce any siderophores; however, *Campylobacter* could utilize different siderophores produced by other microorganisms (Field *et al.*, 1986). Of these, the enterobactin (Ent)-mediated iron scavenging is tightly linked to the ability of *Campylobacter* colonization in the intestine. More importantly, recent research supports a new model of FeEnt acquisition that is different from the conserved process in many other Gram-negative bacteria. In addition, many recent findings discussed in this chapter are from the studies using both *C. jejuni* and *C. coli*, the two closely related *Campylobacter* species, which provide new insights into the evolution of FeEnt acquisition in *Campylobacter*.

Iron sources in the intestine: an ecological view

Because the intestine is the major colonization site by *Campylobacter*, understanding iron sources in the intestine is important for us to understand the evolution and adaptation. As a human pathogen, *Campylobacter* has natural reservoirs in various animals, such as poultry, swine, and cattle. Potential iron sources in the intestine of human and various animal hosts include mucosal secretions and cells, food components, and iron-binding compounds produced by gut microbiota and host (Stintzi *et al.*, 2008). Most of these iron sources, such as iron-binding glycoproteins from the mucosal secretions (e.g. lactoferrin and transferrin), haem and ferritin (released from intestinal and/or blood cells during inflammation), consistently appear in the intestine of various hosts. However, the high-affinity siderophores produced by gut microorganisms may display high diversity in the intestine of various animal hosts.

The low molecular mass (500–1000 Da) iron chelators with high affinity for ferric iron ($K_{aff} > 10^{30}$) that are synthesized by microorganisms have been collectively termed siderophores (Greek 'iron bearer') (Neilands, 1981). Production and utilization of siderophores has been described in various environmental and pathogenic microorganisms including fungi. Siderophores have highly divergent structures but can be classified into four major groups (catecholate, hydroxamate, phenolate and carboxylate), which is based on the characteristics of the moieties donating the oxygen atoms for coordinating Fe^{3+} ions. There are also 'hybrid' siderophores containing at least two distinct Fe^{3+}-binding moieties. The structure of the representative siderophores for each groups are presented in Fig. 10.1. Despite the extensive information on the biochemistry of different siderophores, there is a significant knowledge gap

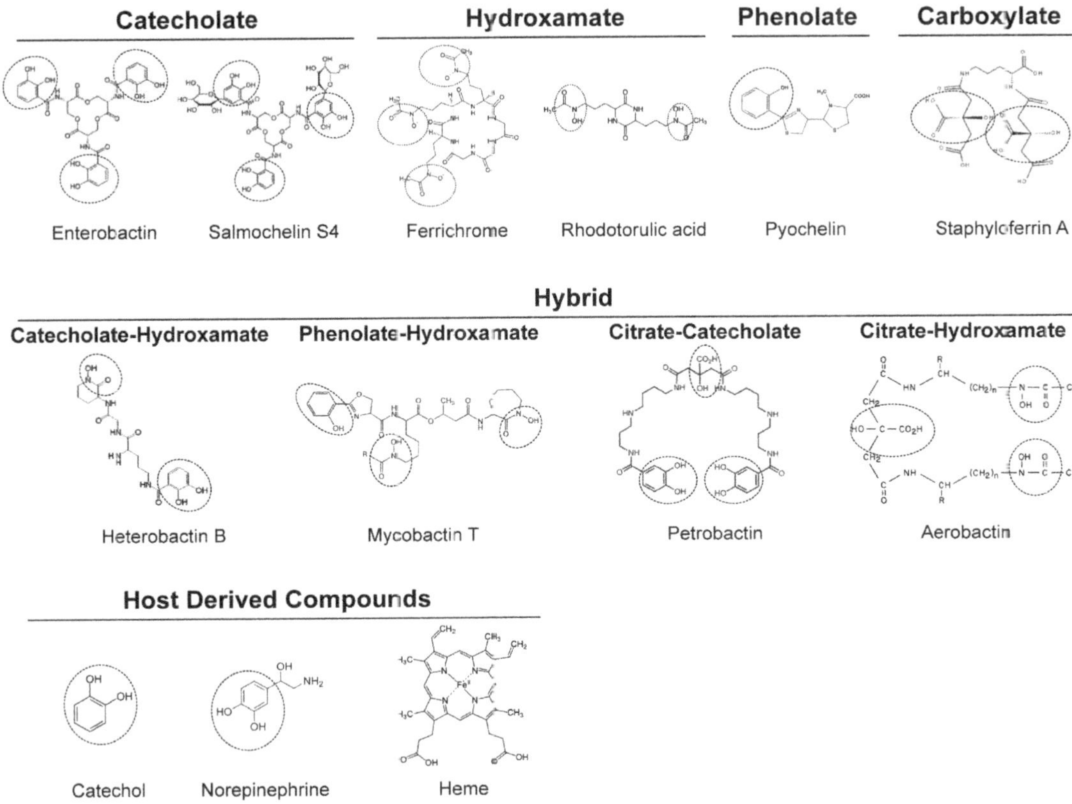

Figure 10.1 Structure of representative siderophores of each category and the iron-binding compounds produced by animal hosts. Catecholate and hydroxamate are the two main groups of siderophore produced by bacteria. The specific moieties involved in iron coordination are highlighted by dotted ovals. The porphyrin in haem is responsible for binding to ferric iron.

on the siderophores and their corresponding producers in the intestine. Little information exists concerning the diversity and relative abundance of siderophores in the intestine. Presence of a specific siderophore in the gut has been primarily based on logical speculation. For example, Ent, a siderophore that has exceptionally high affinity for ferric iron of any natural siderophore compounds tested in neutral and alkaline conditions (Raymond *et al.*, 2003), is produced by a number of bacteria belonging to the family of Enterobacteriaceae. Thus, Ent is likely to be produced in significant amounts by commensal bacteria in the intestine and ferric Ent (FeEnt) may be a significant source of iron to enteric pathogens including *C. jejuni* during colonization in intestine (Stintzi *et al.*, 2008; Wooldridge and van Vliet, 2005).

With the aid of culture-independent molecular approaches and high-throughput pyrosequencing technologies, significant progress has been made in the understanding of microbial ecology of gut microbiota in humans and different animal species (Danzeisen *et al.*, 2011; Jeraldo *et al.*, 2012; Kim *et al.*, 2012; Spor *et al.*, 2011). Establishment and maintenance of intestinal microbiota (> 1000 species) is a complex process that is influenced by environment, and interplay between diet, host and microbes (Backhed *et al.*, 2005; Mackie *et al.*, 1999; McCracken and Lorenz, 2001). Unlike in other microbial ecosystems, such as in soils and oceans, bacterial diversity is low in the intestine. Only 8 of the 55 known bacterial divisions have been identified to date, and of these, two divisions (the Bacteroidetes and the Firmicutes) are dominant in the intestine of humans and animals (Backhed *et al.*, 2005; Zoetendal *et al.*, 2006). In humans, mice and swine, the Firmicutes phylum accounts about 60–80% of 16S rDNA sequences while Bacteriodetes accounts 20–40% of 16S rDNA sequences (Kim *et al.*, 2012; Spor *et al.*, 2011). In the chicken intestine, Firmicutes constitute approximately 90% of bacterial populations with no Bacteriodetes detected using 16S rDNA tag pyrosequencing and total DNA shotgun metagenomic pyrosequencing (Danzeisen *et al.*, 2011). Compared to the relatively low diversity at the division level, gut microbiota is highly diverse at the lower taxonomic ranks (e.g. genus and species) and the composition and diversity of the intestinal microbiota are quite different among different animal species (Backhed *et al.*, 2005; Cho and Blaser, 2012; Danzeisen *et al.*, 2011; Jeraldo *et al.*, 2012; Kim *et al.*, 2012; Spor *et al.*, 2011; Zoetendal *et al.*, 2006). Using high-resolution 16S rDNA tag sequencing for analysing microbiome of three different domesticated vertebrates (chicken, swine and cattle), Jeraldo *et al.* (2012) showed that these microbiomes have been assembled through a process that involves a significant non-neutral (niche) contribution. Such enormous diversity of bacterial populations in the gut clearly would influence the ecology of siderophores in the intestine, which may differ significantly among different animal species. In addition, to date little is known about the abundance and diversity of the fungal microbiota in the intestine, which greatly impedes our understanding on the fungi-produced siderophores in the gut. A recent study (Scupham *et al.*, 2006) showed that diverse fungal taxa are present in the murine intestinal microbial community. Using a unique phospholipid fatty acid biomarker analysis, we recently also observed that abundant viable fungal populations (10–25% of total microbial biomass) were present in the chicken microbiota (Lin *et al.*, 2013). Together, to better understand how intestinal siderophores influence the adaptation and survival of enteric pathogens, examination of the ecology of intestinal siderophores and the corresponding producers are urgently needed.

The representative siderophores that are produced by the microorganisms potentially present in the intestine are summarized in Table 10.1. *Campylobacter* could utilize Ent (a catecholate siderophore) and ferrichrome (a hydroxamate siderophore) produced by other microorganisms (Field *et al.*, 1986; Guerry *et al.*, 1997). Recently, rhodotorulic acid, another hydroxamate siderophore, was observed to be utilized by *C. jejuni* (Stintzi *et al.*, 2008). Since both ferrichrome and rhodotorulic acid are primarily produced by certain soil fungi, Ent is the only known siderophore produced by enteric microorganisms for *Campylobacter* iron acquisition during colonization.

In addition to the siderophores produced by microorganisms, some host-derived iron-binding compounds, such as catechol, norepinephrine (NE), and haem, are also potentially present in the intestine and could be utilized by bacteria including *Campylobacter* (Table 10.1). Utilization of haem by *Campylobacter* is mediated by a prevalent operon of *chuABCDZ* (Miller *et al.*, 2009; Stintzi *et al.*, 2008; Wooldridge and van Vliet, 2005). Obviously, both catechol and NE has the same iron-binding moiety as the catecholate siderophores (Fig. 10.1). Although stress hormone NE is usually at a nanomolar level in sera, the level of NE in the intestine is much higher due to the existence of the enteric nervous system rich in adrenergic neurons (Hughes and Sperandio, 2008). The enteric nervous system, consisting of about 100 million neurons, can secrete NE into the gastrointestinal lumen with concentrations up to the millimolar level or even higher in a pathologic situation (Eldrup and Richter, 2000). Thus, NE is a likely iron source for invading enteric pathogens. Catechol displays high affinity to ferric iron (Bao *et al.*, 2010). The catechols in the intestine are primarily from host metabolism, dietary intake of editable plant (Booth *et al.*, 1960; Shearer and Walton, 2010), and microbiota-mediated conversion of plant polyphenols to simple catechols (Bone *et al.*, 1976; Martin, 1982). It has been shown that dietary catechols can enhance the growth of enteric pathogens under iron-limited conditions (Coulanges *et al.*, 1998; Freestone *et al.*, 2007).

Enterobactin-mediated iron acquisition

Enterobactin (Ent)-mediated iron scavenging is an archetype for iron acquisition and has been well studied in *E. coli* (Raymond *et al.*, 2003). FeEnt acquisition requires coordination of multiple components from outer membrane to cytoplasm,

Table 10.1 Potential intestinal siderophores and other iron-binding compounds produced by animal hosts

Compounds	Producer		Utilization by *Campylobacter*
	Organism	Phylum	
Siderophores			
I. Catecholate			
Enterobactin	Enteric bacteria (O'Brien *et al.*, 1970; Pollack and Neilands, 1970), *Streptomyces* spp. (Fiedler *et al.*, 2001)	Proteobacteria Actinobacteria	Yes (Palyada *et al.*, 2004; Zeng *et al.*, 2009)
Salmochelin S4	*Salmonella* spp., uropathogenic *Escherichia coli* (Hantke *et al.*, 2003)	Proteobacteria	Yes (unpublished data)
Vibriobactin	*Vibrio cholerae* (Griffiths *et al.*, 1984)	Proteobacteria	?
Bacillibactin	*Bacillus subtilis* (May *et al.*, 2001), *Bacillus anthracis* (Koppisch *et al.*, 2005)	Firmicutes	?
Protochelin	*Azobacter vinelandii* (Cornish and Page, 1995)	Proteobacteria	?
Azotochelin	*Azobacter vinelandii* (Corbin and Bulen, 1969)	Proteobacteria	?
Aminochelin	*Azobacter vinelandii* (Page and Tigerstrom, 1988)	Proteobacteria	?
Agrobactin	*Agrobacterium tumefaciens* (Ong *et al.*, 1979)	Proteobacteria	?
Fluvibactin	*Vibrio fluvialis* (Yamamoto *et al.*, 1993)	Proteobacteria	?
II. Hydroxamate			
Ferrichrome	*Ustilago sphaeuogena* (Garibaldi and Neilands, 1955), *Penicillium chrysogenum* (Charlang *et al.*, 1981)	Basidiomycota Ascomycota	Yes (Field *et al.*, 1986; Guerry *et al.*, 1997)
Rhodotorulic acid	*Rhodotorula pilimanae* (Atkin *et al.*, 1970)	Basidiomycota	Yes (Stintzi *et al.*, 2008)
Alcaligin	*Alcaligenes denitrificans* (Nishio *et al.*, 1988), *Bordetella pertussis* and *Bordetella brochiseptica* (Moore *et al.*, 1995)	Proteobacteria	?
Desferrioxamine B	*Streptomyces pilosus* (Bickel *et al.*, 1960)	Actinobacteria	No
III. Phenolate			
Yersiniabactin	*Yersinia pestis* (Perry *et al.*, 1999), *Yersinia enterocolitica* (Chambers *et al.*, 1996), *Pseudomonas syringae*, some *Escherichia coli* (Bultreys *et al.*, 2006)	Proteobacteria Proteobacteria	?
Pyochelin	*Pseudomonas aeruginosa* (Cox *et al.*, 1981)	Proteobacteria	?
IV. Carboxylate			
Staphyloferrin A	*Staphylococcus* spp. (Konetschny-Rapp *et al.*, 1990)	Firmicutes	?
Achromobactin	*Erwinia chrysanthemi* (Munzinger *et al.*, 2000), *Pseudomonas syringae* (Owen and Ackerley, 2011)	Proteobacteria	?
V. Hybrid			
i. Catecholate-Hydroxamate			
Heterobactin B	*Rhodococcus erythropolis* (Carran *et al.*, 2001)	Actinobacteria	?
ii. Phenolate-Hydroxamate			
Mycobactin T	*Mycobacterium tuberculosis* (Snow, 1965)	Actinobacteria	?
iii. Citrate-Catecholate			
Petrobactin	*Bacillus anthracis* (Koppisch *et al.*, 2005), *Bacillus cereus* (Wilson *et al.*, 2006), *Marinobacter hydrocarbonoclasticus* (Barbeau *et al.*, 2002)	Firmicutes Proteobacteria	?

Table 10.1 (Continued)

Compounds	Producer Organism	Phylum	Utilization by *Campylobacter*
iv. Citrate-Hydroxamate			
Aerobactin	*Enterobacter* spp. (Gibson and Magrath, 1969), *Escherichia coli* (Warner *et al*., 1981), *Shigella flexneri* (Payne, 1980)	Proteobacteria	?
Animal host-derived compounds			
Catechol	Plants, mammals (Dirmikis and Darbre, 1974)		?
Noradrenaline	Vertebrate (Von Euler and Hillarp, 1956)		Yes (Zeng *et al*., 2009)
Haem	Vertebrate		Yes (Ridley *et al*., 2006)

such as specific receptors in the outer membrane, TonB-ExbB-ExbD energy transduction systems, ABC transporter system, and cytoplasmic trilactone esterase (Miethke and Marahiel, 2007; Raymond *et al.*, 2003). The molecular, functional and evolutional features of FeEnt acquisition systems in *Campylobacter* were not well understood until recent characterization of the two FeEnt receptors (CfrA and CfrB) and the novel periplasmic enterobactin esterase Cee (Palyada *et al.*, 2004; Xu *et al.*, 2010; Zeng *et al.*, 2009). A review of recent findings on this significant iron scavenging system in *Campylobacter* is given below.

FeEnt receptors CfrA and CfrB

FeEnt (molecular mass of 719 Da) is too large to pass through the porin channels of the bacterial outer membrane. Like all other siderophores, FeEnt requires a specific iron-regulated outer membrane protein (OMP) to enable its passage across the outer membrane and into the periplasm (Miethke and Marahiel, 2007). Thus, FeEnt receptors function as a 'gatekeeper' for FeEnt acquisition and have been extensively studied in different organisms (e.g. FepA in *E. coli*) (Klebba, 2003). Although CfrA was identified as an iron-regulated OMP in *C. coli* as early as 1997 (Guerry *et al.*, 1997), its function as a FeEnt receptor was not established until 2004 by using *C. jejuni* NCTC 11168 strain (Palyada *et al.*, 2004). Strikingly, inactivation of the *cfrA* gene alone not only impaired Ent-mediated iron assimilation in *C. jejuni* but also completely abolished colonization of *C. jejuni* in chickens even if the parent strain colonized all chickens (Palyada *et al.*, 2004). This finding indicates that other iron-uptake system(s) in *C. jejuni* cannot compensate the function of CfrA and CfrA play an essential role in colonization of in chickens. Therefore, the surface-exposed CfrA is likely a promising vaccine candidate against *C. jejuni*. To test this, Zeng *et al.* (2009) further examined molecular, antigenic, and functional characteristics of CfrA in *C. jejuni.*

CfrA has been shown to be induced under iron-restricted conditions and is prevalent in *Campylobacter* strains from various sources and geographic locations (Zeng *et al.*, 2009). The sera from *C. jejuni*-infected chickens can react with CfrA, indicating that CfrA is expressed and immunogenic in chicken hosts. Specifically, purified CfrA antibodies may block the function of CfrA and diminish FeEnt-mediated growth promotion under iron-restricted conditions (Zeng *et al.*, 2009). Recently, studies have shown that a family of stress-related neuroendorine hormones, such as NE, can influence growth and modulate of virulence-associated properties in *C. jejuni* (Cogan *et al.*, 2007; Hughes and Sperandio, 2008). Interestingly, Zeng *et al.* (2009) observed that CfrA is involved in the NE-mediated growth promotion in *C. jejuni*. Given the presence of catechol structure (Fig. 10.1), NE might function as an iron chelator that could be recognized by CfrA for iron assimilation in *C. jejuni*. Together, these findings provide useful molecular, antigenic and functional evidence suggesting that CfrA may be a promising subunit vaccine candidate for controlling *C. jejuni* in humans and animal reservoirs.

In light of the CfrA study by Zeng *et al.* (2009), a new FeEnt receptor (designated CfrB) was identified and characterized using both *in vitro* and *in vivo* systems in *Campylobacter* (Xu *et al.*, 2010). CfrB, a homologue of *C. jejuni* NCTC 11168 pseudogene Cj0444, shares approximately 34% of amino acid identity with CfrA. Sequence alignment showed that CfrB displayed a high level of homology among various *Campylobacter* strains; however, phylogenetic analysis resulted in two clusters for CfrB. The levels of aa identity for CfrB within each cluster ranged from 94% to 99% while the CfrB sequences between cluster 1 (all derived from *C. coli*) and cluster 2 (all derived from *C. jejuni*) share lower homology with aa identity ranging from 84% to 87% (Xu *et al.*, 2010). In terms of prevalence, CfrB is produced in the majority of *C. coli* (41 out of 45) and in a small population of *C. jejuni* (8 out of 32). Notably, all CfrB-producing *C. coli* strains also produce CfrA but CfrB dominates in FeEnt acquisition. However, in *C. jejuni*, CfrB is very rarely co-produced with CfrA although CfrB alone functions as efficiently as CfrA for FeEnt acquisition. Interestingly, examination of currently available *C. jejuni* genomes (http://www.xbase.ac.uk/campydb/) showed that the *cfrB* gene commonly exists as a pseudogene in CfrA-producing *C. jejuni*. This suggests that the existence of the two FeEnt acquisition systems (CfrA- and CfrB-dependent) in a single *C. jejuni* strain may be unnecessary. Although a recent study showed that *C. coli* and *C. jejuni*, the two closely related zoonotic pathogenic species, are exchanging large amounts of DNA (Sheppard *et al.*, 2008), CfrA and CfrB clearly exhibit considerable differences in their distribution and relative contribution to FeEnt acquisition in *C. jejuni* and *C. coli*. Chicken studies indicated CfrB also plays a critical role in *in vivo* colonization and is another attractive subunit vaccine candidate against *Campylobacter* (Xu *et al.*, 2010)

TonB-ExbB-ExbD energy transduction system

Following binding, FeEnt is transported through the receptor channel using the TonB-ExbB-ExbD energy transduction system, in which TonB is an inner membrane-anchored periplasm-spanning protein and ExbB and ExbD proteins are embedded in inner membrane (Miethke and Marahiel, 2007; Raymond *et al.*, 2003). The TonB-ExbB-ExbD system transduces the proton motive force energy to the receptor to allow FeEnt translocation. Typically, TonB interacts with the FeEnt receptor via its C-terminal domain and with both ExbB and ExbD via its N-terminal domain (Miethke and Marahiel, 2007; Raymond *et al.*, 2003). Interestingly, unlike many Gram-negative bacteria that have one type of TonB complex, there are up to three TonB-ExbB-ExbD systems in *Campylobacter*. All the three TonB complexes are subject to iron-regulation in *C. jejuni* NCTC 11168 (Palyada *et al.*, 2004). The *tonB1* (*Cj0181*) and *tonB2* (*Cj1630*) are organized into the same operon with their corresponding *exbB*/*exbD* genes (*Cj0179*/*Cj0180* and *Cj1628*/*Cj1629*, respectively). The *tonB3* (*Cj0753c*) is an 'orphan' gene with no adjacent *exbB*/*exbD* genes. However, *cfrA* (*Cj0755*) is next to the *tonB3* gene in an opposite orientation, suggesting that TonB3 is involved in CfrA-dependent FeEnt acquisition. There is a third pair of *exbB*/*exbD* (*Cj0109*/*Cj0110*) in NCTC 11168 genome, designated *exbB3*/*exbD3* despite their distant location with the *tonB3*. In *C. coli*, *exbB1*/*exbD1*/*tonB1* and *tonB3*/*cfrA* are highly conserved while *tonB2* is missing from *exbB2*/*exbD2* locus in all 12 *C. coli* genomes that have been deposited in Integrated Microbial Genomes (IMG) system. The redundancy of multiple TonB complexes in *Campylobacter* may allow this organism to utilize different iron sources. Recently, we have determined the cognate TonB-ExbB-ExbD systems of CfrA- and CfrB-dependent FeEnt acquisition using genomic analysis, site-directed mutagenesis, complementation, and random transposon mutagenesis (data not shown). Based on the preliminary work, CfrA appears to interact with the TonB3-ExbB2-ExbD2 system in *Campylobacter*. The ExbB2 and ExbD2 also appear to be essential for CfrB-dependent pathway; however, CfrB could interact with either TonB2 or TonB3 for efficient iron uptake. In another words, both TonB2 and TonB3 could actively interact with the ExbB2/ExbD2 complex for CfrB-dependent FeEnt acquisition. Other components, including

TonB1, ExbB1-ExbD1 and ExbB3-ExbD3, are not involved in FeEnt acquisition in *Campylobacter*. However, these components are likely to be involved in other iron acquisition systems and/or other physiological functions. For example, the genes adjacent to *exbB3/exbD3* (*exbB3-exbD3-Cj0111-tolB-pal-Cj0114*) appear to encode orthologues related to the Tol-Pal system in other Gram-negative bacteria, which are involved in maintaining the integrity of the cell envelope (Cascales *et al.*, 2002; Dubuisson *et al.*, 2005; Lazzaroni *et al.*, 2002).

Enterobactin esterase Cee

It has been widely accepted that an efficient cytoplasmic trilactone esterase is required for intracellular hydrolytic iron release from FeEnt, the last and essential step in Ent-mediated iron acquisition (Miethke and Marahiel, 2007; Raymond *et al.*, 2003). Our recent identification and characterization of a periplasmic esterase Cee revealed unique features of FeEnt acquisition in *Campylobacter* and also suggested the presence of reductive iron release mechanisms for FeEnt acquisition (Zeng *et al.*, 2013).

The Cee gene (Cj1376, 269 aa) is homologous to the trilactone esterases identified in other bacteria with a closer phylogenetic relationship to the periplasmic IroE than Fes and IroD, the two efficient Ent esterases localized in cytoplasm (Zeng *et al.*, 2013). IroE is a trilactone esterase with low efficiency for hydrolysis of Ent as well as its analogue salmochelin in some *E. coli* and *S. enterica* strains (Bäumler *et al.*, 1996; Lin *et al.*, 2005; Zhu *et al.*, 2005). Enzymatic analysis by Zeng *et al.* (2013) indicated that the Cee and its homologues Fes and IroD, displayed high efficiency and similar dynamics for hydrolysing both *apo* and ferric forms of Ent while IroE catalysed Ent inefficiently. Genetic manipulation of diverse *Campylobacter* strains demonstrated that Cee is essential for CfrB-dependent FeEnt acquisition; however, Cee also could utilize CfrA-transported FeEnt when the CeuBCDE ABC transporter is not functional. Notably, the Cee, which appears to be the sole trilactone esterase in *C. jejuni*, is localized in the periplasm. Therefore, these findings strongly support a new pathway for CfrB-dependent FeEnt acquisition: after passing through outer membrane, FeEnt is hydrolysed by the periplasmic Cee, which leads to immediate iron release/transfer in the periplasm.

Prevalence of *cee* varies between *C. jejuni* and *C. coli* (Zeng *et al.*, 2013). The *cee* gene is present in all examined *C. coli* strains, which is consistent with the finding that Cee is essential in CfrB-dependent pathway and the CfrB-dependent pathway plays a dominant role in FeEnt acquisition in *C. coli*. In *C. jejuni*, *cee* is present in about half of the strains examined, which is also consistent with the finding that *C. jejuni* usually has only one functional system (CfrA- or CfrB-dependent) for FeEnt acquisition. Since Cee could utilize the CfrA-transported FeEnt, expression of a functional Cee may allow *C. jejuni* to utilize FeEnt during evolution. Zeng *et al.* (2013) also showed that Cee is required for optimal *in vivo* colonization of *C. jejuni* in a chicken model system. However, the definitive role of Cee in *Campylobacter* pathogenesis is still not clear due to the difference in the prevalence of *cee* in *C. jejuni* and *C. coli*.

In light of the discovery of Cee, an intracellular reductive iron release mechanism is likely to be used by *Campylobacter* for CfrA-dependent FeEnt utilization. For most siderophores that have higher redox potential than Ent, reduction of ferric–siderophore complex by reductase is a common route for iron acquisition (Miethke and Marahiel, 2007). However, little compelling evidence exists showing the intracellular reduction of FeEnt followed by release of Fe^{2+}. Recently, Miethke *et al.* (Miethke *et al.*, 2011) demonstrated that the siderophore-interacting protein YqiH functions as a ferric reductase to catalyse the release of iron from Ent in *E. coli*. Zeng *et al.* (2013) demonstrated that inactivation of Cee did not affect the CfrA-mediated FeEnt acquisition in NCTC 11168 although Cee appears to be the sole Ent esterase in *Campylobacter*, which suggests that the CfrA-transported FeEnt is subjected to intracellular reductive iron release. The FeEnt reduction mechanism may be important for *Campylobacter* that only utilize exogenous Ent because ferric siderophore reduction provides the advantage of reutilizing intact siderophores (Miethke *et al.*, 2011). A recent study has supported the close relationship among riboflavin biosynthesis, assimilatory ferric reduction, and iron acquisition in *C. jejuni* (Crossley *et al.*, 2007); but the identity of the putative ferric reductase is still unknown. Thus, identification and characterization of ferric reductases in *Campylobacter* requires more study.

New FeEnt acquisition model in *Campylobacter*

The recent studies described above led to a new FeEnt acquisition model in *Campylobacter* (Fig. 10.2). In this new model, Cee is a periplasmic Ent esterase that is involved in both CfrA- and CfrB-dependent Ent utilization systems. Cee is an essential player in CfrB-dependent pathway but also could hydrolyse CfrA-transported FeEnt when the CeuBCDE system is not functional. In periplasm, the degraded products of FeEnt by Cee may follow two routes. First, immediate release/transfer of Fe^{3+} from the low-affinity DHBS monomers and/or dimers, and second, transport of linear Fe^{3+}-$(DHBS)_3$ into cytoplasm for utilization. In CfrA-dependent pathway, FeEnt is likely to be subjected to reductive iron release by an uncharacterized ferric reductase system in the cytoplasm. Future validation studies for the proposed model will provide new insights into the role of FeEnt acquisition in the evolution and pathogenesis of *Campylobacter* and other bacteria.

Salmochelin-mediated iron acquisition

Salmochelin, the glucosylated analogue of Ent, has been isolated from *Salmonella enterica* and *E. coli* strains (Hantke *et al.*, 2003). Salmochelin

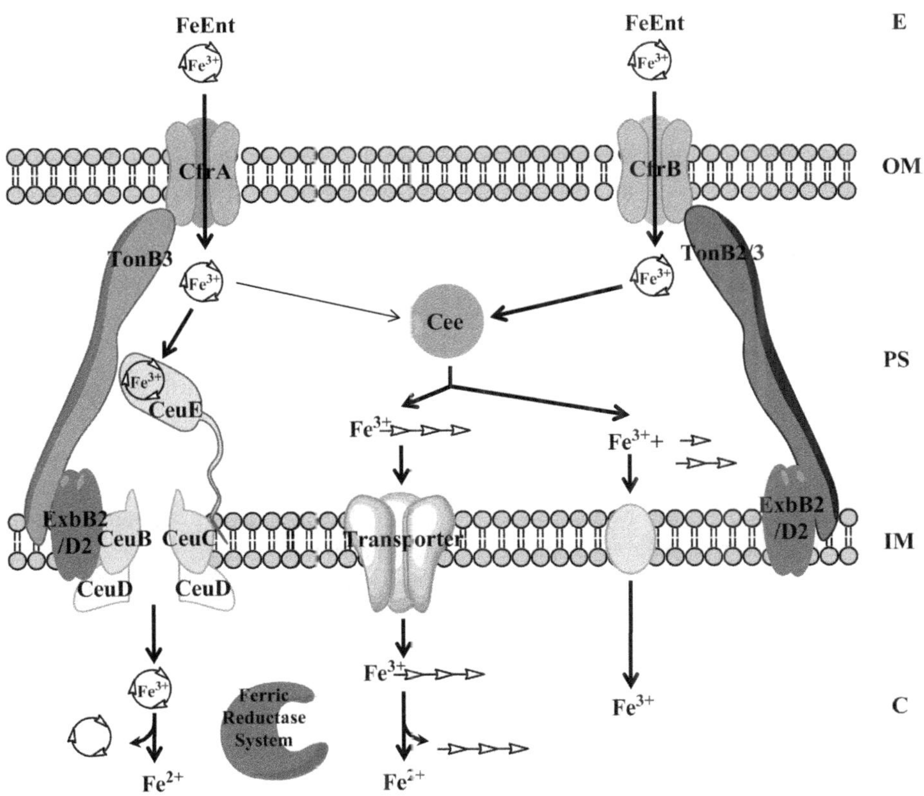

Figure 10.2 The proposed new model of FeEnt acquisition system in *Campylobacter.* , Ent, or cyclic trimer of DHBS; , linear trimer of DHBS; , dimer of DHBS; , DHBS monomer. The thinner arrow line indicates an alternative pathway for utilizing CfrA-transported FeEnt when CeuBCDE system is not functional. E, extracellular environment; OM, outer membrane; PS, periplasmic space; IM, inner membrane; C, cytoplasm (this figure was cited from Zeng *et al.* (2013) with slight modification). A colour version of this figure is available in the plate section at the back of the book.

may be another physiologically relevant, high-affinity siderophore present in the intestine under either symptomatic conditions, resulting from polymicrobial *Salmonella* infection, or from asymptomatic conditions with heavy a commensal *Salmonella* load in the natural animal hosts of *Campylobacter*, such as poultry and swine. Unlike Ent that can be sequestered by host innate immunity protein lipocalins, salmochelin has low affinity for lipocalins and thus the use of salmochelin could confer an *in vivo* growth advantage to bacteria by resisting lipocalins, particularly in the inflamed intestine (Flo *et al.*, 2004; Raffatellu *et al.*, 2009). Both IroD and IroE could hydrolyse salmochelin and play a role in uptake or export of salmochelin products (Lin *et al.*, 2005; Zhu *et al.*, 2005). IroD is likely the only trilactone esterase that hydrolyses ferric forms of salmochelins to release iron in cytoplasm (Lin *et al.*, 2005). Given the functional similarity of Cee and IroD for hydrolysis of trilactone Ent (Zeng *et al.*, 2013), it is possible that *Campylobacter* utilizes salmochelin via Cee. Recently, we have demonstrated that *Campylobacter* could utilize different salmochelins (mono-, di- and triglucosyl Ent) in a Cee-dependent manner (data not shown). Further studies on the salmochelin-mediated iron acquisition in *Campylobacter* will greatly improve our understanding of the role of this iron scavenging mechanism in *Campylobacter* infection.

Hydroxamate siderophore-mediated iron acquisition

Campylobacter can utilize ferrichrome and rhodotorulic acid, the two hydroxamate siderophores primarily produced by certain soli fungi (Field *et al.*, 1986; Guerry *et al.*, 1997; Stintzi *et al.*, 2008). However, the cognate components for the acquisition of these hydroxamate siderophores in *Campylobacter* are still largely unknown (Miller *et al.*, 2009; Stintzi *et al.*, 2008). Galindo *et al.* (2001) identified the *cfhuABD* operon in *C. jejuni* M129 strain that may encode a ferrichrome utilization system but direct evidence for the role of CfhuABD in ferrichrome utilization is still lacking. In addition, the *cfhuABD* operon was not found in the published *C. jejuni* genomes (Miller *et al.*, 2009; Stintzi *et al.*, 2008). It has been proposed that ferrichrome-mediated iron acquisition is likely to be important for *Campylobacter* survival outside of the intestine (Stintzi *et al.*, 2008). However, we cannot rule out the possibility that ferrichrome siderophores may be present in significant amounts in the intestine due to the presence of corresponding fungal producers in gut microbiota. This speculation needs to be examined in future studies.

Conclusion and future directions

High-affinity siderophore-mediated iron acquisition systems are critical for bacterial pathogenesis and thus have been proposed as attractive targets for vaccine and antimicrobial development (Lin *et al.*, 2002; Miethke and Marahiel, 2007). Previous studies on siderophore-mediated iron scavenging in *Campylobacter* were primarily focused on FeEnt acquisition, a highly efficient and conserved iron scavenging system in Gram-negative bacteria. Recent characterization of CfrA, CfrB as well as Cee not only reveals unique features of this system in *Campylobacter* but also suggests that *Campylobacter* may serve as an ideal model organism to examine FeEnt acquisition due to its small genome (1.6–1.8 Mbp), inability to synthesize any siderophores, low redundancy of iron acquisition systems, and enormous strain diversity. Despite greatly improved knowledge on the molecular and immunologic characteristics of the FeEnt system in *Campylobacter*, several issues still need to be addressed to validate the proposed model (Fig. 10.2) and to better illustrate the role of FeEnt acquisition in the evolution and pathogenesis of *Campylobacter*. For example, identification of the ferric reductase required for intracellular iron release from FeEnt in *Campylobacter* would fill a significant knowledge gap concerning mechanisms of iron release from triscatecholate siderophores in bacteria.

Characterization of CfrA and CfrB has suggested that these two surface-exposed OMPs are promising vaccine candidates against *Campylobacter* colonization in human and animal reservoirs. Appropriate CfrA/CfrB-based vaccination strategies, such as intranasal vaccine and live attenuated *Salmonella*-vectored vaccine, should be developed and their efficacy to trigger protective mucosal immunity should be determined. Previous studies have also revealed the complexity of two FeEnt systems (CfrA- and CfrB-dependent) in *C. jejuni* and *C. coli*. With the aid of new generation of high-throughput sequencing technologies, more comprehensive comparative genomics and functional studies are needed to understand the delicate molecular interaction of these two FeEnt systems and to explain why the prevalence of some key components (e.g. CfrA, CfrB, Cee and TonB) differ significantly in *C. jejuni* and *C. coli*.

In addition to Ent, utilization of other intestinal siderophores by *Campylobacter* provide an interesting subject for future study. Recently, we have demonstrated that salmochelin is another physiologically relevant siderophore that could be utilized by *Campylobacter*. Given its unique role in immune evasion, examination of salmochelin-mediated iron acquisition in *Campylobacter* should provide novel information on the delicate interaction between *Campylobacter* and its animal hosts. Finally, diverse and abundant siderophores may exist in the intestine, which could serve as a significant iron source for *Campylobacter* and other enteric pathogens. However, there is still a significant knowledge gap in the identity of siderophores and the corresponding producers present in the intestine. Genomic, metagenomic and metabolomic examination of gut microbiome provide possibilities for understanding the ecology of siderophore production in the intestine.

Acknowledgements

This work is supported by NIH grant AI072551 and AI090095 and by AFRI Food Safety Challenge Grant No. 2012-68003-19679 to J.L.

References

Andrews, S.C., Robinson, A.K., and Rodriguez-Quinones, F. (2003). Bacterial iron homeostasis. FEMS Microbiol. Rev. *27*, 215–237.

Atkin, C.L., Neilands, J.B., and Phaff, H.J. (1970). Rhodotorulic acid from species of *Leucosporidium, Rhodosporidium, Rhodotorula, Sporidiobolus,* and *Sporobolomyces,* and a new alanine-containing ferrichrome from *Cryptococcus melibiosum*. J. Bacteriol. *103*, 722–733.

Backhed, F., Ley, R.E., Sonnenburg, J.L., Peterson, D.A., and Gordon, J.I. (2005). Host-bacterial mutualism in the human intestine. Science *307*, 1915–1920.

Bao, G., Clifton, M., Hoette, T.M., Mori, K., Deng, S.X., Qiu, A., Viltard, M., Williams, D., Paragas, N., Leete, T., *et al.* (2010). Iron traffics in circulation bound to a siderocalin (Ngal)–catechol complex. Nat. Chem. Biol. *6*, 602–609.

Barbeau, K., Zhang, G., Live, D.H., and Butler, A. (2002). Petrobactin, a photoreactive siderophore produced by the oil-degrading marine bacterium *Marinobacter hydrocarbonoclasticus*. J. Am. Chem. Soc. *124*, 378–379.

Bäumler, A.J., Tsolis, R.M., van der Velden, A.W., Stojiljkovic, I., Anic, S., and Heffron, F. (1996). Identification of a new iron regulated locus of *Salmonella typhi*. Gene *183*, 207–213.

Bickel, H., Bosshardt, R., Gaumann, E., Reusser, P., Vischer, E., Voser, W., Wettstein, A., and Zahner, H. (1960). Uber die Isolierung and Charakterisierung der Ferrioxamine A-F, neuer Wuchsstoffe der Sideramingruppe. Helv. Chim. Acta *43*, 2119 – 2128.

Black, R.E., Levine, M.M., Clements, M.L., Hughes, T.P., and Blaser, M.J. (1988). Experimental *Campylobacter jejuni* infection in humans. J. Infect. Dis. *157*, 472–479.

Bone, E., Tamm, A., and Hill, M. (1976). The production of urinary phenols by gut bacteria and their possible role in the causation of large bowel cancer. Am. J. Clin. Nutr. *29*, 1448–1454.

Booth, A.N., Robbins, D.J., Masri, M.S., and De, E.F. (1960). Excretion of catechol after ingestion of quinic and shikimic acids. Nature *187*, 691.

Braun, V., Hantke, K., and Koster, W. (1998). Bacterial iron transport: mechanisms, genetics, and regulation Metal ions biol. Syst. *35*, 67–145.

Bultreys, A., Gheysen, I., and de Hoffmann, E. (2006) Yersiniabactin production by *Pseudomonas syringae* and *Escherichia coli*, and description of a second yersiniabactin locus evolutionary group. Appl. Environ. Microbiol. *72*, 3814–3825.

Carran, C.J., Jordan, M., Drechsel, H., Schmid, D.G., and Winkelmann, G. (2001). Heterobactins: A new class of siderophores from *Rhodococcus erythropolis* IGTS8 containing both hydroxamate and catecholate donor groups. Biometals *14*, 119–125.

Cascales, E., Bernadac, A., Gavioli, M., Lazzaroni, J.C., and Lloubes, R. (2002). Pal lipoprotein of *Escherichia coli* plays a major role in outer membrane integrity. J. Bacteriol. *184*, 754–759.

Chambers, C.E., McIntyre, D.D., Mouck, M., and Sokol, P.A. (1996). Physical and structural characterization of yersiniophore, a siderophore produced by clinical isolates of *Yersinia enterocolitica*. Biometals *9*, 157–167.

Charlang, G., Ng, B., Horowitz, N.H., and Horowitz, R.M. (1981). Cellular and extracellular siderophores of *Aspergillus nidulans* and *Penicillium chrysogenum*. Mol. Cell. Biol. *1*, 94–100.

Cho, I., and Blaser, M.J. (2012). The human microbiome: at the interface of health and disease. Nat. Rev. Genet. *13*, 260–270.

Cogan, T.A., Thomas, A.O., Rees, L.E., Taylor, A.H., Jepson, M.A., Williams, P.H., Ketley, J., and Humphrey, T.J. (2007). Norepinephrine increases the pathogenic potential of *Campylobacter jejuni*. Gut *56*, 1060–1065.

Corbin, J.L., and Bulen, W.A. (1969). Isolation and identification of 2,3-dihydroxybenzoic acid and.

N2,N6-di(2,3-dihydroxybenzoyl)-L-lysine formed by iron-deficient *Azotobacter vinelandii*. Biochemistry *8*, 757–762.

Cornish, A.S., and Page, W.J. (1995). Production of the triacetecholate siderophore protochelin by *Azotobacter vinelandii*. Biometals *8*, 332–338.

Coulanges, V., Andre, P., and Vidon, D.J. (1998). Effect of siderophores, catecholamines, and catechol compounds on *Listeria* spp. Growth in iron-complexed medium. Biochem. Biophys. Res. Commun. *249*, 526–530.

Cox, C.D., Rinehart, K.L., Jr., Moore, M.L., and Cook, J.C., Jr. (1981). Pyochelin: novel structure of an iron-chelating growth promoter for *Pseudomonas aeruginosa*. Proc. Natl. Acad. Sci. U.S.A. *78*, 4256–4260.

Crossley, R.A., Gaskin, D.J., Holmes, K., Mulholland, F., Wells, J.M., Kelly, D.J., van Vliet, A.H., and Walton, N.J. (2007). Riboflavin biosynthesis is associated with assimilatory ferric reduction and iron acquisition by *Campylobacter jejuni*. Appl. Environ. Microbiol. *73*, 7819–7825.

Danzeisen, J.L., Kim, H.B., Isaacson, R.E., Tu, Z.J., and Johnson, T.J. (2011). Modulations of the chicken cecal microbiome and metagenome in response to anticoccidial and growth promoter treatment. PloS One *6*, e27949.

Dirmikis, S.M., and Darbre, A. (1974). Gas–liquid chromatography of simple phenols for urinalysis. J. Chromatogr. *94*, 169–187.

Dubuisson, J.F., Vianney, A., Hugouvieux-Cotte-Pattat, N., and Lazzaroni, J.C. (2005). Tol-Pal proteins are critical cell envelope components of *Erwinia chrysanthemi* affecting cell morphology and virulence. Microbiology *151*, 3337–3347.

Eldrup, E., and Richter, E.A. (2000). DOPA, dopamine, and DOPAC concentrations in the rat gastrointestinal tract decrease during fasting. Am. J. Physiol. Endocrinol. Metab. *279*, E815–822.

Fiedler, H.P., Krastel, P., Muller, J., Gebhardt, K., and Zeeck, A. (2001). Enterobactin: the characteristic catecholate siderophore of *Enterobacteriaceae* is produced by *Streptomyces* species. (1). FEMS Microbiol. Lett. *196*, 147–151.

Field, L.H., Headley, V.L., Payne, S.M., and Berry, L.J. (1986). Influence of iron on growth, morphology, outer membrane protein composition, and synthesis of siderophores in *Campylobacter jejuni*. Infect. Immun. *54*, 126–132.

Flo, T.H., Smith, K.D., Sato, S., Rodriguez, D.J., Holmes, M.A., Strong, R.K., Akira, S., and Aderem, A. (2004). Lipocalin 2 mediates an innate immune response to bacterial infection by sequestrating iron. Nature *432*, 917–921.

Freestone, P.P., Walton, N.J., Haigh, R.D., and Lyte, M. (2007). Influence of dietary catechols on the growth of enteropathogenic bacteria. Int. J. Food Microbiol. *119*, 159–169.

Friedman, C.R., Neimann, J., Wegener, H.C., and Tauxe, R.V. (2000). Epidemiology of *Campylobacter jejuni* infections in the United States and other industrialized nations. In *Campylobacter*, Nachamkin, I., and Blaser, M.J., eds. (ASM Press, Washington, DC), pp. 121–138.

Galindo, M.A., Day, W.A., Raphael, B.H., and Joens, L.A. (2001). Cloning and characterization of a *Campylobacter jejuni* iron-uptake operon. Curr. Microbiol. *42*, 139–143.

Garibaldi, J.A., and Neilands, J.B. (1955). Isolation and properties of ferrichrome A. J. Am. Chem. Soc. *77*, 2429–2430.

Gibson, F., and Magrath, D.I. (1969). The isolation and characterization of a hydroxamic acid (aerobactin) formed by *Aerobacter aerogenes* 62-I. Biochim. Biophys. Acta *192*, 175–184.

Griffiths, G.L., Sigel, S.P., Payne, S.M., and Neilands, J.B. (1984). Vibriobactin, a siderophore from *Vibrio cholerae*. J. Biol. Chem. *259*, 383–385.

Guerry, P., Perez-Casal, J., Yao, R., McVeigh, A., and Trust, T.J. (1997). A genetic locus involved in iron utilization unique to some *Campylobacter* strains. J. Bacteriol. *179*, 3997–4002.

Hantke, K., Nicholson, G., Rabsch, W., and Winkelmann, G. (2003). Salmochelins, siderophores of *Salmonella enterica* and uropathogenic *Escherichia coli* strains, are recognized by the outer membrane receptor IroN. Proc. Natl. Acad. Sci. U.S.A. *100*, 3677–3682.

Hughes, D.T., and Sperandio, V. (2008). Inter-kingdom signalling: communication between bacteria and their hosts. Nat. Rev. Microbiol. *6*, 111–120.

Jeraldo, P., Sipos, M., Chia, N., Brulc, J.M., Dhillon, A.S., Konkel, M.E., Larson, C.L., Nelson, K.E., Qu, A., Schook, L.B., *et al.* (2012). Quantification of the relative roles of niche and neutral processes in structuring gastrointestinal microbiomes. Proc. Natl. Acad. Sci. U.S.A. *109*, 9692–9698.

Kim, H.B., Borewicz, K., White, B.A., Singer, R.S., Sreevatsan, S., Tu, Z.J., and Isaacson, R.E. (2012). Microbial shifts in the swine distal gut in response to the treatment with antimicrobial growth promoter, tylosin. Proc. Natl. Acad. Sci. U.S.A. *109*, 15485–15490.

Klebba, P.E. (2003). Three paradoxes of ferric enterobactin uptake. Front. Biosci. *8*, s1422–1436.

Konetschny-Rapp, S., Jung, G., Meiwes, J., and Zahner, H. (1990). Staphyloferrin A: a structurally new siderophore from *staphylococci*. FEBS *191*, 65–74.

Koppenol, W.H. (2001). The Haber–Weiss cycle – 70 years later. Redox Report: comm. Free radical res. *6*, 229–234.

Koppisch, A.T., Browder, C.C., Moe, A.L., Shelley, J.T., Kinkel, B.A., Hersman, L.E., Iyer, S., and Ruggiero, C.E. (2005). Petrobactin is the primary siderophore synthesized by *Bacillus anthracis* str. *Sterne* under conditions of iron starvation. Biometals *18*, 577–585.

Koster, W. (2005). Cytoplasmic membrane iron permease systems in the bacterial cell envelope. Front. Biosci. *10*, 462–477.

Lazzaroni, J.C., Dubuisson, J.F., and Vianney, A. (2002). The Tol proteins of *Escherichia coli* and their involvement in the translocation of group A colicins. Biochimie *84*, 391–397.

Lin, H., Fischbach, M.A., Liu, D.R., and Walsh, C.T. (2005). *In vitro* characterization of salmochelin and enterobactin trilactone hydrolases IroD, IroE, and Fes. J. Am. Chem. Soc. *127*, 11075–11084.

Lin, J., Huang, S., and Zhang, Q. (2002). Outer membrane proteins: key players for bacterial adaptation in host niches. Microbes Infect. *4*, 325–331.

Lin, J., Hunkapiller, A.A., Layton, A.C., Chang, Y.J., and Robbins, K.R. (2013). Response of intestinal microbiota to antibiotic growth promoters in chickens. Food. Pathog. Dis. *10*, 331–337.

McCracken, V.J., and Lorenz, R.G. (2001). The gastrointestinal ecosystem: a precarious alliance among epithelium, immunity and microbiota. Cell. Microbiol. *3*, 1–11.

Mackie, R.I., Sghir, A., and Gaskins, H.R. (1999). Developmental microbial ecology of the neonatal gastrointestinal tract. Am. J. Clin. Nutr. *69*, 1035S–1045S.

Martin, A.K. (1982). The origin of urinary aromatic compounds excreted by ruminants. 3. The metabolism of phenolic compounds to simple phenols. Br. J. Nutr. *48*, 497–507.

May, J.J., Wendrich, T.M., and Marahiel, M.A. (2001). The dhb operon of *Bacillus subtilis* encodes the biosynthetic template for the catecholic siderophore 2,3-dihydroxybenzoate-glycine-threonine trimeric ester bacillibactin. J. Biol. Chem. *276*, 7209–7217.

Mentzing, L.O. (1981). Waterborne outbreaks of *Campylobacter* enteritis in central Sweden. Lancet *2*, 352–354.

Miethke, M., and Marahiel, M.A. (2007). Siderophore-based iron acquisition and pathogen control. Microbiol. Mol. Biol. Rev. *71*, 413–451.

Miethke, M., Hou, J., and Marahiel, M.A. (2011). The Siderophore-Interacting Protein YqjH Acts as a Ferric Reductase in Different Iron Assimilation Pathways of *Escherichia coli*. Biochemistry *50*, 10951–10964.

Miller, C.E., Williams, P.H., and Ketley, J.M. (2009). Pumping iron: mechanisms for iron uptake by *Campylobacter*. Microbiology *155*, 3157–3165.

Moore, C.H., Foster, L.A., Gerbig, D.G., Jr., Dyer, D.W., and Gibson, B.W. (1995). Identification of alcaligin as the siderophore produced by *Bordetella pertussis* and *B. bronchiseptica*. J. Bacteriol. *177*, 1116–1118.

Munzinger, M., Budzikiewicz, H., Expert, D., Enard, C., and Meyer, J.M. (2000). *Achromobactin*, a new citrate siderophore of *Erwinia chrysanthemi*. Zeitschrift fur Naturforschung C, J. Biosci. *55*, 328–332.

Neilands, J.B. (1981). Microbial iron compounds. Annu. Rev. Biochem. *50*, 715–731.

Nishio, T., Tanaka, N., Hiratake, J., Katsube, Y., Ishida, Y., and Oda, J. (1988). Isolation and structure of the novel dihydroxamate siderophore alcaligin. J. Am. Chem. Soc. *110*, 8733–8734.

O'Brien, I.G., Cox, G.B., and Gibson, F. (1970). Biologically active compounds containing 2,3-dihydroxybenzoic acid and serine formed by *Escherichia coli*. Biochim. Biophys. Acta *201*, 453–460.

Olson, C.K., Ethelberg, S., Pelt, W.V., and Tauxe, R.V. (2008). Epidemiology of *Campylobacter jejuni* infections in industrialized nations. In *Campylobacter*, Nachamkin, I., Szymanski, C.M., and Blaser, M.J., eds. (ASM Press, Washington, DC), pp. 163–189.

Ong, S.A., Peterson, T., and Neilands, J.B. (1979). Agrobactin, a siderophore from *Agrobacterium tumefaciens*. J. Biol. Chem. *254*, 1860–1865.

Owen, J.G., and Ackerley, D.F. (2011). Characterization of pyoverdine and achromobactin in *Pseudomonas syringae* pv. *phaseolicola* 1448a. BMC Microbiol. *11*, 218.

Page, W.J., and Tigerstrom, M.V. (1988). Aminochelin, a Catecholamine Siderophore Produced by *Azotobacter vinelandii*. J. Gen. Microbiol. *134*, 453–460.

Palyada, K., Threadgill, D., and Stintzi, A. (2004). Iron acquisition and regulation in *Campylobacter jejuni*. J. Bacteriol. *186*, 4714–4729.

Payne, S.M. (1980). Synthesis and utilization of siderophores by *Shigella flexneri*. J. Bacteriol. *143*, 1420–1424.

Perry, R.D., Balbo, P.B., Jones, H.A., Fetherston, J.D., and DeMoll, E. (1999). Yersiniabactin from *Yersinia pestis*: biochemical characterization of the siderophore and its role in iron transport and regulation. Microbiology *145(Pt 5)*, 1181–1190.

Pollack, J.R., and Neilands, J.B. (1970). Enterobactin, an iron transport compound from *Salmonella typhimurium*. Biochem. Biophys. Res. Commun. *38*, 989–992.

Raffatellu, M., George, M.D., Akiyama, Y., Hornsby, M.J., Nuccio, S.P., Paixao, T.A., Butler, B.P., Chu, H., Santos, R.L., Berger, T., *et al.* (2009). Lipocalin-2 resistance confers an advantage to *Salmonella enterica* serotype Typhimurium for growth and survival in the inflamed intestine. Cell. Host Microbe *5*, 476–486.

Raymond, K.N., Dertz, E.A., and Kim, S.S. (2003). Enterobactin: an archetype for microbial iron transport. Proc. Natl. Acad. Sci. U.S.A. *100*, 3584–3588.

Ridley, K.A., Rock, J.D., Li, Y., and Ketley, J.M. (2006). Heme utilization in *Campylobacter jejuni*. J. Bacteriol. *188*, 7862–7875.

Scupham, A.J., Presley, L.L., Wei, B., Bent, E., Griffith, N., McPherson, M., Zhu, F., Oluwadara, O., Rao, N., Braun, J., *et al.* (2006). Abundant and diverse fungal microbiota in the murine intestine. Appl. Environ. Microbiol. *72*, 793–801.

Shearer, N., and Walton, N.J. (2010). Dietary Catechols and their Relationship to Microbial Endocrinology. In Microbial Endocrinology, Lyte, M., and Freestone, P.P.E., eds. (Springer, New York), pp. 69–87.

Sheppard, S.K., McCarthy, N.D., Falush, D., and Maiden, M.C. (2008). Convergence of *Campylobacter* species: implications for bacterial evolution. Science *320*, 237–239.

Snow, G.A. (1965). Isolation and structure of mycobactin T, a growth factor from *Mycobacterium tuberculosis*. Biochem. J. *97*, 166–175.

Spor, A., Koren, O., and Ley, R. (2011). Unravelling the effects of the environment and host genotype on the gut microbiome. Nat. Rev. Microbiol. *9*, 279–290.

Stintzi, A., Vliet, A.H.M.V., and Ketley, J.M. (2008). Iron Metabolism, Transport, and Regulation. In *Campylobacter*, Nachamkin, I., Szymanski, C.M., and Blaser, M.J., eds. (ASM Press, Washington, DC), pp. 591–610.

Tauxe, R.V. (2002). Emerging foodborne pathogens. Int. j. food microbiol. *78*, 31–41.

Von Euler, U.S., and Hillarp, N.A. (1956). Evidence for the presence of noradrenaline in submicroscopic structures of adrenergic axons. Nature *177*, 44–45.

Wandersman, C., and Delepelaire, P. (2004). Bacterial iron sources: from siderophores to hemophores. Annu. Rev. Microbiol. *58*, 611–647.

Warner, P.J., Williams, P.H., Bindereif, A., and Neilands, J.B. (1981). ColV plasmid-specific aerobactin synthesis by invasive strains of *Escherichia coli*. Infect. Immun. *33*, 540–545.

Wilson, M.K., Abergel, R.J., Raymond, K.N., Arceneaux, J.E., and Byers, B.R. (2006). Siderophores of *Bacillus anthracis*, *Bacillus cereus*, and *Bacillus thuringiensis*. Biochem. Biophys. Res. Commun. *348*, 320–325.

Wooldridge, K.G., and van Vliet, A.H.M. (2005). Iron transport and regulation. In *Campylobacter*: molecular and cellular biology, Ketley, J.M., and Konkel, M.E., eds. (Norfolk, UK, Horizon Bioscience), pp. 293–310.

Wooldridge, K.G., and Williams, P.H. (1993). Iron uptake mechanisms of pathogenic bacteria. FEMS Microbiol. Rev. *12*, 325–348.

Xu, F., Zeng, X., Haigh, R.D., Ketley, J.M., and Lin, J. (2010). Identification and characterization of a new ferric enterobactin receptor, CfrB, in *Campylobacter*. J. Bacteriol. *192*, 4425–4435.

Yamamoto, S., Okujo, N., Fujita, Y., Saito, M., Yoshida, T., and Shinoda, S. (1993). Structures of two polyamine-containing catecholate siderophores from *Vibrio fluvialis*. J. Biochem. *113*, 538–544.

Young, K.T., Davis, L.M., and Dirita, V.J. (2007). *Campylobacter jejuni*: molecular biology and pathogenesis. Nat. Rev. Microbiol. *5*, 665–679.

Zeng, X., Xu, F., and Lin, J. (2009). Molecular, antigenic, and functional characteristics of ferric enterobactin receptor CfrA in *Campylobacter jejuni*. Infect. Immun. *77*, 5437–5448.

Zeng, X., Mo, Y., Xu, F., and Lin, J. (2013). Identification and characterization of a periplasmic trilactone esterase, Cee, revealed unique features of ferric enterobactin acquisition in *Campylobacter*. Mol. Microbiol. *87*, 594–608.

Zhu, M., Valdebenito, M., Winkelmann, G., and Hantke, K. (2005). Functions of the siderophore esterases IroD and IroE in iron-salmochelin utilization. Microbiology *151*, 2363–2372.

Zoetendal, E.G., Vaughan, E.E., and de Vos, W.M. (2006). A microbial world within us. Mol. Microbiol. *59*, 1639–1650.

Genetic Mechanisms Involved in *Campylobacter jejuni* Survival Under Oxidative Stress Conditions

11

Ozan Gundogdu, Brendan W. Wren and Nick Dorrell

Abstract

Campylobacter jejuni is the major cause of human bacterial gastroenteritis. As a microaerophilic bacterium, *C. jejuni* will be exposed to reactive oxygen species (ROS) during the course of normal bacterial metabolism as well as during colonization or infection, from the host immune system. *C. jejuni* contains a number of different mechanisms for countering the effects of oxidative stress and the control of the *C. jejuni* oxidative stress response is complex involving multiple inter-linked levels of regulation, with two new regulators of the oxidative stress response recently identified. In this chapter, we cover both the mechanisms of *C. jejuni* oxidative stress defence and the current understanding of the increasingly complex regulation of this oxidative stress response.

The *C. jejuni* survival conundrum

Human infections with *Campylobacter* species are estimated to cause 5–14% of diarrhoeal cases worldwide (Coker *et al.*, 2002). Each year *Campylobacter* infections are estimated to be responsible for 400 million human cases of gastroenteritis worldwide, making this the leading cause of bacterial food borne disease and a major causative agent of traveller's diarrhoea (Friedman *et al.*, 2000; Walker, 2005). The human intestinal pathogen *C. jejuni* is a microaerophilic bacterium which grows optimally in an atmosphere containing 5–10% O_2 and 5–12% CO_2 and at a temperature between 37°C and 42°C (Fitzgerald *et al.*, 2008; Liu *et al.*, 2013; Park, 2002). Historically, several factors have hindered the study of *Campylobacter* species compared to other enteric pathogens such as *Escherichia coli* and *Salmonella* species. Important advances were the identification of the microaerobic growth requirements of *Campylobacter* species and the development of appropriate isolation and selection methods for bacterial culture (Debruyne *et al.*, 2008; Skirrow, 1977). In the last two decades, microaerobic cabinets have greatly assisted in the routine growth and study of *Campylobacter* species. Additionally, the lack of a convenient animal model has also hindered the understanding of the pathogenesis of *Campylobacter* species (Dorrell and Wren, 2007; Mansfield *et al.*, 2008). This paucity of knowledge about this important human pathogen was one of the driving forces for the original *C. jejuni* NCTC11168 genome sequencing project (Parkhill *et al.*, 2000). Important genetic loci identified included those encoding proteins involved in the biosynthesis of the capsular polysaccharide (referred to as surface polysaccharide at the time), flagella modification and lipooligosaccharide. Since then, multiple *C. jejuni* genomes have been sequenced which have allowed further insights into this microorganism (Hofreuter *et al.*, 2006; Miller *et al.*, 2000; Poly *et al.*, 2007).

Given the physiological characteristics of *C. jejuni*, one of the major questions regarding this pathogen is how a microaerophilic bacterium is able to survive under unfavourable environmental conditions? The ability of *C. jejuni* to respond to oxidative stresses is believed to be paramount during *in vivo* survival (Atack and Kelly, 2008; Fang, 2004; Palyada *et al.*, 2009; Zaki *et al.*, 2005). The availability of *C. jejuni* genome sequences has allowed a number of genes to be identified which encode proteins that have a direct role in countering oxidative stresses. In addition, multiple oxidative stress regulators that control the expression of the oxidative stress responses have been identified. In 2009, Atack and Kelly wrote a comprehensive review covering the *C. jejuni* oxidative stress response that highlighted the fact that the regulation of the oxidative stress response was still poorly understood (Atack and Kelly, 2009). Since then two new regulators of the *C. jejuni* oxidative stress response have been characterized; the OmpR-like regulator CosR and the MarR-like regulator Cj1556 (Gundogdu *et al.*, 2011; Hwang *et al.*, 2011). A network consisting of multiple inter-linked levels of regulation and control is emerging.

Oxidative stress overview

Reactive oxygen species (ROS) is a collective term that describes the chemical species generated upon incomplete reduction of oxygen (Imlay, 2003, 2008). ROS include the superoxide anion (O_2^-), hydrogen peroxide (H_2O_2) and the hydroxyl radical (HO·) (D'Autreaux and Toledano, 2007). The accumulation of these toxic compounds in the bacterial cytoplasm and periplasm results in damage to proteins, nucleic acids and membranes. As *C. jejuni* is microaerophilic, the bacteria require a small amount of free oxygen for growth (Sellars *et al.*, 2002). As such *C. jejuni* will be exposed to ROS during the course of normal metabolism (Imlay, 2003). Therefore the ability to neutralize these toxic compounds is vital for bacterial survival both in the environment and also within a host where ROS are produced by the host immune system (Atack and Kelly, 2009; Imlay, 2008). Biological systems have a significant role to play in such reactions. The Fenton reaction, where ferric iron (Fe^{3+}) is reduced by O_2^- to produce ferrous iron (Fe^{2+}) that can then react with H_2O_2 to yield HO· is one example. Thus, the level of iron concentration within bacterial cells is important in controlling oxidative stress (Atack and Kelly, 2009; Imlay, 2006; van Vliet *et al.*, 2002). Additionally, O_2^- is a major cause of oxidation leading to damage of iron–sulfur clusters in proteins (Imlay, 2006). This damage causes the release of free iron into the cell that can inhibit enzymes and lead to the formation of HO• via the Fenton reaction (Atack and Kelly, 2009; Imlay, 2006).

The effects of ROS have been studied in a number of different microorganisms and have been shown to display varying targets. Experiments in *E. coli* have shown that high O_2^- concentrations can lead to decreased dehydratase activity, impaired growth and significant DNA damage (Gort and Imlay, 1998). H_2O_2 readily breaks down into two HO• molecules or alternatively into hydrogen and a hydroperoxyl radical (HO_2). H_2O_2 is produced both endogenously within cells that use O_2 as a terminal electron acceptor and exogenously by, for example, certain types of host cells when pathogens induce an immune response (Atack and Kelly, 2009). H_2O_2 can damage proteins through the oxidation of cofactors, such as iron–sulfur clusters (Imlay, 2006, 2008), damage membranes or DNA, and can also react with O_2^- to form HO• using iron via the Fenton reaction (Atack and Kelly, 2009; Imlay, 2008). Studies in *E. coli* have shown that mutants unable to neutralize H_2O_2 suffer significant DNA and metabolic pathway damage (Atack and Kelly, 2009; Jang and Imlay, 2007; Park *et al.*, 2005).

C. jejuni mechanisms for countering oxidative stress

In order to protect against the damaging effects of ROS, bacteria produce a variety of oxidative stress defence proteins. Analysis of the *C. jejuni* genomes published to date indicates that genes that encode proteins directly involved in countering oxidative stress are generally conserved amongst *C. jejuni* strains. Here we give a summary of those identified so far:

Catalase (KatA)

Catalase converts H_2O_2 to H_2O and O_2. Catalases can be classified into three types: monofunctional, bi-functional catalase-peroxidases and manganese-containing (non-haem) catalases (Atack and Kelly, 2009; Chelikani *et al.*, 2004). *C. jejuni* has been shown to contain a single catalase that is induced by H_2O_2 and O_2^- (Garenaux *et al.*, 2009; Grant and Park, 1995). The *C. jejuni* catalase is encoded by *katA* (*Cj1385* in NCTC11168). A *C. coli katA* mutant strain (CK100 Kat⁻) displayed increased sensitivity to 1 mM H_2O_2 *in vitro* compared to the wild-type strain (UA585 Kat[+]) (Grant and Park, 1995). A separate study investigated the effect of paraquat on *katA* expression by performing RT-PCR (Garenaux *et al.*, 2008). Transcriptomic data showed *katA* expression was 13-fold higher in paraquat-stressed cells and thus indicated that catalase was induced by this type of oxidative stress (Garenaux *et al.*, 2008). A 15 min exposure to paraquat resulted in significant increase in catalase expression whilst a 30 min exposure led to an expression level that was 30 times above the basal level. After 45 and 60 min, the expression level decreased and stabilized at about 10 times above the basal level. However expression of SodB (see below) remained unaffected by exposure to paraquat and this was confirmed by proteomic analysis (Garenaux *et al.*, 2008).

Recently the gene (*Cj1386*) directly downstream of *katA* was shown to encode an ankyrin-containing protein that contributes to bacterial catalase activity (Flint *et al.*, 2012). *Cj1386* was shown to be up-regulated in response to oxidative stress, as was *katA* (Palyada *et al.*, 2009). A NCTC11168 *Cj1386* mutant was shown to exhibit a similar level of increased sensitivity to H_2O_2 as both a *katA* mutant and a *katA Cj1386* double mutant (Flint *et al.*, 2012). The *Cj1386* mutant has identical KatA levels compared to the wild-type strain, but reduced catalase activity (Flint *et al.*, 2012). This reduction in catalase activity was demonstrated to be due to a significant decrease in haem concentration associated with KatA, suggesting a role for Cj1386 in haem trafficking to KatA and promoting optimal levels of catalase activity (Flint *et al.*, 2012).

Although *C. jejuni* does not appear to survive in macrophages (Wassenaar *et al.*, 1997; Watson and Galán, 2008), in comparison to the M129N wild-type strain, a *katA* mutant displayed a further reduction in survival within two different macrophage cell lines (murine and porcine peritoneal) (Day *et al.*, 2000). However, experiments using epithelial HEp-2 cells showed no difference in survival (Day *et al.*, 2000). Inhibition of macrophage nitric oxide synthase or NADPH oxidase allowed the recovery of a *C. jejuni katA* mutant to survival levels similar to that of the wild-type strain (Day *et al.*, 2000).

Superoxide dismutase (SodB)

Superoxide dismutase (SOD) functions by converting O_2^- to H_2O_2 and O_2 (Hassan, 1988). Bacteria typically contain two distinct types of SOD; a manganese-cofactored SOD (MnSOD/SodA) or an iron-cofactored SOD (FeSOD/SodB), however *C. jejuni* possesses a single SOD (SodB) with an iron cofactor (Pesci *et al.*, 1994). This was the first oxidative stress response mechanism identified in *C. jejuni* (Pesci *et al.*, 1994). Unlike *E. coli*, which expresses three distinct types of SOD, there is no biochemical evidence to support the existence of other classes of SOD in *C. jejuni* apart from SodB (Purdy *et al.*, 1999). All *C. jejuni* strains contain a *sodB* gene (encoded by *Cj0169* in NCTC11168) and SodB activity was originally thought to be the main mechanism of resistance against O_2^-. Early studies showed a 81-176 *sodB* mutant to have significantly reduced ability (12-fold) to invade INT 407 cells compared to the wild-type strain, indicating a potential role for SodB in *C. jejuni* intracellular survival (Pesci *et al.*, 1994). Also *C. coli sodB* mutants have shown a decreased ability to colonize 1-day-old chicks, further linking SOD functionality to bacterial survival *in vivo* (Purdy *et al.*, 1999). More recent studies have demonstrated that *C. jejuni* NCTC11168 expresses KatA, not SodB, when exposed to O_2^- (Garenaux *et al.*, 2009). However a basal level of SOD production has been shown to be important as *sodB* mutants are overly sensitive to oxidative stress (Pesci *et al.*, 1994).

Peroxiredoxins

Peroxiredoxins are abundant cellular antioxidant proteins that help to control intracellular peroxide levels (Hall *et al.*, 2009). These non-haem proteins have been shown to confer resistance to H_2O_2 and organic hydroperoxides through breakdown to H_2O and the corresponding alcohol (Wood *et al.*, 2003). Peroxiredoxins use active site cysteine residues in the catalytic cycle rather than metal cofactors like SOD and catalase (Wang *et al.*, 2006). On the basis of the presence or absence of the resolving cysteine residue, peroxiredoxins are classified into 2-Cys, atypical 2-Cys, and 1-Cys peroxiredoxin subfamilies (Chae *et al.*, 1994; Rhee *et al.*, 2005; Rhee and Woo, 2011). All peroxiredoxins share the same basic catalytic mechanism, in which an active site cysteine (the peroxidatic cysteine) is oxidized to a sulfenic acid by the peroxide substrate. The recycling of the sulfenic acid back to a thiol also distinguishes the peroxiredoxin subfamily (Wood *et al.*, 2003).

Many bacteria possess an alkyl hydroperoxidase reductase (Ahp) consisting of the AhpC catalytic subunit and the AhpF flavoprotein, which reduces the oxidized AhpC to reactivate the Ahp enzyme (Poole *et al.*, 2000). *C. jejuni* strains contain an *ahpC* gene (*Cj0334* in NCTC11168) that encodes the catalytic domain, however lack an *ahpF* gene encoding the flavoprotein (Parkhill *et al.*, 2000). Therefore, it is likely that *C. jejuni* utilizes an alternative system for the reduction of oxidized AhpC, possibly via the thioredoxin reductase TrxB (Holmes *et al.*, 2005; Palyada *et al.*, 2004). *C. jejuni* 81116 *ahpC* mutants were shown to be hypersensitive to killing by cumene hydroperoxide and aerobic stress (Baillon *et al.*, 1999). However, *C. jejuni* AhpC has never been over-expressed and studied *in vitro*, so the exact substrate range of this alkyl hydroperoxidase reductase remains unclear (Atack and Kelly, 2009). Interestingly, although catalase is the most widely studied H_2O_2 detoxification mechanism in the cytoplasm (Chelikani *et al.*, 2004), it may not represent the main way in which H_2O_2 is degraded by bacteria at low concentrations (Seaver and Imlay, 2001). It is possible that this could be the function of AhpC (Atack and Kelly, 2009; Seaver and Imlay, 2004).

Along with AhpC, the thiol peroxidase Tpx (encoded by *Cj0779* in NCTC11168) is a member of another group of peroxiredoxins that catalyse the breakdown of peroxides, specifically H_2O_2. Studies have shown that *C. jejuni* Tpx is located in the bacterial cytoplasm and not within the periplasm (as is the case in *E. coli*) and also uses the cytoplasmically located thioredoxin system as an electron donor (Atack *et al.*, 2008). It has been hypothesized that in *C. jejuni*, Tpx could act as the major H_2O_2 scavenger, similar to AhpC in *E. coli* (Atack *et al.*, 2008; Day *et al.*, 2000). Alternatively in *C. jejuni*, AhpC and Tpx may act in concert with other H_2O_2 detoxification systems, such as catalase, in order to provide maximum resistance to H_2O_2 (Atack and Kelly, 2009). *C. jejuni* NCTC11168 *tpx* mutants were no more sensitive than the wild-type strains in disc diffusion assays with hydrogen peroxide, organic peroxides, superoxide or nitrosative stress agents (Atack *et al.*, 2008). However, quantitative cell viability and cellular lipid peroxidation assays indicated the *tpx* mutant exhibited some increased sensitivity to peroxide stress (Atack *et al.*, 2008). Analysis of the peroxidase activity of a purified recombinant *C. jejuni* Tpx showed surprisingly that Tpx reduced only H_2O_2 as a substrate and immunoblotting of wild-type cell extracts with Tpx-specific antibodies showing increased abundance of Tpx under high aeration compared to that under microaerobic growth conditions (Atack *et al.*, 2008). Taken together, the results suggest that Tpx is a partially redundant antioxidant enzyme that plays an important role in protection of *C. jejuni* against oxygen-induced oxidative stress (Atack *et al.*, 2008; Atack and Kelly, 2009).

The other *C. jejuni* peroxiredoxin is the bacterioferritin comigratory protein Bcp (encoded by *Cj0271* in NCTC11168). Bcp is able to reduce H_2O_2, organic peroxides and lipid peroxides (Atack *et al.*, 2008). As such Bcp appears to play a role as a non-specific peroxide reductase in *C. jejuni*, in addition to the dedicated H_2O_2 reductase activity of Tpx (Atack and Kelly, 2009). A *C. jejuni* NCTC11168 *bcp* mutant exhibited a similar phenotype to that of a NCTC11168 *tpx* mutant as both mutants exhibited an increased sensitivity to organic peroxides, but not to H_2O_2 and O_2^-, showed a marked aerobic growth defect and suffered oxidative protein damage when challenged with peroxides or grown aerobically (Atack *et al.*, 2008; Atack and Kelly, 2009). A double *tpx*/*bcp* mutant exhibits an exacerbated phenotype when compared with the single mutants and was sensitive to both H_2O_2 and O_2^- (Atack *et al.*, 2008). As is the case with Tpx, the data suggest that Bcp plays a partially redundant antioxidant role protecting *C. jejuni* against oxygen-induced oxidative stress (Atack *et al.*, 2008; Atack and Kelly, 2009).

Cytochrome c peroxidases

Cytochrome c peroxidases (CCPs) detoxify H_2O_2 by reduction to H_2O (Atack and Kelly, 2007). In prokaryotes, CCPs are periplasmic proteins containing two *c*-type haems. *C. jejuni* contains two CCPs, encoded by *Cj0020c* and *Cj0358* in NCTC11168 (Bingham-Ramos and Hendrixson, 2008). Though the exact function of *C. jejuni* CCPs has yet to be fully elucidated, *C. jejuni* 81-176 mutants lacking the first CCP (termed DocA) were shown to have a significantly reduced ability to colonize chicks when compared to the wild-type strain (Bingham-Ramos and Hendrixson, 2008; Hendrixson and DiRita, 2004). Additionally, studies with a *C. jejuni* 81-176 mutant lacking the second CCP (encoded by *Cjj0382* in 81-176) was also shown to exhibit slight chick colonization defects, but not to the same extent as the 81-176 *docA* mutant (Hendrixson and DiRita, 2004). Currently unpublished studies in *C. jejuni* NCTC11168 suggest that both Cj0020 and Cj0358 are required for full resistance to peroxides as single mutants appear to be more sensitive to H_2O_2 than the wild-type strain (Atack and Kelly, 2009). It is still unknown why *C. jejuni* requires two CCPs, though it has been speculated that they may have differing roles (Atack and Kelly, 2009). Further work is required to define the role of both CCPs in the *C. jejuni* oxidative stress response.

Methionine sulfoxide reductases

In *C. jejuni* NCTC11168, *Cj0637c* and *Cj1112c* encode MsrA and MsrB respectively and the function of these methionine sulfoxide reductases (Msrs) has been confirmed using a thioredoxin-coupled assay (Atack and Kelly, 2008). Msrs convert methionine sulfoxides (Met-SO) back to methionine using thioredoxin as an electron donor. The oxidation of methionine residues by ROS is a major source of oxidative damage in cells, as the formation of Met-SO can lead to protein conformational changes and inactivation (Atack *et al.*, 2008; Moskovitz, 2005), as well as causing new proteins to fold incorrectly, as Met-SO is a much poorer substrate for methionine aminopeptidase than methionine (Atack *et al.*, 2008; Chang *et al.*, 1989). MsrA and MsrB are specific for the Met-SO S and R stereoisomer forms (Atack and Kelly, 2008; Ezraty *et al.*, 2005). *C. jejuni* studies have shown that both *msrA* and *msrB* are expressed during stationary phase (Atack and Kelly, 2008). When actively growing cells are challenged with peroxide stress, a *msrA* mutant, a *msrB* mutant and a double *msrA msrB* mutant were all killed at a higher rate than the wild-type strain (Atack and Kelly, 2008). However during stationary phase challenge, only the double *msrA msrB* mutant showed an increased sensitivity to peroxide stress, indicating that the presence of just one active Msr is sufficient to allow cells to survive during this phase of bacterial growth (Atack and Kelly, 2008). In contrast, superoxide stress assays resulted in increased killing of single mutants compared to the wild-type strain and the double mutant exhibited an enhanced sensitivity to superoxide stress than either of the single mutants (Atack and Kelly, 2008). *C. jejuni* NCTC11168 expresses KatA but not SodB when exposed to O_2^- (Garenaux *et al.*, 2009), so basal levels of SOD activity within *C. jejuni* may not be sufficient to prevent O_2^--mediated oxidation of methionine residues to such an extent as to result in an increased level of sensitivity when either MsrA or MsrB are non-functional (Atack and Kelly, 2009).

Additional oxidative stress responses

In *C. jejuni* NCTC11168, *Cj1545c* encodes MdaB, an orthologue of which has been shown to be an NADPH-specific quinone reductase in *E. coli* (Chatterjee and Sternberg, 1995). The *E. coli* MdaB has been shown to have reductase activity towards menadione (Hayashi *et al.*, 1996). *Helicobacter pylori* MdaB was shown to be important for oxidative stress protection and also colonization ability within mice (Wang and Maier, 2004). In *C. jejuni* NCTC11168, *mdaB* appears to be upregulated in response to oxidative stress (Palyada *et al.*, 2009). Further studies are required to definitively characterize the role of *C. jejuni* MdaB in the oxidative stress response.

In *H. pylori*, NapA has been shown to play a role in neutrophil activation and also in the oxidative stress response (Evans *et al.*, 1995). The *H. pylori* NapA has amino acid similarity to the Dps family of iron-binding proteins (Zanotti *et al.*, 2002). Dps family proteins have been shown to bind DNA and act as a shield against ROS (Wolf *et al.*, 1999). *C. jejuni* contains a Dps/bacterioferritin family protein orthologue (encoded by *Cj1534c* in NCTC11168) that protects bacterial cells against ROS by sequestering free iron (Ishikawa *et al.*,

2003). A *C. jejuni dps* mutant displayed increased sensitivity to oxidative stress and subsequently it was shown that Dps binds iron, but not DNA (Ishikawa *et al.*, 2003). In *C. jejuni* NCTC11168, *dps* appears to be up-regulated in response to oxidative stress (Palyada *et al.*, 2009). Further work is required to fully understand the role of Dps within *C. jejuni* (Atack et al., 2008).

There is some evidence that sulfate-reducing bacteria possess an alternative oxidative stress protection mechanism to superoxide dismutase-catalase system involving two non-haem iron proteins rubredoxin oxidoreductase (Rbo) and rubrerythrin (Rbr) (Lumppio *et al.*, 2001). The *C. jejuni* Rbo/Rbr-like protein (Rrc) is a non-haem iron-containing 27 kDa protein encoded by *Cj0012* in NCTC11168 and a *rrc* mutant was found to be more sensitive to ROS than the wild-type strain (Yamasaki *et al.*, 2004). The function is unclear though as orthologues of Rrc have been shown to be important in oxidative stress defence in anaerobes via redox reactions using iron cofactors (Lumppio *et al.*, 2001). This highlights again the importance of further study of the role of iron and non-haem iron proteins in *C. jejuni* oxidative stress defence, especially as iron metabolism and oxidative stress defence overlap so closely (Atack *et al.*, 2008). In *C. jejuni* NCTC11168, *rrc* appears to be up-regulated in response to oxidative stress (Palyada *et al.*, 2009).

NCTC11168 Cj1371 is a periplasmic protein identified with a potential role in oxidative stress response (Garenaux *et al.*, 2008). Mutant studies showed an increased sensitivity to O_2^- stress compared to the wild-type strain (Garenaux *et al.*, 2008). Again, further studies are required to fully elucidate the functionality of Cj1371 in the oxidative stress response.

Oxidative stress regulation in *C. jejuni*

A number of classical bacterial stress response regulatory proteins found in other enteric bacteria are absent from *C. jejuni* strains. These include the SoxRS and OxyR regulons (Blanchard *et al.*, 2007; Imlay, 2008; Pomposiello *et al.*, 2001). *E. coli* SoxRS has been shown to control genes involved in countering the harmful effects of superoxide and nitric oxide and mutations in *soxR* or *soxS* result in a lack of induction of the SoxRS regulon (Amabile-Cuevas and Demple, 1991; Farr and Kogoma, 1991). The only identified target of activated SoxR is the *soxS* gene (Pomposiello *et al.*, 2001). Increased SoxS levels have been shown to induce expression of more than 100 genes (Blanchard *et al.*, 2007). Different components of the SoxRS regulon combat the toxic effects of oxidants at various levels, including ROS scavenging, replacement of sensitive targets by resistant counterparts and protecting against DNA damage repair (Imlay, 2008; Krapp *et al.*, 2011). In *E. coli*, OxyR has been shown to regulate genes involved in the defence against H_2O_2 (Zheng and Storz, 2000). OxyR is a transcription factor that senses oxidative stress through disulfide bond formation and the conformational change this produces in OxyR results in transcription of the OxyR regulon (Storz *et al.*, 1990; Toledano *et al.*, 1994; Zheng *et al.*, 1998). OxyR is considered a global regulator and *H. influenzae oxyR* mutants are unable to respond to oxidative stress (Maciver and Hansen, 1996). However, both the SoxRS and OxyR oxidative stress response systems are absent from *C. jejuni*. The functions of the important regulators of the *C. jejuni* oxidative stress response are summarized in Table 11.1 and Fig. 11.1 and covered in detail below.

Peroxide-sensing regulator PerR

As stated above *C. jejuni* lacks an OxyR orthologue, which regulates KatA and AhpC expression in response to oxidative stress in many enteric bacteria such as *Salmonella* species and *E. coli* (Cabiscol *et al.*, 2000). In *C. jejuni*, the peroxide sensing regulator PerR (encoded by *Cj0322* in NCTC11168) was demonstrated to repress *ahpC* and *katA* transcription in an iron-dependent manner, thus making PerR a functional, but non-homologous, substitute for OxyR (van Vliet *et al.*, 1999). A *C. jejuni* NCTC11168 *perR* mutant constitutively expressed both *katA* and *ahpC*, and also exhibited enhanced resistance to peroxide stress (van Vliet *et al.*, 1999). A later study confirmed this observation, demonstrating that a *perR* mutant was more resistant to both H_2O_2 and cumene hyrdroperoxide oxidative stresses than the wild-type strain, yet also showed that

Table 11.1 Overview of *C. jejuni* oxidative stress regulators

Regulator	Oxidative stress defence mechanism (repression)	Oxidative stress defence mechanism (activation)	Regulator interaction	Reference(s)
PerR	AhpC KatA Dps MdaB	SodB	CprRS	Kim *et al.* (2011), Palyada *et al.* (2009), van Vliet *et al.* (1999)
Fur	Rrc	AhpC KatA SodB Dps MdaB Tpx	PerR CprRS	Holmes *et al.* (2005), Palyada *et al.* (2004, 2009), Stintzi *et al.* (2008), van Vliet *et al.* (1998, 2000, 2002)
CosR	SodB Dps Rrc LuxS	AhpC KatA	–	Garenaux *et al.* (2008), Hwang *et al.* (2011, 2012), Raphael *et al.* (2005)
Cj1556	–	KatA	PerR CpsRS	Gundogdu *et al.* (2007, 2011)
CpsRS	AhpC KatA TrxB	SodB Rrc	CosR	Palyada *et al.* (2009), Svensson *et al.* (2009)
RacRS	Cj0358 (CCP)	–	–	Apel *et al.* (2012), Bras *et al.* (1999)

the *perR* mutant was more sensitive to menadione stress than the wild-type strain (Palyada *et al.*, 2009). Microarray analysis of gene expression in a NCTC11168 *perR* mutant in comparison to the wild-type strain identified 104 genes belonging to the PerR regulon (Palyada *et al.*, 2009). Both PerR-repressed and PerR-activated genes were identified. Genes repressed by PerR included both *katA* and *ahpC*, but also *dps* and *mdaB* (Palyada *et al.*, 2009). Our reanalysis of the microarray data from this study (Palyada *et al.*, 2009) suggests that *sodB* is also a PerR-activated gene, as expression of *sodB* was down-regulated in the *perR* mutant. This study also highlighted the close interlink between iron metabolism and oxidative stress, which has been extended by a more recent study. Using a *perR* promoter–*lacZ* fusion assay, it was demonstrated that whilst other metal ions did not affect *perR* transcription, this was significantly reduced in the presence of iron (Kim *et al.*, 2011). The level of *perR* transcription was increased in a *perR* mutant whilst wild-type levels were restored by complementation with *perR* in *trans*, indicating autoregulation of *perR* transcription (Kim *et al.*, 2011). Binding of PerR to the *perR* promoter was demonstrated using electrophoretic mobility shift assays (EMSAs) and a PerR binding site identified using DNase I footprinting assays (Kim *et al.*, 2011). Comparison of the PerR binding sequence between the promoters of *perR*, *katA*, *ahpC* and *dps* revealed a 16bp consensus PerR binding sequence (Kim *et al.*, 2011). In *Bacillus subtilis*, PerR has been shown to sense H_2O_2 by Fe^{2+}-catalysed oxidation of two histidines within the PerR binding site, leading to Fe^{2+} release and derepression of PerR repressed genes (Lee and Helmann, 2006). Further studies are required to investigate the sensing of peroxide stress by *C. jejuni* PerR.

Ferric iron uptake regulator Fur

The ferric uptake regulator (Fur) is an iron-responsive regulator controlling the expression of genes encoding iron uptake systems, directly repressing iron acquisition once the intracellular iron concentration becomes too high and thus preventing accumulation of iron to toxic concentrations in the cytoplasm (Stintzi *et al.*, 2008). As already stated, the control of iron metabolism and the response to oxidative stress are closely linked,

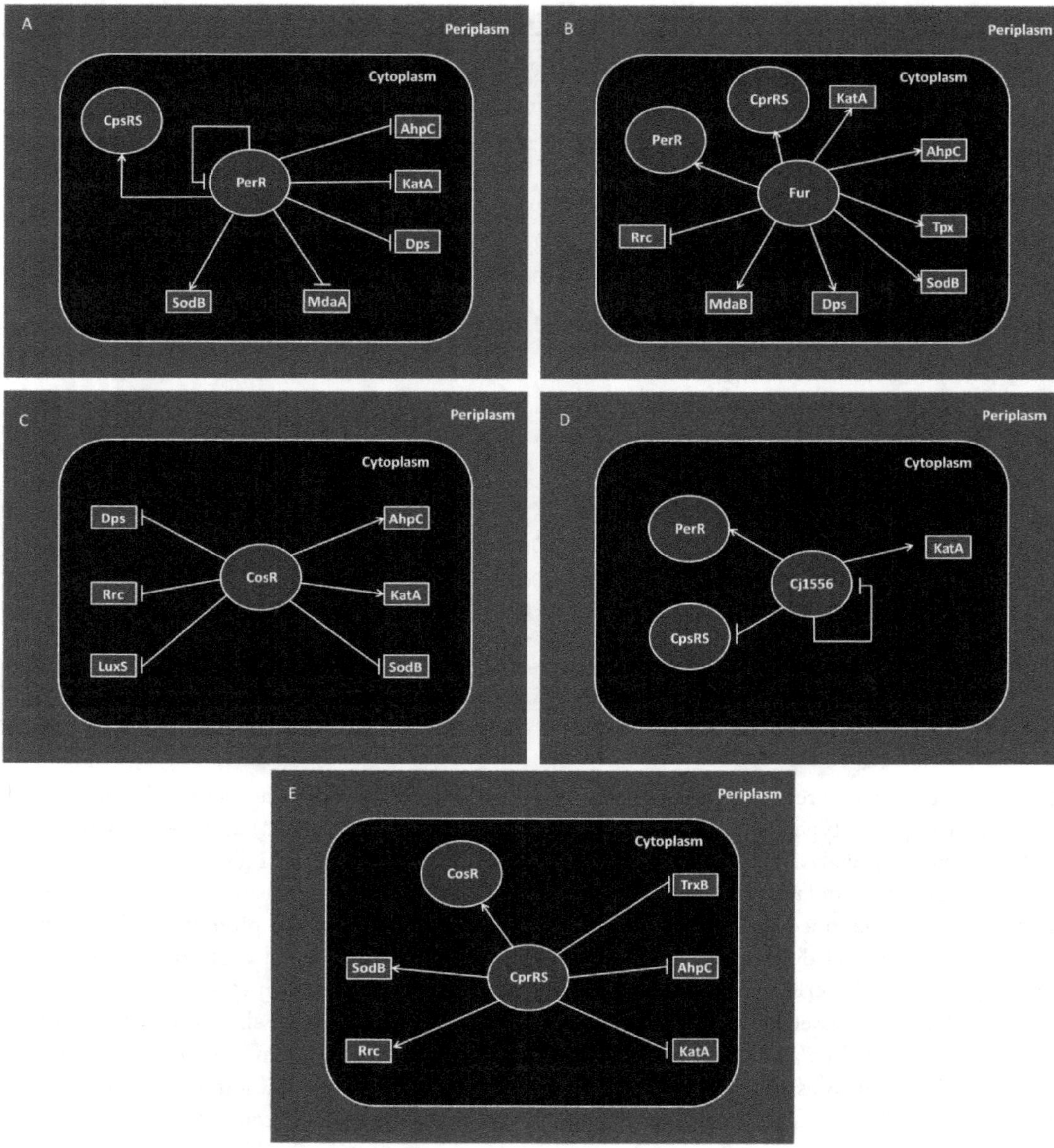

Figure 11.1 *C. jejuni* oxidative stress regulators. (A) PerR; (B) Fur; (C) CosR; (D) Cj1556; (E) CprRS. Proteins involved in countering ROS within *C. jejuni* are displayed in grey boxes. Regulatory proteins are displayed in pink boxes. Arrows indicate activation of gene expression whilst a halt indicates suppression of gene expression. This is determined from published microarray expression studies. If mutation of a gene encoding a regulator results in up-regulation of gene expression, it is assumed that the regulator normally suppresses expression of that gene in the wild-type strain. The opposite is true for genes down-regulated in microarray expression studies. A colour version of this figure is available in the plate section at the back of the book.

as iron in combination with oxygen mediates the generation of ROS (van Vliet *et al.*, 2002). A *C. jejuni* NCTC11168 *fur* mutant displayed a decreased growth rate when compared to the wild-type strain under both iron-restricted and iron-sufficient conditions, but as mutagenesis of *fur* might also affect transcription of the downstream *lysS* and *glyA* genes, the growth defect cannot yet be attributed solely to the absence of Fur (van Vliet *et al.*, 1998, 2000, 2002). Fur (encoded by *Cj0400* in NCTC11168) has also been linked to a role in oxidative stress defence. Fur directly regulates expression of *katA*, as both *fur* and *perR* must be mutated to completely abolish iron-dependent *katA* expression (Palyada *et al.*, 2009; van Vliet *et al.*, 1999). Genome-wide transcriptional analyses have also revealed that *C. jejuni* Fur regulates the expression of a number of genes encoding proteins with a role in oxidative stress defence. These include *perR, katA, ahpC, tpx* and *sodB* (Palyada *et al.*, 2004). Under iron limited conditions, *trxB, trxA* and *tpx* were expressed at increased levels (Holmes *et al.*, 2005). A NCTC11168 *fur* mutant was shown to be more resistant to H_2O_2 oxidative stress than the wild-type strain, yet was more sensitive to menadione and cumene hydroperoxide oxidative stresses than the wild-type strain (Palyada *et al.*, 2009). Our reanalysis of the microarray data from this study (Palyada *et al.*, 2009) suggests that *katA, ahpC, tpx* and *sodB* are Fur-activated genes but also *dps* and *mdaB*, as expression of these genes was down-regulated in the *fur* mutant. This reanalysis of the microarray data also suggests that *rrc* is a Fur-repressed gene as expression was up-regulated in the *fur* mutant. PerR is a Fur orthologue and studies have demonstrated that Fur is able to bind the *perR* binding sequence in *C. jejuni* (van Vliet *et al.*, 1999). This raised the question as to which regulatory protein is actually controlling expression of specific oxidative stress genes in *C. jejuni* (Atack and Kelly, 2009). A *C. jejuni* NCTC11168 *fur* mutant showed different expression patterns of PerR-controlled genes, whereas *perR* mutants did not affect any Fur controlled genes (van Vliet *et al.*, 1999). Thus, it is thought that *perR* boxes are a subset of *fur* boxes and Fur can bind both *fur* and *perR* boxes whereas PerR can only bind *perR* boxes (Atack and Kelly, 2009; van Vliet *et al.*, 2002).

OmpR-type response regulator CosR

C. jejuni also contains the recently identified *Campylobacter* oxidative stress regulator CosR (Hwang *et al.*, 2011). In NCTC11168, CosR is encoded by *Cj0355c* and is a member of the OmpR family of transcriptional regulators (Hwang *et al.*, 2011). CosR is essential for the viability of *C. jejuni*, as it has not been possible to construct a *cosR* mutant indicating that the mutation is lethal (Garenaux *et al.*, 2008; Hwang *et al.*, 2011; Raphael *et al.*, 2005). The inability to construct a defined mutant has hampered attempts to investigate the function of CosR, until antisense peptide nucleic acid (PNA) technology was used to repress the expression of CosR (Hwang *et al.*, 2011). CosR was shown to be responsible for the negative regulation of the oxidative stress response proteins SodB, Dps, Rrc and LuxS. Interestingly CosR also positively controls expression of AhpC. Binding of CosR to the promoter regions of *ahpC, sodB, dps* and *luxS* was demonstrated using EMSAs. A 21 bp CosR binding sequence was identified within the *sodB* and *ahpC* promoters using DNase I footprinting assays. CosR expression was significantly reduced in the presence of paraquat (a superoxide generator), but not by H_2O_2 (Hwang *et al.*, 2011). Consistent with the overall negative regulation of *C. jejuni* oxidative stress defence proteins by CosR, knockdown of CosR resulted in bacteria more resistant to oxidative stress whilst bacteria overexpressing CosR were more sensitive to oxidative stress (Hwang *et al.*, 2011). Further studies have since identified 93 genes whose transcriptional levels were altered more than 2-fold due to the repression of CosR expression by antisense technology (Hwang *et al.*, 2012). A significant finding was that CosR also regulates the transcription of *katA*. Further characterization demonstrated that CosR positively controls *katA* transcription and catalase activity by direct interactions with the *katA* promoter. As such CosR has an important role in regulating oxidative stress defence in *C. jejuni* (Hwang *et al.*, 2012).

MarR-type response regulator Cj1556

Re-annotation of the *C. jejuni* NCTC11168 genome sequence identified Cj1556 (originally annotated as a hypothetical protein) as a MarR family transcriptional regulator and further analysis indicated a role in regulating the oxidative stress response (Gundogdu *et al.*, 2007, 2011). A *C. jejuni* 11168H *Cj1556* mutant exhibited increased sensitivity to oxidative and aerobic (O_2) stress, decreased ability for intracellular survival in Caco-2 human intestinal epithelial cells and

J774A.1 mouse macrophages and a reduction in virulence in the *Galleria mellonella* infection model. Microarray analysis of gene expression changes in the *Cj1556* mutant indicated negative autoregulation of *Cj1556* expression and down-regulation of genes associated with oxidative and aerobic (O_2) stress responses, such as *katA* and *perR*. EMSAs confirmed the binding of recombinant Cj1556 to the promoter region upstream of the *Cj1556* gene. *cprS* was also up-regulated in the *Cj1556* mutant and subsequent studies showed that this mutant had a reduced ability to form biofilms (Gundogdu *et al.*, 2011). It is interesting to note that Cj1556 appears to positively regulate KatA and PerR expression, whilst CosR appears to positively regulate KatA and AhpC expression and negatively regulates SodB, Dps, Rrc and LuxS expression. The occurrence of cross-talk between the different regulatory systems is becoming increasingly apparent.

Re-annotation of the *C. jejuni* NCTC11168 genome sequence also identified a second MarR family transcriptional regulator Cj1546 (Gundogdu *et al.*, 2007, 2011). Preliminary studies indicate that a 11168H *Cj1546* mutant also exhibits increased sensitivity to oxidative stress (Gundogdu and Dorrell, unpublished data). This suggests that yet another regulatory protein may play a role in *C. jejuni* oxidative stress resistance.

Global post-transcriptional regulator CsrA

CsrA (encoded by *Cj1103* in NCTC11168) has been identified within *C. jejuni* as a post-transcriptional regulator (Fields and Thompson, 2008). The *H. pylori* CsrA orthologue is responsible for the activation or repression of many important bacterial processes, including oxidative stress resistance (Barnard *et al.*, 2004). A *C. jejuni* 81-176 *csrA* mutant displayed increased sensitivity to various stresses including atmospheric oxygen and H_2O_2 (Fields and Thompson, 2008). Expression of CsrA in *E. coli* is regulated by small non-coding RNAs *csrB* and *csrC*, but orthologues have not yet been identified in *C. jejuni* (Babitzke and Romeo, 2007; Fields and Thompson, 2008). In *H. pylori*, CsrA has been shown to regulate *ahpC* and *fur* expression (Barnard *et al.*, 2004), however similar studies have not been performed for *C. jejuni*. It is becoming increasingly apparent that the control of oxidative stress resistance in *C. jejuni* is more complex than was originally envisaged.

CprRS two-component regulatory system

C. jejuni CprRS (encoded by *Cj1227c and Cj1226c* in NCTC11168) has also been shown to be important for oxidative stress resistance (Svensson *et al.*, 2009). The response regulator CprR appears to be essential as mutations in *cprR* were lethal, however the sensor kinase CprS is not essential and a 81-176 *cprS* mutant exhibited decreased oxidative stress tolerance to t-butyl-hydroperoxide, paraquat and H_2O_2 compared to the wild-type strain (Svensson *et al.*, 2009). Proteomic analysis identified a number of differences between the wild-type strain and the *cprS* mutant in the expression of oxidative stress defence proteins. KatA, TrxB and AhpC were all up-regulated in the *cprS* mutant whilst SodB and Rrc were down-regulated (Svensson *et al.*, 2009). The *cprS* mutant also showed a decreased colonization ability of 1-day-old chicks when compared to the wild-type strain, but showed increased motility and secretion ability (Svensson *et al.*, 2009). Intriguingly the proteomic analysis also identified down-regulation of CosR in the *cprS* mutant (Svensson *et al.*, 2009). In *C. jejuni* NCTC11168, *cprS* appears to be down-regulated in response to oxidative stress and also down-regulated in both a *perR* and a *fur* mutant (Palyada *et al.*, 2009).

RacRS two-component regulatory system

Other regulatory proteins have also been identified which have been shown to have oxidative stress countering roles. RacR (encoded by *Cj1261* in NCTC11168) has been designated a temperature-responsive regulator and has been shown to play a role in the regulation of the CCP encoded by *Cj0358* in NCTC11168 (Bras *et al.*, 1999). The RacRS two-component regulatory system (reduced ability to colonize) is involved in a temperature-dependent signalling system (Bras *et al.*, 1999). A *C. jejuni* 81116 *racR* mutant was shown to have reduced ability to colonize the chicken intestinal tract and resulted in temperature-dependent changes in its protein profile along with growth characteristics (Bras *et al.*, 1999). A more recent study further investigated the role of RacRS and revealed roles in the heat shock response, motility, and maintenance of cell length homogeneity (Apel *et al.*, 2012). However, the authors did not note any further link to the oxidative stress response.

Quorum sensing

Bacterial quorum sensing systems have been shown to play significant roles in a variety of processes, including oxidative stress (Plummer, 2012). *C. jejuni* does not possess an acyl-homoserine lactone autoinducer (AI-1) quorum sensing system, according to the genome sequences published to date (Plummer, 2012), but does possess a AI-2 quorum sensing system where the AI-2 is produced by the *S*-ribosylhomocysteinase LuxS encoded by *Cj1198* in NCTC11168 (Elvers and Park, 2002). The role of LuxS in the *C. jejuni* response to oxidative stress was initially investigated using a well diffusion assay to compare the sensitivity of a NCTC11168 *luxS* mutant to paraquat and H_2O_2 to that of the wild-type strain. These studies did not observe any differences in sensitivity (Elvers and Park, 2002). However a later study using both disk inhibition assays and minimum inhibitory concentration (MIC) assays showed that a 81-176 *luxS* mutant exhibited significant increased sensitivity to cumene hydroperoxide and H_2O_2 compared to the wild-type strain (He *et al.*, 2008). Further evidence for a role for LuxS in regulating the *C. jejuni* oxidative stress response was provided when the expression of both *aphC* and *tpx* were shown to be down-regulated in the 81-176 *luxS* mutant (He *et al.*, 2008). Further studies are required to characterize the exact role of LuxS in the *C. jejuni* oxidative stress response, especially in light of the indication that the effects of the *luxS* mutation on the transcriptome are related to the consequential loss of function in the activated methyl cycle (Holmes *et al.*, 2009).

Cross-talk between different regulatory systems?

The activation of expression of oxidative stress defence genes appears to be both oxidant-specific and dependent on the activity of a number of different regulators (Palyada *et al.*, 2009). A NCTC11168 *perR* mutant was more resistant to both H_2O_2 and cumene hyrdroperoxide oxidative stresses than the wild-type strain, yet was more sensitive to menadione oxidative stress whilst a NCTC11168 *fur* mutant was shown to be more resistant to H_2O_2 oxidative stress than the wild-type strain, yet was more sensitive to menadione and cumene hydroperoxide oxidative stresses than the wild-type strain (Palyada *et al.*, 2009). Intriguingly, a NCTC11168 *perR fur* double mutant was more resistant to both H_2O_2 and cumene hydroperoxide oxidative stresses than the wild-type strain (though not as resistant as the *perR* single mutant), yet was more sensitive to menadione oxidative stress (though not as sensitive as the *perR* single mutant) (Palyada *et al.*, 2009). It is now clear that the control of the *C. jejuni* oxidative stress response is a complex process consisting of multiple inter-linked levels of both sensing and regulation. The response to H_2O_2 is also dependent on the level of iron (Palyada *et al.*, 2009). The control of the expression of oxidative stress defence genes is multi-factorial with different regulatory proteins acting as repressors or activators and many regulators appear to affect the expression of other regulatory proteins. Repression of CosR results in up-regulation of LuxS, whilst over expression of CosR down-regulates LuxS expression (Hwang *et al.*, 2011). Proteomic analysis identified down-regulation of both CosR and LuxS in a *cprS* mutant (Svensson *et al.*, 2009). Microarray analysis of gene expression changes in the *Cj1556* mutant indicated that *perR* was down-regulated and *cprS* was up-regulated (Gundogdu *et al.*, 2011), whilst *cprS* appears to be down-regulated in both *perR* and *fur* mutants (Palyada *et al.*, 2009). This regulatory system appears to be of central importance to *C. jejuni* as both CosR and CprR regulators are essential. Further studies are required to fully elucidate the different aspects of both sensing of oxidative stresses and the regulation of the oxidative stress response to these different stresses.

Oxidative stress regulators with multifactorial roles

The list of regulatory proteins involved in oxidative stress response is extensive. Importantly many of these regulators have additional roles often relating to virulence. PerR has been shown to be up-regulated under acid-stress conditions indicating there may be overlap of oxidative and other stress response regulons (Reid *et al.*, 2008). *C. jejuni* NCTC11168 *perR* mutants displayed a significant reduction in motility when compared to the wild-type strain (Palyada *et al.*, 2009). Colonization of the chick caeca by the NCTC11168 *perR* mutant strain was reduced from the wild-type level of 2×10^7 to 2×10^4 CFU/g of caecal content respectively (Palyada *et al.*, 2009). However this reduction in the colonization ability of the *perR* mutant may in part be due to the lack of motility. A *C. jejuni* NCTC11168 *fur* mutant displayed a decreased growth rate when compared with the wild-type strain under both iron-restricted and iron-sufficient conditions, though it has not been ruled out that the *fur* mutation affects transcription of the downstream *lysS* and *glyA* genes, so this growth defect may not yet be attributed solely to the absence of Fur (van Vliet *et al.*, 1998, 2000, 2002). Colonization of the chick caeca by the NCTC11168 *fur* mutant was reduced from the wild-type level of 2×10^7 to 5×10^3 CFU/g of caecal content respectively (Palyada *et al.*, 2009). CosR has been shown to negatively regulate *cmeABC* expression, which affects the extrusion of toxic compounds out of the bacterial cell (Hwang *et al.*, 2012). Cj1556 has been shown to have multiple phenotypes relating to virulence. A *C. jejuni* 11168H *Cj1556* mutant exhibited increased sensitivity to oxidative and aerobic stress, decreased ability for intracellular survival within Caco-2 human intestinal epithelial cells and J774A.1 mouse macrophages, a reduction in virulence in the *G. mellonella* infection model and a reduction in biofilm formation compared to the wild-type strain (Gundogdu *et al.*, 2011). CsrA has been shown to play a role in the activation of biofilm formation, motility, and interactions with host cells *in vitro* (Fields and Thompson, 2008). Intriguingly a 81-176 *csrA* mutant exhibits a reduction in adherence to INT 407 cells, but a 20-fold increase in invasion (Fields and Thompson, 2008). CprS has been shown to play a role in biofilm formation, chick colonization and stress tolerance (Svensson *et al.*, 2009). A 81-176 *cprS* mutant displayed a decrease in osmotolerance and colonization ability of 1-day-old chicks when compared to the wild-type strain, but showed increased motility and secretion ability (Svensson *et al.*, 2009). The 81-176 *cprS* mutant displayed an apparent growth defect and formed biofilms at an enhanced rate compared to the wild-type strain (Svensson *et al.*, 2009). RacRS is involved in a temperature-dependent signalling system (Bras *et al.*, 1999). A 81116 *racR* mutant was shown to have reduced ability to colonize the chicken intestinal tract and mutation of *racR* resulted in temperature-dependent changes in protein expression and growth (Bras *et al.*, 1999). A more recent study further investigated the role of RacRS in the *C. jejuni* 81-176 strain, revealing roles in the heat shock response, motility and maintenance of cell length homogeneity (Apel *et al.*, 2012). The fact that many *C. jejuni* oxidative stress response regulators also play important roles in other aspects of bacterial metabolism, physiology and pathogenesis indicates both the complexity of the regulatory process and also the central importance of these regulatory systems in *C. jejuni*.

Discussion and future perspectives

C. jejuni contains a large repertoire of genes encoding proteins that allow the bacterium to tolerate and survive oxidative stresses. In addition, a complex network of regulatory proteins is involved controlling the *C. jejuni* oxidative stress response dependent on both the oxidant and other physiological conditions. It has been stated that oxidative stress regulation is poorly understood within *C. jejuni* (Atack and Kelly, 2009). PerR is arguably the best studied regulatory system within *C. jejuni*. However, many aspects of sensing and regulation are still poorly understood, mainly due to overlap with the ferric iron uptake regulator system. Other regulators such as CosR and Cj1556 have only recently been identified and appear to have pleiotropic functions, so it is currently difficult to accurately ascertain what the primary function of each regulator is. This number of regulatory proteins involved in the oxidative stress response is now almost comparable in size to the number of actual proteins

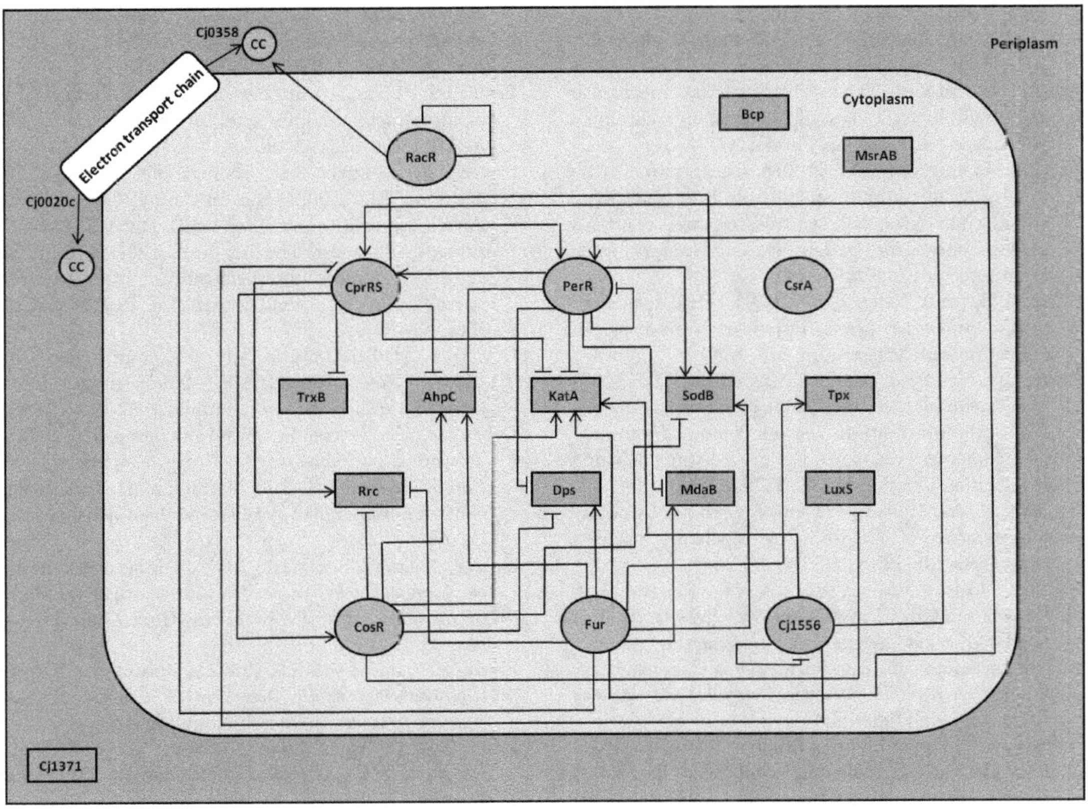

Figure 11.2 The regulation of the *C. jejuni* oxidative stress response. Proteins involved in countering ROS within *C. jejuni* are displayed in grey boxes. Regulatory proteins are displayed in pink boxes. CC indicates cytochrome C peroxidase. Arrows indicate activation of gene expression whilst a halt indicates suppression of gene expression. This is determined from published microarray expression studies. If mutation of a gene encoding a regulator results in up-regulation of gene expression, it is assumed that the regulator normally suppresses expression of that gene in the wild-type strain. The opposite is true for genes down-regulated in microarray expression studies. A colour version of this figure is available in the plate section at the back of the book.

involved in the neutralization of ROS by *C. jejuni*. A complex network for sensing and regulating the *C. jejuni* oxidative stress response has emerged. However, it is still difficult to state the functionally, mechanisms and limitations of each regulator. For instance, we do not know whether these same regulators control expression of more than one oxidative response gene or even other genes, since many of the mutant regulator experiments have shown multiple phenotypes. In addition, we do not know if these systems function simultaneously, whether in concert or in conflict, or individually. Also, the role of non-coding RNAs in the control of the *C. jejuni* oxidative stress response has yet to be investigated. Our current understanding of the regulation of the *C. jejuni* oxidative stress response is summarized in Fig. 11.2. The control of the expression of many oxidative stress response proteins by multiple regulators certainly warrants further investigation and we predict that our understanding of the hierarchical nature of the regulatory system will start to be elucidated in the near future.

References

Amabile-Cuevas, C.F., and Demple, B. (1991). Molecular characterization of the *soxRS* genes of *Escherichia coli*: two genes control a superoxide stress regulon. Nucleic Acids Res. *19*, 4479–4484.

Apel, D., Ellermeier, J., Pryjma, M., Dirita, V.J., and Gaynor, E.C. (2012). Characterization of *Campylobacter jejuni* RacRS reveals roles in the heat shock response, motility, and maintenance of cell length homogeneity. J. Bacteriol. *194*, 2342–2354.

Atack, J.M., and Kelly, D.J. (2007). Structure, mechanism and physiological roles of bacterial cytochrome c peroxidases. Adv. Microb. Physiol. *52*, 73–106.

Atack, J.M., and Kelly, D.J. (2008). Contribution of the stereospecific methionine sulfoxide reductases MsrA and MsrB to oxidative and nitrosative stress resistance in the food-borne pathogen *Campylobacter jejuni*. Microbiology *154*, 2219–2230.

Atack, J.M., and Kelly, D.J. (2009). Oxidative stress in *Campylobacter jejuni*: responses, resistance and regulation. Fut. Microbiol. *4*, 677–690.

Atack, J.M., Harvey, P., Jones, M.A., and Kelly, D.J. (2008). The *Campylobacter jejuni* thiol peroxidases Tpx and Bcp both contribute to aerotolerance and peroxide-mediated stress resistance but have distinct substrate specificities. J. Bacteriol. *190*, 5279–5290.

Babitzke, P., and Romeo, T. (2007). CsrB sRNA family: sequestration of RNA-binding regulatory proteins. Curr. Opin. Microbiol. *10*, 156–163.

Baillon, M.L., van Vliet, A.H., Ketley, J.M., Constantinidou, C., and Penn, C.W. (1999). An iron-regulated alkyl hydroperoxide reductase (AhpC) confers aerotolerance and oxidative stress resistance to the microaerophilic pathogen *Campylobacter jejuni*. J. Bacteriol. *181*, 4798–4804.

Barnard, F.M., Loughlin, M.F., Fainberg, H.P., Messenger, M.P., Ussery, D.W., Williams, P., and Jenks, P.J. (2004). Global regulation of virulence and the stress response by CsrA in the highly adapted human gastric pathogen *Helicobacter pylori*. Mol. Microbiol. *51*, 15–32.

Bingham-Ramos, L.K., and Hendrixson, D.R. (2008). Characterization of two putative cytochrome c peroxidases of *Campylobacter jejuni* involved in promoting commensal colonization of poultry. Infect. Immun. *76*, 1105–1114.

Blanchard, J.L., Wholey, W.Y., Conlon, E.M., and Pomposiello, P.J. (2007). Rapid changes in gene expression dynamics in response to superoxide reveal SoxRS-dependent and independent transcriptional networks. PLoS One *2*, e1186.

Bras, A.M., Chatterjee, S., Wren, B.W., Newell, D.G., and Ketley, J.M. (1999). A Novel *Campylobacter jejuni* Two-Component Regulatory System Important for Temperature-Dependent Growth and Colonization. J. Bacteriol. *181*, 3298–3302.

Cabiscol, E., Tamarit, J., and Ros, J. (2000). Oxidative stress in bacteria and protein damage by reactive oxygen species. Int. Microbiol. *3*, 3–8.

Chae, H.Z., Robison, K., Poole, L.B., Church, G., Storz, G., and Rhee, S.G. (1994). Cloning and sequencing of thiol-specific antioxidant from mammalian brain: alkyl hydroperoxide reductase and thiol-specific antioxidant define a large family of antioxidant enzymes. Proc. Natl. Acad. Sci. U.S.A. *91*, 7017–7021.

Chang, S.Y., McGary, E.C., and Chang, S. (1989). Methionine aminopeptidase gene of *Escherichia coli* is essential for cell growth. J. Bacteriol. *171*, 4071–4072.

Chatterjee, P.K., and Sternberg, N.L. (1995). A general genetic approach in *Escherichia coli* for determining the mechanism(s) of action of tumoricidal agents: application to DMP 840, a tumoricidal agent. Proc. Natl. Acad. Sci. U.S.A. *92*, 8950–8954.

Chelikani, P., Fita, I., and Loewen, P.C. (2004). Diversity of structures and properties among catalases. Cell. Mol. Life Sci. *61*, 192–208.

Coker, A.O., Isokpehi, R.D., Thomas, B.N., Amisu, K.O., and Obi, C.L. (2002). Human campylobacteriosis in developing countries. Emerg. Infect. Dis. *8*, 237–244.

D'Autreaux, B., and Toledano, M.B. (2007). ROS as signalling molecules: mechanisms that generate specificity in ROS homeostasis. Nat. Rev. Mol. Cell Biol. *8*, 813–824.

Day, W.A., Jr., Sajecki, J.L., Pitts, T.M., and Joens, L.A. (2000). Role of catalase in *Campylobacter jejuni* intracellular survival. Infect. Immun. *68*, 6337–6345.

Debruyne, L., Gevers, D., and Vandamme, P. (2008). Taxonomy of the family *Campylobacteraceae*. In *Campylobacter*, 3rd edn, Nachmkin, I., Szymanski, C.M., and Blaser, M.J., eds. (ASM Press, Washington, DC), pp. 3–25.

Dorrell, N., and Wren, B.W. (2007). The second century of *Campylobacter* research: recent advances, new opportunities and old problems. Curr. Opin. Infect. Dis. *20*, 514–518.

Elvers, K.T., and Park, S.F. (2002). Quorum sensing in *Campylobacter jejuni*: detection of a luxS encoded signalling molecule. Microbiology *148*, 1475–1481.

Evans, D.J., Jr., Evans, D.G., Takemura, T., Nakano, H., Lampert, H.C., Graham, D.Y., Granger, D.N., and Kvietys, P.R. (1995). Characterization of a *Helicobacter pylori* neutrophil-activating protein. Infect. Immun. *63*, 2213–2220.

Ezraty, B., Aussel, L., and Barras, F. (2005). Methionine sulfoxide reductases in prokaryotes. Biochim. Biophys. Acta *1703*, 221–229.

Fang, F.C. (2004). Antimicrobial reactive oxygen and nitrogen species: concepts and controversies. Nat. Rev. Microbiol. *2*, 820–832.

Farr, S.B., and Kogoma, T. (1991). Oxidative stress responses in *Escherichia coli* and *Salmonella typhimurium*. Microbiol. Rev. *55*, 561–585.

Fields, J.A., and Thompson, S.A. (2008). *Campylobacter jejuni* CsrA mediates oxidative stress responses, biofilm formation, and host cell invasion. J. Bacteriol. *190*, 3411–3416.

Fitzgerald, C., Whichard, J., and Nachamkin, I. (2008). Diagnosis and antimicrobial susceptibility of *Campylobacter* species. In *Campylobacter*, 3rd edn, Nachmkin, I., Szymanski, C.M., and Blaser, M.J., eds. (ASM Press, Washington, DC), pp. 227–243.

Flint, A., Sun, Y.Q., and Stintzi, A. (2012). Cj1386 is an ankyrin-containing protein involved in heme trafficking to catalase in *Campylobacter jejuni*. J. Bacteriol. *194*, 334–345.

Friedman, C.R., Neimann, J., Wegener, H.C., and Tauxe, R.V. (2000). Epidemioogy of *Camylobacter jejuni* infections in the United States and other industrialized nations. In *Campylobacter*, 2nd edn, Nachamkin, I., and Blaser, M.J., eds. (ASM Press, Washington, DC) pp. 121–138.

Garenaux, A., Guillou, S., Ermel, G., Wren, B., Federighi, M., and Ritz, M. (2008). Role of the Cj1371 periplasmic protein and the Cj0355c two-component regulator in the *Campylobacter jejuni* NCTC 11168 response to oxidative stress caused by paraquat. Res. Microbiol. *159*, 718–726.

Garenaux, A., Ritz, M., Jugiau, F., Rama, F., Federighi, M., and de Jonge, R. (2009). Role of oxidative stress in *C. jejuni* inactivation during freeze–thaw treatment. Curr. Microbiol. *58*, 134–138.

Gort, A.S., and Imlay, J.A. (1998). Balance between endogenous superoxide stress and antioxidant defenses. J. Bacteriol. *180*, 1402–1410.

Grant, K.A., and Park, S.F. (1995). Molecular characterization of katA from *Campylobacter jejuni* and generation of a catalase-deficient mutant of *Campylobacter coli* by interspecific allelic exchange. Microbiology *141(Pt 6)*, 1369–1376.

Gundogdu, O., Bentley, S.D., Holden, M.T., Parkhill, J., Dorrell, N., and Wren, B.W. (2007). Re-annotation and re-analysis of the *Campylobacter jejuni* NCTC11168 genome sequence. BMC Genomics *8*, 162.

Gundogdu, O., Mills, D.C., Elmi, A., Martin, M.J., Wren, B.W., and Dorrell, N. (2011). The *Campylobacter jejuni* transcriptional regulator Cj1556 plays a role in the oxidative and aerobic stress response and is important for bacterial survival *in vivo*. J. Bacteriol. *193*, 4238–4249.

Hall, A., Karplus, P.A., and Poole, L.B. (2009). Typical 2-Cys peroxiredoxins--structures, mechanisms and functions. FEBS J. *276*, 2469–2477.

Hassan, H.M. (1988). Biosynthesis and regulation of superoxide dismutases. Free Radic. Biol. Med. *5*, 377–385.

Hayashi, M., Ohzeki, H., Shimada, H., and Unemoto, T. (1996). NADPH-specific quinone reductase is induced by 2-methylene-4-butyrolactone in *Escherichia coli*. Biochim. Biophys. Acta *1273*, 165–170.

He, Y., Frye, J.G., Strobaugh, T.P., and Chen, C.Y. (2008). Analysis of AI-2/LuxS-dependent transcription in *Campylobacter jejuni* strain 81-176. Food. Pathog. Dis. *5*, 399–415.

Hendrixson, D.R., and DiRita, V.J. (2004). Identification of *Campylobacter jejuni* genes involved in commensal colonization of the chick gastrointestinal tract. Mol. Microbiol. *52*, 471–484.

Hofreuter, D., Tsai, J., Watson, R.O., Novik, V., Altman, B., Benitez, M., Clark, C., Perbost, C., Jarvie, T., Du, L., *et al.* (2006). Unique features of a highly pathogenic *Campylobacter jejuni* strain. Infect. Immun. *74*, 4694–4707.

Holmes, K., Mulholland, F., Pearson, B.M., Pin, C., McNicholl-Kennedy, J., Ketley, J.M., and Wells, J.M. (2005). *Campylobacter jejuni* gene expression in response to iron limitation and the role of Fur. Microbiology *151*, 243–257.

Holmes, K., Tavender, T.J., Winzer, K., Wells, J.M., and Hardie, K.R. (2009). AI-2 does not function as a quorum sensing molecule in *Campylobacter jejuni* during exponential growth in vitro. BMC Microbiol. *9*, 214.

Hwang, S., Kim, M., Ryu, S., and Jeon, B. (2011). Regulation of oxidative stress response by CosR, an essential response regulator in *Campylobacter jejuni*. PLoS One *6*, e22300.

Hwang, S., Zhang, Q., Ryu, S., and Jeon, B. (2012). Transcriptional regulation of the CmeABC multidrug efflux pump and the KatA catalase by CosR in *Campylobacter jejuni*. J. Bacteriol. *194*, 6883–6891.

Imlay, J.A. (2003). Pathways of oxidative damage. Annu. Rev. Microbiol. *57*, 395–418.

Imlay, J.A. (2006). Iron–sulfur clusters and the problem with oxygen. Mol. Microbiol. *59*, 1073–1082.

Imlay, J.A. (2008). Cellular defenses against superoxide and hydrogen peroxide. Annu. Rev. Biochem. *77*, 755–776.

Ishikawa, T., Mizunoe, Y., Kawabata, S., Takade, A., Harada, M., Wai, S.N., and Yoshida, S. (2003). The iron-binding protein Dps confers hydrogen peroxide stress resistance to *Campylobacter jejuni*. J. Bacteriol. *185*, 1010–1017.

Jang, S., and Imlay, J.A. (2007). Micromolar intracellular hydrogen peroxide disrupts metabolism by damaging iron–sulfur enzymes. J. Biol. Chem. *282*, 929–937.

Kim, M., Hwang, S., Ryu, S., and Jeon, B. (2011). Regulation of perR expression by iron and PerR in *Campylobacter jejuni*. J. Bacteriol. *193*, 6171–6178.

Krapp, A.R., Humbert, M.V., and Carrillo, N. (2011). The soxRS response of *Escherichia coli* can be induced in the absence of oxidative stress and oxygen by modulation of NADPH content. Microbiology *157*, 957–965.

Lee, J.W., and Helmann, J.D. (2006). The PerR transcription factor senses H_2O2 by metal-catalysed histidine oxidation. Nature *440*, 363–367.

Liu, Y.W., Denkmann, K., Kosciow, K., Dahl, C., and Kelly, D.J. (2013). Tetrathionate stimulated growth of *Campylobacter jejuni* identifies a new type of bi-functional tetrathionate reductase (TsdA) that is widely distributed in bacteria. Mol. Microbiol. *88*, 173–188.

Lumppio, H.L., Shenvi, N.V., Summers, A.O., Voordouw, G., and Kurtz, D.M., Jr. (2001). Rubrerythrin and rubredoxin oxidoreductase in *Desulfovibrio vulgaris*: a novel oxidative stress protection system. J. Bacteriol. *183*, 101–108.

Maciver, I., and Hansen, E.J. (1996). Lack of expression of the global regulator OxyR in *Haemophilus influenzae* has a profound effect on growth phenotype. Infect. Immun. *64*, 4618–4629.

Mansfield, L.S., Schauer, D.B., and Fox, J.G. (2008). Animal models of *Campylobacter* jejuni infections. In *Campylobacter*, 3rd edn, Nachmkin, I., Szymanski, C.M., and Blaser, M.J., eds. (ASM Press, Washington, DC), pp. 367–379.

Miller, W.G., Bates, A.H., Horn, S.T., Brandl, M.T., Wachtel, M.R., and Mandrell, R.E. (2000). Detection on surfaces and in Caco-2 cells of *Campylobacter jejuni* cells transformed with new *gfp*, *yfp*, and *cfp* marker plasmids. Appl. Environ. Microbiol. *66*, 5426–5436.

Moskovitz, J. (2005). Methionine sulfoxide reductases: ubiquitous enzymes involved in antioxidant defense, protein regulation, and prevention of aging-associated diseases. Biochim. Biophys. Acta *1703*, 213–219.

Palyada, K., Threadgill, D., and Stintzi, A. (2004). Iron acquisition and regulation in *Campylobacter jejuni*. J. Bacteriol. *186*, 4714–4729.

Palyada, K., Sun, Y.Q., Flint, A., Butcher, J., Naikare, H., and Stintzi, A. (2009). Characterization of the oxidative stress stimulon and PerR regulon of *Campylobacter jejuni*. BMC Genomics *10*, 481.

Park, S., You, X., and Imlay, J.A. (2005). Substantial DNA damage from submicromolar intracellular hydrogen peroxide detected in Hpx- mutants of *Escherichia coli*. Proc. Natl. Acad. Sci. U.S.A. *102*, 9317–9322.

Park, S.F. (2002). The physiology of *Campylobacter* species and its relevance to their role as foodborne pathogens. Int. J. Food Microbiol. *74*, 177–188.

Parkhill, J., Wren, B.W., Mungall, K., Ketley, J.M., Churcher, C., Basham, D., Chillingworth, T., Davies, R.M., Feltwell, T., Holroyd, S., *et al.* (2000). The genome sequence of the food-borne pathogen *Campylobacter jejuni* reveals hypervariable sequences. Nature *403*, 665–668.

Pesci, E.C., Cottle, D.L., and Pickett, C.L. (1994). Genetic, enzymatic, and pathogenic studies of the iron superoxide dismutase of *Campylobacter jejuni*. Infect. Immun. *62*, 2687–2694.

Plummer, P.J. (2012). LuxS and quorum-sensing in *Campylobacter*. Front. Cell. Infect. Microbiol. *2*, 22.

Poly, F., Read, T., Tribble, D.R., Baqar, S., Lorenzo, M., and Guerry, P. (2007). Genome sequence of a clinical isolate of *Campylobacter jejuni* from Thailand. Infect. Immun. *75*, 3425–3433.

Pomposiello, P.J., Bennik, M.H., and Demple, B. (2001). Genome-wide transcriptional profiling of the *Escherichia coli* responses to superoxide stress and sodium salicylate. J. Bacteriol. *183*, 3890–3902.

Purdy, D., Cawthraw, S., Dickinson, J.H., Newell, D.G., and Park, S.F. (1999). Generation of a superoxide dismutase (SOD)-deficient mutant of *Campylobacter coli*: evidence for the significance of SOD in *Campylobacter* survival and colonization. Appl. Environ. Microbiol. *65*, 2540–2546.

Raphael, B.H., Pereira, S., Flom, G.A., Zhang, Q., Ketley, J.M., and Konkel, M.E. (2005). The *Campylobacter jejuni* response regulator, CbrR, modulates sodium deoxycholate resistance and chicken colonization. J. Bacteriol. *187*, 3662–3670.

Reid, A.N., Pandey, R., Palyada, K., Naikare, H., and Stintzi, A. (2008). Identification of *Campylobacter jejuni* genes involved in the response to acidic pH and stomach transit. Appl. Environ. Microbiol. *74*, 1583–1597.

Rhee, S.G., and Woo, H.A. (2011). Multiple functions of peroxiredoxins: peroxidases, sensors and regulators of the intracellular messenger H(2)O(2), and protein chaperones. Antioxid. Redox Signal *15*, 781–794.

Rhee, S.G., Chae, H.Z., and Kim, K. (2005). Peroxiredoxins: a historical overview and speculative preview of novel mechanisms and emerging concepts in cell signaling. Free Radic. Biol. Med. *38*, 1543–1552.

Seaver, L.C., and Imlay, J.A. (2001). Alkyl hydroperoxide reductase is the primary scavenger of endogenous hydrogen peroxide in *Escherichia coli*. J. Bacteriol. *183*, 7173–7181.

Seaver, L.C., and Imlay, J.A. (2004). Are respiratory enzymes the primary sources of intracellular hydrogen peroxide? J. Biol. Chem. *279*, 48742–48750.

Sellars, M.J., Hall, S.J., and Kelly, D.J. (2002). Growth of *Campylobacter jejuni* supported by respiration of fumarate, nitrate, nitrite, trimethylamine-*N*-oxide, or dimethyl sulfoxide requires oxygen. J. Bacteriol. *184*, 4187–4196.

Skirrow, M.B. (1977). Campylobacter enteritis: a 'new' disease. Br. Med. J. *2*, 9–11.

Stintzi, A., van Vliet, A.H., and Ketley, J.M. (2008). Iron metabolism, transport, and regulation. In *Campylobacter*, 3rd edn, Nachmkin, I., Szymanski, C.M., and Blaser, M.J., eds. (ASM Press, Washington, DC), pp. 591–610.

Storz, G., Tartaglia, L.A., and Ames, B.N. (1990). Transcriptional regulator of oxidative stress-inducible genes: direct activation by oxidation. Science *248*, 189–194.

Svensson, S.L., Davis, L.M., MacKichan, J.K., Allan, B.J., Pajaniappan, M., Thompson, S.A., and Gaynor, E.C. (2009). The CprS sensor kinase of the zoonotic pathogen *Campylobacter jejuni* influences biofilm formation and is required for optimal chick colonization. Mol. Microbiol. *71*, 253–272.

Toledano, M.B., Kullik, I., Trinh, F., Baird, P.T., Schneider, T.D., and Storz, G. (1994). Redox-dependent shift of OxyR-DNA contacts along an extended DNA-binding site: a mechanism for differential promoter selection. Cell *78*, 897–909.

van Vliet, A.H., Wooldridge, K.G., and Ketley, J.M. (1998). Iron-responsive gene regulation in a *Campylobacter jejuni fur* mutant. J. Bacteriol. *180*, 5291–5298.

van Vliet, A.H., Baillon, M.L., Penn, C.W., and Ketley, J.M. (1999). *Campylobacter jejuni* contains two fur homologs: characterization of iron-responsive regulation of peroxide stress defense genes by the PerR repressor. J. Bacteriol. *181*, 6371–6376.

van Vliet, A.H., Rock, J.D., Madeleine, L.N., and Ketley, J.M. (2000). The iron-responsive regulator Fur of *Campylobacter jejuni* is expressed from two separate promoters. FEMS Microbiol. Lett. *188*, 115–118.

van Vliet, A.H., Ketley, J.M., Park, S.F., and Penn, C.W. (2002). The role of iron in *Campylobacter* gene regulation, metabolism and oxidative stress defense. FEMS Microbiol. Rev. *26*, 173–186.

Walker, R.I. (2005). Considerations for development of whole cell bacterial vaccines to prevent diarrheal diseases in children in developing countries. Vaccine *23*, 3369–3385.

Wang, G., and Maier, R.J. (2004). An NADPH quinone reductase of *Helicobacter pylori* plays an important role in oxidative stress resistance and host colonization. Infect. Immun. *72*, 1391–1396.

Wang, G., Alamuri, P., and Maier, R.J. (2006). The diverse antioxidant systems of *Helicobacter pylori*. Mol. Microbiol. *61*, 847–860.

Wassenaar, T.M., Engelskirchen, M., Park, S., and Lastovica, A. (1997). Differential uptake and killing potential of *Campylobacter jejuni* by human peripheral monocytes/macrophages. Med. Microbiol. Immunol. (Berl.) *186*, 139–144.

Watson, R.O., and Galán, J.E. (2008). Interaction of *Campylobacter jejuni* with host cells. In *Campylobacter*, 3rd edn, Nachmkin, I., Szymanski, C.M., and Blaser, M.J., eds. (ASM Press, Washington, DC), pp. 289–296.

Wolf, S.G., Frenkiel, D., Arad, T., Finkel, S.E., Kolter, R., and Minsky, A. (1999). DNA protection by stress-induced biocrystallization. Nature *400*, 83–85.

Wood, Z.A., Schroder, E., Robin Harris, J., and Poole, L.B. (2003). Structure, mechanism and regulation of peroxiredoxins. Trends Biochem. Sci. *28*, 32–40.

Yamasaki, M., Igimi, S., Katayama, Y., Yamamoto, S., and Amano, F. (2004). Identification of an oxidative stress-sensitive protein from *Campylobacter jejuni*, homologous to rubredoxin oxidoreductase/rubrerythrin. FEMS Microbiol. Lett. *235*, 57–63.

Zaki, M.H., Akuta, T., and Akaike, T. (2005). Nitric oxide-induced nitrative stress involved in microbial pathogenesis. J. Pharmacol. Sci. *98*, 117–129.

Zanotti, G., Papinutto, E., Dundon, W., Battistutta, R., Seveso, M., Giudice, G., Rappuoli, R., and Montecucco, C. (2002). Structure of the neutrophil-activating protein from *Helicobacter pylori*. J. Mol. Biol. *323*, 125–130.

Zheng, M., and Storz, G. (2000). Redox sensing by prokaryotic transcription factors. Biochem. Pharmacol. *59*, 1–6.

Zheng, M., Aslund, F., and Storz, G. (1998). Activation of the OxyR transcription factor by reversible disulfide bond formation. Science *279*, 1718–1721.

Oxidative Stress Survival During *Campylobacter* Transmission and Infection

12

Rebecca Handley, Mark Reuter and Arnoud H.M. van Vliet

Abstract

Illness from *Campylobacter jejuni* infection is a widespread burden in the developed world. The spread and high incidence of this pathogen is in no small part due to its ability to survive exposure to exogenous stresses. This provides *C. jejuni* with the opportunity to spread to and colonize a variety of niches that lack the conditions in which *C. jejuni* can grow. In this chapter we will describe the common transmission routes of *C. jejuni*, followed by a discussion of the systems and mechanisms that are part of the protective responses *C. jejuni* has at its disposal. Finally, we will discuss the challenges faced by *C. jejuni* as a microaerophilic organism, which has to survive in an aerobic world.

Campylobacter jejuni transmission and aerotolerance

The members of the genus *Campylobacter* are both microaerophiles and capnophiles. They have a limited tolerance to aerobic environments but still require oxygen as a terminal electron acceptor for respiration (Hoffman *et al.*, 1979a,b). They also require increased carbon dioxide concentrations, although the reasons for this are unclear. The thermophilic *Campylobacter* species *C. jejuni* and *C. coli* are now firmly established as the most common causal agents of acute bacterial enteritis worldwide (Tauxe, 1992). They are also associated with auto-inflammatory diseases such as reactive arthritis and Guillain-Barré syndrome (Berden *et al.*, 1979; Nachamkin *et al.*, 1998). *C. jejuni* and *C. coli* differ in many aspects, including transmission routes. Most information on aerotolerance and oxidative stress resistance is available for *C. jejuni* and hence we will focus on *C. jejuni* in this chapter.

C. jejuni is a prolific pathogen, shown by the high number of infections seen each year (predicted to be 371,000 cases of campylobacteriosis in the UK per year (2010). Infections can arise from a variety of sources (Fig. 12.1) however in the developed world, animals for consumption are considered the primary source of *Campylobacter* infections (Adak *et al.*, 1995; Harris *et al.*, 1986). *C. jejuni* is a fastidious organism, it has an optimum growth temperature of 42°C, and has specific requirements for growth, such as a combination of complex media and defined atmospheric conditions (5–10% v/v oxygen, 5–10% carbon dioxide) (Atack and Kelly, 2009b). These conditions are thought to resemble those occurring in the natural habitat of *C. jejuni*, the avian caecum (Atack, 2010; Jacobs-Reitsma *et al.*, 2000).

C. jejuni is exposed to atmospheric oxygen concentrations during transmission, which results in an oxidative stress burden upon this microaerophilic organism. Further oxidative stress burdens are experienced by *C. jejuni* during host invasion, due to the host immune response. Therefore it is not surprising that bacterial species such as *C. jejuni* that have to cope with drastic transitions in oxygen availability and oxidative stress, possess an arsenal of cellular defences to combat the damaging effects of oxidative stress (Purdy *et al.*, 1999). Given the demands of life in an environment where oxygen levels may widely fluctuate, it is surprising that *C. jejuni* possesses a compact genome of only 1.7 Mbp, which is about one-third

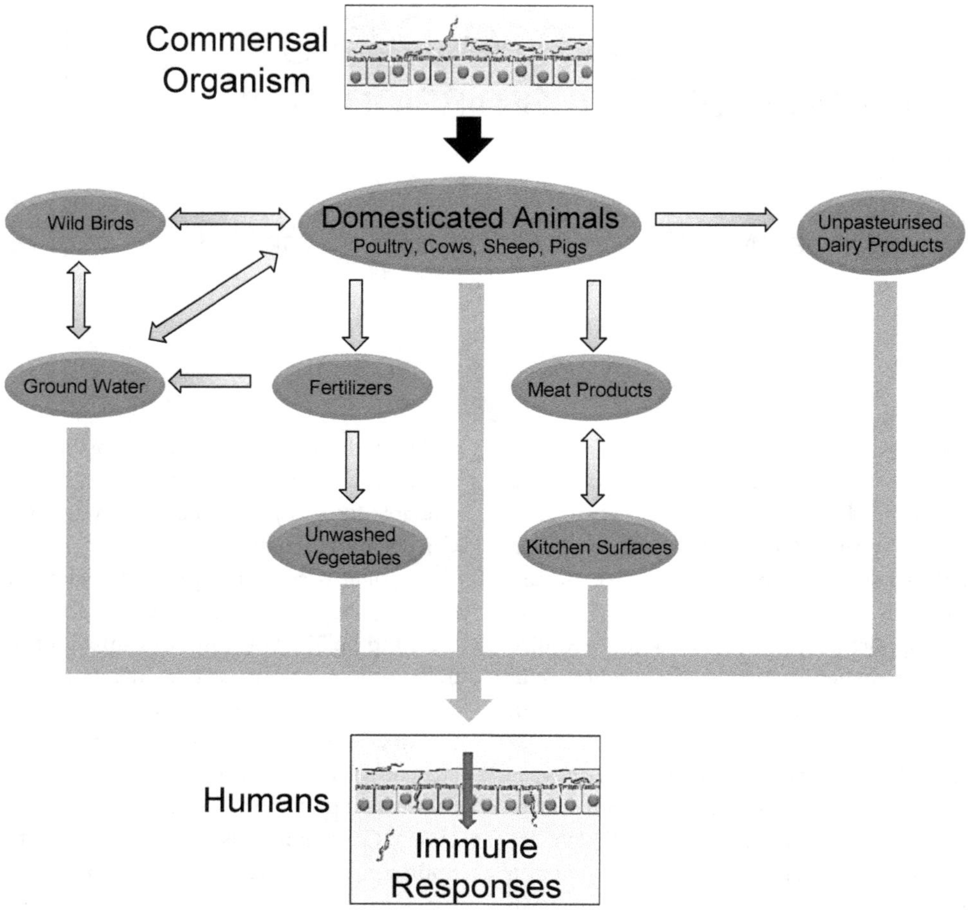

Figure 12.1 Sources and distribution of *C. jejuni* leading to human infection: *C. jejuni* is a commensal organism in poultry and spreads quickly through bird flocks. *C. jejuni* multiplies rapidly in infected birds and is excreted in high numbers. The faecal–oral route allows *C. jejuni* to spread to other farm animals and to wild birds – a potential source of contamination between flocks. The use of animal manure as a fertilizer promotes the further spread of *C. jejuni*, many vegetables are not cooked or heated, and may be consumed unwittingly by humans who are unaware they may carry *C. jejuni* on their surface. Surface water run-off from animal farms can contaminate ground water supplies (such as ponds and streams) with *C. jejuni*. Growing research provides evidence for the planktonic survival of the *Campylobacter* species in water, whether via interaction with protozoa or the formation of biofilms. The most common sources of *C. jejuni* infection are undercooked meat products and unpasteurized dairy, respectively. Contaminated meat may be a source of further outbreaks, and spread *C. jejuni* to other foods in food preparation areas when food hygiene standards are not adhered to.

of the size of that of well-studied enteric pathogens like *E. coli* and *Salmonella enterica*. This difference in genome size is reflected in the lower number of oxidative stress defence systems that have been identified in *C. jejuni*. For example, there are no orthologues of the well-characterized oxidative stress regulatory systems such as OxyRS, RpoS and SoxRS, which control the expression of defence systems in response to oxidative stress (Parkhill *et al.*, 2000). Similarly, there is a relative scarcity of genes encoding known proteins for degrading or inactivating reactive oxygen species. However it is apparent that despite this defensive shortfall there is still a high incidence of *C. jejuni*-related illness. Within the next section we will discuss the *C. jejuni* effector proteins that are directly involved in detoxifying oxidative stress generated during *C. jejuni* transmission, with particular emphasis on aerotolerance and surviving aerobic stress (as shown in Figs. 12.1 and 12.2).

As of yet, no dedicated pathway for aerobic survival or aerotolerance has been identified in *Campylobacter*, yet many different genes have been implicated in the aerotolerance of *C. jejuni*, with mutants typically showing a reduced ability to survive aerobic exposure. Here we will discuss the generation of oxidative stress during the transmission route of *C. jejuni*, and some of the major detoxification pathways present.

Reactive oxygen species

Aerobic metabolism generates reactive oxygen species (ROS) within the bacterial cell. ROS are generated during the incomplete reduction of oxygen to water and are also produced by a step by step reduction of molecular oxygen to more reactive compounds (Fig. 12.3) (Atack and Kelly, 2009a; van Vliet *et al.*, 1999). These reactive compounds include the superoxide anion (O_2^-), hydrogen peroxide (H_2O_2) and the hydroxyl radicals ($OH^•$) (Atack and Kelly, 2009a; Imlay, 2003). *C. jejuni* will also encounter ROS as part of host defences and as part of natural processes. ROS are produced both inside and outside *C. jejuni* animal hosts and are a component of the lysosome, which *Campylobacter* will be exposed to during phagocytosis (Svensson *et al.*, 2008). H_2O_2 and O_2^- are also formed by the auto-oxidation of respiratory dehydrogenases (Imlay, 2008).

ROS may also be formed by reactions of oxygen with cellular iron, this will be discussed later in a later section. ROS produced from a variety of sources still have the same detrimental effects on

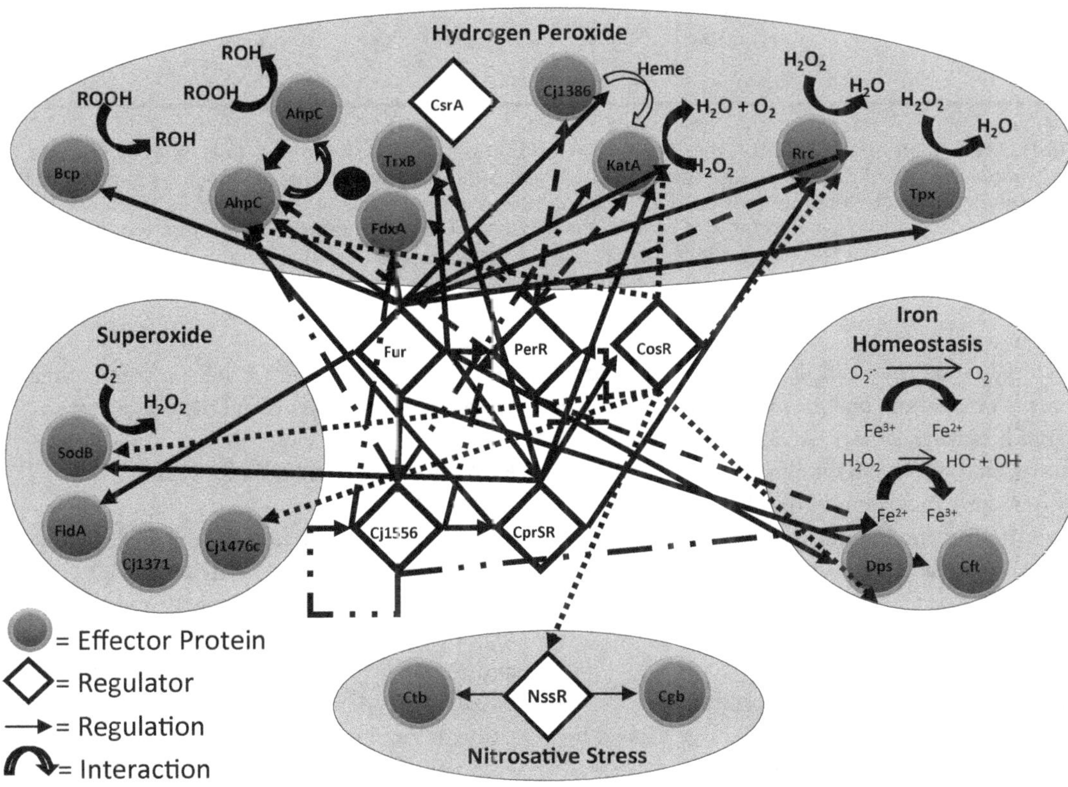

Figure 12.2 Function and regulation of *C. jejuni* major stress defence systems that aid *C. jejuni* survival during transmission. Effector proteins are segregated based on their proposed function, and discussed in more detail below. The regulation of effector proteins is indicated by the presence of arrows, which vary for each regulator; Fur (black arrow), PerR (black, dashed arrow), CosR (black, dotted arrow), Cj1556 (Grey, dotted and dashed arrow) and CprSR (grey arrow).

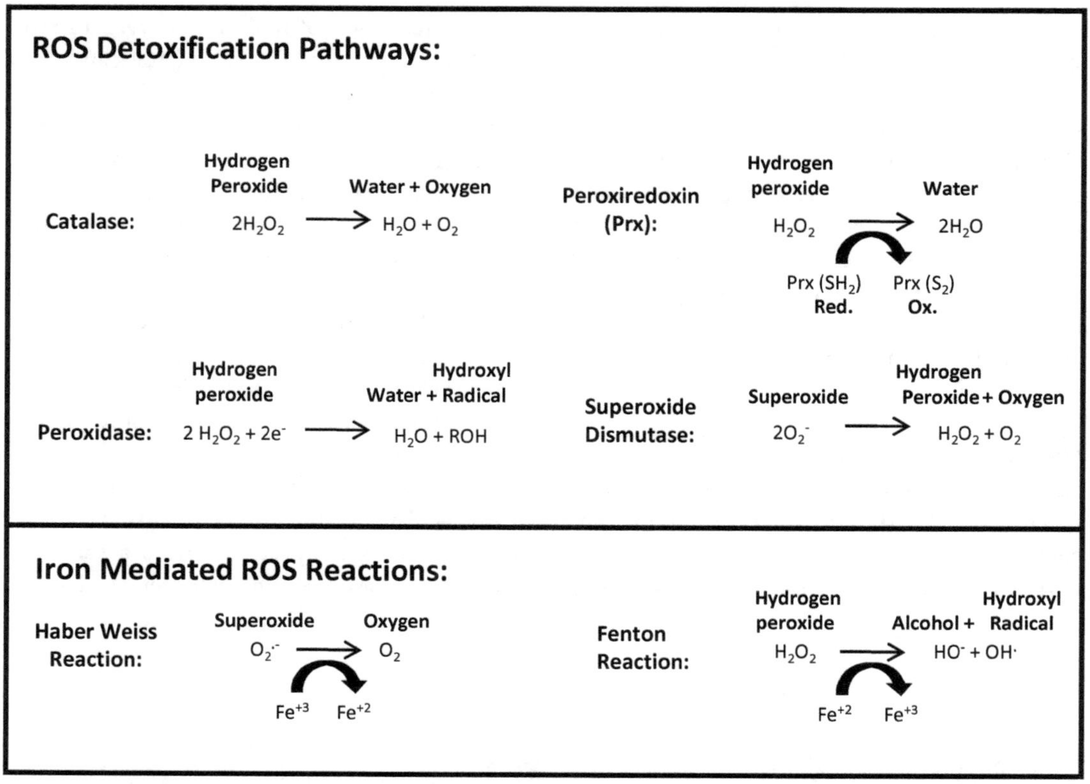

Figure 12.3 Major pathways for the generation and break down of the reactive oxygen species in *Campylobacter jejuni*. *C. jejuni* possesses an assortment of proteins capable of detoxifying members of the ROS, some of which generate further ROS – for example superoxide dismutase catalyses the break down of superoxide yet forms hydrogen peroxide in the process. The hydrogen peroxide must then be further detoxified by other oxidative stress effector proteins. Processes such as these may explain the apparent co-regulation of so many oxidative stress effector proteins in *C. jejuni*.

bacterial cells. They can damage DNA and proteins via oxidation, and also induce peroxidation of lipids causing damage to cell membranes (Svensson *et al.*, 2008). Methionine residues oxidized by ROS are a major source of cell damage; also the formation of methionine sulfoxide can produce conformational changes in proteins (Skaar *et al.*, 2002; Weissbach *et al.*, 2002).

Aerobic stress effector proteins

To actively resist or avoid the damage caused by ROS, protective proteins are constantly expressed in low levels by bacterial cells to protect from, and repair any cellular damage (Traore *et al.*, 2006). Characteristically, these proteins are negatively controlled by regulators creating an inducible, flexible response system (Mongkolsuk and Helmann, 2002). Fig. 12.2 gives a brief summary of major known effector proteins and their regulators.

Peroxide resistance effector proteins

Peroxides are generated at various stages throughout *C. jejuni* transmission and can damage *C. jejuni* both intra- and extracellularly. Peroxides form a major part of the ROS stress burden on *C. jejuni*, and as result the bacterium has evolved a number of different detoxification pathways. *C. jejuni* contains multiple detoxification pathways for combating general hydroperoxides and more specifically hydrogen peroxide, with systems proposed to function in different cell locations, such as the periplasm or cytoplasm. Some of these major peroxide detoxification systems, as shown in Fig. 12.2, will be discussed here.

Alkyl hydroperoxide reductase, ferredoxin and thioredoxin reductase

In *C. jejuni* an alkyl hydroperoxide reductase (AhpC, Cj0334) has been linked to increased aerotolerance (Baillon *et al.*, 1999). AhpC proteins are also referred to as Thiol-specific antioxidants (TSA) or peroxiredoxins (see Fig. 12.3). When reduced they degrade alkylhydroperoxides to the corresponding alcohols whilst becoming oxidized in the process ($ROOH + AhpC^{red} \rightarrow ROH + AhpC^{ox}$). Oxidized AhpC proteins can be recycled back to the reduced protein by a recycling partner, similar to the AhpF protein in *E. coli*. AhpF re-reduces AhpC by transferring electrons from NAD(P)H to the oxidized AhpC (Poole, 1996; Poole and Ellis, 1996). However in *C. jejuni* and the closely related pathogen *Helicobacter pylori*, no AhpF homologues have been found – indicating an alternative method of AhpC recycling (Tomb *et al.*, 1997). Proteins such as ferredoxin (FdxA) or thioredoxin reductase (TrxB) have been suggested as potential recycling partners in *C. jejuni*, their roles as potential recycling partners for AhpC will be discussed below.

The role of AhpC in *C. jejuni* aerotolerance was noted by Baillion *et al.* (1999), who showed that in stationary phase cells, an *ahpC* mutant had a significantly reduced ability to survive in an aerobic environment (aerotolerance) compared to the wild-type strain. Alkyl hydroperoxide reductases are able to detoxify peroxidated molecules and destroy hydroperoxide intermediates generated by reactive oxygen species, which typically challenge *C. jejuni* during aerobic exposure (Baillon *et al.*, 1999). Ferredoxins (FdxA) are reducing agents (the *C. jejuni fdxA* gene encodes a 4Fe–4S ferredoxin) and *fdxA* is divergently located from *ahpC* in the *C. jejuni* genome, which is why it is a tempting candidate as an AhpC recycling partner (van Vliet *et al.*, 2001). However, both *ahpC* and *fdxA* have differential regulation under iron depletion. Whereas *ahpC* expressed is induced in low iron conditions under the control of PerR and CosR, *fdxA* is repressed and regulated by Fur (Baillon *et al.*, 1999; Holmes *et al.*, 2005; Hwang *et al.*, 2011; Palyada *et al.*, 2009a; van Vliet 2001). In *C. jejuni* strain 81116, an *fdxA* mutant showed reduced aerotolerance and increased sensitivity to hydrogen peroxide and cumene hydroperoxide, indicating a potential role during *C. jejuni* survival on food surfaces and in the environment (van Vliet *et al.*, 2001; van Vliet and Ketley, 2001).

Thioredoxin reductase (TrxB) is a flavoprotein that reduces disulfide containing substrates via electron transfer from pyrimidine nucleotides by a flavin carrier. In the *C. jejuni* genome, it is located alongside *trxA*, which encodes a thioredoxin that undergoes reversible oxidation and reduction mediated by TrxB and NADPH (Arner and Holmgren, 2000; Cohen *et al.*, 1993). Under iron limitation both TrxA and TrxB have increased expression, owing to TrxB's strong reductase activity and similar regulation to AhpC, it makes a plausible candidate as a recycling partner (Holmes *et al.*, 2005). However both TrxB and FdxA are regulated by Fur whereas AhpC is not, indicating further levels of complexity in the search for a recycling partner for AhpC.

Bcp (Cj0271)

Bacterioferritin Comigratory protein (Bcp) is a thiol peroxidase with a broad substrate range in *C. jejuni*, although in *E. coli* and *H. pylori* Bcp uses linoleic acid hydroperoxide as a preferred substrate. In *C. jejuni* Bcp provides important protection against protein oxidation, aerobic exposure and lipid peroxidation. Lipid peroxidation is particularly important when host inflammatory responses are triggered during infection where membrane lipids will be oxidized (Atack *et al.*, 2008; Jeong *et al.*, 2000; Wang *et al.*, 2005).

Catalase and Cj1386

The catalase enzyme is ubiquitous in bacteria, several different types of catalase enzymes exist, these show intricate patterns of regulation and substrate specificity. The catalase enzyme rapidly dismutates hydrogen peroxide (H_2O_2) to water (H_2O) and oxygen (O_2), removing the hydrogen peroxide generated by the breakdown of superoxide by SodB (Grant and Park, 1995; van Vliet and Ketley, 2001). *C. jejuni* contains a single haem-cofactored catalase that is important for *C. jejuni* survival against aerobic and oxidative stresses in the environment and against macrophage killing by oxidative bursts – therefore it may also contribute to cell survival during infection (Grant and Park, 1995). However, in *C. coli*, *katA* mutants did not show a reduced ability to survive aerobically on chicken skin and skimmed milk (Purdy *et al.*, 1999; Stead and Park, 2000). This hints at the presence of alternative hydrogen peroxide detoxification pathways, for example AhpC (see above) or Rrc encoded by the *Cj0012c* gene, a recently characterized unique desulforubrerythrin, which possesses a significant hydrogen peroxide reductase activity (Pinto *et al.*, 2011; Yamasaki *et al.*, 2004).

In the *C. jejuni* genome directly downstream from catalase is KapA (Cj1386) and despite not being co-transcribed with catalase, it is believed to function within the same hydrogen peroxide detoxification pathway. KapA may function in haem trafficking to catalase, haem is required for catalase to breakdown hydrogen peroxide, however the exact role that haem plays is still not understood (Flint *et al.*, 2012).

Tpx (Cj0779)

The *C. jejuni* thiol peroxidase (Tpx) belongs to a group of peroxiredoxins that are commonly found throughout bacterial species (Cha *et al.*, 2004; Olczak *et al.*, 2003; Rho *et al.*, 2006). In other bacteria Tpx is capable of breaking down a variety of peroxides, similar to Bcp. However in *C. jejuni* Tpx is a dedicated hydrogen peroxide reductase and shows no activity against other hydroperoxides. A *C. jejujni tpx* null mutant shows poor growth capabilities in aerobic conditions and *tpx* also has increased expression when *C. jejuni* cells are in a biofilm (Atack *et al.*, 2008). Both of these findings are indicative of a role for Tpx in the survival of *C. jejuni* during transmission throughout aerobic environments.

Rrc (Cj0012c)

The rubredoxin oxidoreductase-rubreythrin like protein of *C. jejuni* (Rrc), also termed desulforubreythin (DRbr), is a non-haem containing iron protein. It was first recognized in *C. jejuni* for its apparent degradation after *C. jejuni* cells were exposed to hydrogen peroxide (Yamasaki *et al.*, 2004). Rrc is a novel multidomain metalloprotein and the first characterized example of a protein containing three distinct structural domains; an N-terminal desulforedoxin like domain, a four helix bundle domain and a C-terminal rubredoxin domain. Rrc is conserved in all sequenced *Campylobacter* species and has a significant NADH-linked hydrogen peroxide reductase activity (Pinto *et al.*, 2011). Rrc is also unique in that currently it is the only oxidative stress effector protein to be jointly regulated by all four major oxidative stress regulation systems in *C. jejuni*, incorporating PerR, Fur, CosR and CprSR regulation (Holmes *et al.*, 2005; Hwang *et al.*, 2011; Palyada *et al.*, 2009a; Svensson *et al.*, 2009). This tight regulation hints that Rrc has a significant role in oxidative stress detoxification, one that has not yet been fully recognized.

Superoxide effector proteins

Superoxide is a harmful member of the ROS and the sole *C. jejuni* superoxide dismutase was the first ROS responsive gene identified in *C. jejuni*. It is responsible for catalysing the dismutation of superoxide to H_2O_2 and dioxygen (Svensson *et al.*, 2008). SOD genes are classified by their metal co-factors, commonly copper-zinc, nickel, manganese or iron. However, unlike *E. coli*, which expresses three SOD proteins (Fe, Mn and CuZn-cofactored), *sodB* encodes the sole SOD present in *C. jejuni* and requires iron as a cofactor (Lynch and Kuramitsu, 2000; Pesci *et al.*, 1994; Purdy and Park, 1994). SodB and its role during *C. jejuni* transmission are discussed here, along with other proteins that have been shown to function in response to paraquat, a superoxide generating compound.

Superoxide dismutase

The superoxide dismutase enzyme is ubiquitous in all kingdoms of life, and catalyses the dismutation of superoxide into hydrogen peroxide and oxygen ($2O_2^- + 2H^+ \rightarrow H_2O_2 + O_2$). The hydrogen peroxide generated during this process is then further detoxified by catalase (KatA). Superoxide dismutase enzymes utilize metal cofactors for their function, with copper-zinc being the common form in eukaryotes, while bacteria most commonly use manganese and iron-cofactored SOD enzymes. *Campylobacter* species have a single SOD enzyme (Cj0169), which contains iron as co-factor (Purdy and Park 1995. Pesci *et al* 1995); this plays important roles in the protection of *C. jejuni* and *C. coli* against superoxides and other oxidative stresses. In *C. coli*, *sodB* null mutants showed reduced aerotolerance in stationary phase cells, had a reduced ability to survive freeze–thawing, and were less able to survive on foods such as chicken skin and skimmed milk (Purdy *et al.*, 1999; Stead and Park, 2000)

Other superoxide effector proteins: flavodoxin (FldA), Cj1371, Cj1476c

Garenaux *et al* (2008) looked at the responses of *C. jejuni* to paraquat, a superoxide anion generating compound. Using global proteomic profiling they observed differential regulation of a number of *C. jejuni* genes in response to paraquat exposure, some of these were known oxidative stress response effector proteins, whereas other proteins identified were less well characterized. These proteins include FldA, Cj1476c, Cj1371 and the orphan response regulator CosR.

Flavodoxin and Cj1476c, a pyruvate-flavodoxin oxidoreductase were overexpressed in response to paraquat, both having roles in oxidative stress responses or redox status. *C. jejuni* flavodoxin is an electron acceptor for the pyruvate oxidoreductase complex (POR) responsible for catalysing NADPH production by oxidative decarboxylation. Flavodoxin is iron repressed and located in an operon (*Cj1384c-Cj1383c-fldA*) divergently transcribed from catalase (*katA*). In comparison, little is known about Cj1371, it is co-ordinately expressed with flagellar proteins although not directly involved in motility (Garenaux *et al.*, 2008; Palyada *et al.*, 2004). In other bacteria such as *Pseudomonas putida*, homologues of Cj1371 have been shown to function in response to other sources of oxidative stress, namely phenol induced oxidative stress (Santos *et al.*, 2004).

Nitrosative stress effector proteins

Nitric oxide and its redox intermediates are intricately linked with the ROS. Nitrosative stress forms part of the protective response of humans, which *C. jejuni* will encounter during infection and invasion. The stomach is a source of nitrosative stress, generating nitric oxide during the digestion of food. Nitrosating compounds also form part of the protective inducible defence systems of intestinal epithelial cells, which will be encountered by *C. jejuni* during cell invasion (Dykhuizen *et al.*, 1996; Elvers *et al.*, 2004).

Haemoglobins (Ctb, Cgb), NssR

C. jejuni expresses two putative haemoglobins, a single domain globin, known as Cgb (*Campylobacter* globin, Cj1586), and a truncated globin, Ctb (Cj0465c). Cgb has been implicated in a protective response to nitrosative stress, being up-regulated when *C. jejuni* is challenged with nitrosative stress. Mutants lacking Cgb are hypersensitive to nitrosating agents, which may indirectly link with oxygen-related stresses. *cgb* mutants also show a reduced ability to survive exposure to aerobic environments (Elvers *et al.*, 2004; Wainwright *et al.*, 2005). The *C. jejuni* truncated haemoglobin Ctb is also thought to function in the nitrosative stress response regulon, which is controlled by the regulatory protein NssR (Cj0466), although it may also play a role in in moderating O_2 flux in *C. jejuni* (Elvers *et al.*, 2005; Wainwright *et al.*, 2005)

Iron homeostasis effector proteins

Iron homeostasis is carefully monitored on many levels including uptake, utilization, and storage. To date, nine iron uptake and transport systems have been identified in *C. jejuni* (van Vliet *et al.*, 2002). When excess iron and ROS are both present there is a risk of cell damage (Palyada *et al.*, 2009b). Harmful radicals are formed during reactions such as the Haber–Weiss reaction and Fenton chemistry (Fig. 12.3). Ferrous iron in the form of cellular iron, catalyses electron transfer to H_2O_2, aiding the formation of the hydroxyl radical, a strong aqueous oxidant (Imlay, 2003; Storz and Imlay, 1999; Touati, 2000). It is for this reason that iron homeostasis is often paired with oxidative stress responses within the cell, as oxidative stress response proteins use iron cofactored prosthetic groups. Also excess iron may lead to iron toxicity in the cell (Storz and Imlay, 1999; Touati, 2000; van Vliet *et al.*, 2002).

Iron homeostasis proteins that are also known to function during aerobic stress are illustrated in Fig. 12.2, these include *Campylobacter* ferritin (Cft), a Dps homologue (Dps) and the ferric uptake regulator (Fur).

Ferritin (Cft)

Campylobacter ferritin (Cft) has a significant role in *C. jejuni*, functioning in both iron storage and intracellular iron homeostasis. Iron homeostasis has a dual role, by limiting intracellular free iron concentrations ferritin also protects *C. jejuni* from iron mediated oxidative stress, such as the Fenton and Haber Weiss reactions (Garenaux *et al.*, 2008). In other species, overexpression of ferritin has been shown to increase aerotolerance by removing free iron from the cell. This prevents the Fenton reaction from occurring and allows bacteria to survive in an aerobic environment for much longer (Touati *et al.*, 1995; Wai *et al.*, 1996).

Dps

The Dps protein is encoded by the *Cj1534c* gene, it derives its name from the *E. coli* counterpart (DNA binding protein from starved cells) and is important for oxidative stress resistance. It may play a role similar to bacterioferritin which protects DNA by aspecific binding whilst sequestering iron, thus effectively reducing the opportunity of free metal ions to produce harmful Fenton Chemistry intermediates (Ishikawa *et al.*, 2003).

The *C. jejuni* Dps protein shows high affinity DNA binding in the presence of Fe^{2+} or hydrogen peroxide under acidic conditions, as shown with both electrophoretic mobility shift assays and surface plasmon resonance (Huergo *et al*, 2013). Moreover, Dps–DNA complexes are protected from DNase I degradation and damage from hydroxyl radicals, suggesting a direct role for *C. jejuni* Dps in protecting DNA from oxidative damage. Two studies have shown that inactivating the *dps* gene in *C. jejuni* results in increased sensitivity to hydrogen peroxide (Huergo *et al*, 2013; Ishikawa *et al*, 2003). Direct DNA protection may account for, or contribute to, this phenotype. The fact that pH conditions below neutral pH increased DNA-binding point to a role in protecting DNA at acidic pH, which may explain why *dps* mutants failed to colonize both piglet and poultry models (Theoret *et al.*, 2011, 2012).

Regulation of aerobic stress responses

PerR and Fur

Gene regulation in *C. jejuni* is complex and control of oxidative stress effector proteins is no exception. PerR – the peroxide regulator, and Fur – the ferric uptake regulator, are two major transcriptional regulators in *C. jejuni*, both belonging to the Fur family of metalloproteins. A genome wide transcriptional profile of the PerR regulon revealed a total of 143 genes that are differentially regulated by PerR in response to oxidative stress (Palyada *et al.*, 2009b). This shows PerR to be a major regulator of oxidative stress responses in *C. jejuni*, and as seen in Fig. 12.2 both Fur and PerR are known to regulate many of the major known effector proteins involved in the protective responses of *C. jejuni* to aerobic and oxidative stress that are encountered during transmission. The PerR and Fur regulons have some functional overlap (approximately 50 co-regulated genes in *C. jejuni*), this is perhaps due to the interplay iron uptake and oxidative stress (Palyada *et al.*, 2009b; van Vliet *et al.*, 1999). As well as Fur and PerR, several other regulators have been recognized as controlling regulation of oxidative stress effector proteins, these are summarized below.

CprSR

The *Campylobacter* planktonic growth regulation system *cprSR* (*cj1227–1226*) encodes a two component regulatory system, which consists of a membrane-associated signal-transducing histidine kinase (CprS) and a cytoplasmic response regulator (CprR). *C. jejuni* cells lacking CprS show colonization defects in chickens, but enhanced motility and biofilm formation (Svensson *et al.*, 2009). The CprSR system also influences transcription levels of genes encoding different proteins involved in oxidative stress responses, such as *katA*, *trxB* and *ahpC* (repressed) and *sodB* and *rrc* (induced).

Cj1556

The *C. jejuni* genome encodes several putative regulatory proteins for which no specific function can be predicted. One of these genes (*cj1556*) encodes a MarA family transcriptional regulator, when inactivated results in a reduced ability to form biofilms, decreased infection of the *Galleria melonella* invertebrate virulence model and increased sensitivity to aerobic and oxidative stress (Gundogdu *et al.*, 2011).

CosR

The *Campylobacter* oxidative stress regulator is an essential OmpR type response regulator in *C. jejuni* that controls expression of approximately 93 proteins, including some of the major oxidative stress effector proteins (SodB, Rrc, Dps, AhpC and KatA) (Hwang *et al.*, 2011, 2012). The *cosR* gene in *C. jejuni* is conserved and cannot be inactivated. In order to study genes under its regulation, expression of *cosR* was suppressed using antisense peptide nucleic acids (PNA).

CosR differentially regulates expression of a number of genes important to the *C. jejuni* protective response to oxidative stress, including *dps*, *rrc*, *ahpC*, *sodB*, *luxS* and *katA*. LuxS itself has been shown to regulate *ahpC* and *tpx* in the presence of hydrogen peroxide in *C. jejuni* (He *et al.*, 2008). CosR acts a repressor of *sodB*, *dps*, *rrc* and *luxS*, and an activator of *katA* and *ahpC* (Hwang *et al.*, 2012). Expression of *cosR* in *C. jejuni* can be manipulated by the presence of oxidative stress. Expression was reduced in response to challenging *C. jejuni* with paraquat, a super-oxide generating compound, which is in keeping with its role in repressing oxidative stress responses (Garenaux *et al.*, 2008). However expression was not altered by the addition of hydrogen peroxide, indicating CosR has more functions in response to superoxide stress, which is implicated by its regulation of superoxide dismutase (Hwang *et al.*, 2011).

CsrA

CsrA is a homologue of the *E. coli* carbon storage regulator and a putative transcriptional regulator in *C. jejuni*. In *E. coli* CsrA is an RNA-binding global regulator that mediates gene expression by post transcriptionally regulating translation, decaying specific mRNA targets (Liu *et al.*, 1997; Romeo, 1996). More specifically, CsrA inhibits the ribosome from binding the ribosome binding site by itself binding the 5′ untranslated region of target mRNAs. This prevents the initiation of translation and may increase or decrease the half-life of the mRNA (Baker *et al.*, 2002; Romeo, 1998). CsrA also potentially has many roles in *C. jejuni*, it has been identified as having regulatory roles in motility, adherence and oxidative stress responses. In *C. jejuni* strain 81-176, a *csrA* mutant had a reduced ability to survive oxidative stress, including hydrogen peroxide stress and reduced ability to survive aerobic exposure compared to the wild-type. A *csrA* mutant also had reduced motility and biofilm formation, indicating an activating role for CsrA in these processes. Interestingly, the *csrA* mutant had an increased ability to invade human cells, this suggests that CsrA may have important functions during *C. jejuni* pathogenesis. At present little is known about *C. jejuni* CsrA, such as details of its mRNA targets and during what stages of the *C. jejuni* growth phase it is active. Further work is required to understand this regulator; however, it is clear it has important functions in *C. jejuni* survival after exposure to stress (Fields and Thompson, 2008).

Oxidative stress responses during transmission of *C. jejuni*

Studies on risk factors and prevalence of *Campylobacter* species pathogenic to humans have shown that the organisms are found in many places, both inside the gut of birds and other domesticated and wild animals, but there are also environmental sources such as water (Doorduyn *et al.*, 2010). *C. jejuni* is known to rapidly spread within poultry populations via the faecal–oral route, and its association with food necessitates survival outside the intestine in atmospheric oxygen conditions during transmission. In general, host-associated bacteria must be able to survive the environmental stresses encountered during transmission, which in the case of *Campylobacter* means survival in conditions that are growth-limiting and stressful. Therefore, the successful transmission of *C. jejuni* will be dependent on the organism detecting and responding to environmental stresses such as oxygen, temperature and nutrient shortage if the bacterium is to survive. As previously mentioned, *C. jejuni* has a fastidious nature, which would be expected to limit the bacteria's ability to survive outside normal growth conditions. However, owing to the vast arsenal of stress defences, *C. jejuni* is remarkably resilient and able to survive transmission through a variety of environments, and these will be discussed below.

Oxidative stress and transmission of *C. jejuni* in an avian host

Poultry are considered to be the primary source of *C. jejuni* infection, responsible for up to 70% of campylobacteriosis in humans (Harris *et al.*, 1986; Stern, 1992). *C. jejuni* are largely regarded as commensal organisms forming part of the natural gut flora of chickens (Stern, 1992). The bacteria colonize the intestinal tract of poultry, specifically the mucosal layer overlying the epithelial cells of the small intestine and caeca (Newell and Fearnley, 2003). The immune systems of chickens provide little resistance to colonization by *Campylobacter*, which is illustrated by the absence of clinical illness and the relatively low infection dose required for colonization (40 CFU) (Cawthraw *et al.*, 1996; Newell and Fearnley, 2003). *C. jejuni* must pass through the avian digestive tract prior to colonization, but it has sufficient acid resistance mechanisms to survive exposure to the acidic proventriculus (glandular stomach) in chickens (Avila *et al.*, 1986; Le *et al.*, 2012). Studies indicate that young chicks become colonized with *C. jejuni* between 2 and 4 weeks of age, not from birth, suggesting that transmission occurs via an environmental source (Berndtson *et al.*, 1996; Evans and Sayers, 2000; Jacobs-Reitsma *et al.*, 1995). The initial source of *Campylobacter* is likely to vary from flock to flock, but all studies agree that once an initial bird is infected transmission throughout the whole flock occurs rapidly (Evans and Sayers, 2000; Jacobs-Reitsma *et al.*, 1995). *C. jejuni* are shed from colonized animals in high numbers and bird-to-bird contamination occurs via the faecal–oral route (Wesley *et al.*, 2000). The level of *Campylobacter* on a colonized chicken ranges between log 5–9 CFU/g in the small intestine and caeca reducing to between log 2–6 on a chicken carcass after slaughter and processing (Rosenquist *et al.*, 2006). It is likely that oxidative stress during freezing accounts for some reduction in *C. jejuni* numbers of chicken carcasses. The freeze–thaw cycle generates superoxide, in *C. coli* SodB had an important role in aiding bacterial survival during oxidative stress throughout the freeze thaw cycle. *C. coli* cells lacking SodB had reduced survival after freezing and thawing in the presence of oxygen (Sears *et al.*, 2011; Stead and Park, 2000).

Oxidative stress survival of *C. jejuni* outside the host

Large numbers of *Campylobacter* are continually excreted into the environment from avian hosts. Once out in the environment *C. jejuni* are exposed to atmospheric levels of oxygen, which is toxic to the bacteria. This highlights aerotolerance as a key survival feature for *C. jejuni*, which is required for successful transmission in and across avian hosts (Newell and Fearnley, 2003). In addition to aerobic stress, *Campylobacter* must also combat other environmental pressures, such as osmotic stress, nutrient starvation, non growth temperatures, pH variations and oxidative stress.

Once shed from a reservoir, transmission of *C. jejuni* can occur to many places in the food processing environment (see Fig. 12.1). Typically *C. jejuni* will make its way into domestic or commercial food preparation areas by the purchasing of contaminated poultry food products. Unless sufficient food hygiene standards are adhered too, cross-contamination of work surfaces and utensils may occur – spreading *C. jejuni* to other food products and surfaces (Cogan *et al.*, 1999). Cross-contamination of other food stuffs may also occur at earlier stages during poultry processing. Chicken manure has prominent uses as a fertilizer, giving *C. jejuni* an opportunity to contaminate food crops. *Campylobacter* species also affect other farm animals such as cattle, sheep and pigs, as well as small wild rodents (Manser and Dalziel, 1985). *C. jejuni is* also prominent in water supplies, especially rivers or standing water in close vicinity to poultry farms and the consumption of un-chlorinated water has been linked to infection by *C. jejuni* (Doorduyn *et al.*, 2010; Thomas *et al.*, 1998).

Oxidative stress survival during infection

To successfully infect humans, *C. jejuni* must survive the route through the human digestive system. Once ingested, *C. jejuni* is exposed to the harsh acidic environment of the stomach. The human stomach can reach pH levels ranging from 2–7 depending on the acid output of the subject, the stomach contents and food is often retained in the stomach for 30–60 min before continuing along the digestive tract (Dressman *et al.*, 1990; Siegel *et al.*, 1988). *C. jejuni* is capable of surviving exposure to ~ pH 4 for 30 min, potentially due to increased motility and increased expression of catalase, heat shock and general stress response proteins when challenged with a low pH (Le *et al.*, 2012). As well as acid stress, the stomach is also a source of nitrosative stress, generated by the break down of food-borne nitrate to nitrite (Dykhuizen *et al.*, 1996). As previously mentioned, Cgb has a role in the protective response to nitrosative stress and may aid *C. jejuni* survival through the stomach (Elvers *et al.*, 2004).

Once through the stomach, *C. jejuni* must begin its route though the small intestine, competing with natural gut flora for space and nutrients whilst also avoiding host immune responses and combating low oxygen levels, iron limitation, bile salts and hyperosmolarity (Park, 2005). Oxygen levels vary greatly within the entire gastrointestinal tract due to variations in blood flow and metabolism that cause shifts in local oxygen partial pressure. But particularly, there is a large oxygen gradient from the lumen of the intestine, which is regarded as anaerobic, compared to the sub-epithelium that is highly vascularized (Colgan and Taylor, 2010). Blood flow and the diffusion of oxygen from capillaries at the tips of villi generates a layer of oxygenation (approximately 70 μM) near to the mucosal surface. This oxygenated layer may act as a signal for aerotaxis or bacterial invasion and adhesion, however more research is required (Marteyn *et al.*, 2011). Initiation of the oxidative burst immune response from polymorphonuclear cells also uses large quantities of oxygen and results in depletion of oxygen and local hypoxia (Colgan and Taylor, 2010).

This oxygen deficiency presents a problem for *C. jejuni*, which cannot grow in an anaerobic environment. However a number of genes have been identified in *C. jejuni* that indicate the ability to utilize alternative electron acceptors in low oxygen conditions, such as fumarate, nitrate and nitrite (Kelly, 2005). Exposure to acid shock in the stomach may prime *C. jejuni* for invasion of the oxygenated epithelial, to avoid low oxygen and growth limiting environments (Colgan and Taylor, 2010; Le *et al.*, 2012).

Upon reaching the epithelial cells, *C. jejuni* interacts with the surface of the epithelial mucosal lining triggering host signal transduction events. These events lead to cytoskeletal rearrangement and eventually bacterial cell uptake, although epithelial cell invasion by *C. jejuni* is still not fully understood. However, it is clear that *C. jejuni* is able to adhere to and penetrate the epithelial cell barrier and reach the lamina propria, causing inflammation (Ketley, 1997; Kopecko *et al.*, 2001; van Alphen *et al.*, 2008). The mode of transport across the epithelial cell barrier is currently unclear, whether this occurs transcellularly through M cells or enterocytes, or paracellularly (Hu and Kopecko, 2008). Adherence and invasion are essential virulence factors but once through the epithelial mucosa *C. jejuni* must combat an array of internal host defences.

Conclusions and future trends

Oxidative stress is a key feature of the immune response of humans and a wide variety of animals. The investigation of oxidative stress responses holds great merit, not only allowing us to unravel how *C. jejuni* survives and persists in the environment, but also helping our understanding of how *C. jejuni* can delay detection and removal by immune responses. Overall, the picture of how *C. jejuni* survives oxidative stress is being constantly improved, but it is evident that oxidative stress defence is a highly coordinated process, with genes often under the transcriptional control of multiple regulators. This may give the organism a regulatory versatility, adapted to cope with the variety of sources of oxidative stress (Hwang *et al.*, 2011; Palyada *et al.*, 2009a).

Many of the oxidative stress effector proteins have now been discovered, but further work is required to unravel regulation of these responses. With several regulators of oxidative stress now having been identified in *C. jejuni*, it is important to unravel their interplay with one another and their significance in response to environmental stimuli. Further work is required to detail modes of regulation activation, and it is likely that structural insights into these regulatory proteins will prove enlightening. It is important to decipher the key defence mechanisms used by *C. jejuni* during environmental transmission, as this information can then be used to focus on combating *C. jejuni*, with a long term goal of reducing the amount of *C. jejuni* in the human food chain.

Acknowledgements

The author(s) gratefully acknowledge the support of the Norwich Research Park through a PhD-studentship to RH, and the Biotechnology and Biological Sciences Research Council (BBSRC) through the Institute Strategic Programmes IFR/08/3 (Biology and Complexity of Food-borne Bacterial Pathogens) and BB/J004529/1 (Gut Health and Food Safety).

References

Adak, G.K., Cowden, J.M., Nicholas, S., and Evans, H.S. (1995). The Public Health Laboratory Service national case–control study of primary indigenous sporadic cases of campylobacter infection. Epidemiol. Infect. *115*, 15–22.

van Alphen, L.B., Bleumink-Pluym, N.M., Rochat, K.D., van Balkom, B.W., Wosten, M.M., and van Putten, J.P. (2008). Active migration into the subcellular space precedes *Campylobacter jejuni* invasion of epithelial cells. Cell. Microbiol. *10*, 53–66.

Arner, E.S., and Holmgren, A. (2000). Physiological functions of thioredoxin and thioredoxin reductase. Eur. J. Biochem. *267*, 6102–6109.

Atack, J.M., and Kelly, D.J. (2009a). Oxidative stress in *Campylobacter jejuni*: responses, resistance and regulation. Future Microbiol. *4*, 677–690.

Atack, J.M., and Kelly, D.J. (2009b). Oxidative stress in *Campylobacter jejuni*: responses, resistance and regulation. Future Microbiol. *4*, 677–690.

Atack, J.M., Harvey, P., Jones, M.A., and Kelly, D.J. (2008). The *Campylobacter jejuni* thiol peroxidases Tpx and Bcp both contribute to aerotolerance and peroxide-mediated stress resistance but have distinct substrate specificities. J. Bacteriol. *190*, 5279–5290.

Atack, J.R. (2010). Preclinical and clinical pharmacology of the GABAA receptor alpha5 subtype-selective inverse agonist alpha5IA. Pharmacol. Ther. *125*, 11–26.

Avila, R.E., Semar, M.E., and de Fabro, S.P. (1986). Ultrastructural differentiation of glandular stomach (proventriculus) in chick embryo. Folia Histochem. Cytobiol. *24*, 227–231.

Baillon, M.L., van Vliet, A.H., Ketley, J.M., Constantinidou, C., and Penn, C.W. (1999). An iron-regulated alkyl hydroperoxide reductase (AhpC) confers aerotolerance and oxidative stress resistance to the microaerophilic pathogen *Campylobacter jejuni*. J. Bacteriol. *181*, 4798–4804.

Baker, C.S., Morozov, I., Suzuki, K., Romeo, T., and Babitzke, P. (2002). CsrA regulates glycogen biosynthesis by preventing translation of glgC in *Escherichia coli*. Mol. Microbiol. *44*, 1599–1610.

Berden, J.H., Muytjens, H.L., and van de Putte, L.B. (1979). Reactive arthritis associated with *Campylobacter jejuni* enteritis. Br. Med. J. *1*, 380–381.

Berndtson, E., Danielsson-Tham, M.L., and Engvall, A. (1996). *Campylobacter* incidence on a chicken farm and the spread of *Campylobacter* during the slaughter process. Int. J. Food. Microbiol. *32*, 35–47.

Cawthraw, S.A., Wassenaar, T.M., Ayling, R., and Newell, D.G. (1996). Increased colonization potential of *Campylobacter jejuni* strain 81116 after passage through chickens and its implication on the rate of transmission within flocks. Epidemiol. Infect. *117*, 213–215.

Cha, M.K., Kim, W.C., Lim, C.J., Kim, K., and Kim, I.H. (2004). *Escherichia coli* periplasmic thiol peroxidase acts as lipid hydroperoxide peroxidase and the principal antioxidative function during anaerobic growth. J. Biol. Chem. *279*, 8769–8778.

Cogan, T.A., Bloomfield, S.F., and Humphrey, T.J. (1999). The effectiveness of hygiene procedures for prevention of cross-contamination from chicken carcases in the domestic kitchen. Lett. Appl. Microbiol. *29*, 354–358.

Cohen, G., Yanko, M., Mislovati, M., Argaman, A., Schreiber, R., Av-Gay, Y., and Aharonowitz, Y. (1993). Thioredoxin-thioredoxin reductase system of *Streptomyces clavuligerus*: sequences, expression, and organization of the genes. J. Bacteriol. *175*, 5159–5167.

Colgan, S.P., and Taylor, C.T. (2010). Hypoxia: an alarm signal during intestinal inflammation. Nat. Rev. Gastroenterol. Hepatol. 7, 281–287.

Doorduyn, Y., Van Den Brandhof, W.E., Van Duynhoven, Y.T., Breukink, B.J., Wagenaar, J.A., and Van Pelt, W. (2010). Risk factors for indigenous *Campylobacter jejuni* and *Campylobacter coli* infections in the Netherlands: a case–control study. Epidemiol. Infect. *138*, 1391–1404.

Dressman, J.B., Berardi, R.R., Dermentzoglou, L.C., Russell, T.L., Schmaltz, S.P., Barnett, J.L., and Jarvenpaa, K.M. (1990). Upper gastrointestinal (GI) pH in young, healthy men and women. Pharm. Res. 7, 756–761.

Dykhuizen, R.S., Frazer, R., Duncan, C., Smith, C.C., Golden, M., Benjamin, N., and Leifert, C. (1996). Antimicrobial effect of acidified nitrite on gut pathogens: importance of dietary nitrate in host defense. Antimicrob. Agents Chemother. *40*, 1422–1425.

EFSA Panel on Biological Hazards (BIOHAZ) (2010). Scientific Opinion on Quantification of the risk posed by broiler meat to human campylobacteriosis in the EU. EFSA *8*, 1437.

Elvers, K.T., Wu, G., Gilberthorpe, N.J., Poole, R.K., and Park, S.F. (2004). Role of an inducible single-domain hemoglobin in mediating resistance to nitric oxide and nitrosative stress in *Campylobacter jejuni* and *Campylobacter coli*. J. Bacteriol. *186*, 5332–5341.

Elvers, K.T., Turner, S.M., Wainwright, L.M., Marsden, G., Hinds, J., Cole, J.A., Poole, R.K., Penn, C.W., and Park, S.F. (2005). NssR, a member of the Crp-Fnr superfamily from *Campylobacter jejuni*, regulates a nitrosative stress-responsive regulon that includes both a single-domain and a truncated haemoglobin. Mol. Microbiol. *57*, 735–750.

Evans, S.J., and Sayers, A.R. (2000). A longitudinal study of *campylobacter* infection of broiler flocks in Great Britain. Prev. Vet. Med. *46*, 209–223.

Fields, J.A., and Thompson, S.A. (2008). *Campylobacter jejuni* CsrA mediates oxidative stress responses, biofilm formation, and host cell invasion. J. Bacteriol. *190*, 3411–3416.

Flint, A., Sun, Y.Q., and Stintzi, A. (2012). Cj1386 is an ankyrin-containing protein involved in heme trafficking to catalase in *Campylobacter jejuni*. J. Bacteriol. *194*, 334–345.

Garenaux, A., Guillou, S., Ermel, G., Wren, B., Federighi, M., and Ritz, M. (2008). Role of the Cj1371 periplasmic protein and the Cj0355c two-component regulator in the *Campylobacter jejuni* NCTC 11168 response to oxidative stress caused by paraquat. Res. Microbiol. *159*, 718–726.

Grant, K.A., and Park, S.F. (1995). Molecular characterization of katA from *Campylobacter jejuni* and generation of a catalase-deficient mutant of *Campylobacter coli* by interspecific allelic exchange. Microbiology *141(Pt 6)*, 1369–1376.

Gundogdu, O., Mills, D.C., Elmi, A., Martin, M.J., Wren, B.W., and Dorrell, N. (2011). The *Campylobacter jejuni* transcriptional regulator Cj1556 plays a role in the oxidative and aerobic stress response and is important for bacterial survival *in vivo*. J. Bacteriol. *193*, 4238–4249.

Harris, N.V., Weiss, N.S., and Nolan, C.M. (1986). The role of poultry and meats in the etiology of *Campylobacter jejuni/coli* enteritis. Am. J. Public Health *76*, 407–411.

He, Y., Frye, J.G., Strobaugh, T.P., and Chen, C.Y. (2008). Analysis of AI-2/LuxS-dependent transcription in *Campylobacter jejuni* strain 81-176. Food. Pathog. Dis. *5*, 399–415.

Hoffman, P.S., George, H.A., Krieg, N.R., and Smibert, R.M. (1979a). Studies of the microaerophilic nature of *Campylobacter fetus* subsp. *jejuni*. II. Role of exogenous superoxide anions and hydrogen peroxide. Can. J. Microbiol. *25*, 8–16.

Hoffman, P.S., Krieg, N.R., and Smibert, R.M. (1979b). Studies of the microaerophilic nature of *Campylobacter fetus* subsp. *jejuni*. I. Physiological aspects of enhanced aerotolerance. Can. J. Microbiol. *25*, 1–7.

Holmes, K., Mulholland, F., Pearson, B.M., Pin, C., McNicholl-Kennedy, J., Ketley, J.M., and Wells, J.M. (2005). *Campylobacter jejuni* gene expression in response to iron limitation and the role of Fur. Microbiology *151*, 243–257.

Hu, L., and Kopecko, D.J. (2008). Cell Biology of Human Host Cell Entry by *Campylobacter jejuni*. In *Campylobacter*, Nachamkin, I., Szymanski, C.M., and Blaser, M.J., eds. (ASM Press, Washington, DC), pp. 297–314.

Huergo, L.F., Rahman, H., Ibrahimovic, A., Day, C.J., and Korolik, V. (2013). *Campylobacter jejuni* Dps Protein Binds DNA in the Presence of Iron or Hydrogen Peroxide. J. Bacteriol. *195*, 1970–1978.

Hwang, S., Kim, M., Ryu, S., and Jeon, B. (2011). Regulation of oxidative stress response by CosR, an essential response regulator in *Campylobacter jejuni*. PLoS One *6*, e22300.

Hwang, S., Zhang, Q., Ryu, S., and Jeon, B. (2012). Transcriptional Regulation of the CmeABC Multidrug Efflux Pump and the KatA Catalase by CosR in *Campylobacter jejuni*. J. Bacteriol. *194*, 6883–6891.

Imlay, J.A. (2003). Pathways of oxidative damage. Annu. Rev. Microbiol. *57*, 395–418.

Imlay, J.A. (2008). Cellular defenses against superoxide and hydrogen peroxide. Annu. Rev. Biochem. *77*, 755–776.

Ishikawa, T., Mizunoe, Y., Kawabata, S., Takade, A., Harada, M., Wai, S.N., and Yoshida, S. (2003). The iron-binding protein Dps confers hydrogen peroxide stress resistance to *Campylobacter jejuni*. J. Bacteriol. *185*, 1010–1017.

Jacobs-Reitsma, W.F., van de Giessen, A.W., Bolder, N.M., and Mulder, R.W. (1995). Epidemiology of *Campylobacter* spp. at two Dutch broiler farms. Epidemiol. Infect. *114*, 413–421.

Jacobs-Reitsma, W., Lyhs, U., and Wagenaar, J. (2000). *Campylobacter* in the Food Supply. In *Campylobacter*, Nachamkin, I., Szymanski, C.M., and Blaser, M.J., eds. (ASM Press, Washington, DC), pp. 467–482.

Jeong, W., Cha, M.K., and Kim, I.H. (2000). Thioredoxin-dependent hydroperoxide peroxidase activity of bacterioferritin comigratory protein (BCP) as a new member of the thiol-specific antioxidant protein (TSA)/Alkyl hydroperoxide peroxidase C (AhpC) family. J. Biol. Chem. *275*, 2924–2930.

Kelly, D.J. (2005). Metabolism, Electron Transport and Bioenergetics of *Campylobacter jejuni*: Implications for Understanding Life in the Gut and Survival in the Environment. In *Campylobacter* Molecular and Cellular Biology, Ketley, J.M., and Konkel, M.E., eds. (Horizon Bioscience, Norwich, UK), pp. 275–293.

Ketley, J.M. (1997). Pathogenesis of enteric infection by *Campylobacter*. Microbiology *143(Pt 1)*, 5–21.

Kopecko, D.J., Hu, L., and Zaal, K.J. (2001). *Campylobacter jejuni*–microtubule-dependent invasion. Trends Microbiol. *9*, 389–396.

Le, M.T., Porcelli, I., Weight, C.M., Gaskin, D.J.H., Carding, S.R., and van Vliet, A.H.M. (2012). Acid shock of *Campylobacter jejuni* induces flagellar gene expression and host cell invasion. Eur. J. Microbiol. Immunol. *2*, 12–19.

Liu, M.Y., Gui, G., Wei, B., Preston, J.F., 3rd, Oakford, L., Yuksel, U., Giedroc, D.P., and Romeo, T. (1997). The RNA molecule CsrB binds to the global regulatory protein CsrA and antagonizes its activity in *Escherichia coli*. J. Biol. Chem. *272*, 17502–17510.

Lynch, M., and Kuramitsu, H. (2000). Expression and role of superoxide dismutases (SOD) in pathogenic bacteria. Microbes Infect. *2*, 1245–1255.

Manser, P.A., and Dalziel, R.W. (1985). A survey of *Campylobacter* in animals. J. Hyg. (Lond.) *95*, 15–21.

Marteyn, B., Scorza, F.B., Sansonetti, P.J., and Tang, C. (2011). Breathing life into pathogens: the influence of oxygen on bacterial virulence and host responses in the gastrointestinal tract. Cell Microbiol. *13*, 171–176.

Mongkolsuk, S., and Helmann, J.D. (2002). Regulation of inducible peroxide stress responses. Mol Microbiol. *45*, 9–15.

Nachamkin, I., Allos, B.M., and Ho, T. (1998). *Campylobacter* species and Guillain-Barré syndrome. Clin. Microbiol. Rev. *11*, 555–567.

Newell, D.G., and Fearnley, C. (2003). Sources of *Campylobacter* colonization in broiler chickens. Appl. Environ. Microbiol. *69*, 4343–4351.

Olczak, A.A., Seyler, R.W., Jr., Olson, J.W., and Maier, R.J. (2003). Association of Helicobacter pylori antioxidant activities with host colonization proficiency. Infect. Immun. *71*, 580–583.

Palyada, K., Threadgill, D., and Stintzi, A. (2004). Iron acquisition and regulation in *Campylobacter jejuni*. J. Bacteriol. *186*, 4714–4729.

Palyada, K., Sun, Y.Q., Flint, A., Butcher, J., Naikare, H., and Stintzi, A. (2009a). Characterization of the oxidative stress stimulon and PerR regulon of *Campylobacter jejuni*. BMC Genomics *10*, 481.

Palyada, K., Sun, Y.Q., Flint, A., Butcher, J., Naikare, H., and Stintzi, A. (2009b). Characterization of the oxidative stress stimulon and PerR regulon of *Campylobacter jejuni*. BMC Genomics *10*, 481.

Park, S.F. (2005). *Campylobacter jejujni* Stress Responses During Survival in the Food Chain and Colonization. In *Campylobacter* Molecular and Cellular Biology, Ketley, J.M., and Konkel, M.E., eds. (Horizon Bioscience, Norwich, UK), pp. 311–331.

Parkhill, J., Wren, B.W., Mungall, K., Ketley, J.M., Churcher, C., Basham, D., Chillingworth, T., Davies, R.M., Feltwell, T., Holroyd, S., *et al.* (2000). The genome sequence of the food-borne pathogen *Campylobacter jejuni* reveals hypervariable sequences. Nature *403*, 665–668.

Pesci, E.C., Cottle, D.L., and Pickett, C.L. (1994). Genetic, enzymatic, and pathogenic studies of the iron superoxide dismutase of *Campylobacter jejuni*. Infect. Immun. *62*, 2687–2694.

Pinto, A.F., Todorovic, S., Hildebrandt, P., Yamazaki, M., Amano, F., Igimi, S., Romao, C.V., and Teixeira, M. (2011). Desulforubrerythrin from *Campylobacter jejuni*, a novel multidomain protein. J. Biol. Inorg. Chem. *16*, 501–510.

Poole, L.B. (1996). Flavin-dependent alkyl hydroperoxide reductase from Salmonella typhimurium. 2. Cystine disulfides involved in catalysis of peroxide reduction. Biochemistry *35*, 65–75.

Poole, L.B., and Ellis, H.R. (1996). Flavin-dependent alkyl hydroperoxide reductase from *Salmonella typhimurium*. 1. Purification and enzymatic activities of overexpressed AhpF and AhpC proteins. Biochemistry *35*, 56–64.

Purdy, D., and Park, S.F. (1994). Cloning, nucleotide sequence and characterization of a gene encoding superoxide dismutase from *Campylobacter jejuni* and *Campylobacter coli*. Microbiology *140(Pt 5)*, 1203–1208.

Purdy, D., Cawthraw, S., Dickinson, J.H., Newell, D.G., and Park, S.F. (1999). Generation of a superoxide dismutase (SOD)-deficient mutant of *Campylobacter coli*: evidence for the significance of SOD in *Campylobacter* survival and colonization. Appl. Environ. Microbiol. *65*, 2540–2546.

Rho, B.S., Hung, L.W., Holton, J.M., Vigil, D., Kim, S.I., Park, M.S., Terwilliger, T.C., and Pedelacq, J.D. (2006). Functional and structural characterization of a thiol peroxidase from *Mycobacterium tuberculosis*. J. Mol. Biol. *361*, 850–863.

Romeo, T. (1996). Post-transcriptional regulation of bacterial carbohydrate metabolism: evidence that the gene product CsrA is a global mRNA decay factor. Res. Microbiol. *147*, 505–512.

Romeo, T. (1998). Global regulation by the small RNA-binding protein CsrA and the non-coding RNA molecule CsrB. Mol. Microbiol. *29*, 1321–1330.

Rosenquist, H., Sommer, H.M., Nielsen, N.L., and Christensen, B.B. (2006). The effect of slaughter operations on the contamination of chicken carcasses with thermotolerant *Campylobacter*. Int. J. Food Microbiol. *108*, 226–232.

Santos, P.M., Benndorf, D., and Sá-Correia, I. (2004). Insights into *Pseudomonas putida* KT2440 response to phenol-induced stress by quantitative proteomics. Proteomics *4*, 2640–2652.

Sears, A., Baker, M.G., Wilson, N., Marshall, J., Muellner, P., Campbell, D.M., Lake, R.J., and French, N.P. (2011). Marked campylobacteriosis decline after interventions aimed at poultry, New Zealand. Emerg. Infect. Dis. *17*, 1007–1015.

Siegel, J.A., Urbain, J.L., Adler, L.P., Charkes, N.D., Maurer, A.H., Krevsky, B., Knight, L.C., Fisher, R.S., and Malmud, L.S. (1988). Biphasic nature of gastric emptying. Gut *29*, 85–89.

Skaar, E.P., Tobiason, D.M., Quick, J., Judd, R.C., Weissbach, H., Etienne, F., Brot, N., and Seifert, H.S. (2002). The outer membrane localization of the *Neisseria gonorrhoeae* MsrA/B is involved in survival against reactive oxygen species. Proc. Natl. Acad. Sci. U.S.A. *99*, 10108–10113.

Stead, D., and Park, S.F. (2000). Roles of Fe superoxide dismutase and catalase in resistance of *Campylobacter coli* to freeze–thaw stress. Appl. Environ. Microbiol. *66*, 3110–3112.

Stern, N.J. (1992). Reservoirs for Campylobacter jejuni and Approaches for Intervention in Poultry. In *Campylobacter jejuni* Current Status and Future Trends, Nachamkin, I., Blaser, M.J., and Tompkins, L.S., eds. (American Society for Microbiology, Washington, DC), pp. 49–61.

Storz, G., and Imlay, J.A. (1999). Oxidative stress. Curr. Opin. Microbiol. *2*, 188–194.

Svensson, S.L., Frirdich, E., and Gaynor, E.C. (2008). Survival Strategies of *Campylobacter jejuni*: Stress Responses, the Viable but Nonculturable state, and Biofilms. In *Campylobacter*, 3rd edn, Nachamkin, I., Szymanski, C.M., and Blaser, M.J., eds. (ASM Press, Washington, DC), pp. 571–590.

Svensson, S.L., Davis, L.M., MacKichan, J.K., Allan, B.J., Pajaniappan, M., Thompson, S.A., and Gaynor, E.C. (2009). The CprS sensor kinase of the zoonotic pathogen *Campylobacter jejuni* influences biofilm formation and is required for optimal chick colonization. Mol. Microbiol. *71*, 253–272.

Tauxe, R.V. (1992). Epidemiology of *Campylobacter jejuni* Infections in the United States and Other Industrialized Nations. In *Campylobacter jejuni: Current Staus and Future Trends*, Nachamkin, I., Blaser, M.J., and Tompkins, L.S., eds. (American Society for Microbiology, Washington, DC), pp. 9–19.

Theoret, J.R., Cooper, K.K., Glock, R.D., and Joens, L.A. (2011). A *Campylobacter jejuni* Dps homolog has a role in intracellular survival and in the development of campylobacterosis in neonate piglets. Food. Pathog. Dis. *8*, 1263–1268.

Theoret, J.R., Cooper, K.K., Zekarias, B., Roland, K.L., Law, B.F., Curtiss, R., 3rd and Joens, L.A. (2012). The *Campylobacter jejuni* Dps homologue is important for in vitro biofilm formation and cecal colonization of poultry and may serve as a protective antigen for vaccination. Clin. Vaccine Immunol. *19*, 1426–1431.

Thomas, C., Gibson, H., Hill, D.J., and Mabey, M. (1998). *Campylobacter* epidemiology: an aquatic perspective. J. Appl. Microbiol. *85(Suppl 1)*, 168S–177S.

Tomb, J.F., White, O., Kerlavage, A.R., Clayton, R.A., Sutton, G.G., Fleischmann, R.D., Ketchum, K.A., Klenk, H.P., Gill, S., Dougherty, B.A., *et al.* (1997). The complete genome sequence of the gastric pathogen *Helicobacter pylori*. Nature *388*, 539–547.

Touati, D. (2000). Iron and oxidative stress in bacteria. Arch. Biochem. Biophys. 373, 1–6.

Touati, D., Jacques, M., Tardat, B., Bouchard, L., and Despied, S. (1995). Lethal oxidative damage and mutagenesis are generated by iron in delta fur mutants of *Escherichia coli*: protective role of superoxide dismutase. J. Bacteriol. *177*, 2305–2314.

Traore, D.A., El Ghazouani, A., Ilango, S., Dupuy, J., Jacquamet, L., Ferrer, J.L., Caux-Thang, C., Duarte, V., and Latour, J.M. (2006). Crystal structure of the apo-PerR-Zn protein from *Bacillus subtilis*. Mol. Microbiol. *61*, 1211–1219.

van Vliet, A.H., and Ketley, J.M. (2001). Pathogenesis of enteric *Campylobacter* infection. Symp. Ser. Soc. Appl. Microbiol. 45S–56S.

van Vliet, A.H., Baillon, M.L., Penn, C.W., and Ketley, J.M. (1999). *Campylobacter jejuni* contains two fur homologs: characterization of iron-responsive regulation of peroxide stress defense genes by the PerR repressor. J. Bacteriol. *181*, 6371–6376.

van Vliet, A.H., Baillon, M.A., Penn, C.W., and Ketley, J.M. (2001). The iron-induced ferredoxin FdxA of *Campylobacter jejuni* is involved in aerotolerance. FEMS Microbiol. Lett. *196*, 189–193.

van Vliet, A.H., Ketley, J.M., Park, S.F., and Penn, C.W. (2002). The role of iron in *Campylobacter* gene regulation, metabolism and oxidative stress defense. FEMS Microbiol. Rev. *26*, 173–186.

Wai, S.N., Nakayama, K., Umene, K., Moriya, T., and Amako, K. (1996). Construction of a ferritin-deficient mutant of *Campylobacter jejuni*: contribution of ferritin to iron storage and protection against oxidative stress. Mol. Microbiol. *20*, 1127–1134.

Wainwright, L.M., Elvers, K.T., Park, S.F., and Poole, R.K. (2005). A truncated haemoglobin implicated in oxygen metabolism by the microaerophilic food-borne pathogen *Campylobacter jejuni*. Microbiology *151*, 4079–4091.

Wang, G., Olczak, A.A., Walton, J.P., and Maier, R.J. (2005). Contribution of the *Helicobacter pylori* thiol peroxidase bacterioferritin comigratory protein to oxidative stress resistance and host colonization. Infect. Immun. 73, 378–384.

Weissbach, H., Etienne, F., Hoshi, T., Heinemann, S.H., Lowther, W.T., Matthews, B., St John, G., Nathan, C., and Brot, N. (2002). Peptide methionine sulfoxide reductase: structure, mechanism of action, and biological function. Arch. Biochem. Biophys. *397*, 172–178.

Wesley, I.V., Wells, S.J., Harmon, K.M., Green, A., Schroeder-Tucker, L., Glover, M., and Siddique, I. (2000). Fecal shedding of *Campylobacter* and *Arcobacter* spp. in dairy cattle. Appl. Environ. Microbiol. *66*, 1994–2000.

Yamasaki, M., Igimi, S., Katayama, Y., Yamamoto, S., and Amano, F. (2004). Identification of an oxidative stress-sensitive protein from *Campylobacter jejuni*, homologous to rubredoxin oxidoreductase/rubrerythrin. FEMS Microbiol. Lett. *235*, 57–63.

The Role of the Flagellum in *Campylobacter jejuni* Colonization and Disease

13

Jason L. O'Loughlin and Michael E. Konkel

Abstract

Campylobacter jejuni causes acute human enterocolitis via mechanisms requiring colonization of the host intestine, adherence to, and subsequent invasion of epithelial cells. The *C. jejuni* flagellum is responsible for the motility necessary for colonization and contributes to adherence and internalization of host cells. The flagellum also secretes non-flagellar (i.e. effector) proteins that enhance internalization by modifying host cell behaviour. Additionally, the flagellum is the only type three secretion system (T3SS) encoded by *C. jejuni* and its structure is similar to the injectisome of a classical virulence-associated T3SS common to other enteric pathogens. Numerous proteins secreted by the flagellum have been identified and characterized, with several proteins delivered to the cytosol of host cells. Furthermore, the flagellum of *C. jejuni* contributes to biofilm formation. Taken together, flagellar function, including the secretion of effector proteins and the characterization of their effects on the host, is an important area of *C. jejuni* research.

Introduction

Campylobacter jejuni is a motile, Gram-negative pathogen and a leading bacterial cause of human gastrointestinal disease worldwide (Fig. 13.1) (Samuel *et al.*, 2004). *C. jejuni* causes acute human enterocolitis (campylobacteriosis). Symptoms of *C. jejuni*-mediated disease generally occur 2–5 days following ingestion of bacteria and are characterized by diarrhoea, possibly with blood and leucocytes in the stool, severe abdominal pain, and fever (Allos, 2001). Furthermore, campylobacteriosis may be fatal, particularly in young children in developing countries. The infectious dose is as few as 500 organisms for humans (Black *et al.*, 1988). *C. jejuni* is a commensal organism of animals, including poultry, and human infection often occurs by consuming food products that are undercooked or cross-contaminated with the bacterium (Black *et al.*, 1988; Friedman *et al.*, 2004; van Gerwe *et al.*, 2009). However, *Campylobacter* spp. can also be acquired from unpasteurized milk, contaminated water, and from pets with diarrhoea (Ketley, 1997). Reducing the incidence of *C. jejuni*-mediated disease requires an increased understanding of the unique pathogenic mechanisms used by the bacterium.

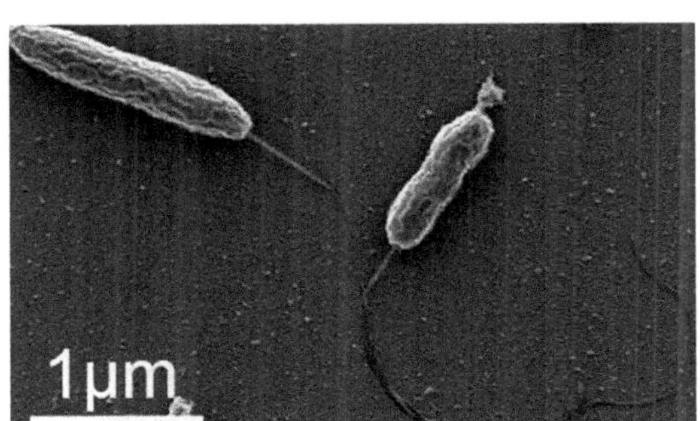

Figure 13.1 Scanning electron micrograph of *C. jejuni* with flagellum. Image courtesy of D.R. Samuelson (unpublished). A colour version of this figure is available in the plate section at the back of the book.

C. jejuni-mediated enteritis

The flagellum of *C. jejuni* is important for the characteristic darting motility and plays an integral role in our model of *C. jejuni* pathogenesis (Fig. 13.2). Initially, *C. jejuni* in the presence of host factors initiates *de novo* synthesis of the proteins we have termed the *Campylobacter* invasion antigens (Cia) (Konkel and Cieplak, 1992; Konkel *et al.*, 1993). *C. jejuni* then disrupts the epithelial barrier allowing for bacterial migration across the intestinal epithelium via a paracellular route of translocation (Allos, 2001; Baqar *et al.*, 2001; Larson *et al.*, 2008a; Wassenaar and Blaser, 1999). The recruitment of innate phagocytic cells, including macrophages and polymorphonuclear leucocytes, may contribute to barrier disruption. At the basolateral surface of host cells, *C. jejuni* then binds to extracellular matrix molecules, including fibronectin, and invades epithelial cells (Buelow *et al.*, 2011; Flanagan *et al.*, 2009; Konkel *et al.*, 1992b, 1997, 1999, 2004; Konkel and Joens, 1989; Monteville and Konkel, 2002). The majority of adherent bacteria are observed at cellular junctions (tight junctions) with few binding at the apical surface (Hu *et al.*, 2008; Konkel *et al.*, 1992b). The flagellar secretion apparatus is used for the secretion of proteins delivered to the cytosol of host cells (Konkel *et al.*, 1992b; Larson *et al.*, 2008b). Bacterial uptake by host cells requires stimulation of host cell signalling events, facilitated in part by secreted proteins that modulate Rho GTPase activation (Rac1) responsible for the rearrangement of the host cell cytoskeleton, that create the characteristic membrane ruffling observed during *C. jejuni* infection (Eucker and Konkel, 2012; Krause-Gruszczynska *et al.*, 2007; Monteville *et al.*, 2003). Finally, *C. jejuni* multiplies within a vacuole, where it ultimately causes cell lysis (Buelow *et al.*, 2011; Konkel *et al.*, 1992a). In summary, multiple flagellar functions, including motility and protein secretion, are necessary for *C. jejuni* pathogenesis.

Flagellar structure and gene expression

The expression of genes in the flagellar biosynthetic pathway of enteric organisms is tightly regulated in a sequential genetic hierarchy. Expression begins with the flagellar genes necessary for encoding structures spanning the inner membrane and ends with those most distal to the cell surface. The ordered flagellar assembly of the enteric microorganisms *Escherichia coli* and *Salmonella enterica* serotype Typhimurium has been extensively characterized, with over 60 genes grouped into a three class transcriptional hierarchy (Aldridge and Hughes, 2002; Kutsukake *et al.*, 1990). Bioinformatic analysis of the *C. jejuni* genome revealed 36 open reading frames that may encode proteins involved in the export and assembly of the flagellum (Parkhill *et al.*, 2000).

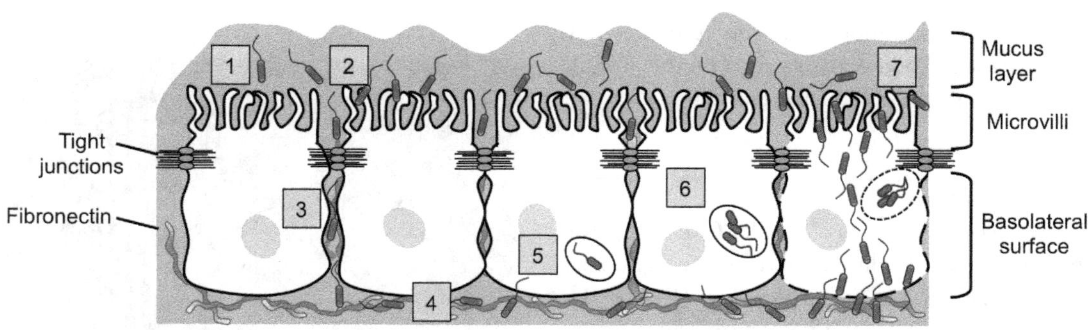

Figure 13.2 Model of *C. jejuni* pathogenesis. *C. jejuni* engages in a novel synthetic response (effector protein synthesis) (Step 1), delivers effector proteins (i.e. Cia proteins) to host cells to induce an inflammatory response, ultimately resulting in barrier disruption (Step 2), migrates across the intestinal cell barrier via a paracellular route of translocation (Step 3), binds to fibronectin (Step 4), invades an epithelial cell (Step 5), and multiplies within a vacuole to high numbers intracellularly (Step 6), and escapes from the vacuole (Step 7). The dashed lines indicate host cell death. A colour version of this figure is available in the plate section at the back of the book.

Completion of flagellar synthesis in *Campylobacter* results in flagellum at one or both poles, although some strains have flagella at both poles. The flagellum in *C. jejuni* is composed of three central components (the basal body, hook, and filament) and flagellar biosynthesis occurs in three stages (class I, II and III genes, discussed below) (Table 13.1). However, the environmental cues responsible for flagellar synthesis and the signalling molecules responsible for Class I activation (i.e. a protein analogous to the master regulator FlhDC in *Yersinia* (Kapatral *et al.*, 2004)) have not been identified in *C. jejuni*. Once the basal body and hook structure are formed, monomers of the flagellin proteins are translocated through the newly created hollow core and polymerize at the distal end to form the filament (Yonekura *et al.*, 2000). Flagellin is likely unfolded as it is secreted through the channel to the tip before refolding and contributing to the growing filament (Samatey *et al.*, 2001). The flagellar assembly process, from initiation of transcription to the formation of a complete and functional flagellum, requires at least one bacterial generation (Karlinsey *et al.*, 2000).

Class I gene expression is σ^{70}-dependent (*rpoD*) and the encoded protein products are necessary for assembly of the basal body and export apparatus (Carrillo *et al.*, 2004). Class II gene expression is dependent on class I gene products and σ^{54} (*rpoN*). Class III gene expression is also dependent on Class I gene products and σ^{28} (*fliA*) (Carrillo *et al.*, 2004). FliA also functions to repress expression of σ^{54}-dependent genes. Mutants deficient in σ^{54}-regulated genes demonstrate a reduction in the colonization of chicks compared to a wild-type strain (Carrillo *et al.*, 2004). In *C. jejuni*, *flgM* encodes a negative repressor of σ^{28}, although it is not as efficient at repressing gene expression as compared to a homologue in *Helicobacter pylori* (Hendrixson and DiRita, 2003). The FlgM represser is exported once a functional export apparatus is completed, allowing for filament formation via σ^{28} expression. Class II and III genes are necessary for complete flagellar assembly, including completion of the basal body, hook, and filament. The flagellar export apparatus is dependent on the coordination of gene expression, and specifically the expression of σ^{54}-regulated genes, including the two component system FlgSR (Hendrixson and DiRita, 2003). A *C. jejuni* strain with a single point mutation in FlgSR is less virulent, suggesting that content, expression, and regulation of flagellar genes contribute to pathogenicity (Malik-Kale *et al.*, 2007). Furthermore, a *Campylobacter flgE* mutant (encodes a hook protein) does not form flagellum, although the expression of other flagellar genes appears unaffected (Kinsella *et al.*, 1997). Thus, flagellar assembly is partially controlled by an ordered hierarchy of gene expression components comprising the basal body, hook, and filament.

The flagellar filament is comprised of two homologous proteins with predicted molecular masses of 59 kDa: the σ^{28}-regulated FlaA and the σ^{54}-regulated FlaB (Nuijten *et al.*, 1990). The two proteins are encoded by adjacent genes that share 95% sequence identity within a single strain, although there is sequence variation between strains of *C. jejuni*, and there is evidence that the encoded proteins undergo phase variation and antigenic variation (Caldwell *et al.*, 1985; Guerry *et al.*, 1990; Hendrixson, 2006; Nuijten *et al.*, 1990; Park *et al.*, 2000). FlaA is the major component of the filament, as the *flaA* gene is expressed at a higher level than *flaB*, and FlaA is readily incorporated into the filament (Neal-McKinney *et al.*, 2010). Furthermore, a *flaA* mutant does not form a full-length flagellar filament (and is therefore non-motile) and shows a reduction of invasion (Guerry *et al.*, 1991; Neal-McKinney *et al.*, 2010; Wassenaar *et al.*, 1991, 1994). In comparison, a *flaB* deletion displays no significant phenotype. It is still unclear what advantage regulating *flaA* and *flaB* by separate sigma factors is to the bacterium.

The flagellin proteins of *Salmonella* and *E. coli* contain highly conserved regions necessary for filament export and formation. This conserved region on the filament is known as a pathogen associated molecular pattern (PAMP) and is recognized by the host innate Toll-Like Receptor (TLR) 5. The ultimate result of TLR5 activation is the induction of a cytokine storm at the site of incidence, including interleukin 8 (IL-8), leading to acute inflammation (Smith *et al.*, 2003). Flagellin residues 89–96 of *Salmonella* are recognized by TLR5 and homologous residues are lacking

Table 13.1 *C. jejuni* genes encoding flagellar components and secreted proteins

Protein structure	Gene name[a]	σ factor[b]	Function
Basal body	Cj0882c *(flhA)*	σ^{70}	Biosynthesis
	Cj0335 *(flhB)*	σ^{70}	Biosynthesis
	Cj0318 *(fliF)*	σ^{70}	MS-ring
	Cj0319 *(fliG)*	σ^{70}	Motor switch
	Cj0320 *(fliH)*	σ^{70}	Assembly
	Cj0195 *(fliI)*	σ^{70}	Flagellum-specific ATP synthase
	Cj0060c *(fliM)*	σ^{70}	Motor switch
	Cj0351 *(fliN)*	σ^{70}	Motor switch
	Cj0820c *(fliP)*		Biosynthesis
	Cj1675 *(fliQ)*	σ^{70}	Biosynthesis
	Cj1179c *(fliR)*		Biosynthesis
	Cj0337c *(motA)*		Motor
	Cj0336c *(motB)*		Motor
	Cj0062c-Cj0061c *(fliA)*	σ^{70}	Integral membrane protein, biosynthesis sigma factor
	Cj0697-Cj0698 *(flgG2/G)*	σ^{54}	Basal body rod
	Cj0687c *(flgH)*	σ^{54}	Basal body L-ring
	Cj1462 *(flgI)*	σ^{54}	Basal body P-ring
	Cj0528c-Cj0526c *(flgB/C/FliE)*	σ^{54}	Basal body rod proteins, hook- basal body protein
	Cj0042-Cj0043 *(flgD/E)*	σ^{54}	Basal body rod modification, hook
Hook	Cj1729c *(flgE2)*	σ^{54}	Hook
	Cj1466 *(flgK)*	σ^{54}	Hook-associated
	Cj0887c *(flgL)*	σ^{54}	Hook-associated
Filament	Cj1339c *(flaA)*	σ^{28}	Flagellin
	Cj1338c *(flaB)*	σ^{54}	Flagellin
	Cj0547-Cj0549 *(flaG/fliD/fliS)*	σ^{28}	Flagellar proteins and cap
	Cj1464 *(flgM)*	σ^{28}	Hypothetical flagellar protein
Putative effectors	Cj0914c *(ciaB)*[c]		Invasion
	Cj1242c *(ciaC)*[c]	σ^{54}	Invasion through host cell modification
	Cj1450 *(ciaI)*[c]	σ^{28}	Intracellular survival
	Cj0859c *(fspA)*	σ^{28}	Induces host cell death
Secreted proteins	Cj0977	σ^{54}	Invasion
	Cj0045c *(fedA)*	σ^{28}	Invasion
	CJ0391c *(fedB)*	σ^{28}	Putative invasion protein
	Cj1034 *(fedC)*	σ^{28}	Putative invasion protein
	Cj1656c *(fedD)*	σ^{28}	Putative invasion protein
	Cj0720c *(flaC)*		Invasion, adhesin
Other	Cj0079c-Cj0077c *(cdtA-C)*	σ^{28}, σ^{70}	Cytolethal distending toxin
	Cj1316c *(pseA)*	σ^{70}	Pseudaminic acid biosynthesis

[a] Locus names refer to *C. jejuni* NCTC11168.
[b] Known σ factors are indicated.
[c] Indicates the secreted protein product is delivered to host cells using the ACD-reporter assay.

in *C. jejuni*, thus allowing *C. jejuni* to evade TLR5 detection (Andersen-Nissen *et al.*, 2005). The quaternary structure of the *C. jejuni* flagellar filament consists of seven protofilaments compared with the 11 protofilament structure found in *Salmonella*, which may also explain why the *C. jejuni* flagellin does not contain the PAMP (Galkin *et al.*, 2008). The absence of TLR5 recognition sites may be a common strategy by select enteric pathogens, including *H. pylori*, to limit immune detection (Andersen-Nissen *et al.*, 2005). The potential for immune activation by flagellin was evaluated by injecting mice with flagellin (Watson and Galan, 2005). The flagellin of *Salmonella* (FliC) elicited a strong innate immune response, as measured by high serum levels of tumour necrosis factor (TNF) α and IL-6. In contrast, the flagellin from *C. jejuni* (FlaA) demonstrated poor immune activation.

The *Campylobacter* flagellum is glycosylated and many genes appear to play a role in this post-translational modification, including the *ptmAB* genes (Guerry *et al.*, 1996), the *pgl* locus (Szymanski *et al.*, 1999), and the *neuB2/B3* genes (Linton *et al.*, 2000). The glycosylation of flagellin proteins is necessary for filament formation in *C. jejuni* and *H. pylori* (Guerry *et al.*, 2006; Josenhans *et al.*, 2002). Furthermore, a *Campylobacter pseA* mutant, which is deficient for post-translational modification of flagellin with pseudaminic acid, is less virulent in the ferret model of disease, even though the strain remained motile (Guerry *et al.*, 2006). Reintroduction of a wild-type copy of the gene restored virulence. Thus, while *Campylobacter* flagellar gene expression and structure share characteristics with other enteric pathogens, it also has several unique aspects that merit further investigation.

The role of flagellum in adherence and invasion of host cells

The *Campylobacter* flagellum was one of the first identified components involved in adherence (Yao *et al.*, 1994). For example, non-motile *C. jejuni* strains are unable to colonize suckling mice (Morooka *et al.*, 1985). It was further shown that flagellar synthesis and motility are necessary for maximal cell adhesion and invasion by comparing the phenotypes of various flagellar mutants to a *C. jejuni* wild-type strain (Yao *et al.*, 1994). More recent experimental evidence using both tissue culture cells and *ex vivo* human intestinal tissue suggests that the *C. jejuni* flagellum is necessary for the bacterium to interact with the host epithelium (Haddock *et al.*, 2010). In addition to the flagellum, *C. jejuni* encodes adhesins that are necessary for host cell binding based on the observation that mutants deficient in adhesins demonstrate a reduction in binding and invasion (Konkel *et al.*, 1997, 2010). Multiple adhesive proteins play a role in host cell adherence, including CadF, CapA, FlpA, HtrA, VirK, and possibly JlpA, although there are differences between strains. Researchers have also hypothesized that colonization may be due, in part, to the organisms motility through the mucus barrier covering the intestinal cells (Lee *et al.*, 1986; Newell *et al.*, 1985; Szymanski *et al.*, 1995). Thus, adhesion to host cells and subsequent colonization is dependent on numerous factors, including the flagellum.

Once *C. jejuni* has bound to an epithelial cell, the bacterium can then invade the cell. *C. jejuni* invasion is dependent on a metabolically active bacterium, as treatment with a bacterial protein synthesis inhibitor (i.e. chloramphenicol) reduces *C. jejuni* invasion (Konkel and Cieplak, 1992). In addition, a functional flagellum is necessary for maximal invasion of an epithelial cell (Morooka *et al.*, 1985; Nachamkin *et al.*, 1993; Takata *et al.*, 1992; Wassenaar *et al.*, 1993). Using the piglet model of infection, a flagellar mutant of *C. jejuni* induced less severe diarrhoea than a motile wild-type strain (Malik-Kale *et al.*, 2007). The necessity of the flagellum for internalization was demonstrated using a *C. jejuni flgL* mutant (Neal-McKinney and Konkel, 2012). FlgL is required for the formation of the flagellar hook–filament junction. A *flgL* mutant does not induce membrane ruffling, which is a marker of internalization, as visualized by scanning electron microscopy (SEM). Additionally, a *flgL* mutant does not invade even when bacteria-host cell contact is promoted by centrifugation (unpublished). Reintroducing a wild-type copy of *flgL* into the mutant restored motility and the ability of the bacteria to invade epithelial cells. In summary, the flagellum is necessary for *C. jejuni* to efficiently bind to and invade host cells.

C. jejuni engages in a novel synthetic response when cultured with epithelial cells leading to protein secretion

C. jejuni cultured in the presence of host cells or *in vivo*-like conditions, including physiological concentrations of the bile acid deoxycholate (DOC), engage in a novel synthetic response. *C. jejuni* up-regulates 156 genes in response to DOC, as judged by microarray analysis (Malik-Kale *et al.*, 2008). Furthermore, *C. jejuni* secretes at least 18 newly synthesized proteins during co-culture with epithelial cells or host cell components (Biswas *et al.*, 2007; Konkel and Cieplak, 1992; Konkel *et al.*, 1999). We have collectively termed these secreted proteins the *Campylobacter* invasion antigens, or Cia proteins. Up-regulation of gene expression and secretion of proteins into the medium indicate that *C. jejuni* secretes putative effector proteins. An effector protein is defined as a protein delivered from a pathogen to a host cell that ultimately functions to alter host cell behaviour. Flagellar secretion of the Cia proteins is also contact dependent (Rivera-Amill and Konkel, 1999). Importantly, bioinformatic and biochemical analyses indicate that *C. jejuni* has only one type three export apparatus capable of protein secretion, which is the flagellum (Konkel *et al.*, 2001).

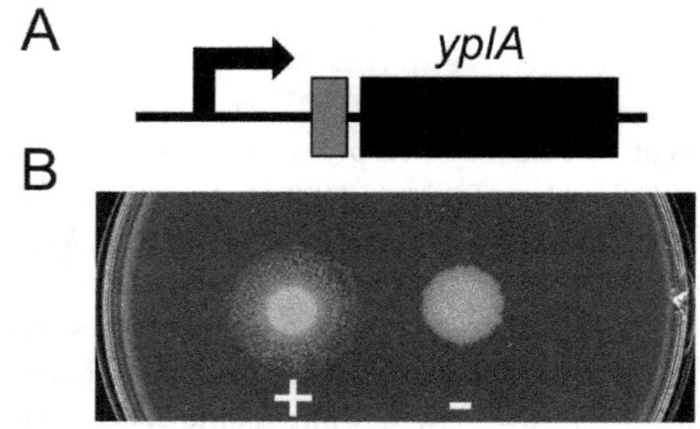

Figure 13.3 Identification of proteins secreted by the flagellum. (A) Schematic of the secretion reporter construct with the constitutive chloramphenicol acetyltransferase (*cat*) promoter (black arrow), a restriction site used for cloning *C. jejuni* fragments (or other pathogens) with potential flagellar secretion signals (grey), and the *yplA* reporter gene. (B) *Y. enterocolitica* harbouring a plasmid with the reporter construct plated on a phospholipase indicator plate. A zone of haze indicates secretion of the YplA phospholipase enzyme. A colour version of this figure is available in the plate section at the back of the book.

Identification of genes encoding the Cia virulence proteins

In order to identify the *C. jejuni* proteins secreted from the flagellum, an assay was developed to screen for proteins containing a type three secretion system (T3SS) export sequence. This screen identified 42 putative *C. jejuni* secreted proteins (Christensen *et al.*, 2009). YplA was used as a reporter to identify the *C. jejuni* flagellar secreted proteins, as YplA is a phospholipase known to be secreted from the flagellum in *Y. enterocolitica* (Fig. 13.3) (Berring *et al.*, 2004; Schmiel *et al.*, 1998; Warren and Young, 2005). We engineered a fusion of *yplA* with 321 *C. jejuni* genes suspected to encode an amino-terminal sequence required for type three secretion. The constructs were transformed into *Y. enterocolitica* and secretion was assessed visually following growth on agar containing Tween 80. Strains positive for secretion of YplA resulted in a characteristic fatty acid precipitate, or halo, surrounding the YplA secretion-competent colonies. This screen has been used to identify several new Cia proteins, two of which have been published (CiaI and CiaC) (Neal-McKinney and Konkel, 2012; Buelow *et al.*, 2011; Christensen *et al.*, 2009). The phospholipase screen has applications beyond *C. jejuni*, aiding in the identification of putative type three secreted proteins from other Gram-negative bacteria (Christensen *et al.*, 2009).

Identification and characterization of the Cia proteins

To date, three Cia secreted proteins have been characterized from the 18 originally identified in the phosphorimage and confirmed using the phospholipase screen (Fig. 13.4). The first Cia protein identified is a 73 kDa protein termed CiaB

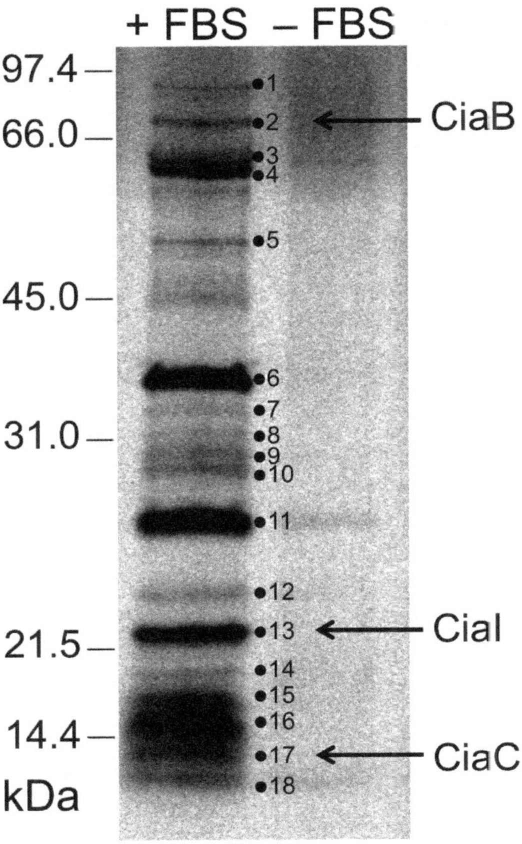

Figure 13.4 Profile of secreted proteins with the three identified Cia proteins indicated. *C. jejuni* incubated in medium containing an inducing signal (+ fetal bovine serum, FBS) or without an inducing signal (–FBS). The secreted proteins are numbered 1–18. Phosphorimage of radioactively labelled proteins from (Larson *et al.*, 2008a), with permission.

(Konkel *et al.*, 1999). A *ciaB* mutant is motile but deficient in Cia protein secretion. The mutant is also deficient in invasion of epithelial cells. The importance of the Cia proteins in *C. jejuni* pathogenesis was demonstrated using a piglet model (Konkel *et al.*, 2001). Inoculation of piglets with a *C. jejuni* wild-type strain resulted in diarrhoea within 24 h and the intestine exhibited severe histological lesions, including shortening of the villi and production of an exudate in the lumen. In contrast, piglets inoculated with a *C. jejuni ciaB* mutant did not develop diarrhoea until 3 days post-infection and only exhibited minor histological lesions. Introducing a wild-type copy of the *C. jejuni ciaB* gene into the mutant restored the strain's virulence. In addition, strains that do not secrete the Cia proteins are poor colonizers of chickens (Biswas *et al.*, 2007). Interestingly, sera isolated from chickens inoculated with *C. jejuni* recognized the Cia proteins by immunoblot, indicating that these proteins are immunogenic (Biswas *et al.*, 2007).

Using the YplA screen, a Cia protein termed CiaI was identified that plays a role in pathogenesis. CiaI is a σ^{28}-regulated gene and encodes a 22.4 kDa protein that is secreted from the flagellum (Buelow *et al.*, 2011). The CiaI protein was recently demonstrated to be delivered to the cytosol of host cells (INT 407) (Neal-McKinney and Konkel, 2012). In contrast to a wild-type strain, a *ciaI* mutant is reduced in intracellular survival in cultured epithelial cells, as judged by a gentamicin survival assay. Following uptake by epithelial cells, the *C. jejuni ciaI* mutant co-localized with a lysosomal marker in the *Campylobacter* containing vacuole when viewed by fluorescence microscopy. Furthermore, a punctate phenotype was observed in cells ectopically expressing *ciaI* and this phenotype was lost upon mutation of the dileucine motif. The dileucine motif is necessary for transport to endosomes. Although a mechanism has yet to be elucidated, CiaI contributes to the survival of *C. jejuni* within epithelial cells. The CiaI protein has also been shown to contribute to colonization of chicks (Barrero-Tobon and Hendrixson, 2012). Although the chicken colonization model is informative, the role of CiaI needs to be evaluated in a *Campylobacter* virulence model.

An additional Cia protein was identified by the YplA screen and is termed CiaC. CiaC is encoded by a σ^{54}-regulated gene and the 12.4 kDa protein plays a role in internalization through modification of host cell signalling (Christensen *et al.*, 2009). The evidence that CiaC is an effector protein is supported by the observation that a *ciaC* mutant is less invasive and induces less host cell membrane ruffling compared to a wild-type strain. Additionally, the CiaC protein is delivered to the cytosol of host cells (Neal-McKinney and Konkel, 2012). Flagellar delivery of CiaC is not dependent on internalization, as CiaC is delivered even when invasion is blocked by treatment with the inhibitor TAE226. TAE226 targets the host cell protein

Focal Adhesion Kinase (FAK), which is required for invasion. Epithelial cells pretreated with the FAK inhibitor results in limited membrane ruffling and reduced *C. jejuni* internalization (Eucker and Konkel, 2012). Noteworthy is that the *ciaC* mutant displays the same growth rate, motility, cell adherence phenotype, and secretion profile (except for the absence of CiaC) to that of the wild-type strain. In summary, CiaC is a secreted effector protein that modifies the host to promote internalization.

Putative roles of the Cia proteins

A model of host cell signalling events following flagellar secretion and delivery of effector proteins is beginning to emerge (Fig. 13.5). The binding of *C. jejuni* to host cells results in the activation of $\alpha_5\beta_1$ integrin receptors, leading to activation of the epidermal growth factor (EGF) receptor and FAK signalling pathways (Eucker and Konkel, 2012). These signalling components are termed the *C. jejuni* invasion complex (CIC). The CIC consists of many proteins, including FAK, paxillin, c-Src, PI3-kinase (PI3K), DOCK180, Rac1 and Cdc42. Integrin signalling results in the activation of FAK. Activated FAK associates with c-Src, which further phosphorylates FAK as well as paxillin and the adaptor protein p130Cas. CrkII is recruited to p130Cas and binds DOCK180, which associates with Elmo. Finally, the DOCK180/Elmo complex activates the Rho monomeric G protein Rac1. These events result in the rearrangement of the actin cytoskeleton, forming the characteristic membrane ruffling that is necessary for invasion. Interestingly, a *C. jejuni ciaC* mutant demonstrated reduced internalization by host cells *in vitro* (Eucker and Konkel, 2012). Furthermore, a *C. jejuni ciaC* mutant is deficient in Rac1 activation, indicating that CiaC modulates host cell behaviour upstream of Rac1. However, the

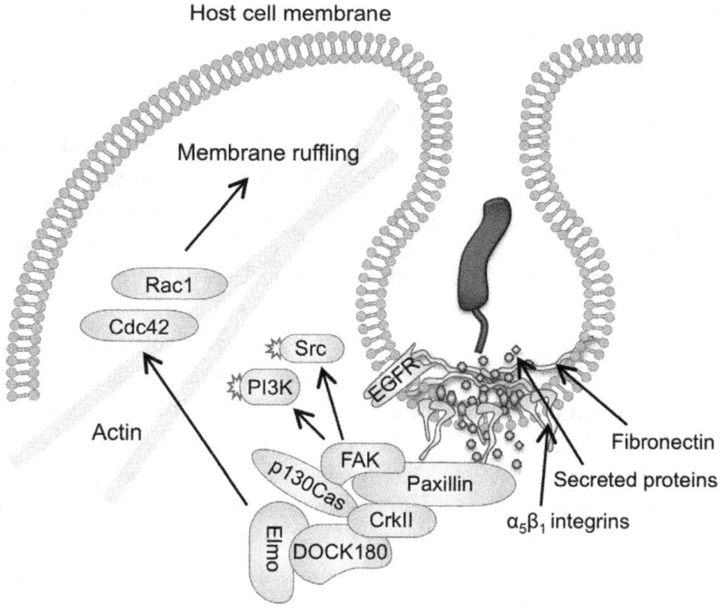

Figure 13.5 Model of host cell signalling. *C. jejuni* secretes proteins that function to modify host cell processes. Initial binding of the bacterium to fibronectin induces $\alpha_5\beta_1$ integrin occupancy and clustering, ultimately resulting in activation. The integrins activate the epidermal growth factor receptor (EGFR) and FAK. Activated FAK associates with and phosphorylates c-Src and PI3K, which contributes to the activation of the focal complex (i.e. FAK, paxillin, p130Cas, CrkII, DOCK180, ELMO). The activated focal complex results in the activation of Rho G proteins including Rac1 and Cdc42. The ultimate result is increased actin cytoskeletal rearrangement leading to membrane ruffling, which is necessary for maximal invasion of host cells by *C. jejuni*. A colour version of this figure is available in the plate section at the back of the book.

phenotype can be rescued by adding excess EGF (Eucker and Konkel, 2012). Thus, CiaC contributes to cytoskeletal rearrangement through the modification of host cell components.

Evidence that the *C. jejuni* flagellum is necessary for protein export

The flagellar export apparatus shares structural and functional properties with the virulence-dedicated T3SS (injectisome) (Fig. 13.6) (Blocker *et al.*, 2003). The hook and basal body complex of the flagellum morphologically resembles the virulence-dedicated needle complex (Macnab, 2003). Moreover, at least eight flagellar components demonstrate sequence homology or similar membrane topology to injectisome proteins and many proteins demonstrate functional conservation (similar phenotypes in assembly or function when knocked out) despite having no significant sequence homology (Blocker *et al.*, 2003; Van Gijsegem *et al.*, 1995). The injectisome and flagellum each span both bacterial membranes and the peptidoglycan to secrete proteins through a 2–3 nm channel. Moreover, many bacteria use environmental cues to regulate each system and ensure appropriate expression of the flagellar and injectisome components (Kutsukake *et al.*, 1990). In spite of both systems being highly similar in structure and function, they do not appear to share a common ancestor in phylogenetic studies (Gophna *et al.*, 2003).

Young and colleagues first demonstrated export of non-flagellar proteins by the flagellar export apparatus in *Y. enterocolitica* (Young *et al.*,

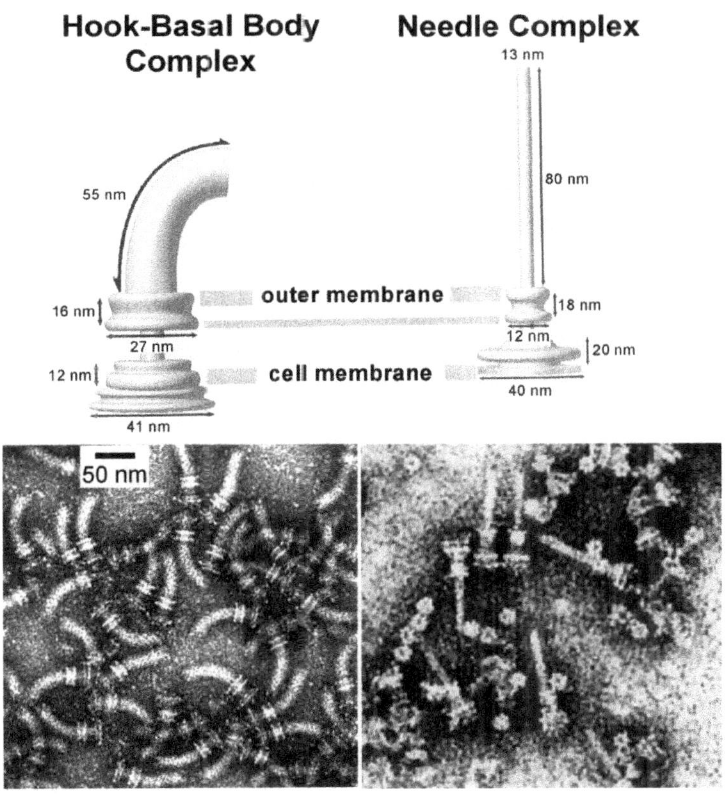

Figure 13.6 Structural similarity of flagellar hook and basal body complexes and the injectisome needle complexes. Top panel: schematic of the hook and basal body complex and the needle complex. Left lower panel: Electron micrograph (EM) of purified flagellar hook and basal body complexes. Right lower panel: EM of purified needle complexes. Figure reproduced from Macnab (2003) with *upper panel* from Shin-Ichi Aizawa, *lower panel*, *left* from Jones *et al.* (1990) and *right* from Kubori *et al.* (1998), with permission.

1999a). These non-flagellar proteins were termed flagellar outer proteins (Fops) and one was found to be the YplA phospholipase. Substrate reciprocity between the injectisome and the flagellar export apparatus has been demonstrated by inappropriate secretion of effector proteins from the flagellar export apparatus in *S. enterica* Typhimurium (Lee and Galan, 2004). Experimental evidence in both *Yersinia* and *Salmonella* demonstrates that the flagellar export apparatus is used to secrete non-flagellar proteins, including virulence factors (Schmiel *et al.*, 1998; Young *et al.*, 1999b). Additionally, flagellin can be delivered by the injectisome (Sun *et al.*, 2007). These results indicate that the flagellum, independent of motility, influences host–pathogen interactions via effector protein export in some bacteria.

The minimal export apparatus for the secretion of non-flagellar (virulence) proteins from *C. jejuni* has been determined. The hook and basal body are essential for secretion, including the hook protein FlgE and the proteins comprising the hook–filament junction, FlgK and FlgL (Neal-McKinney and Konkel, 2012). Proteins that make up the filament, including the flagellar cap protein FliD, are not necessary for secretion. It was previously shown that flagellin proteins were necessary for secretion, as a *C. jejuni flaA flaB* double mutant is unable to secrete proteins but strains with single mutations of either flagellin remained secretion competent (Grant *et al.*, 1993; Konkel *et al.*, 2004). However, we recently determined that the technical nature of these deletions were responsible for the observed lack of protein secretion (Neal-McKinney and Konkel, 2012). The previous *flaA flaB* deletion mutant possessed a chromosomal 5′ portion of the *flaA* gene; however, when the genes were completely deleted in three separate *C. jejuni* strains, protein secretion was observed. Effector protein secretion requires a minimum flagellar structure, as mutants including *flgB, flgC, flgE, flgK, flgL, flhB* and *fliI* that encode basal body and hook components, are deficient in Cia secretion (Konkel *et al.*, 2004; Neal-McKinney and Konkel, 2012). Distinguishing between the contributions of flagellar and non-flagellar (i.e. effector) protein secretion using mutants will further aid our understanding of *C. jejuni* pathogenesis.

The *C. jejuni* flagellar export apparatus is necessary for Cia protein delivery to host cells

It was recently determined that the Cia proteins are delivered into host epithelial cells. The delivery assay used *C. jejuni* genes cloned in-frame with the adenylate cyclase domain (ACD) of the *Bordetella pertussis* CyaA protein (Neal-McKinney and Konkel, 2012). The ACD is only activated in the presence of calmodulin in the host cytosol and catalyses the production of cAMP from ATP. By measuring increased levels of cAMP in host cells, it was determined that the previously identified Cia proteins including CiaB (unpublished), CiaI, and CiaC, were delivered to the cytosol of host cells (Fig. 13.7). A negative control consisting of the cytoplasmic *S*-adenosylmethionine synthetase gene (*metK*) driven by its native promoter did not result in increased cAMP levels. Similar

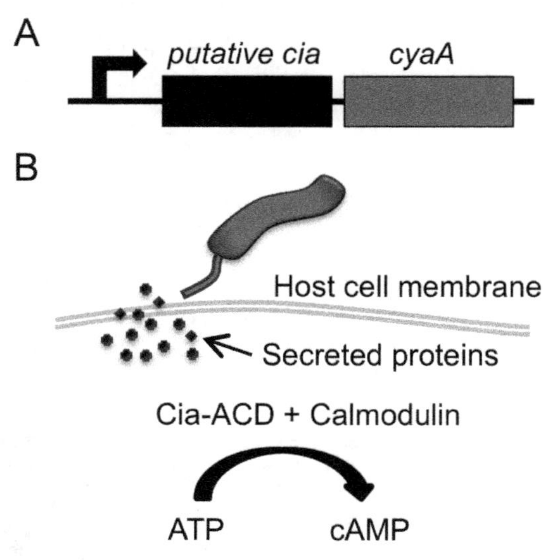

Figure 13.7 The adenylate cyclase delivery assay is used to determine the proteins that are delivered to epithelial cells. (A) Schematic of delivery constructs, including the native promoter (black arrow) with gene of interest (i.e. *cia*) in black and *cyaA* (adenylate cyclase domain, ACD) in red. (B) *C. jejuni* harbouring plasmid encoding reporter construct secretes and delivers the protein to cells (red and black dots). The ACD of the fusion protein is activated in the presence of calmodulin in the cytosol of host cells and catalyses cAMP from ATP, which is measured by ELISA. A colour version of this figure is available in the plate section at the back of the book.

to observations with the injectisome, delivery of secreted proteins is dependent on a functional basal body and hook complex, along with bacteria-host cell contact.

Mechanism of flagellar effector secretion and delivery

Although protein secretion by the injectisome of numerous enteric pathogens has been widely observed, the mechanism of secretion remains debatable after two decades of research. Recently, Wolf-Watz and colleagues proposed that T3SS-dependent secretion and delivery of effector proteins is different than the generally accepted microinjection mechanism (Edgren *et al.*, 2012). The microinjection model implies that effectors are secreted from the basal body through the hollow channel of the needle complex and into the host cell, using the needle complex as a conduit to effectively pass through the host cell membrane. The new Wolf-Watz model incorporates recent findings in *Yersinia*, demonstrating that numerous T3SS-dependent effector proteins are distributed on the surface of the bacterium prior to host cell contact (Akopyan *et al.*, 2011). In support of this model, effector proteins of *Shigella* have been observed to be surface-localized prior to target cell contact (Watarai *et al.*, 1995). The investigators have suggested that the needle complex provides a sensory role for host cell contact, followed by the surface-localized effectors being translocated to the target cell membrane. Effector delivery is then assisted by translocator proteins, such as YopE identified in *Yersinia* (Rosqvist *et al.* 1991). Furthermore, YopB and YopD in *Yersinia* are known to create a pore in the host cell membrane, facilitating delivery of effector proteins (Montagner *et al.*, 2011). This new model of secretion may have implications in *Campylobacter* (Fig. 13.8). Secreted proteins may include effector proteins and translocons that associate with the outer surface of the bacterium and are delivered to the host cell. However, pore forming translocons have yet to be identified in *C. jejuni*. In summary, although many questions remain unanswered in terms of synthesis and delivery of flagellar effector proteins, it is clear that *Campylobacter* secretes proteins that are delivered to host cells in a flagellar-dependent manner.

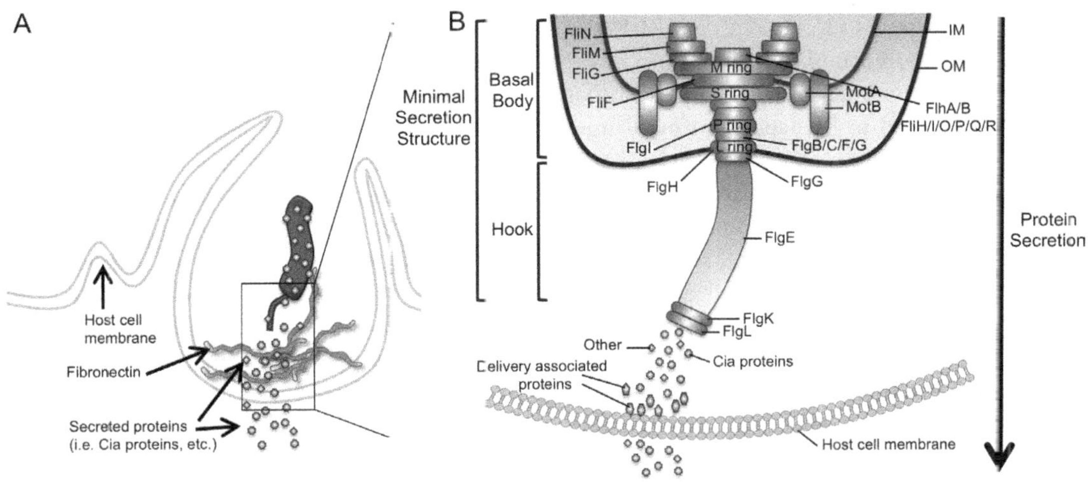

Figure 13.8 Model of *Campylobacter* protein secretion. (A) The proteins secreted by the flagellum are surface-exposed and are delivered to the host cytosol in a mechanism that is not fully defined. (B) Boxed inlet demonstrating the minimal secretion structure of *C. jejuni*. The secreted proteins include the Cia proteins and putative delivery associated proteins. The flagellar filament (not shown) may still be involved during the secretion of effector proteins, although it is not essential. IM, inner membrane; OM, outer membrane. A colour version of this figure is available in the plate section at the back of the book.

Identification of additional non-flagellar proteins exported from the flagellum

There are numerous reports of *C. jejuni* proteins that are secreted from the flagellum. In 2004, FlaC was found to be secreted and shares homology to the flagellin proteins (FlaA and FlaB) (Song *et al.*, 2004). FlaC may influence invasion, as a *flaC* mutant demonstrated reduced invasion. Furthermore, secreted or purified recombinant FlaC binds to host epithelial cells, although the biological role of this observation is not clear as binding is not reduced in a *flaC* mutant as compared to a wild-type strain. FlaC is regulated differently than the Cia proteins, as it is not induced upon culturing *C. jejuni* with bile salts. Another *C. jejuni* secreted protein is the σ^{28}-regulated Cj0977 (Goon *et al.*, 2006). A Cj0977 mutant is motile, but displays reduced invasion of epithelial cells and is attenuated in the ferret model of disease. The protein structure of the Cj0977 gene product has homology to coenzyme A binding proteins (Yokoyama *et al.*, 2008). However, in the *C. jejuni* 81-176 strain, the Cj0977 deletion mutant (referred to as *0996*) showed a defect in motility in liquid medium (Novik *et al.*, 2010). In 2007, Guerry and colleagues identified secreted proteins they termed Fsps for flagellar secreted proteins. FspA is a 15.5 kDa protein and has been found to promote rapid apoptosis in cultured epithelial cells when an excess of recombinant protein was added (Poly *et al.*, 2007). Interestingly, the secretion of FspA was reported to be dependent on a functional flagellum. The investigators also vaccinated mice with two variants, FspA1 and FspA2, to test their protective efficacy. They found that FspA1 and FspA2, when combined with an adjuvant, protected 63.8% and 47.2% of mice from *C. jejuni*-mediated disease, respectively (Baqar *et al.*, 2008). More recently, proteins termed Feds for flagellar coexpressed determinants, including FedA, FedB, FedC, and FedD, were described that play a role in chicken colonization (Barrero-Tobon and Hendrixson, 2012). All of the genes are σ^{28}-regulated and all *fed* mutants are motile. FedA is the best characterized, and was reported to contribute to invasion of cultured epithelial cells and to the colonization of chicks. At this point, more work is required to determine how the FlaC, Cj0977, FspA, and the Fed proteins are secreted from *C. jejuni* and whether they are delivered to host cells. Notwithstanding, there is considerable evidence that *C. jejuni* secretes numerous proteins that contribute to pathogenesis.

Biofilm formation is dependent on flagellum

C. jejuni has the ability to form microcolonies and biofilms (Joshua *et al.*, 2006; Kalmokoff *et al.*, 2006; Lehtola *et al.*, 2006; Reeser *et al.*, 2007). *C. jejuni* biofilm formation *in vitro* is largely dependent on flagellar activity and autoagglutination (Asakura *et al.*, 2007; Guerry *et al.*, 2002, 2006, 2007; Hanning *et al.*, 2009; Kalmokoff *et al.*, 2006; Svensson *et al.*, 2009). Moreover, using *ex vivo* colon sections, it was visually determined by microscopy that *C. jejuni* adhered to host cells initially as microcolonies and later as a biofilm (Haddock *et al.*, 2010). Biofilm formation on host cells is dependent on flagellum, as *flaA* or *flaB* mutants did not form a biofilm (Haddock *et al.*, 2010). Autoagglutination is largely mediated by glycosylation of the flagellar filament at numerous serine and threonine residues with pseudaminic acid and its derivatives (Thibault *et al.*, 2001). Furthermore, based on *in vitro* binding assay, a *C. jejuni* wild-type strain adhered to cells in microcolonies whereas a glycan synthesis mutant (*pseA*) only attached as single cells (Guerry *et al.*, 2006). Thus, glycans decorating the surface of flagellin contribute to the number of bacteria associated with epithelial cells and may result in an increase in the concentration of secreted proteins in microcolonies and biofilms. This notion is supported by attenuation of a *C. jejuni pseA* mutant in ferrets compared to a *C. jejuni* wild-type strain (Guerry *et al.*, 2006). Chromosomal insertion of a wild-type copy of *pseA* into the mutant functionally complemented the phenotype. Additionally, biofilm formation by *C. jejuni* resulted in increased tolerance to stresses including antimicrobials (Svensson *et al.*, 2009). It is possible that biofilm formation is also necessary for *C. jejuni* survival in the environment (Ica *et al.*, 2012). For example, the watering system in poultry houses is often contaminated with *C. jejuni* biofilms on the drinking apparatus and this is a possible source of continued infection for poultry (Hanning *et al.*, 2008).

Conclusions and future perspectives

Since the initial observation that non-motile strains of *C. jejuni* were unable to colonize suckling mice (Morooka *et al.*, 1985), investigators have demonstrated that the *C. jejuni* flagellum contributes to pathogenesis. The flagellum is recognized for therapeutic potential and has been included in a *C. jejuni* subunit vaccine in mice (Lee *et al.*, 1999). The role of the flagellum in the virulence of *C. jejuni* is multifactorial, and the elucidation of the molecular mechanisms governing the expression and regulation of the genes encoding flagellar and effector proteins will further aid in understanding this unique pathogen. Interesting questions remain regarding the mechanism necessary for switching between flagellar protein secretion and non-flagellar protein (i.e. Cia) secretion. Continued research will increase our understanding of *C. jejuni* virulence determinants and offer targets for therapeutic intervention to reduce the incidence and severity of campylobacteriosis.

Acknowledgements

We wish to thank Dr Jason Neal-McKinney, Dr Chris Gourley, Derrick Samuelson and Tyson Eucker for reading this manuscript and providing constructive comments. This work is supported from funds awarded to Dr Konkel from the National Institutes of Health (NIH, Award Number R56 AI088518-01A1). Dr O'Loughlin is supported by the NIH T32 Training program in Infectious Diseases and Microbial Immunology (Award Number 5 T32 AI 7025-33). The content is solely the responsibility of the authors and does not necessarily represent the official views of the National Institute of General Medical Sciences or the NIH.

References

Akopyan, K., Edgren, T., Wang-Edgren, H., Rosqvist, R., Fahlgren, A., Wolf-Watz, H., and Fallman, M. (2011). Translocation of surface-localized effectors in type III secretion. Proc. Natl. Acad. Sci. U.S.A. *108*, 1639–1644.

Aldridge, P., and Hughes, K.T. (2002). Regulation of flagellar assembly. Curr. Opin. Microbiol. *5*, 160–165.

Allos, B.M. (2001). *Campylobacter jejuni* Infections: update on emerging issues and trends. Clin. Infect. Dis. *32*, 1201–1206.

Andersen-Nissen, E., Smith, K.D., Strobe, K.L., Barrett, S.L., Cookson, B.T., Logan, S.M., and Aderem, A. (2005). Evasion of Toll-like receptor 5 by flagellated bacteria. Proc. Natl. Acad. Sci. U.S.A. *102*, 9247–9252.

Asakura, H., Yamasaki, M., Yamamoto, S., and Igimi, S. (2007). Deletion of *peb4* gene impairs cell adhesion and biofilm formation in *Campylobacter jejuni*. FEMS Microbiol. Lett. *275*, 278–285.

Baqar, S., Rice, B., Lee, L., Bourgeois, A.L., El Din, A.N., Tribble, D.R., Heresi, G.P., Mourad, A.S., and Murphy, J.R. (2001). *Campylobacter jejuni* enteritis. Clin. Infect. Dis. *33*, 901–905.

Baqar, S., Applebee, L.A., Gilliland, T.C., Jr., Lee, L.H., Porter, C.K., and Guerry, P. (2008). Immunogenicity and protective efficacy of recombinant *Campylobacter jejuni* flagellum-secreted proteins in mice. Infect. Immun. *76*, 3170–3175.

Barrero-Tobon, A.M., and Hendrixson, D.R. (2012). Identification and analysis of flagellar coexpressed determinants (Feds) of *Campylobacter jejuni* involved in colonization. Mol. Microbiol. *84*, 352–369.

Berring, E., Brancato, S., Grant, K., Schaper, E., Kadavil, S., Smagin, H., Hatic, S.O., 2nd, Picking, W., and Serfis, A.B. (2004). Destabilization of phospholipid model membranes by YplA, a phospholipase A2 secreted by *Yersinia enterocolitica*. Chem. Phys. Lipids *131*, 135–149.

Biswas, D., Fernando, U.M., Reiman, C.D. Willson, P.J., Townsend, H.G., Potter, A.A., and Allan, B.J. (2007). Correlation between *in vitro* secretion of virulence-associated proteins of *Campylobacter jejuni* and colonization of chickens. Curr. Microbiol. *54*, 207–212.

Black, R.E., Levine, M.M., Clements, M.L., Hughes, T.P., and Blaser, M.J. (1988). Experimental *Campylobacter jejuni* infection in humans. J. Infect. Dis. *157*, 472–479.

Blocker, A., Komoriya, K., and Aizawa, S. (2003). Type III secretion systems and bacterial flagella: insights into their function from structural similarities. Proc. Natl. Acad. Sci. U.S.A. *100*, 3027–3030.

Buelow, D.R., Christensen, J.E., Neal-McKinney, J.M., and Konkel, M.E. (2011). *Campylobacter jejuni* survival within human epithelial cells is enhanced by the secreted protein CiaI. Mol. Microbiol. *80*, 1296–1312.

Caldwell, M.B., Guerry, P., Lee, E.C., Burans, J.P., and Walker, R.I. (1985). Reversible expression of flagella in *Campylobacter jejuni*. Infect. Immun. *50*, 941–943.

Carrillo, C.D., Taboada, E., Nash, J.H., Lanthier, P., Kelly, J., Lau, P.C., Verhulp, R., Mykytczuk, O., Sy, J., Findlay, W.A., *et al.* (2004). Genome-wide expression analyses of *Campylobacter jejuni* NCTC11168 reveals coordinate regulation of motility and virulence by *flhA*. J. Biol. Chem. 279, 20327–20338.

Christensen, J.E., Pacheco, S.A., and Konkel, M.E. (2009). Identification of a *Campylobacter jejuni*-secreted protein required for maximal invasion of host cells. Mol. Microbiol. 73, 650–662.

Edgren, T., Forsberg, A., Rosqvist, R., and Wolf-Watz, H. (2012). Type III secretion in *Yersinia*: injectisome or not? PLoS Pathog. *8*, e1002669.

Eucker, T.P., and Konkel, M.E. (2012). The cooperative action of bacterial fibronectin-binding proteins and secreted proteins promote maximal *Campylobacter jejuni* invasion of host cells by stimulating membrane ruffling. Cell. Microbiol. *14*, 226–238.

Flanagan, R.C., Neal-McKinney, J.M., Dhillon, A.S., Miller, W.G., and Konkel, M.E. (2009). Examination of *Campylobacter jejuni* putative adhesins leads to the identification of a new protein, designated FlpA, required for chicken colonization. Infect. Immun. *77*, 2399–2407.

Friedman, C.R., Hoekstra, R.M., Samuel, M., Marcus, R., Bender, J., Shiferaw, B., Reddy, S., Ahuja, S.D., Helfrick, D.L., Hardnett, F., *et al.* (2004). Risk factors for sporadic *Campylobacter* infection in the United States: a case–control study in FoodNet sites. Clin. Infect. Dis. *38(Suppl 3)*, S285–296.

Galkin, V.E., Yu, X., Bielnicki, J., Heuser, J., Ewing, C.P., Guerry, P., and Egelman, E.H. (2008). Divergence of quaternary structures among bacterial flagellar filaments. Science *320*, 382–385.

van Gerwe, T., Miflin, J.K., Templeton, J.M., Bouma, A., Wagenaar, J.A., Jacobs-Reitsma, W.F., Stegeman, A., and Klinkenberg, D. (2009). Quantifying transmission of *Campylobacter jejuni* in commercial broiler flocks. Appl. Environ. Microbiol. *75*, 625–628.

Goon, S., Ewing, C.P., Lorenzo, M., Pattarini, D., Majam, G., and Guerry, P. (2006). A sigma28-regulated nonflagella gene contributes to virulence of *Campylobacter jejuni* 81-176. Infect. Immun. *74*, 769–772.

Gophna, U., Ron, E.Z., and Graur, D. (2003). Bacterial type III secretion systems are ancient and evolved by multiple horizontal-transfer events. Gene *312*, 151–163.

Grant, C.C., Konkel, M.E., Cieplak, W., Jr., and Tompkins, L.S. (1993). Role of flagella in adherence, internalization, and translocation of *Campylobacter jejuni* in nonpolarized and polarized epithelial cell cultures. Infect. Immun. *61*, 1764–1771.

Guerry, P., Logan, S.M., Thornton, S., and Trust, T.J. (1990). Genomic organization and expression of *Campylobacter* flagellin genes. J. Bacteriol. *172*, 1853–1860.

Guerry, P., Alm, R.A., Power, M.E., Logan, S.M., and Trust, T.J. (1991). Role of two flagellin genes in *Campylobacter* motility. J. Bacteriol. *173*, 4757–4764.

Guerry, P., Doig, P., Alm, R.A., Burr, D.H., Kinsella, N., and Trust, T.J. (1996). Identification and characterization of genes required for post-translational modification of *Campylobacter coli* VC167 flagellin. Mol. Microbiol. *19*, 369–378.

Guerry, P., Szymanski, C.M., Prendergast, M.M., Hickey, T.E., Ewing, C.P., Pattarini, D.L., and Moran, A.P. (2002). Phase variation of *Campylobacter jejuni* 81-176 lipooligosaccharide affects ganglioside mimicry and invasiveness *in vitro*. Infect. Immun. *70*, 787–793.

Guerry, P., Ewing, C.P., Schirm, M., Lorenzo, M., Kelly, J., Pattarini, D., Majam, G., Thibault, P., and Logan, S. (2006). Changes in flagellin glycosylation affect *Campylobacter* autoagglutination and virulence. Mol. Microbiol. *60*, 299–311.

Guerry, P., Ewing, C.P., Schoenhofen, I.C., and Logan, S.M. (2007). Protein glycosylation in *Campylobacter jejuni*: partial suppression of *pglF* by mutation of *pseC*. J. Bacteriol. *189*, 6731–6733.

Haddock, G., Mullin, M., MacCallum, A., Sherry, A., Tetley, L., Watson, E., Dagleish, M., Smith, D.G., and Everest, P. (2010). *Campylobacter jejuni* 81-176 forms distinct microcolonies on *in vitro*-infected human small intestinal tissue prior to biofilm formation. Microbiology *156*, 3079–3084.

Hanning, I., Jarquin, R., and Slavik, M. (2008). *Campylobacter jejuni* as a secondary colonizer of poultry biofilms. J. Appl. Microbiol. *105*, 1199–1208.

Hanning, I., Donoghue, D.J., Jarquin, R., Kumar, G.S., Aguiar, V.F., Metcalf, J.H., Reyes-Herrera, I., and Slavik, M. (2009). *Campylobacter* biofilm phenotype exhibits reduced colonization potential in young chickens and altered *in vitro* virulence. Poult. Sci. *88*, 1102–1107.

Hendrixson, D.R. (2006). A phase-variable mechanism controlling the *Campylobacter jejuni* FlgR response regulator influences commensalism. Mol. Microbiol. *61*, 1646–1659.

Hendrixson, D.R., and DiRita, V.J. (2003). Transcription of sigma54-dependent but not sigma28-dependent flagellar genes in *Campylobacter jejuni* is associated with formation of the flagellar secretory apparatus. Mol. Microbiol. *50*, 687–702.

Hu, L., Tall, B.D., Curtis, S.K., and Kopecko, D.J. (2008). Enhanced microscopic definition of *Campylobacter jejuni* 81-176 adherence to, invasion of, translocation across, and exocytosis from polarized human intestinal Caco-2 cells. Infect. Immun. *76*, 5294–5304.

Ica, T., Caner, V., Istanbullu, O., Nguyen, H.D., Ahmed, B., Call, D.R., and Beyenal, H. (2012). Characterization of mono- and mixed-culture *Campylobacter jejuni* biofilms. Appl. Environ. Microbiol. *78*, 1033–1038.

Jones, C.J., Macnab, R.M., Okino, H., and Aizawa, S. (1990). Stoichiometric analysis of the flagellar hook-(basal-body) complex of *Salmonella typhimurium*. J. Mol. Biol. *212*, 377–387.

Josenhans, C., Vossebein, L., Friedrich, S., and Suerbaum, S. (2002). The *neuA/flmD* gene cluster of *Helicobacter pylori* is involved in flagellar biosynthesis and flagellin glycosylation. FEMS Microbiol. Lett. *210*, 165–172.

Joshua, G.W., Guthrie-Irons, C., Karlyshev, A.V., and Wren, B.W. (2006). Biofilm formation in *Campylobacter jejuni*. Microbiology *152*, 387–396.

Kalmokoff, M., Lanthier, P., Tremblay, T.L., Foss, M., Lau, P.C., Sanders, G., Austin, J., Kelly, J., and Szymanski, C.M. (2006). Proteomic analysis of *Campylobacter jejuni* 11168 biofilms reveals a role for the motility complex in biofilm formation. J. Bacteriol. *188*, 4312–4320.

Kapatral, V., Campbell, J.W., Minnich, S.A., Thomson, N.R., Matsumura, P., and Pruss, B.M. (2004). Gene array analysis of *Yersinia enterocolitica* FlhD and FlhC: regulation of enzymes affecting synthesis and degradation of carbamoylphosphate. Microbiology *150*, 2289–2300.

Karlinsey, J.E., Tanaka, S., Bettenworth, V., Yamaguchi, S., Boos, W., Aizawa, S.I., and Hughes, K.T. (2000). Completion of the hook-basal body complex of the *Salmonella typhimurium* flagellum is coupled to FlgM secretion and *fliC* transcription. Mol. Microbiol. 37, 1220–1231.

Ketley, J.M. (1997). Pathogenesis of enteric infection by *Campylobacter*. Microbiology *143(Pt 1)*, 5–21.

Kinsella, N., Guerry, P., Cooney, J., and Trust, T.J. (1997). The *flgE* gene of *Campylobacter coli* is under the control of the alternative sigma factor sigma54. J. Bacteriol. *179*, 4647–4653.

Konkel, M.E., and Cieplak, W., Jr. (1992). Altered synthetic response of *Campylobacter jejuni* to cocultivation with human epithelial cells is associated with enhanced internalization. Infect. Immun. *60*, 4945–4949.

Konkel, M.E., and Joens, L.A. (1989). Adhesion to and invasion of HEp-2 cells by *Campylobacter* spp. Infect. Immun. *57*, 2984–2990.

Konkel, M.E., Hayes, S.F., Joens, L.A., and Cieplak, W., Jr. (1992a). Characteristics of the internalization and intracellular survival of *Campylobacter jejuni* in human epithelial cell cultures. Microb. Pathog. *13*, 357–370.

Konkel, M.E., Mead, D.J., Hayes, S.F., and Cieplak, W., Jr. (1992b). Translocation of *Campylobacter jejuni* across human polarized epithelial cell monolayer cultures. J. Infect. Dis. *166*, 308–315.

Konkel, M.E., Mead, D.J., and Cieplak, W., Jr. (1993). Kinetic and antigenic characterization of altered protein synthesis by *Campylobacter jejuni* during cultivation with human epithelial cells. J. Infect. Dis. *168*, 948–954.

Konkel, M.E., Garvis, S.G., Tipton, S.L., Anderson, D.E., Jr., and Cieplak, W., Jr. (1997). Identification and molecular cloning of a gene encoding a fibronectin-binding protein (CadF) from *Campylobacter jejuni*. Mol. Microbiol. *24*, 953–963.

Konkel, M.E., Kim, B.J., Rivera-Amill, V., and Garvis, S.G. (1999). Bacterial secreted proteins are required for the internalization of *Campylobacter jejuni* into cultured mammalian cells. Mol. Microbiol. *32*, 691–701.

Konkel, M.E., Monteville, M.R., Rivera-Amill, V., and Joens, L.A. (2001). The pathogenesis of *Campylobacter jejuni*-mediated enteritis. Curr. Issues Intest. Microbiol. 2, 55–71.

Konkel, M.E., Klena, J.D., Rivera-Amill, V., Monteville, M.R., Biswas, D., Raphael, B., and Mickelson, J. (2004). Secretion of virulence proteins from *Campylobacter jejuni* is dependent on a functional flagellar export apparatus. J. Bacteriol. *186*, 3296–3303.

Konkel, M.E., Larson, C.L., and Flanagan, R.C. (2010). *Campylobacter jejuni* FlpA binds fibronectin and is required for maximal host cell adherence. J. Bacteriol. *192*, 68–76.

Krause-Gruszczynska, M., Rohde, M., Hartig, R., Genth, H., Schmidt, G., Keo, T., Konig, W., Miller, W.G., Konkel, M.E., and Backert, S. (2007). Role of the small Rho GTPases Rac1 and Cdc42 in host cell invasion of *Campylobacter jejuni*. Cell. Microbiol. *9*, 2431–2444.

Kubori, T., Matsushima, Y., Nakamura, D., Uralil, J., Lara-Tejero, M., Sukhan, A., Galan, J.E., and Aizawa, S.I. (1998). Supramolecular structure of the *Salmonella typhimurium* type III protein secretion system. Science *280*, 602–605.

Kutsukake, K., Ohya, Y., and Iino, T. (1990). Transcriptional analysis of the flagellar regulon of *Salmonella typhimurium*. J. Bacteriol. *172*, 741–747.

Larson, C., Christensen, J., Pacheco, S., Minnich, S., and Konkel, M. (2008a). *Campylobacter jejuni* Secretes Proteins via the Flagellar Type III Secretion System that Contribute to Host cell Invasion and Gastroenteritis. In *Campylobacter*, 3rd edn, Nachamkin, I., Szymanski, C.M., and Blaser, M.J., eds. (ASM Press, Washington, DC), pp. 315–332.

Larson, C.L., Shah, D.H., Dhillon, A.S., Call, D.R., Ahn, S., Haldorson, G.J., Davitt, C., and Konkel, M.E. (2008b). *Campylobacter jejuni* invade chicken LMH cells inefficiently and stimulate differential expression of the chicken CXCLi1 and CXCLi2 cytokines. Microbiology *154*, 3835–3847.

Lee, A., O'Rourke, J.L., Barrington, P.J., and Trust, T.J. (1986). Mucus colonization as a determinant of pathogenicity in intestinal infection by *Campylobacter jejuni*: a mouse cecal model. Infect. Immun. *51*, 536–546.

Lee, L.H., Burg, E., 3rd, Baqar, S., Bourgeois, A.L., Burr, D.H., Ewing, C.P., Trust, T.J., and Guerry, P. (1999). Evaluation of a truncated recombinant flagellin subunit vaccine against *Campylobacter jejuni*. Infect. Immun. *67*, 5799–5805.

Lee, S.H., and Galan, J.E. (2004). *Salmonella* type III secretion-associated chaperones confer secretion-pathway specificity. Mol. Microbiol. *51*, 483–495.

Lehtola, M.J., Pitkanen, T., Miebach, L., and Miettinen, I.T. (2006). Survival of *Campylobacter jejuni* in potable water biofilms: a comparative study with different detection methods. Water Sci. Technol. *54*, 57–61.

Linton, D., Karlyshev, A.V., Hitchen, P.G., Morris, H.R., Dell, A., Gregson, N.A., and Wren, B.W. (2000). Multiple N-acetyl neuraminic acid synthetase (*neuB*) genes in *Campylobacter jejuni*: identification and characterization of the gene involved in sialylation of lipo-oligosaccharide. Mol. Microbiol. 35, 1120–1134.

Macnab, R.M. (2003). How bacteria assemble flagella. Annu. Rev. Microbiol. *57*, 77–100.

Malik-Kale, P., Raphael, B.H., Parker, C.T., Joens, L.A., Klena, J.D., Quinones, B., Keech, A.M., and Konkel, M.E. (2007). Characterization of genetically matched isolates of *Campylobacter jejuni* reveals that mutations in genes involved in flagellar biosynthesis alter the organism's virulence potential. Appl. Environ. Microbiol. 73, 3123–3136.

Malik-Kale, P., Parker, C.T., and Konkel, M.E. (2008). Culture of *Campylobacter jejuni* with sodium deoxycholate induces virulence gene expression. J. Bacteriol. *190*, 2286–2297.

Montagner, C., Arquint, C., and Cornelis, G.R. (2011). Translocators YopB and YopD from *Yersinia enterocolitica* form a multimeric integral membrane complex in eukaryotic cell membranes. J. Bacteriol. *193*, 6923–6928.

Monteville, M.R., and Konkel, M.E. (2002). Fibronectin-facilitated invasion of T84 eukaryotic cells by

Campylobacter jejuni occurs preferentially at the basolateral cell surface. Infect. Immun. *70*, 6665–6671.

Monteville, M.R., Yoon, J.E., and Konkel, M.E. (2003). Maximal adherence and invasion of INT 407 cells by *Campylobacter jejuni* requires the CadF outer-membrane protein and microfilament reorganization. Microbiology *149*, 153–165.

Morooka, T., Umeda, A., and Amako, K. (1985). Motility as an intestinal colonization factor for *Campylobacter jejuni*. J. Gen. Microbiol. *131*, 1973–1980.

Nachamkin, I., Yang, X.H., and Stern, N.J. (1993). Role of *Campylobacter jejuni* flagella as colonization factors for three-day-old chicks: analysis with flagellar mutants. Appl. Environ. Microbiol. 59, 1269–1273.

Neal-McKinney, J.M., Christensen, J.E., and Konkel, M.E. (2010). Amino-terminal residues dictate the export efficiency of the *Campylobacter jejuni* filament proteins via the flagellum. Mol. Microbiol. *76*, 918–931.

Neal-McKinney, J.M., and Konkel, M.E. (2012). The *Campylobacter jejuni* CiaC virulence protein is secreted from the flagellum and delivered to the cytosol of host cells. Front. Cell. Infect. Microbiol. 2, 31.

Newell, D.G., McBride, H., and Dolby, J.M. (1985). Investigations on the role of flagella in the colonization of infant mice with *Campylobacter jejuni* and attachment of *Campylobacter jejuni* to human epithelial cell lines. J. Hyg. (Lond.) *95*, 217–227.

Novik, V., Hofreuter, D., and Galan, J.E. (2010). Identification of *Campylobacter jejuni* genes involved in its interaction with epithelial cells. Infect. Immun. *78*, 3540–3553.

Nuijten, P.J., van Asten, F.J., Gaastra, W., and van der Zeijst, B.A. (1990). Structural and functional analysis of two *Campylobacter jejuni* flagellin genes. J. Biol. Chem. *265*, 17798–17804.

Park, S.F., Purdy, D., and Leach, S. (2000). Localized reversible frameshift mutation in the *flhA* gene confers phase variability to flagellin gene expression in *Campylobacter coli*. J. Bacteriol. *182*, 207–210.

Parkhill, J., Wren, B.W., Mungall, K., Ketley, J.M., Churcher, C., Basham, D., Chillingworth, T., Davies, R.M., Feltwell, T., Holroyd, S., *et al.* (2000). The genome sequence of the food-borne pathogen *Campylobacter jejuni* reveals hypervariable sequences. Nature *403*, 665–668.

Poly, F., Ewing, C., Goon, S., Hickey, T.E., Rockabrand, D., Majam, G., Lee, L., Phan, J., Savarino, N.J., and Guerry, P. (2007). Heterogeneity of a *Campylobacter jejuni* protein that is secreted through the flagellar filament. Infect. Immun. *75*, 3859–3867.

Reeser, R.J., Medler, R.T., Billington, S.J., Jost, B.H., and Joens, L.A. (2007). Characterization of *Campylobacter jejuni* biofilms under defined growth conditions. Appl. Environ. Microbiol. 73, 1908–1913.

Rivera-Amill, V., and Konkel, M.E. (1999). Secretion of *Campylobacter jejuni* Cia proteins is contact dependent. Adv. Exp. Med. Biol. *473*, 225–229.

Rosqvist, R., Forsberg, A., and Wolf-Watz, H. (1991). Intracellular targeting of the *Yersinia* YopE cytotoxin in mammalian cells induces actin microfilament disruption. Infect. Immun. *59*, 4562–4569.

Samatey, F.A., Imada, K., Nagashima, S., Vonderviszt, F., Kumasaka, T., Yamamoto, M., and Namba, K. (2001). Structure of the bacterial flagellar protofilament and implications for a switch for supercoiling. Nature *410*, 331–337.

Samuel, M.C., Vugia, D.J., Shallow, S., Marcus, R., Segler, S., McGivern, T., Kassenborg, H., Reilly, K., Kennedy, M., Angulo, F., *et al.* (2004). Epidemiology of sporadic *Campylobacter* infection in the United States and declining trend in incidence, FoodNet 1996–1999. Clin. Infect. Dis. *38(Suppl 3)*, S165–174.

Schmiel, D.H., Wagar, E., Karamanou, L., Weeks, D., and Miller, V.L. (1998). Phospholipase A of *Yersinia enterocolitica* contributes to pathogenesis in a mouse model. Infect. Immun. *66*, 3941–3951.

Smith, M.F., Jr., Mitchell, A., Li, G., Ding, S., Fitzmaurice, A.M., Ryan, K., Crowe, S., and Goldberg, J.B. (2003). Toll-like receptor (TLR) 2 and TLR5, but not TLR4, are required for *Helicobacter pylori*-induced NF-kappa B activation and chemokine expression by epithelial cells. J. Biol. Chem. *278*, 32552–32560.

Song, Y.C., Jin, S., Louie, H., Ng, D., Lau, R., Zhang, Y., Weerasekera, R., Al Rashid, S., Ward, L.A., Der, S.D., *et al.* (2004). FlaC, a protein of *Campylobacter jejuni* TGH9011 (ATCC43431) secreted through the flagellar apparatus, binds epithelial cells and influences cell invasion. Mol. Microbiol. *53*, 541–553.

Sun, Y.H., Rolan, H.G., and Tsolis, R.M. (2007). Injection of flagellin into the host cell cytosol by *Salmonella enterica* serotype Typhimurium. J. Biol. Chem. *282*, 33897–33901.

Svensson, S.L., Davis, L.M., MacKichan, J.K., Allan, B.J., Pajaniappan, M., Thompson, S.A., and Gaynor, E.C. (2009). The CprS sensor kinase of the zoonotic pathogen *Campylobacter jejuni* influences biofilm formation and is required for optimal chick colonization. Mol. Microbiol. *71*, 253–272.

Szymanski, C.M., King, M., Haardt, M., and Armstrong, G.D. (1995). *Campylobacter jejuni* motility and invasion of Caco-2 cells. Infect. Immun. *63*, 4295–4300.

Szymanski, C.M., Yao, R., Ewing, C.P., Trust, T.J., and Guerry, P. (1999). Evidence for a system of general protein glycosylation in *Campylobacter jejuni*. Mol. Microbiol. 32, 1022–1030.

Takata, T., Fujimoto, S., and Amako, K. (1992). Isolation of nonchemotactic mutants of *Campylobacter jejuni* and their colonization of the mouse intestinal tract. Infect. Immun. *60*, 3596–3600.

Thibault, P., Logan, S.M., Kelly, J.F., Brisson, J.R., Ewing, C.P., Trust, T.J., and Guerry, P. (2001). Identification of the carbohydrate moieties and glycosylation motifs in *Campylobacter jejuni* flagellin. J. Biol. Chem. *276*, 34862–34870.

Van Gijsegem, F., Gough, C., Zischek, C., Niqueux, E., Arlat, M., Genin, S., Barberis, P., German, S., Castello, P., and Boucher, C. (1995). The *hrp* gene locus of *Pseudomonas solanacearum*, which controls the production of a type III secretion system, encodes eight proteins related to components of the bacterial flagellar biogenesis complex. Mol. Microbiol. *15*, 1095–1114.

Warren, S.M., and Young, G.M. (2005). An amino-terminal secretion signal is required for YplA export by the Ysa, Ysc, and flagellar type III secretion systems of *Yersinia enterocolitica* biovar 1B. J. Bacteriol. *187*, 6075–6083.

Wassenaar, T.M., and Blaser, M.J. (1999). Pathophysiology of *Campylobacter jejuni* infections of humans. Microbes Infect. *1*, 1023–1033.

Wassenaar, T.M., Bleumink-Pluym, N.M., and van der Zeijst, B.A. (1991). Inactivation of *Campylobacter jejuni* flagellin genes by homologous recombination demonstrates that *flaA* but not *flaB* is required for invasion. EMBO J. *10*, 2055–2061.

Wassenaar, T.M., van der Zeijst, B.A., Ayling, R., and Newell, D.G. (1993). Colonization of chicks by motility mutants of *Campylobacter jejuni* demonstrates the importance of flagellin A expression. J. Gen. Microbiol. *139(Pt 6)*, 1171–1175.

Wassenaar, T.M., Bleumink-Pluym, N.M., Newell, D.G., Nuijten, P.J., and van der Zeijst, B.A. (1994). Differential flagellin expression in a *flaA flaB*+ mutant of *Campylobacter jejuni*. Infect. Immun. *62*, 3901–3906.

Watarai, M., Tobe, T., Yoshikawa, M., and Sasakawa, C. (1995). Contact of *Shigella* with host cells triggers release of Ipa invasins and is an essential function of invasiveness. EMBO J. *14*, 2461–2470.

Watson, R.O., and Galan, J.E. (2005). Signal transduction in *Campylobacter jejuni*-induced cytokine production. Cell. Microbiol. *7*, 655–665.

Yao, R., Burr, D.H., Doig, P., Trust, T.J., Niu, H., and Guerry, P. (1994). Isolation of motile and non-motile insertional mutants of *Campylobacter jejuni*: the role of motility in adherence and invasion of eukaryotic cells. Mol. Microbiol. *14*, 883–893.

Yokoyama, T., Paek, S., Ewing, C.P., Guerry, P., and Yeo, H.J. (2008). Structure of a sigma28-regulated nonflagellar virulence protein from *Campylobacter jejuni*. J. Mol. Biol. *384*, 364–376.

Yonekura, K., Maki, S., Morgan, D.G., DeRosier, D.J., Vonderviszt, F., Imada, K., and Namba, K. (2000). The bacterial flagellar cap as the rotary promoter of flagellin self-assembly. Science *290*, 2148–2152.

Young, G.M., Schmiel, D.H., and Miller, V.L. (1999a). A new pathway for the secretion of virulence factors by bacteria: the flagellar export apparatus functions as a protein-secretion system. Proc. Natl. Acad. Sci. U.S.A. *96*, 6456–6461.

Young, G.M., Smith, M.J., Minnich, S.A., and Miller, V.L. (1999b). The *Yersinia enterocolitica* motility master regulatory operon, *flhDC*, is required for flagellin production, swimming motility, and swarming motility. J. Bacteriol. *181*, 2823–2833.

14 Does Biofilm Formation Aid Colonization and Infection in *Campylobacter*?

Ben Pascoe, Hilary Lappin-Scott, Samuel K. Sheppard and Hiroshi Asakura

Abstract

Biofilm formation can promote the colonization of host and non-host environments for some bacteria. In some species cell aggregation and attachment are key pathogenic determinants. However, in *Campylobacter*, many questions remain about the importance of biofilms for surface attachment in order to establish an infection in animals or humans. In this chapter, we explore the factors that influence biofilm formation in *Campylobacter*, and discuss this in the context of invasion, colonization and the onset of disease.

Introduction

Biofilm formation can be an important component of bacterial growth and survival. Broadly, this involves the attachment and accumulation of bacteria encased within a matrix of extracellular polymeric substances (EPS) which can provide protection from potentially harmful environmental conditions (Hall-Stoodley *et al.*, 2004; Stewart and Costerton, 2001). It is likely that biofilm formation is an ancient adaptive trait that has evolved by several different mechanisms to help overcome external stresses (Donlan and Costerton, 2002) and bacteria within biofilms are usually more resistant to disinfectants and antimicrobials than their free-living counterparts (Fux *et al.*, 2005). *Campylobacter* biofilms allow the organism to survive up to twice as long under atmospheric conditions (Asakura *et al.*, 2007; Joshua *et al.*, 2006) or even longer in water systems (Lehtola *et al.*, 2006). *Campylobacter* biofilm formation has been demonstrated *in vitro*, under laboratory conditions and from poultry-rearing facilities (Bull *et al.*, 2006; Pearson *et al.*, 1993; Trachoo and Frank, 2002; Zimmer *et al.*, 2003). However, unlike in some other pathogenic species where biofilm formation is thought to be essential for survival in certain environments (Costerton *et al.*, 1978, 1995; Hall-Stoodley *et al.*, 2004), there is little evidence linking *Campylobacter* surface attachment to infection. Therefore the question remains: is biofilm formation important for *Campylobacter* infection?

Attachment and adhesins

For bacteria to establish a biofilm they must first adhere and attach. There is evidence of attachment of *Campylobacter* to each other and non-biological surfaces (Buswell *et al.*, 1998; Somers *et al.*, 1994). Some *Campylobacter* species are capable of forming a mono-species biofilm (Joshua *et al.*, 2006; Kalmokoff *et al.*, 2006) or can colonize a pre-existing biofilm (Hanning *et al.*, 2008). Motility is also important in *Campylobacter* adhesion and flagella have been shown to have a role in establishing biofilm micro-colonies (Table 14.1; Bacon *et al.*, 2001; Golden and Acheson, 2002; Jeon *et al.*, 2003), which is in contrast to biofilm formation for many other bacteria (Sherlock *et al.*, 2005). A common factor in biofilms in other bacterial species is binding to fibronectin, a component of the extracellular matrix of epithelial cells. For example, this can enable well known biofilm forming pathogens such as *Staphylococcus epidermidis* to adhere and initiate biofilm development within human and mammalian hosts (Mack *et al.*, 2013; Otto, 2013). *Campylobacter* also has a suite of surface exposed fibronectin-binding proteins, and it

Table 14.1 Motility genes and potential adhesins influencing *Campylobacter* biofilm formation and development

Gene ID	Gene name	Function
Flagellar assembly		
Cj1338	*flaB*	Flagellin B
Cj1339c	*flaA*	Flagellin A
Cj0041	*fliK*	Flagellar hook–length control protein
Cj0042	*flgD*	Flagellar basal body rod modification protein
Cj0043	*flgE*	Flagellar hook protein
Cj0059c	*fliY*	Flagellar motor switch protein
Cj0060c	*fliM*	Flagellar motor switch protein
Cj0064c	*flhF*	Flagellar biosynthetic protein
Cj0195	*fliI*	Flagellum-specific ATP synthase
Cj0318	*fliF*	Flagellar MS-ring protein
Cj0319	*fliG*	Flagellar motor switch protein
Cj0320	*fliH*	Flagellar assembly protein
Cj0335	*flhB*	Flagellar biosynthetic protein
Cj0336c	*motB*	Flagellar motor protein
Cj0337c	*motA*	Flagellar motor protein
Cj0351	*fliN*	Flagellar motor switch protein
Cj0352	*fliO*	Putative transmembrane protein
Cj0526c	*fliE*	Flagellar hook-basal body protein
Cj0527c	*flgC*	Flagellar hook-basal body protein
Cj0528c	*flgB*	Flagellar basal body rod protein
Cj0547	*flaG*	Flagella export protein
Cj0548	*fliD*	Flagellar capping protein
Cj0549	*fliS*	Flagellin-specific chaperone
Cj0687c	*flgH*	Flagellin-specific chaperone
Cj0697	*flgG2*	Flagellar basal-body rod protein
Cj0698	*flgG*	Flagellar basal-body rod protein
Cj0769c	*flgA*	Flagellar basal-body rod protein
Cj0820c	*fliP*	Flagellar biosynthetic protein
Cj0882c	*flhA*	Flagellar biosynthetic protein
Cj0887c	*flgL*	Flagellar hook-associated protein
Cj1075	*fliW*	flagellar assembly protein
Cj1179c	*fliR*	flagellar biosynthetic protein
O-linked glycosylation sites (Cj1293–1342) essential for flagellar glycosylation		
Cj1293	*pseB*	UDP-GlcNAc-specific C4,6 dehydratase/C5 epimerase
Cj1294	*pseC*	C4 aminotransferase
Cj1311	*pseF*	acylneuraminate cytidylytransferase
Cj1312	*pseG*	PseC nucleotidase
Cj1313	*pseH*	PseC N-acetyltransferase
Cj1317	*pseI*	Pse synthetase
Cj1318	*maf1*	Motility accessory factor 1

Gene ID	Gene name	Function
Cj1337	*pseE (maf5)*	Glycan transfer to flagellin
Flagellar regulator, chemotaxis, and others		
Cj0061c	*fliA*	sigma28
Cj0670	*rpoN*	RNA polymerase sigma 54
Cj1001	*rpoD*	RNA polymerase sigma 70
Cj0256	*pEtN*	phosphoethanolamine, flagellar length regulator
Cj0284c	*cheA*	Chemotaxis histidine kinase
Cj0285c	*cheV*	Chemotaxis protein
Cj1118c	*cheY*	Chemotaxis regulatory protein
Cj1335/Cj1336	*maf4*	Motility accessory factor
Cj0062c	Hypothetical	Probable integral membrane protein
Cj0248	Hypothetical	Motility-associated protein
Cj0883c	Hypothetical	Motility-associated protein
Cj1026c	Hypothetical	Motility-associated protein
Adhesins		
Cj0596	*peb4*	Major antigenic peptide PEB4/cell binding factor 2 (CBF2)
Cj0921c	*peb1A*	ABC-type amino acid transporter/periplasmic solute-binding protein
Cj0720c	*flaC*	Putative flagellin, affecting cell adhesion
Cj1478c	*cadF*	Outer membrane fibronectin-binding protein
–	ciaC	*Campylobacter* invasion antigen C

is possible that adhesion of *Campylobacter* to host epithelial cells has some similarities to surface attachment in *S. epidermidis* where this has been studied in more detail.

The host cell binding proteins Peb1A and Peb4 assist with adhesion to intestinal epithelial cells, with mutant species demonstrating significantly lower adhesion and reduced biofilm formation (Asakura *et al.*, 2007; Kalmokoff *et al.*, 2006; Leon-Kempis Mdel *et al.*, 2006; Pei *et al.*, 1998). These mutants were also less virulent in a mouse infection model, suggesting a role for biofilms in colonization and infection. Several other biofilm associated proteins have been identified in *Campylobacter*. The surface lipoproteins *Campylobacter* adhesion protein A (CapA), (Ashgar *et al.*, 2007; Flanagan *et al.*, 2009), and *C. jejuni* lipoprotein A (JlpA) (Jin *et al.*, 2001, 2003), the outer membrane fibronectin-binding proteins, Cj1279c and Cj1349c (Flanagan *et al.*, 2009), the secreted *Campylobacter* invasion antigen C (CiaC) (Barrero-Tobon and Hendrixson, 2012; Christensen *et al.*, 2009), and the outer membrane fibronectin-binding protein *Campylobacter* adhesin (CadF) (Eucker and Konkel, 2012; Hirayama *et al.*, 2009; Konkel *et al.*, 1993, 2005; Krause-Gruszczynska *et al.*, 2007), all have a role in adhesion and invasion of epithelial cells.

Role of flagella and motility in adhesion and invasion

Campylobacter flagella differ from the previously established model for flagella structure documented in *Salmonella* (Galkin *et al.*, 2008). They have seven protofilaments rather than the 11 observed in *Salmonella* species. *Campylobacter* flagella are often bipolar (Balaban *et al.*, 2009) with elongated flagellins, FlaA and FlaB that are secreted through the flagellar secretion complex and are responsible for the distinctive darting movement (Guerry, 2007). FlaA and FlaB can facilitate *Campylobacter* adhesion and a *flaAB* mutant has been shown to exhibit reduced biofilm formation (Reeser *et al.*, 2007). Correct flagella assembly is vital in motility and surface attachment and mutants defective in flagellar modification (*cj1337*) and assembly (*fliS*) genes adhere less well to glass surfaces (Joshua *et al.*, 2006). Investigation of expressed proteins from biofilm-grown cells has shown increased levels of motility-associated proteins, including FlaA, FlaB, FliD, FlgG, and FlgG2 (Kalmokoff *et al.*, 2006). Expression of FlaC and Cj0391c, which share the same regulator protein, may also be up-regulated in biofilms of *C. jejuni* (Guerry, 2007). *Campylobacter* flagella not only have a role in motility, but also act as the bacterium's secretion system, through which the flagellins, FlaA and FlaB are secreted (Guerry, 2007). Furthermore, *Campylobacter* show extensive glycosylation of their surface exposed proteins. Both of these traits enhance surface interactions and may assist attachment, invasion and colonization of the host cells.

What is the matrix?

During biofilm development, bacteria commonly secrete proteins or fatty acids to help provide a matrix sticking the bacteria together. In the *Staphylococci*, systems have been identified in which proteinaceous material and oligosaccharides are involved in biofilm formation, typically polysaccharide intercellular adhesin (PIA; Rohde *et al.*, 2010), accumulation-associated protein (Aap; O'Gara, 2007) and other surface proteins (Arciola *et al.*, 2012). Secretion of fatty acids via the flagella secretion system also facilitates agglutination (Guerry *et al.*, 2006).

Lipooligosaccharides are secreted on the surface of many mucosal pathogens providing a sticky exterior to aid adhesion. The lipooligosaccharide (LOS) gene cluster drives secretion of fatty acids that interact with the host immune system and are essential in colonization and infection. The host response includes the release of cytokines (Al-Amri *et al.*, 2008; Al-Salloom *et al.*, 2003) and activation of macrophages (Heikema *et al.*, 2010; Huizinga *et al.*, 2012). Truncation of LOS polysaccharides can reduce the ability of *Campylobacter* to adhere to and invade host cells (Table 14.2; Holden *et al.*, 2012; Naito *et al.*, 2010). Capsular polysaccharides (CPS) also contribute to this sticky exterior and have been shown in other bacterial species to be involved in external interactions, including quorum sensing, bacteriophage resistance and osmotic stress. However, as yet there is no evidence that they are also involved in *Campylobacter* biofilms (Day *et al.*, 2012; Richards *et al.*, 2013).

Increasingly, extracellular DNA is being found to give structure to bacterial biofilm matrices (Arciola *et al.*, 2012; Campoccia *et al.*, 2011; Costerton *et al.*, 2005; Montanaro *et al.*, 2011; Ravaioli *et al.*, 2011). Extracellular DNA shields the biofilm from antimicrobial peptide attack leading to increased antimicrobial resistance in *Pseudomonas aeruginosa* (Lewenza, 2013). It is likely that the extracellular DNA also interacts with exopolysaccharides to strengthen the extracellular matrix of biofilms as in *Myxococcus xanthus* (Hu *et al.*, 2012). *Campylobacter* appears to be similar, with expression of extracellular DNA in *C. rectus* periodontal biofilms (Palmer *et al.*, 2012) and evidence that application of DNase can be sufficient to resuspend autoagglutinated *Campylobacter* cells (Ritchie *et al.*, 1983). A *C. jejuni cprS* mutant showing enhanced biofilm formation was also sensitive to DNase digestion (Svensson *et al.*, 2009).

Effect of oxygen on biofilm formation

Following accumulation and agglutination of bacteria, a physical barrier to the external environment is created within the biofilm and an oxygen gradient is generated (Park *et al.*, 2010; Skolimowski *et al.*, 2010; Snider *et al.*, 2012). Under aerobic conditions, or mild aerobic stress, *Campylobacter* biofilm formation is increased (Reuter *et al.*, 2010). This is consistent with the idea that biofilms in *Campylobacter* may have a role in reducing exposure to oxygen. Adhesion of *C. jejuni* to inert surfaces and swarming motility can also be enhanced in oxygen-acclimatised cells, presumably due to the higher abundance of membrane proteins involved in adhesion and biofilm formation. Specifically, over-expression of the fibronectin-binding CadF in the outer membrane

Table 14.2 Surface proteins influencing *Campylobacter* biofilm formation and development

Gene ID	Gene name	Function
Cj1131c	*gne*	UDP-GlcNAc/Glc 4-epimerase
Cj1133	*waaC*	Heptosy transferase I
Cj1134	*htrB*	Lipid A biosynthesis lauroyl acyltransferase
Cj1135	*lgtF*	Probable two-domain glycosyltransferase
Cj1136	Hypothetical	Probable glycosyltransferase
Cj1137/1138	Hypothetical	Probable glycosyltransferase
Cj1139c	*wlaN*	Beta-1,3-galactosyltransferase
Cj1140	*cstIII*	Alpha-2,3 sialyltransferase (transfer sialic acid from CMP-NeuAc to LOS)
Cj1141	*neuB1*	Sialic acid synthase (probable N-acetylneuraminic acid synthetase)
Cj1142	*neuC1*	Sialic acid synthase (probable N-acetylglycosamine-6-phosphate 2-epimerase)
Cj1143	*neuA1(cgtA)*	β-1,4-N-Acetylgalactosaminyltransferase/CMP-Neu5Ac synthase
Cj1146c	*waaV*	Glucosyltransferase
Cj1148	*waaF*	Heptosyltransferase II
Cj1149	*gmhA*	Phosphoheptose isomerase
Cj1150c	*hldE*	D-Beta-D-heptose 7-phosphate kinase
Cj1151c	*hldD*	ADP-glyceromanno-heptose 6-epimerase
Cj1152c	*gmhB*	D,D-heptose 1,7-bisphosphate phosphatase
Cj1148	*waaF*	LOS synthesis
Cj1135	*lgtF*	LOS synthesis
CJJ81176_1157	*cstII*	LOS synthesis
CJJ81176_1165	*galT*	LOS synthesis
Cj1272c	*spoT*	LOS truncation (Penta-phosphate guanosine-3′-pyrophosphohydrolase, RelA/SpoT)

of oxygen-acclimatised cells supports its putative role in adhesion along with the observation of a reduction in the adherence of a *C. jejuni* 81-176 mutant lacking *cadF* (Sulaeman *et al.*, 2012).

Trace metals may also have an indirect influence on biofilm formation through established links between oxidative stress and iron uptake (Palyada *et al.*, 2009; van Vliet *et al.*, 1999). The putative MarR family regulator protein (non-specific multiple antibiotic resistance), Cj1556 binds to the promoter region upstream of the *cj1556* locus, thereby regulating the transcription of genes for oxidative and aerobic stress responses such as *katA*, *perR*, and *hspR*. A *cj1556* mutant has shown reduced biofilm formation (Gundogdu *et al.*, 2011). Indirectly, the LuxS/autoinducer-2 (AI-2) mediated quorum sensing system (Plummer, 2012) may also influence biofilm formation through interactions with the oxidative stress, motility and surface protein expression systems. Studies using *Campylobacter* mutants that lacked LuxS and were inhibited in their expression of the *flaAB* flagellins, were restricted in biofilm formation (Reeser *et al.*, 2007) and unable to autoagglutinate (Guerry *et al.*, 2006).

Phase switching in biofilm formation

Phase switching as a result of mutations that change protein expression levels can induce a reversible change in the phenotype. The stochastic nature of these genetic rearrangements means that different phenotypes can coexist within the same population (Chia *et al.*, 2008). In *Staphylococcus epidermidis* phase variants that produce a polysaccharide intercellular adhesin (PIA), important in the formation of thick multilayered biofilms, were

Table 14.3 Regulatory proteins influencing *Campylobacter* biofilm formation and development

Gene ID	Gene name	Function
Metabolic and protein synthesis		
Cj1400c	*FabI*	Enoyl-[acyl-carrier-protein] reductase
CJM1_1307	*FbpA*	Ferric binding protein
Cj1364c	*FumC*	Fumarate hydratase
Cj0597	*Fba*	Fructose-bisphosphotase aldolase
Cj1023c	*Asd*	Aspartate-semialdehyde dehydrogenase
Cj0534	*SucD*	Succinyl-CoA synthetase alpha chain
Cj0765c	*HisS*	Histidyl-tRNA synthatase
Cj0893c	*RpsA*	30S ribosomal protein S1
Cj1181c	*EF-Ts*	Translation elongation factor TS
Cj0551	*EF-P*	Translation elongation factor P
Cj0470	*EF-Tu*	Translation elongation factor Tu
Cj0493	*EF-G*	Translation elongation factor G
Cj0193c	*Tig*	Trigger factor
Stress response		
Cj0334	*AhpC*	Antioxidant, AhpC/Tsa family
Cj1385	*KatA*	Catalase
Cj0012c	*Rrc*	Non-haem iron protein
Cj0146c	*TrxB*	Thioredoxin reductase
Cj0169	*SodB*	Superoxide dismutase
Phosphate metabolism, quorum sensing (QS), and respiratory response		
Cj0578c	*tatC*	Twin arginine transport (TAT) system
Cj0145	*phoX*	Poly-P synthesis, TAT pathway signal sequence domain protein
Cj1556	hypothetical	MarR family transcriptional regulator
Cj0604	*ppk2*	Inorganic polyphosphate (polyP) synthesis
Cj0688	*pta*	Phosphate acetyltransferase
Cj1359	*ppk1*	Inorganic polyphosphate (polyP) synthesis
Cj1198	*luxS*	S-Ribosylhomocysteinase (homoserine lactone, HSL)
Cj1267c	*hydA*	Ni/Fe-hydrogenase small chain, HydA
Cj1266c	*hydB*	Ni/Fe-hydrogenase large subunit, HydB
Cj0889/Cj0890	*phosS/phosR*	Phosphate sensor/phosphate response regulator
Cj0332c	*ndk*	Probable nucleoside diphosphate kinase
Cj0414	–	Putative oxidoreductase subunit
Cj0780	*napA*	Periplasmic nitrate reductase, NapA
Other regulatory proteins		
Cj1259	*porA*	Major outlet membrane protein (MOMP)
Cj0706	*Cj0706*	Hypothetical
Cj0092	*Cj0092*	Putative periplasmic protein
Cj0998	*Cj0998*	Putative periplasmic protein, YceI-like
Cj0355c	*Cj0355c*	Two-component response regulator
Cj1226	*cprR*	*Campylobacter* planktonic growth regulator system
Cj1227c	*cprS*	Sensor kinase

found to have arisen from an insertion sequence that inactivates genes involved in PIA regulation. On excision, this insertion was reversible and the original biofilm phenotype was restored (Ziebuhr *et al.*, 2001). Therefore, even a biofilm grown from a single microbe is capable of producing a community consisting of several variant phenotypes, a phenomenon that is an important component for the survival and persistence of biofilms in general (Kiem *et al.*, 2004; Tormo *et al.*, 2007). The genome sequence of *C. jejuni* suggests a high potential for phase variation (Parkhill *et al.*, 2000) and, in particular, proteins involved in surface interactions, like the LOS genes (Guerry *et al.*, 2002; Prendergast *et al.*, 2004) and exposed motility genes (Guerry *et al.*, 1990; Karlyshev *et al.*, 2002; Logan *et al.*, 1989; Park *et al.*, 2000). This is reflected in observations of biofilm phenotype variation in *Campylobacter* strains (Guerry *et al.*, 2002).

Environmental phosphate

Environmental phosphate may also play a role in biofilm formation as the ability to obtain extracellular phosphate is central to core metabolic responses (Drozd *et al.*, 2011). In *C. jejuni*, the alkaline phosphatase (PhoX) acts as a substrate of the twin argentine transport (TAT) system and disruption of the TAT system results in reduced stress tolerance and colonization, but increased biofilm formation (Table 14.3). This suggests that *phoX* has a role in sensing nutrient stress and inducing biofilm formation (Drozd *et al.*, 2011). Another polyphosphate kinase (*ppk2*) has also been found to be critical for *C. jejuni* survival, adaptation, and persistence in host environments. A *ppk2* deficient mutant was more susceptible to osmotic, nutrient, aerobic, and antimicrobial stresses, while also displaying an enhanced ability to form biofilm (Gangaiah *et al.*, 2010).

Campylobacter planktonic growth regulation complex (CprRS)

Two-component regulatory systems are often used by prokaryotes to respond to environmental changes (Eppinger *et al.*, 2004). Environmental conditions are sensed via interaction between two proteins, a membrane-bound sensor histidine kinase and its cytoplasmic DNA-binding response regulator. Activation of the regulator protein initiates changes in protein expression that help adaptation to the current challenge. Investigation in *Campylobacter* genetics has identified one such regulatory system with effects on biofilm formation (Svensson *et al.*, 2009).

A study into how *C. jejuni* respond to epithelial cells (Gaynor *et al.*, 2005) identified a two-component regulatory system encoded by *cj1226c* and *cj1227c* that may contribute to adaptation during host invasion. This was termed the *Campylobacter* planktonic growth regulation complex (Cpr). Mutation experiments by Svensson and colleagues (2009) found that the response regulator is essential and *cprS* sensor kinase mutants demonstrated accelerated biofilm formation, as well as enhanced motility and secretion. The mutant also exhibited decreased tolerance to osmotic and oxidative stresses, increased expression of stress tolerance genes, as well as expression of the biofilm-associated major outer membrane protein (MOMP, encoded by *porA*) and *flaA*.

Concluding remarks

Understanding of the mechanisms of biofilm formation has been enhanced in recent years but questions remain about the role of this important phenotype in *Campylobacter* ecology and evolution. In particular, how it might be involved in survival outside of the host and transmission through the food chain. Differences in the capacity to form biofilms have been noted between *Campylobacter* strains. For example, in one study ST-22 and ST-4526 isolates were shown to form biofilms (Asakura *et al.*, 2012) while other isolates tested in the same study (ST-1947) did not (Revez *et al.*, 2011). This hints at the possibility that strains may adopt different strategies in relation to transmission, adhesion, epithelial cell invasion, intestinal colonization and infection. Increased understanding of the genes involved in biofilm formation and the availability genome data for large numbers of isolates have the potential to further unravel the factors involved in this complex phenotype.

References

Al-Amri, A.I., Botta, G.A., Tabbara, K.S., Ismaeel, A.Y., Al-Mahmeed, A.E., Qareeballa, A.Y., Bin Dayna, K.M., and Bakhiet, M.O. (2008). *Campylobacter jejuni* induces diverse kinetics and profiles of cytokine genes in INT-407 cells. Saudi Med. J. *29*, 514–519.

Al-Salloom, F.S., Al Mahmeed, A., Ismaeel, A., Botta, G.A., and Bakhiet, M. (2003). *Campylobacter*-stimulated INT407 cells produce dissociated cytokine profiles. J. Infect. *47*, 217–224.

Arciola, C.R., Campoccia, D., Speziale, P., Montanaro, L., and Costerton, J.W. (2012). Biofilm formation in *Staphylococcus* implant infections. A review of molecular mechanisms and implications for biofilm-resistant materials. Biomaterials *33*, 5967–5982.

Asakura, H., Yamasaki, M., Yamamoto, S., and Igimi, S. (2007). Deletion of *peb4* gene impairs cell adhesion and biofilm formation in *Campylobacter jejuni*. FEMS Microbiol. Lett. *275*, 278–285.

Asakura, H., Bruggemann, H., Sheppard, S.K., Ekawa, T., Meyer, T.F., Yamamoto, S., and Igimi, S. (2012). Molecular evidence for the thriving of *Campylobacter jejuni* ST-4526 in Japan. PLoS One *7*, e48394.

Ashgar, S.S., Oldfield, N.J., Wooldridge, K.G., Jones, M.A., Irving, G.J., Turner, D.P., and Ala'Aldeen, D.A. (2007). CapA, an autotransporter protein of *Campylobacter jejuni*, mediates association with human epithelial cells and colonization of the chicken gut. J. Bacteriol. *189*, 1856–1865.

Bacon, D.J., Szymanski, C.M., Burr, D.H., Silver, R.P., Alm, R.A., and Guerry, P. (2001). A phase-variable capsule is involved in virulence of *Campylobacter jejuni* 81-176. Mol. Microbiol. *40*, 769–777.

Balaban, M., Joslin, S.N., and Hendrixson, D.R. (2009). FlhF and its GTPase activity are required for distinct processes in flagellar gene regulation and biosynthesis in *Campylobacter jejuni*. J. Bacteriol. *191*, 6602–6611.

Barrero-Tobon, A.M., and Hendrixson, D.R. (2012). Identification and analysis of flagellar coexpressed determinants (Feds) of *Campylobacter jejuni* involved in colonization. Mol. Microbiol. *84*, 352–369.

Bull, S.A., Allen, V.M., Domingue, G., Jorgensen, F., Frost, J.A., Ure, R., Whyte, R., Tinker, D., Corry, J.E., Gillard-King, J., *et al.* (2006). Sources of *Campylobacter* spp. colonizing housed broiler flocks during rearing. Appl. Environ. Microbiol. *72*, 645–652.

Buswell, C.M., Herlihy, Y.M., Keevil, C.W., Marsh, P.D., and Leach, S.A. (1998). Carbon load in aquatic ecosystems affects the diversity and biomass of water biofilm consortia and the persistence of the pathogen *Campylobacter jejuni* within them. J. Appl. Microbiol. *85(Suppl 1)*, 161S–167S.

Campoccia, D., Montanaro, L., Ravaioli, S., Pirini, V., Cangini, I., and Arciola, C.R. (2011). Exopolysaccharide production by *Staphylococcus epidermidis* and its relationship with biofilm extracellular DNA. Int. J. Artif. Organs *34*, 832–839.

Chia, N., Woese, C.R., and Goldenfeld, N. (2008). A collective mechanism for phase variation in biofilms. Proc. Natl. Acad. Sci. U.S.A. *105*, 14597–14602.

Christensen, J.E., Pacheco, S.A., and Konkel, M.E. (2009). Identification of a *Campylobacter jejuni*-secreted protein required for maximal invasion of host cells. Mol. Microbiol. *73*, 650–662.

Costerton, J.W., Geesey, G.G., and Cheng, K.J. (1978). How bacteria stick. Sci. Am. *238*, 86–95.

Costerton, J.W., Lewandowski, Z., Caldwell, D.E., Korber, D.R., and Lappin-Scott, H.M. (1995). Microbial biofilms. Annu. Rev. Microbiol. *49*, 711–745.

Costerton, J.W., Montanaro, L., and Arciola, C.R. (2005). Biofilm in implant infections: its production and regulation. Int. J. Artif. Organs *28*, 1062–1068.

Day, C.J., Semchenko, E.A., and Korolik, V. (2012). Glycoconjugates play a key role in *Campylobacter jejuni* infection: interactions between host and pathogen. Front. Cell. Infect. Microbiol. *2*, 9.

Donlan, R.M., and Costerton, J.W. (2002). Biofilms: survival mechanisms of clinically relevant microorganisms. Clin. Microbiol. Rev. *15*, 167–193.

Drozd, M., Gangaiah, D., Liu, Z., and Rajashekara, G. (2011). Contribution of TAT system translocated PhoX to *Campylobacter jejuni* phosphate metabolism and resilience to environmental stresses. PLoS One *6*, e26336.

Eppinger, M., Baar, C., Raddatz, G., Huson, D.H., and Schuster, S.C. (2004). Comparative analysis of four Campylobacterales. Nat. Rev. Microbiol. *2*, 872–885.

Eucker, T.P., and Konkel, M.E. (2012). The cooperative action of bacterial fibronectin-binding proteins and secreted proteins promote maximal *Campylobacter jejuni* invasion of host cells by stimulating membrane ruffling. Cell. Microbiol. *14*, 226–238.

Flanagan, R.C., Neal-McKinney, J.M., Dhillon, A.S., Miller, W.G., and Konkel, M.E. (2009). Examination of *Campylobacter jejuni* putative adhesins leads to the identification of a new protein, designated FlpA, required for chicken colonization. Infect. Immun. *77*, 2399–2407.

Fux, C.A., Shirtliff, M., Stoodley, P., and Costerton, J.W. (2005). Can laboratory reference strains mirror 'real-world' pathogenesis? Trends Microbiol. *13*, 58–63.

Galkin, V.E., Yu, X., Bielnicki, J., Heuser, J., Ewing, C.P., Guerry, P., and Egelman, E.H. (2008). Divergence of quaternary structures among bacterial flagellar filaments. Science *320*, 382–385.

Gangaiah, D., Liu, Z., Arcos, J., Kassem, II, Sanad, Y., Torrelles, J.B., and Rajashekara, G. (2010). Polyphosphate kinase 2: a novel determinant of stress responses and pathogenesis in *Campylobacter jejuni*. PLoS One *5*, e12142.

Gaynor, E.C., Wells, D.H., MacKichan, J.K., and Falkow, S. (2005). The *Campylobacter jejuni* stringent response controls specific stress survival and virulence-associated phenotypes. Mol. Microbiol. *56*, 8–27.

Golden, N.J., and Acheson, D.W. (2002). Identification of motility and autoagglutination *Campylobacter jejuni* mutants by random transposon mutagenesis. Infect. Immun. *70*, 1761–1771.

Guerry, P. (2007). *Campylobacter* flagella: not just for motility. Trends Microbiol. *15*, 456–461.

Guerry, P., Logan, S.M., Thornton, S., and Trust, T.J. (1990). Genomic organization and expression of

Campylobacter flagellin genes. J. Bacteriol. *172*, 1853–1860.

Guerry, P., Szymanski, C.M., Prendergast, M.M., Hickey, T.E., Ewing, C.P., Pattarini, D.L., and Moran, A.P. (2002). Phase variation of *Campylobacter jejuni* 81-176 lipooligosaccharide affects ganglioside mimicry and invasiveness in vitro. Infect. Immun. *70*, 787–793.

Guerry, P., Ewing, C.P., Schirm, M., Lorenzo, M., Kelly, J., Pattarini, D., Majam, G., Thibault, P., and Logan, S. (2006). Changes in flagellin glycosylation affect *Campylobacter* autoagglutination and virulence. Mol. Microbiol. *60*, 299–311.

Gundogdu, O., Mills, D.C., Elmi, A., Martin, M.J., Wren, B.W., and Dorrell, N. (2011). The *Campylobacter jejuni* transcriptional regulator Cj1556 plays a role in the oxidative and aerobic stress response and is important for bacterial survival in vivo. J. Bacteriol. *193*, 4238–4249.

Hall-Stoodley, L., Costerton, J.W., and Stoodley, P. (2004). Bacterial biofilms: from the natural environment to infectious diseases. Nat. Rev. Microbiol. *2*, 95–108.

Hanning, I., Jarquin, R., and Slavik, M. (2008). *Campylobacter jejuni* as a secondary colonizer of poultry biofilms. J. Appl. Microbiol. *105*, 1199–1208.

Heikema, A.P., Bergman, M.P., Richards, H., Crocker, P.R., Gilbert, M., Samsom, J.N., van Wamel, W.J., Endtz, H.P., and van Belkum, A. (2010). Characterization of the specific interaction between sialoadhesin and sialylated *Campylobacter jejuni* lipooligosaccharides. Infect. Immun. *78*, 3237–3246.

Hirayama, J., Sekizuka, T., Tazumi, A., Taneike, I., Moore, J.E., Millar, B.C., and Matsuda, M. (2009). Structural analysis of the full-length gene encoding a fibronectin-binding-like protein (CadF) and its adjacent genetic loci within *Campylobacter lari*. BMC Microbiol. *9*, 192.

Holden, K.M., Gilbert, M., Coloe, P.J., Li, J., and Fry, B.N. (2012). The role of WlaRG, WlaTB and WlaTC in lipooligosaccharide synthesis by *Campylobacter jejuni* strain 81116. Microb. Pathog. *52*, 344–352.

Hu, W., Li, L., Sharma, S., Wang, J., McHardy, I., Lux, R., Yang, Z., He, X., Gimzewski, J.K., Li, Y., *et al.* (2012). DNA builds and strengthens the extracellular matrix in *Myxococcus xanthus* biofilms by interacting with exopolysaccharides. PLoS One *7*, e51905.

Huizinga, R., Easton, A.S., Donachie, A.M., Guthrie, J., van Rijs, W., Heikema, A., Boon, L., Samsom, J.N., Jacobs, B.C., Willison, H.J., *et al.* (2012). Sialylation of *Campylobacter jejuni* lipo-oligosaccharides: impact on phagocytosis and cytokine production in mice. PLoS One *7*, e34416.

Jeon, B., Itoh, K., Misawa, N., and Ryu, S. (2003). Effects of quorum sensing on *flaA* transcription and autoagglutination in *Campylobacter jejuni*. Microbiol. Immunol. *47*, 833–839.

Jin, S., Joe, A., Lynett, J., Hani, E.K., Sherman, P., and Chan, V.L. (2001). JlpA, a novel surface-exposed lipoprotein specific to *Campylobacter jejuni*, mediates adherence to host epithelial cells. Mol. Microbiol. *39*, 1225–1236.

Jin, S., Song, Y.C., Emili, A., Sherman, P.M., and Chan, V.L. (2003). JlpA of *Campylobacter jejuni* interacts with surface-exposed heat shock protein 90alpha and triggers signalling pathways leading to the activation of NF-kappaB and p38 MAP kinase in epithelial cells. Cell. Microbiol. *5*, 165–174.

Joshua, G.W., Guthrie-Irons, C., Karlyshev, A.V., and Wren, B.W. (2006). Biofilm formation in *Campylobacter jejuni*. Microbiology *152*, 387–396.

Kalmokoff, M., Lanthier, P., Tremblay, T.L., Foss, M., Lau, P.C., Sanders, G., Austin, J., Kelly, J., and Szymanski, C.M. (2006). Proteomic analysis of *Campylobacter jejuni* 11168 biofilms reveals a role for the motility complex in biofilm formation. J. Bacteriol. *188*, 4312–4320.

Karlyshev, A.V., Linton, D., Gregson, N.A., and Wren, B.W. (2002). A novel paralogous gene family involved in phase-variable flagella-mediated motility in *Campylobacter jejuni*. Microbiology *148*, 473–480.

Kiem, S., Oh, W.S., Peck, K.R., Lee, N.Y., Lee, J.Y., Song, J.H., Hwang, E.S., Kim, E.C., Cha, C.Y., and Choe, K.W. (2004). Phase variation of biofilm formation in *Staphylococcus aureus* by IS 256 insertion and its impact on the capacity adhering to polyurethane surface. J. Korean Med. Sci. *19*, 779–782.

Konkel, M.E., Mead, D.J., and Cieplak, W., Jr. (1993). Kinetic and antigenic characterization of altered protein synthesis by *Campylobacter jejuni* during cultivation with human epithelial cells. J. Infect. Dis. *168*, 948–954.

Konkel, M.E., Christensen, J.E., Keech, A.M., Monteville, M.R., Klena, J.D., and Garvis, S.G. (2005). Identification of a fibronectin-binding domain within the *Campylobacter jejuni* CadF protein. Mol. Microbiol. *57*, 1022–1035.

Krause-Gruszczynska, M., van Alphen, L.B., Oyarzabal, O.A., Alter, T., Hanel, I., Schliephake, A., Konig, W., van Putten, J.P., Konkel, M.E., and Backert, S. (2007). Expression patterns and role of the CadF protein in *Campylobacter jejuni* and *Campylobacter coli*. FEMS Microbiol. Lett. *274*, 9–16.

Lehtola, M.J., Pitkanen, T., Miebach, L., and Miettinen, I.T. (2006). Survival of *Campylobacter jejuni* in potable water biofilms: a comparative study with different detection methods. Water Sci. Technol. *54*, 57–61.

Leon-Kempis Mdel, R., Guccione, E., Mulholland, F., Williamson, M.P., and Kelly, D.J. (2006). The *Campylobacter jejuni* PEB1a adhesin is an aspartate/glutamate-binding protein of an ABC transporter essential for microaerobic growth on dicarboxylic amino acids. Mol. Microbiol. *60*, 1262–1275.

Lewenza, S. (2013). Extracellular DNA-induced antimicrobial peptide resistance mechanisms in *Pseudomonas aeruginosa*. Front Microbiol. *4*, 21.

Logan, S.M., Guerry, P., Rollins, D.M., Burr, D.H., and Trust, T.J. (1989). In vivo antigenic variation of *Campylobacter* flagellin. Infect. Immun. *57*, 2583–2585.

Mack, D., Davies, A., Harris, L., Jeeves, R., Pascoe, B., Knobloch, J.M., Rohde, H., and Wilkinson, T. (2013). *Staphylococcus epidermidis* in Biomaterial-Associated Infections. In Biomaterials Associated Infection, Moriarty, T.F., Zaat, S.A.J., and Busscher, H.J., eds. (Springer, New York), pp. 25–56.

Montanaro, L., Poggi, A., Visai, L., Ravaioli, S., Campoccia, D., Speziale, P., and Arciola, C.R. (2011). Extracellular DNA in biofilms. Int. J. Artif. Organs *34*, 824–831.

Naito, M., Frirdich, E., Fields, J.A., Pryjma, M., Li, J., Cameron, A., Gilbert, M., Thompson, S.A., and Gaynor, E.C. (2010). Effects of sequential *Campylobacter jejuni* 81-176 lipooligosaccharide core truncations on biofilm formation, stress survival, and pathogenesis. J. Bacteriol. *192*, 2182–2192.

O'Gara, J.P. (2007). ica and beyond: biofilm mechanisms and regulation in *Staphylococcus epidermidis* and *Staphylococcus aureus*. FEMS Microbiol. Lett. *270*, 179–188.

Otto, M. (2013). Staphylococcal infections: mechanisms of biofilm maturation and detachment as critical determinants of pathogenicity. Annu. Rev. Med. *64*, 175–188.

Palmer, L.J., Chapple, I.L., Wright, H.J., Roberts, A., and Cooper, P.R. (2012). Extracellular deoxyribonuclease production by periodontal bacteria. J. Periodontal Res. *47*, 439–445.

Palyada, K., Sun, Y.Q., Flint, A., Butcher, J., Naikare, H., and Stintzi, A. (2009). Characterization of the oxidative stress stimulon and PerR regulon of *Campylobacter jejuni*. BMC Genomics *10*, 481.

Park, S.F., Purdy, D., and Leach, S. (2000). Localized reversible frameshift mutation in the *flhA* gene confers phase variability to flagellin gene expression in *Campylobacter coli*. J. Bacteriol. *182*, 207–210.

Park, S., Bae, W., and Rittmann, B.E. (2010). Multi-species nitrifying biofilm model (MSNBM) including free ammonia and free nitrous acid inhibition and oxygen limitation. Biotechnol. Bioeng. *105*, 1115–1130.

Parkhill, J., Wren, B.W., Mungall, K., Ketley, J.M., Churcher, C., Basham, D., Chillingworth, T., Davies, R.M., Feltwell, T., Holroyd, S., *et al.* (2000). The genome sequence of the food-borne pathogen *Campylobacter jejuni* reveals hypervariable sequences. Nature *403*, 665–668.

Pearson, A.D., Greenwood, M., Healing, T.D., Rollins, D., Shahamat, M., Donaldson, J., and Colwell, R.R. (1993). Colonization of broiler chickens by waterborne *Campylobacter jejuni*. Appl. Environ. Microbiol. *59*, 987–996.

Pei, Z., Burucoa, C., Grignon, B., Baqar, S., Huang, X.Z., Kopecko, D.J., Bourgeois, A.L., Fauchere, J.L., and Blaser, M.J. (1998). Mutation in the peb1A locus of *Campylobacter jejuni* reduces interactions with epithelial cells and intestinal colonization of mice. Infect. Immun. *66*, 938–943.

Plummer, P.J. (2012). LuxS and quorum-sensing in *Campylobacter*. Front. Cell. Infect. Microbiol. *2*, 22.

Prendergast, M.M., Tribble, D.R., Baqar, S., Scott, D.A., Ferris, J.A., Walker, R.I., and Moran, A.P. (2004). *In vivo* phase variation and serologic response to lipooligosaccharide of *Campylobacter jejuni* in experimental human infection. Infect. Immun. *72*, 916–922.

Ravaioli, S., Campoccia, D., Visai, L., Pirini, V., Cangini, I., Corazzari, T., Maso, A., Poggio, C., Pegreffi, F., Montanaro, L., *et al.* (2011). Biofilm extracellular-DNA in 55 *Staphylococcus epidermidis* clinical isolates from implant infections. Int. J. Artif. Organs *34*, 840–846.

Reeser, R.J., Medler, R.T., Billington, S.J., Jost, B.H., and Joens, L.A. (2007). Characterization of *Campylobacter jejuni* biofilms under defined growth conditions. Appl. Environ. Microbiol. *73*, 1908–1913.

Reuter, M., Mallett, A., Pearson, B.M., and van Vliet, A.H. (2010). Biofilm formation by *Campylobacter jejuni* is increased under aerobic conditions. Appl. Environ. Microbiol. *76*, 2122–2128.

Revez, J., Rossi, M., Ellstrom, P., de Haan, C., Rautelin, H., and Hanninen, M.L. (2011). Finnish *Campylobacter jejuni* strains of multilocus sequence type ST-22 complex have two lineages with different characteristics. PLoS One *6*, e26880.

Richards, V.P., Lefebure, T., Pavinski Bitar, P.D., and Stanhope, M.J. (2013). Comparative characterization of the virulence gene clusters (lipooligosaccharide [LOS] and capsular polysaccharide [CPS]) for *Campylobacter coli*, *Campylobacter jejuni* subsp. *jejuni* and related *Campylobacter* species. Infect. Genet. Evol. *14*, 200–213.

Ritchie, A.E., Bryner, J.H., and Foley, J.W. (1983). Role of DNA and bacteriophage in *Campylobacter* autoagglutination. J. Med. Microbiol. *16*, 333–340.

Rohde, H., Frankenberger, S., Zahringer, U., and Mack, D. (2010). Structure, function and contribution of polysaccharide intercellular adhesin (PIA) to *Staphylococcus epidermidis* biofilm formation and pathogenesis of biomaterial-associated infections. Eur. J. Cell. Biol. *89*, 103–111.

Sherlock, O., Vejborg, R.M., and Klemm, P. (2005). The TibA adhesin/invasin from enterotoxigenic *Escherichia coli* is self recognizing and induces bacterial aggregation and biofilm formation. Infect. Immun. *73*, 1954–1963.

Skolimowski, M., Nielsen, M.W., Emneus, J., Molin, S., Taboryski, R., Sternberg, C., Dufva, M., and Geschke, O. (2010). Microfluidic dissolved oxygen gradient generator biochip as a useful tool in bacterial biofilm studies. Lab. Chip. *10*, 2162–2169.

Snider, R.M., Strycharz-Glaven, S.M., Tsoi, S.D., Erickson, J.S., and Tender, L.M. (2012). Long-range electron transport in *Geobacter sulfurreducens* biofilms is redox gradient-driven. Proc. Natl. Acad. Sci. U.S.A. *109*, 15467–15472.

Somers, E.B., Schoeni, J.L., and Wong, A.C. (1994). Effect of trisodium phosphate on biofilm and planktonic cells of *Campylobacter jejuni*, *Escherichia coli* O157: H7, *Listeria monocytogenes* and *Salmonella typhimurium*. Int. J. Food. Microbiol. *22*, 269–276.

Stewart, P.S., and Costerton, J.W. (2001). Antibiotic resistance of bacteria in biofilms. Lancet *358*, 135–138.

Sulaeman, S., Hernould, M., Schaumann, A., Coquet, L., Bolla, J.M., De, E., and Tresse, O. (2012). Enhanced adhesion of *Campylobacter jejuni* to abiotic surfaces is mediated by membrane proteins in oxygen-enriched conditions. PLoS One 7, e46402.

Svensson, S.L., Davis, L.M., MacKichan, J.K., Allan, B.J., Pajaniappan, M., Thompson, S.A., and Gaynor, E.C. (2009). The CprS sensor kinase of the zoonotic pathogen *Campylobacter jejuni* influences biofilm formation and is required for optimal chick colonization. Mol. Microbiol. *71*, 253–272.

Tormo, M.A., Ubeda, C., Marti, M., Maiques, E., Cucarella, C., Valle, J., Foster, T.J., Lasa, I., and Penades, J.R. (2007). Phase-variable expression of the biofilm-associated protein (Bap) in *Staphylococcus aureus*. Microbiology *153*, 1702–1710.

Trachoo, N., and Frank, J.F. (2002). Effectiveness of chemical sanitizers against *Campylobacter jejuni*-containing biofilms. J. Food. Prot. *65*, 1117–1121.

van Vliet, A.H., Baillon, M.L., Penn, C.W., and Ketley, J.M. (1999). *Campylobacter jejuni* contains two fur homologs: characterization of iron-responsive regulation of peroxide stress defense genes by the PerR repressor. J. Bacteriol. *181*, 6371–6376.

Ziebuhr, W., Loessner, I., Krimmer, V., and Hacker, J. (2001). Methods to detect and analyze phenotypic variation in biofilm-forming Staphylococci. Methods Enzymol. *336*, 195–205.

Zimmer, M., Barnhart, H., Idris, U., and Lee, M.D. (2003). Detection of *Campylobacter jejuni* strains in the water lines of a commercial broiler house and their relationship to the strains that colonized the chickens. Avian Dis. *47*, 101–107.

Alternative Infection Models for *Campylobacter jejuni*

15

Nicola J. Senior, Sok Kiang Lau, Roberto M. La Ragione and Olivia L. Champion

Abstract

Research using experimental hosts enables the lifecycle, metabolism and virulence mechanisms of *Campylobacter jejuni* to be interpreted in a contextual manner. However, the ability of *C. jejuni* to cause disease appears to be dependent not only on the bacterial strain but also on the host; *C. jejuni* is able to cause disease in certain hosts yet it remains avirulent in other host species. Mammalian models of infection are the most widely used and best characterized models of human disease; yet there are ethical, social and scientific limitations of these models. These concerns have driven work to reduce, refine and replace mammalian models of infection and this chapter considers the use of alternative infection models for *C. jejuni* research. This chapter ends with a review of the metabolism of *C. jejuni* in the context of host adaptation.

Introduction

Although *Campylobacter jejuni* is the leading bacterial cause of diarrhoeal disease, its lifecycle in the environment and its mechanisms of virulence have been poorly characterized. Research into mechanisms of virulence of *C. jejuni* underpins work to devise improved control measures for this important zoonotic pathogen. However, the ability of *C. jejuni* to cause disease is implicitly dependent on the host, and *C. jejuni* strain variation may also play a role. A clear example of this is the ability of *C. jejuni* to cause disease in humans whilst avirulently colonizing the gastrointestinal tract of avian species. Mammalian models of infection continue to be the most widely used and best-characterized models of human disease. Murine infection models, whilst used extensively for research into other enteric pathogens such as *E. coli* and *Salmonella enterica* Typhimurium, are not amenable to infection with *Campylobacter*. However, in recent years transgenic murine models have been developed that can be used to model *in vivo* colonization of *C. jejuni*. Similarly, the chick colonization and persistence model can be used to model *C. jejuni* colonization in poultry. However, this model is not suitable to study virulence mechanisms of *C. jejuni*, which is avirulent in poultry. The colostrum deprived piglet model and ferret infection models have been successfully used to study pathogenesis following challenge with *C. jejuni*.

Although mammalian models of infection do have a clear value to the study of *C. jejuni* pathogenicity, it is also important to understand the limitations of these models from ethical, social and scientific viewpoints. The use of mammalian models of infection is regulated in most countries, and the use of mammals for experimental purposes is dependent on the appropriate training of staff, the availability of suitable experimental facilities and the rigorous justification of the work to be undertaken. Particular ethical and social concerns are associated with the use of mammalian infection models. Finally, although mammals can often serve as valuable models of human infectious disease, the mammalian species in question may lack the molecular features of humans, such as cell surface receptors, which means that the pathogenesis of disease in these species is different to that in humans. Collectively these concerns have driven work to reduce, refine and replace mammalian models of infection.

Table 15.1 Examples of alternative infection models for *C. Jejuni*, as described in the literature

Alternative infection model	*Campylobacter* investigation	Outcome of challenge	Reference
HeLa cells (human)	Mutations of CDT genes	Decreased invasion	Biswas *et al.* (2006)
	Comparison of non-inflammatory diarrhoea strains with colitis strains	Decreased invasion	Everest *et al.* (1992)
	Comparison of mean toxicity levels of isolates from chickens and humans	Chicken isolates had decreased toxicity levels	Gilbert and Slavik (2004)
	CdtB mutants	Decreased toxicity	Purdy *et al.* (2000)
	Response to Cj infection	Elevated interleukin production	MacCallum *et al.* (2006)
INT 407 cells (human)	HtrA mutant	Decreased adherence and invasion	Brøndsted *et al.* (2005)
	CsrA mutant	Decreased adhesion, elevated invasion	Fields and Thompson (2008)
	Pgp1 mutant	No defects in invasion	Frirdich *et al.* (2012)
	Fla mutants	No defects in adhesion, decreased invasion	Grant *et al.* (1993)
	Cj1461 mutant	Increased adhesion, decreased invasion	Kim *et al.* (2008)
	Comparison of human and poultry isolates	Differences in levels of invasiveness	Malik-Kale *et al.* (2007)
	Strains lacking genes involved in ganglioside mimicry	Decreased invasion	Poly *et al.* (2008)
	DksA-like protein mutant	Decreased invasion	Yun *et al.* (2008)
	Pkp1 mutant	Decreased invasion	Candon *et al.* (2007)
	Response to Cj infection	Elevated interleukin production	MacCallum *et al.* (2006)
Caco-2 cells (human)	Investigation of less motile strain	Reduced adhesion and invasion	Carrillo *et al.* (2004)
	Comparison of non-inflammatory diarrhoea strains with colitis strains	Decreased invasion	Everest *et al.* (1992)
	Pgp1 mutant	No defects in invasion	Frirdich *et al.* (2012)
	Strains expressing different classes of LOS	Differences in levels of invasiveness	Habib *et al.* (2009)
	Comparison of strains producing non-sialylated and sialylated LOS	Increased invasion	Louwen *et al.* (2008)
	Responses to environmental stresses	Differences in adhesion and invasion	Mihaljevic *et al.* (2007)
	Strains lacking genes involved in ganglioside mimicry	Decreased invasion	Poly *et al.* (2008)
	Interaction with culturable and non-culturable cells	Differences in adhesion and invasion	Verhoeff-Bakkenes *et al.* (2009)
	Response to Cj infection	No difference in interleukin production	MacCallum *et al.* (2006)
Hep-2 cells (human)	Cellular events during Cj infection	Inability to survive intracellularly for extended periods	De Melo *et al.* (1989)
	Prevalence of CDT genes in Cj	Differences in cytotoxicity	Ripabelli *et al.* (2009)

Alternative infection model	*Campylobacter* investigation	Outcome of challenge	Reference
28SC cells (human)	Effect of Cj CDT on human monocytes	Intracellular survival	Hickey *et al.* (2005)
T84 cells (human)	Characterization of a highly pathogenic strain	No defects in invasion or intracellular survival	Hofreuter *et al.* (2006)
	Comparison of strains producing non-sialylated and sialylated LOS	Increased invasion	Louwen *et al.* (2008)
	Comparison of human and porcine cell lines	No differences in invasion	Murphy *et al.* (2011)
	VirK mutant	No differences in invasion	Novik *et al.* (2009)
	Investigation of Cj infection mechanisms	Intracellular survival	Watson and Galán (2008)
Henle-407 cells (human)	Investigation of CDT	Intoxication of cells	Lara-Tejero and Galán (2001)
LMH cells (chicken)	Investigation of adhesins	Decreased adhesion (CadF, CapA, Cj1279c and Cj1349 mutants)	Flanagan *et al.* (2009)
	Comparison of Cj invasion into poultry and human cells	Decreased invasion of chicken cells	Larson *et al.* (2008)
Chinese hamster ovary (CHO) cells	Different host factors for CDTs	Intoxication of cells	Eshragi *et al.* (2010)
	Comparison of non-inflammatory diarrhoea strains with colitis strains	Intoxication of cells	Everest *et al* (1992)
	PorA protein of Cj	Lack of cytotoxicity	Khan *et al.* (2005)
J774A.1 cells (mouse)	KatA mutant	Intracellular survival	Day *et al.* (2000)
	Cj1556 mutant	Decreased intracellular survival	Gundogdu *et al.* (2011)
	Responses to environmental stresses	Differences in intracellular survival	Pogačar *et al.* (2009)
	Comparison of murine and insect cell line	Intracellular survival	Senior *et al.* (2011)
RAW 264.7 cells (mouse)	Pgp1 mutant	No defects in invasion or intracellular survival	Frirdich *et al.* (2012)
IPEC-1 cells (pig)	Comparison of human and porcine cell lines	No differences in invasion	Murphy *et al.* (2011)
Vero cells (monkey)	PorA protein of Cj	No intoxication of cells	Khan *et al.* (2005)
COS-7 cells (monkey)	VirK mutant	No differences in invasion	Novik *et al.* (2009)
SF9 cells (insect)	Comparison of murine and insect cell line	Intracellular survival	Senior *et al.* (2011)
Galleria mellonella model	Cj1556 mutant	Attenuation of virulence	Gundogdu *et al.* (2011)
	Characterization of model	Attenuation of virulence in different strains	Senior *et al.* (2011)
	O-methylphosphoramidate	Attenuation of virulence in mutants and strains	Champion *et al.* (2010)

Studies on *C. jejuni* mechanisms of virulence and their interpretation in an appropriate contextual manner require a suitable experimental host. This chapter considers the use of alternative *in vitro* and *in vivo* infection models for investigating mechanisms of virulence of *C. jejuni*. A summary of these alternative models is provided in Table 15.1. This chapter also reviews the metabolic traits of *C. jejuni* with respect to the lifecycle and host adaptation of the pathogen.

In vitro cultured cell infection models

Cultured cell infection models have played a central role in the interrogation of host–pathogen interactions. Robust cell culture models provide a cost effective, high throughput experimental platform that facilitates the interrogation of *C. jejuni* interactions with a single cell type. Different cell lines have advantages and disadvantages in terms of their relevance and use as an infection model. Cultured mammalian cells include both primary and immortalized cell lines. Primary cells are derived from a host and have a short life span but may be more representative of an *in vivo* response, for example bone marrow derived macrophages. Immortalized cell lines proliferate indefinitely through random mutation or genetic modification. For example, immortalized cell lines represent mammalian professional phagocytic cells such as the murine macrophage cell line J774A.1 or human monocyte cell line THP-1. Monkey and canine derived COS-1 and MDCK cell lines are readily transfected with genetic material. However, HeLa and Hep-2 cells are derived from human tissue and may be more representative of clinical infection. CaCo-2 and T84 cells are derived from human intestinal epithelium tissue, and may be more physiologically relevant for the analysis of gastrointestinal pathogens such as *C. jejuni* that must breach the gut wall but these cell lines are difficult to transfect (Hurley and McCormick, 2003). Stem cells are amenable to genetic manipulation and can differentiate into any cell type making them an attractive model in which to dissect host pathogen interactions (Yu *et al.*, 2009).

Cultured cell infections also include single cell organisms such as amoebae including water borne *Acanthamoeba castellanii* and soil dwelling *Dictyostelium discodium* (Kurz and Ewbank, 2007; Salah *et al.*, 2009). In the environment, amoebae predate upon bacteria and phagocytose them into a phagosome that undergoes fusion with lysosomes, eradicating bacteria that do not possess intracellular defence mechanisms. Amoebae have therefore been proposed as a model for mammalian professional phagocytes.

Hypotheses about host pathogen interactions developed using a reductionist cell culture approach should be verified in a whole system to validate the data in the context of an infection.

Human cell lines

In vitro systems used to dissect *Campylobacter* virulence mechanisms

Two pathogenesis mechanisms have been proposed through which *C. jejuni* establishes infection in humans; the first is that adhesion of the pathogen to the intestine is followed by production of toxin that alters the intestine fluid absorption capability, leading to secretory diarrhoea (Butzler and Oosterom, 1991; Janssen *et al.*, 2008). The other mechanism proposed involves invasion and proliferation of the pathogen within the intestinal mucosa, which would lead to inflammatory responses and eventually bloody diarrhoea (Janssen *et al.*, 2008). *In vitro* cell culture assays that mimic the different stages of human *Campylobacter* infection have been used to gain understanding about the molecular mechanisms through which *C. jejuni* establishes infection. Several types of human cell-lines have been used in these studies. Table 15.2 outlines the origin of human cell lines commonly used in *Campylobacter* research.

Adhesion and invasion

After gaining entry into the body system via the gastrointestinal (GI) tract, *Campylobacter* must survive the acidic stomach environment (Butzler and Oosterom, 1991; Janssen *et al.*, 2008; van der Sar *et al.*, 2003) to reach and colonize the lower intestinal tract (Dasti *et al.*, 2009). Inside the

Table 15.2 Examples of cell lines of human origin used in the study of campylobacteriosis

Cell line	Morphology	Source organ	Additional notes	Type of studies conducted using this cell line	Source of information on cell line
HeLa	Epithelial	Cervix		Adhesion, invasion, cytotoxicity	ATCC® Number: CCL-2™
INT 407	Epithelial	Embryonic intestine	HeLa derivative	Adhesion, invasion, intracellular survivability	Health Protection Agency Culture Collection Catalogue no.: 85051004
Caco-2	Epithelial	Colon		Adhesion, invasion, intracellular survivability	ATCC® Number: HTB-37™
HEp-2	Epithelial	HeLa contaminant	Contain human papillomavirus	Adhesion, invasion, intracellular survivability	ATCC® Number: CCL-23™
28SC	Monocyte/ macrophages	Peripheral blood		Intracellular survivability	ATCC® Number: CRL-9855™
Thp-1	Monocyte	Peripheral blood		Invasion, intracellular survivability	ATCC® Number: TIB-202™
T84	Epithelial	Colon		Invasion, intracellular survivability	ATCC® Number: CCL-248™
Henle-407	Epithelial	HeLa contaminant	Contain human papillomavirus (HPV-18)	Cytotoxicity	ATCC® Number: CCL-6™

GI tract, the mucus layer forms the first line of defence and in order to establish itself, *C. jejuni* harbours traits such as motility and corkscrew shape to facilitate its penetration through the mucus layer and interaction with the epithelial cells (Young *et al.*, 2007). Following adhesion, *C. jejuni* invades the epithelial cells via microtubule polymerization and some strains require microfilament polymerization as well (cited in Young *et al.*, 2007). The process of epithelial cells extending out pseudopods and enveloping the pathogen has been reported (Biswas *et al.*, 2000). After internalization, the pathogen is contained within a vacuole, however, its fate and the activities involved in pathogenesis are still unknown (Young *et al.*, 2007).

Epithelial cell-lines of both human and animal origin have been used to study *Campylobacter* host–pathogen interaction. For example, *Campylobacter* invasion antigens (Cia) proteins (Konkel *et al.*, 1999) and their flagellar export apparatus were identified through co-culture of *C. jejuni* with INT 407 colonic epithelial cells (Konkel *et al.*, 2004). Invasion assays using INT 407 cells also indicated that some *C. jejuni* strains invaded the host cells while some other strains did not (Malik-Kale, 2007). This finding led to further investigations to look for differences between hyper invasive and low invasive *C. jejuni* strains using INT 407 cells. A single nucleotide difference in the *flgR* gene that resulted in a change in amino acid at the receiver domain of FlgR protein was observed in two of the strains studied (Malik-Kale *et al.*, 2007). The amino acid change resulted in non-motility for the strain that was unable to invade INT 407 cells (Malik-Kale *et al.*, 2007). All findings indicated that flagella play an important role in invasion of epithelial cells.

The epithelial cell line, INT 407, was also used to show that flagella were not required for adherence but could play a role in internalization (Grant *et al.*, 1993). Epithelial cell lines including INT 407, CaCo-2 and T84 as well as a monocyte cell line Thp-1 were used to show that a peptidoglycan peptidase recently found in *C. jejuni* does not play a significant role in cell invasion and intracellular survival (Frirdich *et al.*, 2012).

Intracellular survival

The invasion of cells by *C. jejuni* is well-established. However, the fate of the bacteria after this event is less certain. During invasion of the host, *C. jejuni* encounter professional phagocytic cells; survival within these may allow the bacteria to be transported to different parts of the body. Once established, intracellular survival within non-phagocytic cells may provide a niche with little microbial competition in which the organism can become established (Day *et al.*, 2000). *In vitro* cultured cell infection models have been used to study the interaction of *C. jejuni* with both professional phagocytic and non-phagocytic cells.

An initial investigation by Kiehlbauch *et al.* (1985) looked at the phagocytosis of *C. jejuni*, and its ability to survive intracellularly. Using primary human peripheral blood monocytes, it was observed that *C. jejuni* were readily internalized by these cells, and could be recovered from within the macrophage for up to 6 days. They hypothesized that phagocytosis may actually promote *C. jejuni* survival.

However, a later study suggested that *C. jejuni* invaded HEp-2 cells slowly before being internalized into vacuoles (De Melo *et al.*, 1989). These phagosomes then fused with lysosomes, resulting in a decrease in the viability of the bacteria after 9 h. In contrast, Day *et al.* (2000) observed an initial death phase between 6 and 48 h for *C. jejuni* within HEp-2 cells, which was followed by an increase in bacteria recovered. This increase was short-lived, with numbers recovered dropping away by 96 h post infection. A similar effect was observed in INT 407 cells infected with *C. jejuni* and cultured in medium containing gentamicin; the greatest number of bacteria was recovered at 6 h post infection before dropping off until 48 h. In the interim, the antibiotic was removed, and from 48 h bacterial numbers increased within the cells, reaching a maximum between 60 and 80 h. After this, recoverable bacteria declined rapidly, although this may have been related to the degeneration of the cell monolayers (Konkel *et al.*, 1992). When *C. jejuni* was internalized by human 28SC monocyte cells, an increase in bacteria recovered peaked at 48 h post infection before dropping away, although viable *C. jejuni* were recovered up to day 7 (Hickey *et al.*, 2005).

During these studies, it was noted that the internalized bacteria often underwent a physiological change from spiral to coccoid cells (Kiehlbauch *et al.*, 1985; De Melo *et al.*, 1989). Watson and Galán (2008) showed intracellular survival of *C. jejuni* within human intestinal epithelial T84 cells, with significant numbers recovered at 4 h post infection followed by an approximate 500-fold decrease in CFU by 24 h. However, they sought to distinguish viable and non-viable bacteria within the cells, and found no decrease in viability of intracellular *C. jejuni* over 24 h. They hypothesized that the bacteria undergo physiological changes triggered by a low oxygen environment, which subsequently impeded their recovery from lysed cells.

Cytotoxicity assay

One proposed mechanism that is used by *Campylobacter* during the course of infection suggests that adhesion of the pathogen to the intestine is followed by production of toxin(s) that alters the intestine fluid absorption capability, leading to secretory diarrhoea (Butzler and Oosterom, 1991; Janssen *et al.*, 2008). Cytolethal distending toxin (CDT) in *C. jejuni* can be either a secreted product, present in culture supernatant, or cell-associated (Jinadasa *et al.*, 2011). Cytotoxicity assays have been performed by exposing HeLa cells to *C. jejuni* culture supernatant (Gilbert and Slavik, 2004; Purdy *et al.*, 2000). HeLa cells exposed to culture supernatant from wild-type *C. jejuni* showed cytopathic effects whereas those exposed to culture supernatant from CDT-deficient *C. jejuni* did not (Purdy *et al.*, 2000). It has been reported that activity of CDT in *Campylobacter* and other Gram-negative pathogens requires three functional proteins; CdtA, CdtB and CdtC (Lara-Tejero and Galan, 2001; Pickett *et al.*, 1994; Scott and Kaper, 1994). Henle-407 cells were used to demonstrate that in *C. jejuni*, the CdtB protein is the enzymatically active subunit of three proteins while CdtA and CdtC are involved in the delivery of CdtB (Lara-Tejero and Galan, 2001).

In vitro systems allow interactions of *C. jejuni* with specific target cell types to be studied, which would not be possible in an animal model. One drawback in current *in vitro* tissue culture techniques for *Campylobacter* studies is that the assays have been conducted under aerobic conditions, which is neither favourable for *Campylobacter* nor a true representation of the oxygen content in intestinal lumen environment. The aerobic environment, to which the pathogen was exposed during *in vitro* assays, could be a contributing factor towards low levels of *Campylobacter* interacting with and/or invading host cells. A recent study conducted by Mills *et al.* (2012) demonstrated, with the use of a vertical diffusion chamber (VDC) model system, that the numbers of *C. jejuni* interacting with or residing within the intestinal epithelial cell (IEC) had increased. The VDC model allows co-culturing of *C. jejuni* and IECs under microaerobic conditions, where the apical surfaces of IECs were exposed to microaerobic conditions while the basolateral surfaces of the cells were under aerobic conditions. The increase in the number of *C. jejuni* interacting with or residing within culture cells recovered more materials for downstream work, such as transcriptomics and/or proteomics analysis.

Non-human cell lines

Campylobacter is not restricted exclusively to humans. Strains of *C. jejuni* have been isolated from pigs, cows and sheep, and the bacteria is known to be part of the normal avian flora, with poultry in particular providing a major reservoir of infection (Janssen *et al.*, 2008). The use of non-human cell lines, for example, the chicken-derived continuous cell line LMH hepatocellular carcinoma cells, offers an ideal model for studying interactions such as colonization. As a result, non-human cell lines have proved invaluable in the study of adhesion and invasion, the effect of *Campylobacter* toxin and in determining the potential roles of genes associated with colonization and pathogenicity.

Non-human cell lines are not only sourced from poultry. There are a number of macrophage-like continuous cell lines derived from mice allowing for the assessment of *Campylobacter* intracellular survival. These murine cells, being mammalian in origin, will have similarities to human macrophages. Further mammalian cell lines in use include Chinese hamster ovary cells and monkey Vero cells, which are standard assays for intoxication of cells, and porcine IPEC-1 cells, which mimic human gastrointestinal physiology.

Adhesion and invasion assays

Adhesion assays are designed to enumerate bacteria that bind to cells whereas invasion assays look for the number of microbes internalized by the cells. Both begin with culturing of monolayers of the appropriate cells at a specific density. In an adhesion assay, the cells are inoculated with bacteria at a known multiplicity of infection (MOI); there is a short incubation period to allow the bacteria to adhere after which the cells are thoroughly rinsed to remove non-adherent cells and lysed, either chemically or mechanically. Dilutions are made of the samples and plated out to determine the number of adherent bacteria. In an invasion assay, cell monolayers are overlaid with a solution containing gentamicin, which does not cross into the cells and is thus used to inactivate extracellular bacteria. This is also known as a gentamicin assay.

Larson *et al.* (2008) used chicken-derived LMH cells to investigate the difference between invasion of avian and human cells, their hypothesis being that *C. jejuni* would be less prone to invading chicken cells, and thus trigger the release of host cell cytokines, due to the commensal relationship between the bacteria and the birds. It was demonstrated that *C. jejuni* strains were significantly less invasive in LMH cells when compared to the human epithelial cell line INT 407; the number of internalized *C. jejuni* strains in chicken LMH cells ranged from 7×10^3 to 1.6×10^4, whereas in human INT 407 cells numbers ranged from 2.8×10^5 to 6.2×10^5. The difference in ability of *C. jejuni* to invade chicken-derived cells was further highlighted by the observation that *C. jejuni* strains bound to LMH and INT 407 cells in similar numbers, and this meant that when invasion was expressed as a percentage of the bound cells, there was more than a 20 times reduction in chicken cells compared to human cells. In addition, LMH cells inoculated with *C. jejuni* did not express chicken IL-8 orthologues (chCXCLi2 and chCXCLi1), which stimulate the host inflammatory response, although human IL-8 was expressed in the human cell line inoculated with the same strains. These data indicate that *C. jejuni* invade and elicit an inflammatory response in human cells that is not observed in chicken cells.

Another study used chicken LMH cells to identify proteins required for chicken colonization (Flanagan *et al.*, 2009). Adherence to the host cells is essential to prevent clearance of the bacteria by the host. This adherence seems to be dependent on multiple surface-exposed adhesins; in *C. jejuni*, these include CadF, CapA, JlpA, PEB1, Cj1279c (FlpA) and Cj1349c. Mutants were created in the genes encoding these putative adhesins by insertional mutagenesis, and these were used to perform cell adherence assays with LMH cells. A significant reduction in adherence to LMH cells was observed for the *cadF*, *capA*, *flpA* and *cj1349c* mutants in comparison to a wild-type strain, and FlpA was identified as a novel adhesin. The results were consistent with the hypothesis that the adherence of *C. jejuni* to host epithelial cells requires more than one protein.

However, Larson *et al.* (2008) proposed that LMH cells do not behave in the same way as *in vivo* cells, due to differences between their results and other published data obtained from primary cells. In contrast to their observation that *C. jejuni* is less invasive in chicken cells than in human cells, *C. jejuni* isolates invade chicken primary cells and human cells at comparable levels (Smith *et al.*, 2005; Byrne *et al.*, 2007). Van Deun *et al.* (2008) observed that *C. jejuni* invades primary chicken caecal epithelial cells at significantly lower levels than human T84 epithelial cells. It was suggested that the specific cell type may influence the invasion phenotype more than the virulence of any particular *C. jejuni* strain.

Chicken cell lines are useful to study *C. jejuni* colonization mechanisms in this host. However, cell lines from susceptible hosts have been used to study the response to invading *Campylobacter*. Pigs are susceptible to *C. jejuni* challenge and infection reflects that observed in humans, therefore a porcine cell line has been used for assaying invasion. IPEC-1 cells are a continuous cell line derived from fetal pig intestinal epithelium; since pigs and humans share a similar gastrointestinal physiology, the response of human and porcine intestinal cells to *Campylobacter* was compared by Murphy *et al.* (2011). Two strains were used in the study; *C. jejuni* 81116 was originally isolated from a human clinical case, whilst *C. coli* 1669 was originally isolated from a pig. These were tested in gentamicin assays performed with a range of MOIs in both IPEC-1 cells and human T84 cells. Values were calculated for the minimum MOI at which maximum invasion was achieved, and the maximum number of bacteria invaded per cell. No significant differences were found between T84 and IPEC-1 cells for either bacterial strain indicating that both *C. jejuni* and *C. coli* invaded both cell lines at a similar level. It was also shown that infection of both cell types resulted in significant increases in IL-8 release. The study concluded that pigs may mount a similar innate immune response to *Campylobacter* infection as humans.

Campylobacter toxicity

Chinese hamster ovary (CHO) cells have been used to test for toxin production in *Campylobacter* strains (Everest *et al.*, 1992). A collection of clinical isolates, known to cause either colitis or watery diarrhoea, were added to CHO cells. Cells which were elongated, defined as a length: width ratio of > 3:1, demonstrated enterotoxin activity by the bacteria. In a separate assay, cells that were rounded and dead demonstrated cytotoxic activity by the bacteria. It was shown that enterotoxin production was more prevalent in *C. jejuni* strains as opposed to *C. coli* strains, although there was no significant association between enterotoxin production and the clinical symptoms. Cytotoxic activity was reported in all of the colitis strains and the majority of those strains causing watery diarrhoea, suggesting a role for cytotoxins in disease.

Khan *et al.* (2005) used CHO cells, along with Vero cells, which are a continuous cell line derived from African green monkey kidney epithelial cells, to investigate a *C. jejuni* outer membrane protein, PorA, reported to exhibit cytotoxicity in mammalian cells. Whilst the bacterial strain from which the protein was isolated demonstrated cytotoxicity in the CHO cells, the isolated protein itself did not. In the Vero cells, neither the strain nor the isolated protein had any cytotoxic effect.

Mutated CHO cells, along with a range of mouse cell lines, were used to demonstrate how cytolethal distending toxins (CDT) from a number of bacteria including *C. jejuni* vary in their ability to interact with host cells (Eshraghi *et al.*, 2012). *C. jejuni* CDTs were shown to display distinct host cell preferences for murine 3T3 (embryonic fibroblast) cells and Y-1 (epithelial) cells. It was concluded that CDTs derived from different bacteria intoxicate host cells in very specific, unique ways, which may be caused by bacterial species having different receptors.

The toxic activity of CDT was examined in COS-1 cells, which is a fibroblast-like cell line derived from African green monkey kidney. It was known that the toxin was comprised of three subunits, CdtA, CdtB and CdtC. Lara-Tejero and Galan (2000) transfected COS-1 cells with plasmids expressing separate subunits of *C. jejuni* CDT. Only cells infected with plasmids expressing CdtB exhibited signs of intoxication, with fragmented nuclei and changes in chromatin, even as far as total collapse of the latter. It was shown through amino acid sequencing that CdtB was very similar to deoxyribonuclease (DNase) I-like proteins, and induces cell cycle arrest at the G_2/M phase.

Intracellular survival

Campylobacter is cited to be primarily an extracellular pathogen, but there is evidence to suggest that it can survive within cells (Hickey *et al.*, 2005). Studies with human *in vitro* models have reached differing conclusions, with some groups reporting *C. jejuni* survival within epithelial cells (De Melo *et al.*, 1989; Day *et al.*, 2000) but not within macrophages (Watson and Galan, 2008). Other studies indicate that *C. jejuni* can survive within phagocytes; for example, Kiehlbauch *et al.* (1985) used J774G8, a murine peritoneal macrophage primary cell line, to examine *C. jejuni* phagocytosis alongside human monocytes, and found that the bacteria were readily internalized but not rapidly killed. They supposed that phagocytosis may actually promote *C. jejuni* survival, and observed that the phagocytosed cells survived longer than cells in preparations without phagocytes.

This enhanced survival was examined by Day *et al.* (2000) in a murine macrophage-like continuous cell line, J774A.1. This cell line is known to have retained the capability for a respiratory burst. Seventy-two hours post-infection, a *C. jejuni* catalase mutant was unculturable from the cell line whereas strains that still produced catalase were recovered in significant numbers, indicating that they were able to survive and multiply within the cells. The catalase was hypothesized to counteract the effects of both nitric oxide synthesis and the respiratory burst, facilitating intramacrophage survival. A gene indicated as playing a potential role in regulating the oxidative stress response, *Cj1556*, was studied by Gundogdu *et al.* (2011), who observed that a *C. jejuni Cj1556* mutant was less able to survive intracellularly in J774A.1 cells, thus emphasizing the importance of this gene for *in vivo* persistence.

Another murine macrophage cell line, RAW 264.7, was used to investigate whether a gene involved in the maintenance of the characteristic shape of *C. jejuni*, *pgp 1*, also affected intracellular survival (Frirdich *et al.*, 2012). No differences were observed when the cell line was infected with *C. jejuni pgp 1* mutants.

In our laboratory we were interested in whether *Campylobacter* could survive within insect cells, as part of a characterization of *Galleria mellonella* as an infection model for *C. jejuni*. We used an insect-derived cell line, SF9, alongside the J774A.1 cell line to gauge responses to challenge with *C. jejuni*. Cell lines were challenged with bacteria at an MOI of 10, and were then monitored 4 h and 24 h post infection (Senior *et al.*, 2011). Although bacterial numbers declined 100-fold in J774A.1 between infection and the 4 h time point, they then maintained that level through to the 24 h time point. Within SF9, the initial decline was approximately 1000-fold, but by the 24 h time point there had been a significant increase in bacterial numbers of about 10-fold. We hypothesized that this showed not only survival within the insect line, but also replication. The role of insects in the life cycle of *C. jejuni* may therefore warrant further investigation.

Campylobacter infects and survives within a range of animals, many of which are domesticated, and, as such, act as reservoirs of infection for humans. Understanding how the bacteria interact with these different hosts will inform biocontrol strategies and public health measures.

In vivo alternative infection models

Galleria mellonella

Larvae of the Greater Waxmoth, *Galleria mellonella*, have increasingly been used as models to assess the virulence of human pathogens; in the year 2000, the number of published studies using this model was less than 10, but by 2010, it had risen to nearly 80 (Desbois and Coote, 2012). These papers have included studies on the bacterial pathogens *Acinetobacter baumannii* (Peleg *et al.*, 2009), *Yersinia pseudotuberculosis* (Champion *et al.*, 2009), *Burkholderia pseudomallei, B. thailandensis, B. oklahomensis* (Wand *et al.*, 2011), *Pseudomonas aeruginosa, Bacillus anthracis, Listeria monocytogenes* and *Enterococcus faecalis* (Fedhila *et al.*, 2010). Given the lack of *Campylobacter* infection models and the demand to reduce, refine and replace mammalian infection models we investigated whether the alternative infection model *G. mellonella* was susceptible to challenge with *C. jejuni* (Champion *et al.*, 2010; Senior *et al.*, 2011).

There are a number of advantages to using *G. mellonella* as an infection model. The use of insect infection models is ethically more acceptable than mammalian models, requiring no Home Office license. Larvae are commercially available and inexpensive, without the need for additional costs associated with maintained facilities or specialist staff. *G. mellonella* can be kept for up to 1 month on woodchips in a 15°C incubator. *G. mellonella* used for infection assays tend to be between 0.2 and 0.3 g, and around 20 mm in length. Their size thus renders them easier to handle than other alternative infection models such as *Drosophila melanogaster*, which must be inoculated using a microinjector; *G. mellonella* can be infected under the naked eye using a syringe. With inoculation as the route of infection, the ability to deliver a precise dose is preferable to the feeding method, routinely used in the *Caenorhabditis elegans* model, where it is difficult to establish how much inoculum the insect has consumed. *G. mellonella* are inoculated in their prolegs, which provide a well-defined inoculation site, and, furthermore, seal themselves post-injection so as to minimize the risk of the larva dying through loss of hemocoel. Furthermore, the larvae will withstand multiple injections, which allows for the *in vivo* testing of antimicrobial compounds against specific pathogens; for example, McKenney *et al.* (2012) inoculated *G. mellonella* with *Francisella novicida*, then 2 h later injected infected larvae with one of a number of antibiotics. Untreated larvae had a mean time to death of 59 h, whilst the mean times of death for treated larvae ranged from 93 to 103 h, depending on the compound administered. There are positive correlations between the results of virulence assays in *G. mellonella* and mice that are lacking for other insect models. Wand *et al.* (2011) observed that *B. pseudomallei* isolates with high, intermediate and low levels of virulence showed the same virulence patterns in *G. mellonella*. In addition, *G. mellonella* model discriminated between *B. pseudomallei* and clinical isolates of *B. thailandensis* and *B. oklahomensis* on the basis of their virulence, Similarly, a *C. jejuni* lipooligosaccharide (LOS) mutant was shown to be attenuated in *G. mellonella*, which correlates with observations of the importance of LOS in invasion of cultured cells (Champion *et al.*, 2010).

G. mellonella can be kept at 37°C to mimic the physiological conditions in humans. Finally, whereas previously there was a lack of genomic information for the larvae compared to other insect models, a recent study has developed extensive transcriptomic resources for *G. mellonella* (Vogel *et al*, 2011).

Some physiological comparisons can be drawn between insects and humans. *G. mellonella* larvae possess a cuticle, analogous to skin, which acts as a primary barrier against microbial infection as it is impervious to environmental exposure (Kemp and Massey, 2007). Internally, the larvae are able to mount a humoral response to pathogens that includes the concurrent release of at least 18 known or putative antimicrobial peptides (Brown *et al.*, 2009). Furthermore, in parallel with this, insects produce a cellular response involving the circulating hemocytes within the larval hemolymph, which are able to phagocytose microbial cells and kill them by production of superoxide (Bergin *et al.*, 2005). While insects lack an adaptive immune system, *G. mellonella* can raise an immune response following an initial challenge which provides protection against subsequent infections, although this usually declines within 24 h of the challenge; for example, when exposed to a variety of non-lethal doses of *Aspergillus fumigatus*, larvae were more resistant to a subsequent lethal challenge through differential activation of both cellular and humoral responses. The mechanisms for this have not yet been determined (Fallon *et al.*, 2011).

G. mellonella assay

The most common route of infection for the *G. mellonella* model is via injection. Cultured *Campylobacter* strains are adjusted to a known concentration in phosphate-buffered saline before 10 µl are inoculated into a specified proleg for each larva, providing a precise challenge dose.

G. mellonella are generally infected with ten replicates per condition. A group of 10 is set aside as an uninfected control group, and another group of 10 is inoculated with 10 µl of PBS to act as control for the mechanical injury. All groups are then incubated at 37°C and their progress monitored at defined time points.

The larvae are assayed normally through their survival. A dead larva is one that does not move in response to stimuli, and, when turned *G. mellonella* cannot right themselves. However, a more refined pathological scoring system may include taking observations about the health of the *G. mellonella*. Uninfected larvae remain cream-coloured and move freely in response to stimuli. Infected larvae become less motile, often accompanied by melanisation, which may be brown, grey or black.

Melanisation can be correlated to the presence of bacteria in the body cavity; we found a significant association between colour change and bacteria in the body cavity (Senior *et al.*, 2011). Melanisation occurs around nodules composed of aggregations of hemocytes produced in response to bacterial invasion (Harding *et al.*, 2012); the hemocytes produce prophenoloxidase, which induces the production of melanin to isolate microbes from the circulation and kill them (Shao *et al.*, 2012). This process in insect hosts possibly acts as protection against systemic damage for endogenous tissues within the body during phagocytosis and encapsulation (Nappi and Christensen, 2005).

Identification of virulence-associated genes

The ability to use *G. mellonella* as an inexpensive, large-scale alternative infection model allows for the investigation of virulence factors, as large numbers of mutants can be assayed to identify genes with a role in bacterial virulence. Screening assays would be unethical and prohibitively expensive in mammalian infection model.

We have used the *G. mellonella* infection model to screen *C. jejuni* strains with gene insertional inactivations. The screens showed that it was possible to identify attenuated mutants, and also to discriminate between degrees of attenuation. Attenuation data in *G. mellonella* was validated by mammalian infection model data; a mutant in *cj0079*, which produces cytolethal distending toxin (CDT), was attenuated in a susceptible mouse model (Champion *et al.*, 2010). As would be expected, mutants in *cj1321* and *cj1324*, which form part of the O-linked glycosylation locus that is important for the colonization of chickens, were not attenuated in the *G. mellonella* model as this model bypasses the colonization step in infection with bacteria being injected directly into the larval body cavity (Champion *et al.*, 2010).

A lipooligosaccharide mutant (*cj1132–1152*) was shown to be completely attenuated in the larvae. This was significant as it was the first direct evidence that LOS is essential for *C. jejuni* virulence (Champion *et al.*, 2010). The same study also demonstrated attenuation in an aflagellate *C. jejuni* mutant (*cj1339*); this may be linked to the secretion of CDT, or it may be a result of the bacteria's reduced motility. The mutant *cj0511* was noticeably attenuated; this gene encodes a protease that is not a characterized virulence factor, but the results in the *G. mellonella* model suggest it warrants further study.

Furthermore, it is known that the capsule of *C. jejuni* is a virulence factor, but Champion *et al.* (2010) were able to focus on a particular gene, *cj1416*, which is responsible for an O-methyl phosphoramidate modification of the capsule, present in many isolates associated with post-infection complications such as enteritis and Guillain–Barré syndrome. A Cj1416 mutant was attenuated in *G. mellonella*, and the virulence could be restored via complementation. This particular modification has structural similarities with the active structures of organophosphate pesticides, and it was hypothesized that this may account for its lethal effect on the larvae.

The transcriptional regulator Cj1556 was originally identified as a hypothetical protein, but further annotation of the *C. jejuni* NCTC11168 genome identified it as a MarR family transcriptional regulator, with potential for regulating the oxidative stress response. An insertion mutant for this gene demonstrated significant attenuation in the *G. mellonella* model, whilst a complemented strain restored the wild-type phenotype. The same mutant displayed reduced intracellular survival in murine J774 cells (Gundogdu *et al.*, 2011).

Summary

G. mellonella has been used increasingly as an infection model for bacterial pathogens and has been shown to be a useful tool for studying *C. jejuni* virulence. The *G. mellonella* infection model could be used to assess the virulence of veterinary or animal-derived isolates of *Campylobacter*. It could support epidemiological studies, in which the virulence of isolates from different locations could be compared. There is potential for mechanisms of virulence to be identified, thus allowing the development of vaccines or evaluation of other approaches – such as niche competition – to control *Campylobacter* infection. Moreover, large panels of isolates could be screened so that potential virulence determinants or differences in virulence at 37°C could be studied.

Zebra fish embryo as a model for *Campylobacter* studies

Zebra fish (*Danio rerio*) embryos (ZFE) have been widely used in developmental studies and have recently gained attention as an infection model. ZFE develop innate immunity mediated through pathogen recognition receptors, a system that is highly conserved in eukaryotes (Kimbrell and Beutler, 2001; Akira *et al.*, 2006). Adaptive immunity comprises specific pathogen antigen recognition, immunological memory of pathogens and pathogen specific antibodies. ZFE and mammals both possess phagocytic cells, Toll-like receptors, complement, proinflammatory cytokines, interferon dependent defence functions, acute phase response proteins, leucocytes, antigen processing cells, T cells, B cells and immunoglobulins (Trede *et al.*, 2001; Yoder *et al.*, 2002; Zarkadis *et al.*, 2001; Lin *et al.*, 2007; Meijer *et al.*, 2004; Sieger *et al.*, 2009; van der Sar *et al.*, 2006). Utilizing these similarities, the immunological response of zebra fish to challenge with bacterial pathogens may be used to model the mammalian response. The availability of genetic manipulation tools for zebra fish increases the value of this model as specific host genes may be knocked down to address the roles played by a host gene of interest during host–pathogen interaction. Genetic tools available for the zebra fish system include the creation of target selected homozygous mutations (Wienholds *et al.*, 2002, 2003). In addition, morpholinos can be used to transiently knockdown specific genes (Shultz, 1991). These techniques provide a platform for large scale, economical forward and reverse genetic screens.

ZFE also have many practical advantages over mammalian infection models; they are small, economical, prolific and easily confined and maintained. ZFE lend themselves to the study of host pathogen interactions as they are transparent for the first three weeks of development, forming an ideal system to monitor real time infections and fluorescent transgenic immune cells (Davis *et al.*, 2002). However, ZFE as an infection model has some limitations, including inability of zebra fish embryos to survive at 37°C. This prevents the study of temperature regulated bacterial virulence genes. In addition, ZFE may not possess the mammalian target organ or receptor and they may not exhibit the complete mammalian immune response. However, no infection model is perfect, but must be selected based on the research question being asked. ZFE as an infection model has been used to elucidate complex host – pathogen interactions that could not be answered using traditional animal infection models (Hamm *et al.*, 2006; Volkman *et al.*, 2010). A key unanswered question about *C. jejuni* pathogenesis is the nature of its interaction with monocytes *in vivo* (Young *et al.*, 2007) and ZFE could be a useful tool to study this interaction.

Utilization of zebra fish embryos in understanding pathogenesis

Although currently ZFE have not been reported as a model for Campylobacteriosis studies, zebra fish have been used as a model organism to gain understanding of other enteric pathogens including *Salmonella* Typhimurium (van der Sar *et al.*, 2003), *Listeria monocytogenes* (Levraud *et al.*, 2009) and *E. coli* (Wiles *et al.*, 2009). For example, zebra fish embryos have been used to image

L. monocytogenes capture by macrophages in *vivo* and in real time. Virulence and disease progression of *L. monocytogenes* in ZFE was shown to be similar to that observed in mammals, with the formation of actin comet tails and protrusions indicating cell to cell spread (Levraud *et al.*, 2009). In *Mycobacterium* studies, disease progression in murine models differs to that seen in human infection, as granulomas are not observed (Wang, 2010). However, infected ZFE formed granulomas similar to that in human tuberculosis (Pozos and Ramakrishnan, 2004) making it an attractive model for study on granuloma formation.

Transgenic lines with fluorescently tagged immune cells, such as GFP-expressing neutrophils and mCherry-expressing macrophages, are available (Grey *et al.*, 2011) and could be used to track interaction between pathogens of interest with the immune cells in real-time. For example, GFP-neutrophils were used to demonstrate interaction with *Staphylococcus aureus*, 2 h after being injected into 30 hpf ZFE (Prajsnar *et al.*, 2008). Such a system could be applied to investigate the interaction of *C. jejuni* with professional macrophages *in vivo*.

ZFE manipulation

Microinjection is commonly used to introduce the pathogen of interest into ZFE. Common sites of microinjection include the blood stream (e.g. axial vein [van der Sar *et al.*, 2003]), yolk sac circulation valley (Clatworthy *et al.*, 2009) or yolk sac (Prajsnar *et al.*, 2008). The low cost and maintenance of ZFE as a model system facilitates economical screening of numerous strains and mutants. However, the labour intensive and time consuming procedure to inoculate embryos through microinjection is one of the drawbacks of this system. As it is a living system, the embryos continue to develop throughout handling and manipulation. Thus it is crucial to complete inoculating the ZFE within a short time frame in order to ensure that all embryos are infected at similar developmental stage. This restriction limits the number of embryos that could be inoculated and the number of strains or mutants to be screened each time. This could be overcome through the use of automated robotic microinjection system as such systems can inoculate 15 ZFE per minute with a survival rate of 98% (Wang *et al.*, 2007). Alternative routes of infection such as immersion of embryos in media containing the pathogen of interest have also been explored. A study using *Edwardsiella tarda* was conducted to compare static immersion and intravenous injection. This study revealed that ZFE immersed in media containing the pathogen were resistant to the infection whereas those inoculated intravenously were susceptible (van Soest *et al.*, 2011). Taking into consideration the numerous advantages of ZFE, it would be worthwhile exploring the use of this infection model for *Campylobacter jejuni*.

Campylobacter metabolism and host adaptation

Campylobacter jejuni is able to colonize the intestinal tract of a diverse range of hosts and in order to do this it must compete against the resident intestinal microbiota. To perform this effectively, it must be able to efficiently acquire nutrients from its environment to support its survival and growth in the intestine. Recent studies on *C. jejuni* metabolism have facilitated a greater understanding of how *C. jejuni* acquires and metabolizes key nutrients (Velayudhan *et al.*, 2004; Leon-Kempis Mdel *et al.*, 2006; Hofreuter *et al.*, 2008; Stahl *et al.*, 2011).

It is well established that different isolates of the same pathogen often display significant diversity within their repertoire of virulence factors. However, less is known about diversity within basic metabolic traits, which are often considered more stable attributes of bacterial pathogens. Pathogens that as part of their normal ecology inhabit different environments with different available nutrients are also likely to be under strong selective pressure to diversify and evolve their basic metabolic traits. Despite this, the exact role of metabolism and nutrient acquisition in the pathobiology of *C. jejuni* remains unclear. A better understanding of the metabolism of *Campylobacter* may aid the development of rationale control strategies including vaccines (Hofreuter *et al.*, 2008).

Interestingly, *C. jejuni* lacks many of the common metabolic pathways necessary for metabolizing glucose, galactose and other carbohydrates that a number of microbes commonly utilize. However, it is able to utilize citric acid cycle intermediates and various amino acids. *C. jejuni* readily metabolizes the amino acids aspartate, glutamate, serine, and proline. Interestingly, some strains also possess additional pathways enabling them to utilize glutamine and asparagine. Recent studies have revealed that some *C. jejuni* strains can metabolize L-fucose. *C. jejuni* also possesses diverse mechanisms for the acquisition of various transition metals that are required for metabolic activities. Furthermore, *C. jejuni* is unique in possessing both molybdate and tungsten co-factored proteins (Stahl *et al.*, 2012).

The presence of multiple respiratory mechanisms in *C. jejuni* and their differential expressions under various conditions may be advantageous for the bacteria to deal with the changes in oxygen availability in the *in vivo* environment. Evidence also exists for both oxygen-dependent and oxygen-independent pathways being activated to ensure the survival of *C. jejuni* in the chicken caecum. In the caecum, *C. jejuni* grows under oxygen-limited conditions, utilizing a range of electron acceptors and electron donors for respiration, exploiting specific amino acids for growth, and consuming haemin as a primary source of iron (Woodhall *et al.*, 2005).

Furthermore, some *C. jejuni* sequence types (STs) such as ST-21 appear to use mechanisms such as inter-strain recombination, metabolic and surface-associated genetic modules and contingency genes to permit a generalist lifestyle and the flexible expression of fitness factors in diverse environments (Gripp *et al.*, 2011).

Carbon and nitrogen utilization

C. jejuni lacks the ability to use many common carbohydrates such as carbon sources (Parkhill *et al.*, 2000). Furthermore, apart from the absence of the appropriate transporters to take up sugars such as glucose or galactose, *Campylobacter* also lacks numerous enzymes within the glycolytic pathway. Although *C. jejuni* possesses the complete set of enzymes necessary for the synthesis of glucose-6-phosphate from pyruvate via gluconeogenesis, it lacks any form of glucokinase to phosphorylate extracellular glucose (Velayudhan and Kelly, 2002). In addition the 6-phosphofructokinase enzyme required for the irreversible phosphorylation of fructose-6-phosphate to fructose-1,6-diphosphate during glycolysis is also absent from *C. jejuni*. Interestingly, *C. jejuni* still possesses pyruvate kinase (Velayudhan and Kelly, 2002).

Amino acids utilization

Several amino acids (Asp, Glu, Ser, Asn, and Pro) are utilized by *C. jejuni* (Guccione *et al.*, 2008; Leach *et al.*, 1997; Mohammed *et al.*, 2004), and mutations in the genes necessary for their metabolism attenuate *C. jejuni*'s in vivo colonization potential (Guccione *et al.*, 2008; Velayudhan *et al.*, 2004). The current understating of *C. jejuni*'s in vivo metabolism is focused on amino acids to support growth and persistence in the host intestines. Although metabolism has not traditionally been considered a virulence factor, the abilities to survive and persist in the environment and to cause disease are intimately linked to nutrient availability. Thus, metabolic diversity within pathogens may be directly linked to host specificity and virulence potential.

It is well characterized that *C. jejuni* is able to acquire a limited number of amino acids while inhabiting the intestine. *C. jejuni* is also able to utilize available amino acids in a sequential manner, with serine, aspartate, asparagine, and glutamate being preferred in that order, although this may vary between different *C. jejuni* strains. Proline can also be metabolized, but only after other nutrients have been exhausted (Wright *et al.*, 2009). Furthermore, *C. jejuni* 81-176 mutants unable to utilize serine were shown to be severely defective for colonization of the intestine and systemic tissues in a mouse model of infection. In contrast, *C. jejuni* 81-176 mutants unable to utilize proline were only defective for intestinal colonization. These results further emphasize the importance of amino acid utilization in *C. jejuni* colonization of various tissues (Hofreuter *et al.*, 2012).

Interestingly, serine, aspartate, glutamate and proline make up some of the most common amino acids found in chicken faeces, perhaps explaining why they play such a central role in *C. jejuni* metabolism (Parsons, 1984).

Citric acid cycle intermediate utilization

C. jejuni depends heavily on the citric acid cycle for its energy needs. All of the aforementioned pathways produce pyruvate, fumarate, oxaloacetate, or 2-oxoglutarate, each of which is fed directly into the citric acid cycle. In addition to this, *C. jejuni* is capable of transporting several of the citric acid cycle intermediates, and using them directly as nutrient sources. For example 2-oxoglutarate can be transported by the permease KgtP. In addition, succinate, fumarate, and malate can all be transported by the C4-dicarboxylate transporters, DcuA, and DcuB (Guccione *et al.*, 2008). The putative C4-dicarboxylate transporter DctA may also play a minor role in the uptake of both aspartate and CAC intermediates (Leon-Kempis Mdel *et al.*, 2006; Guccione *et al.*, 2008). Many strains also possess a citrate transporter, allowing for the acquisition of citrate (Parkhill *et al.*, 2000). *C. jejuni* is known to utilize pyruvate as a primary carbon source while it is present in the growth medium (Velayudhan and Kelly, 2002). However, no pyruvate transporters have been identified or studied in *C. jejuni*, thus the mechanism by which it transports pyruvate into the cell remains elusive.

Short-chain fatty acid utilization

Short-chain fatty acids (SCFAs) are produced as metabolic by-products by a number of bacterial species found in the intestinal tract. The host or other bacteria subsequently metabolize the SCFA as energy sources. Interestingly, the SCFA butyrate, which is produced as a by-product of bacterial acetate metabolism (Duncan *et al.*, 2004), is important in general intestinal function. Thus, butyrate is readily taken up by enterocytes and provides a primary energy source for the cells in the intestine (Louis *et al.*, 2010). Acetate is the most significant SCFA found within the intestine (Duncan *et al.*, 2004), with smaller levels of lactate, propionate, and butyrate also present (Belenguer *et al.*, 2011). However, studies suggest that *C. jejuni* only possess the capacity to transport and metabolize acetate and lactate (Wright *et al.*, 2009; Thomas *et al.*, 2011).

Molybdate and tungstate utilization

Campylobacter jejuni is unusual amongst bacteria in that it possesses both molybdenum (Moco) and tungsten (Woco) co-factored proteins (Stahl *et al.*, 2012). *C. jejuni* formate dehydrogenase, a key enzyme for energy metabolism using formate as an electron donor, has been shown to require Woco for proper function, while nitrate reductase, sulfite oxidase, and SN oxide reductase require Mocos for their activity (Smart *et al.*, 2009; Taveirne *et al.*, 2009). This peculiar metabolic requirement likely explains the unusual presence in *C. jejuni* of specific high affinity ABC transporters for both metals. ModABC and TupABC transporters import either molybdate or tungstate, respectively, across the inner membrane in an ATP-dependent manner, but both are regulated together by a single ModE-like regulatory protein (Taveirne *et al.*, 2009). Interestingly, the *C. jejuni* ModE protein represses the mod operon in the presence of excess molybdate or tungstate while the tup operon is only repressed in response to tungsten. Moreover, the *C. jejuni* ModE protein does not appear to contain a metal binding domain and instead contains a coiled–coiled domain that is indicative of protein to protein interactions (Taveirne *et al.*, 2009).

Copper, iron, zinc and other metal utilization

The zinc homeostasis system remains largely uncharacterized in *C. jejuni*. There are no obvious candidates for the regulation of zinc uptake or efflux despite the almost universal presence of these proteins in other bacteria (Hantke, 2001). *C. jejuni* does possess a homologue of the high-efficiency *E. coli* ZnuABC zinc uptake system that is also present in many other bacteria (Hantke, 2001). Recent work by Davis *et al.* (2009) has demonstrated that this system is functional in *C. jejuni* and is required for both survival in low zinc environments and for successful chick colonization.

In the absence of sufficient iron sources, the growth of *C. jejuni* is significantly reduced, if not completely halted (Palyada *et al.*, 2004). Thus, *C. jejuni*, and indeed other bacteria, have developed a wide variety of mechanisms for iron acquisition

(Stintzi *et al.*, 2008; Miller *et al.*, 2009; Butcher *et al.*, 2010). However, there is a paucity of information regarding copper and zinc homeostasis in *C. jejuni*. Interestingly, copper chelation results in slowed *C. jejuni* growth and this growth inhibition can be partially relieved by deleting non-essential copper containing proteins such as the p19 iron acquisition protein (Chan *et al.*, 2010). *C. jejuni* does not possess any annotated homologues of the copper regulators CueR or CusR. However, it does harbour CopA and CeuO homologues (Hall *et al.*, 2008). CopA proteins are transmembrane ATPases that are involved in pumping excess cytosolic copper into the periplasmic space (Osman and Cavet, 2008). Cuprous ions in the periplasmic space are oxidized to the less toxic cupric form by CeuO (Osman and Cavet, 2008). It is currently unclear whether *C. jejuni* harbours specific copper transporters for copper import or export from the cell.

In addition to those described above, *C. jejuni* also makes use of a number of other metals for either enzymatic or structural purposes, such as cobalt and nickel. Although there are no obvious transporters for these metals in the *C. jejuni* genome, there are several examples where these metals are required for protein function.

References

Akira, S., Uematsu, S., and Takeuchi, O. (2006). Pathogen recognition and innate immunity. Cell *124*, 783–801.

Belenguer, A., Holtrop, G., Duncan, S.H., Anderson, S.E., Calder, A.G., Flint, H.J., and Lobley, G.E. (2011). Rates of production and utilization of lactate by microbial communities from the human colon. FEMS Microb. Ecol. *77*, 107–119.

Bergin, D., Reeves, E.P., Renwick, J., Wientjes, F.B., and Kavanagh, K. (2005). Superoxide production in *Galleria mellonella* hemocytes: Identification of proteins homologous to the NADPH oxidase complex of human neutrophils. Infect. Immun. *73*, 4161–4170.

Biswas, D., Itoh, K., and Sasakawa, C. (2000). Uptake pathways of clinical and healhty animal isolates of *Campylobacter jejuni* into INT-407 cells. FEMS Immunol. Med. Microbiol. *29*, 203–211.

Biswas, D., Fernando, U., Reiman, C., Willson, P., Potter, A., and Allan, B. (2006). Effect of cytolethal distending toxin of *Campylobacter jejuni* on adhesion and internalization in cultured cells and in colonization of the chicken gut. Avian Dis. *50*, 586–593.

Brøndsted, L., Thorup Andersen, M., Parker, M., Jørgensen, K., and Ingmer, H. (2005). The HtrA protease of *Campylobacter jejuni* is required for heat and oxygen tolerance and for optimal interaction with human epithelial cells. Appl. Environ. Microbiol. *71*, 3205–3212.

Brown, S.E., Howard, A., Kasprzak, A.B., Gordon, K.H., and East, P.D. (2009). A peptidomics study reveal the impressive antimicrobial peptide arsenal of the wax moth *Galleria mellonella*. Insect Biochem. Mol. Biol. *39*, 792–800.

Butcher, J., Flint, A., Stahl, M., and Stintzi, A. (2010). *Campylobacter* Fur and PerR regulons. In Iron Uptake and Homeostasis in microorganisms, Cornelis, P., and Andrews, S.C., eds. (Norfolk, UK, Caister Academic Press), pp. 167–202.

Butzler, J.P., and Oosterom, J. (1991). *Campylobacter*: pathogenicity and significance in foods. Int. J. Food Microbiol. *12*, 1–8.

Byrne, C.M., Clyne, M., and Bourke, B. (2007). *Campylobacter jejuni* adhere to and invade chicken intestinal epithelial cells in vitro. Microbiology *153*, 561–569.

Candon, H.L., Allan, B.J., Fraley, C.D., and Gaynor, E.C. (2007). Polyphosphate kinase 1 is a pathogenesis determinant in *Campylobacter jejuni*. J. Bacteriol. *189*(22), 8099–8108.

Carrillo, C.D., Taboada, E., Nash, J.H.E., Lanthier, P., Kelly, J., Lau, P.C., Verhulp, R., Mykytczuk, O., Sy, J., Findlay, W.A., *et al.* (2004). Genome-wide expression analyses of *Campylobacter jejuni* NCTC11168 reveals coordinate regulation of motility and virulence by flhA. J. Biol. Chem. *279*, 20327–20338.

Champion, O.L., Cooper, I.A., James, S.L., Ford, D., Karlyshev, A.V., Duffield, M., Oyston, P.C., and Titball, R.W. (2009). *Galleria mellonella* as an alternative infection model for *Yersinia pseudotuberculosis*. Microbiology *155*. 1516–1522.

Champion, O.L., Karlyshev, A.V., Senior, N.J., Woodward, M.J., La Ragione, R.M., Howard, S.L., Wren, B.W., and Titball, R.W. (2010). Insect infection model for *Campylobacter jejuni* reveals that O-methyl phosphoramidate has insecticidal activity. J. Infect. Dis. *201*, 776–782.

Chan, A.C.K., Doukov, T.I., Scofield, M., Tom-Yew, S.A.L., Ramin, A.B., MacKichan, J.K., Gaynor, E.C., and Murphy, M.E.P. (2010). Structure and function of P19, a high-affinity ion transporter of the human pathogen *Campylobacter jejuni*. J. Mol. Biol. *401*, 590–604.

Chen, Y.H., Poly, F., Pakulski, Z., Guerry, P., and Monteiro, M. (2008). The chemical structure and genetic locus of *Campylobacter jejuni* CG8486 (serotype HS:4) capsular polysaccharide: the identification of 6-deoxy-D-ido-heptopyranose. Carbohydrate Res. *343*, 1034–1040.

Clatworthy, A.E., Lee, J.S., Leibman, M., Kostun, Z., Davidson, A.J., and Hung, D.T. (2009). *Pseudomonas aeruginosa* infection of zebrafish involves both host and pathogen determinants. Infect. Immun. 77(4), 1293–1303.

Dasti, J.I., Tareen, A.M., Lugert, R., Zautner, A.E., and Gross, U. (2009). *Campylobacter jejuni*: a brief overview on pathogenicity-associated factors and disease-mediating mechanisms. Int. J. Med. Microbiol. *300*, 205–211.

Davis, J.M., Clay, H., Lewis, J.L., Ghon, N., Herbomel, P., and Ramakrishnan, L. (2002). Real-time visualization of *Mycobacterium*-macrophage interactions leading to initiation of granuloma formation in zebrafish embryos. Immunity *17*, 693–702.

Davis, L.M., Kakuda, T., and DiRita, V.J. (2009) A *Campylobacter jejuni* znuA orthologue is essential for growth in low-zinc environments and chick colonization. J. Bacteriol. *191*(5), 1631–1640.

Day, W.A.J., Sajecki, J.L., Pitts, T.M., and Joens, L.A. (2000). Role of catalase in *Campylobacter jejuni* intracellular survival. Infect. Immun. *68*, 6337–6345.

De Melo, M.A., Gabbiani, G., and Pechère, J.C. (1989). Cellular events and intracellular survival of *Campylobacter jejuni* during infection of Hep-2 cells. Infect. Immun. *57*, 2214–2222.

Desbois, A.P., and Coote, P.J. (2012). Utility of Greater Waxmoth (*Galleria mellonella*) for evaluating the toxicity and efficacy of new antimicrobial agents. Adv. Appl. Microbiol. *78*, 25–53.

Duncan, S.H., Holtrop, G., Lobley, G.E., Calder, A.G., Stewart, C.S., and Flint, H.J. (2004). Contribution of acetate to butyrate formation by human faecal bacteria. Br. J. Nutr. *91*, 915–923.

Eshraghi, A., Maldonado-Arocho, F.J., Gargi, A., Cardwell, M.M., Prouty, M.G., Blanke, S.R., and Bradley, K.A. (2010). Cytolethal distending toxin family members are differentially affected by alterations in host glycans and membrance cholesterol. J. Biol. Chem. *285*, 18199–18207.

Everest, P.H., Goossens, H., Butzler, J.P., Lloyd, D., Knutton, S., Ketley, J.M., and Williams, P.H. (1992). Differentiated Caco-2 cells as a model for enteric invasion by *Campylobacter jejuni* and *C. coli*. J. Med. Microbiol. 37, 319–325.

Fallon, J.P., Troy, N., and Kavanagh, K. (2011). Pre-exposure of *Galleria mellonella* larvae to different doses of *Aspergillus fumigatus conidia* causes differential activation of cellular and humoral immune responses. Virulence 2, 413–421.

Fedhila, S., Buisoon, C., Dusserget, O., Serror, P., Glomski, I.J., Liehl, P., Lereclus, D., and Nielsen-LeRoux, C. (2010). Comparative analysis of the virulence of invertebrate and mammalian pathogenic bacteria in the oral insect infection model *Galleria mellonella*. J. Invertebrate Pathol. *103*, 24–29.

Fields, J.A., and Thompson, S.A. (2008). *Campylobacter jejuni* CsrA mediates oxidative stress responses, biofil formation, and host cell invasion. J. Bacteriol *190*, 3411–3416.

Flanagan, R.C., Neal-McKinney, J.M., Dhillon, A.S., Miller, W.G., and Konkel, M.E. (2009). Examination of *Campylobacter jejuni* putative adhesins leads to the identification of a new protein, designated FlpA, required for chicken colonization. Infect. Immun. 77, 2399–2407.

Frirdich, E., Biboy, J., Adams, C., Lee, J., Ellermeier, J., Davis Gielda, L., DiRita, V.J., Girardin, S.E., Vollmer, W., and Gaynor, E.C. (2012). Peptidoglycan-modifying enzyme Pgp1 is required for helical cell shape and pathogenicity traits in *Campylobacter jejuni*. PLOS Pathogens *8*(3).

Gilbert, C., and Slavik, M. (2004). Determination of toxicity of *Campylobacter jejuni* isolated from humans and from poultry carcasses acquired at various stages of production. J. Appl. Microbiol. *97*, 347–353.

Grant, C.C.R., Konkel, M.E., Cieplak, W. Jr., and Tompkins, L.S. (1993). Role of flagella in adherence, internalization, and translocation of *Campylobacter jejuni* in nonpolarized and polarized epithelial cell cultures. Infect. Immun. *61*, 1764–1771.

Gray, C., Loynes, C.A., Whyte, M.K.B., Crossman, D.C., Renshaw, S.A., and Chico, T.J.A. (2011). Simultaneous intravital imaging of macrophage and neutrophil behaviour during inflammation using a novel transgenic zebrafish. Thrombos. Haemostas. *105*, 743–932.

Gripp, E., Hlahla, D., Didelot, X., Kops, F., Maurischat, S., Tedin, K., Ellerbroek, L., Schreiber, K., Schomberg, D., Janssen, T., *et al.* (2011). Closely related *Campylobacter jejuni* strains from different sources reveal a generalist rather than a specialist lifestyle. BMC Genomics *12*.

Guccione, E., Leon-Kempis Mdel, R., Pearson, B.M., Hitchin, E., Mulholland, F., van Diemen, P.M., Stevens, M.P., and Kelly, D.J. (2008). Amino acid-dependent growth of *Campylobacter jejuni*: key roles for aspartase (AspA) under microaerobic and oxygen-limited conditions and identifacation of AspB (Cj0762), essential for growth on glutamate. Mol. Microbiol. *69*, 77–93.

Gundogdu, O., Mills, D.C., Elmi, A., Martin, M.J., Wren, B.W., and Dorrell, N. (2011). The *Campylobacter jejuni* transcriptional regulator Cj1556 plays a role in the oxidative and aerobic stress response and is important for bacterial survival *in vivo*. J. Bacteriol. *19*, 4238–4249.

Habib, I., Louwen, R., Uyttendaele, M., Houf, K., Vandenberg, O., Nieuwenhuis, E.E., Miller, W.G., van Belkum, A., and De Zutter, L. (2009). Correlation between genotypic diveristy, lipooligosaccharide gene locus class variation, and Caco-2 cell invasion potential of *Campylobacter jejuni* isoaltes from chicken meat and humans: contributing to virulotyping. Appl. Environ. Microbiol. *75*, 4277–4288.

Hall, S.J., Hitchcock, A., Butler, C.S., and Kelly, D.J. (2008). A multicopper oxidase (Cj1516) and a CcpA homologue (Cj1161) are major components of the copper homeostasis system of *Campylobacter jejuni*. J. Bacteriol. *190*, 8075–8085.

Hamm, E.E., Voth, D.E., and Ballard, J.D. (2006). Identification of *Clostridium difficile* toxin B cardiotoxicity using a zebrafish embryo model of intoxication. Proc. Natl. Acad. Sci. U.S.A. *103*, 14176–14181.

Hantke, K. (2001). Bacterial zinc transporters and regulators. Biometals *14*, 239–249.

Harding, C.R., Schroeder, G.N., Reynolds, S., Kosta, A., Collins, J.W., Mousnier, A., and Frankel, G. (2012). *Legionella pneomophila* pathogenesis in the *Galleria mellonella* infection model. Infect. Immun. *80,* 2780–2790.

Hickey, T.E., Majam, G., and Guerry, P. (2005). Intracellular survival of *Campylobacter jejuni* in human monocytic cells and induction of apoptotic death by cytolethal distending toxin. Infect. Immun. *73,* 5194–5197.

Hofreuter, D., Tsai, J., Watson, R.O., Novik, V., Altman, B., Benitez, M., Clark, C., Perbost, C., Jarvie, T., Du, L., *et al.* (2006). Unique features of a highly pathogenic *Campylobacter jejuni* strain. Infect. Immun. *74,* 4694–4707.

Hofreuter, D., Novik, V., and Galán, J.E. (2008). Metabolic diveristy in *Campylobacter jejuni* enhances specific tissue colonization. Cell Host Microbe 4(5), 425–433.

Hofreuter, D., Mohr, J., Wensel, O., Rademacher, S., Schreiber, K., Schomburg, D., Gao, B., and Galán, J.E. (2012). Contribution of amino acid catabolism to the tissue specific persistence of *Campylobacter jejuni* in a murine colonization model. PLoS One 7(11), 1–14.

Hurley, B.P., and McCormick, B.A. (2003). Translating tissue culture results into animal models: the case of *Salmonella typhimurium*. Trends Microbiol. *11,* 562–569.

Jander, G., Rahme, L.G., and Ausubel, F.M. (2000). Positive correlation between virulence of *Pseudomonas aeruginosa* mutants in mice and insects. J. Bacteriol. *182,* 3843–3845.

Janssen, R., Krogfelt, K.A., Cawthraw, S.A., van Pelt, W., Wagenaar, J.A., and Owen, R.J. (2008). Host–pathogen interactions in *Campylobacter* infections: the host perspective. Clin. Microbiol. Rev. *21,* 505–518.

Jinadasa, R.N., Bloom, S.E., Weiss, R.S., and Duhamel, G.E. (2011). Cytolethal distending toxin: a conserved bacterial genotoxin that blocks cell cycle progression, leading to apoptosis of a broad range of mammalian cell lineages. Microbiology *157,* 1851–1875.

Kemp, M.W., and Massey, R.C. (2007). The use of insect models to study human pathogens. Drug Discov. Today: Dis. Models 4(3), 105–110.

Khan, I., Adler, B., Haridas, S., and Albert, M.J. (2005). PorA protein of *Campylobacter jejuni* is not a cytotoxin mediating inflammatory diarrhoea. Microb. Infect. *7,* 853–859.

Kiehlbauch, J.A., Albach, R.A., Baum, L.L., and Chang, K.-P. (1985). Phagocytosis of *Campylobacter jejuni* and its intracellular survival in mononuclear phagocytes. Infect. Immun. *48,* 446–451.

Kim, J.S., Li, J., Barnes, I.H.A., Baltzegar, D.A., Pajaniappan, M., Cullen, T.W., Trent, M.S., Burns, C.M., and Thompson, S.A. (2008). Role of the Campylobacter jejuni Cj1461 DNA methyltransferase in regulating virulence characteristics. J. Bacteriol. *190,* 6524–6529.

Kimbrell, D.A., and Beutler, B. (2001). The evolution and genetics of innate immunity. Nature Rev. Genet. *2,* 256–267.

Konkel, M.E., Hayes, S.F., Joens, L.A., and Cieplak Jr, W. (1992). Characteristics of the internalization and intracellular survival of *Campylobacter jejuni* in human epithelial cell cultures. Microb. Pathog. *13,* 357–370.

Konkel, M.E., Kim, B.J., Rivera-Amill, V., and Garvis, S.G. (1999). Bacterial secreted proteins are required for the internalization of *Campylobacter jejuni* into cultured mammalian cells. Mol. Microbiol. *32,* 691–701.

Konkel, M.E., Klena, J.D., Rivera-Amill, V., Monteville, M.R., Biswas, D., Raphael, B.H., and Mickelson, J. (2004). Secretion of virulence proteins from *Campylobacter jejuni* is dependent on a functional flagellar export apparatus. J. Bacteriol. *186,* 3296–3303.

Kurz, C.L., and Ewbank, J.J. (2007). Infection in a dish: high-throughput analyses of bacterial pathogenesis. Curr. Opin. Microbiol. *10,* 10–16.

Lara-Tejero, M., and Galán, J.E. (2000). A bacterial toxin that controls cell cycle progression as a deoxyribonuclease I-like protein. Science *290,* 354–357.

Lara-Tejero, M., and Galán, J.E. (2001). CdtA, CdtB, and CdtC form a tripartite complex that is required for cytolethal distending toxin activity. Infect. Immun. *69*(7), 4358–4365.

Larson, C.L., Shah, D.H., Dhillon, A.S., Call, D.R., Ahn, S., Haldorson, G.J., Davitt, C., and Konkel, M.E. (2008). *Campylobacter jejuni* invade chicken LMH cells inefficiently and stimulate differential expression of the chicken CXCLi1 and CXCLi2 cytokines. Microbiology *154,* 3835–3845.

Leach, S.A., Harvey, P., and Wait, R. (1997). Changes with growth rate in the membrane lipid composition of and amino acid utilization by continuous cultures of *Campylobacter jejuni*. J. Appl. Microbiol. *82,* 631–640.

Leon-Kempis Mdel, R., Guccione, E., Mulholland, F., Williamson, M.P., and Kelly, D.J. (2006). The *Campylobacter jejuni* PEB1a adhesin is an aspartate/glutamate binding protein of an ABC transporter essential for microaerobic growth on dicarboxylic amino acids. Mol. Microbiol. *60,* 1262–1275.

Levraud, J.P., Disson, O., Kissa, K., Bonne, I., Cossart, P., Herbomel, P., and Lecuit, M. (2009). Real-time observation of Listeria monocytogenes–phagocyte interactions in living zebrafish larvae. Infect. Immun. *77,* 3651–3660.

Lin, B., Chen, S., Cao, Z., Lin, Y., Mo, D., Zhang, H., Gu, J., Dong, M., Liu, Z., and Xu, A. (2007). Acute phase response in zebrafish upon *Aeromonas salmonicida* and *Staphylococcus aureus* infection: Striking similarities and obvious differences with mammals. Mol. Immun. *44,* 295–301.

Louis, P., Young, P., Holtrop, G., and Flint, H.J. (2010). Diveristy of human colonic butyrate-producing bacteria revealed by analysis of the butyryl-CoA:acetate CoA-transferase gene. Environ. Microbiol. *12,* 304–314.

Louwen, R., Heikema, A., van Belkum, A., Ott, A., Gilbert, M., Ang, W., Endtz, H.P., Bergman, M.P., and Nieuwenhuis, E.E. (2008). The sialylated lipooligosaccharide outer core in *Campylobacter jejuni* is an important determinant for epithelial cell invasion. Infect. Immun. *76,* 4431–4438.

MacCallum, A.J., Harris, D., Haddock, G., and Everest, P.H. (2006). *Campylobacter jejuni*-infected human epithelial cell lines vary in their ability to secrete inerleukin-8 compared to in vitro-infected primary human intestinal tissue. Microbiology *152*, 3661–3665.

McKenney, E.S., Sargent, M., Khan, H., Uh, E., Jackson, E.R., San Jose, G., Couch, R.D., Dowd, C.S., and Van Hoek, M.L. (2012). Lipophilic prodrugs of FR900098 are antimicrobial against *Francisella novicida in vivo* and *in vitro* and show GlpT independent activity. PLoS One 7(10).

Malik-Kale, P., Raphael, B.H., Parker, C.T., Joens, L.A., Klena, J.D., Quinones, B., Keech, A.M., and Konkel, M.E. (2007). Characterization of genetically matched isolates of *Campylobacter jejuni* reveals that mutations in genes involved in flagellar biosynthesis alter the organism's virulence potential. Appl. Environ. Microbiol. 73, 3123–3136.

Meijer, A.H., Krens, S.F.G., Medina Rodriguez, I.A., He, S., Bitter, W., Snaar-Jagalska, B.E., and Spaink, H.P. (2004). Expresion analysis of the Toll-like receptor and TIR domain adaptor families of zebrafish. Mol. Immunol. *40*, 773–783.

Mihaljevic, R.R., Sikic, M., Klancnik, A. Brumini, G., Mozina, S.S., and Abram, M. (2007). Environmental stress factors affecting survival and virulence of *Campylobacter jejuni*. Microb. Pathog. *43*, 120–125.

Miller, C.E., Williams, P.H., and Ketley, J.M. (2009). Pumping iron: mechanisms for iron uptake by *Campylobacter*. Microbiology *155*, 3157–3165.

Mills, D.C., Gundogdu, O., Elmi, A., Bajaj-Elliott, M., Taylor, P.W., Wren, B.W., and Dorrell, N. (2012). Increase in *Campylobacter jejuni* invasion of intestinal epithelial cells under low-oxygen coculture conditions that reflect the in vivo environment. Infect. Immun. *80*, 1690–1698.

Mohammed, K.A.S., Miles, R.J., and Halablab, M.A. (2004). The pattern and kinetics of substrate metabolism of *Campylobacter jejuni* and *Campylobacter coli*. Lett. Appl. Microbiol. 39, 261–266.

Murphy, H., Cogan, T., Hughes, R., and Humphrey, T. (2011). Porcine intestinal epithelial responses to *Campylobacter* infection. Comp. Immunol. Microbiol. Infect. Dis. 34, 489–495.

Nappi, A.J., and Christensen, B.M. (2005). Melanogenesis and associated cytotoxic reactions: Applications to insect innate immunity. Insect Biochem. Mol. Biol. 35, 443–459.

Novik, V., Hofreuter, D., and Galán, J.E. (2009). Characterization of a *Campylobacter jejuni* VirK protein homolog as a novel virulence determinant. Infect. Immun. 77, 5428–5436.

Osman, D., and Cavet, J.S. (2008). Copper homeostasis in bacteria. Adv. Appl. Microbiol. *65*, 217–247.

Palyada, K., Threadgill, D., and Stintzi, A. (2004). Iron acquisition and regulation in *Campylobacter jejuni*. J. Bacteriol. *186*, 4714–4729.

Parkhill, J., Wren, B.W., Mungall, K., Ketley, J.M., Churcher, C., Basham, D., Chillingworth, T., Davies, R.M., Feltwell, T., Holroyd, S., *et al.* (2000). The genome sequence of the food-borne pathogen *Campylobacter jejuni* reveals hypervariable sequences. Nature *403*, 665–668.

Parsons, C.M. (1984). Influence of caecectomy and source of dietary fibre or starch on excretion of endogenous amino acids by laying hens. Br. J. Nutr. *51*, 541–548.

Peleg, A.Y., Jara, S., Monga, D., Eliopoulos, G.M., Moellering, R.C. Jr., and Mylonakis, E. (2009). *Galleria mellonella* as a model system to study *Acinetobacter baumannii* pathogenesis and therapeutics. Antimicrob. Agents Chemother. 53, 2605–2609.

Pickett, C.L., Cottle, D.L., Pesci, E.C., and Bikah, G. (1994). Cloning, sequencing, and expression of the *Escherichia coli* cytolethal distending toxin genes. Infect. Immun. *62*, 1046–1051.

Pogacar, M.S., Mihaljevic, R.R., Klancnik, A., Brumini, G., Abram, M., and. Mozina, S.S. (2009). Survival of stress exposed *Campylobacter jejuni* in the murine macrophage J774 cell line. Int. J. Food Microbiol. *129*, 68–73.

Poly, F., Read, T.D., Chen, Y.H., Monteiro, M., Serichantalergs, O., Pootong, P., Bodhidatta, L., Mason, C.J., Rockabrand, D., Baqar, S., *et al.* (2008). Characterization of two *Campylobacter jejuni* strains for use in volunteer experimental-infection studies. Infect. Immun. *76*, 5655–5667.

Pozos, T.C., and Ramakrishnan, L. (2004). New models for the study of *Mycobacterium*–host interactions. Curr. Opin. Immunol. *16*, 499–505.

Prajsnar, T.K., Cunliffe, V.T., Foster, S.J., and Renshaw, S.A. (2008). A novel vertebrate model of *Staphylococcus aureus* infection reveals phagovyte-dependent resistance of zebrafish to non-host specialized pathogens. Cell. Microbiol. *10*, 2312–2325.

Purdy, D., Buswell, C.M., Hodgson, A.E., McAlpine, K., Henderson, I., and Leach, S.A. (2000). Characterisation of cytolethal distending toxin (CDT) mutants of *Campylobacter jejuni*. J. Med. Microbiol. *49*, 473–479.

Ripabelli, G., Tamburro, M., Minelli, F., Leone, A., and Sammarco, M.L. (2009). Prevalence of virulence-associated genes and cytolethal distending toxin production in *Campylobacter* spp. isolated in Italy. Comp. Immunol. Microbiol. Infect. Dis.

Salah, I.B., Ghigo, E., and Drancourt, M. (2009). Free-living amoebae, a training field for macrophage resistance of *mycobacteria*. Clin. Microbiol. Infect. *15*(10), 894–905.

van der Sar, A.M., Musters, R.J.P., van Eeden, F.J.M., Appelmelk, B.J., Vandenbroucke-Grauls, C.M.J.E., and Bitter, W. (2003). Zebrafish embryos as a model host for the real time analysis of *Salmonella typhimurium* infections. Cell. Microbiol. 5(9), 601–611.

van der Sar, A.M., Stockhammer, O.W., van der Laan, C., Spaink, H.P., Bitter, W., and Meijer, A.H. (2006). MyD88 innate immune fiunction in a zebrafish embryo model. Infect. Immun. 74(4), 2436–2441.

Scott, D.A., and Kaper, J.B. (1994). Cloning and sequencing of the genes encoding *Escherichia coli* cytolethal distending toxin. Infect. Immun. 62(1), 244–251.

Senior, N.J., Bagnall, M.C., Champion, O.L., Reynolds, S.E., La Ragione, R.M., Woodward, M., Salguero, F.J., and Titball, R.W. (2011). *Galleria mellonella* as an infection model for *Campylobacter jejuni* virulence. J. Med. Microbiol. *60*, 661–669.

Shao, Q., Yang, B., Xu, Q., Li, X., Lu, Z., Wang, C., Huang, Y., Söderhäll, K., and Ling, E. (2012). Hindgut innate immunity andregualtion of fecal microbiota through melanisation in insects. J. Biol. Chem. *287*, 14270–14279.

Shultz, L.D. (1991). Immunological mutants of the mouse. Am. J. Anat. *191*, 303–311.

Sieger, D., Stein, C., Neifer, D., van der Sar A.M., and Leptin, M. (2009). The role of gamma interferon in innate immunity in the zebrafish embryo. Dis. Models Mech. *2*(11–12), 571–581.

Smart, J.P., Cliff, M.J., and Kelly, D.J. (2009). A role for tungsten in the biology of *Campylobacter jejuni*: tungstate stimulates formate dehydrogenase activity and is transported via an ultra-high afiinity ABC system distinct from the molybdate transporter. Mol. Microbiol. *74*(3), 742–757.

Smith, C.K., Kaiser, P., Rothwell, L., Humphrey, T., Barrow, P.A., and Jones, M.A. (2005). *Campylobacter jejuni*-induced cytokine responses in avian cells. Infect. Immun. *73*(4), 2094–2100.

van Soest, J.J., Stockhammer, O.W., Ordas, A., Bloemberg, G.V., Spaink, H.P., and Meijer, A.H. (2011). Comparison of static immersion and intravenous injection systems for exposure of zebrafish embryos to the natural pathogen *Edwardsiella tarda*. BMC Immunol. *12*, 58.

Stahl, M., Butcher, J., and Stintzi, A. (2012). Nutrient acquisition and metabolism by *Campylobacter jejuni*. Front. Cell. Infect. Microbiol. *2*.

Stahl, M., Friis, L.M., Nothaft, H., Liu, X., Li, J., Szymanski, C., and Stintzi, A. (2011). L-Fucose utilization provides *Campylobacter jejuni* with a competitive advantage. Proc. Natl. Acad. Sci. U.S.A. *108*(17), 7194–7199.

Stintzi, A., van Vliet, A.H.M., and Ketley, J.M. (2008). Iron metabolism, transport and regulation. In *Campylobacter*, Nachamkin, I., Szymanski, C., and Blaser, M.J., eds. (ASM Press, Washington, DC): p. 591.

Taveirne, M.E., Sikes, M.L., and Olson, J.W. (2009). Molybdenum and tungsten in *Campylobacter jejuni*: their physiological role and identification of separate transporters regulated by a single ModE-like protein. Mol. Microbiol. *74*(3), 758–771.

Thomas, M.T., Shepherd, M., Poole, R.K., van Vliet, A.H.M., Kelly, D.J., and Pearson, B.M. (2011). Two respiratory enzyme systems in *Campylobacter jejuni* NCTC 11168 contribute to growth on L-lactate. Environ. Microbiol. *13*(1), 48–61.

Trede, N.S., Zapata, A., and Zon, L.I. (2001). Fishing for lymphoid genes. Trends Immunol. *22*(6), 302–307.

Van Deun, K., Pasmans, F., Ducatelle, R., Flahou, B., Vissenberg, K., Martel, A., Van den Broeck, W., Van Immerseel, F., and Haesebrouck, F. (2008). Colonization strategy of *Campylobacter jejuni* results in persistent infection of the chicken gut. Vet. Microbiol. *130*, 285–297.

Velayudhan, J., and Kelly, D.J. (2002). Analysis of gluconeogenic and anaplerotic enzymes in *Campylobacter jejuni*: an essential role for phosphoenolpyruvate carboxykinase. Microbiology *148*, 685–694.

Velayudhan, J., Jones, M.A., Barrow, P.A., and Kelly, D.J. (2004). L-Serine catabolism via an oxygen-labile L-serine dehydratase is essential for colonization of the avian gut by *Campylobacter jejuni*. Infect. Immun. *72*(1), 260–268.

Verhoeff-Bakkenes, L., Hazeleger, W.C., de Jonge, R., and Zwietering, M.H. (2009). *Campylobacter jejuni*: a study on environmental conditions affecting culturability and in vitro adhesion/invasion. J. Appl. Microbiol. *106*, 924–931.

Vogel, H., Altincicek, B., Glöckner, G., and Vliconskas, A. (2011). A comprehensive transcriptome and immune-gene repertoire of the lepidopteran model host *Galleria mellonella*. BMC Genomics *12*.

Volkman, H.E., Pozos, T.C., Zheng, J., Davis, J.M., Rawls, J.F., and Ramakrishnan, L. (2010). *Tuberculous granuloma* induction via interaction of a bacterial secreted protein with host epithelium. Science *327*(5964), 466–469.

Wand, M.E., Mueller, C.M., Titball, R.W., and Michell, S.L. (2011). Macrophage and *Galleria mellonella* infection models reflect the virulence of naturally occurring isolates of *B. pseudomallei*, *B. thailandensis* and *B. oklahomensis*. BMC Microbiol. *11*.

Wang, L. (2010). Zebrafishing for tuberculosis infection. Protein Cell *1*, 309–311.

Wang, W., Liu, X., Gelinas, D., Ciruna, B., and Sun, Y. (2007). A fully automated robotic system for microinjection of zebrafish embryos. PLoS One *2*(9).

Watson, R.O., and Galán, J.E. (2008). *Campylobacter jejuni* survives within epithelial cells by avoiding delivery to lysosomes. PLOS Pathogens *4*(1).

Wienholds, E., Schulte-Merker, S., Walderich, B., and Plasterk, R.H.A. (2002). Target-selected inactivation of the zebrafish rag1 gene. Science *297*(5578), 99–102.

Wienholds, E., van Eeden, F., Kosters, M., Mudde, J., Plasterk, R.H.A., and Cuppen, E. (2003). Efficient target-selected mutagenesis in zebrafish. Genome Res. *13*(12), 2700–2707.

Wiles, T.J., Bower, J.M., Redd, M.J., and Mulvey, M.A. (2009). Use of zebrafish to probe the divergent virulence potentials and toxin requirements of extraintestinal pathogenic *Escherichia coli*. PLoS Pathogens *5*(12).

Woodall, C.A., Jones, M.A., Barrow, P.A., Hinds, J., Marsden, G.L., Kelly, D.J., Dorrell, N., Wren, B.W., and Maskell, D.J. (2005). *Campylobacter jejuni* gene expression in the chick cecum: Evidence for adaptation to a low-oxygen environment. Infect. Immun. *73*(8), 5278–5285.

Wright, J.A., Grant, A.J., Hurd, D., Harrison, M., Guccione, E., Kelly, D.J., and Maskell, D.J. (2009). Metabolite and transcriptome analysis of *Campylobacter jejuni in vitro* growth reveals a stationary-phase physiological switch. Microbiology *155*, 80–94.

Yoder, J.A., Nielsen, M.E., Amemiya, C.T., and Litman, G.W. (2002). Zebrafish as an immunological model system. Microb. Infect. *4*, 1469–1478.

Young, K.T., Davis, L.M., and DiRita, V.J. (2007). *Campylobacter jejuni*: molecular biology and pathogenesis. Nature Rev. Microbiol. *5*: 665–679.

Yu, J., Rossi, R., Hale, C., Goulding, D., and Dougan, G. (2009). Interaction of enteric bacterial pathogens with murine embryonic stem cells. Infect. Immun. *77*, 585–597.

Yun, J., Jeon, B., Barton, Y.-W., Plummer, P., Zhang, Q., and Ryu, S. (2008). Role of the DksA-like protein in the pathogenesis and divrse metaolic actvity of *Campylobacter jejuni*. J. Bacteriol. *190*(13), 4512–4520.

Zarkadis, I.K., Mastellos, D., and Lambris, J.D. (2001). Phylogenetic aspects of the complement system. Develop. Comp. Immunol. *25*, 745–762.

Part III

Molecular Epidemiology

Exploring the Population Dynamics of Multiple *Campylobacter* Strains in Poultry

16

Frances M. Colles and Martin C.J. Maiden

Abstract

In order to reduce the burden of human campylobacteriosis, there is an urgent need to better understand the natural ecology of *Campylobacter* amongst poultry flocks, so that effective on-farm intervention strategies may be designed. Broiler flocks contaminated early in the production cycle, as well as older birds, may be colonized by multiple strains of *Campylobacter*. There is evidence that *Campylobacter* strains are unevenly distributed amongst individual birds within a flock and that faecal samples are efficient in capturing diversity at a flock level. *Campylobacter* strains are not necessarily cleared from a flock and may persist for some time, meaning that the diversity of *Campylobacter* strains that can be isolated from a flock may increase over time. Groups of wild birds are typically colonized by a high diversity of *Campylobacter* strains, which may be a consequence of diversifying selection generated by host–pathogen interactions. However, studies of *Campylobacter* population dynamics over more than a few weeks are hindered by conventional sampling methods and the short life cycle of broiler chicken flocks. The precise mechanisms controlling the dynamics of *Campylobacter* colonization are likely to be multifactorial and remain to be fully understood.

Introduction

Although *Campylobacter* can be isolated from a variety of animal and environmental sources, genetic attribution models have focussed attention on the poultry industry as they estimate that up to 40–78% of human campylobacteriosis in industrialized countries results from consumption of contaminated chicken meat (Sheppard *et al.*, 2009). Biosecurity measures are currently the only means by which *Campylobacter* prevalence is targeted on-farm, yet, in a challenging environment these are prone to breaking down, with an average of 71.2% of European Union broiler flocks being positive for the presence of *Campylobacter* at slaughter (EFSA, 2010). An improved understanding of the natural ecology of *Campylobacter* within chicken flocks, is therefore important for the development of alternative and effective on-farm intervention strategies, with the ultimate aim of reducing human disease.

There are a number of inherent problems that must be overcome in the investigation of naturally occurring *Campylobacter* populations in chickens. There is evidence that both poultry flocks and individual birds within flocks can be colonized by multiple *Campylobacter* strains, including both *C. jejuni* and *C. coli* species, which vary in a number of different attributes (Calderon-Gomez *et al.*, 2009; Colles *et al.*, 2008b; Denis *et al.*, 2008; El-Shibiny *et al.*, 2005, 2007; Fearnley *et al.*, 2008; Jorgensen *et al.*, 2011; Rivoal *et al.*, 2005a; Wittwer *et al.*, 2005). There is little opportunity to investigate the dynamics of the colonization of poultry by *Campylobacter* over time, since conventionally reared broiler flocks are typically slaughtered at an immature age of just 5–6 weeks. In addition, conventional sampling strategies requiring culture of the caecal contents are inappropriate for repeated study of individuals since they require birds to be sacrificed. These challenges must be met to address fundamental questions including: how transmission of *Campylobacter* strains occurs within flocks; why the lag phase of colonization occurs in young chicks; why there is geographical variation in contamination rates of broiler flocks; and whether the prevalence and attributes of particular genotypes can be related to those found in human disease (Conlan *et al.*, 2007; Jonsson *et al.*, 2010; Jorgensen *et al.*, 2011).

Here, we present new data, and examine existing data, investigating the natural ecology of multiple *Campylobacter* strains amongst broiler chicken flocks, and discuss some of the issues regarding sampling strategies for future investigations.

Multiple carriage of *Campylobacter* strains within flocks

We followed the turnover of *Campylobacter* strains amongst two different free range broiler flocks on the same farm in Oxfordshire, UK (Fig. 16.1). Where repeated sampling of a flock has been undertaken previously, *Campylobacter* strains have appeared to replace each other in waves. However, over an extended period of sampling in our study there was evidence that *Campylobacter* strains were not necessarily cleared from the flock during their life cycle, but that they co-existed with other strains, with the diversity of strains present in the flock increasing over time (Calderon-Gomez *et al.*, 2009; Colles *et al.*, 2008b; El-Shibiny *et al.*, 2007; Messens *et al.*, 2009; Wallace *et al.*, 1998). There is likely to be a combination of mechanisms controlling the prevalence of different *Campylobacter* strains. To date, research has concentrated on differences in colonization and invasion potential of strains, competitive exclusion of different strains at protein binding sites, changing immunity and microbiota of the chicken host, and strain specific phage activity (Calderon-Gomez *et al.*, 2009; El-Shibiny *et al.*, 2005, 2007; Coote *et al.*, 2007; Fearnley *et al.*, 2008; Konkel *et al.*, 2007; Scott *et al.*, 2007; Skanseng *et al.*, 2007). In common with findings from a Norwegian study, the flock that was colonized at an earlier age had highest diversity of genotypes at the time of slaughter (Johnsen *et al.*, 2006).

A more substantial study of a free range broiler breeder flock over a year found the diversity index of *Campylobacter* strains to be related primarily to age of the flock rather than climatic factors, forming an asymptotic trend (Colles *et al.*, 2011). The results were consistent with the flock being exposed to an increasing number of *Campylobacter* types over time and also with changes in host immunity as the birds matured. The high diversity of genotypes was similar to that seen amongst wild birds and was perhaps the reason that *Campylobacter* colonization remained relatively stable overtime (Colles *et al.*, 2008a; Skanseng *et al.*, 2007). It is possible that dynamics of *Campylobacter* within individual birds influenced dynamics of flock colonization, for example, some birds may become chronic shedders allowing particular strains to persist within a flock (Achen *et al.*, 1998). In the example shown (Fig. 16.1), only one *C. coli* genotype, ST-814, was common to both flocks but the pattern of isolation was different, it being identified at a much earlier age and on repeated occasions in the second flock. These results are consistent with those of an experimental study that found transmission of *Campylobacter* strains amongst a small group of birds to be associated with a number of stochastic events, and thus dominating strains were unpredictable (Grant *et al.*, 2005).

Studies of commercial broiler flocks report between 40% and 77% of chickens to be colonized by more than one *Campylobacter* genotype, with one UK study finding 2.8% (4/147) to be co-colonized with the different species *C. jejuni* and *C. coli* (Jorgensen *et al.*, 2011; Messens *et al.*, 2009). *C. coli*, in particular, has been associated with colonization later in the chicken growth cycle, and with older birds such as free-range flocks (Colles *et al.*, 2008b; De Cesare *et al.*, 2008; Denis *et al.*, 2008; El-Shibiny *et al.*, 2005; Jorgensen *et al.*, 2011). Between one and six *Campylobacter* strains have been recovered from flocks by sampling live birds, which may increase post-slaughter when examining carcass rinses, reflecting cross- contamination in the abattoir (De Cesare *et al.*, 2008; Denis *et al.*, 2008; Ridley *et al.*, 2011; Rivoal *et al.*, 2005b; Wittwer *et al.*, 2005).

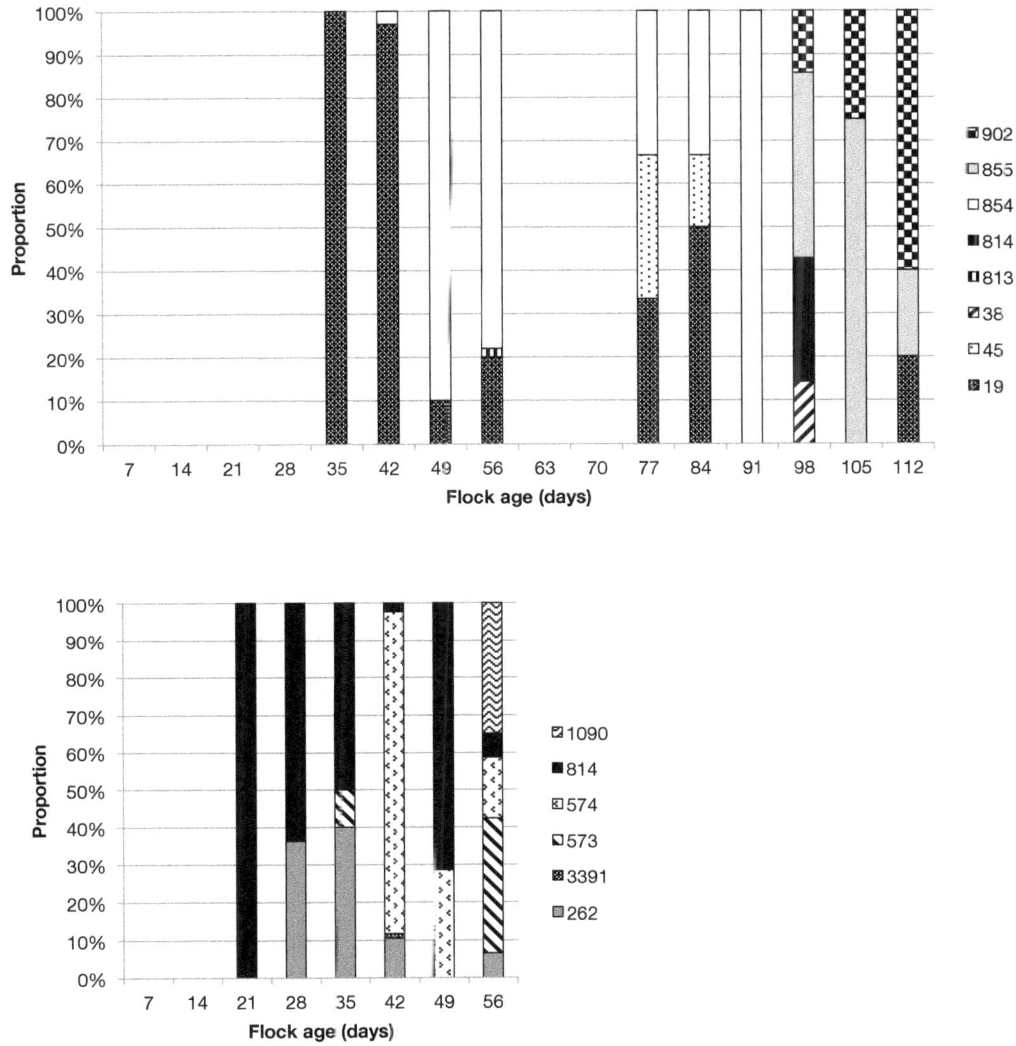

Figure 16.1 The *Campylobacter* MLST types (STs) isolated over time from two commercial free-range broiler flocks using faecal samples up to 28 days of age and swabs of the cloacal opening from 35 days of age are shown. One hundred samples were collected and cultured individually on mCCDA from both flocks on days 28, 42 and 56; 10 swabs were collected on the remainder of the sample dates (no samples were collected from flock 1 on days 63 and 70, and none were collected from flock 57 after 56 days of age). The free-range flocks, consisting of 1340 birds, were raised on the same farm but separated by six months, flock 1 being sampled in January to April and flock 57 being sampled during October and November, 2003. Flock 1 was reduced in size to approximately 100 birds at 56 days of age and moved to an indoor barn. Further details of the study are available (Colles *et al*., 2008b). Different STs are represented by different shading.

Multiple carriage of *Campylobacter* strains amongst individual birds

The diversity of *Campylobacter* strains recovered when up to 30 colonies typed from individual broiler breeder birds aged 57 weeks was investigated. These data showed that majority of birds (14/19, 73.7%) were colonized by multiple (between two and seven) *Campylobacter* strains, often including both *C. jejuni* and *C. coli* species, or a hybrid of the two containing introgressed alleles that could not be separated upon repeated subculture (Fig. 16.2) (Sheppard *et al*., 2011). The issue of multiple carriage of *Campylobacter* strains

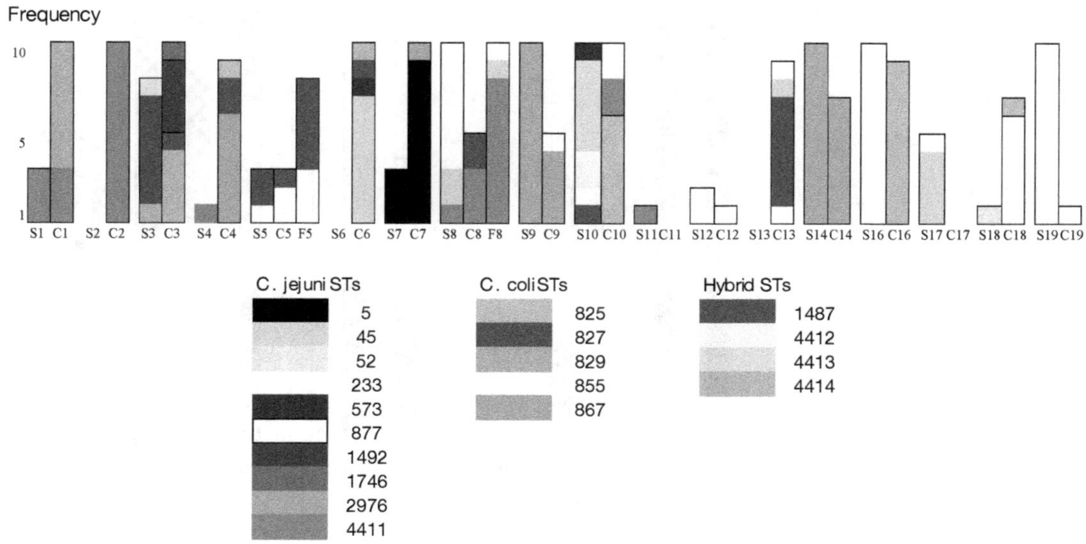

Figure 16.2 The *Campylobacter* MLST types isolated from swab of the cloacal opening (S), caecal contents (C) and faecal samples (F), grouped by individual bird are shown. (No *Campylobacter* was recovered from Bird 15). The birds were from the same commercial free-range broiler breeder flock aged 57 weeks, detailed in Table 16.1. Following direct culture onto mCCDA and depending on growth, up to ten *Campylobacter* colonies were sub-cultured and typed. Different STs are represented by different colours. A colour version of this figure is available in the plate section at the back of the book.

amongst individual birds within commercial broiler flocks has rarely been addressed; however, other studies in Italy and the UK testing between five and ten colonies from individual samples, identified between two and five different *Campylobacter* strains from caecal samples and between four and eight strains from carcass rinse samples (Colles *et al.*, 2010; De Cesare *et al.*, 2008).

Variation in the number and type of colonizing strains amongst individual birds was observed; the same *Campylobacter* strain was not always isolated from different birds in the same proportions. In another study, examining only *Campylobacter* load, large variation was observed among individual birds within the same flock, with loads varying by 6.3 log cfu/g caecal contents (Hansson *et al.*, 2010). The longitudinal study of multiple *Campylobacter* strains amongst individual birds on commercial farms has yet to be fully evaluated, however, there is evidence that single, presumed to be dominant, *Campylobacter* strains show a quick turnover amongst wild birds captured on multiple occasions (Colles *et al.*, 2009; Waldenstrom *et al.*, 2007). Individual broiler chickens in experimental conditions have also been observed to shed *Campylobacter* loads in cyclic patterns (Achen *et al.*, 1998).

Multiple carriage of *Campylobacter* strains and implications for sampling

Faecal shedding and coprophagy are thought to be important factors in the transmission of *Campylobacter* amongst birds in a flock (Newell and Fearnley, 2003). Thus, the possible differential survival of *Campylobacter* strains outside the host means that the *Campylobacter* population in the caecum does not necessarily reflect that which is transmissible to other birds in a flock. A comparison of different sampling methods demonstrated that for a free-range broiler breeder flock the greatest number of genotypes, of both *C. jejuni* and *C. coli*, was recovered using fresh faecal samples rather than caecal contents or swabs of the cloacal opening, although the diversity index for each method was not significantly different, ranging from $D=0.89$ to $D=0.91$ (Table 16.1). A separate study of 40 broiler flocks in the UK found that enrichment culture of faecal

Table 16.1 The frequency of species and MLST types identified amongst *Campylobacter* isolates collected from a 57-week-old commercial free-range broiler breeder flock, using different sampling methods, is shown. Swabs of the cloacal opening and caecal contents were taken from the same 19 individual birds; two faecal samples were also collected from two of the individual birds, and another 18 were collected from the flock of 500 birds as a whole, sampling separate regions throughout the house. Following direct culture onto mCCDA and depending on growth, up to ten *Campylobacter* colonies were subcultured and typed giving a total of 89 isolates from swabs, 117 isolates from caecal contents and 151 isolates from faecal samples

			Number of chickens colonized		
Species	Clonal complex	Sequence type	Swab	Caecal contents	Faeces
C. jejuni	45	45	2	2	2
		233	6	6	5
	52	52	1	0	4
	283	267	0	0	1
		2975	0	0	1
		5167	0	0	1
	353	5	1	1	4
		2976	1	3	2
	443	51	0	0	1
	573	573	0	1	0
		1746	0	1	1
	Unassigned	877	1	1	7
		1492	2	5	2
		4411	1	3	3
C. coli	828	825	0	4	1
		827	0	2	0
		829	1	1	1
		855	0	1	0
		867	0	1	1
		2273	0	1	1
		3616	0	0	1
		5171	0	0	1
	Unassigned	3095	0	0	1
Hybrid	1150	1487	2	5	2
		4412	1	0	0
		4413	1	0	0
		4414	0	1	0
Total number of different STs			12	17	21

samples was the most sensitive method for the detection of *C. coli*, and boot swabs used by farmers whilst checking flocks (and thus providing the ultimate in convenience), being most sensitive for detection of *C. jejuni*, in comparison with caecal contents samples (Vidal *et al.*, 2012). Boot swabs or fresh faecal samples provide a non-invasive and practical way of sampling large flocks of birds for *Campylobacter* on a longitudinal basis.

Culture of caecal contents is considered to be the gold standard sampling technique, since this is the region of the gut found to contain the highest numbers of *Campylobacter*; however, this renders longitudinal sampling of individuals impossible, since birds are required to be sacrificed (Achen *et al.*, 1998; Wallace *et al.*, 1998). Results comparing *Campylobacter* strains recovered from swabs of the cloacal opening (and thus repeatable over time) and from caecal contents of the same individual birds demonstrate that at least one *Campylobacter* strain, as defined by MLST type, was recovered from both sample types for half (9 of 18) of the birds tested (Fig. 16.2). It was not possible to recover all STs using one method, and whilst swabs may be perceived as less sensitive, culture of caecal contents was also problematic since the heavy growth of coliform bacteria may have prevented the isolation of *Campylobacter* on at least two occasions. Given the uneven distribution of *Campylobacter* strains within a flock, together with practical and financial constraints, the optimal sampling strategy to obtain broad representation of *Campylobacter* subtypes at flock level is therefore taking samples from more birds but with fewer colony isolations per flock, rather than few birds with many colony isolations.

Future directions

Investigating the epidemiology, transmission and population dynamics of *Campylobacter* amongst poultry is very complex when multiple strains are considered; however, the usefulness of this approach has been demonstrated in the investigation of outbreak sources of human disease (Forbes *et al.*, 2009; Mullner *et al.*, 2010). Conventional culture-based microbiology methods have imposed limits on the number of bacterial colonies on an agar plate, or number of samples per flock that can efficiently be tested, but it cannot be certain that a greater diversity of strains will not be discovered if they are looked for. This may be particularly relevant in older birds or in large flocks of 20,000 birds or more. Rapidly improving molecular methods and metagenomics approaches mean that broader investigation is becoming increasingly feasible. MLST data are compatible with next generation sequencing, and indeed, 'virtual' MLST using full genome sequence and any number of 'schemes' made up of genes of interest, is becoming increasingly efficient and cost effective (Colles and Maiden, 2012). Depending on the question to be answered, the size and scale of the poultry industry may mean very large sampling efforts are required to effectively evaluate the *Campylobacter* population at a particular point in time. However, it is sometimes the rare or unusual strains arising from such investigations that give most benefit in comparative studies downstream.

Longitudinal and population sampling of flocks has often been neglected and is difficult to perform on a sufficiently large scale, and there remain many fundamental questions to be answered that will contribute to the modelling *Campylobacter* colonization of chickens. Although *Campylobacter* strains have been found to vary in colonization potential and virulence in the laboratory, it is not known how strains interact in the field, for example, it is not known whether or not the same strains always become dominant within a flock, on a farm or within a company. Further questions include whether *Campylobacter* strains consistently vary with respect to surviving different stages in the processing plant or whether differing strain prevalence can be correlated with seasonal and geographical factors. More detailed provenance information for specimens provides added value for any experimental system using *Campylobacter* strains in the laboratory.

Conclusion

There is evidence that while some housed flocks are colonized by a single *Campylobacter* strain by the age of slaughter, between 40% and 77% of flocks are colonized by multiple strains, and some may also be co-colonized by both *C. jejuni* and *C. coli.* Flocks that are colonized earlier in the production cycle, as well as older birds such as those in a free-range system, are more likely to be colonized by multiple strains of *Campylobacter.* Longitudinal sampling of flocks gives benefits in understanding the population dynamics of different *Campylobacter* strains, and non-invasive sampling methods using faecal material that are necessarily required, show good efficiency in capturing the diversity at flock level. *Campylobacter* strains are not always cleared from a flock, with many persisting over time, and both host and bacterial mechanisms controlling multiple carriage remain to be fully elucidated. Detailed population sampling using sequence based typing methods is urgently needed in order that experimental and modelling approaches are better informed and the ultimate goal of reducing human infection is reached most efficiently.

References

Achen, M., Morishita, T.Y., and Ley, E.C. (1998). Shedding and colonization of *Campylobacter jejuni* in broilers from day-of-hatch to slaughter age. Avian Dis. *42*, 732–737.

Calderon-Gomez, L.I., Hartley, L.E., McCormack, A., Ringoir, D.D., and Korolik, V. (2009). Potential use of characterised hyper-colonising strain(s) of *Campylobacter jejuni* to reduce circulation of environmental strains in commercial poultry. Vet. Microbiol. *134*, 353–361.

Colles, F.M., and Maiden, M.C. (2012). *Campylobacter* sequence typing: applications and future prospects. Microbiology *158*, 2695–2709.

Colles, F.M., Dingle, K.E., Cody, A.J., and Maiden, M.C. (2008a). Comparison of *Campylobacter* populations in wild geese with those in starlings and free-range poultry on the same farm. Appl. Environ. Microbiol. *74*, 3583–3590.

Colles, F.M., Jones, T.A., McCarthy, N.D., Sheppard, S.K., Cody, A.J., Dingle, K.E., Dawkins, M.S., and Maiden, M.C. (2008b). *Campylobacter* infection of broiler chickens in a free-range environment. Environ. Microbiol. *10*, 2042–2050.

Colles, F.M., McCarthy, N.D., Howe, J.C., Devereux, C.L., Gosler, A.G., and Maiden, M.C. (2009). Dynamics of *Campylobacter* colonization of a natural host, *Sturnus vulgaris* (European starling). Environ. Microbiol. *11*, 258–267.

Colles, F.M., McCarthy, N., Sheppard, S.K., Layton, R., and Maiden, M.C.J. (2010). Comparison of *Campylobacter* populations isolated from a free-range broiler flock before and after slaughter. Int. J. Food Microbiol *137*, 259–264.

Colles, F.M., McCarthy, N.D., Layton, R., and Maiden, M.C. (2011). The prevalence of *Campylobacter* amongst a free-range broiler breeder flock was primarily affected by flock age. PLoS One *6*, e22825.

Conlan, A.J., Coward, C., Grant, A.J., Maskell, D.J., and Gog, J.R. (2007). *Campylobacter jejuni* colonization and transmission in broiler chickens: a modelling perspective. J. R. Soc. Interface *4*, 819–829.

Coote, J.G., Stewart-Tull, D.E., Owen, R.J., Bolton, F.J., Siemer, B.L., Candlish, D., Thompson, D.H., Wardlaw, A.C., On, S.L., Candlish, A., *et al.* (2007). Comparison of virulence-associated in vitro properties of typed strains of *Campylobacter jejuni* from different sources. J. Med. Microbiol. *56*, 722–732.

De Cesare, A., Parisi, A., Bondioli, V., Normanno, G., and Manfreda, G. (2008). Genotypic and phenotypic diversity within three *Campylobacter* populations isolated from broiler ceca and carcasses. Poult. Sci. *87*, 2152–2159.

Denis, M., Rose, V., Huneau-Salaun, A., Balaine, L., and Salvat, G. (2008). Diversity of pulsed-field gel electrophoresis profiles of *Campylobacter jejuni* and *Campylobacter coli* from broiler chickens in France. Poult. Sci. *87*, 1662–1671.

EFSA (2010). Analysis of the baseline survey on the prevalence of *Campylobacter* in broiler batches and of *Campylobacter* and *Salmonella* on broiler carcasses in the EU, 2008. EFSA Journal *8*, 1522–1654.

El-Shibiny, A., Connerton, P.L., and Connerton, I.F. (2005). Enumeration and diversity of campylobacters and bacteriophages isolated during the rearing cycles of free-range and organic chickens. Appl. Environ. Microbiol. *71*, 1259–1266.

El-Shibiny, A., Connerton, P.L., and Connerton, I.F. (2007). *Campylobacter* succession in broiler chickens. Vet. Microbiol. *125*, 323–332.

Fearnley, C., Manning, G., Bagnall, M., Javed, M.A., Wassenaar, T.M., and Newell, D.G. (2008). Identification of hyperinvasive *Campylobacter jejuni* strains isolated from poultry and human clinical sources. J. Med. Microbiol. *57*, 570–580.

Forbes, K.J., Gormley, F.J., Dallas, J.F., Labovitiadi, O., MacRae, M., Owen, R.J., Richardson, J., Strachan, N.J., Cowden, J.M., Ogden, I.D., *et al.* (2009). *Campylobacter* immunity and coinfection following a large outbreak in a farming community. J. Clin. Microbiol. *47*, 111–116.

Grant, A.J., Coward, C., Jones, M.A., Woodall, C.A., Barrow, P.A., and Maskell, D.J. (2005). Signature-tagged transposon mutagenesis studies demonstrate the dynamic nature of cecal colonization of 2-week-old chickens by *Campylobacter jejuni*. Appl. Environ. Microbiol. *71*, 8031–8041.

Hansson, I., Pudas, N., Harbom, B., and Engvall, E.O. (2010). Within-flock variations of *Campylobacter* loads in caeca and on carcasses from broilers. Int. J. Food Microbiol. *141*, 51–55.

Johnsen, G., Kruse, H., and Hofshagen, M. (2006). Genetic diversity and description of transmission routes for *Campylobacter* on broiler farms by amplified-fragment length polymorphism. J. Appl. Microbiol. *101*, 1130–1139.

Jonsson, M.E., Norstrom, M., Sandberg, M., Ersbøll, A.K., and Hofshagen, M. (2010). Space-time patterns of *Campylobacter* spp. colonization in broiler flocks, 2002–2006. Epidemiol. Infect. *138*, 1336–1345.

Jorgensen, F., Ellis-Iversen, J., Rushton, S., Bull, S.A., Harris, S.A., Bryan, S.J., Gonzalez, A., and Humphrey, T.J. (2011). Influence of season and geography on *Campylobacter jejuni* and *C. coli* subtypes in housed broiler flocks reared in Great Britain. Appl. Environ. Microbiol. *77*, 3741–3748.

Konkel, M.E., Christensen, J.E., Dhillon, A.S., Lane, A.B., Hare-Sanford, R., Schaberg, D.M., and Larson, C.L. (2007). *Campylobacter jejuni* strains compete for colonization in broiler chicks. Appl. Environ. Microbiol. *73*, 2297–2305.

Messens, W., Herman, L., De Zutter, L., and Heyndrickx, M. (2009). Multiple typing for the epidemiological study of contamination of broilers with thermotolerant *Campylobacter*. Vet. Microbiol. *138*, 120–131.

Mullner, P., Collins-Emerson, J.M., Midwinter, A.C., Carter, P., Spencer, S.E.F., van der Logt, P., Hathaway,

S., and French, N.P. (2010). Molecular Epidemiology of *Campylobacter jejuni* in a Geographically Isolated Country with a Uniquely Structured Poultry Industry. Appl. Environ. Microbiol. *76*, 2145–2154.

Newell, D.G., and Fearnley, C. (2003). Sources of *Campylobacter* colonization in broiler chickens. Appl. Environ. Microbiol. *69*, 4343–4351.

Ridley, A.M., Morris, V.K., Cawthraw, S.A., Ellis-Iversen, J., Harris, J.A., Kennedy, E.M., Newell, D.G., and Allen, V.M. (2011). Longitudinal molecular epidemiological study of thermophilic campylobacters on one conventional broiler chicken farm. Appl. Environ. Microbiol. *77*, 98–107.

Rivoal, K., Ragimbeau, C., Salvat, G., Colin, P., and Ermel, G. (2005a). Genomic diversity of *Campylobacter coli* and *Campylobacter jejuni* isolates recovered from free-range broiler farms and comparison with isolates of various origins. Appl. Environ. Microbiol. *71*, 6216–6227.

Rivoal, K., Ragimbeau, C., Salvat, G., Colin, P., and Ermel, G. (2005b). Genomic diversity of *Campylobacter coli* and *Campylobacter jejuni* isolates recovered from free-range broiler farms and comparison with isolates of various origins. Appl. Environ. Microbiol. *71*, 6216–6227.

Scott, A.E., Timms, A.R., Connerton, P.L., El-Shibiny, A., and Connerton, I.F. (2007). Bacteriophage influence *Campylobacter jejuni* types populating broiler chickens. Environ. Microbiol. *9*, 2341–2353.

Sheppard, S.K., Dallas, J.F., Strachan, N.J., MacRae, M., McCarthy, N.D., Wilson, D.J., Gormley, F.J., Falush, D., Ogden, I.D., Maiden, M.C., *et al.* (2009). *Campylobacter* genotyping to determine the source of human infection. Clin. Infect. Dis. *48*, 1072–1078.

Sheppard, S.K., McCarthy, N.D., Jolley, K.A., and Maiden, M.C.J. (2011). Introgression in the genus *Campylobacter*: generation and spread of mosaic alleles. Microbiology *157*, 1066–1074.

Skanseng, B., Trosvik, P., Zimonja, M., Johnsen, G., Bjerrum, L., Pedersen, K., Wallin, N., and Rudi, K. (2007). Co-infection dynamics of a major food-borne zoonotic pathogen in chicken. PLoS Pathog. *3*, e175.

Vidal, A.B., Rodgers, J., Arnold, M., and Clifton-Hadley, F. (2012). Comparison of different sampling strategies and laboratory methods for the detection of *C. jejuni* and *C. coli* from Broiler Flocks at Primary Production. Zoonoses Pub. Health *60*, 412–425.

Waldenstrom, J., On, S.L., Ottvall, R., Hasselquist, D., and Olsen, B. (2007). Species diversity of campylobacteria in a wild bird community in Sweden. J. Appl. Microbiol. *102*, 424–432.

Wallace, J.S., Stanley, K.N., and Jones, K. (1998). The colonisation of turkeys by thermophilic campylobacters. J. Appl. Microbiol. *85*, 224–230.

Wittwer, M., Keller, J., Wassenaar, T.M., Stephan, R., Howald, D., Regula, G., and Bissig-Choisat, B. (2005). Genetic diversity and antibiotic resistance patterns in a *Campylobacter* population isolated from poultry farms in Switzerland. Appl. Environ. Microbiol. *71*, 2840–2847.

Evolution of *Campylobacter* Species in New Zealand

17

Nigel French, Shoukai Yu, Patrick Biggs, Barbara Holland, Paul Fearnhead, Barbara Binney, Andrew Fox, Dai Grove-White, Jessica W. Leigh, William Miller, Petra Muellner and Philip Carter

Abstract

New Zealand is an isolated archipelago in the South-West Pacific with a unique fauna and flora, a feature partly attributable to it being the last sizable land mass to be colonized by man. In this chapter we test the hypothesis that different periods in the history of New Zealand – from pre-history to post-Polynesian/pre-European arrival and post-European – have left a signature on the evolution of *Campylobacter* spp. that is detectable in the extant population today. In order to explore and test this hypothesis we describe a brief, relevant history of New Zealand, followed by three studies: (1) an examination of recently discovered populations of new *Campylobacter* spp. in endemic wild birds and environmental water; (2) a comparison of isolates from humans, ruminants and poultry in the United Kingdom and New Zealand and; (3) an examination of the phylogeny of samples of multilocus sequence typing (MLST) genotype ST-474, the strain responsible for a multi-year poultry-associated epidemic in New Zealand.

History of New Zealand, including domestic and wild animal introductions, from pre-human to post European colonization

In order to understand some of the key features that may have shaped the evolution and population structure of *Campylobacter* spp. in New Zealand it is important to be aware of its history and geographical isolation and the pattern of animal introductions that have led to the domestic and wild animal populations that are hosts to the *Campylobacter* genus in the present day. An estimated 82–85 million years ago the ancient landmass of Zealandia separated from the combined coast of Australia-Antarctica that formed part of the supercontinent Gondwana, eventually forming an isolated archipelago that is now modern New Zealand. The subsequent long period of isolation led to the evolution of a unique flora and fauna and created an opportunity for endemic flightless species of birds, such as the takahē and kiwi to evolve (Trewick, 1997; Wallis and Trewick, 2009; Trewick and Gibb, 2010).

When humans arrived about 800 years ago there were no terrestrial mammals, with the exception of two bat species, although an early mammalian fossil has been found (Worthy *et al.*, 2006; Jolivet and Verma, 2010). These Polynesian voyagers arrived and introduced a species of rat (the kiore, *Rattus exulans*) and the dog (the kuri, *Canis familiaris* Linnaeus) (Anderson and Clark, 2001; Anderson, 2009). It is not thought that either pigs or poultry arrived with the Polynesians prior to European contact (Simmons, 1969; Gongora *et al.*, 2004; Storey and Ladefoged, 2008). Early European explorers, starting with Captain James Cook introduced sheep and other domesticated animals to New Zealand (Diez-Tascon *et al.*, 2000; Parkes, 2001). From 1840 onwards, European settlers accelerated the rate of livestock introduction with the advent of a system of pastoral agriculture (Fairweather, 1985; Patterson, 1991).

Australia was the main source of early livestock introductions, which led to the genesis of the modern New Zealand agricultural system (Evans,

1969; Peden, 2002). The pattern of introduction for sheep and cattle was similar: very large numbers were imported in the 1860s, but relatively few in subsequent decades (Fig. 17.1). The records show that from 1860 to 1959 approximately 650,000 sheep were imported, compared to approximately 63,000 cattle imported from 1860–1979. Current day estimates indicate the population of sheep is approximately 31.1 million, which is much lower than the peak of over 70 million in 1982, and the current cattle population is approximately 10 million.

Poultry were introduced to New Zealand from the time of James Cook (Thomson, 1922), but in contrast to cattle and sheep, the numbers of poultry introduced were relatively low until the 1960s. Between 1880 and 1939, approximately 21,000 poultry importations were recorded, compared to approximately 290,000 in the 1960s and 1.3 million in the 1970s, when the industry changed from small scale production to a modern high production system (Stone, 1972) (Fig. 17.1).

During the 1860s to 1880s regional 'acclimatisation societies' and some individuals introduced many exotic wild birds and game animals into New Zealand (Thomson, 1922). In total some 82 species of mammal, bird and fish have established populations in New Zealand since 1840, as well as numerous accidentally introduced invertebrates (Atkinson and Cameron, 1993). Not all introductions have extant populations; for example, of the more than 130 introduced wild bird species only about a quarter are believed to be surviving in the wild today (Atkinson and Cameron, 1993; Green, 1997). Europeans introduced 54 mammalian species for a variety of reasons including livestock for agricultural production, meat, fur, domestic pet and game hunting; of these, 32 species have established populations lasting 25 years or more (King and Barrett, 2005).

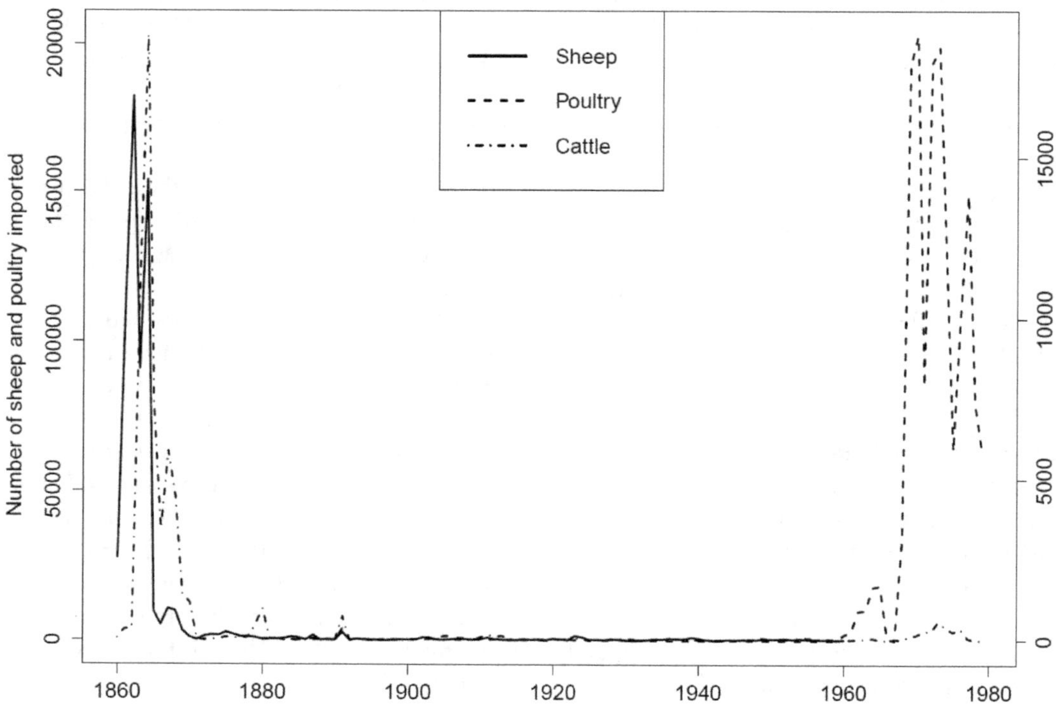

Figure 17.1 Numbers of cattle, sheep and poultry introduced into New Zealand between 1860 and 1980.

Theoretical effect of isolation and introduction on the population of *Campylobacter* spp.

The recent estimated divergence times of the members of the genus *Campylobacter* indicate a phylogenetic split around 19,000–62,600 years ago at the root of the tree, where the non-thermotolerant species *C. fetus* branches away from other members of the genus containing thermotolerant *Campylobacter* spp. such as *C. jejuni*, *C. coli*, *C. lari*, *C. helveticus*, *C. upsaliensis* and *C. insulaenigrae* (Wilson *et al.*, 2009; Sheppard *et al.*, 2010b). This is relatively recent compared to the geological isolation and natural history of New Zealand. It is likely that early introductions of descendants of the common ancestor of the thermotolerant campylobacters were from migratory wild birds, after the split between non-thermotolerant and thermotolerant species, and these may have colonized populations of native wild birds and evolved into new species through a process of genetic drift and/or natural selection. In this chapter we hypothesize that existing descendants of these early introductions manifest themselves as host-adapted, highly divergent lineages/distinct species that are found exclusively in endemic, non-migratory (e.g. flightless) wild birds. We explore this by studying populations of divergent, previously unreported *Campylobacter* spp. lineages isolated from endemic members of the rail family in New Zealand (Study 1). Similarly, native wild bird species that have become established in New Zealand post-human colonization, but without the help of man, may be expected to be colonized with more recently diverged species of *Campylobacter* and, furthermore, with variants of these species that reflect their time of introduction. Other studies have shown strong evidence of genetic structuring and host-association of *C. jejuni* in wild birds (Sheppard *et al.*, 2011). We explore this by considering populations of *C. jejuni* – a species that was previously estimated to show genetic variation that dates back just 400 years (Wilson *et al.*, 2009) – in recently introduced native wild birds.

Migratory wild birds are likely to have continued to introduce campylobacters into New Zealand to the present day, and potentially export evolved strains to other parts of the world. However, post human colonization (primarily post-European colonization), the introduction of domestic and wild animal populations is likely to have introduced substantial new populations of livestock-associated *Campylobacter*, particularly in the 19th century. These relatively isolated and rapidly expanding animal populations are likely to have played host to an equally rapidly expanding population of campylobacters, that have subsequently diverged from their ancestral, founder populations to form new strains and lineages that, provided they are not subsequently re-exported to their source destinations, will be unique to New Zealand. We examine this by comparing populations of *C. jejuni* from the same hosts in New Zealand and the United Kingdom, isolated over the same time period and test the null hypothesis that the rapid movement of campylobacters around the world in domestic animals, and possibly man, has swamped the effect of evolution to result in a homogeneous 'everything is everywhere' distribution of genotypes (Study 2). We also examine, in detail, the population genetics of a poultry-associated strain of *C. jejuni* that was responsible for a high proportion of human cases of campylobacteriosis in New Zealand, but is relatively rare elsewhere in the world, and question whether the genetic variation observed in this strain was generated in New Zealand and, if so, over what time period (Study 3)?

Molecular epidemiology of human campylobacteriosis in New Zealand

Although the focus of this chapter is on the evolution of *Campylobacter* spp. in multiple hosts in New Zealand, it is also important to consider this in the context of the molecular epidemiology of human campylobacteriosis. New Zealand has historically had a high rate of campylobacteriosis notifications (Baker *et al.*, 2006, 2007). Results from a molecular-based surveillance programme in a sentinel site provided strong evidence that a large proportion of human cases were linked to poultry meat consumption (Mullner *et al.*, 2009), adding to existing evidence implicating poultry as the main contributor to the human disease burden (Eberhart-Phillips *et al.*, 1997). Regulatory and voluntary control strategies were implemented along the poultry supply chain. This resulted in a 50% reduction in disease incidence in 2008 compared with the previous high level between 2002 and 2006 (Mullner *et al.*, 2010b; Sears *et al.*, 2011). Furthermore, in the New Zealand sentinel surveillance site, a dominant poultry-associated MLST sequence type of *C. jejuni* (ST-474) was identified that, to date, has rarely been reported in other countries. Before the implementation of the poultry interventions, ST-474 accounted for 30% of human cases in the sentinel surveillance site (Mullner *et al.*, 2009), but in 2010–11 was isolated from fewer than 5% of cases (Muellner *et al.*, 2013).

Study 1: population genetics of *Campylobacter* spp. isolated from native non-migratory wild bird species in New Zealand

In this section we describe the population genetics of campylobacters recently isolated from surface water and native members of the rail family in New Zealand. We ask the question: do these represent ancient lineages that have evolved in relative isolation in New Zealand? In order to test this we use single locus, multilocus and full genome data, and describe the phylogeny using phylogenetic network methods. Further, we estimate the divergence times of these proposed new species in the context of the history of New Zealand using Bayesian evolutionary analyses (Drummond and Rambaut, 2007).

Phylogeny based on housekeeping genes

The first indication that previously unidentified lineages of *Campylobacter* spp. were present in New Zealand followed the isolation of unusual campylobacters, originally of unknown origin, in a study of environmental surface waters (Carter *et al.*, 2009). Parallel studies of the carriage of *Campylobacter* spp. in multiple species of wild birds and domestic animals revealed a collection of atypical isolates recovered only from river water and three native members of the Rallidae family (takahē, weka and pukeko) that clustered to form a distinct group based on the sequences of housekeeping genes used in the MLST scheme developed for *C. jejuni/C. coli* (Dingle *et al.*, 2001). These are provisionally called *C.* sp. nov. 1 and *C.* sp. nov. 2 and are illustrated in the *glnA* phylogeny shown in Fig. 17.2, which, interestingly, places this group relatively close to the common ancestor of *C. jejuni* and *C. coli*. *C.* sp. nov. 1 to date has only been isolated from water ($n=4$), takahē ($n=90$), weka ($n=8$) and pukeko ($n=10$), *C.* sp. nov. 2 has only been isolated from weka ($n=4$). The takahē (*Porphyrio hochstetteri*) is an 'endangered' flightless endemic rail that was thought to be extinct until a remnant population was discovered in 1940, the weka (*Gallirallus australis*) is a 'vulnerable' flightless endemic rail and the pukeko (*Porphyrio porphyrio melanotus*) is a 'non-threatened' member of the same genus as the takahē, and is the same subspecies as the Australian purple swamp hen. The limited subfossil evidence suggests that the pukeko may be a recent self introduction, probably from Australia within the last 500 years – all fossil occurrences are in sites younger than 400 years, and there is no evidence of pukeko in mainland New Zealand prior to the arrival of Maori (Holdaway *et al.*, 2001; Trewick and Worthy, 2001.) A third cluster of isolates provisionally named *C.* sp. nov. 3, was isolated from water ($n=2$) and European starlings ($n=4$) in the city of Palmerston North, North Island.

These initial studies were based on a multiple isolates and a small number of housekeeping genes. The availability of full genome sequencing has enabled us to examine a small subset of these more closely, exploring the relationship between these proposed new 'species' and other campylobacters, including *C. jejuni* and *C. coli* isolated from multiple hosts in New Zealand.

Phylogeny of three proposed new species based on full genome sequencing

A total of 64 *Campylobacter* spp. isolates from humans, farm animals and wildlife in New Zealand were grown to generate high quality intact genomic DNA. This was used to make randomly sheared paired end libraries that were sequenced using Illumina technology (GA_{IIx} and a MiSeq). The resulting fastq files were quality trimmed

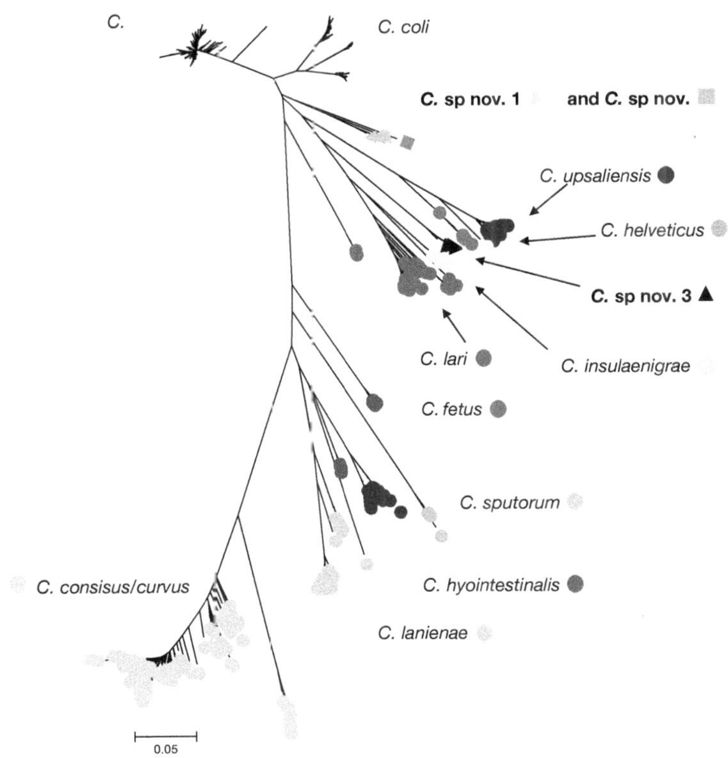

Figure 17.2 Neighbour-Joining phylogeny of the *glnA* locus including isolates from water and wild birds in New Zealand and the proposed new species: *C.* sp nov. 1, 2 and 3. A colour version of this figure is available in the plate section at the back of the book.

using SolexaQA (Cox *et al.*, 2010) and assembled with the *de novo* assembler Velvet (Zerbino and Birney, 2008) to generate contigs. After a process of contig checking, the best contigs from a given assembly parameter were used in a Glimmer (Salzberg *et al.*, 1998; Delcher *et al.*, 1999, 2007) gene prediction analysis. The resultant genes from the 64 isolates were then clustered using OrthoMCL (Li *et al.*, 2003; Chen *et al.*, 2006), including the genes from the isolates of *C. lari* and *C. upsaliensis* deposited in GenBank. Custom Perl scripts were used to extract the orthologous cluster information for those clusters that had a member of each isolate under investigation, and where the length range of the genes in a given cluster was fewer than or equal to 20 amino acids. Genes were aligned individually using Muscle (Edgar, 2004), and then concatenated for visualization using NeighborNets (Bryant and Moulton, 2004) inside SplitsTree (Huson *et al.*, 2010). For the ribosomal MLST gene (Jolley and Maiden, 2010; Jolley *et al.*, 2012) set containing the 52 genes (out of 53) present in the Campylobacterales, individual orthologous clusters of the genes were identified, aligned, concatenated and visualized as above in a separate NeighborNet.

A detailed reconstruction of the phylogeny using 393 variable length (≤ 20 amino acids) genes present in all isolates, and the 52 ribosomal genes (Fig. 17.3) confirm the presence of a distinct cluster or 'phylogroup' comprising three members of our proposed new species provisionally named *C.* sp. nov. 1 in Fig. 17.3. *C.* sp. nov. 2 (a single isolate from a weka) forms a more distinct taxon in Fig. 17.3, compared to the single gene phylogeny shown in Fig. 17.2. In common with the single gene analysis shown in Fig. 17.2, the multilocus, high coverage analysis locates *C.* sp. nov. 1 and *C.* sp. nov. 2 relatively close to the common ancestor of the two major human pathogens (*C. jejuni and C. coli*). Our representative of *C.* spp. nov. 3, isolated from a European starling, forms a distinct

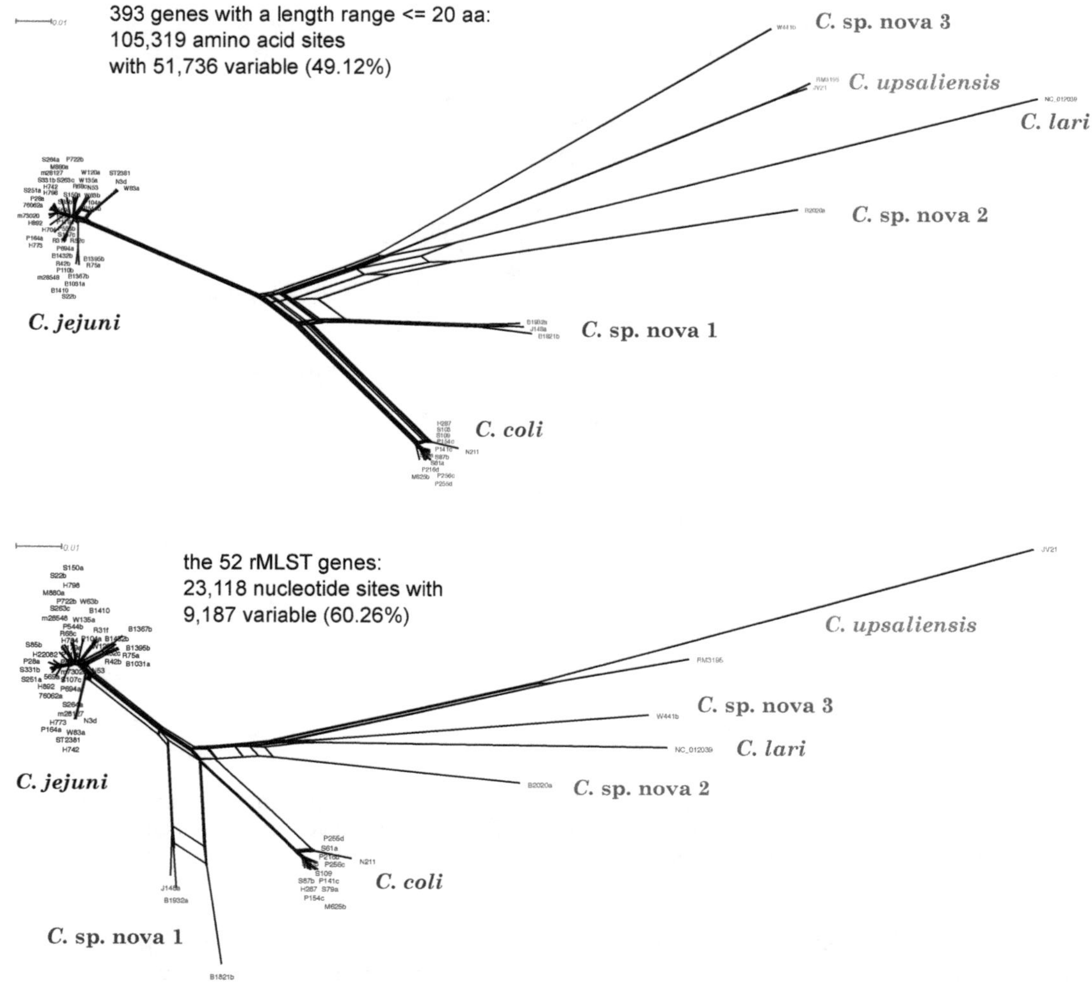

Figure 17.3 NeighbourNet plots showing the relatedness of isolates from New Zealand and *C. lari* and *C. upsaliensis* reference strains, based on full genome analysis. The top plot shows the phylogeny based on 393 concatenated genes, the bottom plot is based on a concatenation of the 52 ribosomal genes. A colour version of this figure is available in the plate section at the back of the book.

taxon located closer to, but distinct from, *C. lari* and *C. upsaliensis*. This is consistent with its location in the *glnA* NJ tree shown in Fig. 17.2.

Estimating divergence times of lineages isolated from New Zealand

A reconstruction of the phylogeny of *Campylobacter* spp. was conducted using BEAST (Drummond and Rambaut, 2007), comparing the newly identified *C.* nov. spp. with strains of *C. jejuni* and *C. coli* isolated from New Zealand and representatives of other species (*C. fetus* ST-4, *C. helveticus* ST-2, *C. insulaenigrae* ST-12, *C. lari* ST-6, and *C. upsaliensis* ST-25) used in a previous study (Wilson *et al.*, 2009; Sheppard *et al.*, 2010b). The approach taken was similar to Wilson *et al.* (2009), using the same loci (*glnA*, *glyA*, *tkt*, and *uncA*), first testing for evidence of recombination using a Sawyer's runs test (Sawyer, 1989) using START2 (Jolley *et al.*, 2001). The Markov Chain Monte Carlo (MCMC) chain was run for 10,000,000 iterations, and parameters were stored every 1000 iterations. Two chains were run, and convergence was compared and visually checked. The smallest effective sample size for posterior distribution was 407; the edge

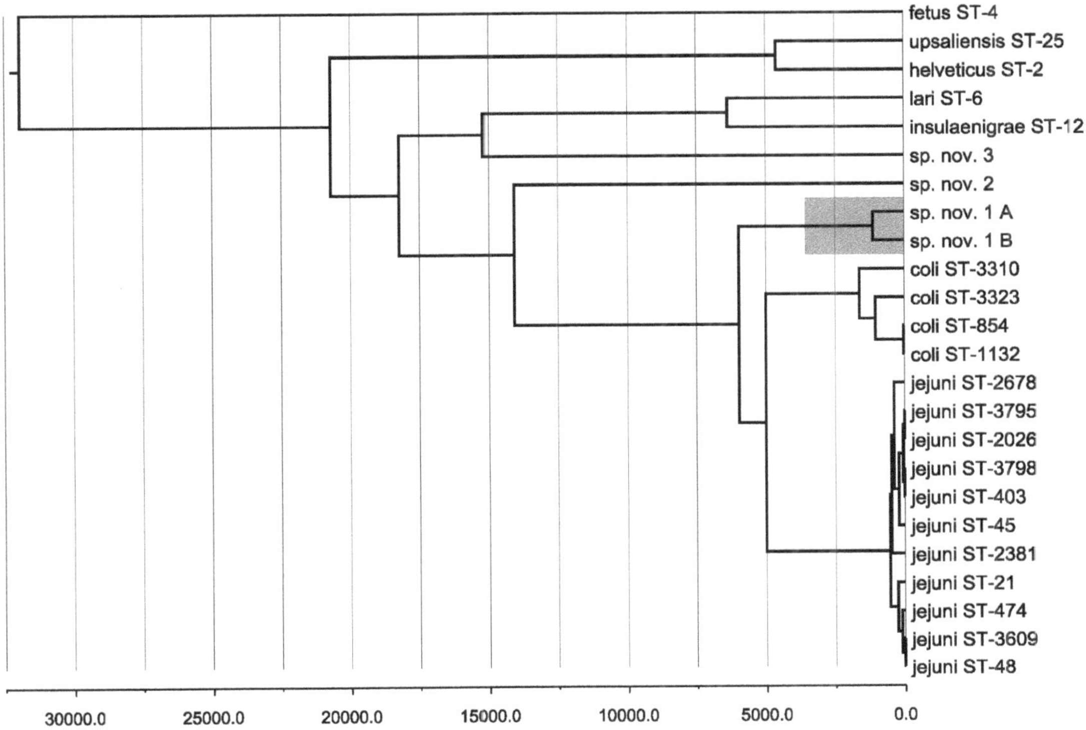

Figure 17.4 Estimated phylogeny and mean split times (in years on x axis) for New Zealand *C. jejuni*, *C. coli* and *C.* sp. nov. isolates, compared to reference strains of *C. lari*, *C. upsaliensis*, *C. helveticus*, *C. insulaenegrae* and *C. fetus*. The two clades of *C.* sp. nov. 1 are highlighted in the shaded area.

lengths in the BEAST trees were scaled according to time.

The BEAST analysis confirms the presence of previously unrecognised, highly divergent lineages in New Zealand water and native wild birds (*C.* sp. nov. 1 and 2), and introduced wild birds (*C.* sp. nov. 3) (Fig. 17.4). The analysis also provides evidence of at least two clades of *C.* sp. nov. 1 that diverged prior to the colonization of New Zealand by man (i.e. over 1000 years ago). The estimated divergence times of the New Zealand isolates in this study are comparable to the estimates based on isolates from other countries (Wilson *et al.*, 2009; Sheppard *et al.*, 2010b), showing, for example, a similar tree topology and divergence times within *C. jejuni* and between *C. jejuni* and *C. coli* (Fig. 17.4 and Table 17.1). In addition to the species that diverged pre-history, the very recent evolution of *C. jejuni*, is apparent and shows the ST-21 and ST-45 associated lineages. Some STs, such as ST-3609, are estimated to have diverged from a common ancestor with ST-48 as recently as the last decade. This is consistent with the history of isolation of this genotype: ST-3609 was isolated frequently from one poultry supplier and no other source in 2005, and has not been recovered since.

C. jejuni in rails and surface water in New Zealand

In the previous section we examined the phylogeny of divergent lineages of *Campylobacter* that, to date, have only been isolated in New Zealand water and wild birds. In this section we focus on the species *C. jejuni* and consider the question: is there evidence of evolving lineages of *C. jejuni* in New Zealand that were introduced pre-European colonization and before the importation of domestic livestock populations? We explore this by considering the phylogeny of an unusual group of closely related *C. jejuni* isolates, which, like *C.* sp. nov. 1 and *C.* sp. nov. 2 considered in the previous section, have only been isolated from environmental water and rails in New Zealand. Again, this group is of particular interest given

Table 17.1 Estimated mean phylogenetic split times for the *Campylobacter* genus using isolates from New Zealand and reference strains of *C. lari*, *C. upsaliensis*, *C. helveticus*, *C. insulaenegrae* and *C. fetus*. HPD stands for highest posterior density and is similar to the credible interval in Bayesian theory

Split name	Time (unit years)	95% HPD lower	95% HPD upper
fetus – *jejuni*	32000	14000	54000
lari – *jejuni*	21000	9000	36000
sp. nov. 3 – *jejuni*	19000	8000	32000
sp. nov. 3 – *lari*	16000	6000	29000
sp. nov. 2 – sp. nov. 1	14000	7000	24000
insulaenigrae – *lari*	6500	2700	11000
sp. nov. 1 – *jejuni*	6000	2600	9800
jejuni-coli	5100	2400	8500
helveticus – *upsaliensis*	4700	2100	7900
coli splits	1600	700	2700
sp. nov. 1 splits	1100	600	1900
coli ST-3323 – ST-854	1100	500	1800
jejuni splits	540	230	870
jejuni ST-48 – ST-2381	530	220	860
jejuni ST-2678 – ST-45	390	170	640
jejuni ST-474 – ST-21	260	110	440
jejuni ST-3795 – ST-45	220	90	370
jejuni ST-48 – ST-474	120	40	210
jejuni ST-3798 – ST-3795	62	19	113
coli ST-1132 – ST-854	36	8	71
jejuni ST-3795 – ST-2026	12	0	31
jejuni ST-3798 – ST-403	10	0	25
jejuni ST-48 – ST-3609	8	0	21

the natural history of the two rails in question: the endemic, rare flightless takahē, and the more recently introduced pukeko.

In the recent study of surface waters in both North and South Island, the most commonly isolated sequence types of *C. jejuni* were ST-2381, and single locus variants (SLVs) of ST-2381 (Carter *et al.*, 2009). However, after culturing 2185 human, 5218 farm animal, 822 domestic pets and 2256 wildlife samples, the only host species that ST-2381 has been isolated from to date are two species of rail, namely the pukeko and takahē (Fig. 17.5). Despite the high prevalence of ST-2381 and its SLVs in New Zealand, members of this lineage and the alleles that comprise ST-2381 have rarely been reported elsewhere. Only 27 of the 6000+ STs in the current PubMLST share alleles with ST-2381 and, including ST-2381, 23 of these have, to-date, only been isolated from water and rails in NZ. Only one, ST-3673, has been isolated in both NZ and another country (ST-3673 is a five locus variant of ST-2381 isolated from a poultry sample in the USA). The remaining four share a single allele with ST-2381 (*uncA* 102), and these have been isolated from cattle in the USA, surface water in Canada and a human case in Luxembourg. A haplotype (integer neighbour-joining) network of these isolates is provided in Fig. 17.5, which shows a distinct cluster of water and rail-associated STs that are allelic variants of ST-2381, indicating that the allelic profile of ST-2381 may represent an ancestral or 'founder' state. The interrelationship between other STs isolated from NZ water and rails, and the small number of internationally isolated STs, is more

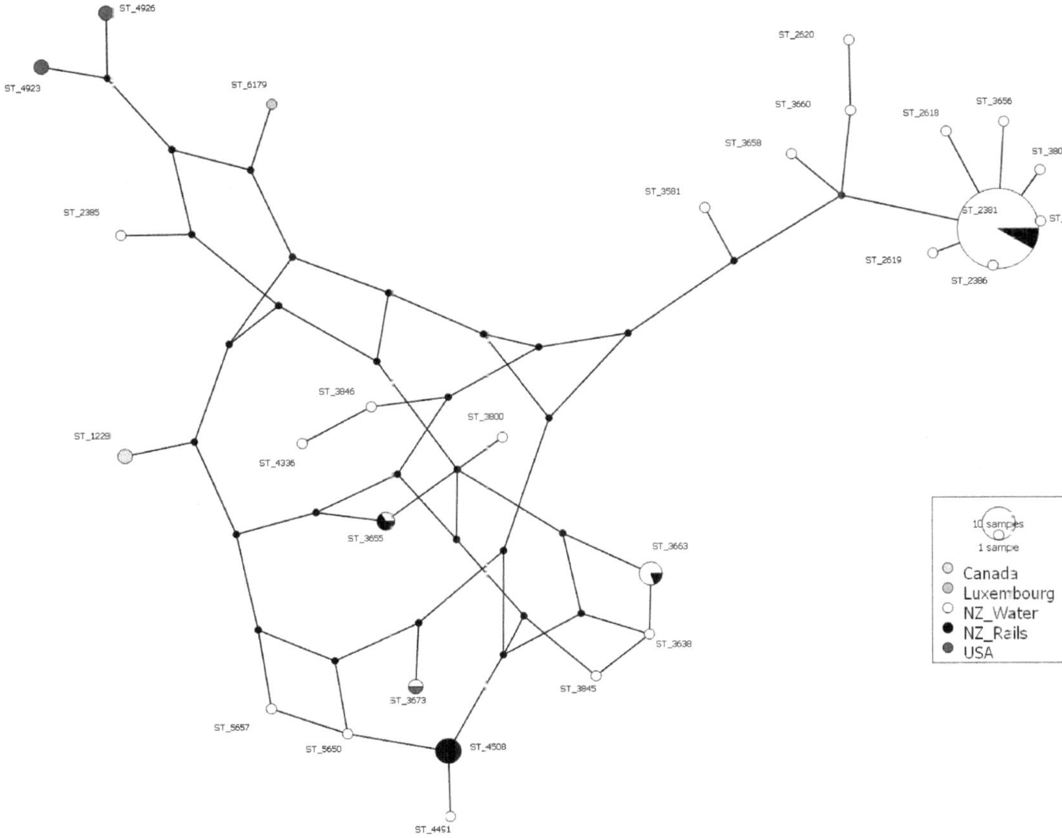

Figure 17.5 Integer neighbour–joining network of *C. jejuni* ST-2381 and related sequence types. ST-2381 has been isolated from 55 samples of river water, 3 pukeko and 2 takahē samples, ST-4508 has been isolated from 6 pukeko samples, and ST-3655 has been isolated from 2 pukeko and 5 river water samples. ST-3663 has been isolated from 1 takahē and 4 water samples. The remaining isolates from NZ are single isolates from river water. ST-2381 and SLVs of ST-2381 are highlighted in the box. Integer NJ starts with an NJ tree topology, then uses integer linear programming to select integer edge lengths such that the total length of the tree is minimized, with the constraint that distances between leaves in the tree are no shorter than distances between sequences. Then, for pairs of leaves whose distance, d_G is greater than the corresponding sequence distance, d_S, extra edges may be added to correct this distance: if a candidate edge increases the length of the network by less than $d_G - d_S + \varepsilon$, it is added. The term ε is known as the 'reticulation tolerance'. A reticulation tolerance value of 1 was used.

complex and may represent patterns of recombination between ST-2381 and other lineages. The pattern of alleles in these variants suggests recombination has occurred with members of the ST-179 (e.g. *aspA* 1, *glnA* 6 and *uncA* 3) and ST-45 (e.g. *glyA* 4 and *tkt* 7) clonal complexes, both of which are frequently isolated from wild birds.

The estimated divergence time of ST-2381 from other members of the *C. jejuni* species is approximately 500 years ago, placing it close to the divergence of the two major lineages of *C. jejuni* represented by ST-21 and ST-45 (Table 17.1 and Fig. 17.4). Although these estimates are associated with wide confidence intervals and heavily dependent on the conversion from coalescent units to years, it is interesting to note that this divergence time is similar to the estimated time of introduction of the pukeko into New Zealand from Australia (Trewick, 1997; Trewick and Gibb, 2010). A full genome evolutionary analysis of *C. jejuni* and *C. coli* isolates from Europe and the USA including a New Zealand ST-2381 isolate is also described in (Sheppard *et al.*, 2013).

Study 2: population structure of *Campylobacter jejuni* in New Zealand compared with the UK

In this analysis the effect of geographical isolation on the evolution of *C. jejuni* is examined by comparing MLST data from humans, poultry and ruminants in New Zealand, with similar data from the UK. More specifically, we compare isolates from the Manawatu region of North Island ($n = 869$), New Zealand, with the county of Lancashire in the UK ($n = 1815$), using data from the same time period (2006–7). These regions have equivalent areas, and an equivalent mix of urban and rural populations.

Species richness was standardized and compared by the rarefaction technique (Heck *et al.*, 1975). Rarefaction curves calculate the species richness for a given number of sampled individuals and show the number of unique sequence types (STs) as a function of the number of individuals sampled. This analysis showed that taxonomic richness of isolates from the UK was significantly greater than those sampled in New Zealand for humans, poultry and ruminants, and the number of indistinguishable sequence types that were shared between the two datasets was low (Fig. 17.6).

Yu (2013) also calculated taxonomic diversity using the Chao1-bc estimator (Chao *et al.*, 2005), and this provided evidence of a significantly higher diversity in the UK populations, compared to the New Zealand populations.

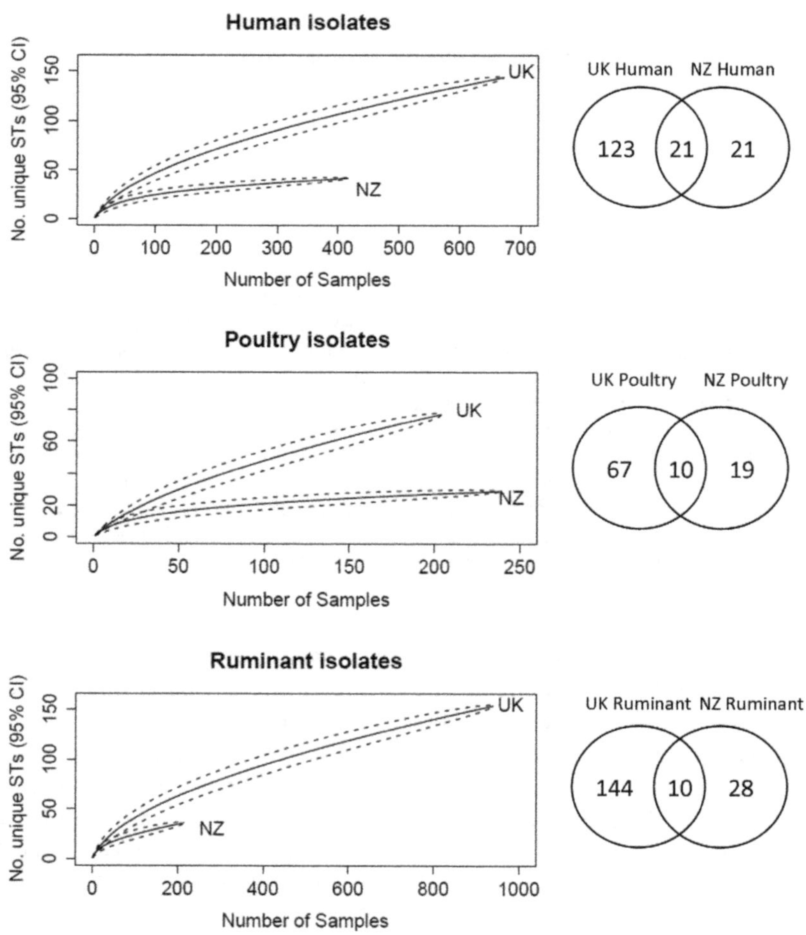

Figure 17.6 Rarefaction curves, with 95% confidence intervals, comparing isolates sampled from populations in New Zealand and the United Kingdom in 2006 and 2007. The Venn diagrams show the number of multilocus sequence types that are shared between populations.

The structure and relatedness of these populations were further compared by computing the proportional similarity or Czekanowski index (PSI) (Kapperud and Rosef, 1983) with 95% bootstrapped confidence intervals (Garrett *et al.*, 2007). The PSI is the area of overlap of two distributions and ranges from 1 (identical genotype distributions) to 0 (no shared sequence types). In this analysis the relatedness of the eight populations of isolates are compared (human, poultry, cattle and sheep from NZ and the UK) and a distance matrix created by calculating 1-PSI for all pairs of populations. The distance matrix is represented as phylogenetic network of all eight populations using the Neighbour-Net method (Bryant and Moulton, 2004) as implemented in SplitsTree (Huson *et al.*, 2010).

The pairs of populations with the highest degree of similarity are the human and poultry populations in NZ (PSI = 0.57, CI = 0.48–0.61) and the UK (PSI = 0.48, CI = 0.39–0.50) and cattle and sheep populations in NZ (PSI = 0.50, CI = 0.37–0.56). Populations with a higher degree of similarity are connected by shorter branch lengths in the Neighbour-Net splits graph (Fig. 17.7). The splits graph also shows splits dividing the two countries (A) and host populations (B).

By comparing the data at three levels of genetic resolution, the same data showed evidence that geographical isolation affects the evolution and diversity of *Campylobacter* genotypes over short time-scales, but that this effect diminishes over longer time-scales. This can be seen by analysing sequence data at different levels of resolution, from the nucleotide level, to the allelic profile level, to the sequence type level and comparing Fst values (Yu, 2013). Fearnhead (2007) demonstrates that this may reflect structure at different time-scales that suggests that the effect of geographical structure is recent and may be short-lived.

A similar study of the effects of host and geographical separation on the population structure of both *C. jejuni* and *C. coil*, which included isolates from wild birds and farmed animals from the UK, the Netherlands, the USA and Senegal, also provided evidence of both host- and location-associated structuring. They concluded that the association with host transcended the effects of geographical location and this was evidence of the effects of international transmission of host-associated lineages (Sheppard *et al.*, 2010a).

The relationship between *C. jejuni* populations from the UK and NZ were further explored using multi-dimensional scaling (MDS) and permutational multivariate analysis of variance (PERMANOVA) (McArdle and Anderson, 2001). These techniques use the pairwise distance matrix generated from the allelic profiles to: (1) represent the relatedness of isolates in a two-dimensional space and (2) test specific hypotheses concerning the contribution of covariates to the variation in pairwise distances and calculate components of variation. PERMANOVA analyses were carried out on 10 randomly selected subsets of the full dataset, each of size 87 (the size of the smallest population, cattle isolates from New Zealand).

PERMANOVA results confirm the results from PSI that both country and host species are significant and independent determinants of population structure and differentiation ($P < 0.0001$ in all ten replicates). Further the interaction between country and species was significant ($P < 0.0001$ in all 10 replicates), supporting evidence from the PSI analysis represented in Table 17.2 and Fig. 17.7. For example, the cattle and sheep populations are more similar to each other in New Zealand, compared to the equivalent populations in the UK. Table 17.3 shows the average estimated coefficients of variation for country, host and the interaction term. Although only 12.1% of the proportion of the variation in allelic profiles can be attributed to these variables, each contributes a similar percentage

Fig. 17.8 shows MDS plots of the relatedness of all STs sampled in the UK/New Zealand comparison. In each subplot, STs are scaled by a bubble that represents the proportion of isolates recovered from each host population. A number of features are evident: the patterns are more similar for within-country comparisons of human and poultry isolates (subplots A and C, and subplots B and D) compared to the ruminant populations. This is in accordance with the PSI estimates in Table 17.2 and Fig. 17.7; and the patterns for the ruminant populations (subplots E and F) are more similar than the equivalent comparisons of human and poultry populations between countries. Such

Table 17.2 and Figure 17.7 Examination of the relatedness of populations of *C. jejuni* isolated from New Zealand (Manawatu) and the United Kingdom (Lancashire) in 2005/6. The table shows the pairwise estimates of 1-PSI (the proportionality index, or area of overlap of genotype distributions). Split A divides the two countries, Split B divides the ruminant populations from the human and poultry populations.

		New Zealand				**United Kingdom**			
		Cattle	Human	Poultry	Sheep	Cattle	Human	Poultry	Sheep
New Zealand	Cattle	1	0.25–0.38	0.21–0.36	**0.37–0.56**	0.15–0.29	0.14–0.24	0.06–0.15	0.17–0.32
	Human	0.32	1	**0.48–0.61**	0.21–0.31	0.15–0.21	0.28–0.35	0.20–0.30	0.21–0.33
	Poultry	0.30	**0.57**	1	0.19–0.31	0.10–0.16	0.23–0.31	0.22–0.34	0.14–0.26
	Sheep	**0.50**	0.27	0.27	1	0.14–0.26	0.07–0.14	0.03–0.10	0.21–0.34
United Kingdom	Cattle	0.23	0.18	0.13	0.20	1	0.20–0.26	0.14–0.22	0.27–0.39
	Human	0.19	0.33	0.28	0.11	0.24	1	**0.39–0.50**	0.17–0.26
	Poultry	0.11	0.26	0.29	0.06	0.19	**0.48**	1	0.09–0.19
	Sheep	0.27	0.28	0.21	0.31	0.34	0.23	0.15	1

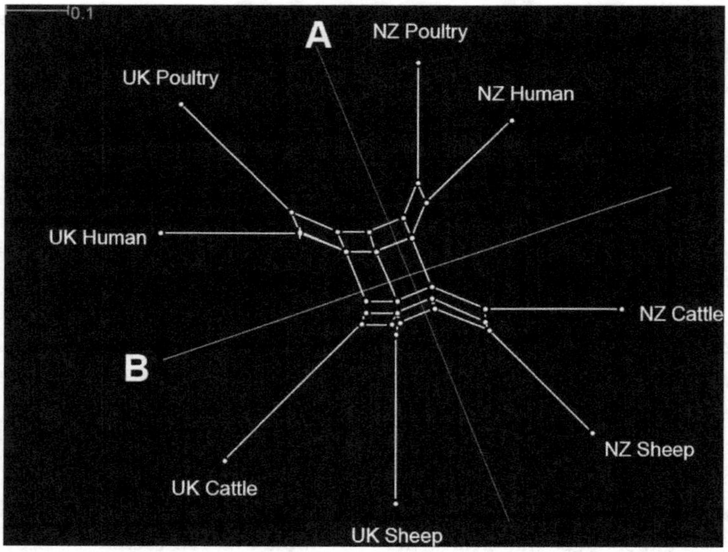

associations between countries (subplots E and F) and between hosts within countries (subplots A and C, and subplots B and D) explain the significant contribution of country, host and the interaction between the two described in Table 17.3.

These patterns of genetic diversity, taxonomic richness and the genetic relationship between isolates from different hosts in the UK and New Zealand are consistent with the presence of an historic barrier to gene flow between these geographical locations following the export of large numbers of animals from the UK to Australia and on to New Zealand in the 19th century. This would have been a permeable barrier, occasionally broken by the importation of smaller numbers of domestic animals – a practice that has continued to the present day – and

Table 17.3 Average estimated components of variation (CoV) from 10 randomly re-sampled subsets each of 87 isolates from 8 populations (human, poultry, cattle and sheep from the UK and NZ)

Variable	Mean CoV	S.D	% CoV	S.D
Country (UK or NZ)	0.75	0.17	4.0	0.8
Host (human, poultry, cattle, sheep)	0.73	0.09	3.7	0.4
Country×host interaction	0.95	0.17	5.0	0.8
Residual	17.10	0.20	87.9	1.1

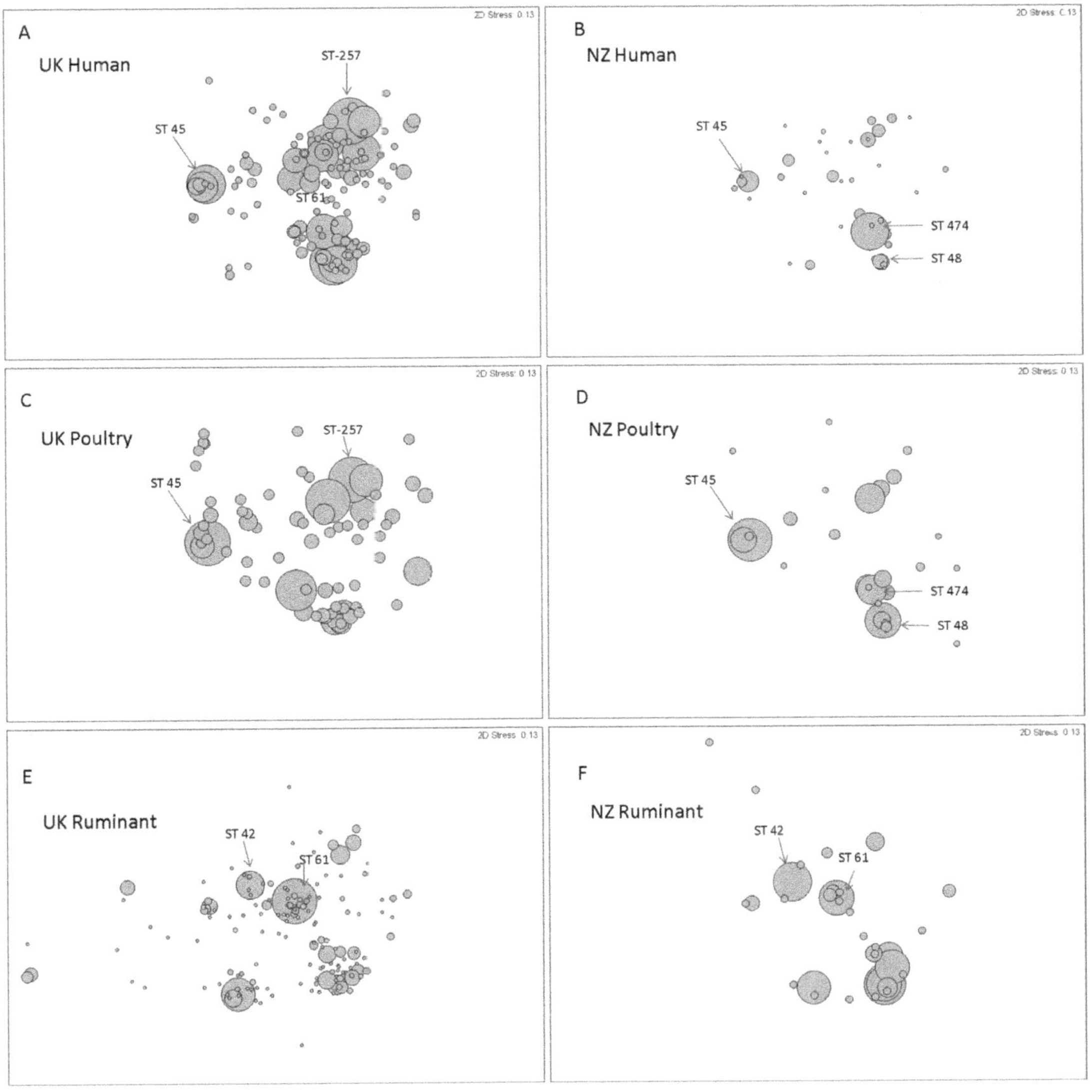

Figure 17.8 Multi-dimensional scaling bubble plots showing 2-D representations of the populations of *C. jejuni* isolated from different host species in the UK and NZ. Each bubble represents a sequence type; its location is determined by the distance matrix generated from allelic profiles. The bubble size is scaled according to the number of isolates from each host–country combination; the smallest bubbles represent single isolates and the largest bubbles are UK human=50, UK poultry=19, UK ruminant=183, NZ human=128, NZ poultry=52, NZ ruminant=31. Common STs are highlighted.

the migration of wild birds and infected humans. In addition to the geographical barriers, between host barriers were introduced by certain farming practices (Sheppard *et al.*, 2008), although the co-grazing of both cattle and sheep would have allowed greater gene flow between these species. The physical separation of host species, and possibly host adaptation by *Campylobacter*, may also have contributed to the population structure observed in this study.

In summary, it is important to consider multiple factors that may have contributed to the pattern observed. These include:

- New Zealand populations of *C. jejuni* have evolved from founder populations that were introduced with the historical introduction of their host species; large numbers of ruminants were introduced in the 19th century, whereas significant poultry importation did not start until the 1960s. The bottleneck resulting from the introduction of a subset of the global population, with subsequent evolution in relative geographical isolation, could have resulted in lower taxonomic richness and diversity that, despite the global movement of people and animals, persists to the present day.
- Poultry products in NZ are sourced exclusively from domestic production, with no imports and a small number of suppliers, in contrast to the UK where there are more suppliers and imported poultry is available. In general, the human population in UK may be exposed to a more diverse range of *C. jejuni* genotypes from multiple sources.
- Bias as a result of sampling from differently structured populations, particularly the ruminant populations, may have created spurious differences between the UK and New Zealand populations. For example, the populations of isolates from cattle and sheep in the UK were sampled from faeces from 15 dairy farms and four sheep farms (Grove-White *et al.*, 2011), whereas the equivalent samples in New Zealand were faecal samples from 4 dairy farms and 3 sheep farms and red meat and offal samples from each species sampled at retail outlets. However, the isolates from NZ food samples are likely to have originated from multiple farms, and would be expected to display greater richness and diversity.
- Even though the methods used in both countries were very similar, bias as a result of subtle differences in microbiological sampling and culture methods may have affected the distribution of genotypes observed.

Study 3: The evolution of ST-474: New Zealand's most important public health-related sequence type

C. jejuni ST-474 was identified as the most prevalent sequence type in sporadic human campylobacteriosis between 2005 and 2007 (Muellner *et al.*, 2013) and in a large nationwide outbreak in 2006 (McTavish *et al.*, 2008). There are a number of important features of this sequence type:

- Although it was first reported in a poultry sample in the Czech Republic in 1998, it has rarely been reported outside of New Zealand (Best *et al.*, 2005; Clark *et al.*, 2007; Wilson *et al.*, 2008).
- It is a double locus variant of the globally distributed ST-21, and a member of clonal complex CC ST-48.
- It has been isolated predominantly from one poultry-supplier (Mullner *et al.*, 2010a), but has also been recovered from other poultry suppliers and cattle and sheep in New Zealand.
- It accounted for up to a third of all human cases at the peak of notifications in 2006 (Mullner *et al.*, 2009), but had subsequently declined to less than 5% following the introduction of regulations in the poultry industry (Muellner *et al.*, 2013).
- It has multiple pulsed- field gel electrophoresis (PFGE) variants (McTavish *et al.*, 2008).
- At the full-genome level, there is evidence of multiple mutation and recombination events within members of ST-474 in New Zealand. A comparison of two ST-474 genomes based on *de-novo* assemblies revealed evidence of at least 83 genes that differed between the two isolates, located in 37 regions (Fig. 17.9 taken from Biggs *et al.*, 2011). Many of the differences

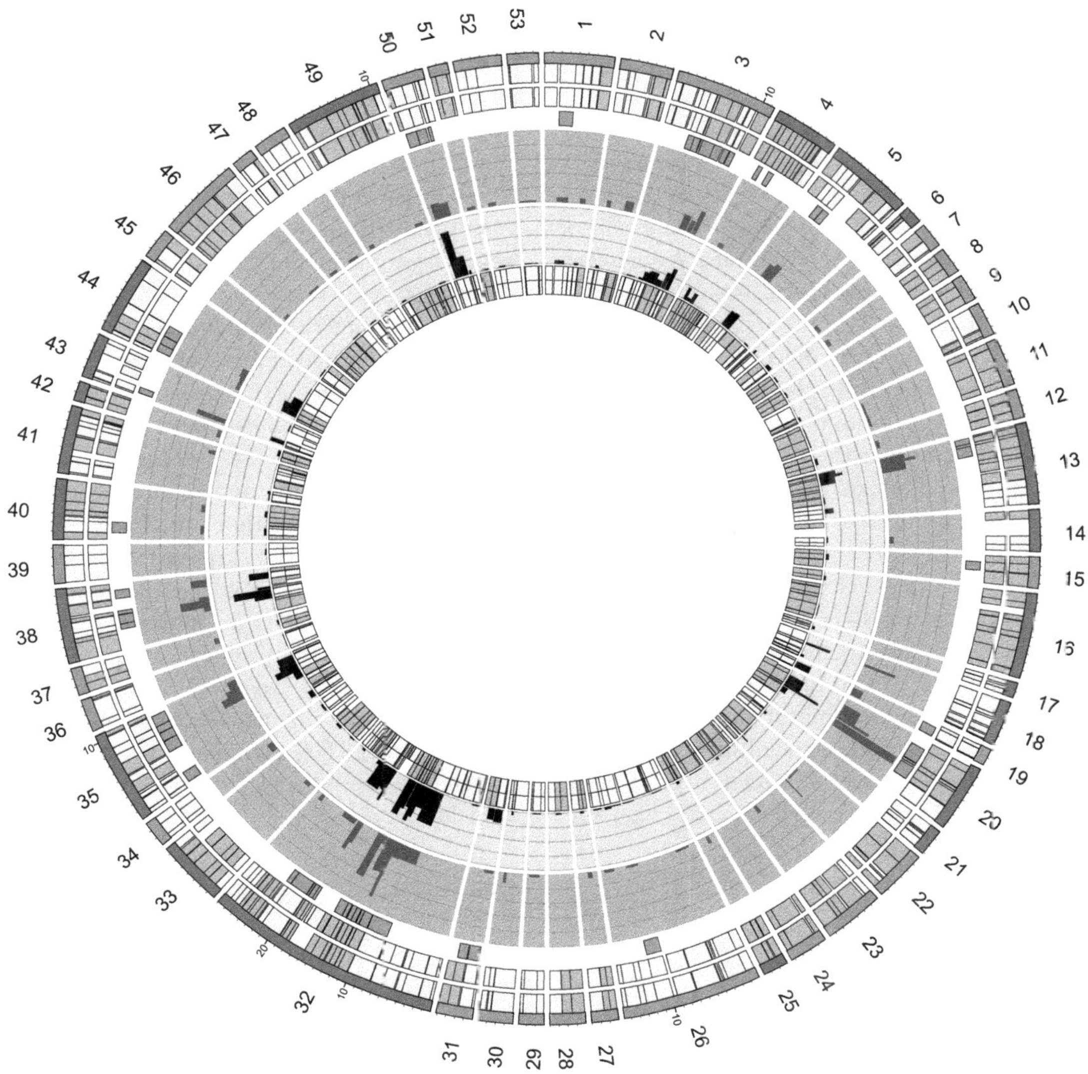

Figure 17.9 Circos plot of orthologous genomic regions for two *C. jejuni* ST-474 genomes isolated in New Zealand where sequence differences were found (from Biggs *et al.*, 2011). The figure shows 37 detectable regions. Orthologous gene pairs in P110b and H22082 were calculated using OrthoMCL with a 'pi_cutoff' value of 0.95. The tracks from outside to inside are; genome region name and size (green, regions of same length in P110b and H22082; blue, the region from P110b is longer; red, the region from H22082 is longer); genes and sizes for P110b; genes and sizes for H22082; genes coloured cyan showing evidence of recombination; histograms for P110b (orange) and H22082 (red) showing the nucleotide coverage from the short reads plotted as an average over the length of the gene, along with the standard deviations for the coverage as round circles; histograms showing the number of sequence differences between the genes at the protein level (dark blue on a light blue background) and DNA level (black on a grey background) as a fraction of protein or gene length respectively; a repeat of the gene sizes and locations in P110b and H22082. Gene orientations are shown in purple and yellow for the forward and reverse strand respectively. The nucleotide coverage histograms have the same scale but a different magnitude, 180 and 120 for P110b and H22082 respectively. Similarly the sequence difference histograms have the same scale but a different magnitude, 0.05 and 0.04 for the protein and the DNA histograms respectively. A colour version of this figure is available in the plate section at the back of the book.

were attributed to recombination with multiple lineages of *C. jejuni*.

The estimated divergence time of ST-474 from its common ancestor with ST-21 was ~260 years ago, and ~120 years ago from its common ancestor with ST-48 (Table 17.1). If these figures are representative of the evolutionary timescale of this ST, then we can speculate that the extant population of ST-474 in New Zealand evolved from a common ancestor of ST-48, which is much more widespread geographically, representing a lineage introduced into the country, possibly by imported livestock, in the 19th Century.

A more detailed exploration of ST-474, using 7 full genome sequences, and a sequence of ST-48 (the putative founder of clonal complex 48 of which ST-474 is a member), was carried out using a comparative analysis of 1667 genes (loci). Each locus was assigned an allele number according to its sequence (i.e. all unique sequences were assigned a different number), and these alleles were used to create an extended allelic profile for each isolate producing a character matrix with 8 taxa and 1667 characters, where each character represented a locus. We then constructed the maximum parsimony (MP) tree using allelic profiles and mapped each character onto the tree using a parsimony approach. There were 274 loci that showed allelic variation, and of these 83 were parsimony informative (i.e. had the potential to disagree with the tree, the remainder would have fitted perfectly on any tree). Of these 274 loci, 44 fit the tree in Fig. 17.10 perfectly and, if the ST-48 (H892) isolate is excluded from the analysis, 34 required multiple changes. These 34 genes are shown in Fig. 17.10 and may represent loci that have recombined between the lineages represented by the 7 ST-474 isolates under investigation, and not undergone subsequent mutation. If recombination usually introduces novel material then the expectation is that all the characters in the allelic profile should be mutually compatible. The presence of many characters in the allelic profile that require parallel changes on the MP tree could be an indicator of recombination moving alleles around within a population of ST-474 strains, or perhaps selection giving some pressure for parallel acquisition of allele types.

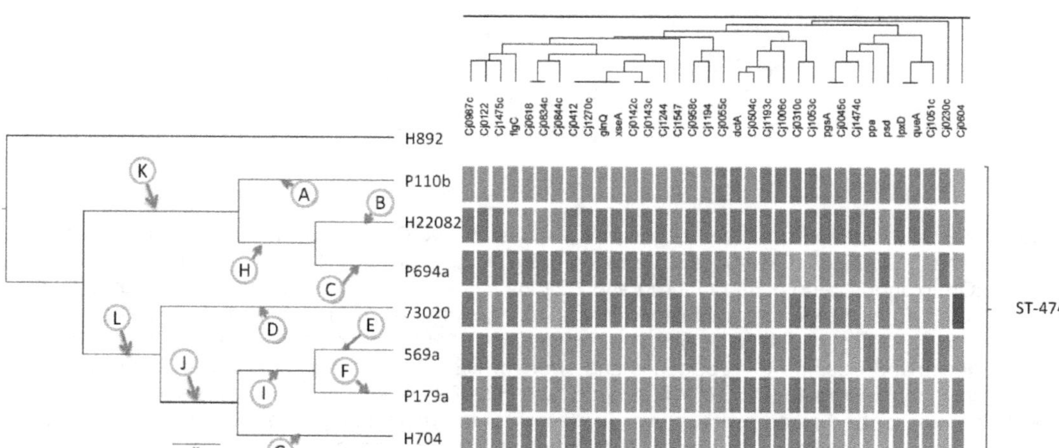

Figure 17.10 Maximum parsimony (MP) cladogram based on seven full-genome sequences of *C. jejuni* ST-474 isolated from humans, poultry and ruminants in New Zealand, and an isolate of ST-48 (horizontal dendrogram), and clustering of loci with similar character states across isolates (vertical dendrogram). The MP tree was drawn based on 1667 loci, of which 274 showed allelic variation between the isolates. 83 of these were parsimony informative. The 34 loci shown in the matrix did not fit perfectly on the tree, requiring more than one change, and are coloured according to their allelic designation (i.e. different colours represent different alleles). The loci are clustered to identify similar patterns of allelic variation across the ST-474 genomes. The circles labelling each branch (A-L) are referred to in Table 17.4. A colour version of this figure is available in the plate section at the back of the book.

Some of these parallel changes appear to involve more than one gene. The clustering of loci that displayed a similar pattern of allelic variation across the ST-474 isolates (vertical dendrogram in Fig. 17.10) revealed some evidence of grouping of loci according to their location on the genome; two pairs of genes (*Cj0142c/Cj0143c* and *lpxD/queA*) are overlapping genes according to their location on reference genome NC_002163. However, none of the remaining loci that shared an identical pattern were located close to each other on the genome, including the largest group comprising *Cj0412, Cj1270c, glnQ* and *xseA*. Of the 274 loci that showed allelic variation, 24 were different in all seven isolates (i.e. there were 7 different alleles). This included members of the hypervariable gene families 617 (*Cj1305*) and 1318 (*Cj1337*) and other hypervariable genes (*Cj0685c, Cj1442c, Cj1295* and *flaA*) (Parkhill *et al.*, 2000).

Allelic variants were mapped onto the MP tree to provide an indication of the extent of variation on each branch, including both the terminal branches (also called pendant edges) and internal branches (Table 17.4). This table counts how many of the 274 variable loci had a state that was unique for each isolate (the number on each terminal branch) and the number of loci that have a unique state that appears at all the taxa below an internal branch and nowhere else. This gives an indication of the number of genes whose nucleotide sequences have changed as a result of either mutation or external recombination (i.e. excluding recombination between lineages represented in the tree), following their divergence from a common ancestor. For example, the 63 allelic variants assigned to branch A represent an estimate of the number of loci that have undergone nucleotide substitutions between the isolation of P110b and its common ancestor with H22082 and P694a. The larger number of unique states on the terminal branches, compared to the internal branches could be explained by recent population expansion within this lineage.

Conclusion

In this chapter we provide evidence for the evolution of *Campylobacter* spp. in the relative isolation of New Zealand's wildlife and farmed livestock. We cannot rule out the effects of sampling bias and the possibility that similar intensive sampling of, for example, water rails in other countries, may reveal the same lineages

Table 17.4 The number of unique states on each branch of the maximum Parsimony tree shown in Fig. 17.10. Positions A-G represent terminal branches, whereas H-L represent internal branches

Position name	Location	Number of allelic variants unique to branch
Terminal branches	*Node*	
A	P110b	63
B	H22082	107
C	P694a	146
D	73020	64
E	569a	44
F	P179a	48
G	H704	100
Internal branches	*Above nodes*	
H	H22082/P694a	31
I	569a/P179a	9
J	569a/179a/H704	3
K	P110b/H22082/P694a	0
L	73020/569a/179a/H704	4

and population structure elsewhere. However, the presence of previously unreported lineages and proposed new species in flightless and non-migratory wild birds, coupled with the distinct patterns of *C. jejuni* genotypes in New Zealand compared to the UK, support the hypotheses that a combination of geographical separation, coupled with the onset of farming post-European colonization (e.g. importation of farmed livestock and mixed and single species farming), have all influenced the evolution of *Campylobacter* spp. and shaped their population structure and biology in New Zealand today.

In summary, a detailed analysis of partial and full genome sequences of isolates from a wide variety of host species in New Zealand, provide evidence that relative geographical isolation, combined with events both pre-history, and post human colonization, have all left an indelible signature on the populations of *Campylobacter* spp. isolated in the last decade. Arguably, this culminated in the emergence and expansion of ST-474 in New Zealand, a *C. jejuni* strain that continued to evolve, following multiple mutation and recombination events, to become associated with one of the largest epidemics of campylobacteriosis in history (Sears *et al.*, 2011).

Acknowledgements

We would like to thank Dr Sam Sheppard for providing the genome sequence for ST-2381 and Matt Balm and Zoe Grange for information on the prevalence of *C.* sp. nov. in rails and Dr. Steve Trewick for his assistance with acquiring samples from rails. We would also like to thank the mEpiLab microbiology and molecular biology teams, led by Drs Anne Midwinter and Julie Collins-Emerson, for the isolation and characterization of *Campylobacter* spp. in New Zealand. We would also like to thank Professor Martin Maiden for his stimulating discussions on the evolution of *Campylobacter* spp. during his visit to New Zealand and Dr Keith Jolley for providing expertise and access to BIGSdb.

This work was funded by the Ministry for Primary Industries (formerly the New Zealand Food Safety Authority) and the Royal Society of New Zealand Marsden Fund.

References

Anderson, A. (2009). The rat and the octopus: initial human colonization and the prehistoric introduction of domestic animals to Remote Oceania. Biol. Invasions *11*, 1503–1519.

Anderson, A.J., and Clark, G. (2001). Advances in New Zealand mammalogy 1990–2000: Polynesian dog or kuri. J. R. Soc. N.Z. *31*, 161–163.

Atkinson, I.A., and Cameron, E.K. (1993). Human influence on the terrestrial biota and biotic communities of New Zealand. Trends Ecol. Evol. *8*, 447–451.

Baker, M., Wilson, N., Ikram, R., Chambers, S., Shoemack, P., and Cook, G. (2006). Regulation of chicken contamination is urgently needed to control New Zealand's serious campylobacteriosis epidemic. N. Z. Med. J. *119*, U2264.

Baker, M.G., Sneyd, E., and Wilson, N.A. (2007). Is the major increase in notified campylobacteriosis in New Zealand real? Epidemiol. Infect. *135*, 163–170.

Best, E.L., Fox, A.J., Frost, J.A., and Bolton, F.J. (2005). Real-time single-nucleotide polymorphism profiling using Taqman technology for rapid recognition of *Campylobacter jejuni* clonal complexes. J. Med. Microbiol. *54*, 919–925.

Biggs, P.J., Fearnhead, P., Hotter, G., Mohan, V., Collins-Emerson, J., Kwan, E., Besser, T.E., Cookson, A., Carter, P.E., and French, N.P. (2011). Whole-genome comparison of two *Campylobacter jejuni* isolates of the same sequence type reveals multiple loci of different ancestral lineage. PLoS One *6*, e27121.

Bryant, D., and Moulton, V. (2004). Neighbor-net: an agglomerative method for the construction of phylogenetic networks. Mol. Biol. Evol. *21*, 255–265.

Carter, P.E., McTavish, S.M., Brooks, H.J., Campbell, D., Collins-Emerson, J.M., Midwinter, A.C., and French, N.P. (2009). Novel clonal complexes with unknown animal reservoir dominate *Campylobacter jejuni* isolated from river water in New Zealand. Appl. Environ. Microbiol. *75*, 6038–6046.

Chao, A., Chazdon, R.L., Colwell, R.K., and Shen, T.-J. (2005). A new statistical approach for assessing similarity of species composition with incidence and abundance data. Ecol. Lett. *8*, 148–159.

Chen, F., Mackey, A.J., Stoeckert, C.J., Jr., and Roos, D.S. (2006). OrthoMCL-DB: querying a comprehensive multi-species collection of ortholog groups. Nucleic Acids Res. *34*, D363–368.

Clark, C.G., Beeston, A., Bryden, L., Wang, G., Barton, C., Cuff, W., Gilmour, M.W., and Ng, L.K. (2007). Phylogenetic relationships of *Campylobacter jejuni* based on porA sequences. Can. J. Microbiol. *53*, 27–38.

Cox, M.P., Peterson, D.A., and Biggs, P.J. (2010). SolexaQA: At-a-glance quality assessment of Illumina second-generation sequencing data. BMC Bioinformatics *11*, 485.

Delcher, A.L., Harmon, D., Kasif, S., White, O., and Salzberg, S.L. (1999). Improved microbial gene identification with GLIMMER. Nucleic Acids Res. *27*, 4636–4641.

Delcher, A.L., Bratke, K.A., Powers, E.C., and Salzberg, S.L. (2007). Identifying bacterial genes and endosymbiont DNA with Glimmer. Bioinformatics *23*, 673–679.

Diez-Tascon, C., Littlejohn, R.P., Almeida, P.A.R., and Crawford, A.M. (2000). Genetic variation within the Merino sheep breed: analysis of closely related populations using microsatellites. Anim. Genet. *31*, 243–251.

Dingle, K.E., Colles, F.M., Wareing, D.R.A., Ure, R., Fox, A.J., Bolton, F.E., Bootsma, H.J., Willems, R.J.L., Urwin, R., and Maiden, M.C.J. (2001). Multilocus sequence typing system for *Campylobacter jejuni*. J. Clin. Microbiol. *39*, 14–23.

Drummond, A.J., and Rambaut, A. (2007). BEAST: Bayesian evolutionary analysis by sampling trees. BMC Evol. Biol. *7*, 214.

Eberhart-Phillips, J., Walker, N., Garrett, N., Bell, D., Sinclair, D., Rainger, W., and Bates, M. (1997). Campylobacteriosis in New Zealand: results of a case–control study. J. Epidemiol. Comm. Health *51*, 686–691.

Edgar, R.C. (2004). MUSCLE: multiple sequence alignment with high accuracy and high throughput. Nucleic Acids Res. *32*, 1792–1797.

Evans, B.L. (1969). A history of agricultural production and marketing in New Zealand. Keeling and Mundy Ltd Palmerston North, NZ.

Fairweather, J.R. (1985). White-Settler Colonial Development: Early New Zealand Pastoralism and the Formation of Estates1. J. Sociol. *21*, 237–257.

Fearnhead, P. (2007). On the choice of genetic distance in spatial-genetic studies. Genetics *177*, 427–434.

Garrett, N., Devane, M.L., Hudson, J.A., Nicol, C., Ball, A., Klena, J.D., Scholes, P., Baker, M.G., Gilpin, B.J., and Savill, M.G. (2007). Statistical comparison of *Campylobacter jejuni* subtypes from human cases and environmental sources. J. Appl. Microbiol. *103*, 2113–2121.

Gongora, J., Fleming, P., Spencer, P.B.S., Mason, R., Garkavenko, O., Meyer, J.-N., Droegemueller, C., Lee, J.H., and Moran, C. (2004). Phylogenetic relationships of Australian and New Zealand feral pigs assessed by mitochondrial control region sequence and nuclear GPIP genotype. Mol. Phylogenet. Evol. 33, 339–348.

Green, R.E. (1997). The Influence of Numbers Released on the Outcome of Attempts to Introduce Exotic Bird Species to New Zealand. J. Animal Ecol. *66*, 25–35.

Grove-White, D.H., Leatherbarrow, A.J., Cripps, P.J., Diggle, P.J., and French, N.P. (2011). Molecular epidemiology and genetic diversity of *Campylobacter jejuni* in ruminants. Epidemiol. Infect. *139*, 1661–1671.

Heck, K.L., Jr., van Belle, G., and Simberloff, D. (1975). Explicit Calculation of the Rarefaction Diversity Measurement and the Determination of Sufficient Sample Size. Ecology *56*, 1459–1461.

Holdaway, R.N., Worthy, T.H., and Tennyson, A.J. (2001). A working list of breeding bird species of the New Zealand region at first human contact. N.Z. J. Zool. *28*, 119–187.

Huson, D.H., Rupp, R., and Scornavacca, C. (2010). Phylogenetic Networks. Concepts, algorithms and applications. Cambridge University Press, New York.

Jolivet, P., and Verma, K.K. (2010). Good morning Gondwana. Refuge *46*, 53–61.

Jolley, K.A., and Maiden, M.C. (2010). BIGSdb: Scalable analysis of bacterial genome variation at the population level. BMC Bioinformatics *11*, 595.

Jolley, K.A., Feil, E.J., Chan, M.S., and Maiden, M.C.J. (2001). Sequence type analysis and recombinational tests (START). Bioinformatics *17*, 1230–1231.

Jolley, K.A., Bliss, C.M., Bennett, J.S., Bratcher, H.B., Brehony, C., Colles, F.M., Wimalarathna, H., Harrison, O.B., Sheppard, S.K., Cody, A.J., and Maiden, M.C., (2012). Ribosomal multilocus sequence typing: universal characterization of bacteria from domain to strain. Microbiology *158*, 1005–1015.

Kapperud, G., and Rosef, O. (1983). Avian wildlife reservoir of *Campylobacter fetus* subsp. *jejuni*, *Yersinia* spp., and *Salmonella* spp. in Norway. Appl. Environ. Microbiol. *45*, 375–380.

King, C., and Barrett, P. (2005). The handbook of New Zealand mammals. Oxford University Press, Auckland.

Li, L., Stoeckert, C.J., Jr., and Roos, D.S. (2003). OrthoMCL: identification of ortholog groups for eukaryotic genomes. Genome Res. *13*, 2178–2189.

McArdle, B., and Anderson, M. (2001). Fitting multivariate models to community data: a comment on distance-based redundancy analysis. Ecology *82*, 290–297.

McTavish, S.M., Pope, C.E., Nicol, C., Sexton, K., French, N., and Carter, P.E. (2008). Wide geographical distribution of internationally rare *Campylobacter* clones within New Zealand. Epidemiol. Infect. *136*, 1244–1252.

Mullner, P., Spencer, S., Wilson, D., Jones, G., Noble, A., Midwinter, A., Collins-Emerson, J., Carter, P., Hathaway, S., and French, N. (2009). Assigning the source of human campylobacteriosis in New Zealand: A comparative genetic and epidemiological approach. Infect. Genet. Evol. *9*, 1311–1319.

Mullner, P., Collins-Emerson, J.M., Midwinter, A.C., Carter, P., Spencer, S.E., van der Logt, P., Hathaway, S., and French, N.P. (2010a). Molecular epidemiology of *Campylobacter jejuni* in a geographically isolated country with a uniquely structured poultry industry. Appl. Environ. Microbiol. *76*, 2145–2154.

Mullner, P., Shadbolt, T., Collins-Emerson, J.M., Midwinter, A.C., Spencer, S.E., Marshall, J., Carter, P.E., Campbell, D.M., Wilson, D.J., Hathaway, S., *et al.* (2010b). Molecular and spatial epidemiology of human campylobacteriosis: source association and genotype-related risk factors. Epidemiol. Infect. *138*, 1372–1383.

Muellner, P., Pleydell, E., Pirie, R., Baker, M., Campbell, D., Carter, P., and French, N. (2013). The contribution of molecular-based surveillance to successful prevention and control of campylobacteriosis in New Zealand – from source attribution to genomic epidemiology. Eurosurveillance *18*(7).

Parkes, J. (2001). Advances in New Zealand mammalogy 1990–2000: Feral livestock. J. R. Soc. N.Z. *31*, 233–241.

Parkhill, J., Wren, B.W., Mungall, K., Ketley, J.M., Churcher, C., Basham, D., Chillingworth, T., Davies, R.M., Feltwell, T., Holroyd, S., *et al.* (2000). The genome sequence of the food-borne pathogen *Campylobacter jejuni* reveals hypervariable sequences. Nature *403,* 665–668.

Patterson, B. (1991). Laagers in the Wilderness. The Origins of Pastoralism in the Southern North Island Districts 1840–55. Stout Centre Rev. *1,* 9.

Peden, R.L. (2002). Sheep farming practice in colonial Canterbury 1843 to 1882. University of Canterbury, Christchurch, NZ., pp. 1–178.

Salzberg, S.L., Delcher, A.L., Kasif, S., and White, O. (1998). Microbial gene identification using interpolated Markov models. Nucleic Acids Res. *26,* 544–548.

Sawyer, S. (1989). Statistical tests for detecting gene conversion. Mol. Biol. Evol. *6,* 526–538.

Sears, A., Baker, M.G., Wilson, N., Marshall, J., Muellner, P., Campbell, D.M., Lake, R.J., and French, N.P. (2011). Marked campylobacteriosis decline after interventions aimed at poultry, new zealand. Emerg. Infect. Dis. *17,* 1007–1015.

Sheppard, S.K., McCarthy, N.D., Falush, D., and Maiden, M.C. (2008). Convergence of *Campylobacter* species: implications for bacterial evolution. Science *320,* 237–239.

Sheppard, S.K., Colles, F., Richardson, J., Cody, A.J., Elson, R., Lawson, A., Brick, G., Meldrum, R., Little, C.L., Owen, R.J., *et al.* (2010a). Host association of *Campylobacter* genotypes transcends geographic variation. Appl. Environ. Microbiol. *76,* 5269–5277.

Sheppard, S.K., Dallas, J.F., Wilson, D.J., Strachan, N.J., McCarthy, N.D., Jolley, K.A., Colles, F.M., Rotariu, O., Ogden, I.D., Forbes, K.J., and Maiden, M.C. (2010b). Evolution of an agriculture-associated disease causing *Campylobacter coli* clade: evidence from national surveillance data in Scotland. PLoS One *5,* e15708.

Sheppard, S.K., Colles, F.M., McCarthy, N.D., Strachan, N.J., Ogden, I.D., Forbes, K.J., Dallas, J.F., and Maiden, M.C. (2011). Niche segregation and genetic structure of Campylobacter jejuni populations from wild and agricultural host species. Mol. Ecol. *20,* 3484–3490.

Sheppard, S.K., Didelot, X., Jolley, K.A., Darling, A.E., Pascoe, B., Meric, G., Kelly, D.J., Cody, A., Colles, F.M., Strachan, N.J., *et al.* (2013). Progressive genome-wide introgression in agricultural *Campylobacter coli.* Mol. Ecol. 22, 1051–1064.

Simmons, D.R. (1969). Economic change in New Zealand prehistory. J. Polynesian Soc. *78,* 3–34.

Stone, W.R. (1972). The poultry industry in New Zealand. N.Z. Vet. J. *20,* 142–145.

Storey, A.A., and Ladefoged, T. (2008). Counting Your Chickens: Density and Distribution of Chicken Remains in Archaeological Sites of Oceania. Int. J. Osteoarchaeol. *261,* 240–261.

Thomson, G.M. (1922). The naturalisation of animals and plants in New Zealand. Cambridge University Press, London.

Trewick, S.A. (1997). Flightlessness and phylogeny amongst endemic rails (Aves:Rallidae) of the New Zealand region. Philos. Trans. R. Soc. Lond. B Biol. Sci. *352,* 429–446.

Trewick, S.A., and Gibb, G. (2010). Vicars, tramps and assembly of the New Zealand avifauna: a review of molecular phylogenetic evidence. Ibis *152,* 226–253.

Trewick, S.A., and Worthy, T.H. (2001). Origins and prehistoric ecology of Takahe based on morphometric, molecular and fossil data. In The Takahe, Fifty Years of Conservation Management and Research, Lee, W.G., and Jamieson, I.G., eds. (Otago University Press, Dunedin), pp. 31–48.

Wallis, G., and Trewick, S. (2009). New Zealand phylogeography: evolution on a small continent. Mol. Ecol. *18,* 3548–3580.

Wilson, D.J., Gabriel, E., Leatherbarrow, A.J., Cheesbrough, J., Gee, S., Bolton, E., Fox, A., Fearnhead, P., Hart, C.A., and Diggle, P.J. (2008). Tracing the source of campylobacteriosis. PLoS Genet. *4,* e1000203.

Wilson, D.J., Gabriel, E., Leatherbarrow, A.J., Cheesbrough, J., Gee, S., Bolton, E., Fox, A., Hart, C.A., Diggle, P.J., and Fearnhead, P. (2009). Rapid evolution and the importance of recombination to the gastroenteric pathogen *Campylobacter jejuni.* Mol. Biol. Evol. *26,* 385–397.

Worthy, T.H., Tennyson, A.J., Archer, M., Musser, A.M., Hand, S.J., Jones, C., Douglas, B.J., McNamara, J.A., and Beck, R.M. (2006). Miocene mammal reveals a Mesozoic ghost lineage on insular New Zealand, southwest Pacific. Proc. Natl. Acad. Sci. U.S.A. *103,* 19419–19423.

Yu, S., Fearnhead, P., Holland, B., Biggs, P., Grove-White, D.H., Carter, P., Fox, A., and French, N., (2013). The evolution of *Campylobacter*. PhD thesis, Massey University, New Zealand.

Zerbino, D.R., and Birney, E. (2008). Velvet: algorithms for de novo short read assembly using de Bruijn graphs. Genome Res. *18,* 821–829.

How Molecular Typing has Changed our Understanding on Sources and Transmission Routes of Campylobacteriosis in Finland

18

Caroline P.A. de Haan, Rauni I. Kivistö, Hilpi Rautelin and Marja-Liisa Hänninen

Abstract

In this chapter we will present a short review on data concerning the analysis of sources/reservoirs important in the epidemiology of campylobacteriosis among Finnish patients who acquired their infection from domestic sources between 1996 and 2006. We describe epidemiological studies on sporadic cases to find important risk factors and studies on water- and milk-borne outbreaks. In addition, we describe strain characterization, which includes Penner serotyping, pulsed-field gel electrophoresis (PFGE) and multilocus sequence typing (MLST). Also, we discuss results of source attribution modelling studies. Our results have shown that the strains circulating in Finland in different hosts are diverse but certain types predominate and some are present persistently in fluctuating proportions. In addition, the chicken flocks were less likely to be positive for *Campylobacter* compared to most other similar studies. This suggests that they may contribute less to the relative burden of human campylobacteriosis compared to estimates in some other European countries. Therefore, other sources, such as water or environmental exposure, may be additional transmission routes in Finland.

Introduction

Campylobacter is the most common reported cause of human bacterial gastroenteritis in Finland (Fig. 18.1). The annual incidence of campylobacteriosis varied between 61 and 76 per 100,000 inhabitants during 2002–2005. The mean incidence was highest (148/100,000) in people aged 25–29 and lowest (range 21–24/100,000) in children aged 5–14 years and the elderly (aged ≥75 years) (Nakari *et al.*, 2010). The annual incidence showed regional differences and was highest in the Helsinki-Uusimaa region (125/100,000 inhabitants, (http://www3.thl.fi/stat/)). This is considerably higher than the EU average of 48.6 cases per 100,000 in 2010. The difference could be due to the fact that Finland has more actual cases or that registration systems between countries differ (European Food Safety Authority, 2012).

A pronounced seasonality of *Campylobacter* infections is seen in many EU countries, including Finland (Nylen *et al.*, 2002; Jore *et al.*, 2010). The highest number of infections is usually detected in the summer, between June and August (Fig. 18.2). According to Nakari *et al.* (2010), almost all isolates from patients in winter were travel-related; whereas 60%–70% of the infections in summer were acquired domestically. Similar strong seasonal distribution is also typical for the other Nordic countries (Nylen *et al.*, 2002; Jore *et al.*, 2010).

The majority of *C. jejuni* infections are sporadic and the sources of these infections are often unknown. Subtyping has been utilized to characterize *C. jejuni* isolates and has made the examination of *C. jejuni* isolates from different sources possible. For *C. jejuni* the most commonly used subtyping methods in Finnish studies have been Penner serotyping, PFGE and MLST.

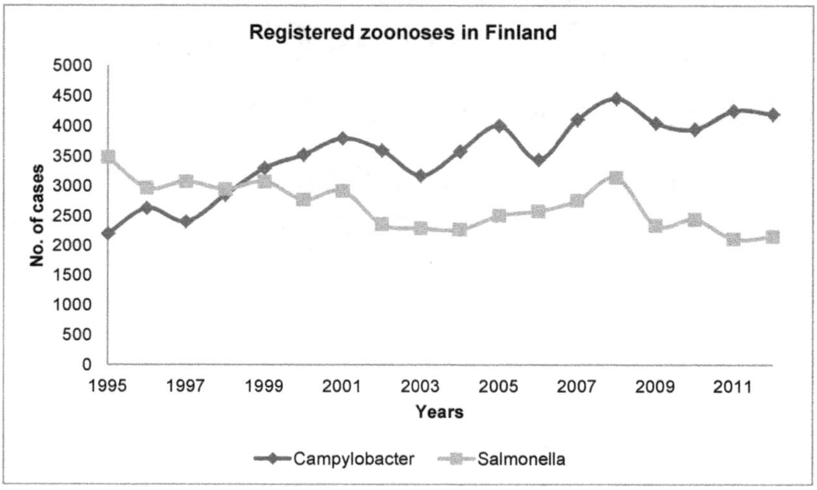

Figure 18.1 Registered laboratory-confirmed campylobacteriosis and salmonellosis cases in Finland from 1995 to 2011 (http://www3.thl.fi/stat/).

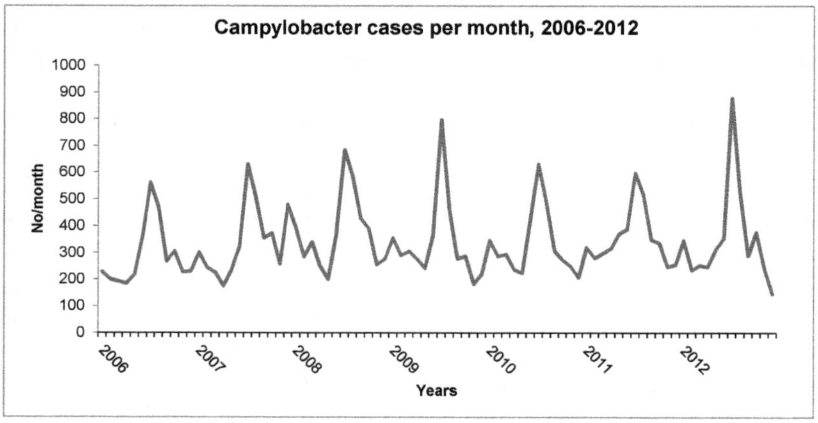

Figure 18.2 *Campylobacter* cases per month in Finland, 2000–2012 (http://www3.thl.fi/stat/).

Epidemiology of campylobacteriosis in Finland

The clinical picture of campylobacteriosis may vary from mild diarrhoea to severe symptoms requiring hospitalization. Furthermore, post-infection complications such as reactive arthritis (ReA) and Guillain-Barré syndrome (GBS) can develop in some patients (Havelaar *et al.*, 2000). Because campylobacteriosis among Finnish patients is frequently acquired while travelling abroad, Feodoroff *et al.* (2010) performed a study comparing symptoms of the patients, as well as characteristics of the infecting strains, with the origin of the infection. In this study, bloody diarrhoea was more common among patients with domestically acquired infection than among those who had travelled abroad in the age group of 30 to 59 years (Feodoroff *et al.*, 2010). *Campylobacter* infection-associated ReA is relatively common in Finland and has been reported in 4–7% of *Campylobacter* patients (Hannu *et al.*, 2002; Schönberg-Norio *et al.*, 2004). This contrasts with GBS, which is rarely detected among Finnish *Campylobacter* patients (Feodoroff *et al.*, 2011).

The most obvious difference between domestically acquired and travel-associated

C. jejuni isolates among Finnish patients is the differing antimicrobial susceptibility patterns. Most domestic strains (> 99%) are susceptible to fluoroquinolones compared to travel-associated strains, which can be up to 100% resistant depending on the country of origin (Rautelin and Hänninen, 2000; Vierikko *et al.*, 2004; Schönberg-Norio *et al.*, 2006; Feodoroff *et al.*, 2009). This finding is related to the controlled usage of antimicrobials in Finnish food animal production and low resistance levels seen among isolates from production animals (Finnish Food Safety Authority Evira *et al.*, 2011).

A Finnish case–control study performed in 2002 during the seasonal peak in three different geographical regions showed that eating undercooked meat (either beef or chicken), swimming in natural sources of water as well as drinking well water were independent risk factors for *Campylobacter* infections (Schönberg-Norio *et al.*, 2004). To support and extend these findings we analysed sources and transmission routes using other methods. In our MLST studies including between 298 (Kärenlampi *et al.*, 2007) and 454 (de Haan *et al.*, 2010a) human *C. jejuni* isolates, the most common sequence types were ST-45, ST-50, ST-677, ST-137 and ST-230. The most common clonal complexes (CCs) were ST-45, ST-21 and ST-677 accounting for a total of 75% of the isolates in both studies (Table 18.1). Fluctuations in the distributions of STs and CCs between the years were prominent, but were also seen during July, August and September within the same year (Fig. 18.3). We also analysed epidemiological data with MLST types of the patient isolates in 2002 and found that ST-48 CC was significantly associated with eating or tasting raw minced beef meat and ST-52 with having contact with cattle and drinking unpasteurized milk (Kärenlampi *et al.*, 2007).

The ST-677 CC seems to be more common in Finland than elsewhere (de Haan *et al.*, 2010a; McCarthy *et al.*, 2012). In the study by Kärenlampi *et al.* (2007) the ST-677 CC was more common among patients requiring hospitalization and a longer stay in hospital, which may indicate that the isolates of this lineage cause more severe clinical symptoms than others, although further studies are needed to confirm this. Nevertheless, a decline in this lineage has been noted in human gastrointestinal infections in a 10-year period from 1996 to 2006. In contrast, a rise of this CC was seen in chicken isolates in the same time period (de Haan *et al.*, 2010a).

Poultry as a source of domestically acquired infections in Finland

Contamination of chicken flocks with *Campylobacter* is low in Finland, although the highest levels of contamination occur during the seasonal peak (European Food Safety Authority, 2012). The number of human *Campylobacter* infections often peak in July, whereas chicken flock contamination peaks in August (Nylen *et al.*, 2002). Between June and October in 2010 only 6% of the Finnish chicken flocks were positive for *Campylobacter*, which further declined in the winter months (European Food Safety Authority, 2012). These figures indicate one of the lowest levels of *Campylobacter* contamination in the EU. Similarly as in many other countries, the chicken meat production in Finland has risen significantly in the past two decades (Fig. 18.4).

We have studied chicken as a reservoir of *C. jejuni* using several approaches such as PFGE genotype overlap between chicken and patient isolates during a seasonal peak (Hänninen *et al.*, 2000). Further we developed a model to study temporal and spatial overlap between chicken and human Penner serotype and PFGE genotype combinations during a seasonal peak (Kärenlampi *et al.*, 2003). In that study, containing 208 human isolates from domestically acquired infections, the overall percentage of overlapping sero- and genotypes was 46%. However, when the dates of chicken flock slaughter and collection of samples from human infections were taken into account there was only an overlap of 31% between the human and chicken isolates. Quite often a sero/genotype was identified earlier in patients than in chickens at slaughter, suggesting that both chicken flocks and humans might have had some common environmental source(s). Interestingly, among patient isolates, we found a certain PFGE genotype in the south in early July; in central Finland at the end of July and in the north in August.

Table 18.1 Summary of most common CCs/STs in bovine, poultry and human *Campylobacter jejuni* isolates in Finland

CC	ST	Bovine (*n*=120) (%)	Poultry (*n*=211) (%)	Human (*n*=454) (%)
ST-21 CC		49.2	8.5	19.4
	ST-21	3.3	0	0.2
	ST-50	3.3	6.6	13.9
	ST-53	13.3	1.4	0
	ST-451	4.2	0.9	0.4
	ST-883	7.5	0.5	2.2
ST-22 CC		0.8	0	4.9
	ST-22	0	0	2.4
	ST-1947	0	0	2
ST-45 CC		10.8	54	43.6
	ST-11	0	1.4	1.8
	ST-45	4.2	18.3	28
	ST-137	0	0.5	4.6
	ST-230	0.8	0.9	4.2
	ST-538	0	1.4	0.9
	ST-583	0	2.4	0.4
	ST-1326	0	2.4	0
	ST-3805	0	3.8	0
ST-48 CC		5	0	3.3
	ST-48	0	0	2
	ST-475	3	0	0.9
ST-61 CC		16.7	0	0.2
	ST-61	8.3	0	0.2
	ST-618	2.5	0	0
	ST-3351	2.5	0	0
ST-283 CC		0	8.1	2.9
	ST-267	0	8.1	2.4
ST-677 CC		2.5	9.5	11.7
	ST-677	2.5	8.1	9.3
	ST-794	0	1.4	2.4
ST-1034 CC		0	1.4	0.4
	ST-4001		1.4	0
ST-1287 CC	ST-945	0	1.9	1.3
ST-1332 CC	ST-1332	0	1.4	0.2
UA[1]		12.5	11.4	5.7
	ST-58	5	0	0
	ST-586	3.3	0.5	0.2

[1]UA=Unassigned to any clonal complex.

These findings demonstrate the transfer of certain *C. jejuni* types throughout the whole country in a short period of time.

ST-45 has been one of the most common STs detected in poultry (Kärenlampi *et al.*, 2007). There is evidence of a recent decline (de Haan

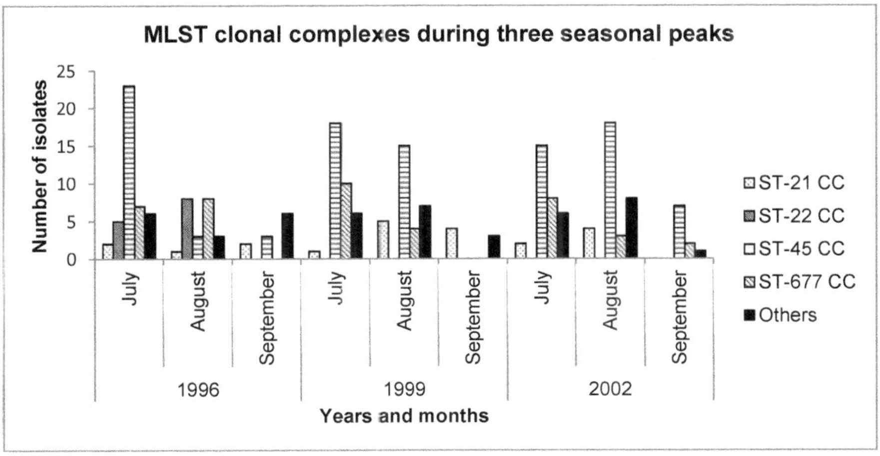

Figure 18.3 *Campylobacter jejuni* MLST clonal complexes during July, August and September in 1996, 1999 and 2002.

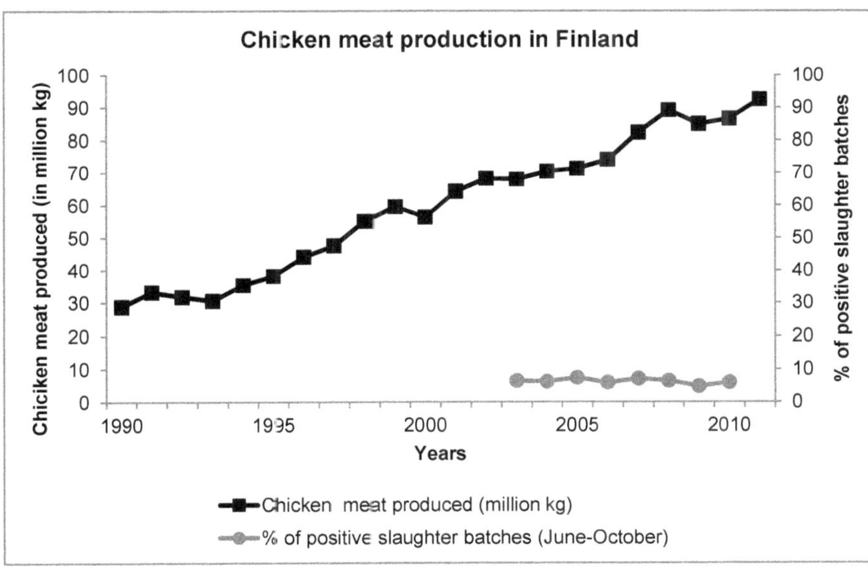

Figure 18.4 Chicken meat production in Finland between 1990 and 2010 and percentage of *Campylobacter*-positive chicken slaughter batches between 2003 and 2010.

et al., 2010a) but the total number of isolates belonging to the ST-45 CC has remained stable. There are many STs within the ST-45 CC in chicken hosts, and the changes in the prevalence of particular STs could indicate microevolution of certain STs within this CC. ST-137 and ST-230, both in the ST-45 CC, were associated with the human isolates (de Haan *et al.*, 2010a). An eBURST analysis (Feil and Spratt, 2006) of STs in the ST-45 CC, and the closely related ST-283 CC among human and chicken isolates, revealed two lineages (Fig. 18.5). Of the chicken isolates in the ST-45 CC, 89% were found in lineage 1. Also, a high proportion of human isolates were found in lineage 1 (72%), mainly due to the high numbers of ST-45 in both hosts and almost all STs in lineage 2 were shared between the two hosts. This makes it difficult to estimate the

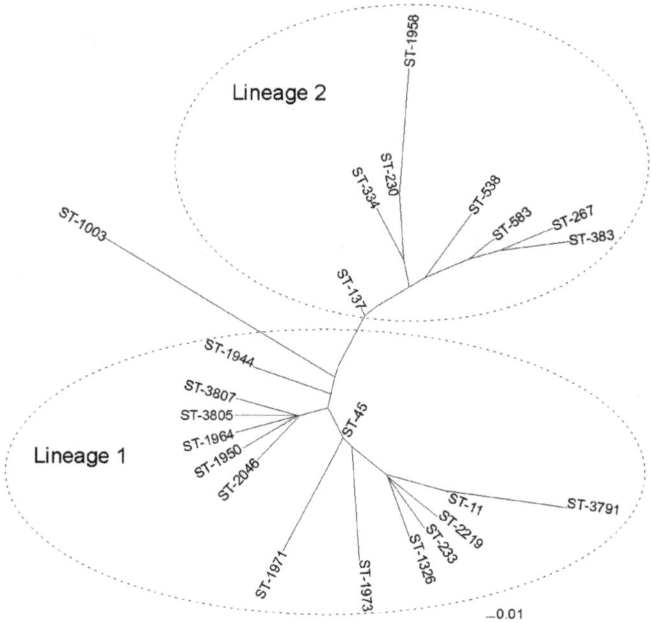

Figure 18.5 eBURST analysis of the ST-45 CC and ST-283 CC among Finnish human and chicken *Campylobacter jejuni* isolates. Unrooted neighbour-joining tree analysis was performed with iTOL (Letunic and Bork, 2007, 2011) at http://itol.embl.de/.

importance of chicken as a source for the ST-45 CC using this approach; particularly as ST-45 is also prevalent in other potential sources (Kärenlampi *et al.*, 2007).

In our longitudinal study (de Haan *et al.*, 2010a) the numbers of ST-883 and ST-50 isolates followed similar trends in both the human and chicken host. Both STs belong to the ST-21 CC. It is plausible that chicken may have been an important source of ST-50 and ST-883 in human infections in the study years. However, frequently trends among other STs did not match perfectly implying that sources other than chicken may be important (de Haan *et al.*, 2010a).

Cattle as a source of domestically acquired infections in Finland

Cattle have been shown to be carriers of *C. jejuni*, although contamination levels of beef are low (Whyte *et al.*, 2004). Nevertheless, indirect transmission via unpasteurized milk and faecal contamination of the environment make cattle a significant reservoir of this pathogen (Hakkinen *et al.*, 2007). In Finland, Hakkinen *et al.* (2007) showed that *Campylobacter* could be detected in 31.1% of the bovine rectal faecal samples and in 3.5% of the bovine carcass surface samples. Beef cattle yielded a higher proportion of *Campylobacter*-positive samples than dairy cattle. The majority of samples were positive for *C. jejuni* (19.5%), but also *C. hyointestinalis* was commonly found among the samples (10.8%). *C. jejuni* was the most common *Campylobacter* species among young cattle (1–3 years old), whereas older cattle (3–7 years old) more commonly carried *C. hyointestinalis*. Unlike in human infections and chicken flock carriage, a clear summer peak was not observed in cattle, although a slight increasing trend was detected towards the end of the summer in August (Hakkinen *et al.*, 2007). Analysis of SmaI/KpnI PFGE profiles has suggested that approximately 31% of the human infections could be attributed to poultry, and 19% to cattle (Hakkinen *et al.*, 2009; Nakari *et al.*, 2010).

We performed MLST on 120 cattle isolates from 2003 (Hakkinen *et al.*, 2007; de Haan *et al.*, 2010b). A summary and comparison of the most common STs and CCs in bovines, poultry and humans from our studies can be found in Table 18.1. Small-scale outbreak investigations are also useful in analysing cattle-associated campylobacter infections. Drinking unpasteurized milk is a well-known cause of campylobacteriosis (Sharp, 1989) and raw milk may be drunk, for example by farmer family members and people visiting cattle farms. We studied a farm-associated outbreak in which a total of five family members acquired the infection. Some of the members had several symptomatic episodes during a period from September to December. SmaI and KpnI PFGE types were identical for the isolates from two family members and tank milk samples at the farm (Schildt *et al.*, 2006). Later studies of the faecal *C. jejuni* isolates from dairy cattle on the farm revealed diverse PFGE patterns but also contained the particular pattern detected from the milk and the patients. This type was later shown to be ST-50. The source of the raw milk contamination was identified as failure in the milking machine and after controlling this no more intestinal symptoms appeared in the family.

Population structure of isolates

MLST has given us an opportunity to use statistical and algorithm-based methodology in studies of *C. jejuni* populations that circulate in humans, chickens and bovines. To facilitate comparison of different populations Bayesian models have been developed to infer the population structure and perform source attribution based on sequence data.

We have performed population genetics studies with Bayesian Analysis of Population Structure (BAPS) (Corander and Marttinen, 2006; Corander and Tang, 2007; Corander *et al.*, 2008; Tang *et al.*, 2009). In our first study, including isolates from domestically acquired human infections, poultry meat and bovine faecal samples all collected in 2003 (de Haan *et al.*, 2010b), we found an almost equal division of the human isolates between the poultry and bovine associated BAPS clusters, with the remaining 10.3% of the human cases unattributed. In our second study we also included, water and wild bird isolates which were typed with MLST (de Haan *et al.*, 2013). In that study, the majority of water (57.9%), poultry (75.5%) and human isolates (65.4%) clustered together (cluster 1), whereas the bovine (49.2%) and wild bird isolates (56.8%) formed a separate cluster (cluster 2). Cluster 1 was marked poultry-associated; despite the high number of natural water isolates present in this cluster. Part of the water isolates (35.7%) formed a separate cluster, mainly consisting of unusual and unassigned STs (de Haan *et al.*, 2013).

Water as a source of domestically acquired infections in Finland

The strict seasonality of *Campylobacter* infections in Finland from July to September (Fig. 18.2) is most probably associated with the climate and the behaviour of people. July is by far the most popular holiday month when many people travel to their cottages in the countryside. Swimming in lakes is a common recreational activity. Another *Campylobacter* risk factor associated with these holidays is the drinking of water from private wells (Hänninen and Kärenlampi, 2004).

Epidemiological studies of outbreaks are a central tool for understanding and managing infectious diseases including campylobacteriosis. Indeed, many of the risks identified for campylobacteriosis in case–control studies have been supported by reports of outbreaks due to contaminated water. In a Finnish case–control study, containing patient data from three regions collected during the seasonal peak in 2002, swimming in lakes, rivers or the sea was shown to be a significant risk factor (Schönberg-Norio *et al.*, 2004). Further analysis of STs of *C. jejuni* strains with the epidemiological data showed that those patients who reported swimming were infected with new and unassigned STs more often than expected (Kärenlampi *et al.*, 2007). This result was further confirmed by de Haan *et al.* (2013) showing that natural water samples collected from pristine areas often had uncommon or unique ST types most probably due to contamination by wildlife. Natural waters may also have been contaminated by human or agricultural sources and thus the composition of STs might reflect more generally the STs common in those sources (ST-21 CC, ST-45 CC). However, the association between swimming and campylobacteriosis requires further study.

Drinking unchlorinated water, but not disinfected water from water plants having treatment processes, was identified as a risk factor in a Finnish case–control study in 2002 (Schönberg-Norio *et al.*, 2004). Similar results were reported in a prospective case–control study of sporadic, domestically acquired *Campylobacter* infections in Norway between 1999 and 2000 (Kapperud *et al.*, 2003). In our study, *C. jejuni* isolates of the patients who reported drinking unchlorinated water or water from natural sources were often associated with CC ST-677 (Kärenlampi *et al.*, 2007). However, if their drinking water had been taken from a well, their isolates (12/45) showed a higher diversity of different STs (e.g. CC-ST-45; ST-48, ST-50), which could suggest contamination from agricultural sources. A private well as a drinking water supply is common in Finland and according to an estimation, approximately 500,000 private wells are in use and these are mostly located in rural areas (www.ymparisto.fi). We have found further support for the role of drinking well water as a source of campylobacteriosis by analysing patient data from July–August 1996 when patients were asked for a personal opinion on the possible source of their infection (Hänninen *et al.*, 1998). Several patients suspected that drinking well water was associated with their illness. In those particular cases the patients were infected with ST-45 CC or ST-677 CC. Wells are often shallow and not well protected from flooding after rains, and could therefore be frequently contaminated by surface water.

Campylobacter has been shown to be associated with waterborne outbreaks all over the world since the 1980s when the first waterborne outbreak was reported in the USA (Craun, 2012). Also in Finland, waterborne outbreaks associated with drinking water have been commonly detected (Hänninen and Kärenlampi, 2004; Kuusi *et al.*, 2004, 2005; Pitkänen *et al.*, 2008; Laine *et al.*, 2010). Most frequently identified microbial agents have been norovirus and *Campylobacter*; the latter was shown to be the causative agent in 19% (16 outbreaks) of the registered outbreaks reported in 1998–2011 (http://www.thl.fi). This indicates that approximately 1 to 2 *Campylobacter* outbreaks occur annually. Also other Nordic countries regularly report outbreaks associated with contaminated drinking water (http://www.euro.who.int/en/home). The reason why waterborne outbreaks are especially common in Finland, Sweden and Norway is that these countries have common practices to use ground water with minimal treatment before distribution to consumers. For example, in Finland, approximately 1500 water plants distribute drinking water and most of these plants are small-sized ground water plants. The number of bigger plants, which usually use surface water and have a full-scale treatment procedure, including disinfection as one of the last steps, is less than 100 (Hänninen *et al.*, 2004; Kuusi *et al.*, 2004, 2005; Pitkänen *et al.*, 2008; Laine *et al.*, 2010). Waterborne outbreaks occur in most cases in rather small communities that use ground water as a source. Data on those Finnish waterborne *Campylobacter* outbreaks, where epidemiological surveys have been performed, indicate heavy rains, flooding and maintenance of pipes, to have an association with the outbreak (Hänninen *et al.*, 2003; Kuusi *et al.*, 2005; Pitkänen *et al.*, 2008; Laine *et al.*, 2010). In Finland, at the end of the 1990s and in the beginning of 2000, when Penner serotyping and PFGE analyses were used as typing methods, three outbreaks were caused by Pen 12 and one by Pen 55 (Kuusi *et al.*, 2005). PFGE has been shown to be a very useful tool for the confirmation of epidemiological association between the patients and drinking water in all studied Finnish outbreaks (Hänninen *et al.*, 2003).

Genetic markers

The use of additional genetic markers can provide further insight into the molecular epidemiology of *Campylobacter*. Our studies on genetic markers have focused on combining MLST and host data with PCR detection of genes encoding a periplasm-secreted form of asparaginase (*ansB*(s)) and *Cj0486*, encoding fucose permease as well as enzymatic detection of gamma-glutamyltranspeptidase (GGT) activity (Hofreuter *et al.*, 2008; Stahl *et al.*, 2011). Among our isolates typed with MLST; we have also studied the distribution of different isoforms of *fspA*, a putative virulence determinant, with FspA1 involved in immunogenicity (Baqar *et al.*, 2008) and FspA2 involved in apoptosis (Poly *et al.*, 2007). Table 18.2 summarizes our findings and includes PCR detection

Table 18.2 PCR detection of *Campylobacter jejuni* LOS locus class, *fspA* isoforms, *AnsB*(s) and *fucP* and enzymatic detection of GGT in selected hosts, multilocus sequence types (ST) and clonal complexes (CC)

Host or ST/CC	LOS locus class	*fspA1*[b], % positive isolates	*fspA2*[b], % positive isolates	*ansB*(s)[c], % positive isolates	*fucP*[c], % positive isolates	GGT[c], % positive isolates
Host						
Human	ND[a]	69.5***	35.4	51.8	28.5	34.4
Poultry	ND	79.2***	20.8	60.6**	18.3	38*
Bovine	ND	25.2	74.8***	15.9	57.1%***	10.[illegible]
ST/CC						
ST-21 CC		2.6	97.4***	1.9	98.7***	1.3
ST-50	C (96.8%)		100***	0	100***	2.6
ST-53	ND		100***	0	95***	0
ST-451	A1		100***	0	88.9***	0
ST-883	A1		100***	10	100***	0
ST-22 CC		100***		100***	0	61.5***[d]
ST-22	A1	100***		100***	0	100***[d]
ST-1947	A1	100***		100***	0	0[d]
ST-45 CC		99.3***	0.7	70***	2.6	58.1***
ST-11	H	100		100***	7.1	92.9***
ST-45	E (59%); H (41%)	99.4***	0.6	59**	2.3	40.5
ST-137	H	95.5**	4.5	95.2***	0	95.2***
ST-230	E	100***		92.9***	7.1	100***
ST-48 CC		44.4	55.6	11.1	88.9***	0
ST-48	B2	100		14.3	100***	0
ST-475	B2		100***	12.5	87.5**	0
ST-52 CC (ST-52)	Not classified		100*	0	100**	0
ST-61 CC	A2		100***	0	0	0
ST-206 CC	B2		100**	0	100*	0
ST-283 CC (ST-267)	E	100***		100***	0	93***
ST-677 CC		100***		98.1***	0	0
ST-677	E	100***		100***	0	0
ST-794	E	100***		100***	0	0
ST-1034 CC	A1		80	0	0	0
ST-1287 CC (ST-945)	N (83.3%)	100		50	0	0
ST-1332 CC (ST-1332)	A1		100*	0	0	0
ST-58	ND		100**	0	16.7	0
ST-586	ND	100		83.3	0	100***

Only positively statistically significant associations are reported.
*$P<0.05$, ** $P<0.01$, *** $P<0.001$.
[a]ND, not done.
[b]de Haan *et al.* (2010c).
[c]de Haan *et al.* (2012).
[d]Revez *et al.* (2012).

of LOS locus classes (de Haan *et al.*, 2012; Revez and Hänninen, 2012).

Selective pressure on *fspA1* seemed to be greater than on *fspA2*; a total of 14 different amino acid sequences were predicted for FspA2, of which at least four were not encoded as full-length proteins (data not shown). Interestingly, the ST-48 CC was divided between *fspA1* (ST-48) and *fspA2* (ST-475) alleles. ST-48 was only found among human isolates, whereas ST-475 was also isolated from bovines.

Clearly, all studied markers, *fspA*, GGT, *ansB*(s) and *fucP* had strong host and MLST associations (de Haan *et al.*, 2012), which is in line with other studies including some of these markers (Gripp *et al.*, 2011; Zautner *et al.*, 2011). Zautner *et al.* (2011) suggested that both GGT and *ansB*(s), together with *dmsA*, could be used to distinguish ST-45 CC and ST-283 CC from ST-21 CC, ST-48 CC and ST-206 CC. However, we showed that ST-45 was rather heterogeneous in the distribution of GGT and *ansB*(s) and a large proportion of the ST-45 strains did not exhibit GGT activity. ST-45 has been isolated from a great number of sources (http://pubmlst.org/campylobacter/) and seems to easily adapt to new niches. However among ST-50, another multihost ST, generally only *fucP* was found. Our results could indicate that none of the studied markers are essential for colonization or establishing infection. However, our findings may indicate different evolutionary histories of the different lineages or dependance of some lineages on certain functions.

Web resources

National Institute for Health and Welfare THL (http://www.thl.fi/fi_FI/web/fi/etusivu)

THL Statistical database (http://www3.thl.fi/stat/)

European Food Safety Authority (EFSA) (http://www.efsa.europa.eu/)

Campylobacter MLST Home Page (http://pubmlst.org/campylobacter/)

eBURST v.3 (http://eburst.mlst.net/)

iTOL Interactive Tree of Life (http://eburst.mlst.net/)

The website of Finland's environmental administration (http://www.ymparisto.fi/)

World Health Organization (http://www.euro.who.int/en/home)

References

Baqar, S., Applebee, L.A., Gilliland, T.C.,Jr, Lee, L.H., Porter, C.K., and Guerry, P. (2008). Immunogenicity and protective efficacy of recombinant *Campylobacter jejuni* flagellum-secreted proteins in mice. Infect. Immun. *76*, 3170–3175.

Corander, J., and Marttinen, P. (2006). Bayesian identification of admixture events using multilocus molecular markers. Mol. Ecol. *15*, 2833–2843.

Corander, J., and Tang, J. (2007). Bayesian analysis of population structure based on linked molecular information. Math. Biosci. *205*, 19–31.

Corander, J., Marttinen, P., Siren, J., and Tang, J. (2008). Enhanced Bayesian modelling in BAPS software for learning genetic structures of populations. BMC Bioinformatics *9*, 539.

Craun, G.F. (2012). The importance of waterborne disease outbreak surveillance in the United States. Ann. Ist. Super. Sanita *48*, 447–459.

European Food Safety Authority (EFSA). (2012). The European Union Summary Report on Trends and Sources of Zoonoses, Zoonotic Agents and Food-borne Outbreaks in 2010. EFSA 3 Journal 2597.

Feil, E., and Spratt, B.G. (2006). eBURST V.3.*2012*,

Feodoroff, B., Ellström, P., Hyytiäinen, H., Sarna, S., Hänninen, M.L., and Rautelin, H. (2010). *Campylobacter jejuni* isolates in Finnish patients differ according to the origin of infection. Gut Pathog. *2*, 22.

Feodoroff, B., Lauhio, A., Ellstrom, P., and Rautelin, H. (2011). A Nationwide Study of *Campylobacter jejuni* and *Campylobacter coli* bacteremia in Finland over a 10-year period, 1998–2007, with special reference to clinical characteristics and antimicrobial susceptibility. Clin. Infect. Dis. *53*, e99-e106.

Feodoroff, F.B., Lauhio, A.R., Sarna, S.J., Hänninen, M.L., and Rautelin, H.I. (2009). Severe diarrhoea caused by highly ciprofloxacin-susceptible *Campylobacter* isolates. Clin. Microbiol. Infect. *15*, 188–192.

Finnish Food Safety Authority Evira, Nuotio, L., Myllyniemi, A.L., and Nykäsenoja, S. (2011). FINRES-Vet 2007–2009; Finnish Veterinary Antimicrobial Resistance Monitoring and Consumption of Antimicrobial Agents (Finnish Food Safety Authority Evira, Helsinki).

Gripp, E., Hlahla, D., Didelot, X., Kops, F., Maurischat, S., Tedin, K., Alter, T., Ellerbroek, L., Schreiber, K., Schomburg, D., *et al.* (2011). Closely related *Campylobacter jejuni* strains from different sources reveal a generalist rather than a specialist lifestyle. BMC Genomics *12*, 584.

de Haan, C.P., Kivistö, R., Hakkinen, M., Rautelin, H., and Hänninen, M.L. (2010a). Decreasing trend of overlapping multilocus sequence types between human and chicken *Campylobacter jejuni* isolates over a decade in Finland. Appl. Environ. Microbiol. *76*, 5228–5236.

de Haan, C.P., Kivistö, R.I., Hakkinen, M., Corander, J., and Hänninen, M.L. (2010b). Multilocus sequence types of Finnish bovine *Campylobacter jejuni* isolates and their attribution to human infections. BMC Microbiol. *10*, 200.

de Haan, C.P., Kivistö, R., and Hänninen, M.L. (2010c). Association of *Campylobacter jejuni* Cj0859c gene (*fspA*) variants with different *C. jejuni* multilocus sequence types. Appl. Environ. Microbiol. *76*, 6942–6943.

de Haan, C.P., Llarena, A.K., Revez, J., and Hänninen, M.L. (2012). Association of *Campylobacter jejuni* metabolic traits with multilocus sequence types. Appl. Environ. Microbiol. *78*, 5550–5554.

de Haan, C.P., Lampén, K., Corander, J., and Hänninen, M.L. (2013). Multilocus sequence types of environmental *Campylobacter jejuni* isolates and their similarities to those of human, poultry and bovine *C. jejuni* isolates. Zoonoses Public. Health. *60*, 125–133.

Hakkinen, M., Heiska, H., and Hänninen, M.L. (2007). Prevalence of *Campylobacter* spp. in cattle in Finland and antimicrobial susceptibilities of bovine *Campylobacter jejuni* strains. Appl. Environ. Microbiol. *73*, 3232–3238.

Hakkinen, M., Nakari, U.M., and Siitonen, A. (2009). Chickens and cattle as sources of sporadic domestically acquired *Campylobacter jejuni* infections in Finland. Appl. Environ. Microbiol. *75*, 5244–5249.

Hänninen, M.L., and Kärenlampi, R. (2004). *Campylobacter* in waterborne epidemics in Finland. Water Sci. Technol. *4*, 39–45.

Hänninen, M.L., Pajarre, S., Klossner, M.L., and Rautelin, H. (1998). Typing of human *Campylobacter jejuni* isolates in Finland by pulsed-field gel electrophoresis. J. Clin. Microbiol. *36*, 1787–1789.

Hänninen, M.L., Perko-Mäkelä, P., Pitkälä, A., and Rautelin, H. (2000). A three-year study of *Campylobacter jejuni* genotypes in humans with domestically acquired infections and in chicken samples from the Helsinki area. J. Clin. Microbiol. *38*, 1998–2000.

Hänninen, M.L., Haajanen, H., Pummi, T., Wermundsen, K., Katila, M.L., Sarkkinen, H., Miettinen, I., and Rautelin, H. (2003). Detection and typing of *Campylobacter jejuni* and *Campylobacter coli* and analysis of indicator organisms in three waterborne outbreaks in Finland. Appl. Environ. Microbiol. *69*, 1391–1396.

Hannu, T., Mattila, L., Rautelin, H., Pelkonen, P., Lahdenne, P., Siitonen, A., and Leirisalo-Repo, M. (2002). *Campylobacter*-triggered reactive arthritis: a population-based study. Rheumatology (Oxford) *41*, 312–318.

Havelaar, A.H., de Wit, M.A., van Koningsveld, R., and van Kempen, E. (2000). Health burden in the Netherlands due to infection with thermophilic *Campylobacter* spp. Epidemiol. Infect. *125*, 505–522.

Hofreuter, D., Novik, V., and Galan, J.E. (2008). Metabolic diversity in *Campylobacter jejuni* enhances specific tissue colonization. Cell. Host Microbe *4*, 425–433.

Jore, S., Viljugrein, H., Brun, E., Heier, B.T., Borck, B., Ethelberg, S., Hakkinen, M., Kuusi, M., Reiersen, J., Hansson, I., *et al.* (2010). Trends in *Campylobacter* incidence in broilers and humans in six European countries, 1997–2007. Prev. Vet. Med. *93*, 33–41.

Kapperud, G., Espeland, G., Wahl, E., Walde, A., Herikstad, H., Gustavsen, S., Tveit, I., Natas, O., Bevanger, L., and Digranes, A. (2003). Factors associated with increased and decreased risk of *Campylobacter* infection: a prospective case–control study in Norway. Am. J. Epidemiol. *158*, 234–242.

Kärenlampi, R., Rautelin, H., Hakkinen, M., and Hänninen, M.L. (2003). Temporal and geographical distribution and overlap of Penner heat-stable serotypes and pulsed-field gel electrophoresis genotypes of *Campylobacter jejuni* isolates collected from humans and chickens in Finland during a seasonal peak. J. Clin. Microbiol. *41*, 4870–4872.

Kärenlampi, R., Rautelin, H., Schönberg-Norio, D., Paulin, L., and Hänninen, M.L. (2007). Longitudinal study of Finnish *Campylobacter jejuni* and *C. coli* isolates from humans, using multilocus sequence typing, including comparison with epidemiological data and isolates from poultry and cattle. Appl. Environ. Microbiol. *73*, 148–155.

Kuusi, M., Klemets, P., Miettinen, I., Laaksonen, I., Sarkkinen, H., Hänninen, M.L., Rautelin, H., Kela, E., and Nuorti, J.P. (2004). An outbreak of gastroenteritis from a non-chlorinated community water supply. J. Epidemiol. Community Health *58*, 273–277.

Kuusi, M., Nuorti, J.P., Hänninen, M.L., Koskela, M., Jussila, V., Kela, E., Miettinen, I., and Ruutu, P. (2005). A large outbreak of campylobacteriosis associated with a municipal water supply in Finland. Epidemiol. Infect. *133*, 593–601.

Laine, J., Huovinen, E., Virtanen, M.J., Snellman, M., Lumio, J., Ruutu, P., Kujansuu, E., Vuento, R., Pitkanen, T., Miettinen, I., *et al.* (2010). An extensive gastroenteritis outbreak after drinking-water contamination by sewage effluent, Finland. Epidemiol. Infect. *15*, 1–9.

Letunic, I., and Bork, P. (2007). Interactive Tree Of Life (iTOL): an online tool for phylogenetic tree display and annotation. Bioinformatics *23*, 127–128.

Letunic, I., and Bork, P. (2011). Interactive Tree Of Life v2: online annotation and display of phylogenetic trees made easy. Nucleic Acids Res. 39, W475–8.

McCarthy, N.D., Gillespie, I.A., Lawson, A.J., Richardson, J., Neal, K.R., Hawtin, P.R., Maiden, M.C., and O'Brien, S.J. (2012). Molecular epidemiology of human *Campylobacter jejuni* shows association between seasonal and international patterns of disease. Epidemiol. Infect. *140*, 2247–2255.

Nakari, U.M., Huovinen, E., Kuusi, M., and Siitonen, A. (2010). Population-based surveillance study of *Campylobacter* infections in Finland. Epidemiol. Infect. *138*, 1712–1718.

Nylen, G., Dunstan, F., Palmer, S.R., Andersson, Y., Bager, F., Cowden, J., Feierl, G., Galloway, Y., Kapperud, G., Megraud, F., *et al.* (2002). The seasonal distribution of *campylobacter* infection in nine European countries and New Zealand. Epidemiol. Infect. *128*, 383–390.

Pitkänen, T., Miettinen, I.T., Nakari, U.M., Takkinen, J., Nieminen, K., Siitonen, A., Kuusi, M., Holopainen, A., and Hänninen, M.L. (2008). Faecal contamination of a municipal drinking water distribution system in association with *Campylobacter jejuni* infections. J. Water. Health. *6*, 365–376.

Poly, F., Ewing, C., Goon, S., Hickey, T.E., Rockabrand, D., Majam, G., Lee, L., Phan, J., Savarino, N.J., and Guerry, P. (2007). Heterogeneity of a *Campylobacter jejuni* protein that is secreted through the flagellar filament. Infect. Immun. *75*, 3859–3867.

Rautelin, H., and Hänninen, M.L. (2000). Campylobacters: the most common bacterial enteropathogens in the Nordic countries. Ann. Med. *32*, 440–445.

Revez, J., and Hänninen, M.L. (2012). Lipooligosaccharide locus classes are associated with certain *Campylobacter jejuni* multilocus sequence types. Eur. J. Clin. Microbiol. Infect. Dis. *31*, 2203–2209.

Schildt, M., Savolainen, S., and Hänninen, M.L. (2006). Long-lasting *Campylobacter jejuni* contamination of milk associated with gastrointestinal illness in a farming family. Epidemiol. Infect. *134*, 401–405.

Schönberg-Norio, D., Takkinen, J., Hänninen, M.L., Katila, M.L., Kaukoranta, S.S., Mattila, L., and Rautelin, H. (2004). Swimming and *Campylobacter* infections. Emerg. Infect. Dis. *10*, 1474–1477.

Schönberg-Norio, D., Hänninen, M.L., Katila, M.L., Kaukoranta, S.S., Koskela, M., Eerola, E., Uksila, J., Pajarre, S., and Rautelin, H. (2006). Activities of telithromycin, erythromycin, fluoroquinolones, and doxycycline against *Campylobacter* strains isolated from Finnish subjects. Antimicrob. Agents Chemother. *50*, 1086–1088.

Sharp, J.C. (1989). Milk-borne infection. J. Med. Microbiol. *29*, 239–242.

Stahl, M., Friis, L.M., Nothaft, H., Liu, X., Li, J., Szymanski, C.M., and Stintzi, A. (2011). L-fucose utilization provides *Campylobacter jejuni* with a competitive advantage. Proc. Natl. Acad. Sci. U.S.A. *108*, 7194–7199.

Tang, J., Hanage, W.P., Fraser, C., and Corander, J. (2009). Identifying currents in the gene pool for bacterial populations using an integrative approach. PLoS Comput. Biol. *5*, e1000455.

Vierikko, A., Hänninen, M.L., Siitonen, A., Ruutu, P., and Rautelin, H. (2004). Domestically acquired *Campylobacter* infections in Finland. Emerg. Infect. Dis. *10*, 127–130.

Whyte, P., McGill, K., Cowley, D., Madden, R.H., Moran, L., Scates, P., Carroll, C., O'Leary, A., Fanning, S., Collins, J.D., *et al.* (2004). Occurrence of *Campylobacter* in retail foods in Ireland. Int. J. Food Microbiol. *95*, 111–118.

Zautner, A.E., Herrmann, S., Corso, J., Tareen, A.M., Alter, T., and Gross, U. (2011). Epidemiological association of different *Campylobacter jejuni* groups with metabolism-associated genetic markers. Appl. Environ. Microbiol. *77*, 2359–2365.

19

Extensive Spatial and Temporal Clustering of *Campylobacter* Infections

Norval J.C. Strachan and Ken Forbes

Abstract

Outbreaks of campylobacteriosis are rare with approximately 99% of cases being individual sporadic human infection, where they are typically unrelated in time and space and presumed to be acquired from different sources. However, the sporadic nature of campylobacteriosis has yet to be evaluated at the strain level using high resolution genotyping. In this chapter we investigated most cases reported in Scotland between July 2005 and September 2006 (N = 5167), where human isolates were typed by sequencing 7 multi-locus sequence typing loci and the flaA locus. Here, general and household outbreaks that were defined using epidemiological data were compared with apparently sporadic cases, with the aim of identifying clusters of cases in time and space by recognizing common strain types that may indicate a common source of infection. Spatial clusters representing 5% of cases occurred in 20 of 247 ST-flaA strains, and temporal clusters representing 12–13% of cases occurred in 67 of 316 ST-flaA strains. Similar to sporadic cases, strains from both temporal and spatial clusters (mainly *C. jejuni*) were attributed mainly to farm ruminants and chicken. *Campylobacter* infections defined as genotypic strains showed up to five times more in clustered cases than was evident from recognized outbreaks using epidemiological data [general outbreaks ($N = 3$) and household outbreaks ($N = 28$)]. These clusters of homogenous genotypes potentially reflect shared origins either via food or the environment. Analysis of isolates characterized by high-resolution genotyping revealed unexpectedly frequent cases in spatial and especially temporal clusters, suggesting that related cases often go unrecognised but are perhaps more common than initially thought.

Introduction

Campylobacter is a zoonotic infection that causes the highest proportion of recognized bacterial gastroenteritis in developed countries (Olson *et al.*, 2008). Since 2005, year on year increases in the number of cases have been observed in the UK (Strachan *et al.*, 2010). Sources of infection include consumption of contaminated retail meats (especially chicken) and raw milk, contact with farm animals and their faeces, and untreated domestic water supplies (Friedman *et al.*, 2004). Incidence is typically 2- to 3-fold higher during summer months, and due to under-reporting community incidences can be 7- to 34-fold higher than those reported (Wheeler *et al.*, 1999; Samuel *et al.*, 2004). *Campylobacter* infection is viewed as predominantly sporadic because identifiable outbreaks are rare. General outbreaks identified from surveillance typically constitute 0.4–1.8% of reported cases (Frost *et al.*, 2002; McTavish *et al.*, 2008), and household outbreaks identified from cases co-occurring in households within a short time window comprise approximately 3% of reported cases (Charlett *et al.*, 2003; Ethelberg *et al.*, 2004; Rotariu *et al.*, 2010). *Campylobacter* outbreaks are viewed as originating from common events or processes surrounding human exposure where they are classified under different terms. Point-source outbreaks are single events (in time and place) where the source of contamination is ingested by a number of individuals for example a restaurant meal (Sails *et al.*, 2003) or the consumption of unpasteurized milk on a 1-day farm field visit, Montrose outbreak (Forbes *et al.*, 2009). Point-source outbreaks tend to be acute, highly localized and have a high level of contamination, with a high attack rate and are relatively easy to recognize through surveillance. Continuous-source outbreaks originate from the distribution (Pearson *et al.*, 2000) or continued sale (O'Leary *et al.*, 2009) of contaminated food product or water (Said *et al.*, 2003). For example, 110 cases of *Campylobacter* occurred over a six week period due to sewage contamination of a groundwater reservoir. The water that was distributed directly without filtration or chlorination to local residents. Isolations from the water were of the same serotype and DNA profile as 29/30 isolates from clinical cases. Continuous outbreaks tend to have a lower levels of contamination with incidences being less local and a lower attack rate than point source outbreaks.

Diffuse outbreaks are often individual cases that occur over a large area (states/countries) and are often linked through a single vehicle of infection such as contaminated food product that has been widely distributed. Diffuse outbreaks are hard to recognize through routine surveillance, where an increase in widely distributed cases is often masked by background sporadic cases. Diffuse outbreaks are often only recognized where an increase in infections is observed within a single region, or from an unusual pathogen or strain, where an increase in a particular sub-type may be recognized. Often diffuse outbreaks are related to fresh products that are widely distributed, again with a generally low level of contamination. Diffuse outbreaks with a relatively large number of cases are easier to recognize than smaller diffuse outbreaks or clusters. Such diffuse outbreaks are recognized for pathogenic strains of *E. coli* (Terajima *et al.*, 2006) and *Salmonella* (Werber *et al.*, 2005) but the extent to which these occur in *Campylobacter* is unclear and there needs to be active laboratory surveillance and communication between reference labs to increase the recognition of diffuse outbreaks.

Campylobacter isolates can be defined as sequence type (ST) strains using multilocus sequence typing (MLST) (Dingle *et al.*, 2001) according to unique allelic profiles of seven housekeeping genes. Many STs can be clustered clonal complexes (CC) or subdivided by adding alleles of highly variable genes, such as *flaA* (Meinersmann *et al.*, 1997). Human *Campylobacter* cases can be attributed to sources by statistical comparison of alleles in human cases and isolates from potential sources. The majority of cases occurring in NW England (Wilson *et al.*, 2008) and Scotland (Sheppard *et al.*, 2009a) are attributable to retail chicken (50–80%) (Strachan *et al.*, 2010) with the majority of the remaining cases acquired from farm ruminants.

Human infections from outbreaks can have distinctive strain compositions (Sintchenko *et al.*, 2007). *Campylobacter* cases from known outbreaks with foodborne (Excoffier *et al.*, 2005; Sails *et al.*, 2003; Pearson *et al.*, 2000) or groundwater-associated (Clark *et al.*, 2005) origins show high proportions of particular strains. If infections from known outbreaks can have increased levels of strain matching then reversing this relationship implies that clusters of particular strains in apparently sporadic cases may indicate potential outbreaks or links to shared infection sources. For example, temporal clusters of ST45 complex strains with putative environmental origins occur in *Campylobacter* cases from north-west England (Sopwith *et al.*, 2008), and molecular strains defined as ST augmented with three highly variable genes, with putative origins in food chains, are evident in cases from Oxfordshire (Dingle *et al.*, 2008). Thus, apparently sporadic *Campylobacter* cases could have clustered incidence patterns at the strain level, but this question has yet to be explored using formal cluster analysis. Spatiotemporal clusters of matching molecular profiles of *Campylobacter* strains are also evident in apparently sporadic clinical cases in Canada and Denmark, potentially indicating outbreaks. Nonetheless, inferring outbreaks according to single-strain matching might fail when the outbreak contains several strains. Chickens, pigs and retail meats often harbour multiple *Campylobacter* strains, as do some human patients.

Cluster analysis is most informative when conducted at large enough scales to encompass typical transmission routes (Sobel *et al.*, 2002), and when any spatiotemporal stratification that could create artifactual clusters is also characterized. In the case of *Campylobacter*, most studies of strains from human infection encompass restricted spatial scales, for example, a single city (Karenlampi *et al.*, 2007) or region (Dingle *et al.*, 2008; Mctavish *et al.*, 2008) and none characterizes strain stratification in any detail. The aims of this study were to (i) characterize spatio-temporal patterns of *Campylobacter* species and strain composition in cases from national surveillance during one annual incidence cycle; (ii) quantify cases occurring in general outbreaks and co-occurring in the same households, and quantify their levels of strain matching; (iii) quantify cases occurring in spatial and temporal clusters within strains; (iv) compare the source attributions of clustered and sporadic cases.

Materials and methods

Campylobacter isolates were submitted from public health bacteriology laboratories in Scotland during July 2005 to September 2006 (Table 19.1) as previously described (Sheppard *et al.*, 2009a). A further six isolates from a reported outbreak were submitted in January 2007, and these were used only for analysing strain composition in general outbreaks. Cases in general outbreaks were identified from ObSurv (HPS, 2009), the surveillance system for all general outbreaks of infectious intestinal disease in Scotland. Cases in household outbreaks were identified from matching addresses and reporting dates within the same 28-day period using data provided by NHS Health Boards. Ethical approval for the collection and use of patient information was obtained.

Isolates were submitted, re-grown for DNA preparation and archived as previously described (Sheppard *et al.*, 2009a). Briefly, isolates were plated onto charcoal cefoperozone deoxycholate agar (CCDA), (CM0739, Oxoid, UK) and incubated for 2 days under microaerophilic conditions, after which presumptive *Campylobacter* cells were identified. The presence of thermophilic *Campylobacter* species was initially confirmed using a latex agglutination test (M46CE, Microgen, UK). Following isolation of a single colony, DNA was extracted using the Chelex-100 resin (142–1253, Bio-Rad Laboratories, USA). All positive isolates were stored in nutrient broth plus 15% glycerol at –80°C.

Molecular typing

Isolates were typed for the seven genes of the *Campylobacter* MLST scheme and for the short

Table 19.1 Numbers and ST-flaA diversity of *Campylobacter* isolates from 12 NHS Health Boards in mainland Scotland

Health boards	N	STf	DI	95% CI of DI
Ayrshire and Arran	248	118	0.977	0.970–0.984
Argyll and Clyde	119	66	0.969	0.957–0.981
Borders	126	70	0.977	0.970–0.983
Dumfries and Galloway	201	102	0.977	0.970–0.983
Fife	257	117	0.979	0.973–0.984
Forth Valley	138	75	0.972	0.963–0.981
Greater Glasgow	862	298	0.980	0.976–0.983
Grampian	777	273	0.980	0.976–0.984
Highland	287	127	0.980	0.976–0.984
Lanarkshire	634	224	0.978	0.974–0.982
Lothian	1022	360	0.985	0.983–0.987
Tayside	496	188	0.980	0.976–0.984
Total	5167	983	0.982	0.981–0.984

N, number of isolates with a ST-flaA profile; STf, number of different ST-flaA profiles; DI, Simpson's diversity index; CI, confidence interval.

variable region sequence (flaA-SVR) of the *flaA* gene as previously described (Gormley *et al.*, 2008; Sheppard *et al.*, 2009a). Sequences were assembled and alleles assigned using the STARS software available on the PubMLST website (http://pubmlst.org/Campylobacter/) and sequence types and clonal complexes were assigned to the array of seven alleles (Samuel *et al.*, 2004). The successful assignment of an allele number to each of the seven loci and subsequently a sequence type (ST) and clonal complex (CC) being allocated, further confirms the species using the pubMLST *C. jejuni*/*C. coli* database (pubMLST, 2009).

A total of 5676 isolates were received from all the laboratories that analyse *Campylobacter* infection in the 15 Health Boards in Scotland, representing 97.3% of cases reported to national surveillance during the study period. Isolates from species other than *C. jejuni* or *C. coli*, mixed strains, typing failures, repeat isolates of the same strain from the same patient within the same 8-week period, and the submissions from three island Health Boards with very low sample sizes were excluded, leaving 5167 isolates with ST-flaA profiles for the main analysis.

Statistical analyses

Values of Simpson's diversity index were calculated using V-DICE(Health Protection Agency, 2009 3424/id The partitioning of variation was analysed using the analysis of molecular variance (AMOVA) framework available in Arlequin 3.11 (Excoffier *et al.*, 2005) Means and confidence intervals were calculated using the bootstrap method from Poptools add-in for Microsoft Excel (http://www.cse.csiro.au/poptools/). Levels of ST-flaA profile matching in known outbreaks were compared with levels in comparison groups of sporadic isolates of the same size as each outbreak group. The sporadic isolates originated from the same Health Board and the same period as the outbreak. In the cases of the outbreaks in Ayrshire and Arran and Tayside Health Boards, the period was two weeks before and after the median date of the outbreak, and in the case of the outbreak in Forth Valley Health Board that occurred after the study period, the period was 1 month before and after the median date of the outbreak from the previous year. Comparisons were made by bootstrap re-sampling (1000 replicates).

Cluster analyses

Spatial and temporal clustering of isolates within strains was analysed using the scan statistic available in SatScan 8.0.1 (Kulldorff *et al.*, 2005), and temporal clustering was also analysed using a hypothesis-free (HF) scan statistic (Cucala *et al.*, 2008) available as scripts for the R statistical package. Strains were analysed at the highest level of discrimination (ST-flaA profile) because supplementing ST with flaA and two other highly variable loci can improve cluster detection (Dingle *et al.*, 2008). The spatial cluster analysis focused on 247 ST-flaA profiles comprising 85% of the isolates with location data and the temporal analysis focused on 316 profiles comprising 87% of the isolates with reporting dates. The remaining isolates had singleton profiles that cannot by definition be clustered.

Isolates were assigned spatial locations, defined as the centroid of the postcode sector of the patient's home postcode, and this information was known for 70.6% of the 5167 isolates with ST-flaA profiles. The census population of each postcode sector was obtained from the General Register of Scotland for mid-2001, and the coordinates of their centroids were obtained from Experian UK. Isolates were assigned dates, defined as either the reporting date or four days before the receipt date in cases when reporting date was unavailable.

Spatial clustering was analysed using the following parameter settings in SatScan: the time precision was day, the maximum cluster size was 50% of the population at risk, the type of analysis was retrospective and purely spatial, the probability model was Poisson, cluster shapes were circles and ellipses with a medium penalty for non-compactness, cluster detection was for high rates only and the time aggregation was 1 day. Temporal clustering was analysed using the same settings as in the spatial analysis except that the maximum cluster size of 90% of the study period was used, the type of analysis was purely temporal, and seasonal changes in relative risk were adjusted using weekly *Campylobacter* incidence values calculated as a rolling average for the focal week and the week before and after. The same values were used to adjust for seasonal changes in relative risk in the HF analysis. Joint spatio–temporal cluster analysis proved to be computationally impractical in SatScan using an appropriate combination of these settings, and is not possible using the HF analysis.

SatScan reports a single (most likely) cluster whereas HF also reports secondary clusters, and SatScan reports both observed and expected numbers of isolates in each cluster whereas HF reports only the observed number. Numbers of isolates in clusters were calculated as the difference between the observed and expected numbers of isolates in each SatScan cluster. The equivalent value for HF clusters was calculated as the observed number of isolates in each cluster multiplied by the ratio of observed minus expected over observed numbers of isolates in the corresponding SatScan cluster. Single isolates in a spatial or temporal cluster cannot be categorized as being observed or expected so cluster size and duration were reported only for the observed isolates.

Sets of *P*-values from analyses involving multiple comparisons were evaluated for significance using the false discovery rate (FDR) framework (Benjamini *et al.*, 1995). Q values were calculated from sets of *P* values using the R package. Statistical source attribution values were calculated for groups of isolates as the product of the number of isolates of each ST and the attribution values obtained using the program STRUCTURE for each ST in a previous analysis (Sheppard *et al.*, 2009b) averaged over all the STs in the group.

Results

Overall spatial and temporal stratification

Each health board yielded isolates with high levels of ST-flaA diversity (Table 19.1), with limited evidence of differences among Health Boards. Lothian Health Board had the highest Simpson's diversity index value, which was distinct from the three lowest values (Argyll and Clyde, Forth Valley and Lanarkshire Health Boards). Stratification of species and strain composition was analysed at the nested levels of among Health Boards, among months within Health Board, and within Health Board-month. There was no evidence for significant partitioning of variation in the composition of cases by species or by CC, ST or ST-flaA strains among health boards (Table 19.2). There was evidence for partitioning of variation at each level, especially in CC, ST and ST-flaA strains, among months within Health Board but these levels accounted for less than 2% of variation. Over 98% of the variation in composition at each categorical level occurred within Health Board-month.

General and household outbreaks

Three general outbreaks were reported during the study period: in Ayrshire and Arran (four cases), Tayside (34 cases; see (Forbes *et al.*, 2009)) and other Health Boards (two cases) and comprising 0.73% of cases. Fifty household outbreaks (range 2–4 cases each) comprising 1.88% of cases were detected. After removing cases in common, these outbreaks together comprised 2.54% of cases during the study period.

Cases from outbreaks were compared with sporadic cases according to their strain diversity. Average pairwise matching (APM) values of ST-flaA profiles among isolates from three general outbreaks were 0.333 for Ayrshire and Arran, 0.221 for Tayside and 0.667 for the outbreak reported after the study period in Forth Valley (six isolates). The comparison resampled groups had APM values of 0.086 (95% CI 0.000–0.500) for Ayrshire and Arran, 0.023 (95% CI 0.014–0.032) for Tayside and 0.047 (95% CI 0.000–0.200) for Forth Valley. Of the 28 household outbreaks for which ST-flaA profiles were available, 22 (78.6%) shared the same profile. In comparison, one pair of isolates randomly resampled 1,000 times from sporadic isolates from the whole study had the same ST-flaA profile in 1.8% of instances.

Spatial and temporal clustering

Spatial clusters were detected for 20 ST-flaA profiles (*Q*-value cut off = 0.05; corresponding *P*-values 0.00001–0.00543; Table 19.3). These clusters contained an average of 9.1 isolates (range 2.0–45.9) that comprised an average of 59.0% of the total isolates in each profile (range 6.6–99.9%). The observed isolates in each cluster were separated by an average pairwise distance of 25.5 km (range 0.0–95.3 km). The percentages of

Table 19.2 Partitioning of variation in the composition of species, CC, ST and ST-flaA profiles of *Campylobacter* isolates by Health Board (HB) and month of reporting

Source of variation	Sum of squares	Variance components	*P*-value	Percentage of variation
1. Species				
Among HB	2.073	0.00021	0.072	0.25
Among months within HB	13.160	0.00052	0.039	0.61
Within HB-month	416.343	0.08547	0.013	99.14
Total	431.576	0.08621		100.00
2. CC				
Among HB	7.665	0.00012	0.178	0.03
Among months within HB	100.828	0.00581	<0.00001	1.28
Within HB-month	2254.090	0.44804	<0.00001	98.69
Total	2362.584	0.45398		100.00
3. ST				
Among HB	6.456	–0.00008	0.594	–0.02
Among months within HB	98.198	0.00413	<0.00001	0.85
Within HB-month	2422.022	0.48142	<0.00001	99.17
Total	2526.675	0.48546		100.00
4. ST-flaA				
Among HB	6.168	–0.00012	0.791	–0.02
Among months within HB	74.323	0.00341	<0.00001	0.69
Within HB-month	2333.025	0.48808	<0.00001	99.30
Total	2413.516	0.49137		100.00

Table 19.3 Spatial and temporal statistical clusters of ST-flaA profiles of clinical *Campylobacter*

	Spatial-SS			Temporal-SS			Temporal-HF		
	Cc	Cj	Total	Cc	Cj	Total	Cc	Cj	Total
No. of isolates analysed	247	2863	3110	343	4156	4499	343	4156	4499
No. of primary clusters	2	18	20	1	66	67	1	55	56
No. of secondary clusters							0	7	7
No. of excess isolates in clusters	15.2	167.3	182.5	8.3	627.3	635.6	8.3	658.2	666.5
Excess isolates in clusters as a percentage of all isolates (incl. singletons)	4.1%	5.1%	5.0%	1.7%	13.4%	12.3%	1.7%	14.1%	12.9%

SS, SatScan cluster analysis; HF, hypothesis-free cluster analysis (see Materials and methods); Cc, *C. coli*; Cj, *C. jejuni*.

isolates in clusters were approximately equal for both *C. jejuni* and *C. coli*. The spatial clusters comprised 5.0% of all isolates with known locations.

Temporal clusters were detected for 67 ST-flaA profiles in a SatScan analysis and for a subset of 56 of the same profiles in an HF analysis (*Q*-value cut-off = 0.015 for both analyses, SatScan *P*-values: 0.00001–0.00786, HF *P*-values: 0.00001–0.00621; Table 19.3). Six of the profiles with an HF cluster also had one secondary cluster and a seventh had two secondary clusters. Of the 56 clusters detected in both analyses, 49 contained exactly the same isolates, three contained all but one of the same isolates and the average overlap of isolates in each corresponding SatScan and HF cluster was 99% (range 58–117%).

The 67 SatScan temporal clusters contained an average of 9.5 isolates (range 2.0–68.7) that comprised an average of 62.3% of all the isolates of its profile (range 7.5–99.9%), and lasted an average of 32.8 days (range 1–320). The 63 HF clusters contained an average of 10.6 isolates (range 2.0–73.3) that comprised an average of 54.8% of all the isolates of its profile (range 0.5–99.9%), and lasted an average of 30.6 days (range 1–241). The isolates in temporal clusters comprised either 12.3% (SatScan) or 12.9% (HF) of all isolates with dates. The proportion of isolates in temporal clusters was approximately eight-fold higher in *C. jejuni* than in *C. coli* (Table 19.3).

Temporal clusters lasting up to 50 days predominated (Fig. 19.1a), all cluster durations contributed to the cumulative percentage of clustered cases (Fig. 19.1b) and clusters occurred throughout the study period (Fig. 19.2). Two of the clusters (Fig. 19.2: black squares) contained 16 of the 34 isolates from the large reported outbreak in Tayside Health Board, and two further clusters (Fig. 19.2: black triangles) corresponded exactly to two household-associated groups of cases separate from that outbreak.

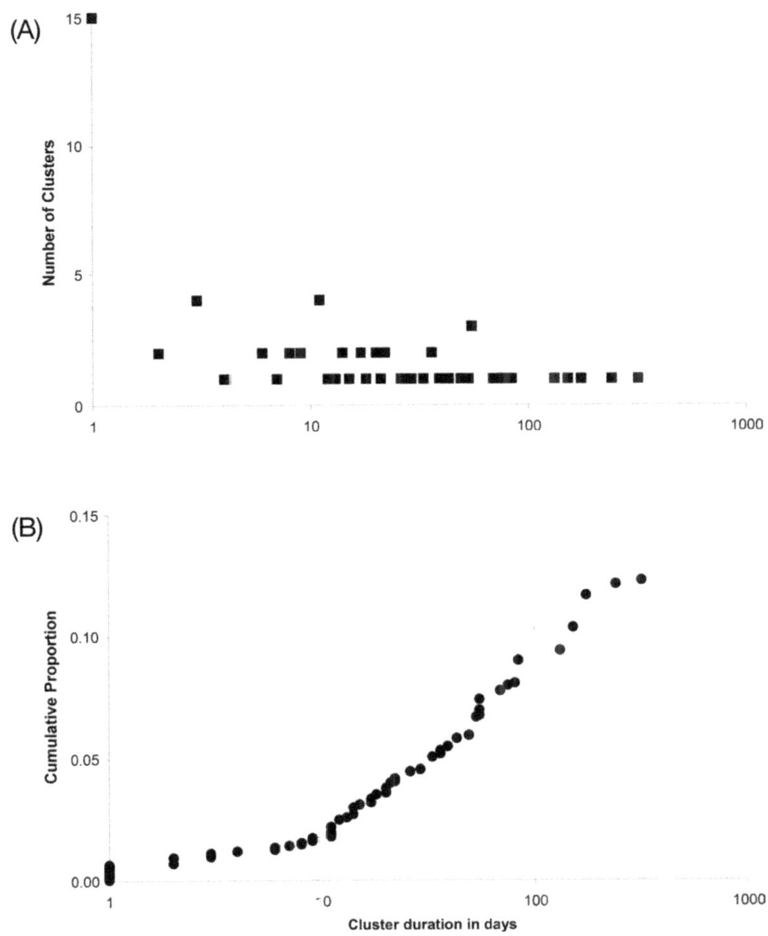

Figure 19.1 Temporal clusters identified in the SatScan analysis. (A) number of clusters of increasing duration. (B) size of clusters of increasing duration as cumulative proportions of all cases.

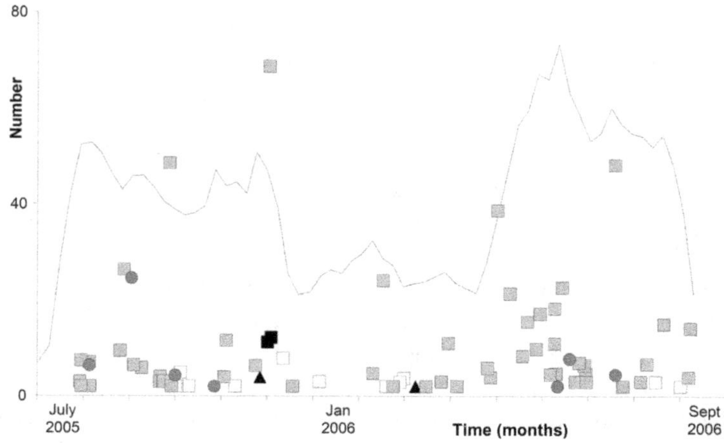

Figure 19.2 Temporal clusters of ST-flaA profiles from clinical *Campylobacter* isolates in 12 mainland Health Boards in Scotland during July 2005–Sept 2006. Each symbol shows one cluster as the median date of its duration plotted against the excess number of isolates it contained. Different types of cluster are indicated as follows: grey square: 56 clusters common to SatScan and HF; white square: 11 SatScan-only clusters; dark grey round: seven secondary HF clusters; black square: SatScan–HF clusters that overlap partly with isolates from a large reported *Campylobacter* outbreak in Tayside (12); black triangle: SatScan–HF clusters that correspond to two household-associated pairs of cases. The line shows the weekly number of total *Campylobacter* isolates as the three-week rolling average reduced by approximately two-fold.

Overlap of outbreaks, spatial clusters and temporal clusters

Cases from the general and household outbreaks occurred in 10 of the spatial clusters, comprising 1.0% of all isolates with known locations, and in 11 of the temporal clusters, comprising 1.1% of all isolates with dates. Seven of the 20 spatial clusters also showed evidence of temporal clustering, having fewer average pairwise days between cases than the lower 95% confidence interval of days among the same number of isolates randomly sampled from all cases with dates. Two of these corresponded to household outbreaks and a further two overlapped with the large general outbreak in the Tayside Health Board area. The total number of cases from the seven clusters comprised 1.5% of all cases with known locations.

Six of the 67 temporal clusters also showed evidence of spatial clustering, having shorter average pairwise distances between cases than the lower 95% confidence interval of distances among the same number of isolates randomly sampled from all the isolates with known locations. Two of the six clusters corresponded to household outbreaks and one overlapped with the large reported outbreak in Tayside Health Board. The total number of cases from the six clusters constituted 0.6% of all cases with known dates.

Source attribution

The isolates in spatial clusters and SatScan temporal clusters had similar average source attribution values (Table 19.4). There was some evidence for non-random attributions to the two minor infection sources, pigs and wild birds but no evidence that the cases in either type of cluster had distinct attribution values for the two major infection sources, ruminant and chicken.

Discussion

Campylobacter cases from national surveillance in mainland Scotland during one annual incidence cycle were homogeneous among Health Boards and almost homogeneous among months at the levels of species, clonal complex, ST and ST-flaA strain. The percentage of cases in general (0.73%) and household outbreaks (1.88%) was low, and these showed significantly increased levels of strain matching. Household outbreaks have been reported to be 30 fold greater than general outbreaks (Rotariu *et al.*, 2010). The numbers of clustered cases were more common than cases in

Table 19.4 Source attributions for isolates in spatial and temporal statistical clusters of ST-flaA profiles of clinical *Campylobacter*

	Spatial clusters		Temporal clusters	
Source	Actual cases	Cases resampled from all cases (95% CI)	Actual cases	Cases resampled from all cases (95% CI)
Ruminant	0.338	0.385 (0.322–0.449)	0.385	0.385 (0.352–0.418)
Pig	0.034	0.012 (0.002–0.024)	0.001	0.012 (0.006–0.018)
Chicken	0.546	0.505 (0.440–0.566)	0.573	0.504 (0.470–0.573)
Wild bird	0.082	0.099 (0.063–0.136)	0.041	0.099 (0.083–0.115)

The temporal clusters were those identified in the SatScan analysis.

household and general outbreaks, with cases in temporal clusters being approximately five times more common and spatial clusters two times more common, with spatial and temporal clusters showing little overlap. Spatial clusters (0.505 chicken, 0.385 ruminants), temporal clusters (0.504 chicken, 0.385 ruminants) and sporadic cases had similar attribution values for the two major sources of infection and the temporal clusters showed an excess of *C. jejuni* strains.

The overall pattern of strain stratification in Scotland resembled a spatially homogeneous system that fluctuated to a limited extent over the course of one seasonal incidence cycle. Spatial homogeneity of CC-defined strains is also evident in cases occurring in two English regions (O'Leary *et al.*, 2009), and spatial heterogeneity is evident in comparisons between geographically distant countries, such as the UK, Australia, Curaçao and New Zealand (Dingle, 2008 *et al.*; McTavish *et al.*, 2008). Temporal heterogeneity is also evident from the seasonal occurrence of ST-45 complex strains in NW England (Sopwith *et al.*, 2008) and macrorestriction profile strains in Helsinki (Hanninen *et al.*, 2000). *Campylobacter* species and strains are typically stratified among the hosts implicated as the major sources of infection (McCarthy *et al.*, 2007; Sheppard *et al.*, 2009a; Wilson *et al.*, 2008; Karenlampi *et al.*, 2007). The spatial homogeneity found here implies that any geographical heterogeneity in exposure to these sources has no strong influence on the species or strain composition of cases aggregated at Health Board level. The spatial and temporal clusters demonstrated in the present study are an unsuspected type of incidence pattern in clinical *Campylobacter* that becomes evident only at the level of genotypically defined strains. The events or processes from which these clusters originate are presently unclear but include foodborne dissemination of specific cohorts of strains (Sobel *et al.*, 2002; Pearson *et al.*, 2000), environmental triggers of incidence peaks in infection sources (Sopwith *et al.*, 2008), or strain-specific immunological responses to human infection, acting singly or in combination.

The extent of clustering suggests that the diffuse outbreak concept is valid for *Campylobacter* as well as *E. coli* and *Salmonella*, implying that outbreak detection technologies based on strain incidence could become a more generic paradigm for surveillance frameworks for foodborne zoonotic pathogens in Europe (European Centre for Disease Prevention and Control *et al.*, 2009) and the USA (Centers for Disease Control and Prevention, 2009) than is presently the case. Furthermore, the proportion of clustered cases implies that interventions (Strachan *et al.*, 2010) targeting the mechanisms producing these could be more effective in reducing *Campylobacter* incidence than the current view of predominantly sporadic transmission suggests.

Acknowledgements

We thank the staff of all the NHS bacteriology laboratories in Scotland for submitting isolates and J.F. Dallas, A. Smith-Palmer, M. Macrae and O. Rotariu for the provision of data and analyses. This study used the *C. jejuni/C. coli* Multilocus Sequence Typing Web site developed by Keith Jolley and Man-Suen Chan at the University of Oxford.

References

Benjamini, Y., and Hochberg, Y. (1995). Controlling the false discovery rate – a practical and powerful approach to multiple testing. J. R. Stat. Soc. Ser. B-Methodol. *57*, 289–300.

Centers for Disease Control and Prevention, PulseNet. Available at: http://www.cdc.gov/PULSENET/index.htm (Accessed 18 December 2011).

Charlett, A., Cowden, J.M., Frost, J.A., Gillespie, I.A., Millward, J., Neal, K.R., O'Brien, S.J., Painter, M.J., Syed, Q., Tompkins, D.S., *et al.* (2003). Point source outbreaks of *Campylobacter jejuni* infection – are they more common than we think and what might cause them? Epidemiol. Infect. *130*, 367–375.

Clark, C.G., Bryden, L., Cuff, W.R., Johnson, P.L., Jamieson, F., Ciebin, B., and Wang, G.H. (2005). Use of the Oxford multilocus sequence typing protocol and sequencing of the flagellin short variable region to characterize isolates from a large outbreak of waterborne *Campylobacter* sp strains in Walkerton, Ontario, Canada. J. Clin. Microbiol. *43*, 2080–2091.

Cucala, L. (2008). A hypothesis-free multiple scan statistic with variable window. Biom. J. *50*, 299–310.

Dabney, A., and Storey, J. Q-VALUE, Available at: http://www.genomics.princeton.edu/storeylab/qvalue/index.html (Accessed 18 December 2011).

Dingle, K.E., Colles, F.M., Wareing, D.R.A., Ure, R., Fox, A.J., Bolton, F.E., Bootsma, H.J., Willems, R.J.L., Urwin, R., and Maiden, M.C.J. (2001). Multilocus sequence typing system for *Campylobacter jejuni*. J. Clin. Microbiol. *39*, 14–23.

Dingle, K.E., McCarthy, N.D., Cody, A.J., Peto, T.E.A., and Maiden, M.C.J. (2008). Extended sequence typing of *Campylobacter* spp, United Kingdom. Emerg. Infect. Dis.*14*, 1620–1622.

Ethelberg, S., Olsen, K.E.P., Gerner-Smidt, P., and Molbak, K. (2004). Household outbreaks among culture-confirmed cases of bacterial gastrointestinal disease. Am. J. Epidemiol. *159*, 406–412.

European Centre for Disease Prevention and Control. Programme on food and water-borne diseases and zoonoses. Available at: http://ecdc.europa.eu/en/activities/diseaseprogrammes/fwd/Pages/index.aspx (Accessed 18 December 2011).

Excoffier, L., Laval, G., and Schneider, S. (2005). Arlequin (version 3.0): An integrated software package for population genetics data analysis. Evol. Bioinform. Online *1*, 47–50. Available at: http://www.ncbi.nlm.nih.gov/pmc/articles/PMC2658868/ (Accessed 18 December 2011).

Forbes, K.J., Gormley, F.J., Dallas, J.F., Labovitiadi, O., MacRae, M., Owen, R.J., Richardson, J., Strachan, N.J.C., Cowden, J.M., Ogden, I.D., and McGuigan, C.C. (2009). *Campylobacter* immunity and coinfection following a large outbreak in a farming community. J. Clin. Microbiol. *47*, 111–116.

Friedman, C.R., Hoekstra, R.M., Samuel, M., Marcus, R., Bender, J., Shiferaw, B., Reddy, S., Ahuja, S.D., Helfrick, D.L., Hardnett, F., *et al.* (2004). Risk factors for sporadic *Campylobacter* infection in the United States: A case–control study in FoodNet sites. Clin. Infect. Dis. *38*, S285–296.

Frost, J.A., Gillespie, I.A., and O'Brien, S.J. (2002). Public health implications of *Campylobacter* outbreaks in England and Wales, 1995–9: epidemiological and microbiological investigations. Epidemiol. Infect. *128*, 111–118.

Gormley, F.J., MacRae, M., Forbes, K.J., Ogden, I.D., Dallas, J.F., and Strachan, N.J.C. (2008). Has retail chicken played a role in the decline of human campylobacteriosis? Appl. Environ. Microbiol. *74*, 383–390.

Hanninen, M.L., Perko-Makela, P., Pitkala, A., and Rautelin, H. (2000). A three-year study of *Campylobacter jejuni* genotypes in humans with domestically acquired infections and in chicken samples from the Helsinki area. J. Clin. Microbiol. *38*, 1998–2000.

Health Protection Agency Bioinformatics. V-DICE: VNTR Diversity and Confidence Extractor. Available at: http://www.hpa-bioinformatics.org.uk/cgi-bin/DICI/DICI.pl (Accessed 18 December 2011).

Health Protection Scotland. Gastrointestinal and zoonoses: outbreaks. Available at: www.hps.scot.nhs.uk/giz/outbreaks.aspx?subjectid=5,23@44 (Accessed 18 December 2011).

Hood, G.M. (2009). PopTools version 3.1.1. Available at: http://www.cse.csiro.au/poptools. (Accessed 18 December 2011).

Jolley, K.A., and Maiden, M.C.J. (2010). BIGSdb: Scalable analysis of bacterial genome variation at the population level. BMC Bioinformat. *11*, 595.

Jonsson, M.E., Heier, B.T., Norstrom, M., and Hofshagen, M. (2010). Analysis of simultaneous space–time clusters of Campylobacter spp. in humans and in broiler flocks using a multiple dataset approach. Int. J. Health Geogr. *9*, 48.

Karenlampi, R., Rautelin, H., Schonberg-Norio, D., Paulin, L., and Hanninen, M.L. (2007). Longitudinal study of Finnish *Campylobacter jejuni* and *C. coli* isolates from humans, using multilocus sequence typing, including comparison with epidemiological data and isolates from poultry and cattle. Appl. Environ. Microbiol. *73*, 148–155.

Kulldorff, M., Heffernan, R., Hartman, J., Assuncao, R., and Mostashari, F. (2005). A space–time permutation scan statistic for disease outbreak detection. PLoS Med. *2*, 216–224.

McCarthy, N.D., Colles, F.M., Dingle, K.E., Bagnall, M.C., Manning, G., Maiden, M.C.J., and Falush, D. (2007). Host-associated genetic import in *Campylobacter jejuni*. Emerg. Infect. Dis. *13*, 267–272.

McTavish, S.M., Pope, C.E., Nicol, C., Sexton, K., French, N., and Carter, P.E. (2008). Wide geographical distribution of internationally rare *Campylobacter* clones within New Zealand. Epidemiol. Infect. *136*, 1244–1252.

Meinersmann, R.J., Helsel, L.O., Fields, P.I., and Hiett, K.L. (1997). Discrimination of *Campylobacter jejuni* isolates by *fla* gene sequencing. J. Clin. Microbiol. *35*, 2810–2814.

O'Leary, M.C., Harding, O., Fisher, L., and Cowden, J. (2009). A continuous common-source outbreak of campylobacteriosis associated with changes to the preparation of chicken liver pate. Epidemiol. Infect. *137*, 383–388.

Olson, C.K., Ethelberg, S., van Pelt, W., and Tauxe, R.V. (2008). Epidemiology of *Campylobacter jejuni* infections in industrialized nations. In *Campylobacter*, 3rd edn., Nachamkin, I., Szymanski, C.M., and Blaser, M.J., eds. (ASM Press, Washington, DC). pp.163–189.

Pearson, A.D., Greenwood, M.H., Donaldson, J., Healing, T.D., Jones, D.M., Shahamat, M., Feltham, R.K.A., and Colwell, R.R. (2000). Continuous source outbreak of campylobacteriosis traced to chicken. J Food Prot. *63*, 309–314.

Rotariu, O., Smith-Palmer, A., Cowden, J., Bessell, P.R., Innocent, G.T., Reid, S.W.J., Matthews, L., Dallas, J., Ogden, I.D., Forbes, K.J., and Strachan, N.J.C. (2010). Putative household outbreaks of campylobacteriosis typically comprise single MLST genotypes. Epidemiol. Infect. *138*, 1744–1747.

Said, B., Wright, F., Nichols, G.L., Reacher, M., and Rutter, M. (2003). Outbreaks of infectious disease associated with private drinking water supplies in England and Wales 1970–2000. Epidemiol. Infect. *130*, 469–479.

Sails, A.D., Swaminathan, B., and Fields, P.I. (2003). Utility of multilocus sequence typing as an epidemiological tool for investigation of outbreaks of gastroenteritis caused by *Campylobacter jejuni*. J. Clin. Microbiol. *41*, 4733–4739.

Samuel, M.C., Vugia, D.J., Shallow, S., Marcus, R., Segler, S., McGivern, T., Kassenborg, H., Reilly, K., Kennedy, M., Angulo, F., and Tauxe, R.V. (2004). Epidemiology of sporadic *Campylobacter* infection in the United States and declining trend in incidence, FoodNet 1996–1999. Clin. Infect. Dis. *38*, S165–174.

Sheppard, S.K., Dallas, J.F., MacRae, M., McCarthy, N.D., Sproston, E.L., Gormley, F.J., Strachan, N.J.C., Ogden, I.D., Maiden, M.C.J., and Forbes, K.J. (2009a). *Campylobacter* genotypes from food animals, environmental sources and clinical disease in Scotland 2005/6. Int. J. Food Microbiol. *134*, 96–103.

Sheppard, S.K., Dallas, J.F., Strachan, N.J.C., MacRae, M., McCarthy, N.D., Wilson, D.J., Gormley, F.J., Falush, D., Ogden, I.D., Maiden, M.C.J., and Forbes, K.J. (2009b). *Campylobacter* genotyping to determine the source of human infection. Clin. Infect. Dis. *48*, 1072–1078.

Sintchenko, V., Iredell, J.R., and Gilbert, G.L. (2007). Opinion – Pathogen profiling for disease management and surveillance. Nat. Rev. Microbiol. *5*, 464–470.

Sobel, J., Griffin, P.M., Slutsker, L., Swerdlow, D.L., and Tauxe, R.V. (2002). Investigation of multistate foodborne disease outbreaks. Pub. Health Rep. *117*, 8–19.

Sopwith, W., Birtles, A., Matthews, M., Fox, A., Gee, S., Painter, M., Regan, M., Syed, Q., and Bolton, E. (2008). Identification of potential environmentally adapted *Campylobacter jejuni* strain, United Kingdom. Emerg. Infect. Dis. *14*, 1769–1773.

Strachan, N.J.C., and Forbes, K.J. (2010). The growing UK epidemic of human campylobacteriosis. Lancet *376*, 665–667.

Terajima, J., Izumiya, H., Iyoda, S., Mitobe, J., Miura, M., and Watanabe, H. (2006). Effectiveness of pulsed-field gel electrophoresis for the early detection of diffuse outbreaks due to Shiga toxin-producing *Escherichia coli* in Japan. Foodborne Pathog. Dis. 3, 68–73.

Werber, D., Dreesman, J., Feil, F., van Treeck, U., Fell, G., Ethelberg, S., Hauri, A.M., Roggentin, P., Prager, R., St Fisher, I., *et al.* (2005). International outbreak of *Salmonella* Oranienburg due to German chocolate. BMC Infect. Dis. 5, 7.

Wheeler, J.G., Sethi, D., Cowden, J.M., Wall, P.G., Rodrigues, L.C., Tompkins, D.S., Hudson, M.J., and Roderick, P.J. (1999). Study of infectious intestinal disease in England: rates in the community, presenting to general practice, and reported to national surveillance. Br Med. J. *318*, 1046–1050.

Wilson, D.J., Gabriel, E., Leatherbarrow, A.J.H., Cheesbrough, J., Gee, S., Bolton, E., Fox, A., Fearnhead, P., Hart, C.A., and Diggle, P.J. (2008). Tracing the source of campylobacteriosis. PLoS Genet. *4*, e1000203.

Anonymous. (2006). Update on multi-state outbreak of *E. coli* O157:H7 infections from fresh spinach. CDC. Available at: http://www.cdc.gov/foodborne/ecolispinach/100606.htm. (Accessed July 2008).

Ecology and Host Associations of *Campylobacter* in Wild Birds

20

Jonas Waldenström and Petra Griekspoor

Abstract

The following review aims to summarize the present knowledge on *Campylobacter jejuni* and other *Campylobacter*-species in wild birds. The question of whether campylobacters present in wildlife, especially those found in wild birds, play a role in the epidemiology of human campylobacteriosis has been around for more than 30 years (Kapperud and Rosef, 1983; Smibert, 1969). This review will summarize what is currently known and point to the gaps in understanding of these organisms and their interplay with hosts and the environment. Here the focus will mainly be on *C. jejuni*, as this species has received most attention, but data for *Campylobacter coli* and *Campylobacter lari* will also be discussed.

Poultry and *Campylobacter*

Human *Campylobacter* infections are generally considered to be food-borne, where the disease is acquired through consumption of contaminated food items. Because the incubation period is approximately 3 days (Skirrow and Blaser, 2000), the food item that caused the infection has often been consumed and the remains thrown away, when the infection has been confirmed by the hospital. This obstructs epidemiological investigations. For this reason, many epidemiological studies have focused on indirect studies of the infection pathway. For example, by comparing the phenotypic and genotypic properties of isolates from different stages in the food chain such as farms, processing plants, and retailers.

Poultry products stand out in the literature as a major cause of human campylobacteriosis (Jacobs-Reitsma, 2000; Newell and Fearnley, 2003) and the fact that the hen, *Gallus gallus*, is the most common bird on the planet. Because of this it is appropriate to begin a discussion of *Campylobacter* in bird hosts with poultry. The proportion of chicken flocks colonized by *C. jejuni* is generally high in most of the world, both in backyard rearing, in organic farming and in high intensity units. However, prevalence shows significant spatial and temporal variation (Newell and Fearnley, 2003), and infection can be reduced with hygiene routines. Chickens exhibit coprophagic behaviour which facilities the spread of campylobacters within flocks and the incidence is normally very high in infected flocks. The infection dose to establish a *C. jejuni* colonization in chickens has been estimated to be as low as 40 CFU in experimental settings (Cawthraw *et al.*, 1996), but may differ depending on strain and host characteristics (Newell and Fearnley, 2003). The growth optimum of 37–42°C for *C. jejuni*, and other thermotolerant campylobacters, may be an adaptation to avian intestines (chickens 41°C). Eggs seem to be free of *Campylobacter*, although faeces-contaminated shell sometimes harbours live bacterial cells, but these are likely to die within a few days. Newly hatched chickens are not colonized by *C. jejuni*, and remain free until about 2–3 weeks of age. This lag-phase has been attributed to the existence of maternal antibodies (Sahin *et al.*, 2001), but other factors such as the existence of commensal organisms in the gut or changes in stock management could perhaps be involved (Newell and Fearnley, 2003).

Poultry usually show no external signs of infection, even when excreting large quantities of *Campylobacter* in their faeces, and often maintain the organisms in their guts throughout the production cycle – at least for the 35-day lifespan of commercially reared broilers. Because of this, *Campylobacter* is often described as a commensal organism in birds. However, *Campylobacter* can be isolated from the liver of chickens and can cause acute pathology in other host species. For example, *C. jejuni* and *C. coli* have been noted in an outbreak of infectious hepatitis in reared ostriches in Australia (Stephens *et al.*, 1998). This suggests the potential for pathogenic effects in poultry.

C. jejuni occurrence in wild birds

Based on the association between *C. jejuni* and poultry, it is not surprising that wild bird carriage of *Campylobacter* has achieved interest from an epidemiological perspective. Wild birds could theoretically transmit campylobacters to humans directly, either through consumption of wild birds as game, or contamination of other foods or water (Hudson *et al.*, 1990; Sacks *et al.*, 1986; Southern *et al.*, 1990; Varslot *et al.*, 1996). Indirect transmission may, however, be of more interest as wild birds could act as sources for colonization of food animals, particularly poultry and cattle (Brown *et al.*, 2004; Colles *et al.*, 2003, 2008a,b; French *et al.*, 2005; Leatherbarrow *et al.*, 2004), which are consumed by humans. Thus, wild birds could in theory be both transient vectors transporting *C. jejuni* between food animals or act as reservoirs for these pathogens, sometimes spilling over to food production systems. Furthermore, wild bird associated genotypes could serve as a genetic reservoir from which new food animal strains may emerge, either by adaptation to the food animal niche, or through transfer of genes by horizontal recombination (Sheppard *et al.*, 2009a, b, 2010, 2011).

Obtaining a comprehensive knowledge of *Campylobacter* populations associated with wild birds is challenging as they constitute a taxonomically, phenotypically and ecologically heterogeneous group of host animals. They differ in factors including size (2 g–140 kg), diet, annual movements between habitats, and association with humans and food animals (Waldenström *et al.*, 2002). Table 20.1 is a compilation of the available literature on the occurrence of *Campylobacter* in different wild bird species. Several of the studies date back to the early years of *Campylobacter* research (Kapperud and Rosef, 1983; Kapperud *et al.*, 1983; Smibert, 1969), others are more recent (Colles *et al.*, 2003, 2008a,c; French *et al.*, 2009; Griekspoor *et al.*, 2013; Hughes *et al.*, 2009; Sippy *et al.*, 2012), meaning that the culture methods and subsequent phenotypic and/or genotypic characterization differ between studies. The sampling techniques also differ. In some studies fresh bird droppings were collected in the field (Colles *et al.*, 2008a; French *et al.*, 2009; Palmgren *et al.*, 2004) and in others faecal swabs were taken from live birds (Broman *et al.*, 2002; Keller *et al.*, 2011; Waldenström *et al.*, 2002), or necroscopic material from natural carcasses or hunted birds (Robino *et al.*, 2010; Wahlstrom *et al.*, 2003). While generally being small, the sample sizes also differ between host species and studies. This means that comparing prevalence rates between studies should be avoided because it is highly dependent on sampling and culturing techniques. Furthermore, older studies, conducted before molecular identification techniques were available, may have been unable to differentiate between organisms which today are treated separately. This is probably most true for the diverse *C. lari* taxon, which in the older nomenclature was termed *Campylobacter laridis* and nalidix acid-resistant thermophilic campylobacters (NARTCs), which may have included isolates that today should have been characterized as UPTC, or other biovars of *C. lari*.

To date at least 11,357 wild birds of 205 species have been screened for the presence of *C. jejuni* (Table 20.1). Of these, 2187 samples were positive either by culture or by PCR. *C. jejuni* have been isolated from wild birds in Europe (Broman *et al.*, 2002, 2004; Brown *et al.*, 2004; Colles *et al.*, 2003, 2008a–c; French *et al.*, 2005; Fricker *et al.*, 1983; Fricker and Metcalfe, 1984; Glunder *et al.*, 1992; Griekspoor *et al.*, 2013; Hughes *et al.*, 2009; Kapperud and Rosef, 1983; Kapperud *et al.*, 1983; McCarthy *et al.*, 2007; Palmgren *et al.*, 2004; Rosef *et al.*, 1983; Sellin *et al.*, 2000; Sheppard *et al.*, 2010, 2011; Sixl *et al.*, 1997; Waldenström *et al.*, 2002, 2005, 2007; Whelan *et al.*, 1988), North America (Keller *et al.*, 2011; Pacha *et al.*, 1988; Quessy and Messier, 1992; Sippy *et al.*, 2012), Asia (Chuma *et al.*, 2000, 2003; Fukuyama *et al.*,

Table 20.1 A checklist of the reported occurrences of *Campylobacter* in wild bird species. For each host species and study, the number of individuals positive for *C. jejuni*, *C. lari*, *C. coli* and unknown *Campylobacter* species are given, as well as the number of sampled individuals (i.e. 5 positive for *C. jejuni* out of 35 sampled: 5/35). *C. lari* include NARTC and UPTC organisms. For all studies *C. jejuni* numbers are given, for other *Campylobacter* only when they have been specifically detected in the published study

Host family	Host species	*C. jejuni*	*C. lari*	*C. coli*	*Campylobacter* spp.	Study
Accipitridae	*Accipiter gentilis*	0/1				Waldenström *et al.* (2002)
	Accipiter nisus	1/32			1/32	Waldenström *et al.* (2002)
		1/–				Petersen *et al.* (2001)
	Accipiter cooperii	0/9				Keller *et al.* (2011)
	Accipiter striatus	0/1				Keller *et al.* (2011)
	Buteo buteo	1/–				Petersen *et al.* (2001)
	Buteo jamaicensis	1/8				Keller *et al.* (2011)
	Buteo lineatus	0/5				Keller *et al.* (2011)
	Circus cyaneus	0/1				Keller *et al.* (2011)
	Haliaeetus leucocephalus	0/12				Keller *et al.* (2011)
	Pandion haliaetus	0/20				Keller *et al.* (2011)
Aegithalidae	*Aegithalos caudatus*	0/8				Waldenström *et al.* (2002)
Alaudidae	*Alauda arvensis*	0/2				Waldenström *et al.* (2002)
Anatidae	*Aix sponsa*	0/9				Keller *et al.* (2011)
	Anas acuta	0/1			1/1	Waldenström *et al.* (2002)
		0/6				Yogasundram *et al.* (1989)
		0/82		2/82		Fukuyama *et al.* (1986)
		15/30				Luechtefeld *et al.* (1980)
	Anas americana	16/38				Luechtefeld *et al.* (1980)
	Anas clypeata	23/35				Luechtefeld *et al.* (1980)
	Anas carolinensis	9/56				Luechtefeld *et al.* (1980)
		1/1				Pacha *et al.* (1988)
	Anas crecca	0/2			1/2	Waldenström *et al.* (2002)
	Anas penelope	0/1				Waldenström *et al.* (2002)
		0/43	1/43			Hughes *et al.* (2009)
	Anas platyrhynchos	4/29				Yogasundram *et al.* (1989)
		82/243				Luechtefeld *et al.* (1980)
		70/89				Pacha *et al.* (1988)
		1/21				Keller *et al.* (2011)
		1/10				Robino *et al.* (2010)
		85/–				Griekspoor *et al.* (2013)
	Anas strepera	0/22				Yogasundram *et al.* (1989)
		4/26				Luechtefeld *et al.* (1980)
	Anas americana	1/3				Yogasundram *et al.* (1989)

Table 20.1 (Continued)

Host family	Host species	*C. jejuni*	*C. lari*	*C. coli*	*Campylobacter* spp.	Study
	Anas clypeata	0/1				Yogasundram *et al.* (1989)
	Anas sp.	7/–		2/–		Kakoyiannis *et al.* (1988)
		4/1				Petersen *et al.* (2001)
	Anser anser	0/1				Yogasundram *et al.* (1989)
	Anser/Branta sp.	166/331		1/331		Colles *et al.* (2008a)
	Aythya americana	11/22				Pacha *et al.* (1988)
	Aythya collaris	0/1				Pacha *et al.* (1988)
	Aythya fuligula	0/2				Waldenström *et al.* (2002)
	Branta bernicla	0/4			1/4	Waldenström *et al.* (2002)[a]
	Branta canadensis	12/105				Wahlstrom *et al.* (2003)
		0/25				Keller *et al.* (2011)
	Branta leucopsis	6/116				Waldenström *et al.* (2007)
	Bucephala clangula	0/1		1/1		Kapperud and Rosef (1983)
	Cygnus olor	0/2				Yogasundram *et al.* (1989)
		0/1				Keller *et al.* (2011)
		0/30		1/30		Hughes *et al.* (2009)
	Somateria mollissima	0/1			1/1	Waldenström *et al.* (2002)
Ardeidae	*Ardea alba*	0/1				Keller *et al.* (2011)
	Ardea herodidas	0/3				Keller *et al.* (2011)
	Bubulcus ibis	0/1				Keller *et al.* (2011)
	Nycticorax nycticorax	0/2				Keller *et al.* (2011)
Caprimulgidae	*Caprimulgus europaeus*	0/1				Waldenström *et al.* (2002)
Cardinalidae	*Cardinalis cardinalis*	0/1				Sacks *et al.* (1986)
		0/3				Sippy *et al.* (2012)
	Pheucticus ludovicianus	1/2				Sippy *et al.* (2012)
Certhiidae	*Certhia familiaris*	0/7				Waldenström *et al.* (2002)
	Certhia americana	0/1				Sippy *et al.* (2012)
Charadriidae	*Charadrius dubius*	0/2				Waldenström *et al.* (2002)
	Charadrius hyaticula	0/6	1/6		1/6	Waldenström *et al.* (2002)
	Charadrius vociferus	0/5				Keller *et al.* (2011)
	Pluvialis squatarola	0/1			1/1	Waldenström *et al.* (2002)
	Vanellus vanellus	29/63	28/63	6/63		Fricker and Metcalfe (1984)
		0/3	3/3			Waldenström *et al.* (2007)

Host family	Host species	*C. jejuni*	*C. lari*	*C. coli*	*Campylobacter* spp.	Study
Columbidae	*Columba palumbus*	0/2				Waldenström *et al.* (2002)
	Columba sp.	1/–				Petersen *et al.* (2001)
					1/5	Smibert (1969)
	Columba livia	2/16				Ito *et al.* (1988)
		51/378				Fukuyama *et al.* (1986)
		3/71				Kapperud and Rosef (1983)
		1/13		2/13		Robino *et al.* (2010)
		2/47				Hughes *et al.* (2009)
	Columbina passerina	0/1				Sacks *et al.* (1986)
	Streptopelia orientalis	1/62				Ito *et al.* (1988)
		0/17				Fukuyama *et al.* (1986)
Corvidae	*Corvus brachyrhynchos*	3/7				Keller *et al.* (2011)
	Corvus monedula	4/4				Waldenström *et al.* (2002)
		1/6				Hughes *et al.* (2009)
	Corvus frugilegus	1/–				Petersen *et al.* (2001)
	Corvus corax	1/–				Petersen *et al.* (2001)
		0/1				Kapperud and Rosef (1983)
	Corvus corone	41/48	1/48	2/48		Kapperud and Rosef (1983)
		39/78		31/78		Robino *et al.* (2010)
	Corvus levaillanti/ corone	30/87				Ito *et al.* (1988)
	Corvus ossifragus	4/10				Keller *et al.* (2011)
	Cyanocitta cristata	5/30				Keller *et al.* (2011)
		0/2				Sippy *et al.* (2012)
	Cyanopica cyana	2/10				Ito *et al.* (1988)
	Garrulus glandarius	1/3				Hughes *et al.* (2009)
	Pica pica	1/–				Petersen *et al.* (2001)
		2/71	3/71			Hughes *et al.* (2009)
Cuculidae	*Cuculus canorum*	0/2				Waldenström *et al.* (2002)
Diomedea	*Diomedea melanophris*	0/40				Broman *et al.* (2000)[b]
	Diomedea chrysostoma	0/100				Broman *et al.* (2000)[b]
Falconidae	*Falco peregrinus*	2/69				Palmgren *et al.* (2004)[b]
	Falco tinnunculus	0/14		1/14		Hughes *et al.* (2009)
Fringillidae	*Carduelis cannabina*	0/2				Waldenström *et al.* (2002)
	Carduelis carduelis	0/5				Waldenström *et al.* (2002)

Table 20.1 (Continued)

Host family	Host species	*C. jejuni*	*C. lari*	*C. coli*	*Campylobacter* spp.	Study
	Carduelis chloris	0/29			1/29	Waldenström *et al*. (2002)
		0/93				Hughes *et al*. (2009)
	Carduelis flammea	0/6				Waldenström *et al*. (2002)
	Carduelis flavirostris	0/15				Waldenström *et al*. (2002)
	Carpodacus erythrinus	0/6				Waldenström *et al*. (2002)
		0/1				Kapperud and Rosef (1983)
	Coccothraustes coccothraustes	0/3				Waldenström *et al*. (2002)
	Emberiza citrinella	1/19				Waldenström *et al*. (2002)
		0/1				Kapperud and Rosef (1983)
	Emberiza hortulana	0/6			1/6	Waldenström *et al*. (2002)
	Emberiza schoeniclus	0/8				Waldenström *et al*. (2002)
		1/5				Kapperud and Rosef (1983)
	Emberiza spodocephala	0/4				Fukuyama *et al*. (1986)
	Emberiza rustica	0/4				Fukuyama *et al*. (1986)
	Fringilla coelebs	0/26				Waldenström *et al*. (2002)
		0/2				Kapperud and Rosef (1983)
		1/84				Hughes *et al*. (2009)
	Fringilla montifringilla	0/9				Waldenström *et al*. (2002)
	Junco hyemalis	0/18				Sippy *et al*. (2012)
	Loxia curvirostra	0/1				Waldenström *et al*. (2002)
	Melospiza americana	0/6				Sippy *et al*. (2012)
	Melospiza melodia	0/4				Sippy *et al*. (2012)
	Serinus serinus	0/7				Waldenström *et al*. (2002)
	Spinus tristis	0/2				Sippy *et al*. (2012)
	Zonotrichia albicollis	4/28		1/28		Sippy *et al*. (2012)
	Zonotrichia querula	0/1				Sippy *et al*. (2012)
Gaviidae	*Gavia immer*	0/2				Keller *et al*. (2011)
Gruidae	*Grus* sp.				71/403	Chuma *et al*. (2003)
Haematopodidae	*Haematopus ostralegus*	36/56	15/56	5/56		Fricker and Metcalfe (1984)
		0/95	6/95			Hughes *et al*. (2009)
Hirundinidae	*Delichon urbica*	0/13				Waldenström *et al*. (2002)
		0/9				Kapperud and Rosef (1983)

Host family	Host species	*C. jejuni*	*C. lari*	*C. coli*	*Campylobacter* spp.	Study
	Hirundo rustica	0/4				Waldenström *et al.* (2002)
		0/2				Kapperud and Rosef (1983)
		0/134				Fukuyama *et al.* (1986)
		0/1				Sippy *et al.* (2012)
Icteridae	*Quiscalus quiscula*	14/34				Sacks *et al.* (1986)
Laniidae	*Lanius collurio*	0/9				Waldenström *et al.* (2002)
	Lanius excubitor	0/1				Waldenström *et al.* (2002)
Laridae	*Larus argentatus*				13/43	Kapperud and Rosef (1983)
		179/942	332/942	92/942		Whelan *et al.* (1988)[c]
		27/27				Glunder *et al.* (1992)[b]
		4/10				Glunder *et al.* (1992)[b]
		96/389	127/389	40/389		Fricker *et al.* (1983)
	Larus a. smithsonianus	0/3				Keller *et al.* (2011)
	Larus atricilla	5/15				Keller *et al.* (2011)
	Larus canus	0/1				Waldenström *et al.* (2002)
					7/37	Kapperud and Rosef (1983)
		0/9				Palmgren *et al.* (1997)
	Larus crassirostris	0/36				Fukuyama *et al.* (1986)
	Larus delawarensis				42/264	Quessy and Messier (1992)
		0/3				Keller *et al.* (2011)
	Larus dominicanus	34/–	43/–	11/–		Kakoyiannis *et al.* (1988)
	Larus fuscus	0/8				Kapperud and Rosef (1983)
	Larus marinus	0/4				Kapperud and Rosef (1983)
	Larus ovaehollandiae	22/–				Griekspoor *et al.* (2013)
	Larus ridibundus	0/3				Waldenström *et al.* (2002)
		26/41				Sixl *et al.* (1997)
					22/88	Kapperud and Rosef (1983)
		0/41				Palmgren *et al.* (1997)
		235/786	8/786	7/786		Broman *et al.* (2002)
		0/13	1/13			Hughes *et al.* (2009)
	Larus sp.	24/180	9/180	12/180	45/180	Kapperud and Rosef (1983)
		2/–				Petersen *et al.* (2001)
		11/104	3/104		2/104	Wahlstrom *et al.* (2003)[a]
					3/–	Kaneko *et al.* (1999)[f]

Table 20.1 (Continued)

Host family	Host species	C. jejuni	C. lari	C. coli	Campylobacter spp.	Study
	Sterna hirundo	1/36		1/36		Kapperud and Rosef (1983)
Mimidae	*Toxostoma rufum*	0/1				Keller *et al*. (2011)
	Dumetella carolinensis	0/2				Sippy *et al*. (2012)
Motacillidae	*Anthus pratensis*	0/9	1/9		2/9	Waldenström *et al*. (2002)
		0/1				Kapperud and Rosef (1983)
	Anthus trivialis	0/12				Waldenström *et al*. (2002)
	Motacilla alba	2/32			8/32	Waldenström *et al*. (2002)
	Motacilla flava	0/3			1/3	Waldenström *et al*. (2002)
Muscicapidae	*Cathartus fuscescens*	0/2				Keller *et al*. (2011)
	Cathartus ustulatus	0/2				Keller *et al*. (2011)
	Erithacus rubecula	0/299				Waldenström *et al*. (2002)
		0/28			1/28	Waldenström *et al*. (2010)
	Ficedula hypoleuca	0/5				Waldenström *et al*. (2002)
		0/2				Kapperud and Rosef (1983)
	Ficedula parva	0/5				Waldenström *et al*. (2002)
	Fidecula albicollis	0/3				Waldenström *et al*. (2002)
	Hylocichla mustelina	0/2				Keller *et al*. (2011)
	Luscinia luscinia	0/11				Waldenström *et al*. (2002)
	Luscinia svecica	0/19				Waldenström *et al*. (2002)
	Muscicapa narcissina	0/6				Fukuyama *et al*. (1986)
	Muscicapa striata	0/11				Waldenström *et al*. (2002)
		0/1				Kapperud and Rosef (1983)
	Oenanthe oenanthe	0/2				Waldenström *et al*. (2002)
		0/1				Kapperud and Rosef (1983)
	Phoenicurus ochruros	0/11				Waldenström *et al*. (2002)
	Phoenicurus phoenicurus	0/44				Waldenström *et al*. (2002)
		0/1				Kapperud and Rosef (1983)
	Saxicola rubetra	0/3				Waldenström *et al*. (2002)
		0/2				Kapperud and Rosef (1983)
	Siala sialis	0/8				Keller *et al*. (2011)
	Turdus chrysolaus	0/6				Fukuyama *et al*. (1986)
	Turdus iliacus	5/10				Waldenström *et al*. (2002)

Host family	Host species	C. jejuni	C. lari	C. coli	Campylobacter spp.	Study
		0/4				Palmgren *et al*. (1997)
	Turdus merula	7/44	4/44	7/44	2/44	Waldenström *et al*. (2002)
		2/32				Palmgren *et al*. (1997)
					1/1	Smibert (1969)
		107/–				Griekspoor *et al*. (2013)
		1/53				Hughes *et al*. (2009)
	Turdus migratorius	4/121				Keller *et al*. (2011)
		0/1				Sippy *et al*. (2012)
	Turdus philomelos	7/25			1/25	Waldenström *et al*. (2002)
		0/27				Palmgren *et al*. (1997)
		84/–				
	Turdus pilaris	2/14				Waldenström *et al*. (2002)
		0/3				Kapperud and Rosef (1983)
		0/5				Palmgren *et al*. (1997)
	Turdus torquatus	0/1				Waldenström *et al*. (2002)
		0/2				Palmgren *et al*. (1997)
	Turdus viscivorus	1/1				Waldenström *et al*. (2002)
Paridae	*Parus ater*	0/2				Waldenström *et al*. (2002)
		0/4				Fukuyama *et al*. (1986)
	Parus caeruleus	0/7				Waldenström *et al*. (2002)
		0/1				Kapperud and Rosef (1983)
	Parus major	0/10				Waldenström *et al*. (2002)
		0/1				Kapperud and Rosef (1983)
		1/154				Hughes *et al*. (2009)
	Poecile atricapillus	0/2				Sippy *et al*. (2012)
Parulidae	*Oreothlypis celata*	0/1				Sippy *et al*. (2012)
	Oreothlypis ruficapilla	0/1				Sippy *et al*. (2012)
Passeridae	*Passer domesticus*	0/4				Waldenström *et al*. (2002)
		0/2				Sacks *et al*. (1986)
		0/7				Kapperud and Rosef (1983)
		0/15				Robino *et al*. (2010)
		2/107		1/107		Sippy *et al*. (2012)
		1/86				Hughes *et al*. (2009)
	Passer montanus	0/11				Waldenström *et al*. (2002)
		0/8				Ito *et al*. (1988)
		3/5				Fukuyama *et al*. (1986)
		0/2				Kapperud and Rosef (1983)
		13/507				Chuma *et al*. (2000)

Table 20.1 (Continued)

Host family	Host species	C. jejuni	C. lari	C. coli	Campylobacter spp.	Study
	Passer sp.				11/15	Smibert (1969)
Phalacrocoracidae	*Phalancrocorax carbo*	0/1				Waldenström *et al.* (2002)
Phasianidae	*Bumbusicola thoracica*	0/34				Ito *et al.* (1988)
	Perdix perdix	0/3				Waldenström *et al.* (2002)
	Phasianus colchicus	0/25				Ito *et al.* (1988)
		1/–				Petersen *et al.* (2001)
Picidae	*Dendrocopus major*	0/1				Kapperud and Rosef (1983)
	Jynx torquilla	0/3				Kapperud and Rosef (1983)
	Picoides pubescens	0/1				Sippy *et al.* (2012)
	Picoides villosus	0/1				Sippy *et al.* (2012)
	Picus viridis	0/5				Waldenström *et al.* (2002)
		0/1				Kapperud and Rosef (1983)
Prunellidae	*Prunella modularis*	1/13				Waldenström *et al.* (2002)
Pycnonotidae	*Hypsipetes amaurotis*	4/36				Ito *et al.* (1988)
Rallidae	*Gallinula chloropus*	1/3				Hughes *et al.* (2009)
	Rallus aquaticus	0/1				Waldenström *et al.* (2002)
Regulidae	*Regulus regulus*	0/200			2/200	Waldenström *et al.* (2002)
	Regulus calendula	0/1				Sippy *et al.* (2012)
Scolopacidae	*Actitis hypoleucos*	0/8	3/8		5/8	Waldenström *et al.* (2002)
		0/1	1/1			Waldenström *et al.* (2007)
	Arenaria interpres	0/4	3/4			Waldenström *et al.* (2002)
		27/49	14/49	8/49		Fricker and Metcalfe (1984)
		0/27	1/27			Hughes *et al.* (2009)
	Calidris acuminata	46/–				Griekspoor *et al.* (2013)
	Calidris alba	0/1				Waldenström *et al.* (2002)
	Calidris alpina	29/313	70/313	6/313	145/313	Griekspoor *et al.* (2013; Waldenström *et al.* (2002)
		0/4	3/4			Waldenström *et al.* (2007)
		1/1				Keller *et al.* (2011)
	Calidris canutus	0/3	2/3			Waldenström *et al.* (2002)
		0/14		1/14		Hughes *et al.* (2009)
	Calidris ferruginea	1/9	6/9		2/9	Waldenström *et al.* (2002)
	Calidris minuta	2/8			5/8	Waldenström *et al.* (2002)
	Calidris temminckii	0/4	2/4			Waldenström *et al.* (2002)
	Gallinaga gallinago	0/1			1/1	Waldenström *et al.* (2002)
		0/1	1/1			Waldenström *et al.* (2007)

Host family	Host species	*C. jejuni*	*C. lari*	*C. coli*	*Campylobacter* spp.	Study
	Limicola falcinellus	4/11	5/11	3/11		Waldenström *et al*. (2002)
	Lymnocryptes minimus	0/1			1/0	Waldenström *et al*. (2002)
	Philomachus pugnax	1/1		1/1		Waldenström *et al*. (2002)
	Scolopax rusticola	0/1			1/1	Waldenström *et al*. (2002)
	Tringa glareola	1/6			1/6	Waldenström *et al*. (2002)
	Tringa nebularia	0/1			1/1	Waldenström *et al*. (2002)
	Tringa ochropus	0/1				Waldenström *et al*. (2002)
	Tringa totanus	26/54	20/54	8/54		Fricker and Metcalfe (1984)
		1/122	105/122			Waldenström *et al*. (2007)
Spheniscidae	*Eudyptes chrysolophus*	3/100				Broman *et al*. (2000; Griekspoor *et al*. (2009)
Strigidae	*Aegolius funereus*	0/1				Waldenström *et al*. (2002)
		0/12				Kapperud and Rosef (1983)
	Asio flammeus	0/2				Waldenström *et al*. (2002)
	Asio otus	10/49				Waldenström *et al*. (2002)
	Strix uralensis	2/2				Kapperud and Rosef (1983)
	Surnia ulula	0/1				Kapperud and Rosef (1983)
Sturnidae	*Sturnus cineraceus*	5/35				Ito *et al*. (1988)
		0/12				Fukuyama *et al*. (1986)
	Sturnus vulgaris	10/31	2/31		2/31	Waldenström *et al*. (2002)
		0/2				Kapperud and Rosef (1983)
		1/31				Palmgren *et al*. (1997)
		9/–				Colles *et al*. (2003)
					1/9	Smibert (1969)
		285/954	60/954	6/954		Colles *et al*. (2008a)
		0/3				Sippy *et al*. (2012)
		1/9				Hughes *et al*. (2009)
Sylviidae	*Acrocephalus palustris*	0/4				Waldenström *et al*. (2002)
	Acrocephalus schoenobaenus	0/14				Waldenström *et al*. (2002)
	Acrocephalus scirpaceus	0/7				Waldenström *et al*. (2002)
	Hippolais icterina	0/8				Waldenström *et al*. (2002)
		0/2				Kapperud and Rosef (1983)
	Locustella naevia	0/1				Waldenström *et al*. (2002)
	Phylloscopus collybita	0/8				Waldenström *et al*. (2002)

Table 20.1 (Continued)

Host family	Host species	C. jejuni	C. lari	C. coli	Campylobacter spp.	Study
	Phylloscopus fuscatus	0/1				Waldenström *et al*. (2002)
	Phylloscopus proregulus	0/2				Waldenström *et al*. (2002)
	Phylloscopus sibilatrix	0/2				Waldenström *et al*. (2002)
	Phylloscopus trochilus	0/107				Waldenström *et al*. (2002)
		0/3				Kapperud and Rosef (1983)
	Sylvia atricapilla	0/10				Waldenström *et al*. (2002)
		0/1				Kapperud and Rosef (1983)
	Sylvia borin	0/4				Waldenström *et al*. (2002)
		0/12				Kapperud and Rosef (1983)
	Sylvia communis	0/20				Waldenström *et al*. (2002)
		0/8				Kapperud and Rosef (1983)
	Sylvia curruca	0/35				Waldenström *et al*. (2002)
	Sylvia nisoria	0/1				Waldenström *et al*. (2002)
Tetraonidae	*Lagopus* spp.	0/37				Kapperud and Rosef (1983)
Troglodytidae	*Troglodytes troglodytes*	1/37	1/37			Waldenström *et al*. (2002)
Tytonidae	*Tytao alba*	1/24				Hughes *et al*. (2009)

[a]Based on pooled samples.
[b]Samples from nestlings.
[c]Numbers of colonized birds calculated from percentages given in the article.
[d]Species not stated in the article, but confirmed by author in email correspondence.
[e]Waldenström *et al*. unpublished.
[f]UPTC.

1986; Ito *et al.*, 1988), Antarctica (Bonnedahl *et al.*, 2005; Broman *et al.*, 2000; Griekspoor *et al.*, 2009), Australia and New Zeeland (French *et al.*, 2009; Griekspoor *et al.*, 2013; Kakoyiannis *et al.*, 1988), suggesting a global distribution of *Campylobacter* in wild birds. There was geographic and taxonomic bias in the datasets with most studies from Europe (Scandinavia and the British isles in particular) and certain bird taxa receiving more interest than others, either because they occur in farm environments, or because that they are easy to sample.

The distribution of *Campylobacter* appears to vary among hosts. Some bird species have high prevalence rates and others have low rates or have not been found to be colonized by *C. jejuni* at all – which also has been noted by some of the original authors (Kapperud and Rosef, 1983; Keller *et al.*, 2011; Sippy *et al.*, 2012; Waldenström *et al.*, 2002). Of the sampled birds, ducks (family *Anatidae*, in particular the dabbling ducks of genus *Anas*), shorebirds (families *Charadriidae*, *Haematopodidae* and *Scopolacidae*), corvids (family *Corvidae*), gulls (family *Laridae*), pigeons and doves (family *Columbidae*), thrushes (genus *Turdus* in the family *Muscicapidae*), and starlings (family *Sturnidae*) were most often colonized (Table 20.2). *C. jejuni* was less commonly detected in passerine birds,

Table 20.2 Distribution of *C. jejuni* positive samples from wild birds per host family

Host family	Number of sampled birds	Number of *C. jejuni*-positive samples
Laridae	3016	666
Anatidae	1486	518
Sturnidae	1086	311
Muscicapidae	1008	220
Passeridae	754	19
Scolopacidae	692	139
Columbidae	608	61
Fringillidae	395	7
Corvidae	360	133
Sylviidae	250	
Regulidae	201	
Paridae	181	1
Hirundinidae	163	
Haematopodidae	151	36
Diomedea	140	
Spheniscidae	100	3
Accipitridae	91	4
Falconidae	83	2
Charadriidae	80	29
Alcidae	76	
Strigidae	67	12
Phasianidae	63	1
Motacillidae	57	2
Tetraonidae	37	
Troglodytidae	37	1
Pycnonotidae	36	4
Icteridae	34	14
Tytonidae	24	1
Prunellidae	13	1
Picidae	12	
Laniidae	10	
Aegithalidae	8	
Certhiidae	8	
Ardeidae	7	
Cardinalidae	6	1
Rallidae	4	1
Mimidae	3	
Alaudidae	2	
Cuculidae	2	
Gaviidae	2	
Parulidae	2	
Caprimulgidae	1	
Phalacrocoracidae	1	

other than thrushes and starlings, and in some passerine families, such as Old World warblers (*Sylvidae* and *Regulidae*), no *C. jejuni* have been found despite fairly large sample sizes (Table 20.2).

The degree of association with humans differs between birds, and thereby the potential for transmission of *Campylobacter* between wild birds and food animals also differs, or vice versa. Generally, bird species associated with human settlements have been overrepresented in terms of *C. jejuni* carriage. For instance, gulls and corvids are often feeding in urban or agricultural areas, often on waste products, and are thereby potentially exposed to *Campylobacter* of human or food animal origin. Other farm-related birds, such as geese, feral pigeons, starlings and some other passerine species (including House sparrow *Passer domesticus*, Tree sparrow *Passer montanus*, White-throated sparrow *Zonotrichia albicollis*) are strongly associated with agricultural areas and often feed (and sometimes breed) in and around farms. Some isolates have been retrieved from birds of prey, which may reflect occurrence of *C. jejuni* in their prey. Two groups of birds that are not as strongly associated with human settlements and still have high prevalence rates of *C. jejuni* are thrushes and shorebirds.

Temporal changes within individual bird hosts can influence that detection of *Campylobacter*. For example, how long an individual bird is colonized, when bacterial cells are shed, and how prevalence varies within populations of hosts. Temporal variation in prevalence has been noted in Black-headed gulls *Larus ridibundus* (Broman *et al.*, 2002), geese (Colles *et al.*, 2008a), thrushes (Griekspoor *et al.*, 2013; Waldenström *et al.*, 2002) and European starlings *Sturnus vulgaris* (Colles *et al.*, 2008c), and it is likely that seasonal changes in gut flora exist in many bird species. Furthermore, work is needed to identify if such changes depend on variations in host biology, such as the proportion of naive juvenile hosts, build-up of immunity, body condition, differences in diet at different times of the year, or environmental conditions affecting survival of bacterial cells outside the host. One study that addresses some of these factors focussed on free-living European starlings in the United Kingdom (Colles *et al.*, 2008c). Both nestlings and adult starlings were susceptible to *C. jejuni* infection, but young birds tended to be more often colonized than older birds, and both shedding rates and prevalence showed temporal patterns. Interestingly, recaptured starlings often shed different *C. jejuni* genotypes, suggesting a rapid turnover of *Campylobacter* in single hosts and no evidence of cross-strain immunity (Colles *et al.*, 2008c). Waldenström *et al.* (Waldenström *et al.*, 2010), studied colonization patterns and putative effects on European Robins *Erithacus rubecula* with an experimental approach. Birds were inoculated orally with *C. jejuni* from either a human patient or from another wild bird species, the song thrush *Turdus philomelos*. These two isolates provoked very different host responses, where the song thrush isolate colonized all challenged birds (colonization lasted 6.8 days on average) and birds infected with the human isolate remained uncolonised. A slight reduction of body mass was detected, implying a cost of infection, and robins infected with the wild bird isolate generated a specific antibody response to *C. jejuni* membrane proteins from the avian isolate, which also was cross-reactive to membrane proteins of the human isolate (Waldenström *et al.*, 2010). Clearly, more studies are needed to establish basic infection biology of *C. jejuni* in wild birds, especially for bird species that live in farm environments and where transmission between food animals and wild birds are most likely to occur.

Epidemiological studies of wild bird campylobacters

Wild birds have been the probable cause in some human campylobacteriosis outbreaks, for instance when 1000 people in Norway were thought to have become infected by *C. jejuni* through faecal contamination of the village's water reservoir by Pink-footed Geese *Anser brachyrhynchus* when drinking water chlorination failed (Varslot *et al.*, 1996). Similarly, a large waterborne outbreak in the USA was suggested to have been caused by malfunctioning water treatment equipment in connection with faecal contamination from wild birds (Sacks *et al.*, 1986). However, in neither study were similar strains found in bird faeces and sick patients, opening the possibility of other contamination sources. One interesting example comes from the UK, where a case–control study showed that Jackdaws (*Corvus monedula*) and Magpies (*Pica pica*) that attacked the foil caps of milk bottles and drank of the milk caused people to become infected with *C. jejuni* (Hudson *et al.*, 1990; Southern *et al.*, 1990). A later study indicated that nearly two thirds of 434 interviewed people would use the milk, even though the top had been pecked (Phillips, 1995).

Over the years, several studies have attempted to compare *Campylobacter* phenotypes/genotypes from wild birds with isolates from human or food animal sources. In a study in Scotland, the Penner serotype distribution of *C. jejuni* and *C. coli* samples from Herring Gulls (*Larus argentatus*) differed from samples from patients of the same region, although some serotypes were common for both groups (Whelan *et al.*, 1988). Furthermore, the proportion of untypable sample with the Penner scheme was much larger in the gull samples (76%) than in the samples from patients (24%) (Whelan *et al.*, 1988). Similarly, in a Japanese study, 34 out of 51 *C. jejuni* strains isolated from free-living feral pigeons were untypable using a slide agglutination technique. The remaining 17 isolates, however, were typed into eight serogroups, two of which were associated with human campylobacter enteritis (Fukuyama *et al.*, 1986). The largest serotyping study, conducted on 374 thermotolerant campylobacters isolated from wild birds, poultry, pigs and flies in Norway, used heat stable antigens and passive hemagglutination for characterization. In this study, poultry and pig isolates had different sets of serotypes (probably many of the pig isolates were *C. coli*) while the wild bird group were grouped into 15 serotypes with 29 samples from gulls (*Larus* sp.), 17 crows (*Corvus corone*), 8 puffins (*Fratercula arctica*), three pigeons (*Columba* sp.) and two owls. Three of the 15 wild bird serotypes were similar to poultry serotypes, and three to pig serotypes. However, nearly 60% of the wild bird isolates were untypeable with the antisera used, compared with 24% and 29% in pigs and poultry, respectively (Rosef *et al.*, 1985).

In a New Zealand study a technique called bacterial restriction endonuclease DNA analysis (BRENDA) – a precursor to PFGE, 34 *C. jejuni* isolates from Kelp gulls (*Larus dominicanus*) and seven from ducks gave rise to 22 and four different restriction profiles, respectively. None of these profiles resembled those of the 316 human isolates, while some of the 98 chicken isolates had profiles matching those from humans (Kakoyiannis *et al.*, 1988). Petersen *et al.* (Petersen *et al.*, 2001) investigated 120 *C. jejuni* isolates from wildlife, humans and broilers in Denmark using a combination of *fla*-typing, PFGE and the Penner serotyping scheme. Thirteen of these isolates came from dead or dying wild birds that had been sent in for necroscopy, while 44 came from Danish patients, 29 from Danish poultry flocks and the remaining from various wild mammals (including hedgehog, fox, badger and seal). Ten of the wild bird isolates gave unique restriction profiles that were not shared by samples from the other reservoirs, and four isolates were untypable with the Penner serotype system. One isolate from a pigeon was indistinguishable from that of a fox in all discrimination techniques applied. Two isolates, one from a Common Buzzard (*Buteo buteo*) and one from a duck, had the O:12 serotype and were similar or identical to isolates from humans, broilers and wild mammals (Petersen *et al.*, 2001).

Using PFGE, Broman *et al.* (Broman *et al.*, 2002) examined the similarities between 76 *C. jejuni* isolates from Black-headed gulls (*Larus ridibundus*) and 36 and 56 isolates from broiler chickens and humans, respectively. Macrorestriction profile similarities were higher between isolates of human and broilers, than between those of wild birds and either of the other two hosts. However, identical profiles were found in two gull isolates and one human isolates, strongly indicating that they may have been colonized by the same clone of *C. jejuni*. The most prevalent profiles of gulls, however, did not occur among isolates of humans or broilers, suggesting the existence of a subpopulation of the bacterium adapted to species-specific colonization or environmental survival. Broman and co-workers have also made another PFGE study on wild bird *C. jejuni* isolates (Broman *et al.*, 2004). In that study, 39 *C. jejuni* isolates from a large collection of campylobacters from migrating birds in Sweden, were compared to 47 domestically acquired patient isolates collected from the same geographical area and time period. Bird strains came from 20 different wild bird species and included raptors, small passerines, corvids and shorebirds. Generally, bird isolates tended to cluster together with other bird isolates, often closest to those from the same host species or of the same feeding guild. The majority of bird isolate macrorestriction profiles could not be identified among the subtypes from humans. However, two bird isolates, one from a European starling and one from a blackbird (*Turdus merula*), had profiles that were very similar to human subtypes. In Sweden, *C. jejuni* has also been isolated from two nestling Peregrine Falcons (*Falco peregrinus*). These isolates were indistinguishable from that of a patient isolate from southern Sweden using PFGE with two different restriction enzymes. It was assumed that the falcons acquired this strain of *C. jejuni* from its diet, possibly from eating gulls or pigeons (Palmgren *et al.*, 2004).

In the last decade, multilocus sequence typing (MLST) has become the method of choice for epidemiologic typing of *C. jejuni* (Dingle *et al.*, 2001, 2002; McCarthy *et al.*, 2007). Colles *et al.*, applied the characterized a collection of 112 isolates from farm animals (turkeys, chickens, cattle and sheep) and wild European starlings. Although the sample size of starlings was as low as nine isolates, some of the wild bird isolates belonged to clonal complexes that are associated with poultry and with human disease (e.g. clonal complexes ST-21 and ST-45 and 257), while some belonged to an environmentally associated clonal complex (Colles *et al.*, 2003). Later more comprehensive studies on starlings from the same team have corroborated these findings, with occurrence of food-animal associated genotypes at low to moderate levels, and the occurrence of a number of sequence types that are uncommon in food animals, including possible starling-associated clonal complexes (Colles *et al.*, 2008a,c). Hughes *et al.* (Hughes *et al.*, 2009) conducted a large-scale screening study of wild bird samples from northern England and used MLST to characterize obtained isolates. Of the resulting 36 *C. jejuni*, some belonged to food animal associated clonal complexes, for instance the ST-48 complex which was found in house sparrows (*Passer domesticus*) and a great tit (*Parus major*) and the ST-45 complex found in moorhens (*Gallinula chloropus*), magpies and a barn owl (*Tyto alba*). Other samples had STs that were unassigned to clonal complexes and unknown, or rare in food animal *C. jejuni* collections (Hughes *et al.*, 2009). Sippy *et al.* (Sippy *et al.*, 2012) studied the occurrence of *Campylobacter* in American birds, and found nine birds positive for *C. jejuni* that gave rise to four different STs, including two ST-806 isolates from House sparrows that also were resistant to fluoroquinolones. Isolates from bird droppings in the field (but where host species is uncertain) have also been used to address involvement of wild birds in *C. jejuni* epidemiology in farmland in UK and playgrounds in New Zealand (French *et al.*, 2005, 2009), again showing similar patterns with a proportion of bird isolates harbouring genotypes associated with human disease and food animals and the remaining isolates with genotypes with unknown epidemiology or associated to other wild bird *C. jejuni* genotypes. Perhaps the most extreme example of human to bird transmission comes from the sub-Antarctic island Bird Island, where visiting researchers or tourists were believed to have transmitted human-associated *C. jejuni* to wild Macaroni Penguins (*Eudyptes chrysolophus*) (Broman *et al.*, 2000; Griekspoor *et al.*, 2009).

The collection of sequence data has enabled studies addressing population genetic structure by phylogenetic analyses and host attribution techniques (Cody *et al.*, 2010; Griekspoor *et al.*, 2010, 2013; McCarthy *et al.*, 2007; Sheppard *et al.*, 2009a,b, 2010, 2011). Generally, these studies corroborate older studies in terms of defining different sets of *C. jejuni* – those that are highly associated to different farm animals, or a general farm animal niche, and environmental wild bird sets that tend to be less common among food animals (Cody *et al.*, 2010; Griekspoor *et al.*, 2010, 2013; McCarthy *et al.*, 2007; Sheppard *et al.*, 2009a,b, 2010, 2011). One example of this comes from the study of Griekspoor *et al.* that compared sets of wild bird *C. jejuni* isolates (in total 928 wild bird isolates) in Europe and Australia with genotypes from different food animals in the same regions (Griekspoor *et al.*, 2013). The *C. jejuni* populations investigated showed strong patterns of genetic subdivision, with distinct genetic subsets associated with particular hosts. *C. jejuni* sets from different food animal hosts (poultry, cattle, sheep) were genetically much more similar to each other than the sets from different wild bird species. Furthermore, the genetic population structure was clearly associated with host taxonomy, with *C. jejuni* populations from taxonomically closely related bird species being most similar to each other. Interestingly, the level of host-associations were far more important than geographical distance as previously shown (Sheppard *et al.*, 2010). For instance, *C. jejuni* from silver gulls (*Larus ovaehollandiae*) sampled in Australia and black-headed gulls sampled in Sweden showed very little genetic subdivision, and similar patterns were seen in comparisons of *C. jejuni* from Australian Sharp-tailed sandpipers (*Calidris acuminata*) and Swedish dunlins (*Calidris alpina*) (Griekspoor *et al.*, 2013).

Concluding remarks

There is no doubt that *Campylobacter* is widespread in animals and in the environment, with isolation from sources as diverse as ground water, insects, wild birds and mammals, and domestic animals including cats, dogs, sheep, cattle and poultry. Because of this the epidemiology is complex. As detailed above, most studies involving wild bird isolates have shown that the range of genotypes found in wild birds is large, and that although there are some genotypes shared between birds and food animals, the majority are not. It is more likely that the birds with *C. jejuni* genotypes overlapping with human and food animal isolates have acquired these bacterial strains through transmission from anthropogenic sources, rather than the reverse. The bird species where food animal associated isolates have been found are mainly those species that often live close to human settlements, and many sample schemes have targeted birds in farm environments where transmission is supposedly largest. Many questions remain and future studies should try to incorporate host ecology, and address key infection biology parameters. Even simple things, for instance the carriage time of *C. jejuni* in different wild bird hosts, is virtually unknown, which makes informed risk assessment difficult.

C. jejuni is one of the most important avian zoonoses, with important consequence for the food industry and public health. Available data suggest that host is an important determinant of the genotypes isolated from wild animal sources but the segregation by host is not strong in food animals. This leads to questions about the whether the population structure observed in food animals is the consequence of more recent introductions and expansions of certain genotypes into the relatively new agricultural niche. The limited host range for some of the genotypes obtained from wild birds suggests host barriers to gene flow, or the existence of host-adapted strains of *C. jejuni* (Griekspoor *et al.*, 2013; McCarthy *et al.*, 2007; Sheppard *et al.*, 2010). With the development of next generation sequence approaches we can now try to identify the genetic elements involved in host adaptation, environmental survival and pathogenicity. By using host associated wild bird *C. jejuni* strains in such comparison, gene content analyses could potentially identify genes that are associated with specific traits. For instance, is it adaptations favouring a broad host range (tolerance for different enteric environments) that predisposes strains to emerge in food animals (Cleaveland *et al.*, 2001)?

References

Bonnedahl, J., Broman, T., Waldenstrom, J., Palmgren, H., Niskanen, T., and Olsen, B. (2005). In search of human-associated bacterial pathogens in Antarctic wildlife: report from six penguin colonies regularly visited by tourists. Ambio. *34*, 430–432.

Broman, T., Bergstrom, S., On, S.L., Palmgren, H., McCafferty, D.J., Sellin, M., and Olsen, B. (2000). Isolation and characterization of *Campylobacter jejuni* subsp. jejuni from macaroni penguins (Eudyptes chrysolophus) in the subantarctic region. Appl. Environ. Microbiol. *66*, 449–452.

Broman, T., Palmgren, H., Bergström, S., Sellin, M., Waldenström, J., Danielsson-Tham, M.L., and Olsen, B. (2002). *Campylobacter jejuni* in Black-Headed Gulls (*Larus ridibundus*): Prevalence, Genotypes, and Influence on *C. jejuni* Epidemiology. J. Clin. Microbiol. *40*, 4594–4602.

Broman, T., Waldenström, J., Dahlgren, D., Carlsson, I., Eliasson, I., and Olsen, B. (2004). Diversities and similarities in PFGE profiles of *Campylobacter jejuni* isolated from migrating birds and humans. J. Appl. Microbiol. *96*, 834–843.

Brown, P.E., Christensen, O.F., Clough, H.E., Diggle, P.J., Hart, C.A., Hazel, S., Kemp, R., Leatherbarrow, A.J., Moore, A., Sutherst, J., *et al.* (2004). Frequency and spatial distribution of environmental Campylobacter spp. Appl. Environ. Microbiol. *70*, 6501–6511.

Cawthraw, S.A., Wassenaar, T.M., Ayling, R., and Newell, D.G. (1996). Increased colonization potential of *Campylobacter jejuni* strain 81116 after passage through chickens and its implication on the rate of transmission within flocks. Epidemiol. Infect. *117*, 213–215.

Chuma, T., Hashimoto, S., and Okamoto, K. (2000). Detection of thermophilic *Campylobacter* from sparrows by multiplex PCR: the role of sparrows as a source of contamination of broilers with *Campylobacter*. J. Vet. Med. Sci. *62*, 1291–1295.

Chuma, T., Kurihara, T., Hashimoto, S., Okamoto, K., Shioya, K., Moriyama, H., Hokazono, T., and Miayazaki, Y. (2003). Antimicrobial susceptibility and fla A PCR-RFLP analysis of *Campylobacter* isolates from migratory cranes, broilers, and sparrows in Kagoshima, Japan. Int. J. Med. Microbiol. *293*, 145–145.

Cleaveland, S., Laurenson, M.K., and Taylor, L.H. (2001). Diseases of humans and their domestic mammals: pathogen characteristics, host range and the risk of emergence. Philos. Trans. R. Soc. Lond. B Biol. Sci. *356*, 991–999.

Cody, A.J., Colles, F.M., Sheppard, S.K., and Maiden, M.C. (2010). Where does *Campylobacter* come from? A molecular odyssey. Adv. Exp. Med. Biol. *659*, 47–56.

Colles, F.M., Jones, K., Harding, R.M., and Maiden, M.C. (2003). Genetic Diversity of *Campylobacter jejuni* Isolates from Farm Animals and the Farm Environment. Appl. Environ. Microbiol. *69*, 7409–7413.

Colles, F.M., Dingle, K.E., Cody, A.J., and Maiden, M.C. (2008a). Comparison of *Campylobacter* populations in wild geese with those in starlings and free-range poultry on the same farm. Appl. Environ. Microbiol. *74*, 3583–3590.

Colles, F.M., Jones, T.A., McCarthy, N.D., Sheppard, S.K., Cody, A.J., Dingle, K.E., Dawkins, M.S., and Maiden, M.C. (2008b). *Campylobacter* infection of broiler chickens in a free-range environment. Environ. Microbiol. *10*, 2042–2050.

Colles, F.M., McCarthy, N.D., Howe, J.C., Devereux, C.L., Gosler, A.G., and Maiden, M.C. (2008c). Dynamics of *Campylobacter* colonization of a natural host, *Sturnus vulgaris* (European Starling). Environ. Microbiol. *11*, 258–267.

Dingle, K.E., Colles, F.M., Wareing, D.R., Ure, R., Fox, A.J., Bolton, F.E., Bootsma, H.J., Willems, R.J., Urwin, R., and Maiden, M.C. (2001). Multilocus sequence typing system for *Campylobacter jejuni*. J. Clin. Microbiol. *39*, 14–23.

Dingle, K.E., Colles, F.M., Ure, R., Wagenaar, J.A., Duim, B., Bolton, F.J., Fox, A.J., Wareing, D.R., and Maiden, M.C. (2002). Molecular characterization of *Campylobacter jejuni* clones: a basis for epidemiologic investigation. Emerg. Infect. Dis. *8*, 949–955.

French, N., Barrigas, M., Brown, P., Ribiero, P., Williams, N., Leatherbarrow, H., Birtles, R., Bolton, E., Fearnhead, P., and Fox, A. (2005). Spatial epidemiology and natural population structure of *Campylobacter jejuni* colonizing a farmland ecosystem. Environ. Microbiol. *7*, 1116–1126.

French, N.P., Midwinter, A., Holland, B., Collins-Emerson, J., Pattison, R., Colles, F., and Carter, P. (2009). Molecular epidemiology of *Campylobacter jejuni* isolates from wild-bird fecal material in children's playgrounds. Appl. Environ. Microbiol. *75*, 779–783.

Fricker, C.R., and Metcalfe, N. (1984). Campylobacters in wading birds (Charadrii): incidence, biotypes and isolation techniques. Zentralbl. Bakteriol. Mikrobiol. Hyg. [B] *179*, 469–475.

Fricker, C.R., Girdwood, R.W., and Munro, D. (1983). A comparison of procedures for the isolation of campylobacters from seagull faeces. J. Hyg.(Lond) *91*, 445–450.

Fukuyama, M., Kamimura, T., Itoh, T., Saito, K., Takahashi, M., Sakai, S., Murata, M., Kohzaki, K., Hara, M., and Shimizu, T. (1986). Distribution of *Campylobacter jejuni* in wild birds and serogroup of isolates by slide agglutination technique. Nippon Juigaku. Zasshi *48*, 487–493.

Glunder, G., Neumann, U., and Braune, S. (1992). Occurrence of *Campylobacter* spp. in young gulls, duration of *Campylobacter* infection and reinfection by contact. Zentralbl. Veterinarmed. [B] *39*, 119–122.

Griekspoor, P., Olsen, B., and Waldenström, J. (2009). *Campylobacter jejuni* in penguins, Antarctica. Emerg. Infect. Dis. *15*, 847–848.

Griekspoor, P., Engvall, E.O., Olsen, B., and Waldenstrîm, J. (2010). Multilocus sequence typing of *Campylobacter jejuni* from broilers. Vet. Microbiol. *140*, 180–185.

Griekspoor, P., Colles, F.M., McCarthy, N.D., Hansbro, P.M., Ashhurt-Smith, C., Olsen, B., Hasselquist, D., Maiden, M.C.J., and Waldenström, J. (2013). Marked host specificity and lack of phylogeographic population structure of *Campylobacter jejuni* in wild birds. Mol. Ecol. *22*, 1463–1472.

Hudson, S.J., Sobo, A.O., Russel, K., and Lightfoot, N.F. (1990). Jackdaws as potential source of milk-borne *Campylobacter jejuni* infection. Lancet *335*, 1160.

Hughes, L.A., Bennett, M., Coffey, P., Elliott, J., Jones, T.R., Jones, R.C., Lahuerta-Marin, A., Leatherbarrow, A.H., McNiffe, K., Norman, D., *et al.* (2009). Molecular epidemiology and characterization of *Campylobacter* spp. isolated from wild bird populations in northern England. Appl. Environ. Microbiol. *75*, 3007–3015.

Ito, K., Kubokura, Y., Kaneko, K., Totake, Y., and Ogawa, M. (1988). Occurrence of *Campylobacter jejuni* in free-living wild birds from Japan. J. Wildl. Dis. *24*, 467–470.

Jacobs-Reitsma, W.F. (2000). *Campylobacter* in the food supply. In *Campylobacter*, Nachamkin, I., and Blaser, M.J., eds. (ASM Press, Washington, DC), pp. 467–481.

Kakoyiannis, C.K., Winter, P.J., and Marshall, R.B. (1988). The relationship between intestinal *Campylobacter* species isolated from animals and humans as determined by BRENDA. Epidemiol. Infect. *100*, 379–387.

Kaneko, A., Matsuda, M., Miyajima, M., Moore, J.E., and Murphy, P.G. (1999). Urease-positive thermophilic strains of *Campylobacter* isolated from seagulls (Larus spp.). Lett. Appl. Microbiol. *29*, 7–9.

Kapperud, G., and Rosef, O. (1983). Avian wildlife reservoir of *Campylobacter fetus* subsp. *jejuni*, *Yersinia* spp., and *Salmonella* spp. in Norway. Appl. Environ. Microbiol. *45*, 375–380.

Kapperud, G., Rosef, O., Rostad, O.W., and Lid, G. (1983). Isolation of *Campylobacter fetus* subsp. *jejuni* from the common puffin (*fratercula arctica*) in Norway. J. Wildl. Dis. *19*, 64–65.

Keller, J.I., Shriver, W.G., Waldenstrom, J., Griekspoor, P., and Olsen, B. (2011). Prevalence of *Campylobacter* in wild birds of the mid-Atlantic region, USA. J. Wildl. Dis. *47*, 750–754.

Leatherbarrow, A.J., Hart, C.A., Kemp, R., Williams, N.J., Ridley, A., Sharma, M., Diggle, P.J., Wright, E.J., Sutherst, J., and French, N.P. (2004). Genotypic and antibiotic susceptibility characteristics of a *Campylobacter coli* population isolated from dairy farmland in the United Kingdom. Appl. Environ. Microbiol. *70*, 822–830.

Luechtefeld, N.A., Blaser, M.J., Reller, L.B., and Wang, W.L. (1980). Isolation of *Campylobacter fetus* subsp. *jejuni* from migratory waterfowl. J. Clin. Microbiol. *12*, 406–408.

McCarthy, N.D., Colles, F.M., Dingle, K.E., Bagnall, M.C., Manning, G., Maiden, M.C., and Falush, D. (2007). Host-associated genetic import in *Campylobacter jejuni*. Emerg. Infect. Dis. *13*, 267–272.

Newell, D.G., and Fearnley, C. (2003). Sources of *Campylobacter* colonization in broiler chickens. Appl. Environ. Microbiol. *69*, 4343–4351.

Pacha, R.E., Clark, G.W., Williams, E.A., and Carter, A.M. (1988). Migratory birds of central Washington as reservoirs of *Campylobacter jejuni*. Can. J. Microbiol. *34*, 80–82.

Palmgren, H., Sellin, M., Bergstrîm, S., and Olsen, B. (1997). Enteropathogenic bacteria in migrating birds arriving in Sweden. Scand. J. Infect. Dis. *29*, 565–568.

Palmgren, H., Broman, T., Waldenstrîm, J., Lindberg, P., Aspan, A., and Olsen, B. (2004). *Salmonella Amager*, *Campylobacter jejuni*, and urease-positive thermophilic *Campylobacter* found in free-flying peregrine falcons (*Falco peregrinus*) in Sweden. J. Wildl. Dis. *40*, 583–587.

Petersen, L., Nielsen, E.M., Engberg, J., On, S.L., and Dietz, H.H. (2001). Comparison of genotypes and serotypes of *Campylobacter jejuni* isolated from Danish wild mammals and birds and from broiler flocks and humans. Appl. Environ. Microbiol. *67*, 3115–3121.

Phillips, C.A. (1995). Incidence of *Campylobacter* and possible modes of transmission. Nutr. Food Sci. *95*, 12–17.

Quessy, S., and Messier, S. (1992). Prevalence of *Salmonella* spp., *Campylobacter* spp. and *Listeria* spp. in ring-billed gulls (*Larus delawarensis*). J. Wildl. Dis. *28*, 526–531.

Robino, P., Tomassone, L., Tramuta, C., Rodo, M., Giammarino, M., Vaschetti, G., and Nebbia, P. (2010). Prevalence of *Campylobacter jejuni*, *Campylobacter coli* and enteric *Helicobacter* in domestic and free living birds in North-Western Italy. Schweizer Archiv fur Tierheilkunde *152*, 425–431.

Rosef, O., Gondrosen, B., Kapperud, G., and Underdal, B. (1983). Isolation and characterization of *Campylobacter jejuni* and *Campylobacter coli* from domestic and wild mammals in Norway. Appl. Environ. Microbiol. *46*, 855–859.

Rosef, O., Kapperud, G., Lauwers, S., and Gondrosen, B. (1985). Serotyping of *Campylobacter jejuni*, *Campylobacter coli*, and *Campylobacter laridis* from domestic and wild animals. Appl. Environ. Microbiol. *49*, 1507–1510.

Sacks, J.J., Lieb, S., Baldy, L.M., Berta, S., Patton, C.M., White, M.C., Bigler, W.J., and Witte, J.J. (1986). Epidemic campylobacteriosis associated with a community water supply. Am. J. Pub. Health *76*, 424–428.

Sahin, O., Zhang, Q., Meitzler, J.C., Harr, B.S., Morishita, T.Y., and Mohan, R. (2001). Prevalence, antigenic specificity, and bactericidal activity of poultry anti-*Campylobacter* maternal antibodies. Appl. Environ. Microbiol. *67*, 3951–3957.

Sellin, M., Palmgren, H., Broman, T., Bergstrom, S., and Olsen, B. (2000). Involving ornithologists in the surveillance of vancomycin-resistant enterococci. Emerg. Infect. Dis. *6*, 87–88.

Sheppard, S.K., Dallas, J.F., MacRae, M., McCarthy, N.D., Sproston, E.L., Gormley, F.J., Strachan, N.J., Ogden, I.D., Maiden, M.C., and Forbes, K.J. (2009a). *Campylobacter* genotypes from food animals, environmental sources and clinical disease in Scotland 2005/6. Int. J. Food Microbiol. *134*, 96–103.

Sheppard, S.K., Dallas, J.F., Strachan, N.J., MacRae, M., McCarthy, N.D., Wilson, D.J., Gormley, F.J., Falush, D., Ogden, I.D., Maiden, M.C., and Forbes, K.J. (2009b). *Campylobacter* genotyping to determine the source of human infection. Clin. Infect. Dis. *48*, 1072–1078.

Sheppard, S.K., Colles, F., Richardson, J., Cody, A.J., Elson, R., Lawson, A., Brick, G., Meldrum, R., Little, C.L., Owen, R.J., *et al.* (2010). Host association of *Campylobacter* genotypes transcends geographic variation. Appl. Environ. Microbiol. *76*, 5269–5277.

Sheppard, S.K., Colles, F.M., McCarthy, N.D., Strachan, N.J., Ogden, I.D., Forbes, K.J., Dallas, J.F., and Maiden, M.C. (2011). Niche segregation and genetic structure of *Campylobacter jejuni* populations from wild and agricultural host species. Mol. Ecol. *20*, 3484–3490.

Sippy, R., Sandoval-Green, C.M., Sahin, O., Plummer, P., Fairbanks, W.S., Zhang, Q., and Blanchong, J.A. (2012). Occurrence and molecular analysis of *Campylobacter* in wildlife on livestock farms. Vet. Microbiol. *157*, 369–375.

Sixl, W., Karpiskova, R., Hubalek, Z., Halouzka, J., Mikulaskova, M., and Salava, J. (1997). *Campylobacter* spp. and *Salmonella* spp. in black-headed gulls (*Larus ridibundus*). Cent. Eur. J. Pub. Health *5*, 24–26.

Skirrow, M.B., and Blaser, M.J. (2000). Clinical aspects of *Campylobacter* infection. In *Campylobacter*, Nachamkin, I., and Blaser, M.J., eds. (ASM Press, Washington, DC), pp. 69–88.

Smibert, R.M. (1969). *Vibrio fetus* var. *intestinalis* isolated from the intestinal content of birds. Am. J. Vet. Res. *30*, 1437–1442.

Southern, J.P., Smith, R.M., and Palmer, S.R. (1990). Bird attack on milk bottles: possible mode of transmission of *Campylobacter jejuni* to man. Lancet *336*, 1425–1427.

Stephens, C.P., On, S.L., and Gibson, J.A. (1998). An outbreak of infectious hepatitis in commercially reared ostriches associated with *Campylobacter coli* and *Campylobacter jejuni*. Vet. Microbiol. *61*, 183–190.

Varslot, M., Resell, J., and Fostad, I.G. (1996). [Waterborne *campylobacter* infection--probably caused by pink-footed geese. Two outbreaks in Nord-Trîndelag, Stjîrtdal in 1994 and Verdal in 1995]. Tidsskr. Nor Laegeforen. *116*, 3366–3369.

Wahlstrom, H., Tysen, E., Olsson, E.E., Brandstrom, B., Eriksson, E., Morner, T., and Vagsholm, I. (2003). Survey of *Campylobacter* species, VTEC O157 and *Salmonella* species in Swedish wildlife. Vet. Rec. *153*, 74–80.

Waldenström, J., Broman, T., Carlsson, I., Hasselquist, D., Achterberg, R.P., Wagenaar, J.A., and Olsen, B. (2002). Prevalence of *Campylobacter jejuni, Campylobacter lari*, and *Campylobacter coli* in Different Ecological Guilds and Taxa of Migrating Birds. Appl. Environ. Microbiol. *68*, 5911–5917.

Waldenström, J., Mevius, D., Veldman, K., Broman, T., Hasselquist, D., and Olsen, B. (2005). Antimicrobial resistance profiles of *Campylobacter jejuni* isolates from wild birds in Sweden. Appl. Environ. Microbiol. *71*, 2438–2441.

Waldenström, J., On, S.L., Ottvall, R., Hasselquist, D., and Olsen, B. (2007). Species diversity of *campylobacteria* in a wild bird community in Sweden. J. Appl. Microbiol. *102*, 424–432.

Waldenström, J., Axelsson-Olsson, D., Olsen, B., Hasselquist, D., Griekspoor, P., Jansson, L., Teneberg, S., Svensson, L., and Ellström, P. (2010). *Campylobacter jejuni* colonization in wild birds: results from an infection experiment. Plos One 5, e9082.

Whelan, C.D., Monaghan, P., Girdwood, R.W., and Fricker, C.R. (1988). The significance of wild birds (Larus sp.) in the epidemiology of *Campylobacter* infections in humans. Epidemiol. Infect. *101*, 259–267.

Yogasundram, K., Shane, S.M., and Harrington, K.S. (1989). Prevalence of *Campylobacter jejuni* in selected domestic and wild birds in Louisiana. Avian Dis. *33*, 664–667.

Part IV

Ecology in Poultry

Epidemiology and Control of *Campylobacter* in Modern Broiler Production

21

Ana B. Vidal, Rob H. Davies, John D. Rodgers, Anne Ridley and Felicity Clifton-Hadley

Abstract

Recent studies continue to indicate that poultry products remain a significant source of foodborne zoonotic *Campylobacter* infections. These organisms have evolved to readily colonize the avian gut and have adapted to achieve high intestinal and faecal concentrations in broiler flocks. The control of *Campylobacter* originating from broiler meat is the most important element of public health strategies aiming at reducing the number of human campylobacteriosis cases. This chapter describes factors related to poultry farm and abattoir environments and processing management practices that are associated with *Campylobacter* contamination at various levels of the broiler production chain, as well as the main available control options. Despite the large body of research in the last 20 years, including the use of molecular typing tools, the ecology and epidemiology of *Campylobacter* in poultry remains poorly understood, possibly as a result of ineffective detection of the organism in key environmental niches. A better understanding of the multiple interactions between this organism, the poultry environment and the chicken host is required in order to help design targeted and cost-effective control strategies in poultry production.

Introduction

Campylobacter is among the most common reported causes of food-borne bacterial gastro-intestinal disease worldwide. It is estimated that there are nine million cases of human campylobacteriosis per year in the EU (EFSA Panel on Biological Hazards (BIOHAZ), 2011). Molecular and epidemiological studies indicate that broiler chickens are an important source for human infections. Handling, preparation and consumption of broiler meat may account for 20% to 30% of the human cases while 50% to 80% may be attributed to the chicken reservoir as a whole suggesting that there are additional routes to infection (EFSA Panel on Biological Hazards (BIOHAZ), 2011). *Campylobacter* are ubiquitous bacteria, frequently associated with the contamination of broiler flocks. *Campylobacter*-colonized broiler chickens can harbour very high numbers of these bacteria in their intestines and are the primary source for contamination of the carcasses during slaughter.

A large body of research carried out in the past 20 years has focused on understanding the sources and routes of colonization for broiler flocks in order to develop targeted interventions that can minimize the risk for carcass contamination and thereby minimize the exposure risk to people. The physiological characteristics of *Campylobacter* suggest that these organisms have evolved to readily colonize the avian gut and adapt to the environment of broiler production, one of the fastest growing and most intensive of livestock sectors. Commercial broiler production has significantly changed over the past decades with potential consequences for the disease risk. Current intensive poultry production involves high throughput animal husbandry in which thousands of animals of similar genotypes are raised in confined, controlled housing environments, with rapid population turnover and slaughter at a young age. A large number of factors related to poultry farm and abattoir environments, and production and processing management practices, have been associated with *Campylobacter* contamination of broiler flocks at various stages of the production chain. These factors are likely to be intimately inter-linked, resulting in a complex epidemiology of *Campylobacter* in broiler flocks that remains poorly understood. A brief description of modern commercial broiler production and a review of current knowledge of the epidemiology of *Campylobacter* at different stages of the broiler production are presented in this chapter. Reliable detection methods for *Campylobacter* are an important tool for understanding the epidemiology in broiler production, but detection offers several challenges and issues that are also discussed in this chapter.

Commercial broiler production, a brief description

Broiler production is a large scale commercial operation in many countries of the world. The world's average stock of chickens is almost 19 billion, or three per person, according to statistics from the UN's Food and Agriculture Organization (Food and Agricultural Organization of the United Nations, 2012). However, the true scale of global chicken population is not easy to measure, as the number of chickens kept in small scale farms and for local or domestic consumption is not easily estimated.

The USA (21.0%), China (13.9%) and Brazil (12.9%) are the main broiler producing countries, collectively accounting for 47.8% of total production, but significant commercial production also occurs in many other countries (Food and Agricultural Organization of the United Nations, 2012). Over 6 billion broiler chickens are slaughtered for meat in the EU each year, producing around 9.4 million tonnes of chicken meat, with an average per capita consumption of 17.4 kg/year (European Commission Directorate-General for Agriculture and Rural Development, 2011). The top six chicken meat-producing countries of the EU account for 67.6% of the total EU production, with the UK being the largest EU producer, followed by Poland, France, Germany and Spain.

Broiler production is frequently undertaken by large integrated companies. Each company normally has a mixture of company farms, where the farm is owned by the company, and contract farmers, who have a commercial agreement with the integrator who provides the farmer with the chickens, services and sells the meat after slaughter. The degree of integration varies between companies. Some companies integrate all the production levels including feed mills, breeding farms, hatcheries and abattoirs. Other companies only integrate part of this, for example the breeding farms and hatchery or only the poultry farms. The number of integrated farms and the degree of integration vary between countries.

The commercial production of broilers is diverse. Most broilers are raised in closed systems (known as 'intensive' or conventional flocks) where birds are kept inside houses with controlled temperature, light and ventilation. Birds are reared on litter (usually based on wood shavings, chopped straw or paper products) and have free access to feed and water. In general, housed broilers are separated from outside contact and biosecurity systems and vaccination programmes are used to help protect birds from disease challenges. Building design varies according to climatic conditions. Naturally lit buildings with windows, static infrared heaters and natural ventilation (mechanically or non-mechanically assisted) are common in southern Europe, whereas in the northern countries more sophisticated buildings are seen with good thermal insulation, dynamic ventilation, artificial light and uniform temperature control within the building. The environment in a poultry house comprises a combination of physical and biological factors (including ventilation, stocking density, litter and air quality, age and health status of the birds), which result in a complex dynamic system of interactions between birds, husbandry system, and the environment. Some alternative poultry management systems also exist, such as free-range (birds have access to the external environment) and organic production on lower stocking density systems, with slower growing birds. In many high broiler producing countries the total number of farm units is small, as the average flock size can be large, with many farms stocking birds at 10,000 to 50,000 animals or more per house on a multiple house holding. Densities in commercial broiler houses are variable from relatively low (11 birds/m^2, 22.5 kg/m^2) to high (20 birds/m^2, 42 kg/m^2) (European Council, 2007). Many poultry companies carry out the practice of 'thinning' for welfare reasons. Broiler houses are stocked with numbers of birds that would be above the recommendations for stocking density if all the birds remained until slaughter weight. To overcome this, at approximately 1–2 weeks before slaughter, a cohort of birds is removed for slaughter, with the remainder being kept until depopulation. Thinning also provides producers with the necessary flexibility to react quickly to the demands of the fresh meat market.

The age at which chickens are slaughtered varies considerably between countries and production systems. Overall, slaughter weight can range from less than 1 kg to more than 3 kg. Age at slaughter varies more or less accordingly, ranging from 21 to 170 days. The growth rate of commercially produced broiler chickens has been increased and now birds can reach 1.5 kg body weight in 30 days whereas 120 days were needed in the 1950s. Simultaneously, the feed conversion ratio was reduced from 4.4 to 1.47, largely as a result of genetic selection (Havenstein *et al.*, 2003). Selection for fast early growth rate has resulted in major changes in the anatomy and physiology of broilers and scientific literature suggests that the selection may have also resulted in reduced resistance to infections (Gross *et al.*, 2002). Standard broiler diets are usually compounded from cereals, such as wheat and maize, as the principal source of energy and protein rich ingredients such as oilseed-meals, pulses and animal proteins (primarily fish meal). Vegetable fats are added as additional sources of energy and diets are routinely supplemented with a range of additives, including minerals, vitamins, synthetic amino acids, and enzymes.

Once birds reach slaughter age, they are removed from the rearing houses and transported to the abattoir in crates, containers or cages, depending on the processing company strategy. At arrival at the plant, birds are placed on the line and stunned using electrical or gas methods followed by bleeding; normally by neck cutting. Variations to normal procedures may be applied in the case of religious slaughter. The carcasses are processed using production line speeds of up to 12,000 carcasses per hour and large plants can process up to two million chicken carcasses per week. Feathers are loosened by 'scalding' in a warm water tank and then mechanically removed by rotating rubber fingers, leaving the skin intact. The water temperature in the scalding tank varies between 52°C and 59°C, depending on whether the end product will be fresh (chilled) or frozen. Various techniques can be used for scalding carcasses, but immersion seems to be the most effective, especially when using a multi-stage facility including counter-current flow to reduce cross-contamination. After removal of the head and feet, the viscera are extracted manually or mechanically. The carcasses are then washed and subsequently chilled to not more than 4°C as rapidly as possible. This chilling operation can be by immersion in cold water or in cold air (dry or with intermittent water sprays) which is the method most commonly used in the EU. Water chilling is mainly used for frozen products while air chilling is used for refrigerated products.

Colonization of broiler flocks by *Campylobacter*

Chickens are frequently colonized with large numbers of thermophilic campylobacters, mainly *C. jejuni* and *C. coli*. Colonization occurs largely in the caeca and is primarily confined to the intestinal mucus layer over the intestinal crypts of the villi (Beery *et al.*, 1988). Translocations to other internal organs such as liver, lymphoid organs, and reproductive tract are also commonly observed (Cox *et al.*, 2002, 2005; Knudsen *et al.*, 2006). In a recent study, *Campylobacter* was cultured from the bloodstream (Richardson *et al.*, 2011), providing insight into the potential for a rapid dissemination to other tissues and organs. After *Campylobacter* colonization, neither associated clinical signs nor pathological lesions are generally observed (Dhillon *et al.*, 2006). Most of the data available indicates that once *Campylobacter* is established within an individual bird and shed in high numbers, transmission occurs rapidly through the flock resulting in the complete colonization of the flock within a few days. Colonized chickens carry high numbers (up to 10^9 cfu/g) of *Campylobacter* in their caeca that are shed in their faeces until the time of slaughter. The mechanism by which a single bird or a group of birds becomes colonized and the time at which this occurs is unknown. Many longitudinal field studies have indicated that broilers are generally free of *Campylobacter* on the day of hatch and that the within-flock prevalence increases with age, with most flocks becoming detectably positive at 2 to 3 weeks of age; this negative period is commonly known as 'lag phase'. Further transmission occurs via the oral-faecal route, probably favoured by the coprophagic activity of broilers during exploratory pecking and preening or contamination of feeding and water delivery systems, although the possibility of transmission through airborne particles has also been suggested (Chinivasagam *et al.*, 2009). Enhanced colonization potential associated with *in vivo* passage through the chicken gut has also been demonstrated (Cawthraw *et al.*, 1996).

In contrast to the evidence provided by field work, experimental studies indicate that day-old chickens and older animals are highly susceptible to colonization. They can be infected by doses as low as 40 cfu/g (Cawthraw *et al.*, 1996), shed high numbers of bacteria in their faeces 2 to 3 days after challenge (Knudsen *et al.*, 2006) and remain colonized, shedding *Campylobacter* in concentrations above 10^6 cfu/g until slaughter age (El-Shibiny *et al.*, 2005).

The reason for the observed 'lag phase' in commercial flocks is not very well understood. Some studies have suggested the role of maternally derived immunity as a potential cause of a delay in the onset of colonization or shedding (Sahin *et al.*, 2003). However, experimental studies show that day-old chickens are highly susceptible to infection. Differences between the artificial inoculation and the natural course of infection, including the immune status of parent birds, and the difficulties in reproducing the natural conditions of colonization and transmission in commercial broiler flocks may be responsible for this discrepancy. Limitations of the standard sampling strategies used for detection of *Campylobacter* in broiler flocks should be also considered as they can only detect colonization above a certain threshold determined by the number of birds sampled and the test sensitivity, and so may not be suitable to detect an initially low within-flock prevalence.

Within-flock transmission

The within-flock prevalence directly relates to the rate of transmission and is therefore a time-dependent phenomenon. It is considered that in a relatively short period, a threshold is reached where the level of contamination of the environment is sufficient to cause extensive colonization in birds as a result of the faecal–oral transmission route. Mathematical modelling of the spread of flock colonization revealed a transmission rate of 2.37 new cases per colonized chicken per day and has estimated that without factors that would delay transmission, a flock will test positive only after 7 days post exposure (Katsma *et al.*, 2007a; van Gerwe *et al.*, 2009). These mathematical models are based on a number of assumptions: (i) all the individuals in the population are identical and have equal susceptibility to infection; (ii) there is only one infectious agent (one strain) being transmitted; (iii) there is homogenous mixing within the population and that the number of contacts between birds leading to transmission is independent of the size of the flock. The population sizes in the studies used by these models are much smaller than the commercial flocks resulting in limited understanding of how transmission scales with flock size, which is crucial to understand population dynamics. Additionally, the models consider an exponential distribution of the latent times within a flock; that is the time required for a broiler to start shedding at an infectious dose. However, it is expected that the period of latency depends on variations in host susceptibility and the magnitude of the infectious dose. Similarly, if there is a requirement for a build-up of environmental contamination, then this should be taken into account and will require information on the levels of viable bacteria and also on how the bacteria survive in the more hostile environment outside the bird (Conlan *et al.*, 2007).

The precise dynamics of *Campylobacter* transmission in poultry flocks is poorly understood and it is likely to be influenced by many factors; some of them not fully investigated. Social behaviours and clustering effect of the birds has been suggested to have a role in the within-flock transmission. Latest modelling approaches considering dose–response data have revealed that an age dependence of transmissibility between birds rather than their susceptibility to colonization is the mechanism behind the lag-phase reported in commercial flocks (Conlan *et al.*, 2011).

Sources, vehicles and routes of infection

Major research efforts have been concentrated on identifying the sources and vehicles of *Campylobacter* infection in poultry flocks in order to introduce targeted interventions to prevent and control flock colonization and disrupt transmission routes.

Vertical transmission

The role of vertical transmission as a source of *Campylobacter* for broiler flocks is still debated (Cox *et al.*, 2012). The majority of the supporting evidence, including extensive molecular epidemiological studies and the fact that it is possible to maintain a high proportion of *Campylobacter*-negative broiler flocks in winter despite on-going infection of all parent flocks, suggests that breeding flocks are not a significant source of contamination for progeny and that vertical transmission occurs rarely, if at all (Callicott *et al.*, 2006). Vertical transmission is usually understood as the internal contamination of the egg within the reproductive tract before completion of shell deposition. However, it can also be interpreted as the transmission of organisms from parent flocks to progeny via routes such as faecal contamination of eggshells (pseudo-vertical transmission). Only a few studies have isolated *Campylobacter* from hatchery samples, eggs and day-old chicks at placement from positive breeder flocks but both the recovery rate and the level of contamination in those samples have been very low (Byrd *et al.*, 2007). Through molecular testing, *Campylobacter* has been detected in hatchery fluff, intestinal tracts of developing embryos and newly hatched chicks (Chuma *et al.*, 1997; Hiett *et al.*, 2002a; Idris *et al.*, 2006). There is evidence of *Campylobacter* entering the eggshell under specific conditions but it has also been demonstrated that campylobacters have poor survival rates in egg albumen (Jones *et al.*, 1991), although they may survive for a period in the moist eggshell membranes (Cox *et al.*, 2012). Overall, a very low incidence of egg contamination can be expected which, combined with an insensitive method of sampling, may explain the rather uncommon recovery from fertile eggs (Byrd *et al.*, 2007). However, a very low transmission rate of 1 in 20,000 could be significant as even a single contaminated bird in a flock could spread *Campylobacter* to adjacent rearing mates within a few weeks. Comparative molecular typing studies have also provided controversial results. Some suggest that the same strains rarely, if at all, colonize both parent and progeny flocks (Petersen and Wedderkopp, 2001; Patriarchi *et al.*, 2011) while others have found parental and progeny flocks containing indistinguishable *Campylobacter* strains (Pearson *et al.*, 1996; Idris *et al.*, 2006).

The lack of an adequate cultural procedure for recovery of *Campylobacter* from different types of samples may explain the difficulties in routinely culturing this organism from eggs. Improvements in laboratory culture methods and the application of recent series of DNA-based assays have been suggested to aid the investigations of possible vertical transmission of *Campylobacter* in broiler flocks (Cox *et al.*, 2012).

Horizontal transmission

The majority of evidence accumulated over the last 15 years indicates that *Campylobacter* is mainly horizontally transmitted to poultry flocks from some external reservoirs, probably via a variety of routes and vehicles. The variety of potential sources and vehicles of infection and their relative importance have been recently reviewed (Adkin *et al.*, 2006; Hermans *et al.*, 2012). Identifying the source of the *Campylobacter* strains that colonize the flock has proved difficult even with modern tracking techniques.

Environmental contamination inside and outside the broiler house

Campylobacter can persist and multiply in many hosts and can survive in the environment under particular conditions. As *Campylobacter* is considered to be unable to grow in the environment, contamination is, presumably, a consequence of faecal contamination of the environment from a living source. Once a flock becomes positive, the poultry house and surrounding farm environment become widely contaminated and this contamination can persist for several weeks. Thus, previous flocks could constitute an important source for subsequent flock infection, and this might be more important with shorter downtimes between flocks (Newell, 2005; Lee and Newell, 2006). Carryover of *Campylobacter* infection from one flock to a subsequent flock in the same house is thought to be limited or negligible, mainly due to the efficacy of cleaning and disinfection as well as the beneficial effect of the down-time between consecutive rearing flocks (Evans and Sayers, 2000; McDowell *et al.*, 2008; Ellis-Iversen *et al.*, 2009; Alter *et al.*, 2011; O'Mahony *et al.*, 2011). Molecular typing studies have identified transmission of house-specific clones between sequential flocks, indicating that carry-over may, at most, account for only a small proportion of flock infections (Shreeve *et al.*, 2002; Alter *et al.*, 2011). However, it has not been clearly established if the presence of similar molecular types in consecutive flocks constitutes a common external source of infection, as samples taken from empty and disinfected houses prior to arrival of a new flock are almost always negative (Messens *et al.*, 2009; Damjanova *et al.*, 2011; Patriarchi *et al.*, 2011), even where PCR approaches have been used (Ridley *et al.*, 2008). The persistence of *Campylobacter* strains in the outside environment and the repeated contamination by out-coming sources such as vehicles or visiting staff could be responsible for repeated infection of consecutive broiler flocks. Various environmental elements around the poultry house, particularly standing water or puddles, are sites from which *Campylobacter* is frequently recovered even before the birds are detectably positive (Hiett *et al.*, 2002b; Herman *et al.*, 2003; Bull *et al.*, 2006; Johnsen *et al.*, 2006; Messens *et al.*, 2009; Ridley *et al.*, 2011b; Ellis-Iversen *et al.*, 2012). Genotypes of isolates from puddles, soil and close surroundings are commonly identical to those later isolated from the flock when it is found to be positive (Bull *et al.*, 2006; Messens *et al.*, 2009; Ridley *et al.*, 2011a), but it is possible that undetected infection in one of the poultry houses is the source of the external contamination and more detailed longitudinal studies are required to investigate this.

Other livestock and animals

Campylobacter has been recovered from the faecal material of other livestock, domestic pets, wild birds and rodents from around the poultry house environment. However, their relative importance as sources of infection for broiler flocks remains unclear. Using molecular epidemiological approaches, strains colonizing target poultry flocks have been found in adjacent livestock, including cattle and pigs (Johnsen *et al.*, 2006; Ridley *et al.*, 2008). This occurrence can be detected prior to the poultry flock being detected positive, indicating that the direction of transmission may be from the livestock to the broilers (Zweifel *et al.*, 2008; Ridley *et al.*, 2011b). Both studies also demonstrated, conversely, that other genotypes initially present in the broiler flocks were subsequently recovered from cattle. However, the majority of strains in adjacent livestock, such as cattle and pigs, are often not recovered subsequently from broilers (Zweifel *et al.*, 2008). Similarly, *Campylobacter* strains from wild bird faeces found on a farm have subsequently been recovered from broiler flocks (Hiett *et al.*, 2002b), albeit infrequently, while other studies showed no significant overlaps in the *Campylobacter* populations from wildlife and broilers, indicating that the role of the wildlife is uncertain (Petersen *et al.*, 2001; Colles *et al.*, 2008). There is no clear evidence of rodents as a source of infection for broiler flocks and their importance is questioned by the low detection rate of *Campylobacter* from captured rodents within the farm environments (Meerburg *et al.*, 2006).

Personnel and equipment

Personnel who regularly move between poultry farms, including field managers, quality auditors and catching crews, and farm equipment such as lorries, forklift trucks, pallets, modules, crates and footwear have been identified as potentially important vehicles for *Campylobacter* being introduced into the poultry house from the external environment (Ramabu *et al.*, 2004). Farm vehicles, even after cleaning, are often contaminated with *Campylobacter* and molecular typing studies have demonstrated that these strains are often subsequently associated with colonization in the concomitant flock (Ridley *et al.*, 2011a). Allen *et al.* (2008) found an association between *C. jejuni* genotypes present on vehicles and crates arriving on a farm at thinning and those subsequently recovered from the birds after slaughter. The spread of a common genotype from one farm to another within the same company through shared bird-catching personnel or vehicles was suggested.

Flies and insects

The role of insects in *Campylobacter* flock colonization remains obscure and is under-investigated in most countries. Crawling insects, arthropods and flies can act as mechanical vectors and passively transfer *Campylobacter* acquired from faecal contamination from the outside to the inside of poultry houses. Some researchers have shown no association between flock positivity and presence of insects in the broiler house, they were unable to culture *Campylobacter* from insects or it was only recovered after the flock became positive (Bates *et al.*, 2004). In contrast, other studies carried out in Denmark showed that a high percentage of flies captured from around the broiler house were positive for *Campylobacter*, particularly if PCR was used as the detection method (Hald *et al.*, 2004; Hald *et al.*, 2008)

Feed, litter and water

Feed, litter, water and air are all potential vehicles of infection for poultry flocks. Both feed and litter are normally free of campylobacters (Zweifel *et al.*, 2008), probably due to the dry environment associated with those sources, which is not suitable for *Campylobacter* survival. *Campylobacter* strains have been found in the feed, water and in the air of broiler houses after flocks were colonized (Bull *et al.*, 2006). The availability of nutrients and the condition of the litter are important factors to facilitate the persistence of *Campylobacter* in broiler houses (Kassem *et al.*, 2010). Acidifying litter treatments can result in a delay of the onset of *Campylobacter* colonization and may help to reduce early horizontal transmission within the flock (Line, 2002, 2006; Line and Bailey, 2006).

Water and water lines can become contaminated either at origin or from the back-tracking of organisms from drinkers contaminated by positive birds. However, attempts to isolate *Campylobacter* from water supplies before flock colonization have failed (Zimmer *et al.*, 2003) and it is believed that water may act as a vehicle for transmission of *Campylobacter* within the flock rather than as a source of infection. Water treatment with organic acids has been shown to be effective at keeping drinking water free of *Campylobacter* and can play a role in a biosecurity strategy for preventing *Campylobacter* spread via drinking water in broiler flocks (Chaveerach *et al.*, 2004). Improved isolation techniques have been described for the detection of *Campylobacter* from water samples (El-Adawy *et al.*, 2012; Jokinen *et al.*, 2012). A PCR-based method was able to detect *Campylobacter* from drinking water samples from a turkey flock at 6 days of age (El-Adawy *et al.*, 2012) in contrast to the longitudinal studies with broilers. Others have recovered identical *Campylobacter* genotypes from borehole well water samples and cloacal swabs from grandparent chickens (Perez-Boto *et al.*, 2010). However, other studies have not been able to show a clear sequence of contamination from the water to the flock or whether the contamination of the water source occurred within or outside the shed (Ogden *et al.*, 2007).

A few studies reported the recovery of low numbers of viable *Campylobacter* from air samples but the methods require sampling of large quantities of air (Chinivasagam *et al.*, 2009). More sophisticated air and dust sampling systems combined with a PCR detection method have been successful in detecting *Campylobacter* colonization in broiler flocks, even prior to detection by the traditional boot-sock samples under normal rearing conditions (Olsen *et al.*, 2009). *Campylobacter* detected from air samples collected outside the broiler house indicate that positive houses are possibly a source of contamination for other flocks through aerosols. However, the *Campylobacter* strains recovered from the air samples have not been recovered from the subsequently colonized flock (Ridley *et al.*, 2011b).

Risk factor analysis studies

Considering the multiple potential sources, vehicles and routes of infection described, it is not surprising that there are also many factors and poultry management practices found to be associated with the detection of *Campylobacter*-positive flocks by risk factor analysis. Some of these factors may be related to the initial colonization of the flock, the subsequent within-flock transmission or with a combination of both.

Age

The trend of increasing numbers of *Campylobacter*-positive flocks with increasing age has been widely described (Hermans *et al.*, 2012). This could be related to an increase of birds susceptibility to infection with age (Sahin *et al.*, 2001; Cawthraw and Newell, 2010) or it could indicate a greater risk of exposure. It is reasonable to assume that the probability of contamination entering a broiler house increases over time (Colles *et al.*, 2011) and the risk is probably related to the increasing number of challenges to the biosecurity barrier and greater inputs of air, water and bedding plus stress related to stocking density. The probability of detecting the infection also increases with age due to an increase in within-flock prevalence.

Season

The prevalence of *Campylobacter*-positive flocks seems to follow a seasonal trend with a peak in the summer (Jore *et al.*, 2010). A recent study carried out in the UK found a strong statistical association between summer months and the *Campylobacter*-positive status of a broiler flock. This study also revealed that the risk for a flock being positive in the summer appears to be more marked in younger birds, which could indicate that in the summer birds are colonized earlier in the life of the flock (Lawes *et al.*, 2012). A parallel survey carried out in broiler abattoirs in the UK showed a higher risk for carcasses to be highly contaminated (> 1000 cfu/g of neck skin) in the summer months (Powell *et al.*, 2012). Interestingly, a seasonal peak of human infections has also been described and it tends to be concurrent or even precede the chicken peak (Nylen *et al.*, 2002; Kovats *et al.*, 2005). The reason for the seasonal trend is unknown but factors influenced by climate may have an effect on the relative importance of some environmental sources such as insects, excretion rates in ruminants or seasonal manure spreading activities as well as unknown transmission routes of *Campylobacter*. Climatic conditions may also affect the ventilation requirements, water consumption rate and water quality in broiler flocks, which may impact on the level of *Campylobacter* flock colonization. The risk factors associated with the additional infections occurring in the summer may be different and this might be reflected in variations in strain types during the seasonal peak. The seasonality in broiler flocks has normally only been examined in relation to *Campylobacter* spp.; however, in the 3-year survey carried out in the UK, it was observed that *C. coli* positive flocks showed a marked seasonal distribution with a clear peak in the summer but this was not so evident for those flocks colonized with *C. jejuni* (Lawes *et al.*, 2012). In contrast to the seasonality of *Campylobacter* colonization observed in housed broiler flocks, a longitudinal study of a free-range broiler breeder flock found no correlation of prevalence or genetic diversity of *Campylobacter* with season, temperature or the amount of rain and sunshine, indicating that the risk of exposure to *Campylobacter* in free-range flocks may remain constant throughout the year (Colles *et al.*, 2011).

Geographic area

Some studies have reported the presence of high risk geographic areas for *Campylobacter*-positive flocks (Jonsson *et al.*, 2010b). Large-scale environmental factors, such as climate, landscape and geography could be associated with a higher risk of *Campylobacter* (Jonsson *et al.*, 2010a). However, factors acting at a more local scale such as high poultry density areas or common poultry management practices may account for the observed spatial patterns of *Campylobacter*-positive flocks (Jonsson *et al.*, 2012). The high density of poultry may be responsible for the wide dissemination of *Campylobacter* within and between farms in some areas. Common poultry management practices may favour the recirculation of these strains between and within farms. The use of common catching crews and other personnel, as well as auditing bodies, across different farms and companies in high density poultry areas may contribute to the transmission of *Campylobacter*. Distances to the nearest broiler farms have been suggested to play a role in introducing *Campylobacter* into a broiler farm (Henry *et al.*, 2011; Chowdhury *et al.*, 2012a; Jonsson *et al.*, 2012). The network between abattoirs and farms, within and between companies, together with the environmental characteristics of the area may explain the dissemination and perpetuation of certain strains in a particular geographical area. Additionally, different *Campylobacter* species or strains may respond differently to certain environmental conditions and management practices which, combined with different survival and invasion characteristics, may explain their particular seasonal and geographical distribution (Jorgensen *et al.*, 2011).

Hygiene and biosecurity practices

Considering the widespread dissemination of *Campylobacter* in the surrounding environment of a broiler house, any management practice that results in a failure of hygiene procedures or in increased opportunities for a breakdown of biosecurity may increase the risk for introduction of *Campylobacter*. The relative risks of the various routes into the house and then to the flock remain unclear, but traffic into the house by farm personnel and particularly visitors is a high risk that should be reduced by best hygiene practices and effective use of hygiene barriers (Newell *et al.*, 2011). The number of visits to the target house has been associated with an increased risk for flock infection (Huneau-Salaun *et al.*, 2007; Chowdhury *et al.*, 2012b). The use of house-specific boots and clothes, the effective management and consistent use of boot dips replenishment, and the use of hand washing/disinfection are all associated with a reduced risk of flock infection (McDowell *et al.*, 2008). The relative efficacy of house-specific boots compared with boot dips remains unclear (Van De Giessen *et al.*, 1998) but both can be combined for increased protection. The presence of hygiene barriers within poultry houses has been considered to be protective in many risk factor studies, although variation in type of hygiene barrier, and the lack of effective use may confuse interpretation of the results (Hald *et al.*, 2000; Lyngstad *et al.*, 2008). In Sweden, an increasing number of boot changes between the house exterior and the occupied section of the house was associated with reduced risk (Hansson *et al.*, 2010a).

Thinning

The practice of thinning or partial depopulation has a number of important public health implications in relation to contamination introduced on-farm by staff and visitors and on crates as well as the deleterious effects of stress within the broiler flock. Thinning has been statistically associated with an increased risk of flock and carcass contamination with *Campylobacter* probably due to the contamination introduced on-farm by personnel and equipment used to remove the birds (Hald *et al.*, 2001; Hue *et al.*, 2010; Lawes *et al.*, 2012; Powell *et al.*, 2012). During this process, personnel and equipment often travel between poultry processing plants and farms without sufficient personal hygiene or biosecurity hence the risk of introducing *Campylobacter* into the broiler house is elevated (Allen *et al.*, 2007, 2008). The recycling of contaminated transport crates that transfer birds between farms and abattoirs can introduce *Campylobacter* into the farm, despite frequently ineffective attempts to clean crates before re-use, and this may result in infection of the remaining birds (Hansson *et al.*, 2005). The process of bird removal may also have deleterious effects of stress and increase the physical mixing of the remaining birds, factors that potentially could increase the within flock transmission of *Campylobacter*, if it is present. Farms often relax biosecurity after thinning, which may also result in increased risk of colonization. Other studies have suggested that there is a strong confounding effect of bird age and season and that in some countries the number of days before final depopulation is not enough to allow the flock to be fully colonized minimizing the importance of this practice as potential risk of flock colonization (Russa *et al.*, 2005).

Other factors

The risk of *Campylobacter* colonization increases with the detection of previously contaminated flocks (Ellis-Iversen *et al.*, 2009; Chowdhury *et al.*, 2012b) and with the number of houses on site (Guerin *et al.*, 2007a; McDowell *et al.*, 2008). This may be a reflection of increased risk of environmental contamination of the poultry house and surrounding farm environment that can persist for several weeks. This risk probably increases with shorter downtimes between flocks, which are more likely on larger farms which may be placing new bedding at one end of the site whilst still depopulating at the other end, thus minimizing the time that the whole site is clear of birds and contaminated wash water. Cleaning and disinfection practices between flocks are considered effective to avoid carry-over of the infection from one flock to the next (Evans and Sayers, 2000). However, some studies have found that cleaning without detergent (Henry *et al.*, 2011) or the use of geothermal water (Guerin *et al.*, 2007b) increased the risk of *Campylobacter*-positive flocks. The presence of wild animals and birds, livestock on farm or within locality, insects, flies, litter beetles, farm surroundings (puddles, soil) and waste disposal on site have also been associated with colonization of broiler flocks (Hansson *et al.*, 2010a; Ellis-Iversen *et al.*, 2012). The presence of rodents during the production cycle has been identified as a risk factor for colonization of broiler flocks, particularly where no rodent control programmes are in place (McDowell *et al.*, 2008).

Despite failures to isolate *Campylobacter* from water supplies in broiler farms, the source (Lyngstad *et al.*, 2008), quality (Sparks, 2009) and chlorination treatment of broiler drinking water (Guerin *et al.*, 2007a) are statistically significant protective factors in some studies. Acidification of drinking water was shown to have an effect on indirect but not on direct transmission of *Campylobacter* in broiler flocks (van Bunnik *et al.*, 2012). The presence of protozoa and biofilms in water lines can delay the decline of *Campylobacter* viability and increase resistance to disinfection (Snelling *et al.*, 2005). The removal of biofilms from water supplies can have a positive impact on the productivity of chickens and the balance of mineral substances (Bobiniene *et al.*, 2012). Ventilation systems, in particular vertical and static systems were also found to be risk factors in some studies (Barrios *et al.*, 2006; Guerin *et al.*, 2007b).

The type of production has also been associated with *Campylobacter*-colonized flocks. Free-range flocks seem to be more frequently colonized with *Campylobacter* (usually close to 100%) than conventionally reared flocks (Heuer *et al.*, 2001). This is partially due to an age effect, as free-range birds are slaughtered at an older age, and in part due to the exposure to the free-range environmental contamination and multi-age production (Colles *et al.*, 2008). Some of these free-range flocks are positive before they are released onto pasture (Huneau-Salaun *et al.*, 2007; Allen *et al.*, 2011) which indicates that other sources and routes of transmission remain important for those flocks, although in these farms the hygiene standards of the brooder accommodation may be lower than in conventional broiler production. A seasonal trend has been reported for free-reared flocks investigated at the end of the indoor rearing period with an increased prevalence in the summer and autumn (Huneau-Salaun *et al.*, 2007; Colles *et al.*, 2010). However, no correlation with season or climatic conditions and the prevalence of *Campylobacter* amongst a free-range breeder flock was observed when followed for 63 weeks (Heuer *et al.*, 2001; Colles *et al.*, 2011). Regarding organic production, a systematic review shows that the prevalence of *Campylobacter* is higher in organic than in conventionally reared chickens at slaughter but no difference is observed in meat chicken products at retail (Young *et al.*, 2009).

Other house-specific risk factors related to husbandry management practices of broiler flocks such as stocking density, flock size, antimicrobial treatment and bird health have been investigated to a lesser extent, with some studies reporting an association with *Campylobacter* flock colonization (Barrios *et al.*, 2006; Guerin *et al.*, 2007b; Bull *et al.*, 2008; Nather *et al.*, 2009).

Risk factors associated with transmission

Broiler management practices could affect not only the risk of flock exposure but also the susceptibility of birds to colonization and the within flock transmission. It has been reported that *Campylobacter* may be found more frequently in birds compromised by poor health. Higher prevalence of *Campylobacter*-positive flocks has been associated with digital dermatitis and rejections at slaughter (Bull *et al.*, 2008) and another study reported an association between *Campylobacter* and *Clostridium perfringens*, which causes necrotic enteritis in broilers (Skanseng *et al.*, 2006). Two risk factor studies carried out in the UK found a positive association of *Campylobacter*-positive flocks and positive carcasses and an increased mortality at 14 days (Lawes *et al.*, 2012; Powell *et al.*, 2012). There are a number of possible reasons for these associations including poor biosecurity and management practices and common environmental effects. Poor condition of the flock and high percentage of rejects may be an indication of poor ventilation and certain environmental conditions within the poultry house that can favour rapid transmission of *Campylobacter* within the flock. Good litter quality for rearing of broilers may be important as wet litter may favour survival and environmental contamination of the house leading to increased transmission of *Campylobacter* within the flock. One study has demonstrated the importance of the relative humidity in the broiler house for the transmission of *Campylobacter* from broiler litter (Line, 2006). Moisture levels may be affected by numerous factors including: litter material, the type of drinkers, water consumption, air temperature and season, stocking density, bird age, diet composition and gut flora. These factors may also influence the water content and viscosity of faeces. The role of health and welfare indicators for *Campylobacter* colonization of broiler flocks requires further studies to clarify the underlying mechanisms that explain these associations.

Transportation and holding

The time and practices associated with removal of birds from the farm, transportation to the slaughterhouse and holding before slaughter have been associated with increased cross-contamination of the birds through direct contact with contaminated birds or contaminated faeces (Hansson *et al.*, 2005; Rasschaert *et al.*, 2007). Crates and modules used to transport the birds are commonly contaminated with *Campylobacter* prior to use and even after washing and sanitizing (Hastings *et al.*, Slader *et al.*, 2002). Washing cages and crates with water and disinfectants can reduce the levels of *Campylobacter* (Berrang and Northcutt, 2005a,c) but it does not completely eliminate the pathogen as cages and crates are difficult to clean and practices have been shown to be relatively ineffective (Berrang *et al.*, 2004). Therefore, crates constitute a potential source of contamination and cross-contamination of the exterior of the birds (Corry *et al.*, 2002; Berrang *et al.*, 2003). Even the design of transport crates (solid *versus* open floor) can influence the level of external contamination on broilers entering the plant, although this did not influence the *Campylobacter* levels of the carcasses in one study (Buhr *et al.*, 2000). Modules that hold the crates are frequently contaminated with impacted litter and are not easily cleaned. Increased duration of water and feed withdrawal prior to slaughter, increased holding time of birds in lairage as well as stress associated with catching and transportation have been shown to increase the shedding of *Campylobacter* in faeces and the concentration of the organism in the caeca (Whyte *et al.*, 2001). Extended feed withdrawal (> 12 h) could cause deterioration in the condition and integrity of the intestinal mucosa and increase the fluidity of gastrointestinal contents, which could increase the potential for faecal contamination of carcasses (Zuidhof *et al.*, 2004). Emptying of the crop during extended feed withdrawal can result in a decrease in the number of lactic acid-producing bacteria and associated increase in crop pH, which may reduce the activity against *Campylobacter* (Hinton and Ingram, 2000; Byrd *et al.*, 2001). It is also possible that hungry birds may attempt to consume contaminated litter.

Travel time between farm and abattoir has been associated with a variation in risk of *Campylobacter* detection in broiler batches. A three-year prevalence study in the UK found that the proportion of positive batches was higher amongst those from farms located within shorter distances or travel time to the abattoir than amongst those travelling longer distances (Lawes *et al.*, 2012). The reason for this association is unknown but it could be related with a longer time for crates and modules to dry when transporting birds for longer distances, a practice that has been shown to be important in the transmission of *Campylobacter* to broiler flocks (Berrang *et al.*, 2004, 2011; Berrang and Northcutt, 2005b). However, the distance travelled could be a confounding factor of the association between *Campylobacter*-positive flocks and particular characteristics of the farms located further away from the abattoir, such as smaller size of the farms or higher proportion of non-company farms. Weather conditions during transportation to the slaughterhouse seem to influence the rate of carcass contamination at the abattoir (Arsenault *et al.*, 2007). Overall, available data are inadequate and the complexity of variables and confounding factors involved make it difficult to assess the effect related with transportation and holding practices on the level of *Campylobacter* contamination.

Slaughter and processing

A positive association between the prevalence of *Campylobacter*-colonized batches and the frequency of *Campylobacter* contamination of the broiler carcasses has been widely reported. Similarly, carcasses from batches of colonized broilers result in carcasses with high numbers of *Campylobacter.* (Rosenquist *et al.*, 2006; Reich *et al.*, 2008; Hue *et al.*, 2011). However, this association varies significantly between countries and among slaughterhouses within countries, indicating that some slaughterhouses are more capable than others of preventing or reducing *Campylobacter* contamination (European Food Safety Authority, 2010b). The slaughterhouse variability may be due to differences in the occurrence of *Campylobacter*-colonized flocks that are delivered to a particular abattoir or it may be related to different slaughter hygiene practices.

Although a colonized broiler flock is more likely to generate contaminated carcasses, in the EU baseline survey there were many contaminated carcasses derived from broiler batches that tested negative (European Food Safety Authority, 2010b). Some of these may be due to limitations of the experimental design used and limited testing sensitivity to detect all *Campylobacter*-colonized batches. However, positive carcasses can be a result of cross-contamination from flocks slaughtered successively or even from flocks slaughtered on different days (Ellerbroek *et al.*, 2010). Processing of *Campylobacter*-colonized flocks can contaminate the abattoir environment and equipment surfaces where the organism can survive, even after cleaning and disinfection, and be a possible source of carcass contamination of subsequent flocks. *Campylobacter* genotypes found in the caeca of broiler batches at the abattoir are frequently found on carcasses from the corresponding batch (Normand *et al.*, 2008). Strains from positive flocks have also been subsequently found in the abattoir environment, contaminating the equipment (Lienau *et al.*, 2007; Peyrat *et al.*, 2008; Damjanova *et al.*, 2011) and can play a role in the spread of contaminating strains to carcasses, particularly for negative flocks (Rasschaert *et al.*, 2006; Elvers *et al.*, 2011). However, the levels of *Campylobacter* contamination and hence, the public health significance of cross-contamination from previous positive flocks or through residual abattoir environmental dissemination is generally low (Johannessen *et al.*, 2007).

Campylobacter carcass contamination can occur throughout the entire slaughter process. There are only a small number of studies that examined *Campylobacter* prevalence or concentration at multiple points during commercial chicken processing. A recent review of those studies found consistent trends of changes in the presence and concentration of *Campylobacter* through different processing stages (Guerin *et al.*, 2010). Specifically, the prevalence decreases after scalding, increases after de-feathering and evisceration and then declines again after chilling. Similarly, the concentration of *Campylobacter* decreases after scalding, increases after de-feathering and decreases again after washing and chilling. No conclusive results can be drawn from the available studies on the effect of different scalding temperatures on *Campylobacte*r prevalence and concentration on broiler carcasses. However, some of the larger decreases in concentration were observed when scalding temperatures were above 55°C. Similarly, air chilling and water immersion chilling are very different methods and although based on results from a limited number of studies, there was evidence of heterogeneity in the effectiveness of these two methods. Water immersion chilling consistently resulted in a greater decrease in the concentration of *Campylobacter* than air chilling, despite the potential for cross-contamination in the chill tank. No consistent trend in the effect of carcass washing on the prevalence of *Campylobacter* is demonstrated by the available data. Some studies found an increase in prevalence after washing which might be due to the possibility that water from the final carcass wash could trap bacteria within skin pockets or the abdominal cavity. By contrast, the de-feathering stage of chicken processing consistently increased the prevalence and level of *Campylobacter* contamination probably due to the escape of faecal material through the cloaca caused by the action of the picker fingers pressing on the abdomen, as well as by vigorous dissemination of surface contamination.

Mitigation measures at different stages of broiler production

Considering the risks associated with *Campylobacter* in broiler flocks and broiler carcasses, several mitigation measures have been suggested at different stages of broiler production. The starting point is to prevent an infectious dose of *Campylobacter* from entering the poultry house through measures implemented on-farm, and biosecurity and best hygiene practice are considered fundamental for this (Katsma *et al.*, 2007b). On-farm interventions would reduce not just the numbers of *Campylobacter* in the food chain but also the numbers of poultry-associated *Campylobacter* in the environment in general. This is considered to give further benefits since between 50% and 80% of human campylobacteriosis is thought to be attributable to the chicken reservoir as a whole (EFSA Panel on Biological Hazards (BIOHAZ), 2011).

However, under practical conditions it has proved extremely difficult to effectively and consistently exclude infection from broiler flocks (Havelaar *et al.*, 2007). The maintenance of hygiene standards depends largely on staff compliance (Newell *et al.*, 2011), which has been shown to be unreliable (Hofshagen and Kruse, 2005; Racicot *et al.*, 2012). The approaches required to deliver and sustain best hygiene practices are currently unknown but they should include staff motivation strategies, such as financial incentives, and continuous education tools (Fraser *et al.*, 2010). Good hygiene practices seem particularly important during thinning or partial depopulation and therefore this is where these should be practised. Crate, module and lorry washing should be properly carried out to ensure these are not contaminated with *Campylobacter* (Allen *et al.*, 2008). Other biosecurity measures in relation to clothing, footwear and hand-washing are also essential. Discontinuing thinning or reducing slaughter age have been suggested as control measures but would have serious economic implications and may not be feasible for the broiler industry in many countries (Newell *et al.*, 2011; Romero-Barrios *et al.*, 2013).

If the introduction of *Campylobacter* into broiler flocks cannot be avoided efforts should be focused on reducing the number of birds that are colonized within a flock or the number of organisms in the bird's gut. This could be achieved by reducing bird susceptibility to infection and/or limiting within flock transmission.

Competitive exclusion, feed and water additives such as short/medium chain fatty acids and organic acids have been shown to reduce the colonization and the number of *Campylobacter* shed in broiler faeces (Santos *et al.*, 2010; Skanseng *et al.*, 2010; Molatova *et al.*, 2011). Probiotic administration, an alternative approach to antibiotic prophylaxis, has been shown in some studies to have marked anti-*Campylobacter* activity when used as a feed additive (Santini *et al.*, 2010; Baffoni *et al.*, 2012; Ghareeb *et al.*, 2012). The use of bacteriocins has been shown to be capable of reducing the concentration of *Campylobacter* in caeca at the end of the production cycle (Svetoch and Stern, 2010). Bacteriophages appear promising under experimental conditions (Connerton *et al.*, 2011) but the rapid development of resistant strains and their distribution into the farm environment are limitations for their use in broilers, apart from as a pre-slaughter intervention. Most current studies are based on limited experimental studies and field validation of *in vivo* trials has not yet been conducted. Large-scale field trials are required to determine not only the effectiveness but also the practicality of these treatments so that the best time and regime for their application in a commercial poultry environment is determined. The use of vaccination or genetic selection is currently hypothetical approaches that could control *Campylobacter* through reducing the susceptibility of the broiler to colonization (Hermans *et al.*, 2011; Calenge and Beaumont, 2012). Vaccination research is on-going to identify suitable candidates to stimulate the immunity of the broiler chick and the advent of high throughput genomics will advance research into the identification of genes responsible for resistance to colonization.

Potential mitigation measures during transportation of the birds to the slaughterhouse may include minimizing the transportation and holding time, optimizing the water and feed withdrawal programmes and improving crate and module washing and disinfection practices. These measures would reduce the contamination of birds before slaughter and the amount of contamination introduced in the slaughter facility. Inadequate available data and the complexity of the factors involved make it difficult to assess the final effect of these interventions during transportation. However, minimizing the transportation time seems to be beneficial for microbiological, product quality, economic and welfare reasons.

At the slaughterhouse, the efficient implementation of hygiene measures, HACCP (Hazard Analysis Critical Control Point) or a Food Safety Management system may lead to low numbers of microorganisms and little variability in microbial counts. A difference in distribution in *Campylobacter* numbers found in poultry products has been shown to be reflected in a difference of the overall slaughter hygiene (Habib *et al.*, 2012b). There are also indications that the lack of process control and effective GHP (good hygiene practice) could result in high prevalence and high *Campylobacter* counts (Habib *et al.*, 2012a,b). Hygienic design of equipment and good operational performance of the different process operations are important for minimizing contamination during processing. Particularly important is the adaptation of the evisceration equipment to the natural variation of carcass sizes to avoid rupture of viscera and subsequent faecal contamination (Rosenquist *et al.*, 2006; Boysen and Rosenquist, 2007). Scheduling of flocks, which implies identifying positive flocks before they are slaughtered and subjecting the carcasses to special treatment such as freezing, is routinely used in some countries. However, in countries with a high prevalence of *Campylobacter*-positive flocks this may not be possible as a high proportion of carcasses would have to be treated, reducing the availability of fresh product (Nauta *et al.*, 2009b). Slaughtering positive flocks after negative flocks ('logistic slaughter') to avoid cross-contamination has also been suggested. However, studies have shown that a rather limited number of carcasses in a negative flock are contaminated by a positive flock slaughtered earlier and the contamination level is generally low (Nauta *et al.*, 2005). Both the 'scheduled' and 'logistic' slaughter options rely on effective early on-farm testing. However, false negative flocks that become detectably positive after the testing limits the applicability of these measures (Havelaar *et al.*, 2005). Improved rapid testing programmes that can facilitate sensitive detection as close to the point of slaughter as possible would facilitate improved application of such measures, but the overall and temporal prevalence of flock infection must be low enough to allow selection of sufficient negative flocks for fresh meat production.

The second type of slaughterhouse control measures are mainly focused on reducing the numbers of organisms on the carcasses. Decontamination can be achieved by chemical or physical treatment and should be considered a supplement and not a substitute for good hygiene practices. Chemical decontamination includes the use of lactic acid, chlorine dioxide, acidified sodium chlorite, trisodium phosphate and peroxyacids. None of these are currently authorized in the EU but some chemicals are used in a number of other countries (e.g. New Zealand and USA). The data available indicate that the efficacy of these chemical products for *Campylobacter* reduction in carcasses ranges from 0.47 to 1.8 $\log_{10}$ cycles at point of application (EFSA Panel on Biological Hazards (BIOHAZ), 2011). The most effective methods that should completely eliminate *Campylobacter* from carcasses are cooking or irradiation. Freezing to −20°C for a few weeks and crust freezing, where the surface of the carcass is temporarily blast frozen, are currently used in some countries (Boysen and Rosenquist, 2007; Rosenquist *et al.*, 2009). Hot water immersion, steam treatment and steam combined with ultrasound are also options although the appearance of carcasses treated with these methods may be compromised (Whyte *et al.*, 2003). These methods can be used in the EU without specific authorization, apart from irradiation, and some are currently in use in some European countries.

Availability of good representative data that

allow an accurate estimation of the effect of specific control options is limited and quite often is based on one or a small number of published or unpublished laboratory experiments or expert opinions. As a consequence, their predicted effects are highly uncertain (Nauta *et al.*, 2009a).

Campylobacter prevalence and monitoring

Detection of *Campylobacter* is not straightforward and differences in methodology and laboratory standards can have a large impact on the test outcome. Over the last 10 years or so, several countries have undertaken national cross-sectional or longitudinal surveillance of *Campylobacter*. These studies clearly indicate that flocks become colonized with *Campylobacter* over time and that at slaughter between 18% and 90% of flocks are *Campylobacter*-positive (Stern *et al.*, 2001; Herman *et al.*, 2003; Hansson *et al.*, 2004; van de Giessen *et al.*, 2006; Lawes *et al.*, 2012). The prevalence of positive flocks varies substantially by country. In general, more northerly European countries, such as Finland, Sweden and Norway, have a lower prevalence (10–30%), indicating that geographical factors are likely to be important but also highlighting potential differences in broiler management practices. However, the lack of standardization in sampling frames, sample collection and transportation, detection methodology and reporting can limit the direct comparison between studies and countries.

To establish a baseline and comparable values for all Member States within the EU, a harmonized baseline survey was carried out in 2008 to determine the prevalence of *Campylobacter* in broiler batches and broiler carcasses at slaughterhouse level. The overall prevalence of *Campylobacter*-colonized broiler batches was 71%, based on testing a pool of ten caecal samples per batch, and that of *Campylobacter*-contaminated broiler carcasses was 76%, based on testing one neck skin sample per batch. This study showed great variation in prevalence between Member States (5–100%). The *Campylobacter* counts on broiler carcasses also varied widely between countries and between slaughterhouses within countries. In general, there was a tendency for high counts on carcasses in countries with high *Campylobacter* prevalence (European Food Safety Authority, 2010a).

Efforts have recently been made to define official monitoring and control programmes to be applied specifically for *Campylobacter* at all stages of the food chain. Some countries have established monitoring programs for *Campylobacter* in broiler production, although the construction of these programmes, including the sampling and testing schemes, varies between countries and data are not directly comparable. In some countries of Northern Europe (i.e. Denmark, Norway, Sweden and Iceland), these strategies are part of official controls or monitoring programmes implemented at a national level (Stern *et al.*, 2003; Hofshagen and Kruse, 2005; Hansson *et al.*, 2007; Rosenquist *et al.*, 2009).

Action plans against *Campylobacter* have been implemented in Denmark, Norway, Iceland, the UK and New Zealand (Hofshagen and Kruse, 2005; Rosenquist *et al.*, 2009; Sears *et al.*, 2011; Tustin *et al.*, 2011). The two main focus areas of the control strategies are on-farm biosecurity to reduce broiler colonization and reduction of *Campylobacter* numbers on broiler meat to reduce consumer exposure to the pathogen. The means for reducing *Campylobacter* numbers at slaughter varies between countries. For example, meat from positive flocks is mandated to be frozen in Iceland and Norway (in Norway, heat treatment is an alternative to freezing). In Denmark, meat from positive flocks should be frozen, to the extent possible (a voluntary strategy which has been adopted by the large slaughter companies). In the UK and New Zealand, the choice of means to reduce *Campylobacter* numbers has been left to the industry.

Performance targets for *Campylobacter* have been adopted in a few countries such as New Zealand and the UK. Targets may concern a specific reduction of cases and/or processing criteria at production level. In New Zealand, mandatory targets have been set for broiler carcasses at the end of processing (on exit from immersion chiller) based on enumeration levels, with escalating regulatory responses if targets are not met (Sears *et al.*, 2011). In the UK, a new voluntary target has recently been set for UK-produced raw chicken, as part of the newly developed *Campylobacter* Risk Management Programme. The aim of the target is to reduce levels of the most contaminated carcasses (> 1000 cfu/g of neck skin) at the end of the slaughter process (post-chill) from 28% in 2008 to 10% by 2015 (Food Standards Agency, 2010a,b).

In 2008, a 5-year action plan against *Campylobacter* was launched by the Danish government (Rosenquist *et al.*, 2009). The action plan aims to reduce the prevalence of *Campylobacter* in Danish broiler flocks and in Danish broiler meat even further. Innovatively, the plan also focuses on reducing the risk of acquiring *Campylobacter* infections from imported broiler meat. In Denmark, testing for *Campylobacter* in every broiler flock at slaughter has been part of the national monitoring programme since 1998. However, in order to be able to channel the production, in 2003, the two major slaughter companies also initiated a private monitoring programme where all flocks are sampled on farm 7–10 days before slaughter. At retail, broiler meat has been monitored since 1995 and from 2004 onwards samples have also been collected at two large slaughterhouses handling more than 98% of Danish broilers.

Detection challenges

Major research efforts have been made to improve the understanding of *Campylobacter* epidemiology in broiler production. Particular focus has been given to the identification of sources and vehicles of infection, colonization, and transmission so that effective and targeted intervention measures can be introduced at different stages of the food chain. Knowledge of the true *Campylobacter* status of broiler flocks is essential for the understanding of the epidemiology and ecology of *Campylobacter* in broiler production and relies on the use of appropriate study designs, sampling strategies and detection methods.

Several investigations have been carried out to determine the best sampling material for the detection of *Campylobacter* in broiler flocks at different stages of the production chain. Caecal contents or faeces have been suggested as good samples for detection of *Campylobacter* at farm level (Rodgers *et al.*, 2010; Vidal *et al.*, 2012). These standard sampling strategies can only reliably detect colonization above a certain threshold of within flock prevalence determined by the number of birds sampled, and may not be suitable to identify recently colonized flocks, where a low within flock prevalence is expected. Recently, the use of boot swabs moistened with appropriate transport media has been suggested as a good alternative for *Campylobacter* detection at flock level, as they will collect faecal material from a large number of birds (Vidal *et al.*, 2012).

The statistical limitations of sampling plans are particularly important in those cases where *Campylobacter* is present in low and variable prevalence and concentrations. It is generally assumed that the within batch prevalence in colonized or contaminated broiler batches is close to 100% and that most carcasses in a positive batch will be contaminated. Consequently the statistical limitations of investigating a limited number of samples from a batch are generally considered negligible. However, a high degree of within-flock variability can be expected (Hansson *et al.*, 2010b). Flocks that are not fully colonized or colonized to a maximal level may both be classified as positive by the standard sampling methods used which could lead to a misinterpretation of results. Interventions can be interpreted as unsuccessful even though the number of colonized chickens or the number of bacteria in the caeca is lower and the overall numbers of campylobacters entering the processing plant would potentially be reduced. The within batch variability of positive birds or carcasses and the within batch variability of counts are important parameters to consider when defining the performance characteristics of sampling plans for *Campylobacter* (Nauta *et al.*, 2012).

Although the sampling site and sample type is of great significance, the culture methods and media used for bacteriological examinations are also important. Numerous culture methods and media types have been developed for isolating *Campylobacter* from a variety of sample types. However, available methods are not necessarily optimal for the recovery of *Campylobacter* from the whole range of sample types (Corry *et al.*, 1995) and optimal culture methods often appear to be sample-type specific. The majority of the methods available were developed for

Campylobacter recovery from faecal samples, in which large populations of *Campylobacter* are expected to be found or from food samples, where the level of competing bacteria is relatively low. However, these methods may not be adequate for recovery of small numbers, sublethally injured, stressed, or viable-non-culturable *Campylobacter*.

Usually a selective medium containing antibiotics is necessary to recover *Campylobacter*. Certain antimicrobial agents present in the media, may be inhibitory to some *Campylobacter* species or strains, resulting in an underestimation of their true prevalence. Additionally, the presence of competitive flora in samples, especially ESBL and AmpC-producing *Enterobacteriaceae* with resistance to extended spectrum cephalosporins may overgrow *Campylobacter*, resulting in false negative samples or in inaccurate counts, particularly when the target organisms are present at low concentrations (Jasson *et al.*, 2009; Moran *et al.*, 2011).

Harmonized methods were recently proposed and used for a baseline survey carried out in the EU to determine the prevalence and concentration of *Campylobacter* in broiler batches and broiler carcasses. The results of the survey indicated that the diagnostic sensitivities of both detection and enumeration tests varied significantly between Member States (European Food Safety Authority, 2010a). Although this could be in part attributed to differences in the microaerobic incubation systems used (Moran *et al.*, 2009), it could also be a reflection of differences inherent in the samples between countries such as the nature of the competitive flora and the *Campylobacter* population within the samples.

The use of different methodologies may result in the recovery of different *Campylobacter* populations (Williams *et al.*, 2012). When considering the species recovered, the test sensitivity estimates are generally lower for *C. coli* than for *C. jejuni* (Vidal *et al.*, 2012). The optimal atmospheric conditions for *C. coli* and *C. jejuni* are known to be different and the detection techniques available are in general more sensitive at detecting *C. jejuni* than *C. coli* (Rosenquist *et al.*, 2007). There is evidence that indicates that the *Campylobacter* populations from colonized flocks frequently comprise several strains and often more than one species. The use of multiple testing approaches has started to reveal an increased reporting of *C. jejuni/C. coli* co-infected flocks and indicate that the prevalence of *C. coli* positive flocks may be largely under-reported (Oyarzabal *et al.*, 2007; Vidal *et al.*, 2012).

Culture methods may influence both the *Campylobacter* species and subtypes recovered; therefore different methods may generate skewed populations (Williams *et al.*, 2012). However, a study carried out to investigate the genetic diversity of *C. jejuni* between and within broiler farms at the end of the rearing period showed that the use of different sampling strategies (caeca, boot swabs and faecal samples) and laboratory methodologies (direct culture versus enrichment) did not result in the recovery of different MLST types, based on one strain characterized per sample. This indicated that differences in sample type and laboratory methodologies had no discernible effect on the genetic profile of the flock (unpublished data).

Broiler flocks are frequently colonized with more than one *Campylobacter* genotype (Messens *et al.*, 2009). Considering the potential diversity of *Campylobacter* strains along the broiler production chain, the representativeness of the strain population is a major issue. Assuming all the strains can grow on the media and under the conditions used, the sensitivity is largely determined by the number of isolates tested. The dynamics of strain diversity in flocks is complex and it may reflect variation in phenotypic properties such as infectious capacity, virulence and stress responses, which could determine survival time, persistence or susceptibility to treatments. Mixed strain infections may lead to even more variety in *Campylobacter* strains as they may exchange DNA leading to recombinant strains thus increasing strain diversity. Some strain diversity also occurs due to genetic instability during colonization as genomic rearrangements can occur enabling strains with varying capacities for colonization to be generated (Taboada *et al.*, 2004).

The genetic diversity of *Campylobacter* in broiler flocks has been associated with the level of colonization. Fully colonized flocks, where all the samples tested were positive, have been shown to have a less diverse population than partially colonized flocks, in which not all the samples tested were positive. This could indicate different times of colonization as by the time flocks are fully colonized competition between types may result in a single predominant type. The presence of a predominant type may also suggest a more stable, adapted and successful type to colonize the flock (unpublished data). The quantification and relative importance of the strain variability and the limitations of the traditional culture methods to capture this variability should be considered. Particularly important are the implications for molecular epidemiology-based studies.

In order to improve the monitoring of *Campylobacter* and ensure freedom from infection in poultry flocks well-designed studies and sensitive sampling methods at individual and at flock level should be used. Culture-based identification of *Campylobacter* is slow and complicated. Developments in rapid detection methods have mainly focussed on the detection of *Campylobacter* proteins or genomic material using antibody- and molecular-based detection, respectively. Laboratory based methodologies such as ELISA (Tissari and Rautelin, 2007), isothermal (Yamazaki *et al.*, 2009) and PCR (Randall *et al.*, 2010) can provide results within a few hours compared to 2–3 days required for culture. The development of portable testing platforms such as lateral flow devices that use antibody specific detection of *Campylobacter* proteins suggest that affordable farm or abattoir-based tests may be soon applicable (Wadl *et al.*, 2009). Portable molecular technologies using PCR are readily available (Whittier *et al.*, 2010) but costs may be prohibitive; however, isothermal testing may provide a simpler low cost alternative (Niemz *et al.*, 2011). The utility of new rapid testing methods will complement research and control programmes, but should not replace traditional culture methods, as it is still important to also establish the viability and concentration of detected organisms (Nocker and Camper, 2009). The possibility of rapidly obtaining accurate information about the colonization status of a broiler flock and the development of simple pen side-tests remain as future challenges.

Future trends

The global production of poultry meat has been growing faster than that of any other meat in both developed and developing countries. This growth pattern can be expected to continue because of the inherent efficiency in feed conversion and the lower production cost associated with intensive poultry production. Despite its many advantages and the positive market outlook the world broiler sector faces increasing challenges and food safety, with *Campylobacter* as top priority, is one of them. The industrialized broiler meat production generates a unique environment and ecosystem that facilitates the evolution of zoonotic pathogens such as *Campylobacter* and their transmission to human populations. Despite the large body of research on the epidemiology and control of this organism, there is little clarity in terms of validated practical control measures. It will be necessary to better understand the multiple interactions between this organism, the poultry environment and the chicken host during transmission along the food chain in order to develop effective preventative and control strategies. To this effect, well designed epidemiological studies of sufficient analytical capacity should be used to generate definitive conclusions. The expense of such studies, the predominance of limited short-term localized funding streams and the high competition for resources mitigate against such definitive work, although the burden to society of *Campylobacter* infection and chronic sequelae continue to be very costly. A more consolidated international task force approach would be likely to meet with greater success.

References

Adkin, A., Hartnett, E., Jordan, L., Newell, D., and Davison, H. (2006). Use of a systematic review to assist the development of *Campylobacter* control strategies in broilers. J. Appl. Microbiol. *100*, 306–315.

Allen, V.M., Ridley, A., Harris, J.A., Morris, V.K., Weaver, H., Emery, J., Sharma, M., Sparks, N., and Edge, S. (2007). Sources and spread of *Campylobacter* spp. during partial depopulation of broiler chicken flocks. Zoo. Pub. Health *54*, 136–136.

Allen, V.M., Ridley, A.M., Harris, J.A., Newell, D.G., and Powell, L. (2011). Influence of production system on the rate of onset of *Campylobacter* colonization in chicken flocks reared extensively in the United Kingdom. Br. Poult. Sci. *52*, 30–39.

Allen, V.M., Weaver, H., Ridley, A.M., Harris, J.A., Sharma, M., Emery, J., Sparks, N., Lewis, M., and Edge, S. (2008). Sources and spread of thermophilic *Campylobacter* spp. during partial depopulation of broiler chicken flocks. J. Food Protect. *71*, 264–270.

Alter, T., Weber, R.M., Hamedy, A., and Gluender, G. (2011). Carry-over of thermophilic *Campylobacter* spp. between sequential and adjacent poultry flocks. Vet. Microbiol. *147*, 90–95.

Arsenault, J., Letellier, A., Quessy, S., and Boulianne, M. (2007). Prevalence and risk factors for *Salmonella* and *Campylobacter* spp. carcass contamination in broiler chickens slaughtered in Quebec, Canada. J. Food Protect. *70*, 1820–1828.

Baffoni, L., Gaggia, F., Di Gioia, D., Santini, C., Mogna, L., and Biavati, B. (2012). A *Bifidobacterium*-based synbiotic product to reduce the transmission of *C. jejuni* along the poultry food chain. Int. J. Food Microbiol. *157*, 156–161.

Barrios, P.R., Reiersen, J., Lowman, R., Bisaillon, J.R., Michel, P., Fridriksdottir, V., Gunnarsson, E., Stern, N., Berke, O., McEwen, S., *et al.* (2006). Risk factors for *Campylobacter* spp. colonization in broiler flocks in Iceland. Prevent. Vet. Med. *74*, 264–278.

Bates, C., Hiett, K.L., and Stern, N.J. (2004). Relationship of *Campylobacter* isolated from poultry and from darkling beetles in New Zealand. Avian Dis. *48*, 138–147.

Beery, J.T., Hugdahl, M.B., and Doyle, M.P. (1988). Colonization of gastrointestinal tracts of chicks by *Campylobacter jejuni*. Appl. Environ Microbiol. *54*, 2365–2370.

Berrang, M.E., and Northcutt, J.K. (2005a). Spray washing and sanitizer immersion to lower *Campylobacter*. numbers on broiler transport cage flooring. Poult. Sci. *84*, 92–92.

Berrang, M.E., and Northcutt, J.K. (2005b). Use of water spray and extended drying time to lower bacterial numbers on soiled flooring from broiler transport coops. Poult. Sci. *84*, 1797–1801.

Berrang, M.E., and Northcutt, J.K. (2005c). Water spray and immersion in chemical sanitizer to lower bacterial numbers on broiler transport coop flooring. J. Appl. Poult. Res. *14*, 315–321.

Berrang, M.E., Northcutt, J.K., Fletcher, D.L., and Cox, N.A. (2003). Role of dump cage fecal contamination in the transfer of *Campylobacter* to carcasses of previously negative broilers. J. Appl. Poult. Res. *12*, 190–195.

Berrang, M.E., Northcutt, J.K., and Cason, J.A. (2004). Recovery of *Campylobacter* from broiler feces during extended storage of transport cages. Poult. Sci. *83*, 1213–1217.

Berrang, M.E., Hofacre, C.L., and Meinersmann, R.J. (2011). Forced hot air to dry feces and kill bacteria on transport cage flooring. J. Appl. Poult. Res. *20*, 567–572.

Bobiniene, R., Miskiniene, M., and Gudaviciute, D. (2012). The impact of a biofilm removal from water supply systems on the productivity of chickens and the balance of mineral substances. Veterinarija ir Zootechnika *59*, 10–14.

Boysen, L., and Rosenquist, H. (2007). Reducing numbers of thermotolerant *Campylobacter* by physical decontamination in broiler processing and optimized hygienic practices. Zoo. Pub. Health *54*, 131–131.

Buhr, R.J., Cason, J.A., Dickens, J.A., Hinton, A., Jr., and Ingram, K.D. (2000). Influence of flooring type during transport and holding on bacteria recovery from broiler carcass rinses before and after defeathering. Poult. Sci. *79*, 436–441.

Bull, S.A., Allen, V.M., Domingue, G., Jorgensen, F., Frost, J.A., Ure, R., Whyte, R., Tinker, D., Corry, J.E.L., Gillard-King, J., *et al.* (2006). Sources of *Campylobacter* spp. colonizing housed broiler flocks during rearing. Appl. Environ. Microbiol. *72*, 645–652.

Bull, S.A., Thomas, A., Humphrey, T., Ellis-Iversen, J., Cook, A.J., Lovell, R., and Jorgensen, F. (2008). Flock health indicators and *Campylobacter* spp. in commercial housed broilers reared in Great Britain. Appl. Environ. Microbiol. *74*, 5408–5413.

van Bunnik, B.A.D., Katsma, W.E.A., Wagenaar, J.A., Jacobs-Reitsma, W.F., and de Jong, M.C.M. (2012). Acidification of drinking water inhibits indirect transmission, but not direct transmission of *Campylobacter* between broilers. Prevent. Vet. Med. *105*, 315–319.

Byrd, J.A., Hargis, B.M., Caldwell, D.J., Bailey, R.H., Herron, K.L., McReynolds, J.L., Brewer, R.L., Anderson, R.C., Bischoff, K.M., Callaway, T.R., *et al.* (2001). Effect of lactic acid administration in the drinking water during preslaughter feed withdrawal on *Salmonella* and *Campylobacter* contamination of broilers. Poult. Sci. *80*, 278–283.

Byrd, J., Bailey, R.H., Wills, R., and Nisbet, D. (2007). Recovery of *Campylobacter* from commercial broiler hatchery trayliners. Poult. Sci. *86*, 26–29.

Calenge, F., and Beaumont, C. (2012). Toward integrative genomics study of genetic resistance to *Salmonella* and *Campylobacter* intestinal colonization in fowl. Front. Genet. *3*, 261–261.

Callicott, K.A., Fridriksdottir, V., Reiersen, J., Lowman, R., Bisaillon, J.-R., Gunnarsson, E., Berndtson, E., Hiett, K.L., Needleman, D.S., and Stern, N.J. (2006). Lack of evidence for vertical transmission of *Campylobacter* spp. in chickens. Appl. Environ. Microbiol. *72*, 5794–5798.

Cawthraw, S.A., and Newell, D.G. (2010). Investigation of the presence and protective effects of maternal antibodies against *Campylobacter jejuni* in chickens. Avian Dis. *54*, 86–93.

Cawthraw, S.A., Wassenaar, T.M., Ayling, R., and Newell, D.G. (1996). Increased colonization potential of *Campylobacter jejuni* strain 81116 after passage through chickens and its implication on the rate of transmission within flocks. Epidemiol. Infect. *117*, 213–215.

Chaveerach, P., Keuzenkamp, D.A., Lipman, L.J.A., and Van Knapen, F. (2004). Effect of organic acids in drinking water for young broilers on *Campylobacter* infection, volatile fatty acid production, gut microflora and histological cell changes. Poult. Sci. *83*, 330–334.

Chinivasagam, H.N., Tran, T., Maddock, L., Gale, A., and Blackall, P.J. (2009). Mechanically ventilated broiler sheds: a possible source of aerosolized *Salmonella*,

Campylobacter, and *Escherichia coli*. Appl. Environ. Microbiol. *75*, 7417–7425.

Chowdhury, S., Sandberg, M., Themudo, G.E., and Ersboll, A.K. (2012a). The effect of presence of infected neighbouring farms for the *Campylobacter* infection status in Danish broiler farms. Spatial Spatio-Temporal Epidemiol. *3*, 311–322.

Chowdhury, S., Sandberg, M., Themudo, G.E., and Ersboll, A.K. (2012b). Risk factors for *Campylobacter* infection in Danish broiler chickens. Poult. Sci. *91*, 2701–2709.

Chuma, T., Yano, K., Omori, H., Okamoto, K., and Yugi, H. (1997). Direct detection of *Campylobacter jejuni* in chicken cecal contents by PCR. J. Vet. Med. Sci. *59*, 85–87.

Colles, F.M., Jones, T.A., McCarthy, N.D., Sheppard, S.K., Cody, A.J., Dingle, K.E., Dawkins, M.S., and Maiden, M.C.J. (2008). *Campylobacter* infection of broiler chickens in a free-range environment. Environ. Microbiol. *10*, 2042–2050.

Colles, F.M., McCarthy, N.D., Sheppard, S.K., Layton, R., and Maiden, M.C.J. (2010). Comparison of *Campylobacter* populations isolated from a free-range broiler flock before and after slaughter. Int. J. Food Microbiol. *137*, 259–264.

Colles, F.M., McCarthy, N.D., Layton, R., and Maiden, M.C.J. (2011). The prevalence of *Campylobacter* amongst a free-range broiler breeder flock was primarily affected by flock age. PLoS ONE, e22825.

Conlan, A.J.K., Coward, C., Grant, A.J., Maskell, D.J., and Gog, J.R. (2007). *Campylobacter jejuni* colonization and transmission in broiler chickens: a modelling perspective. J. R. Soc. Interface *4*, 819–829.

Conlan, A.J.K., Line, J.E., Hiett, K., Coward, C., Van Diemen, P.M., Stevens, M.P., Jones, M.A., Gog, J.R., and Maskell, D.J. (2011). Transmission and dose–response experiments for social animals: a reappraisal of the colonization biology of *Campylobacter jejuni* in chickens. J. R. Soc. Interface *8*, 1720–1735.

Connerton, P.L., Timms, A.R., and Connerton, I.F. (2011). *Campylobacter* bacteriophages and bacteriophage therapy. J. Appl. Microbiol. *111*, 255–265.

Corry, J.E., Post, D.E., Colin, P., and Laisney, M.J. (1995). Culture media for the isolation of campylobacters. Int. J. Food Microbiol. *26*, 43–76.

Corry, J.E.L., Allen, V.M., Hudson, W.R., Breslin, M.F., and Davies, R.H. (2002). Sources of *Salmonella* on broiler carcasses during transportation and processing: modes of contamination and methods of control. J. Appl. Microbiol. *92*, 424–432.

Cox, N.A., Stern, N.J., Wilson, J.L., Musgrove, M.T., Buhr, R.J., and Hiett, K.L. (2002). Isolation of *Campylobacter* spp. from semen samples of commercial broiler breeder roosters. Avian Dis. *46*, 717–720.

Cox, N.A., Hofacre, C.L., Buhr, R.J., Wilson, J.L., Bailey, J.S., Richardson, L.J., Cosby, D.E., Musgrove, M.T., Hiett, K.L., and Russell, S.M. (2005). Attempts to isolate naturally occurring *Campylobacter, Salmonella*, and *Clostridium* perfringens from the ductus deferens, testes, and ceca of commercial broiler breeder roosters. J. Appl. Poult. Res. *14*, 126–129.

Cox, N.A., Richardson, L.J., Maurer, J.J., Berrang, M.E., Fedorka-Cray, P.J., Buhr, R.J., Byrd, J.A., Lee, M.D., Hofacre, C.L., O'Kane, P.M., *et al.* (2012). Evidence for Horizontal and Vertical Transmission in *Campylobacter* Passage from Hen to Her Progeny. J. Food Protect. *75*, 1896–1902.

Damjanova, I., Jakab, M., Farkas, T., Meszaros, J., Galantai, Z., Turcsanyi, I., Bistyak, A., Juhasz, A., Paszti, J., Kiss, I., *et al.* (2011). From farm to fork follow-up of thermotolerant campylobacters throughout the broiler production chain and in human cases in a Hungarian county during a ten-months period. Int. J. Food Microbiol. *150*, 95–102.

Dhillon, A.S., Shivaprasad, H.L., Schaberg, D., Wier, F., Weber, S., and Bandli, D. (2006). *Campylobacter jejuni* infection in broiler chickens. Avian Dis. *50*, 55–58.

EFSA Panel on Biological Hazards (BIOHAZ). (2011). Scientific Opinion on *Campylobacter* in broiler meat production: control options and performance objectives and/or targets at different stages of the food chain. EFSA [Online] *9*, 2105. Available at: http://www.efsa.europa.eu/efsajournal (Accessed 11 March 2013).

El-Adawy, H., Hotzel, H., Tomaso, H., Neubauer, H., and Hafez, H.M. (2012). Elucidation of colonization time and prevalence of thermophilic *Campylobacter* species during turkey rearing using multiplex polymerase chain reaction. Poult. Sci. *91*, 454–459.

El-Shibiny, A., Connerton, P.L., and Connerton, I.F. (2005). Enumeration and diversity of campylobacters and bacteriophages isolated during the rearing cycles of free-range and organic chickens. App. Environ. Microbiol. *71*, 1259–1266.

Ellerbroek, L.I., Lienau, J.A., and Klein, G. (2010). *Campylobacter* spp. in broiler flocks at farm level and the potential for cross-contamination during slaughter. Zoo. Pub. Health *57*, e81-e88.

Ellis-Iversen, J., Jorgensen, F., Bull, S., Powell, L., Cook, A.J., and Humphrey, T.J. (2009). Risk factors for *Campylobacter* colonisation during rearing of broiler flocks in Great Britain. Prevent. Vet. Med. *89*, 178–184.

Ellis-Iversen, J., Ridley, A., Morris, V., Sowa, A., Harris, J., Atterbury, R., Sparks, N., and Allen, V. (2012). Persistent environmental reservoirs on farms as risk factors for *Campylobacter* in commercial poultry. Epidemiol. Infect. *140*, 916–924.

Elvers, K.T., Morris, V.K., Newell, D.G., and Allen, V.M. (2011). Molecular tracking, through processing, of *Campylobacter* strains colonizing broiler flocks. Appl. Environ. Microbiol. *77*, 5722–5729.

European Commission Directorate-General for Agriculture and Rural Development. (2011). Prospects for agricultural markets and income 2011–2020. Available at: http://ec.europa.eu/agriculture/publi/caprep/prospects2011/fullrep_en.pdf (Accessed 11 March 2013).

European Council (2007). Directive 2007/43/EC of 28 June 2007 laying down minimum rules for the protection of chickens kept for meat production. Official Journal of the European Union L 182/19.

Available at: http://faolex.fao.org/docs/pdf/eur72633.pdf (Accessed 28 August 2013).

European Food Safety Authority. (2010a). Analysis of the baseline survey on the prevalence of *Campylobacter* in broiler batches and of *Campylobacter* and *Salmonella* on broiler carcasses in the EU, 2008 – Part A: *Campylobacter* and *Salmonella* prevalence estimates. EFSA [Online] *8*, 1503. Available at: http://www.efsa.europa.eu/efsajournal (Accessed 11 March 2013).

European Food Safety Authority. (2010b). Analysis of the baseline survey on the prevalence of *Campylobacter* in broiler batches and of *Campylobacter* and *Salmonella* on broiler carcasses, in the EU, 2008 – Part B: Analysis of factors associated with *Campylobacter* colonisation of broiler batches and with *Campylobacter* contamination of broiler carcasses; and investigation of the culture method diagnostic characteristics used to analyse broiler carcass samples. EFSA [Online] *8*, 1522. Available at: http://www.efsa.europa.eu/efsajournal (Accessed 11 March 2013).

Evans, S.J., and Sayers, A.R. (2000). A longitudinal study of *Campylobacter* infection of broiler flocks in Great Britain. Prevent. Vet. Med. *46*, 209–223.

Food and Agricultural Organization of the United Nations. (2012). *FAOSTAT*. Available at: http://faostat3.fao.org/home/index.html (Accessed 11 March 2013).

Food Standards Agency. (2010a). The Food Standards Agency's Science and Evidence Strategy 2010–15. Available at: http://www.food.gov.uk/multimedia/pdfs/scistrat.pdf (Accessed 11 March 2013).

Food Standards Agency. (2010b). The joint government and industry target to reduce *Campylobacter* in UK produced chickens by 2015. Available at: http://www.food.gov.uk/multimedia/pdfs/campytarget.pdf (Accessed 11 March 2013).

Fraser, R.W., Williams, N.T., Powell, L.F., and Cook, A.J.C. (2010). Reducing *Campylobacter* and *Salmonella* infection: two studies of the economic cost and attitude to adoption of on-farm biosecurity measures. Zoo. Pub. Health *57*, e109-e115.

van Gerwe, T., Miflin, J.K., Templeton, J.M., Bouma, A., Wagenaar, J.A., Jacobs-Reitsma, W.F., Stegeman, A., and Klinkenberg, D. (2009). Quantifying Transmission of *Campylobacter jejuni* in Commercial Broiler Flocks. Appl. Environ. Microbiol. *75*, 625–628.

Ghareeb, K., Awad, W.A., Mohnl, M., Porta, R., Biarnes, M., Boehm, J., and Schatzmayr, G. (2012). Evaluating the efficacy of an avian-specific probiotic to reduce the colonization of *Campylobacter jejuni* in broiler chickens. Poult. Sci. *91*, 1825–1832.

van de Giessen, A.W., Tilburg, J., Ritmeester, W.S., and Van Der Plas, J. (1998). Reduction of *Campylobacter* infections in broiler flocks by application of hygiene measures. Epidemiol. Infect. *121*, 57–66.

van de Giessen, A.W., Bouwknegt, M., Dam-Deisz, W.D.C., van Pelt, W., Wannet, W.J.B., and Visser, G. (2006). Surveillance of *Salmonella* spp. and *Campylobacter* spp. in poultry production flocks in The Netherlands. Epidemiol. Infect. *134*, 1266–1275.

Gross, W.B., Siegel, P.B., and Pierson, E.W. (2002). Effects of genetic selection for high or low antibody response on resistance to a variety of disease challenges and the relationship of resource allocation. Avian Dis. *46*, 1007–1010.

Guerin, M.T., Martin, W., Reiersen, J., Berke, O., McEwen, S.A., Bisaillon, J.-R., and Lowman, R. (2007a). A farm-level study of risk factors associated with the colonization of broiler flocks with *Campylobacter* spp. in Iceland, 2001–2004. Acta Vet. Scand. *49*, 18.

Guerin, M.T., Martin, W., Reiersen, J., Berke, O., McEwen, S.A., Bisaillon, J.R., and Lowman, R. (2007b). House-level risk factors associated with the colonization of broiler flocks with *Campylobacter* spp. in Iceland, 2001–2004. BMC Vet. Res. *3*, 30.

Guerin, M.T., Sir, C., Sargeant, J.M., Waddell, L., O'Connor, A.M., Wills, R.W., Bailey, R.H., and Byrd, J.A. (2010). The change in prevalence of *Campylobacter* on chicken carcasses during processing: A systematic review. Poult. Sci. *89*, 1070–1084.

Habib, I., Berkvens, D., de Zutter, L., Dierick, K., van Huffel, X., Speybroeck, N., Geeraerd, A.H., and Uyttendaele, M. (2012a). *Campylobacter* contamination in broiler carcasses and correlation with slaughterhouses operational hygiene inspection. Food Microbiol. *29*, 105–112.

Habib, I., de Zutter, L., van Huffel, X., Geeraerd, A.H., and Uyttendaele, M. (2012b). Potential of *Escherichia coli* as a surrogate indicator for postchill broiler carcasses with high *Campylobacter* counts. Food Control. *25*, 96–100.

Hald, B., Wedderkopp, A., and Madsen, M. (2000). Thermophilic *Campylobacter* spp. in Danish broiler production: a cross-sectional survey and a retrospective analysis of risk factors for occurrence in broiler flocks. Avian Pathol. *29*, 123–131.

Hald, B., Rattenborg, E., and Madsen, M. (2001). Role of batch depletion of broiler houses on the occurrence of *Campylobacter* spp. in chicken flocks. Lett. Appl. Microbiol. *32*, 253–256.

Hald, B., Skovgard, H., Dang Duong, B., Pedersen, K., Dybdahl, J., Jespersen, J.B., and Madsen, M. (2004). Flies and *Campylobacter* infection of broiler flocks. Emerg. Infect. Dis. *10*, 1490–1492.

Hald, B., Skovgard, H., Pedersen, K., and Bunkenborg, H. (2008). Influxed insects as vectors for *Campylobacter jejuni* and *Campylobacter coli* in Danish broiler houses. Poult. Sci. *87*, 1428–1434.

Hastings, R., Colles, F.M., McCarthy, N.D., Maiden, M.C., and Sheppard, S.K. (2011). *Campylobacter* genotypes from poultry transportation crates indicate a source of contamination and transmission. J. Appl. Microbiol. *110*, 266–76.

Hansson, I., Engvall, E.O., Lindblad, J., Gunnarsson, A., and Vagsholm, I. (2004). Surveillance programme for *Campylobacter* species in Swedish broilers, July 2001 to June 2002. Vet. Record *155*, 193–196.

Hansson, I., Ederoth, M., Andersson, L., Vagsholm, I., and Engvall, E.O. (2005). Transmission of *Campylobacter* spp. to chickens during transport to slaughter. J. Appl. Microbiol. *99*, 1149–1157.

Hansson, I., Forshell, L.P., Gustafsson, P., Boqvist, S., Lindblad, J., Engvall, E.O., Andersson, Y., and Vagsholm, I. (2007). Summary of the Swedish

Campylobacter program in broilers, 2001 through 2005. J. Food Protect. *70*, 2008–2014.

Hansson, I., Engvall, E.O., Vagsholm, I., and Nyman, A. (2010a). Risk factors associated with the presence of *Campylobacter*-positive broiler flocks in Sweden. Prevent. Vet. Med. *96*, 114–121.

Hansson, I., Pudas, N., Harbom, B., and Engvall, E.O. (2010b). Within-flock variations of *Campylobacter* loads in caeca and on carcasses from broilers. Int. J. Food Microbiol. *141*, 51–55.

Havelaar, A., Evers, E., Nauta, M., and Jacobs-Reitsma, W. (2005). Modelling the effect of logistic slaughter and/or germicidal treatment on the contamination of broiler meat with *Campylobacter* (Beckbergen, the Netherlands, World's Poultry Science Association).

Havelaar, A.H., Mangen, M.-J.J., de Koeijer, A.A., Bogaardt, M.J., Evers, E.G., Jacobs-Reitsma, W.E., van Pelt, W., Wagenaar, J.A., de Wit, G.A., van der Zee, H., *et al.* (2007). Effectiveness and efficiency of controlling *Campylobacter* on broiler chicken meat. Risk Anal. *27*, 831–844.

Havenstein, G.B., Ferket, P.R., and Qureshi, M.A. (2003). Growth, livability, and feed conversion of 1957 versus 2001 broilers when fed representative 1957 and 2001 broiler diets. Poult. Sci. *82*, 1500–1508.

Henry, I., Reichardt, J., Denis, M., and Cardinale, E. (2011). Prevalence and risk factors for *Campylobacter* spp. in chicken broiler flocks in Reunion Island (Indian Ocean). Prevent. Vet. Med. *100*, 64–70.

Herman, L., Heyndrickx, M., Grijspeerdt, K., Vandekerchove, D., Rollier, I., and De Zutter, L. (2003). Routes for *Campylobacter* contamination of poultry meat: epidemiological study from hatchery to slaughterhouse. Epidemiol. Infect. *131*, 1169–1180.

Hermans, D., Van Deun, K., Messens, W., Martel, A., Van Immerseel, F., Haesebrouck, F., Rasschaert, G., Heyndrickx, M., and Pasmans, F. (2011). *Campylobacter* control in poultry by current intervention measures ineffective: Urgent need for intensified fundamental research. Vet. Microbiol. *152*, 219–228.

Hermans, D., Pasmans, F., Messens, W., Martel, A., van Immerseel, F., Rasschaert, G., Heyndrickx, M., van Deun, K., and Haesebrouck, F. (2012). Poultry as a host for the zoonotic pathogen *Campylobacter jejuni*. Vector Borne Zoo. Dis. *12*, 89–98.

Heuer, O.E., Pedersen, K., Andersen, J.S., and Madsen, M. (2001). Prevalence and antimicrobial susceptibility of thermophilic *Campylobacter* in organic and conventional broiler flocks. Lett. Appl. Microbiol. *33*, 269–274.

Hiett, K.L., Cox, N.A., and Stern, N.J. (2002a). Direct polymerase chain reaction detection of *Campylobacter* spp. in poultry hatchery samples. Avian Dis. *46*, 219–223.

Hiett, K.L., Stern, N.J., Fedorka-Cray, P., Cox, N.A., Musgrove, M.T., and Ladely, S. (2002b). Molecular subtype analyses of *Campylobacter* spp. from Arkansas and California poultry operations. Appl. Environ. Microbiol. *68*, 6220–6236.

Hinton, A., and Ingram, K.D. (2000). Use of oleic acid to reduce the population of the bacterial flora of poultry skin. J. Food Protect. *63*, 1282–1286.

Hofshagen, M., and Kruse, H. (2005). Reduction in flock prevalence of *Campylobacter* spp. in broilers in Norway after implementation of an action plan. J. Food Protect. *68*, 2220–2223.

Hue, O., Le Bouquin, S., Laisney, M.J., Allain, V., Lalande, F., Petetin, I., Rouxel, S., Quesne, S., Gloaguen, P.Y., Picherot, M., *et al.* (2010). Prevalence of and risk factors for *Campylobacter* spp. contamination of broiler chicken carcasses at the slaughterhouse. Food Microbiol. *27*, 992–999.

Hue, O., Allain, V., Laisney, M.-J., Le Bouquin, S., Lalande, F., Petetin, I., Rouxel, S., Quesne, S., Gloaguen, P.-Y., Picherot, M., *et al.* (2011). *Campylobacter* contamination of broiler caeca and carcasses at the slaughterhouse and correlation with *Salmonella* contamination. Food Microbiol. *28*, 862–868.

Huneau-Salaun, A., Denis, M., Balaine, L., and Salvat, G. (2007). Risk factors for *Campylobacter* spp. colonization in French free-range broiler-chicken flocks at the end of the indoor rearing period. Prevent. Vet. Med. *80*, 34–48.

Idris, U., Lu, J.R., Maier, M., Sanchez, S., Hofacre, C.L., Harmon, B.G., Maurer, J.J., and Lee, M.D. (2006). Dissemination of fluoroquinolone-resistant *Campylobacter* spp. within an integrated commercial poultry production system. Appl. Environ. Microbiol. *72*, 3441–3447.

Jasson, V., Sampers, I., Botteldoorn, N., Lopez-Galvez, F., Baert, L., Denayer, S., Rajkovic, A., Habib, I., De Zutter, L., Debevere, J., *et al.* (2009). Characterization of *Escherichia coli* from raw poultry in Belgium and impact on the detection of *Campylobacter jejuni* using Bolton broth. Int. J. Food Microbiol. *135*, 248–253.

Johannessen, G.S., Johnsen, G., Okland, M., Cudjoe, K.S., and Hofshagen, M. (2007). Enumeration of thermotolerant *Campylobacter* spp. from poultry carcasses at the end of the slaughter-line. Lett. Appl. Microbiol. *44*, 92–97.

Johnsen, G., Kruse, H., and Hofshagen, M. (2006). Genetic diversity and description of transmission routes for *Campylobacter* on broiler farms by amplified-fragment length polymorphism. J. Appl. Microbiol. *101*, 1130–1139.

Jokinen, C.C., Koot, J.M., Carrillo, C.D., Gannon, V.P.J., Jardine, C.M., Mutschall, S.K., Topp, E., and Taboada, E.N. (2012). An enhanced technique combining pre-enrichment and passive filtration increases the isolation efficiency of *Campylobacter jejuni* and *Campylobacter coli* from water and animal fecal samples. J. Microbiolog. Methods *91*, 506–513.

Jones, F.T., Axtell, R.C., Rives, D.V., Scheideler, S.E., Tarver, F.R., Jr., Walker, R.L., and Wineland, M.J. (1991). A survey of *Campylobacter jejuni* contamination in modern broiler production and processing systems. J. Food Protect. *54*, 259–266.

Jonsson, M.E., Heier, B.T., Norstrom, M., and Hofshagen, M. (2010a). Analysis of simultaneous space–time clusters of *Campylobacter* spp. in humans and in broiler flocks using a multiple dataset approach. Int. J. Health Geograph. *9*, 48.

Jonsson, M.E., Norstrom, M., Sandberg, M., Ersboll, A.K., and Hofshagen, M. (2010b). Space-time patterns of *Campylobacter* spp. colonization in broiler flocks, 2002–2006. Epidemiol. Infect. *138*, 1336–1345.

Jonsson, M.E., Chriel, M., Norstrom, M., and Hofshagen, M. (2012). Effect of climate and farm environment on *Campylobacter* spp. colonisation in Norwegian broiler flocks. Prevent. Vet. Med. *107*, 95–104.

Jore, S., Viljugrein, H., Brun, E., Heier, B.T., Borck, B., Ethelberg, S., Hakkinen, M., Kuusi, M., Reiersen, J., Hansson, I., *et al.* (2010). Trends in *Campylobacter* incidence in broilers and humans in six European countries, 1997–2007. Prevent. Vet. Med. 93, 33–41.

Jorgensen, F., Ellis-Iversen, J., Rushton, S., Bull, S.A., Harris, S.A., Bryan, S.J., Gonzalez, A., and Humphrey, T.J. (2011). Influence of season and geography on *Campylobacter jejuni* and *C. coli* subtypes in housed broiler flocks reared in Great Britain. Appl. Environ. Microbiol. *77*, 3741–3748.

Kassem, I.I., Sanad, Y., Gangaiah, D., Lilburn, M., LeJeune, J., and Rajashekara, G. (2010). Use of bioluminescence imaging to monitor *Campylobacter* survival in chicken litter. J. Appl. Microbiol. *109*, 1988–1997.

Katsma, W.E.A., De Jong, M.C.M., Jacobs-Reitsma, W.F., and Wagenaar, J.A. (2007a). A quantitative study on the direct and indirect *Campylobacter* transmission in broilers under experimental conditions. Zoo. Pub. Health 54, 127–128.

Katsma, W.E.A., de Koeijer, A.A., Jacobs-Reitsma, W.F., Mangen, M.J.J., and Wagenaar, J.A. (2007b). Assessing interventions to reduce the risk of *Campylobacter* prevalence in broilers. Risk Analysis 27, 863–876.

Knudsen, K.N., Bang, D., Andresen, L.O., and Madsen, M. (2006). *Campylobacter jejuni* strains of human and chicken origin are invasive in chickens after oral challenge. Avian Dis. *50*, 10–14.

Kovats, R.S., Edwards, S.J., Charron, D., Cowden, J., D'Souza, R.M., Ebi, K.L., Gauci, C., Gerner-Smidt, P., Hajat, S., Hales, S., *et al.* (2005). Climate variability and *Campylobacter* infection: an international study. Int. J. Biometeorol. *49*, 207–214.

Lawes, J.R., Vidal, A., Clifton-Hadley, F.A., Sayers, R., Rodgers, J., Snow, L., Evans, S.J., and Powell, L.F. (2012). Investigation of prevalence and risk factors for *Campylobacter* in broiler flocks at slaughter: results from a UK survey. Epidemiol. Infect. *140*, 1725–1737.

Lee, M.D., and Newell, D.G. (2006). *Campylobacter* in poultry: Filling an ecological niche. Avian Dis. *50*, 1–9.

Lienau, J.A., Ellerbroek, L., and Klein, G. (2007). Tracing flock-related *Campylobacter* clones from broiler farms through slaughter to retail products by pulsed-field gel electrophoresis. J. Food Protect. *70*, 536–542.

Line, J.E. (2002). *Campylobacter* and *Salmonella* populations associated with chickens raised on acidified litter. Poult. Sci. *81*, 1473–1477.

Line, J.E. (2006). Influence of relative humidity on transmission of *Campylobacter jejuni* in broiler chickens. Poult. Sci. *85*, 1145–1150.

Line, J.E., and Bailey, J.S. (2006). Effect of on-farm litter acidification treatments on *Campylobacter* and *Salmonella* populations in commercial broiler houses in Northeast Georgia. Poult. Sci. *85*, 1529–1534.

Lyngstad, T.M., Jonsson, M.E., Hofshagen, M., and Heier, B.T. (2008). Risk factors associated with the presence of *Campylobacter* species in Norwegian broiler flocks. Poult. Sci. *87*, 1987–1994.

McDowell, S.W.J., Menzies, F.D., McBride, S.H., Oza, A.N., McKenna, J.P., Gordon, A.W., and Neillab, S.D. (2008). *Campylobacter* spp. in conventional broiler flocks in Northern Ireland: Epidemiology and risk factors. Prevent. Vet. Med. *84*, 261–276.

Meerburg, B.G., Jacobs-Reitsma, W.F., Wagenaar, J.A., and Kijlstra, A. (2006). Presence of *Salmonella* and *Campylobacter* spp. in wild small mammals on organic farms. Appl. Environ. Microbiol. *72*, 960–962.

Messens, W., Herman, L., De Zutter, L., and Heyndrickx, M. (2009). Multiple typing for the epidemiological study of contamination of broilers with thermotolerant *Campylobacter*. Vet. Microbiol. *138*, 120–131.

Molatova, Z., Skrivanova, E., Bare, J., Houf, K., Bruggeman, G., and Marounek, M. (2011). Effect of coated and non-coated fatty acid supplementation on broiler chickens experimentally infected with *Campylobacter jejuni*. J. Animal Physiol. Animal Nutrit. *95*, 701–706.

Moran, L., Kelly, C., and Madden, R.H. (2009). Factors affecting the recovery of *Campylobacter* spp. from retail packs of raw, fresh chicken using ISO 10272–1:2006. Lett. Appl. Microbiol. *48*, 628–632.

Moran, L., Kelly, C., Cormican, M., McGettrick, S., and Madden, R.H. (2011). Restoring the selectivity of Bolton broth during enrichment for *Campylobacter* spp. from raw chicken. Lett. Appl. Microbiol. *52*, 614–618.

Nather, G., Alter, T., Martin, A., and Ellerbroek, L. (2009). Analysis of risk factors for *Campylobacter* species infection in broiler flocks. Poult. Sci. *88*, 1299–1305.

Nauta, M., van der Fels-Klerx, I., and Havelaar, A. (2005). A poultry-processing model for quantitative microbiological risk assessment. Risk Analysis *25*, 85–98.

Nauta, M., Hill, A., Rosenquist, H., Brynestad, S., Fetsch, A., van der Logt, P., Fazil, A., Christensen, B., Katsma, E., Borck, B., *et al.* (2009a). A comparison of risk assessments on *Campylobacter* in broiler meat. Int. J. Food Microbiol. *129*, 107–123.

Nauta, M.J., van der Wal, F.J., Putirulan, F.F., Post, J., van de Kassteele, J., and Bolder, N.M. (2009b). Evaluation of the 'testing and scheduling' strategy for control of *Campylobacter* in broiler meat in The Netherlands. Int. J. Food Microbiol. *134*, 216–222.

Nauta, M.J., Sanaa, M., and Havelaar, A.H. (2012). Risk based microbiological criteria for *Campylobacter* in broiler meat in the European Union. Int. J. Food Microbiol. *158*, 209–217.

Newell, D. (2005). Campylobacters: epidemiology, ecology and the potential for control. World Poult. *21*, 26–27.

Newell, D.G., Elvers, K.T., Dopfer, D., Hansson, I., Jones, P., James, S., Gittins, J., Stern, N.J., Davies, R., Connerton, I., *et al.* (2011). Biosecurity-based interventions and strategies to reduce *Campylobacter* spp. on poultry farms. Appl. Environ. Microbiol. 77, 8605–8614.

Niemz, A., Ferguson, T.M., and Boyle, D.S. (2011). Point-of-care nucleic acid testing for infectious diseases. Trends Biotechnol. *29*, 240–250.

Nocker, A., and Camper, A.K. (2009). Novel approaches toward preferential detection of viable cells using nucleic acid amplification techniques. FEMS Microbiol. Lett. *291*, 137–142.

Normand, V., Boulianne, M., and Quessy, S. (2008). Evidence of cross-contamination by *Campylobacter* spp. of broiler carcasses using genetic characterization of isolates. Canad. J. Vet. Res. *72*, 396–402.

Nylen, G., Dunstan, F., Palmer, S.R., Andersson, Y., Bager, F., Cowden, J., Feierl, G., Galloway, Y., Kapperud, G., Megraud, F., *et al.* (2002). The seasonal distribution of campylobacter infection in nine European countries and New Zealand. Epidemiol. Infect. *128*, 383–390.

O'Mahony, E., Buckley, J.F., Bolton, D., Whyte, P., and Fanning, S. (2011). Molecular epidemiology of *Campylobacter* isolates from poultry production units in Southern Ireland. PLoS ONE, e28490.

Ogden, I.D., MacRae, M., Johnston, M., Strachan, N.J.C., Cody, A.J., Dingle, K.E., and Newell, D.G. (2007). Use of multilocus sequence typing to investigate the association between the presence of *Campylobacter* spp. in broiler drinking water and *Campylobacter* colonization in broilers. Appl. Environ. Microbiol. *73*, 5125–5129.

Olsen, K.N., Lund, M., Skov, J., Christensen, L.S., and Hoorfar, J. (2009). Detection of *Campylobacter* bacteria in air samples for continuous real-time monitoring of *Campylobacter* colonization in broiler flocks. Appl. Environ. Microbiol. *75*, 2074–2078.

Oyarzabal, O.A., Backert, S., Nagaraj, M., Miller, R.S., Hussain, S.K., and Oyarzabal, E.A. (2007). Efficacy of supplemented buffered peptone water for the isolation of *Campylobacter jejuni* and *C. coli* from broiler retail products. J. Microbiolog. Methods *69*, 129–136.

Patriarchi, A., Fox, A., Maunsell, B., Fanning, S., and Bolton, D. (2011). Molecular Characterization and Environmental Mapping of *Campylobacter* Isolates in a Subset of Intensive Poultry Flocks in Ireland. Food. Pathog. Dis. *8*, 99–108.

Pearson, A.D., Greenwood, M.H., Feltham, R.K.A., Healing, T.D., Donaldson, J., Jones, D.M., and Colwell, R.R. (1996). Microbial ecology of *Campylobacter jejuni* in a United Kingdom chicken supply chain: Intermittent common source, vertical transmission, and amplification by flock propagation. Appl. Environ. Microbiol. *62*, 4614–4620.

Perez-Boto, D., Garcia-Pena, F.J., Abad-Moreno, J.C., Dolores Hurtado-Pizarro, M., Perez-Cobo, I., and Aurora Echeita, M. (2010). Drinking water as the source of *Campylobacter coli* infection in grandparent heavy breeders. Avian Pathol. *39*, 483–487.

Petersen, L., and Wedderkopp, A. (2001). Evidence that certain clones of *Campylobacter jejuni* persist during successive broiler flock rotations. Appl. Environ. Microbiol. *67*, 2739–2745.

Petersen, L., Nielsen, E.M., Engberg, J., On, S.L.W., and Dietz, H.H. (2001). Comparison of genotypes and serotypes of *Campylobacter jejuni* isolated from Danish wild mammals and birds and from broiler flocks and humans. Appl. Environ. Microbiol. *67*, 3115–3121.

Peyrat, M.B., Soumet, C., Maris, P., and Sanders, P. (2008). Recovery of *Campylobacter jejuni* from surfaces of poultry slaughterhouses after cleaning and disinfection procedures: Analysis of a potential source of carcass contamination. Int. J. Food Microbiol. *124*, 188–194.

Powell, L.F., Lawes, J.R., Clifton-Hadley, F.A., Rodgers, J., Harris, K., Evans, S.J., and Vidal, A. (2012). The prevalence of *Campylobacter* spp. in broiler flocks and on broiler carcases, and the risks associated with highly contaminated carcases. Epidemiol. Infect. *140*, 2233–2246.

Racicot, M., Venne, D., Durivage, A., and Vaillancourt, J.-P. (2012). Evaluation of strategies to enhance biosecurity compliance on poultry farms in Quebec: Effect of audits and cameras. Prevent. Vet. Med. *103*, 208–218.

Ramabu, S.S., Boxall, N.S., Madie, P., and Fenwick, S.G. (2004). Some potential sources for transmission of *Campylobacter jejuni* to broiler chickens. Lett. Appl. Microbiol. *39*, 252–256.

Randall, L., Lemma, F., Rodgers, J., Vidal, A., and Clifton-Hadley, F. (2010). Development and evaluation of internal amplification controls for use in a real-time duplex PCR assay for detection of *Campylobacter coli* and *Campylobacter jejuni*. J. Med. Microbiol. *59*, 172–178.

Rasschaert, G., Houf, K., van Hende, J., and de Zutter, L. (2006). *Campylobacter* contamination during poultry slaughter in Belgium. J. Food Protect. *69*, 27–33.

Rasschaert, G., Houf, K., and de Zutter, L. (2007). External contamination of *Campylobacter*-free flocks after transport in cleaned and disinfected containers. J. Food Protect. *70*, 40–46.

Reich, F., Atanassova, V., Haunhorst, E., and Klein, G. (2008). The effects of *Campylobacter* numbers in caeca on the contamination of broiler carcasses with *Campylobacter*. Int. J. Food Microbiol. *127*, 116–120.

Richardson, L.J., Cox, N.A., Buhr, R.J., and Harrison, M.A. (2011). Isolation of *Campylobacter* from circulating blood of commercial broilers. Avian Dis. *55*, 375–378.

Ridley, A.M., Allen, V.M., Sharma, M., Harris, J.A., and Newell, D.G. (2008). Real-time PCR approach for detection of environmental sources of *Campylobacter* strains colonizing broiler flocks. Appl. Environ. Microbiol. *74*, 2492–2504.

Ridley, A., Morris, V., Gittins, J., Cawthraw, S., Harris, J., Edge, S., and Allen, V. (2011a). Potential sources of *Campylobacter* infection on chicken farms: contamination and control of broiler-harvesting equipment, vehicles and personnel. J. Appl. Microbiol. *111*, 233–244.

Ridley, A.M., Morris, V.K., Cawthraw, S.A., Ellis-Iversen, J., Harris, J.A., Kennedy, E.M., Newell, D.G., and Allen, V.M. (2011b). Longitudinal molecular epidemiological study of thermophilic campylobacters on one conventional broiler chicken farm. Appl. Environ. Microbiol. *77*, 98–107.

Rodgers, J.D., Clifton-Hadley, F.A., Marin, C., and Vidal, A.B. (2010). An evaluation of survival and detection of *Campylobacter jejuni* and *C. coli* in broiler caecal contents using culture-based methods. J. Appl. Microbiol. *109*, 1244–1252.

Romero-Barrios, P., Hempen, M., Messens, W., Stella, P., and Hugas, M. (2013). Quantitative microbiological risk assessment (QMRA) of food-borne zoonoses at the European level. Food Control *29*, 343–349.

Rosenquist, H., Bengtsson, A., and Hansen, T.B. (2007). A collaborative study on a Nordic standard protocol for detection and enumeration of thermotolerant *Campylobacter* in food (NMKL 119, 3. Ed., 2007). Int. J. Food Microbiol. *118*, 201–213.

Rosenquist, H., Boysen, L., Galliano, C., Nordentoft, S., Ethelberg, S., and Borck, B. (2009). Danish strategies to control *Campylobacter* in broilers and broiler meat: facts and effects. Epidemiol. Infect. *137*, 1742–1750.

Rosenquist, H., Sommer, H.M., Nielsen, N.L., and Christensen, B.B. (2006). The effect of slaughter operations on the contamination of chicken carcasses with thermotolerant *Campylobacter*. Int. J. Food Microbiol. *108*, 226–232.

Russa, A.D., Bouma, A., Vernooij, J.C.M., Jacobs-Reitsma, W., and Stegeman, J.A. (2005). No association between partial depopulation and *Campylobacter* spp. colonization of Dutch broiler flocks. Lett. Appl. Microbiol. *41*, 280–285.

Sahin, O., Zhang, Q.J., Meitzler, J.C., Harr, B.S., Morishita, T.Y., and Mohan, R. (2001). Prevalence, antigenic specificity, and bactericidal activity of poultry anti-*Campylobacter* maternal antibodies. Appl. Environ. Microbiol. *67*, 3951–3957.

Sahin, O., Luo, N., Huang, S.X., and Zhang, Q.J. (2003). Effect of *Campylobacter*-specific maternal antibodies on *Campylobacter jejuni* colonization in young chickens. Appl. Environ. Microbiol. *69*, 5372–5379.

Santini, C., Baffoni, L., Gaggia, F., Granata, M., Gasbarri, R., Di Gioia, D., and Biavati, B. (2010). Characterization of probiotic strains: An application as feed additives in poultry against *Campylobacter jejuni*. Int. J. Food Microbiol. *141*, S98-S108.

Sears, A., Baker, M.G., Wilson, N., Marshall, J., Muellner, P., Campbell, D.M., Lake, R.J., and French, N.P. (2011). Marked *Campylobacteriosis* Decline after Interventions Aimed at Poultry, New Zealand. Emerg. Infect. Dis. *17*, 1007–1015.

Shreeve, J.E., Toszeghy, M., Ridley, A., and Newell, D.G. (2002). The carry-over of *Campylobacter* isolates between sequential poultry flocks. Avian Dis. *46*, 378–385.

Skanseng, B., Kaldhusdal, M., and Rudi, K. (2006). Comparison of chicken gut colonisation by the pathogens *Campylobacter jejuni* and *Clostridium perfringens* by real-time quantitative PCR. Mol. Cell. Probes *20*, 269–279.

Skanseng, B., Kaldhusdal, M., Moen, B., Gjevre, A.G., Johannessen, G.S., Sekelja, M., Trosvik, P., and Rudi, K. (2010). Prevention of intestinal *Campylobacter jejuni* colonization in broilers by combinations of in-feed organic acids. J. Appl. Microbiol. *109*, 1265–1273.

Slader, J., Domingue, G., Jorgensen, F., McAlpine, K., Owen, R.J., Bolton, F.J., and Humphrey, T.J. (2002). Impact of transport crate reuse and of catching and processing on *Campylobacter* and *Salmonella* contamination of broiler chickens. Appl. Environ. Microbiol. *68*, 713–719.

Snelling, W.J., McKenna, J.P., Lecky, D.M., and Dooley, J.S.G. (2005). Survival of *Campylobacter jejuni* in waterborne protozoa. Appl. Environ. Microbiol. *71*, 5560–5571.

Solís de los Santos, F., Hume, M., Venkitanarayanan, K., Donoghue, A.M., Hanning, I., Slavik, M.F., Aguiar, V.F., Metcalf, J.H., Reyes-Herrera, I., Blore, P.J., *et al.* (2010). Caprylic acid reduces enteric *Campylobacter* colonization in market-aged broiler chickens but does not appear to alter cecal microbial populations. J. Food Protect. *73*, 251–257.

Sparks, N.H.C. (2009). The role of the water supply system in the infection and control of *Campylobacter* in chicken. Worlds Poult. Sci. J. *65*, 459–473.

Stern, N.J., Fedorka-Cray, P., Bailey, J.S., Cox, N.A., Craven, S.E., Hiett, K.L., Musgrove, M.T., Ladely, S., Cosby, D., and Mead, G.C. (2001). Distribution of *Campylobacter* spp. in selected U.S. poultry production and processing operations. J. Food Protect. *64*, 1705–1710.

Stern, N.J., Hiett, K.L., Alfredsson, G.A., Kristinsson, K.G., Reiersen, J., Hardardottir, H., Briem, H., Gunnarsson, E., Georgsson, F., Lowman, R., *et al.* (2003). *Campylobacter* spp. in Icelandic poultry operations and human disease. Epidemiol. Infect. *130*, 23–32.

Svetoch, E.A., and Stern, N.J. (2010). Bacteriocins to control *Campylobacter* spp. in poultry – a review. Poult. Sci. *89*, 1763–1768.

Taboada, E.N., Acedillo, R.R., Carrillo, C.D., Findlay, W.A., Medeiros, D.T., Mykytczuk, O.L., Roberts, M.J., Valencia, C.A., Farber, J.M., and Nash, J.H.E. (2004). Large-scale comparative genomics meta-analysis of *Campylobacter jejuni* isolates reveals low level of genome plasticity. J. Clin. Microbiol. *42*, 4566–4576.

Tissari, P., and Rautelin, H. (2007). Evaluation of an enzyme immunoassay-based stool antigen test to detect *Campylobacter jejuni* and *Campylobacter coli*. Diagn. Microbiol. Infect. Dis. *58*, 171–175.

Tustin, J., Laberge, K., Michel, P., Reiersen, J., Dadadottir, S., Briem, H., Hardardottir, H., Kristinsson, K., Gunnarsson, E., Fridriksdottir, V., *et al.* (2011). A National Epidemic of Campylobacteriosis in Iceland, Lessons Learned. Zoo. Pub. Health *58*, 440–447.

Vidal, A.B., Rodgers, J., Arnold, M., and Clifton-Hadley, F. 2012. Comparison of Different Sampling Strategies and Laboratory Methods for the Detection of *C. jejuni* and *C. coli* from Broiler Flocks at Primary Production. Zoon. Pub. Health [Online] doi:10.1111/zph.12009 (epub ahead of print). (Accessed 11 March 2013).

Wadl, M., Poelzler, T., Flekna, G., Thompson, L., Slaghuis, J., Koefer, J., Hein, I., and Wagner, M. (2009). Easy-to-Use Rapid Test for Direct Detection of *Campylobacter* spp. in Chicken Feces. J. Food Protect. *72*, 2483–2488.

Whittier, C.A., Cranfield, M.R., and Stoskopf, M.K. (2010). Real-time PCR detection of *Campylobacter* spp. in free-ranging mountain gorillas (*Gorilla beringei beringei*). J. Wild. Dis. *46*, 791–802.

Whyte, P., Collins, J.D., McGill, K., Monahan, C., and O'Mahony, H. (2001). The effect of transportation stress on excretion rates of campylobacters in market-age broilers. Poult. Sci. *80*, 817–820.

Whyte, P., McGill, K., and Collins, J.D. (2003). An assessment of steam pasteurization and hot water immersion treatments for the microbiological decontamination of broiler carcasses. Food Microbiol. *20*, 111–117.

Williams, L.K., Sait, L.C., Cogan, T.A., Jorgensen, F., Grogono-Thomas, R., and Humphrey, T.J. (2012). Enrichment culture can bias the isolation of *Campylobacter* subtypes. Epidemiol. Infect. *140*, 1227–1235.

Yamazaki, W., Taguchi, M., Kawai, T., Kawatsu, K., Sakata, J., Inoue, K., and Misawa, N. (2009). Comparison of Loop-Mediated Isothermal Amplification Assay and Conventional Culture Methods for Detection of *Campylobacter jejuni* and *Campylobacter coli* in Naturally Contaminated Chicken Meat Samples. Appl. Environ. Microbiol. *75*, 1597–1603.

Young, I., Rajic, A., Wilhelm, B.J., Waddell, L., Parker, S., and McEwen, S.A. (2009). Comparison of the prevalence of bacterial enteropathogens, potentially zoonotic bacteria and bacterial resistance to antimicrobials in organic and conventional poultry, swine and beef production: a systematic review and meta-analysis. Epidemiol. Infect. *137*, 1217–1232.

Zimmer, M., Barnhart, H., Idris, U., and Lee, M.D. (2003). Detection of *Campylobacter jejuni* strains in the water lines of a commercial broiler house and their relationship to the strains that colonized the chickens. Avian Dis. *47*, 101–107.

Zuidhof, M.J., McGovern, R.H., Schneider, B.L., Feddes, J.J.R., Robinson, F.E., and Korver, D.R. (2004). Effects of feed withdrawal time on the incidence of fecal spillage and contamination of broiler carcasses at processing. J. Appl. Poult. Res. *13*, 171–177.

Zweifel, C., Scheu, K.D., Keel, M., Renggli, F., and Stephan, R. (2008). Occurrence and genotypes of *Campylobacter* in broiler flocks, other farm animals, and the environment during several rearing periods on selected poultry farms. Int. J. Food Microbiol. *125*, 182–187.

Conditional Commensalism of *Campylobacter* in Chickens

22

Lisa K. Williams, Emma K. Trantham and Tristan A. Cogan

Abstract

Two species of *Campylobacter*, *jejuni* and *coli*, dominate the epidemiology of human infection. Both species are characterized by their broad host range, and spectrum of effects that they have on the hosts that they colonize. In non-immune humans, both cause severe gastrointestinal symptoms, as they can do in companion animals. Farmed animals, and in particular poultry, appear to carry these organisms as intestinal commensals. Given the severe disease seen in some host species this may be because the bacterium is adapted to different hosts. For example, poultry adapted strains may cause severe disease in humans which are an accidental host. Alternatively, the apparent commensalism of *Campylobacter* in poultry may in fact be controlled parasitism, with the host able to control, but not prevent, colonization by the bacterium. Here we review the evidence for commensalism of *Campylobacter* in poultry and ask the questions: are *C. jejuni* and *coli* really avian commensals, or are they just well controlled pathogens?

Colonization of the host

The thermophilic *Campylobacter* spp. are frequently isolated from a variety of animals, and in particular poultry (Denis *et al.*, 2008; Horrocks *et al.*, 2009; Humphrey *et al.*, 2007; Studahl *et al.*, 2000; Rodrigues *et al.*, 2001; Neimann *et al.*, 2003; Wingstrand *et al.*, 2006), pigs (Denis *et al.*, 2008; Horrocks *et al.*, 2009; Zweifel *et al.*, 2008; Little *et al.*, 2008; Oporto *et al.*, 2007), wild and companion animals (Horrocks *et al.*, 2009; Kwan *et al.*, 2008; Meerburg *et al.*, 2006, Flemming 1980; Lee *et al.*, 2004; Petersen *et al.*, 2007; Baker *et al.*, 1999; Hald *et al.*, 2004; Moser *et al.*, 2001; Stanley *et al.*, 1992). *Campylobacter* spp. have also been isolated from cattle (Gilpin *et al.*, 2008; Kwan *et al.*, 2008; Zweifel *et al.*, 2008; Englen *et al.*, 2007) and sheep (Oporto *et al.*, 2007; Sahin *et al.*, 2008; Açik and, Cetinkaya 2006). *C. jejuni* and *C. coli* are the two species most frequently isolated from poultry and cause the majority of cases of human infection. These two species commonly colonize the intestinal tract of birds, often without the animal exhibiting any overt signs of disease. Despite being colonized with up to 10^8 cfu/g gut contents birds typically remain asymptomatic of disease (Dhillion *et al.*, 2006) and because of this *Campylobacter* is often referred to as a commensal in poultry (Dhillion *et al.*, 2006; Macpherson and Uhr 2004). Indeed, the physiology of the organism points to a long evolution in, and adaptation to, avian hosts. The primary clue to this is that *Campylobacter jejuni* and *coli* are well suited to growth at 42°C, the body temperature of chickens. This is in contrast to many of the other *Campylobacter* spp. and may be indicative of a branching out of the ancestral organism into new hosts. The thermophilic species could simplistically be viewed as having a dual existence as commensals in hosts with a body temperature of 42°C and pathogens in hosts with a lower temperature. Although this does appear to be the case in poultry to human transmission where there is no evidence from *in vitro* work that a temperature-dependent virulence switch exists. It appears, therefore, that commensalism may be a product of the host physiology rather than the response of the bacterium itself. Clues to the nature of the interaction of *Campylobacter* with poultry may lie in the detail of colonization of the host.

Invasive behaviour in poultry

It is increasingly recognized that *Campylobacter* can leave the gut and colonize other organs, for example the liver (Knudsen *et al.*, 2006; Van Deun *et al.*, 2007; Sanyal *et al.*, 1984: Jennings *et al.*, 2011; O'Leary *et al.*, 2009). The source of muscle contamination of poultry carcasses has been debated for many years, but it is now clear that spread of *Campylobacter* from the intestine during the scalding and evisceration process is not the only contributor to infected meat. Knudsen *et al.*, (2006) reported that *Campylobacter* can invade the intestinal mucosa and colonize the spleen, liver, lungs and heart. In addition to this Sanyal *et al.* (1984) observed *C. jejuni* was recovered from the spleen in 47% of the chickens analysed and also from the liver and heart blood, indicating that systemic infection occurred.

The appearance of a bacterium in the internal organs is not in itself an indication of an infectious process occurring. Antigen sampling from the intestine and phagocytosis of bacteria in the mucosa will result in the transient presence of bacteria in the organs. Peturbation of the intestinal barrier by damage or inflammation will increase the rate at which the bacterial flora enter the tissues, but an immunocompetent host is able to limit the rate at which bacteria enter the body and eventually degrade such invaders. Pathogens, however, often operate by active invasion rather than passive entry, by which means they can increase their rate of entry and thus their likelihood of establishing an infection. In this respect, the behaviour of *Campylobacter* in poultry is pathogen-like.

It is well known that within the poultry gut *Campylobacter* adhere to epithelial cells, it is thought that adherence is essential for colonization (Hermans *et al.*, 2011) and several adhesins are involved. Konkel *et al.* (2007) reported that different *C. jejuni* strains compete for colonization in broilers. They hypothesized this is due to the sharing of common adhesins among isolates. Using primary intestinal cells from chickens both Van Deun *et al.* (2008) and Byrne *et al.* (2007) showed that *C. jejuni* was able to invade chicken cells, Byrne *et al.* (2007) note this is surprising as *C. jejuni* does not associate with chicken crypt epithelium *in vivo*. However, work in this area is limited by the lack of an immortalized chicken intestinal cell line and complicated handling of primary chicken caecal cells which prevent further investigations towards invasion (and other) mechanisms in chicken caecal cells (Hermans *et al.*, 2011). Despite this obvious invasion, there is little evidence in the literature for disease syndromes primarily associated with *Campylobacter* invasion in poultry. It is, however, becoming increasingly recognized that the organism does in fact alter the state of the host and also have a role in pathological states in birds.

Colonization invokes an immune response

The chicken's immune response to *Campylobacter* colonization is multifaceted. As with many other bacterial species the bacteria are recognized by pattern recognition receptors (PRRs) and this leads to the activation of aspects of an immune response. PRRs activated by *Campylobacter* entry into the gut include TLR4 and TLR21 although the increase in expression of these TLRs in response to *Campylobacter* colonization appears to be short-lived (Shaughnessy, 2009). Glycosylation of the flagellum prevents TLR5 activation. After PRRs are activated, increases in the proinflammatory cytokines IL-6 and IL-1β and the proinflammatory chemokine chicken IL-8 (chIL-8) have been observed (Larson, 2008; Shaughnessy, 2009). There is also release of nitric oxide from macrophages (Linde, 2008). However, after this initial proinflammatory response, proinflammatory cytokine levels decrease suggesting that although the initial colonization event is dealt with by the host's immune system as an attack, the host then reaches a level of active tolerance of *Campylobacter* (Smith, 2008; Hermans, 2012). IL-8 has been shown to attract both heterophils and monocytes to areas of inflammation (Barker, 1993; Poh, 2008), however there is conflicting evidence whether a heterophilia occurs in response to the IL-8 produced with *Campylobacter* colonization. Most work reports no observed heterophilia (Meade, 2008; Van Deun, 2008). However, Smith *et al.* (2008) did observe an increase in the number of heterophils in the caeca and ilia of birds when they were experimentally infected with *Campylobacter* at two weeks of age. Other than this possible heterophilia in response to *Campylobacter* colonization there is very little evidence of other pathology that has been directly attributed to the bacteria. There have been reports of growth retardation and jejunal villus atrophy in experimental models but these symptoms have not been reported in naturally colonized flocks (Sang, 1989; Lamb-Rosteski, 2008). As well as an innate immune response chickens do also mount an adaptive response to colonization with *Campylobacter*, producing antibodies against a number of *Campylobacter* proteins including flagellin (Myszewski and Stern, 1990; Cathraw, 1994; Sahin, 2001, 2002; Smith, 2008).

Jeurissen *et al.* (1998) showed that although the chicken produces a humoral response against T cell-dependent and T cell-independent Type 1 antigens, they produce a very low response against T cell-independent Type 2 antigens. These include the polysaccharide component of the *Campylobacter* capsule and it has been suggested that this may in part contribute to the lack of clearance of *Campylobacter* from the chicken gut.

It is clear that there is a strong immune response by chickens to infection. While a response would be expected to any commensal, the few reports of pathology such as heterophilia indicate that *Campylobacter* may actually be able to cause a disease-like state, and thus behave more as a pathogen than as a commensal. This observation is supported by work on pathological states in which *Campylobacter* is believed to play a role.

Disease

Campylobacter is found in the majority of chickens at slaughter age, regardless of rearing system used (conventional, barn, free-range or organic). For many years *Campylobacter* has been described as a commensal of chickens (Dhillion *et al.*, 2006; Macpherson and Uhr, 2004) as it did not appear to cause any disease in the chickens. Despite this acceptance that in chickens *Campylobacter* is a commensal, more recent studies (Jennings *et al.*, 2011) have shown that *Campylobacter* is associated with vibrionic hepatitis. Vibrionic hepatitis is a disease of poultry that was described in the 1950s (Tudor, 1954) associated with focal liver lesions that are greyish-white in colour and 1–2 mm in size (Burch, 2005; Crawshaw and Young, 2003). It was originally associated with vibrio like organisms which were later identified as *C. jejuni* (Peckham, 1958). Reports of the disease persisted until 1965 when it apparently disappeared from the literature until a re-emergence in the 2000s (Crawshaw and Young, 2003; VLA Surveillance Report, 2005). Although initially linked with the disease (Peckham, 1958) *Campylobacter* has not been shown to cause the focal lesions (Jennings *et al.*, 2011). A causative link to the disease is complicated by the occurrence of *Campylobacter* in the livers of apparently healthy birds (Cox *et al.*, 2006, 2007, 2009). There does appear to be a connection between the bacterium and the disease, though it may be a quantitative rather than qualitative one. Jennings *et al.* (2011) found that significantly fewer healthy livers contained campylobacter compared to those with greyish white spots and using FISH the authors demonstrated that *Campylobacter* was present in higher numbers in the spotty livers. In addition to an antibody response to colonization, the authors of this study observed a T cell response. Although there is no strong causative link to disease in chickens the same cannot be said for other avian species. *Campylobacter* colonization has been associated with focal areas of hepatitis in ostriches and rhea chicks and similarly described lesions (1–2 mm grey-white areas) have been shown to be associated with a higher number of *Campylobacter* organisms in the diseased livers compared to healthy birds from the same flock (Post 1992, Stephens 1998). This extraintestinal spread is of great public health importance and several cases of *Campylobacter* in humans have been linked to the consumption of chicken livers (Medeiros *et al.*, 2008; Cox *et al.*, 2006; Whyte *et al.*, 2006; O'Leary *et al.*, 2009).

Although it is a widely held view that *Campylobacter* is an avian commensal there are multiple reports of *Campylobacter* colonization inducing diarrhoea in chick models (Ruiz-Palacios, 1981; Sanyal, 1984; Sang, 1989). In both experimental

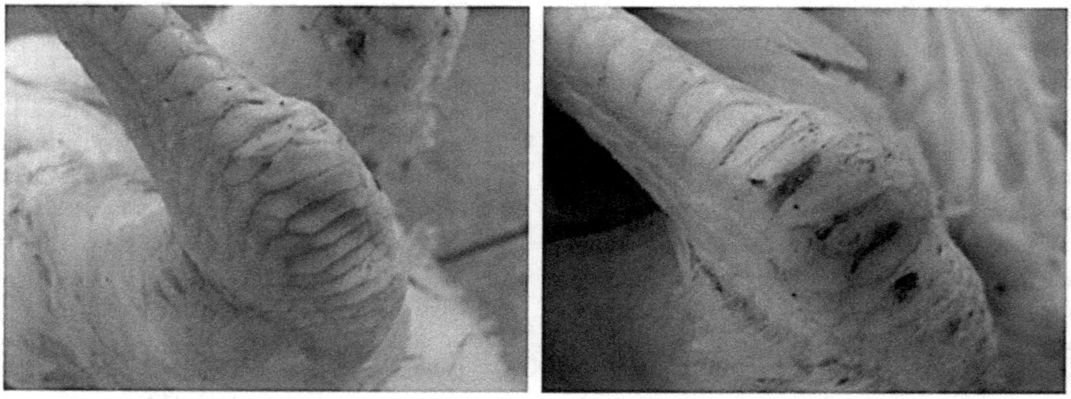

Figure 22.1 Hock marks on broiler chickens. Normal (left) and inflamed (right) hocks. Superficial marks shown here can progress to erosion of the skin as a result of exposure to wet litter. A colour version of this figure is available in the plate section at the back of the book.

and commercial flocks the presence of *Campylobacter* in the flock has been associated with wet litter (Neill, 1984; Williams, 2013). Hock lesions and pododermatitis have been associated both with birds being kept on wet litter and with the presence of *Campylobacter* in the flock (Fig. 22.1) (Dawkins, 2004; Bull, 2008; Colles, 2008). Work by Williams *et al.* (2013) suggests that colonization of broilers with *Campylobacter* is causing these leg lesions indirectly by increasing the looseness of the faeces and so the wetness of the litter. Although there is no obvious diarrhoea induced by infection, this would indicate that infection induces subclinical effects on intestinal transit time or water exchange. These observations taken together place *Campylobacter* as a pathogen, albeit not one that always causes major or readily observable effects on chickens. The reasons for this conditional commensalism are likely to lie in the state of the host.

Host susceptibility to disease

As is evident from the disease that it causes in many hosts, *Campylobacter* has evolved to be pathogen. Enteric and reproductive symptoms and pathology caused by the organism tend to be inflammatory and mediated by a reaction of the host to infection. There is little evidence that the organism itself or the cytolethal distending toxins actually cause damage. It is, therefore, possible that the chicken has evolved to control *Campylobacter* infection and minimize disease symptoms. To explore this, it is necessary to identify the factors that can predispose the host to infection and invasive disease.

Stress is a common predisposing factor for many diseases. Many of the processes involved with modern poultry farming such as thinning and catching evoke stress and fear reactions in chickens (Duncan *et al.*, 1986; Kannan *et al.*, 1997). Simple alteration of the house environment itself can invoke stress responses (Dawkins *et al.*, 2004). Chronic stress in birds results in an elevation of corticosterone. Whilst this promotes adaptive behaviour, and allows the animal to attempt to evade the stressful situation, long-term corticosterone elevation results in increased susceptibility to disease. The hormone decreases responsiveness to IL-2, resulting in a decrease in T-cell function and antibody production and a decrease in intestinal barrier function. In chickens infected with *Campylobacter* this would promote invasive behaviour of the bacterium, colonization and the possibility of development of pathology as discussed above.

Catching, thinning and transport of birds also invoke an acute stress response mediated by noradrenaline. The stress caused by thinning, partial depopulation of the house to prevent overcrowding, is though to be an important risk factor for *Campylobacter* infection of the remaining birds (O'Mahony, 2011) with noradrenaline promoting infectious behaviour of the bacterium as has been shown *in vitro* (Cogan *et al.*, 2006). This stress has been seen to increase colonization and shedding of *Campylobacter* (Whyte *et al.*, 2001). The effects here are likely to be both on bird, with noradrenaline increasing intestinal permeability (Hart and Kamm, 2002), and on the bacterium, with increased iron supply promoting growth and invasive behaviour (Cogan *et al.*, 2006).

These stress responses are an inevitable consequence of the modern process of poultry meat production. To meet demand for chicken meat production has had to become industrialized with different parts of the birds' life cycle occurring in different locations with the need for catching and for transport. Minimizing stressful situations during the rearing life of the chicken will help prevent pathogenic behaviour of *Campylobacter* and the development of attendant pathology. End of life stress during catching and transport to the abattoir will be of public health importance, but occurs over such a short time that it is of little consideration to interaction with the host.

Conclusions

In healthy unstressed birds *Campylobacter* behaves as an apparent commensal. If passage of the bacterium across the intestinal barrier can be limited then pathology and disease will be minimized or prevented entirely. The key clues to the nature of the relationship between host and *Campylobacter* are given by its behaviour when the host is compromised. The development of disease here is conditional upon the health and homeostasis of the host being upset so that invasive behaviour can no longer be controlled. Under these situations *Campylobacter* becomes a pathogen capable of causing disease, either directly as seen with liver disease or indirectly via its effects on litter wetness and attendant leg problems in broilers. It is evident that pathogenesis is normally limited by the host response, but that in the final analysis the relationship between bacterium and bird is a parasitic one and that *Campylobacter* is only a conditional commensal of chickens.

References

Açik, M.N., and Cetinkaya, B. (2006). Random amplified polymorphic DNA analysis of *Campylobacter jejuni* and *Campylobacter coli* isolated from healthy cattle and sheep. J. Med. Microbiol. 55(Pt 3), 331–4.

Baker, J., Barton, M.D., and Lanser, J. (1999). *Campylobacter* species in cats and dogs in South Australia. Aus. Vet J. 77, 662–6.

Bull, S.A., Thomas, A., Humphrey, T., Ellis-Iversen, J., Cook, A.J., Lovell, R., and Jorgensen, F. (2008). Flock health indicators and *Campylobacter* spp. in commercial housed broilers reared in Great Britain. Appl. Environ. Microbiol. 74, 5408–5413.

Burch, D. (2005). Avian vibrionic hepatitis in laying hens. Vet Rec. *157*(17), 528.

Byrne, C.M., Clyne, M., and Bourke, B. (2007). *Campylobacter jejuni* adhere to and invade chicken intestinal epithelial cells *in vitro*. Microbiology *153*(Pt 2), 561–569.

Cawthraw, S., Ayling, R., Nuijten, P., Wassenaar, T., and Newell, D.G. (1994). Isotype, specificity, and kinetics of systemic and mucosal antibodies to *Campylobacter jejuni* antigens, including flagellin, during experimental oral infections of chickens. Avian Dis. *38*, 341–349.

Cogan, T.A., Thomas, A.O., Rees, L.E., Taylor, A.H., Jepson, M.A., Williams, P.H., Ketley, J., and Humphrey, T.J. (2006). Norepinephrine increases the pathogenic potential of *Campylobacter jejuni*. Gut *56*(8), 1060–1065.

Colles, F.M., Jones, T.A., McCarthy, N.D., Sheppard, S.K., Cody, A.J., Dingle, K.E., Dawkins, M.S., and Maiden, M.C. (2008). *Campylobacter* infection of broiler chickens in a free-range environment. Environ. Microbiol. *10*, 2042–2050.

Cox, N.A., Richardson, L.J., Buhr, J.R., Fedorka-Cray, J.P., Bailey, J.S., Wilson, J.L., and Hiett, K.L. (2006). Natural presence of *Campylobacter* spp. in various internal organs of commercial broiler breeder hens. Avian Dis. *50*, 450–453.

Cox, N.A., Richardson, L.J., Buhr, R.J., Northcutt, J.K., Bailey, J.S., and Fedorka-Cray, P.J. (2007). Recovery of *Campylobacter* and *Salmonella* serovars from the spleen, liver/gallbladder, and ceca of six and eight week old commercial broilers. J. Appl. Poult. Res. *16*, 477–480.

Cox, N.A., Richardson, L.J., Buhr, R.J., and Fedorka-Cray, P.J. (2009). *Campylobacter* species occurrence within internal organs and tissues of commercial caged Leghorn laying hens. Poult. Sci. *88*, 2449–2456.

Crawshaw, T., and Young, S. (2003). Increased mortality on a free-range layer site. Vet Rec. *153*(21), 664.

Dawkins, M.S., Donnelly, C.A., and Jones, T.A. (2004). Chicken welfare is influenced more by housing conditions than by stocking density. Nature *427*, 342–344.

Denis, M., Rose, V., Huneau-Salaün, A., Balaine, L., and Salvat, G. (2008). Diversity of pulsed-field gel electrophoresis profiles of *Campylobacter jejuni* and *Campylobacter coli* from broiler chickens in France. Poult. Sci. *87*, 1662–1671.

Dhillon, A.S., Shivaprasad, H.L., Schaberg, D., Wier, F., Weber, S., and Bandli, D. (2006). *Campylobacter jejuni* infection in broiler chickens. Avian Dis. *50*, 55–58.

Duncan, I.J., Slee, G.S., Kettlewell, P., Berry, P., and Carlisle, A.J. (1986). Comparison of the stressfulness of harvesting broiler chickens by machine and by hand. Br. Poult. Sci. 27(1),109–114.

Englen, M.D., Hill, A.E., Dargatz, D.A., Ladely, S.R., and Fedorka-Cray, P.J. (2007). Prevalence and antimicrobial resistance of *Campylobacter* in US dairy cattle. J. Appl. Microbiol. *102*, 1570–1577.

Flemming, M.P. (1980). Incidence of *Campylobacter* infections in dogs. Vet Record. *107*, 202.

Gilpin, B.J., Thorrold, B., Scholes, P., Longhurst, R.D., Devane, M., Nicol, C., Walker, S., Robson, B., and Savill, M. (2008). Comparison of *Campylobacter jejuni* genotypes from dairy cattle and human sources from the Matamata-Piako District of New Zealand. J. Appl. Microbiol. *105*, 1354–1360.

Hald, B., Pedersen, K., Wainø, M., Jørgensen, J.C., and Madsen, M. (2004). Longitudinal study of the excretion patterns of thermophilic *Campylobacter* spp. in young pet dogs in Denmark. J. Clin. Microbiol. *42*, 2003–2012.

Hart, A., and Kamm, M.A. (2002). Mechanisms of initiation and perpetuation of gut inflammation by stress. Aliment Pharmacol. Ther. *16*(12), 2017–2028.

Hermans, D., Van Deun, K., Martel, A., Van Immerseel, F., Messens, W., Heyndrickx, M., Haesebrouck, F., and Pasmans, F. (2011). Colonization factors of *Campylobacter jejuni* in the chicken gut. Vet Res. *42*(1), 82.

Hermans, D., Pasmans, F., Heyndrickx, M., Van Immerseel, F., Martel, A., Van Deun, K., and Haesebrouck, F. (2012). A tolerogenic mucosal immune response leads to persistent *Campylobacter jejuni* colonization in the chicken gut. Crit. Rev. Microbiol. *38*, 17–29.

Horrocks, S.M., Anderson, R.C., Nisbet, D.J., and Ricke, S.C. (2009). Incidence and ecology of *Campylobacter jejuni* and *coli* in animals. Anaerobe *15*, 18–25.

Humphrey, T.J., O'Brien, S., and Madsen, M. (2007). Campylobacters as zoonotic pathogens: A food production perspective. Int. J. Food Microbiol. *117*(3), 237–257.

Jennings, J.L., Sait, L.C., Perrett, C.A., Foster, C., Williams, L.K., Humphrey, T.J., and Cogan, T.A. (2011). *Campylobacter jejuni* is associated with, but not sufficient to cause vibrionic hepatitis in chickens. Vet. Microbiol. *149*, 193–199.

Jeurissen, S.H., Janse, E.M., van Rooijen, N., and Claassen, E. (1998). Inadequate anti-polysaccharide antibody responses in the chicken. Immunobiology *198*, 385–395.

Kannan, G., Heath, J.L., Wabeck, C.J., Souza, M.C., Howe, J.C., and Mench, J.A. (1997). Effects of crating and transport on stress and meat quality characteristics in broilers. Poult. Sci. *76*(3), 523–529.

Knudsen, K.N., Bang, D.D., Andresen, L.O., and Madsen, M. (2006). *Campylobacter jejuni* strains of human and chicken origin are invasive in chickens after oral challenge. Avian Dis. *50*, 10–14.

Konkel, M.E., Christensen, J.E., Dhillon, A.S., Lane, A.B., Hare-Sanford, R., Schaberg, D.M., and Larson, C.L. (2007). *Campylobacter jejuni* strains compete for colonization in broiler chicks. Appl. Environ. Microbiol. *73*, 2297–2305.

Kwan, P.S., Barrigas, M., Bolton, F.J., French, N.P., Gowland, P., Kemp, R., Leatherbarrow, H., Upton, M., and Fox, A.J. (2008). Molecular epidemiology of *Campylobacter jejuni* populations in dairy cattle, wildlife, and the environment in a farmland area. Appl. Environ. Microbiol. *74*, 5130–5138.

Lamb-Rosteski, J.M., Kalischuk, L.D., Inglis, G.D., and Buret, A.G. (2008). Epidermal growth factor inhibits *Campylobacter jejuni*-induced claudin-4 disruption, loss of epithelial barrier function, and *Escherichia coli* translocation. Infect. Immun. *76*, 3390–3398.

Larson, C.L., Shah, D.H., Dhillon, A.S., Call, D.R., Ahn, S., Haldorson, G.J., Davitt, C., and Konkel, M.E. (2008). *Campylobacter jejuni* invade chicken LMH cells inefficiently and stimulate differential expression of the chicken CXCLi1 and CXCLi2 cytokines. Microbiology *154*, 3835–3847.

Lee, M.K., Billington, S.J., and Joens, L.A. (2004). Potential virulence and antimicrobial susceptibility of *Campylobacter jejuni* isolates from food and companion animals. Food. Pathog. Dis. *1*, 223–230.

Linde, A., Ross, C.R., Davis, E.G., Dib, L., Blecha, F., and Melgarejo, T. (2008). Innate immunity and host defense peptides in veterinary medicine. J. Vet. Intern. Med. *22*, 247–265.

Little, C.L., Richardson, J.F., Owen, R.J., de Pinna, E., and Threlfall, E.J. (2008). Prevalence, characterisation and antimicrobial resistance of *Campylobacter* and *Salmonella* in raw poultrymeat in the UK, 2003–2005. Int. J. Environ. Health Res. *18*, 403–414.

Macpherson, A.J., and Uhr, T. (2004). Compartmentalization of the mucosal immune responses to commensal intestinal bacteria. Ann. N.Y. Acad. Sci. *1029*, 36–43.

Meade, K.G., Narciandi, F., Cahalane, S., Reiman, C., Allan, B., and O'Farrelly, C. (2009). Comparative *in vivo* infection models yield insights on early host immune response to *Campylobacter* in chickens. Immunogenetics *61*, 101–110.

Medeiros, D.T., Sattar, S.A., Farber, J.M., and Carrillo, C.D. (2008). Occurrence of Campylobacter spp. in raw and ready-to-eat foods and in a Canadian food service operation. J. Food Prot. *71*, 2087–2093.

Meerburg, B.G., Jacobs-Reitsma, W.F., Wagenaar, J.A., and Kijlstra, A. (2006). Presence of *Salmonella* and *Campylobacter* spp. in wild small mammals on organic farms. Appl. Environ. Microbiol. *72*, 960–962.

Moser, I., Rieksneuwohner, B., Lentzsch, P., Schwerk, P., and Wieler, L.H. (2001). Genomic heterogeneity and O-antigenic diversity of *Campylobacter upsaliensis* and *Campylobacter helveticus* strains isolated from dogs and cats in Germany. J. Clin. Microbiol. *39*, 2548–2557.

Myszewski, M.A., and Stern, N.J. (1990). Influence of *Campylobacter jejuni* cecal colonization on immunoglobulin response in chickens. Avian Dis. *34*, 588–594.

Neimann, J., Engberg, J., Molbak, K., and Wegener, H.C. (2003). A case control study of risk factors for sporadic *Campylobacter* infections in Denmark. Epidemiol. Infect. *130*, 353–366.

O'Leary, M.C., Harding, O., Fisher, L., and Cowden, J. (2009). A continuous common-source outbreak of campylobacteriosis associated with changes to the preparation of chicken liver pâté. Epidemiol. Infect. *137*, 383–388.

O'Mahony, E., Buckley, J.F., Bolton, D., Whyte, P., and Fanning, S. (2011). Molecular epidemiology of *Campylobacter* isolates from poultry production units in southern Ireland. PLoS One *6*(12), e28490.

Oporto, B., Esteban, J.I., Aduriz, G., Juste, R.A., and Hurtado, A. (2007). Prevalence and strain diversity of thermophilic Campylobacters in cattle, sheep and swine farms. J. Appl. Microbiol. *103*(4), 977–984.

Peckham, M.C. (1958). Avian vibrionic hepatitis. Avian Dis. *2*, 348.

Petersen, R.F., Harrington, C.S., Kortegaard, H.E., and On, S.L. (2007). A PCR-DGGE method for detection and identification of *Campylobacter*, *Helicobacter*, *Arcobacter* and related Epsilobacteria and its application to saliva samples from humans and domestic pets. J. Appl. Microbiol. *103*, 2601–2605.

Post, K., Ayers, J.R., Gilmore, W.C., and Raleigh, R.H. (1992). *Campylobacter jejuni* isolated from ratites. J. Vet. Diagn. Invest. *4*, 345–347.

Rodrigues, L.C., Cowden, J.M., Wheeler, J.G., Sethi, D., Wall, P.G., Cumberland, P., Tompkins, D.S., Hudson, M.J., Roberts, J.A., and Roderick, P.J. (2001). The study of infectious intestinal disease in England: risk factors for cases of infectious intestinal disease with *Campylobacter jejuni* infection. Epidemol. Infect. *127*, 185–193.

Ruiz-Palacios, G.M., Escamilla, E., and Torres, N. (1981). Experimental *Campylobacter* diarrhea in chickens. Infect. Immun. *34*, 250–255.

Sahin, O., Zhang, Q., Meitzler, J.C., Harr, B.S., Morishita, T.Y., and Mohan, R. (2001). Prevalence, antigenic specificity, and bactericidal activity of poultry anti-*Campylobacter* maternal antibodies Appl. Environ. Microbiol. *67*, 3951–3957.

Sahin, O., Morishita, T.Y., and Zhang, Q. (2002). *Campylobacter* colonization in poultry: sources of infection and modes of transmission. Anim. Health Res. Rev. 3, 95–105.

Sahin, O., Plummer, P.J., Jordan, D.M., Sulaj, K., Pereira, S., Robbe-Austerman, S., Wang, L., Yaeger, M.J., Hoffman, L.J., and Zhang, Q. (2008). Emergence of a tetracycline-resistant *Campylobacter jejuni* clone associated with outbreaks of ovine abortion in the United States. J. Clin. Microbiol. *46*, 1663–1671.

Sang, F.C., Shane, S.M., Yogasundram, K., Hagstad, H.V., and Kearney, M.T. (1989). Enhancement of *Campylobacter jejuni* virulence by serial passage in chicks. Avian Dis. *33*, 425–430.

Sanyal, S.C., Islam, K.M., Neogy, P.K., Islam, M., Speelman, P., and Huq, M.I. (1984). *Campylobacter jejuni* diarrhea model in infant chickens. Infect. Immun. *43*, 931–936.

Shaughnessy, R.G., Meade, K.G., Cahalane, S., Allan, B., Reiman, C., Callanan, J.J., and O'Farrelly, C. (2009). Innate immune gene expression differentiates the early avian intestinal response between *Salmonella* and *Campylobacter*. Vet. Immunol. Immunopathol. *132*, 191–198.

Smith, C.K., Abuoun, M., Cawthraw, S.A., Humphrey, T.J., Rothwell, L., Kaiser, P., Barrow, P.A., and Jones, M.A. (2008). *Campylobacter* colonization of the chicken induces a proinflammatory response in mucosal tissues. FEMS Immunol. Med. Microbiol. *54*, 114–121.

Stanley, J., Burnens, A.P., Linton, D., On, S.L., Costas, M., and Owen, R.J. (1992). *Campylobacter helveticus* sp. nov., a new thermophilic species from domestic animals: characterization, and cloning of a species-specific DNA probe. J. Gen. Microbiol. *138*, 2293–2303.

Stephens, C.P., On, S.L., and Gibson, J.A. (1998). An outbreak of infectious hepatitis in commercially reared ostriches associated with *Campylobacter coli* and *Campylobacter jejuni*. Vet. Microbiol. *61*, 183–190.

Studahl, A., and Andersson, Y. (2000). Risk factors for indigenous *Campylobacter* infection: a Swedish case control study. Epidemiol. Infect. *125*, 269–275.

Tudor, D.C. (1954). A liver degeneration of unknown origin in chickens. J. Am. Vet. Med. Assoc. *125*(930), 219–220.

Van Deun, K., Haesebrouck, F., Heyndrickx, M., Favoreel, H., Dewulf, J., Ceelen, L., Dumez, L., Messens, W., Leleu, S., Van Immerseel, F., Ducatelle, R., and Pasmans, F. (2007). Virulence properties of *Campylobacter jejuni* isolates of poultry and human origin. J. Med. Microbiol. *56*, 1284–1289.

Van Deun, K., Pasmans, F., Ducatelle, R., Flahou, B., Vissenberg, K., Martel, A., Van den Broeck, W., Van Immerseel, F., and Haesebrouck, F. (2008). Colonization strategy of *Campylobacter jejuni* results in persistent infection of the chicken gut. Vet. Microbiol. 130:285–297.

Veterinary Laboratories Agency. (2005). VLA surveillance report. Vet. Rec. *157*, 399–402.

Whyte, P., Collins, J.D., McGill, K., Monahan, C., and O'Mahony, H. (2001). The effect of transportation stress on excretion rates of campylobacters in market-age broilers. Poult. Sci. *80*(6), 817–820.

Whyte, R., Hudson, J.A., and Graham, C. (2006). *Campylobacter* in chicken livers and their destruction by pan frying. Lett. Appl. Microbiol. *43*, 591–595.

Williams, L.K., Sait, L.C., Trantham, E.K., Cogan, T.A., and Humphrey, T.J. (2013). *Campylobacter* infection has different outcomes in fast and slow growing broiler chickens. Avian Dis. *57*, 238–241.

Wingstrand, A., Neimann, J., Engberg, J., Nielsen, E.M., Gerner-Smidt, P., Wegener, H.C., and Molbak, K. (2006). Fresh chicken as main risk factor for Campylobacteriosis, Denmark. Emerg. Infect. Dis. *12*, 280–285.

Zweifel, C., Scheu, K.D., Keel, M., Renggli, F., and Stephan, R. (2008). Occurrence and genotypes of *Campylobacter* in broiler flocks, other farm animals, and the environment during several rearing periods on selected poultry farms. Int. J. Food Microbiol. *125*, 182–187.

The Chicken Reservoir and Specific Alleles at Conserved Contingency Loci Can Promote Host Colonization and Disease by *Campylobacter jejuni*

23

John E. Linz

Abstract

Campylobacter jejuni ranks among the top foodborne bacterial pathogens in terms of number of disease cases and economic impact. Despite its clear importance as a pathogen, relatively little is known about the specific mechanisms by which *C. jejuni* causes disease. Twenty three or more contingency loci in *C. jejuni* strain 11168 are subject to high frequency mutation within homopolymeric tracts and these genetic changes impact gene expression and/or protein activity. Our laboratory is investigating the role of *C. jejuni* contingency loci in virulence. This chapter reviews recent work in our laboratory that demonstrates that the chicken reservoir promotes phase variation at specific contingency loci and this enhances colonization and disease in a mouse model for disease. A specific allele at contingency locus *Cj0170* for example regulates *C. jejuni* motility and virulence in mice while specific alleles at loci *Cj0045, Cj1420,* and *Cj0685* are strongly associated with colonization and disease. Based on our data, we hypothesize that analysis of the genotype at contingency loci provides a useful tool to predict virulence in mice and humans. The data also prompt us to focus on highly conserved contingency loci as useful targets to prevent disease in humans.

Introduction

Health and economic costs associated with *Campylobacter jejuni* disease

The presence of *Campylobacter jejuni* in poultry products ranks as the number one risk factor for human bacterial foodborne illness (Batz *et al.*, 2011) and the chicken gastrointestinal (GI) tract is the primary reservoir for human disease (Young *et al.*, 2007). On an annual basis in the United States, 75 deaths, 850,000 disease cases, and 1.7 billion dollars in costs are associated with human disease caused by *C. jejuni* (Batz *et al.*, 2011; Center for Disease Control, 2011; Snelling *et al.*, 2005; Young *et al.*, 2007). *C. jejuni* also ranks among the top three foodborne bacterial pathogens in terms of total disease cases and the organism causes similar levels of mortality as *E. coli* O157:H7 (CDC, 2011). *C. jejuni* frequently causes self-limiting gastroenteritis with duration of 5–7 days and affected individuals normally do not require treatment (Deckert *et al.*, 2010). High-risk individuals (pregnant women, infants, immune-suppressed individuals, HIV/AIDS patients) may experience more severe symptoms associated with bacteraemia, systemic infection, recurring infection, or post-infection sequelae (Miller-Fisher and Guillain-Barré syndromes) and they frequently require hospitalization (Humphrey *et al.*, 2007).

C. jejuni contingency loci

Based on the complete genome sequence released in 2000, there are 23 or more contingency loci in *C. jejuni* strain 11168 (Parkhill *et al.*, 2000). The function of very few of these loci has been experimentally determined so we predominantly will use locus designations (e.g. *Cj1420*) in this review; however, we will use gene designations (e.g. *wlaN*, *cipA*) where function is known. Contingency loci carry genes involved predominantly in the synthesis or modification of lipo-oligosaccharide (LOS), flagella, and capsule (Table 23.1) and they are functional only under a specific set of growth conditions, hence the designation 'contingency locus'. Insertions or deletions of one or more bases occur in homopolymeric tracts within *C. jejuni* contingency loci at very high frequency (up to one in every two cells) (Artymovich *et al.*, 2013; Kim *et al.*, 2012; Parkhill *et al.*, 2000). *C. jejuni* strain 11168 appears to lack two or more activities (MutH and MutL) involved in repair of slip strand mutations within homopolymeric tracts consistent with the observed high rate of mutation (Parkhill *et al.*, 2000; Kim *et al.*, 2012). Mutations in promoters can alter gene expression and in the open reading frame (ORF) can alter the activity of the proteins encoded by contingency loci. The high rate of mutation within homopolymeric tracts at contingency loci and the apparent inability to effectively repair these mutations suggests that generation of genetic diversity in *C. jejuni* populations is an important weapon in their evolutionary arsenal to survive and thrive in the environment.

Our group and others demonstrated that phase variation in the expression and function of contingency loci impacts serum response, motility, natural transformation, colonization, invasion of cells in culture, and disease in animal

Table 23.1 Polymorphic contingency loci in *C. jejuni* NCTC11168 and their proposed functions

Region	Locus, gene	HT that gives complete ORF	Proposed gene function [d]
Unspecified	*Cj0031–3* [b]	9G	Putative type IIS restriction/ modification enzyme
	Cj0045c	10G	Putative iron-binding protein
	Cj0170–71[b]	8G	Hypothetical protein
	Cj0685c, *cipA*	9C	Invasion protein CipA
Protein glycosylation and LOS biosynthesis	*Cj1139c*, *wlaN*	8G	Beta-1,3 galactosyltransferase
	Cj1144–45c[b]	19(G+A)	Hypothetical protein
Flagellin modification	*Cj1305c/06c/10c*[c]	9G	Hypothetical proteins
	Cj1318, *maf1/35–36*[b], *maf4*	11G	Motility accessory factor
	Cj1321	10G [a]	Acetyl transferase
	Cj1325–26[b]	9G	Putative methyltransferase
	Cj1342c, *maf7*	9G	Motility accessory factor
Capsule biosynthesis	*Cj1420c*	9G	Putative methyltransferase
	Cj1421c/22c[c]	9G	O-methyl phosphoramidate transferases
	Cj1426c	10G	Putative methyltransferase family protein
	Cj1429c	10G	Hypothetical protein

HT, homopolymeric tract.
[a]Homopolymeric tract located in promoter region rather than ORF.
[b]Genes separated by (–) were merged as a result of the *C. jejuni* NCTC11168 re-annotation (Gundogdu *et al.*, 2007).
[c]Genes separated by (/) are separate genes but share high sequence similarity.
[d]Gene function assigned based on www.ncbi.nlm.nih.gov/nuccore/AL111168.

models and human volunteers (Artymovich *et al.*, 2013; Ashgar *et al.*, 2007; Ewing *et al.*, 2009; Guerry *et al.*, 2002, 2006; Hendrixson, 2006; Howard *et al.*, 2009; Kakuda and DeRita, 2006; Kim *et al.*, 2012; Linton *et al.*, 2000; McNally *et al.*, 2007; Prendergast *et al.*, 2004; Wilson *et al.*, 2010). These studies suggested that phase variation at contingency loci and alterations in surface structure chemistry impact virulence, but little detail was revealed about the underlying mechanisms.

This chapter reviews our recent work on the role of four *C. jejuni* contingency loci in colonization of the chicken reservoir and in virulence in a mouse model for human disease. The work demonstrates the potential importance of the chicken reservoir in promoting human disease and analyzes the specific mechanisms by which contingency loci *Cj0045, Cj0170, Cj0685,* and *Cj1420* contribute to colonization and disease in mice. We also discuss the potential importance of these four loci in human disease and why they became the focus of our current research efforts.

Summary of recent studies in our laboratory

Differential colonization of broiler chickens and mice by *C. jejuni*

Because commercial chickens are the primary reservoir for human illness caused by *C. jejuni*, we analysed the possible impact of passage of this organism through Ross 308 broilers (our model of the chicken reservoir; Kim *et al.*, 2012; Wilson *et al.*, 2010) on disease in C57BL6 IL10-deficient mice (our model for human disease; Mansfield *et al.*, 2007). To initiate these studies, we tested the ability of several different *C. jejuni* strains to colonize broiler chickens or mice. We observed that *C. jejuni* strain NCTC11168 (11168) colonized chickens and mice at high frequency (the number of animals infected per group) and at high levels (measured as CFU per gram of caecal tissue obtained at necropsy at d14 or d21 post inoculation) (Fig. 23.1) (Wilson *et al.*, 2010). In contrast, *C. jejuni* strains 81-176, 33292, KanR4, and CamR2 colonized chickens at lower frequency

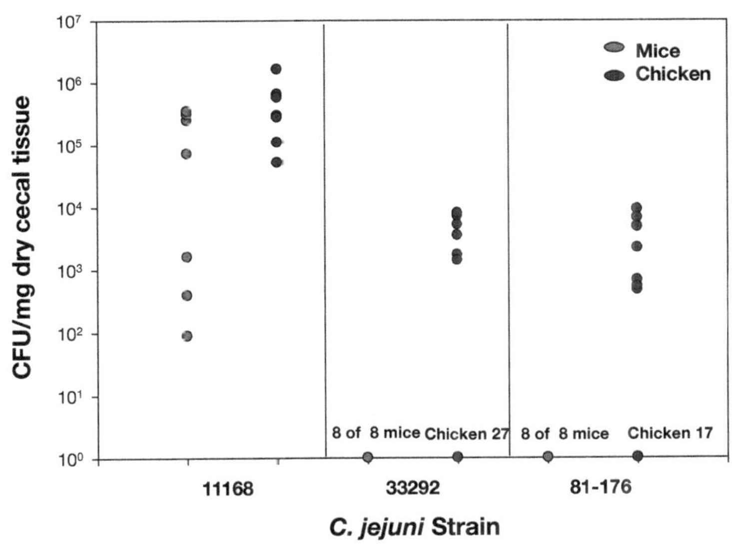

Figure 23.1 *C. jejuni* colonization of C57BL/6J IL10 deficient mice and Ross 308 broiler caecal tissue at 14 day post-inoculation. *C. jejuni* strains 11168, 33292 and 81-176 are represented on the x-axis. Eight mice and eight chickens were sacrificed on day 14 (d14) and caecal tissue homogenized, serially diluted, and plated onto a selective growth medium. Mice inoculated with strains 33292 and 81-176 were culture negative at d14 post-inoculation, with the limit of detection in these experiments ranging from 150 to 300CFUgram^{-1} of dry caecal tissue. Chicken 17 produced a relatively low *C. jejuni* concentration of 5×10^{-1}CFUmg^{-1} dry caecal tissue while chicken 27 did not produce a detectable level of *C. jejuni*. Each animal was housed separately in an individual cage. A colour version of this figure is available in the plate section at the back of the book.

and levels than strain 11168 and only colonized mice at very low frequency (0 of 16 mice with either 81-176 or 33292; 1 of 24 mice with either KanR4 or CamR2) (Figs. 23.1 and 23.2). However, once isolated from colonized mice, KanR4 and CamR2 populations colonized mice at much higher frequency and levels (Fig. 23.2). These data strongly suggest that passage through the GI tract of mice adapts *C. jejuni* populations to subsequent colonization of mice. These early observations also prompted us to speculate that adaptation in the mouse GI tract is closely associated with high frequency genetic changes within contingency loci.

Genetic differences between colonizers and non-colonizers of mice

To identify which genes may be associated with the differential ability of *C. jejuni* strains 11168, 33292, and 81-176 to colonize chickens and mice, we conducted competitive DNA hybridization analysis of a microarray of *C. jejuni* 11168 cDNAs (Wilson *et al.*, 2010). In the analysis, *C. jejuni* 11168 genomic DNA served as the reference probe and genomic DNAs from each of the two other strains served as the test probes. The ability of the test probes to bind to the cDNAs in the microarray in the presence of the reference probe with 100% identity provided a measure of the relative degree of identity between these strains at each gene in the genome. Microarray analyses demonstrated that genetic differences between strain 11168 that colonizes mice and the other two strains that do not colonize mice localized predominantly to gene clusters involved in the synthesis or modification of LOS, capsule, and flagella (Fig. 23.3). Since most of the 23 contingency loci in *C. jejuni* 11168 also map to these clusters and contingency loci are thought to play an important role in surface structure chemistry, we hypothesized that genetic differences in

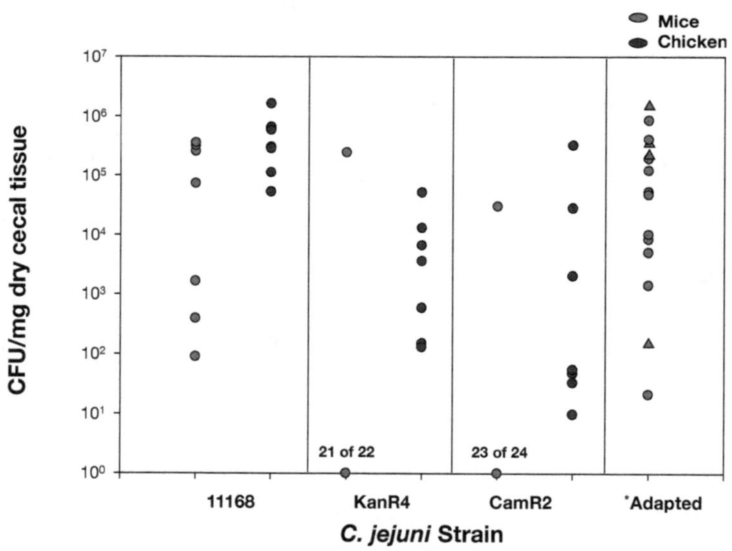

Figure 23.2 Colonization of C57BL/6J IL10 deficient mice and Ross 308 broiler caecal tissue at d14 or d16 post-inoculation by genetically marked *C. jejuni* 11168. Colonization data for wild-type *C. jejuni* 11168 in mouse and chicken caecal tissue are presented for comparison with antibiotic resistant (genetically marked) 11168 variants. 43 of 46 mice inoculated with a *C. jejuni* population prepared under standard laboratory culture conditions (inoculum population) were culture negative for *C. jejuni* at d14 post-inoculation. Mouse-adapted cultures of 11168-Tn5CamR6 (chloramphenicol resistant variant indicated by a red circle in the Adapted column) and 11168-*23S*KanR4 (kanamycin resistant variant indicated by a red triangle in the Adapted column) were recovered from caecal tissue on d 14 post-inoculation from mouse 712 and 682 respectively. The limit of detection in these experiments ranged from 150 to 900CFUgram^{-1} of dry caecal tissue. Each animal was housed separately in an individual cage. A colour version of this figure is available in the plate section at the back of the book.

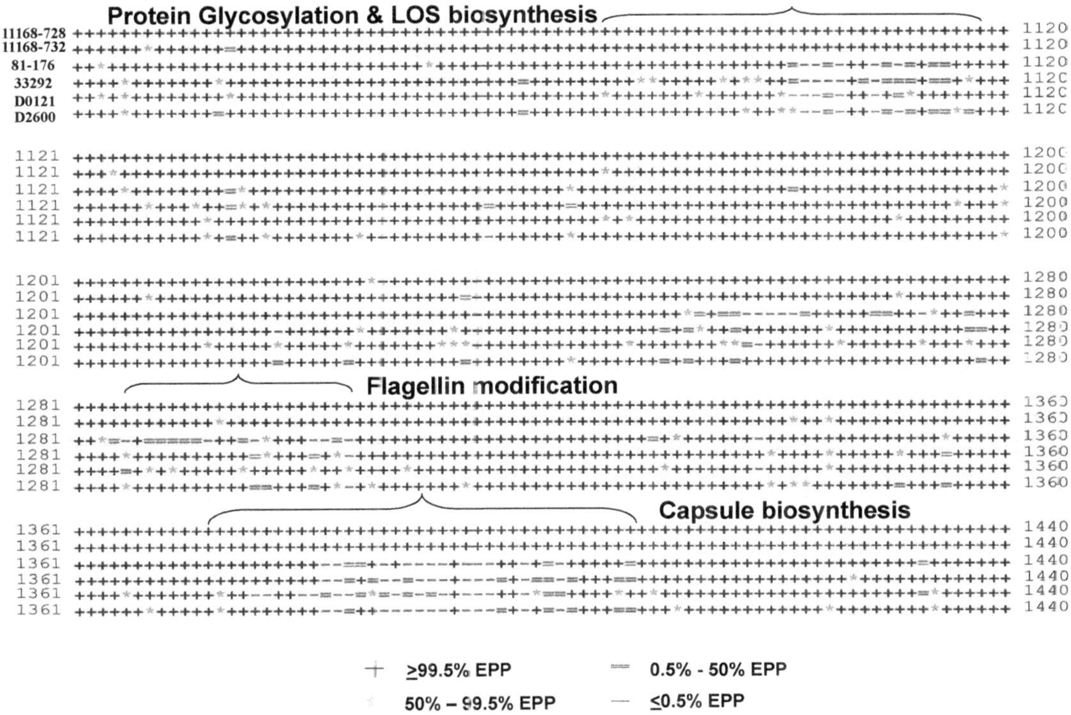

Figure 23.3 Competitive microarray DNA–DNA hybridization of *C. jejuni* chromosomal DNAs. Glass slides carrying several copies of a complete *C. jejuni* 11168 cDNA microarray were hybridized to a chromosomal DNA probe prepared from *C. jejuni* 11168 (reference probe) plus fluorescently labelled chromosomal DNA from either *C. jejuni* strain 33292 or strain 81-176. *C. jejuni* strains D0121 and D2600 are distantly related human isolates and are included here to provide more robust analysis between chromosomes. A GACK value of –0.5 represents a 100% chance of sequence divergence and an Estimated Probability of Presence (EPP) ≤0.5%. A GACK score of 0.5 represents a 100% chance of identical sequence and an EPP ≥99.5%. A colour version of this figure is available in the plate section at the back of the book.

contingency loci could, at least in part, explain differential colonization of chickens and mice.

Genotype at contingency locus Cj1139 (wlaN) is altered by growth conditions

To begin to analyse the potential role of contingency loci in *C. jejuni* colonization of chickens and mice, we first analysed *Cj1139* because the function of the protein encoded by this contingency locus had been determined previously (Linton *et al.*, 2000). Contingency locus *Cj1139* in strain 11168 carries a single gene (*wlaN*) that lies within the LOS gene cluster. *wlaN* encodes a galactosyltransferase (WlaN or CgtB) that catalyses addition of a terminal galactose to LOS. LOS in *C. jejuni* appears to serve as a functional equivalent of LPS in other Gram-negative bacteria, although the chemical structure is significantly less complex (Guerry and Szymanski, 2008).

In order to analyse genotype at contingency loci, we developed 'fragment analysis' based on our modifications to a published protocol (Wassenaar *et al.*, 2002). To conduct fragment analysis, we purified chromosomal DNA from *C. jejuni* populations using standard techniques (Wilson *et al.*, 2010) and used locus-specific primers that flank the homopolymeric tract (HT) (one primer carries a fluorescent tag) to amplify an approximately 200 bp region. Fragment analysis also uses Pfx50, a highly faithful polymerase, to assure that mutations are not introduced into *C. jejuni* DNA during amplification *in vitro*. The labelled fragment population was resolved by capillary electrophoreses, individual fragments detected by fluorescence, and fragment size analysed by

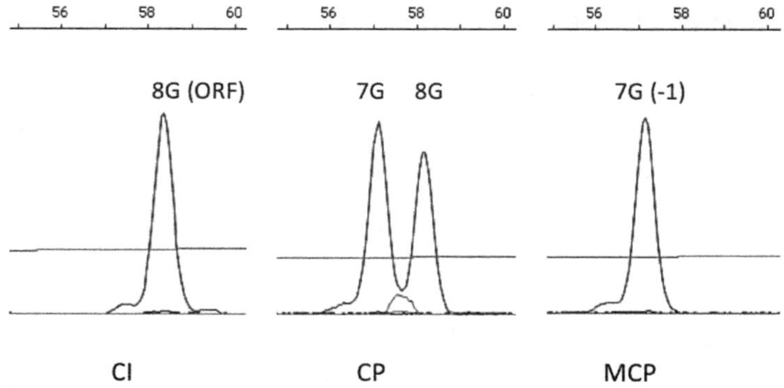

Figure 23.4 Fragment analysis of the homopolymeric tract (HT) at contingency locus *Cj1139 (wlaN)*. DNA samples from an inoculum population prepared under standard laboratory culture conditions (CI), from chicken-passaged populations (CP), and from CP populations passaged through mice (MCP) of strain CamR2 were analysed by fragment analysis. PCR products labelled with 6-FAM fluorescent dye were separated by electrophoresis on capillary columns to resolve alleles. The number of nucleotides in the HT was revealed by parallel sequence analysis of each sample. A colour version of this figure is available in the plate section at the back of the book.

comparison to standard fragments of known size. HT fragments were resolved into peaks that differed by a single nucleotide. Peak area measures the relative abundance of each allele (see Fig. 23.4) and the number of peaks provides the number of alleles.

We also conducted nucleotide sequence analysis to confirm allele frequency data generated by fragment analysis and also to identify the number of nucleotides carried in the HT in each allele. A PCR fragment population was generated from *C. jejuni* genomic DNA samples using a separate set of primers that flanked the region amplified in fragment analysis. The PCR fragments were cloned into a plasmid vector that was transformed into *E. coli.* Twenty or more clones were then sequenced using standard sequencing technology at the Michigan State University Research Technology Support Facility (RTSF) and the approximate allele frequency determined by the number of clones that carried the specific allele divided by the total number of clones analysed. Mutations were not observed in control *C. jejuni* fragments that lacked a HT confirming that mutations did not occur during amplification *in vitro* or during cloning in *E. coli.*

To determine if growth conditions impact genotype at contingency locus *Cj1139,* we grew strain 11168 under standard laboratory culture conditions to generate an inoculum (CI). We then passaged the CI population through Ross 308 broilers (chicken passaged, or CP) or through mice (mouse passaged, or MP). Because we only wished to analyse a single locus in this study, we conducted nucleotide sequence analysis only on the HT region for contingency locus *Cj1139* (Wilson *et al.,* 2010). We purified genomic DNA from *C. jejuni* CI, CP, and MP populations, amplified the region in contingency locus *Cj1139* with primers that flank the HT, and conducted nucleotide sequence analysis on 20 or more individual clones that each carried a representative PCR fragment from the fragment population (Fig. 23.5). We observed that the CI population carried two *wlaN* alleles at locus *Cj1139,* one with an 8G (60% allele frequency) homopolymeric tract and one with a 9G (40% frequency) homopolymeric tract. Subsequent passage of the CI population through chickens (CP) promoted a shift in *wlaN* allele frequency to greater than 80% 9G while passage through mice (MP) promoted a shift to nearly 100% 8G. Based on nucleotide sequence analysis of *wlaN,* we hypothesized that the 8G allele encodes a complete ORF and a functional protein while the 7G and 9G alleles encode truncated and non-functional proteins.

In order to test this hypothesis, we grew *C. jejuni* strain CamR2 under standard laboratory

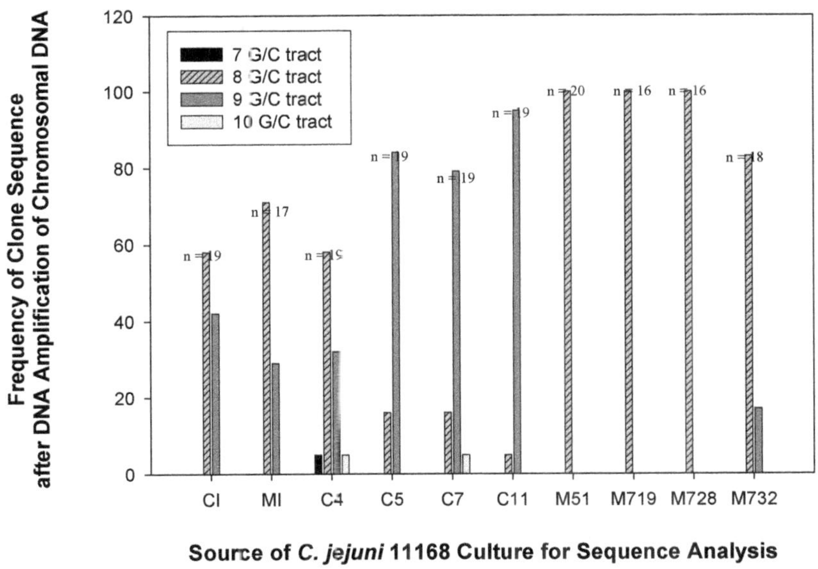

Figure 23.5 DNA sequence analysis of the homopolymeric G/C tract of *wlaN* in *C. jejuni* 11168 cultures recovered from the caecum of C57BL/6J IL10 deficient mice and Ross 308 broilers. We prepared chromosomal DNA extracts for *C. jejuni* 11168 used to inoculate C57BL/6J IL10 deficient mice (MI) and Ross 308 Broilers (CI) as well as from *C. jejuni* isolates recovered from caecal contents of four mice and four chickens. DNA extracts were amplified with Pfx50 high fidelity DNA polymerase and *wlaN* primers. A specific 415bp amplification product containing the *wlaN* homopolymeric G/C tract was purified, ligated and transformed into *E. coli* using the PCR4-TOPO cloning system. Twenty individual *wlaN* clones were selected from each chromosomal DNA amplification and sequenced with either an ABI 3730 Genetic Analyser or an ABI Prism 3700 DNA Analyser using T3 or T7 primers. The relative *wlaN* DNA sequence populations from each culture are reported above (C – Ross 308 Broiler, M=C57BL/6J IL10 deficient mouse). Total sample size for each group is also given.

conditions to generate an inoculum (CI). The CamR2 CI population was inoculated into eight Ross, 308 broilers and this population colonized all 8 birds at day 14 post inoculation. The birds were humanely sacrificed and the *C. jejuni* population was isolated from the caecum by plating on a selective growth medium. The pooled CP population was then inoculated into four mice. Surprisingly, since strain CamR2 normally does not colonize our mouse model, all four mice were colonized at high levels (see below for an explanation of this phenomenon) and the *C. jejuni* population that had been passaged through chickens and then mice (MCP) was isolated on selective growth medium. We analysed genotype at contingency locus *Cj1139* in CI, CP, and MCP populations by nucleotide sequence analysis. The CI population carried the 8G allele (encodes the ORF) at 100% allele frequency, the CP population carried a mixture of the 7G allele (55% allele frequency) and 8G allele (45% allele frequency), and the MCP population carried the 7G allele (100% allele frequency).

In order to test the function of the proteins encoded by the 7G, 8G, and 9G alleles of *wlaN* in the CI, CP, and MCP populations of *C. jejuni* strain CamR2, we conducted a cholera toxin (CT) binding assay using our modifications to a published procedure (Xiang *et al.*, 2006). CT binds to LOS that carries a terminal galactose but does not bind to LOS that lacks galactose (Fig. 23.6). In support of the hypothesis, the CamR2 CI population (100% 8G allele) bound CT at high efficiency and CT binding progressively decreased in the CP (45% 8G allele) and MCP (0% 8G allele) populations as the frequency of the 8G allele decreased. The data demonstrated that the 8G allele encodes a functional WlaN (CgtB) enzyme, that the 7G

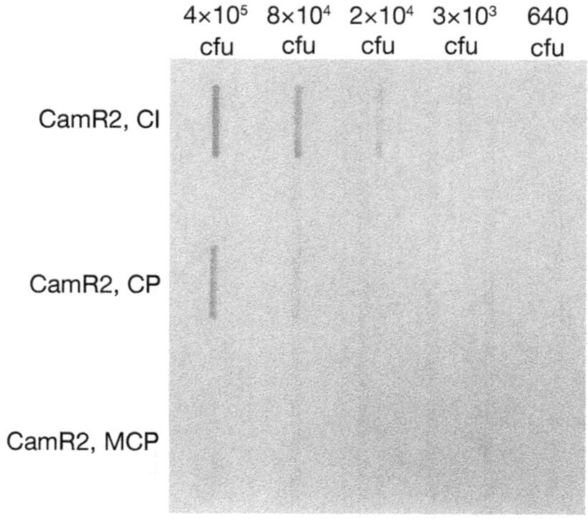

Figure 23.6 Cholera toxin binding assay. A cholera toxin binding assay was used to detect a terminal galactose residue (GM-1-like structure) on LOS of CamR2 CI, CP, and MCP populations. An allele of contingency locus *Cj1139* that carries a homopolymeric tract with 8G encodes an enzyme that produces a GM-1-like LOS structure. The 8G allele of Cj1139 occurred at 94% frequency in the CamR2 CI population. The 8G and 7G allele frequencies in the CP population were 45% and 55%, respectively. The 7G allele frequency was 100% in the MCP population. The designated concentrations of *C. jejuni* cells were boiled for 5 min and cell lysates were loaded onto a PVDF membrane and dried. Then, the terminal galactose residue was detected by peroxidase conjugated to cholera toxin B subunit and a DAB staining kit. A colour version of this figure is available in the plate section at the back of the book.

and 9G alleles encode non-functional enzymes (or enzymes with drastically reduced function), and that animal passage promotes shifts in allele frequency (phase variation) at contingency locus *Cj1139* which can alter WlaN enzyme activity and LOS structure on the cell surface.

Because many contingency loci are involved in synthesis or modification of LOS, flagella, and capsule and these structures are important surface antigens, we hypothesized that phase variation at specific contingency loci would directly impact colonization and disease in the host (mouse) by: (a) altering bacterial attachment to, entry into, or survival in colon epithelial cells in the GI tract; or (b) impacting efficacy of the host innate or adaptive immune response.

Passage of *C. jejuni* through broilers can impact ability to colonize mice

To measure the potential impact of the chicken reservoir on colonization by *C. jejuni* in mice, we grew *C. jejuni* strains 11168, KanR4, CamR2, 81-176, and 33292 under standard laboratory conditions to generate two independent inoculum populations (CI) and these were inoculated by oral gavage into groups of four, Ross 308 broilers in two independent experiments, CM1 and CM2 (Fig. 23.7) (Kim *et al.*, 2012). *C. jejuni* populations were harvested from the chicken caecum at day 14 (CP populations), these populations were pooled from the four birds, and a part of the CI and CP populations for each *C. jejuni* strain were saved for fragment analysis to measure phase variation at contingency loci. We used part of the CP populations of each strain in CM1 and CM2 to inoculate four (CM1) or five (CM2) C57BL6 IL10- deficient mice by oral gavage and measured colonization at day 14 post inoculation. We then froze the remaining portion of the CP populations from CM1 and CM2 for use in experiment CM3 described below. Finally, we purified genomic DNA in *C. jejuni* isolated from colonized mice (MCP populations) for fragment analysis.

Inoculum (CI) populations of *C. jejuni* strains 11168, KanR4, CamR2, 81-176, and 33292 colonized all four chickens in experiments CM1 and CM2. In contrast, only one in two chicken passaged (CP) populations of 33292, KanR4, and CamR2 colonized mice at high frequency in either CM1 or CM2 (Fig. 23.7); the other CP population did not colonize mice at high frequency. The CP population of each strain that colonized mice was designated a 'successful population' whereas the CP population that failed to colonize mice was designated an 'unsuccessful population'. We hypothesized that differences in phase variation at specific contingency loci was responsible for differential colonization of mice by successful and unsuccessful CP populations.

In experiment CM3, we inoculated groups of eight mice with frozen successful CP populations prepared during experiment CM1 or CM2 (Fig. 23.8). We also inoculated groups of eight mice with a fresh inoculum (CI) population for each strain generated specifically for CM3. None of the CI populations for strains KanR4, CamR2, or

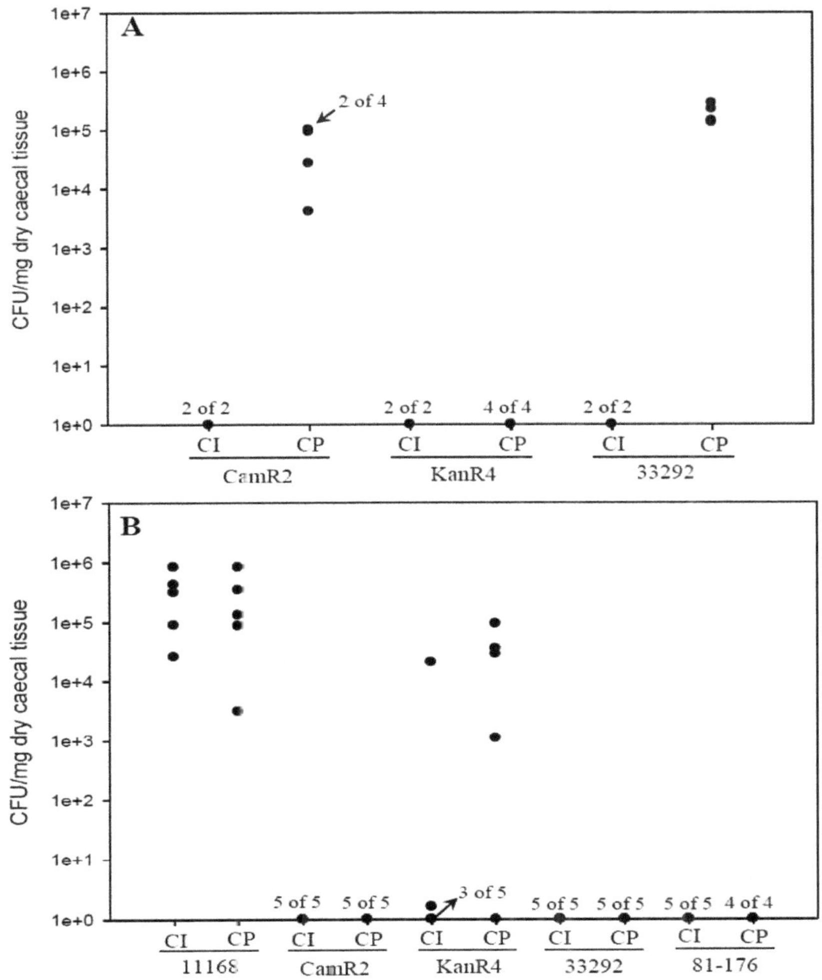

Figure 23.7 *C. jejuni* colonization of caecal tissue in C57BL/6 IL10-deficient mice. In CM1 (A), two mice per group were gavaged orally with *C. jejuni* inoculum (CI), and four mice per group were gavaged with chicken passed (CP) populations. In CM2 (B), five mice per group were gavaged with CI or CP populations. Mice were sacrificed at day 14 post inoculation, and caecal tissue was homogenized, serially diluted, and plated onto selective growth media. Single dots represent colonization in individual mice; in specific cases, numbers adjacent to dots indicate data for multiple mice (e.g. Panel A: the single dot for CamR2 CI represents two mice). CamR2 and 33292 CP populations in CM1 colonized mice at significantly higher levels than CP populations of these same isolates in CM2 (P-values <0.05). The KanR4 CP population in CM2 colonized mice at significantly higher levels than the CP population of this isolate in CM1 (P-value <0.05).

33292 colonized mice at high frequency whereas the frozen successful CP populations for strains KanR4, CamR2, or 33292 colonized mice at high frequency. Differences in mouse colonization between CI and CP populations were statistically significant. These data strongly suggest that passage of *C. jejuni* through the chicken reservoir can promote colonization in the mouse model and this occurred in three different *C. jejuni* strains.

Genotype at specific contingency loci is strongly associated with ability to colonize mice

We analysed allele frequency at 19 contingency loci (placed into 15 groups of individual or genetically similar loci) in *C. jejuni* strains 11168, KanR4, and CamR2, and at five contingency loci in strains 81-176, and 33292 in the inoculum (CI populations), after passage through chickens

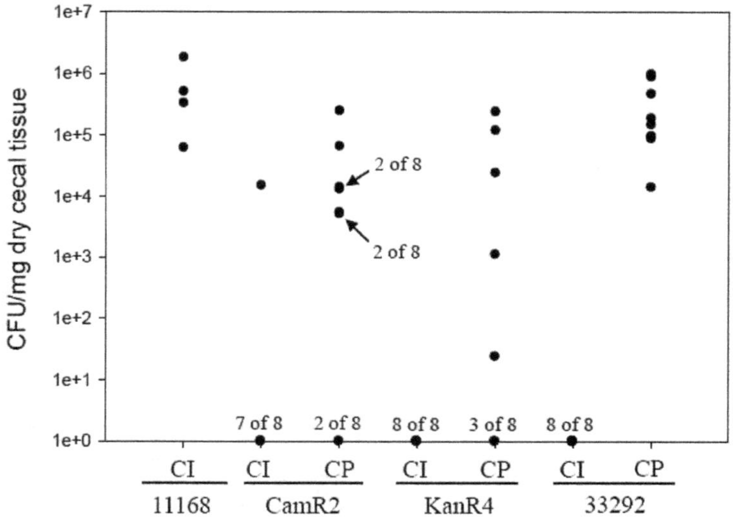

Figure 23.8 Colonization of caecal tissue from C57BL/6 IL 10-deficient mice by *C. jejuni* recovered from frozen chicken caeca samples from CM1 and CM2. Frozen chicken caeca samples were macerated and spread on selective growth media. CamR2 (CM1), KanR4 (CM2), and 33292 (CM1) CP populations were recovered on selective medium. Then, 8 mice per group were gavaged with chicken passed (CP) or chicken inoculum (CI) populations. At 3 weeks post infection, caecal tissue was homogenized, serially diluted, and plated onto selective growth media (see Methods). For all three strains tested, CP populations colonized mice at significantly higher levels than CI populations ($P < 0.05$). Single dots represent colonization in individual mice. Numbers adjacent to dots indicate data for multiple mice.

(CP populations), and after passage through chickens and then mice (MCP populations) using DNA samples prepared from *C. jejuni* populations generated in experiments CM1 and CM2 (Kim *et al.*, 2012) (Fig. 23.9). Overall, there was a strong association between phase variation at 8 contingency loci (*Cj0045*, *Cj0170*, *Cj0685*, *Cj1139*, *Cj11420*, *Cj1421*, *Cj1422*, and *Cj1426*) and the ability of specific *C. jejuni* populations to colonize mice. Below we analyse the observed phase variation in detail and discuss the likely impact of these shifts in allele frequency on the ability to colonize mice by *C. jejuni* populations. To help with the clarity of this discussion, we define a predominant allele as any allele that can be detected at a specific contingency locus in > 50% of the *C. jejuni* population. We define a successful allele as a predominant allele at a specific contingency locus that is closely associated with the ability of that *C. jejuni* population to colonize mice.

Phase variation in successful chicken-passaged (CP) populations but not in unsuccessful CP populations

Large shifts in allele frequency (phase variation) to a successful allele were observed at contingency loci *Cj0685*, *Cj1139*, *Cj1421*, and *Cj1426* in a successful CP population of *C. jejuni* strain CamR2 and these shifts in allele frequency were not observed in an unsuccessful CP population. Of particular significance, phase variation occurred during passage in the chicken reservoir and prior to inoculation of the CP population into mice underscoring a strong association between these four contingency loci and ability to colonize mice. Phase variation to a successful allele was also observed at contingency locus *Cj0045* in a successful CP population of strain 33292 prior to inoculation into mice whereas phase variation was not observed in the unsuccessful population of strain 33292. These data strongly suggest that phase variation at these five contingency loci promotes ability to colonize mice.

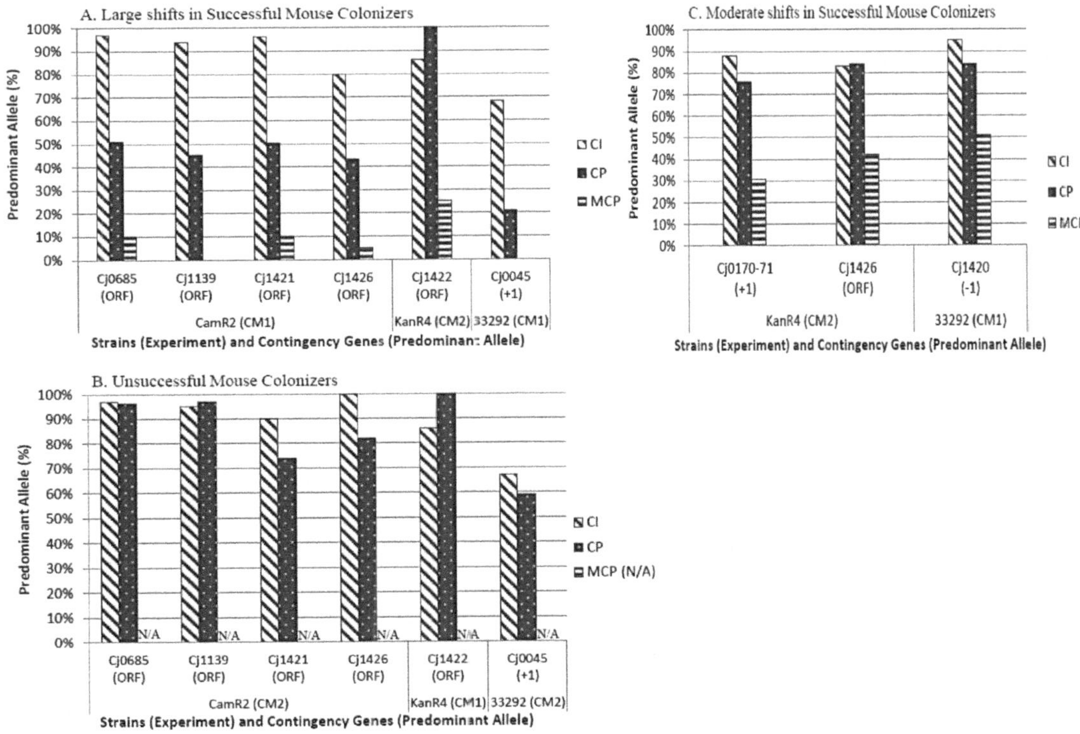

Figure 23.9 Frequency (per cent) of the predominant allele in *C. jejuni* populations. The predominant allele was defined as the highest frequency allele in and inoculum (CI) population grown in culture. Contingency loci were amplified by PCR and fragment and/or sequence analyses were used to determine allele distribution before chicken passage (CI), after chicken passage (CP), and after subsequent mouse passage (MCP). (A) Successful mouse colonizers showed enhanced colonization in mice after passage through chickens. Predominant allele frequencies in CI and CP populations were significantly different (P-values <0.05), except for locus *Cj1422* in KanR4 populations (CM2) (P-value=0.23). Predominant allele frequencies in CP and MCP populations are significantly different (P-values <0.05). (B) Unsuccessful mouse colonizers did not colonize mice after chicken passage. Significant differences in predominant allele frequencies were observed between CI and CP populations (P-values >0.05), except for locus *Cj1426* in CamR2 populations (CM2) (P-value=3.3×10^{-6}). MCP populations were not available (N/A) because mice were not colonized. With the exception of locus *Cj1422*, allele frequencies for all other contingency loci in successful CP populations (A) were significantly different from unsuccessful CP populations (B) (P-value <0.05). (C) MCP populations experienced moderate allele shifts after chicken and mouse passage. Allele frequencies for contingency loci in CI and MCP populations were significantly different (P-value <0.05). Allele frequencies for contingency loci in CI and CP populations were significantly different (P-value <0.05) except for *Cj1426* in KanR4 populations (CM2).

Phase variation to successful alleles during passage in mice (MCP populations)

Phase variation to successful alleles also was observed at four *C. jejuni* contingency loci but the changes occurred after passage in chickens and during passage in mice. Contingency loci *Cj1422, Cj1426,* and *Cj0170* in strain KanR4 and *Cj1420* in strain 33292 belong to this group. These data suggest an association between phase variation at these specific loci and mouse colonization. We confirmed this association for *Cj0170* (see below) and must now determine conclusively if phase variation at the other 3 loci directly contributes to ability to colonize mice.

Phase variation from on to off phase

We observed phase variation to a successful allele at contingency loci *Cj0685 (cipA), Cj1139 (wlaN), Cj1421, Cj1422, Cj1426* and each of these successful alleles is predicted to encode a truncated protein based on nucleotide sequence analysis. These observations strongly suggest that loss of gene function at specific contingency loci may play an important role in ability to colonize mice.

Although these loci encode different enzymes, the available data suggest that all five enzymes modify cell structures that serve as important surface antigens. For example, enzymes encoded by *Cj1421* and *Cj1422* were demonstrated previously to add phosphoramidate to the capsular sugars N-acetylglucosamine and glucoheptopyranose, respectively (McNally *et al.*, 2007). Nucleotide sequence analysis demonstrated a high degree of identity between the CipA protein encoded by contingency locus *Cj0685* (*cipA*) and the phosphoramidate transferases encoded by *Cj1421* and *Cj1422* suggesting that CipA, although not localized in the capsule gene cluster, also is involved in capsule modification. *cipA* was previously demonstrated to play a role in invasion of Caco-2 cells in culture, hence the name *Campylobacter* invasion protein (Poli *et al.*, 2012). Contingency locus *Cj1426* is located in the capsule gene cluster and encodes a methyltransferase that presumably methylates capsule although the details of the reaction catalysed by this enzyme are not known. As described above, *Cj1139* was demonstrated to add galactose to LOS (Linton *et al.*, 2000). Since modification of capsule and LOS can decrease ability of the host immune system to recognize and clear bacterial pathogens, it is reasonable to propose that phase mutations at these five specific contingency loci could reduce modification of capsule and LOS, reduce the innate or adaptive immune response, and block ability to clear bacteria from the mouse GI tract. Alternatively, these changes in surface structure could alter bacterial attachment, invasion, and/or survival in colon epithelial cells.

Phase variation from off to on phase

We observed phase variation from an unsuccessful allele that carries an 8G HT (off phase) to a successful allele that carries a 9G HT (on phase) at contingency locus *Cj1420* in *C. jejuni* strain 33292. Nucleotide sequence analysis and location in the capsule gene cluster suggest that *Cj1420* encodes a methyltransferase involved in capsule modification. Unfortunately, details of the reaction catalysed by this enzyme are not known.

Phase variation from one truncated protein to another

Contingency locus *Cj0170* in *C. jejuni* strain KanR4 experiences phase variation from an unsuccessful allele that carries a 9G HT to a successful allele that carries a 10G HT. This is significant because nucleotide sequence analysis suggests that the 8G allele fuses contingency loci *Cj0170* and *Cj0171* so they encode a complete ORF (Kim *et al.*, 2012). The most recent annotation of the *C. jejuni* 11168 genome (Gundogdu *et al.*, 2007) described *Cj0170/Cj0171* as a single locus whereas the original annotation (Parkhill *et al.*, 2000) described *Cj0170* and *Cj0171* as 2 distinct loci. Based on the recent annotation, the allele of *Cj0170* that carries the 9G allele HT encodes a truncated protein, which is missing the entire *Cj0171* ORF while the 10G allele is missing the entire *Cj0171* ORF plus 6 additional amino acids (YRALWH) from the C terminus of *Cj0170*. These observations suggest that for strain KanR4 (which was derived from the sequenced strain 11168; Parkhill *et al.*, 2000) these two alleles that encode truncated proteins that differ by only 6 amino acids generate phenotypes that are remarkably different. *C. jejuni* KanR4 cells that carry the 10G allele are motile, colonize, and cause disease in mice. Cells that carry 9G allele are nonmotile, and do not colonize or cause disease in mice. We do not know the phenotype of cells that carry predominantly the 8G allele because these cells occur so rarely in the population and we have not isolated them to date.

Phase variation fuses adjacent open reading frames

Most *C. jejuni* strains (including strain 33292) for which the genome sequence has been completed carry contingency locus *Cj0045* in which a successful allele with a 9G HT encodes the complete ORF and a successful 10G allele adds 15 amino acids to the C terminus of *Cj0045* that are derived from the adjacent locus *Cj0044*. Surprisingly, the unsuccessful 11G allele in strain 33292 adds one amino acid to the C-terminus of *Cj0045* but again the phenotype of strains that carry the 11G allele are remarkably different from strains that carry either the 9G or 10G alleles. We hypothesize that the 11G allele is unsuccessful because the translation termination codon for *Cj0045* overlaps the *Cj0044* translation initiation codon reducing its expression (see below proposed gene functions for *Cj0045* and *Cj0044*). Furthermore, we hypothesize that the 9G and 10G alleles of *Cj0045* do not impact expression of *Cj0044*. Testing these hypotheses is one focus of future work.

Genotype at specific contingency loci impacts colonization and disease in mice

Single colony isolates

To more directly test the impact of specific alleles at contingency loci on ability to colonize mice, single colony isolates (strains) of *C. jejuni* KanR4 (Table 23.2) that carry either successful (10G HT; strains SCK1 and SCK7) or unsuccessful (9G HT; SCK18 and SCK21) alleles at contingency locus *Cj0170* as the predominant allele were gavaged into groups of 8 mice and colonization was analysed at d21 post inoculation (Kim *et al.*, 2012). Strains SCK1 and SCK7 that carry a successful 10G allele colonized 11 of 16 mice while strains SCK18 and SCK21 that carry an unsuccessful 9G allele did not colonize mice (0 of 16 mice) (Fig. 23.10). Differences in colonization between strains that carry successful and unsuccessful alleles were statistically significant. Seventeen of 17 strains (including SCK1 and SCK7) that carry the successful 10G allele of *Cj0170* were highly motile while 7 of 7 strains (including SCK18 and SCK21) that carry an unsuccessful 9G allele of *Cj0170* exhibited greatly reduced motility. Of key significance, gene disruption of *Cj0170* in strains DM0170-2 and DM0170-6 eliminated both motility and ability to colonize mice. DM0170-2 and DM0170-6 exhibited a similar phenotype as strains SCK28 and SC21 except motility was reduced to an even greater extent (undetectable).

Strains SC4 and SC13 (isolates of *C. jejuni* strain 33292) that carry the successful 9G allele of contingency locus *Cj0045* and the successful 9G allele of *Cj1420* (Table 23.2), colonized 13 of 16 mice at d21 post inoculation. Strains SC5 and SC9 (also isolates of strain 33292) that carry the unsuccessful 11G allele of *Cj0045* and the unsuccessful 8G allele of *Cj1420* colonized 12 of 16

Table 23.2 Comparison of allele frequencies for contingency loci *Cj0170*, *Cj0045*, and *Cj1420* in single colony isolates prepared from glycerol stocks or prepared for mouse inocula

	Cj0170	Cj0170	Cj0045	Cj0045	Cj1420	Cj1420
Isolate	Glycerol[b]	Inoculum[c]	Glycerol	Inoculum	Glycerol	Inoculum
SCK1 [a] (S)	81 (10G)	78				
SCK7 (S)	84 (10G)	80				
SCK18 (U)	90 (9G)	89				
SCK21 (U)	90 (9G)	90				
SC4 [a] (S)			86 (9G)	85	97 (9G)	100
SC13 (S)			96 (9G)	84	100 (9G)	95
SC5 (U)			72 (11G)	58	96 (8G)	92
SC9 (U)			71 (11G)	53	97 (8G)	95

[a]Single colony isolates SCK1 and SCK7 were derived from a successful population of *C. jejuni* KanR4 and carry the successful allele (S) of *Cj0170* (10G homopolymeric tract). Single colony isolates SCK18 and 21 were derived from the same successful KanR4 population and carry the unsuccessful allele (U) of *Cj0170* (9G homopolymeric tract). Single colony isolates SC4 and SC13 were derived from a successful population of *C. jejuni* 33292 and carry successful alleles (S) of *Cj0045* (9G or 10 G homopolymeric tract) and *Cj1420* (9G homopolymeric tract) Single colony isolates SC5 and SC9 were derived from the same successful population of *C. jejuni* 33292 and carry unsuccessful alleles (U) of *Cj0045* (11G homopolymeric tract) and *Cj1420* (8G homopolymeric tract).
[b]Per cent frequency of the predominant allele in population derived from glycerol stock (see Methods)
[c]Per cent frequency of the predominant allele in population after culture in growth medium during preparation of inocula for oral gavage into mice.

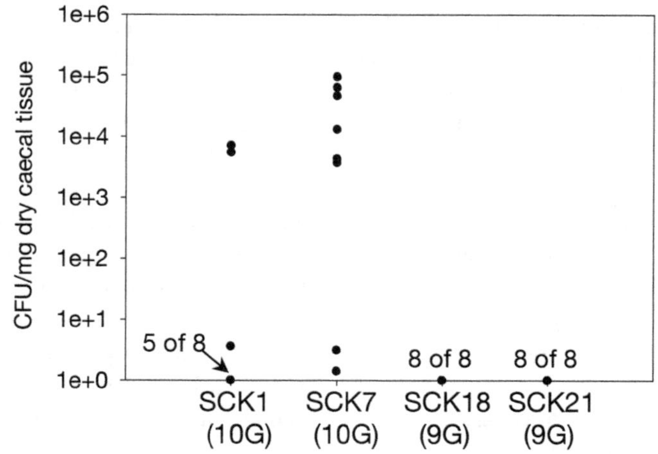

Figure 23.10 Colonization of mouse caeca by KanR4 single colony isolates carrying 2 different *Cj0170* alleles (homopolymeric tract carrying either 9G or 10G). Eight mice were gavaged orally with each KanR4 single-colony isolate. Mice were sacrificed at d 21 post inoculation and caecal tissue was homogenized, serially diluted, and plated onto selective growth media. Single dots represent colonization in individual mice; in specific cases, numbers adjacent to dots indicate data for multiple mice (e.g. the single dot for SCK18 and SCK21 represents 8 mice). The number of mice colonized by SCK1 and SCK7 carrying the 10G allele of *Cj0170* was significantly higher than the number of mice colonized by SCK18 and SCK21 carrying the 9G allele of *Cj0170* (P <0.05).

mice. However, in strains SC4 and SC13, the 9G allele remained the predominate allele at d21 post inoculation. In strains SC5 and SC9, the predominant 11G unsuccessful allele experienced phase variation to either the successful 9G allele or successful 10G allele by d21 post inoculation. Based on these analyses, we could not discern whether phase variation at *Cj0045* contributed directly to mouse colonization in strains SC5 and SC9 or if it occurred as a result of mouse colonization.

To help resolve this issue, we repeated the experiment. We gavaged strain SC13 (with the successful 9G allele as the predominant allele) as well as strains SC5 and SC9 (with the unsuccessful 11G allele as the predominant allele) into mice and conducted fragment analysis on the inoculum and on *C. jejuni* isolated from caecal contents at d21as described above. We also conducted fragment analysis on *C. jejuni* isolated from faecal pellets of colonized mice at d4 post inoculation (Table 23.3). *C. jejuni* strains SC13, SC5, and SC9 each colonized mice at high frequency and significant differences in the numbers of colonized mice were not observed. However, in mice colonized by strain SC13, the 9G successful allele remained the predominant allele throughout the 21-day experiment or it shifted to the successful 10G and/or 7G alleles (Table 23.3). Of particular importance, a shift from the 9G allele to the unsuccessful 11G or 8G alleles was not observed at significant levels so unsuccessful alleles were greatly under-represented in the colonized mice.

In contrast, the 11G predominant unsuccessful allele in strains SC5 and SC9 (59–64% initial allele frequency) nearly disappeared by day 4 post inoculation. In 8 of 12 mice colonized by *C. jejuni* at day 4 post inoculation, the 11G allele shifted to successful 9G or 10G alleles, or to variants of these alleles (6G, 7G, or 12G) and this occurred prior to day 4 post inoculation as the strains established stable colonization. In 4 of 12 mice colonized at day 4, *C. jejuni* also carried the unsuccessful 8G allele (allele frequency ranging from 20–35%). In three of fiur of these mice, the successful 7G allele was present at higher frequency than the unsuccessful 8G allele. In only one mouse was the 8G allele present at greater frequency than the 7G allele. In strains SC5 and SC9, there was a strong selection bias against the unsuccessful 8G and 11G alleles by day 4 post inoculation. These data were confirmed in two biological replicates of the experiment and strongly suggest that successful

Table 23.3 Allele frequencies of *Cj0045* in inoculum (isolate), faecal pellets (d4), and/or caecal samples (d21)

Isolate	6G (–3)	7G (–2)	8G (–1)	9G (ORF)	10G (+1)	11G (+2)	12G (+3)	13G (+4)	14G (+5)
SC13 (S)[a]				90 [b]	10	0			
SC9 (U)[a]				2	34	59	5		
SC5 (U)[a]				0	36	64			
Mouse[e]									
78 d4[c] SC9		38	31	31	0	0	0		
82 d4 SC9		41	16	43	0	0	0		
79 d2[d] SC9		1	6	26	64	3	0		
85 d21 SC9	15		0	52	17	0	6	10	
80 d4 SC9		37	36	27	0	0	0		
80 d21		13	4	80	3	0	0		
81 d4 SC9		42	0	58	0	0	0		
81 d21			0	0	1	6	38	50	5
87 d4 SC5			0	100	0	0			
88 d4 SC5		34	0	66	0	0			
89 d4 SC5		30	23	47	0	0			
90 d4 SC5		19	38	43	0	0			
86 d21 SC5			0	100	0	0			
92 d21 SC5		14	4	78	4	0			
96 d4 SC13		43	15	42	0	0			
93 d21 SC13		12	0	88	0	0			
94 d21 SC13		10	4	81	5	0			
95 d21 SC13	44		0	56	0	0			
98 d4 SC13		55	0	45	0	0			
98 d21		1	4	82	3	0			
100 d4 SC13		47	0	53	0	0			
100 d21	17		0	45	24	0		14	

[a]*Cj0045* allele frequencies in inocula used in oral gavage for isolates SC5, SC9, and SC13. S, successful allele; U, unsuccessful allele.
[b]Numbers represent allele frequencies as a per cent of the total population.
[c]d4. Faecal pellet obtained from mouse at day 4 post inoculation.
[d]d21. Caecal tissue obtained from mouse upon necropsy at d21 post inoculation.
[e]Mice colonized by *C. jejuni* isolates SC5, SC9, and SC13 designated with these isolate numbers. Analysis for faecal pellet and caecal samples from mice 83, 84, 91, 97, and 99 were not completed. Unsuccessful 8G, 11G, and 14G alleles are highlighted in grey. All other alleles are successful alleles.

alleles of *Cj0170* and *Cj0045* enhance ability to colonize mice.

We also observed a statistically significant association between colonization level and gross pathology in the GI tract in mice colonized by single colony isolates of *C. jejuni* strains KanR4 and 33292 that carried successful alleles of *Cj0170* and *Cj0045*, respectively, confirming that successful alleles of *Cj0170* and *Cj0045* impact virulence in mice (Table 23.4). To measure gross pathology, the extent of thickening or enlargement was determined at dix sites (jejunum, caecum, colon, ileo-caeco-colic junction lymph node, mesenteric lymph node, and spleen) in the mouse GI tract. The analysis was conducted in a 'blinded' study by two experienced investigators and an average composite score was generated for individual mice and for each group of mice.

Table 23.4 Association between mouse colonization level and gross pathology score in mice infected with single colony isolates of KanR4 and 33292

Inoculum[a]	CFU/mg dry caecum[b]	Number of mice with level	Number of mice with gross pathology score					Average score[c]
			≤1	2	3	4	≥5	
TSB	0	4	4 (100%)					0.0
11168	>10^3	8	4 (50%)	1 (13%)		1 (13%)	2 (25%)	2.5
KanR4 (S)	0	5	3 (60%)		1 (20%)	1 (20%)		1.8
KanR4 (S)	1–10^3	3	3 (100%)					1.0
KanR4 (S)	>10^3	8	3 (37%)	1 (13%)	4 (50%)			2.1
KanR4 (U)	0	16	10 (62%)	4 (25%)	2 (13%)			1.2
33292	0	7	4 (57%)	1 (14%)	2 (29%)			1.6
33292	1–10^3	2	1 (50%)	1 (50%)				1.0
33292	>10^3	23	6 (26%)	8 (35%)	4 (17%)	1 (4%)	4 (17%)	2.9

[a](S), successful genotypes SCK1 and SCK7; (U), unsuccessful genotypes SCK18 and SCK21.
[b]Colonization level in the caecum at the time of necropsy.
[c]Average gross pathology score.

How does contingency locus Cj0170 impact colonization and disease?

C. jejuni motility is strongly associated with colonization and virulence (Dasti *et al.*, 2001; Guerry, 2007). We demonstrated that phase variation in *Cj0170* impacts both motility and virulence in mice (Artymovich *et al.*, 2013; Kim *et al.*, 2012) prompting us to analyse the mechanisms underlying this functional link. Parkhill *et al.* (2000) and Gundogdu *et al.* (2007) were unable to predict a function for *Cj0170*. However, we searched the NCBI database using the predicted Cj0170 peptide sequence and detected identity between *Cj0170* and methyltransferases from a variety of organisms (e.g. 68% identity, 82% similarity to a DNA adenine [Dam] methyltransferase in *Campylobacter lari*; 31% identity, 49% similarity to a type II methyltransferase of *Aminomonas paucivorans*). Electron microscopy of *C. jejuni* strains that carry successful (10G, SCK1) or unsuccessful (9G, SCK18) alleles of *Cj0170* revealed no detectable differences in gross flagellar structure. These observations prompt us to propose two potential mechanisms by which *Cj0170* could impact motility. First, alteration of gene expression by phase-variable Dam methyltransferases regulate expression of genes involved in motility and chemotaxis in *E. coli* as well as virulence genes in Salmonella and *Yersinia enterocolitica* (Balbontin *et al.*, 2006; Falker *et al.*, 2005; Oshima *et al.*, 2002; and Srikhanta *et al.*, 2010); genes regulated in this fashion belong to a 'phase varion'. Since *Cj0170* does not appear to affect flagellar structure or assembly, genes involved in chemotaxis (e.g. *cheY*), post-translational modification of flagella, and/or flagellar rotation top our list of candidate target genes for activation by the Cj0170 Dam methyltransferase (Yao *et al.*, 1997). Second, posttranslational modification of flagellin subunits by pseudaminic acid (PseAc) is important to proper flagellar assembly and motility (Guerry, 2007; Guerry and Szymanski, 2008). Loci encoding methyltransferase (*Cj1325–26*) and acetyltransferase (*Cj1321*) activities localize in the flagella modification cluster (Guerry, 2007; Guerry and Szymanski, 2008) (see Table 23.1). An acetylated version of PseAc (PseOAc) was identified in flagella supporting a role for the Cj1321 acetyl-transferase in modification of flagellar PseAc. The Cj0170 peptide exhibits 57% identity and 68% similarity to the Cj1325-26 methyltransferase located in the flagella gene cluster suggesting a possible role in methylation of flagellar PseAc.

How does Cj0045 impact colonization and disease?

Cj0045 tentatively was identified as hemerythrin, a non-haem iron/oxygen binding protein (French *et al.*, 2008), in the re-annotation of the *C jejuni* 11168 genome (Gundogdu *et al.*, 2007). In support of this assignment, we observed 99% identity to a putative iron binding protein from another sequenced isolate of *C. jejuni*. Based on these observations, we propose that potential mechanisms by which phase variation in *Cj0045* impacts virulence include competing for iron or modulating or sensing local oxygen levels in the GI tract. However, the *Cj0045* HT lies near the carboxyl terminus of the ORF, so frame shifts induced by phase variation do not generate truncated *Cj0045* proteins; they add 1 or 11 amino acids to the carboxyl terminus. So, we propose that one alternative mechanism by which phase variation in *Cj0045* impacts virulence is by modulating expression of *Cj0044*. We propose that the 9G and 10G alleles of *Cj0045* are successful because the translation termination codon of *Cj0045* does not overlap the translation initiation codon of *Cj0044* allowing normal expression of the Cj0044 protein. Furthermore, we propose that the 11G allele of *Cj0045* is unsuccessful because the translation stop codon of *Cj0045* overlaps the translation initiation codon in *Cj0044* thereby reducing levels of re-initiation at *Cj0044 and* down-regulating synthesis and activity of this protein. Analysis of the *C. jejuni* 11168 sequence (Parkhill *et al.*, 2000; Gondogdu *et al.*, 2007) failed to assign protein function to *Cj0044* and little work has been done on either *Cj0045* or *Cj0044*. The 3 genes immediately upstream from *Cj0044* have been assigned putative roles in flagella assembly including *Cj0041* (motility and invasion), *Cj0042* (*flgD*, hook protein) and *Cj0043* (*flgE* flagellar basal body protein) (Gundogdu *et al.*, 2007; Parkhill *et al.*, 2000), suggesting a possible role for *Cj0044* in motility.

Four contingency loci are highly conserved in *C. jejuni* human disease isolates

Of the 19 *C. jejuni* contingency loci we analyse on a routine basis, *Cj0031*, *Cj0045*, *Cj0170*, *Cj0685*, *Cj1310*, and *Cj1420* are highly conserved (85–100% identity) in all 16 human disease isolates examined to date (Artymovich *et al.*, 2013). Phase variation was observed at significant levels in *Cj0045*, *Cj0170*, *Cj0685*, and *Cj1420* while no significant phase variation was observed in *Cj0031* or *Cj1310* during culture or during passage through chickens and/or mice. Phase variation in *Cj0170* was directly linked with colonization and disease in mice while phase variation in the other 3 conserved contingency loci was strongly associated with colonization and disease. Based on these observations, the 4 highly conserved contingency loci are the major focus of our current research in our laboratory. Of particular interest, a single allele of *Cj0170*, *Cj0685*, and *Cj1420* was enriched in the 16 human disease isolates (Table 23.5); the 9G allele of *Cj0170* and *Cj0685* was the predominant allele (detected in > 50% of the population) in 13 of the 16 isolates and the 9G allele of *Cj1420* was the predominant allele in 14 of the 16 isolates. In *Cj0045*, the unsuccessful 11G allele was detected in 5 of the 16 isolates while successful 9G or 10G alleles were detected in the remaining 11 isolates.

Human disease isolates that carry successful alleles at Cj0045 and Cj1420 colonize mice

To date, we have tested the ability of 7 of the 16 human disease isolates (Table 23.5) to colonize mice. Strains 11168 and E961009 colonized mice without passage through the chicken reservoir (Table 23.6). Genotype analysis at the 4 highly conserved contingency loci revealed common features between these 2 strains; they each carried a 10G successful allele at *Cj0045*, a 9G successful allele at *Cj1420*, and a 9G unsuccessful allele at *Cj0685* (Artymovich *et al.*, 2013). 11168 carried the 9G unsuccessful allele at *Cj0170* and E961009 carried the 10G successful allele at *Cj0170*. Strain 33292 initially was not successful in colonizing mice and it carried a successful allele at *Cj1420* and unsuccessful alleles at the other three loci. However, after passage of strain 33292 through the chicken GI tract, the successful population colonized mice and the unsuccessful 11G allele at *Cj0045* experienced phase variation to either a 9G or 10G successful allele. The other five human disease isolates did not colonize mice and they carried a successful allele at either *Cj1420* or *Cj0045* but not both. Based on these analyses, we hypothesize that successful alleles at both *Cj1420* and *Cj0045* promote colonization in the mouse model. We will test this hypothesis in future work.

Table 23.5 Genotypes at 4 conserved loci in 16 *C. jejuni* human disease isolates analysed to date

Locus	*Cj0045*		*Cj0170*		*Cj0685*		*Cj1420*		
C. jejuni isolate	HPT (G) length		HPT (G) length		HPT (C) length		HPT (G) length		Predicted colonization
4300	10	S	9	U	8	S	10	**U**	No
4239928	10	S	9	U	9	U	9	S	Yes
E961009 (3)	10	S	10	S	9	U	9	S	Yes
E972653	9	S	9	U	9	U	9	S	Yes
E972796	10	S	NC	NC	9	U	9	S	Yes
E9795412 (2)	9	S	10	S	9	U	8	**U**	No
3128 (1)	11	**U**	9	U	9	U	9	S	No
3136	11	**U**	9	U	8	S	9	S	No
33291 (1)	11	**U**	9	U	9	U	9	S	No
43429	12	S	9	U	9	U	9	S	Yes
43433	10	S	9	U	10	U	9	S	Yes
43470	NC	NC	9	U	9	U	9	S	?
49349	9	S	9	U	NC	NC	9	S	Yes
33292 (1)	11	**U**	9	U	9	U	9	S	NO
11168 (2)	10	S	9	U	9	U	9	S	Yes
81-176 (1)	11	**U**	9	U	9	U	9	S	NO

Abbreviations: HT, homopolymeric tract; S, successful allele; U, unsuccessful allele; NC, not completed. Underlined isolates were gavaged into C57BL6 IL10-deficient mice to analyse colonization (Table 23.4). Numbers in parentheses next to isolate indicate the number of loci with successful alleles. Successful alleles of *Cj0045* and *Cj1420* are underlined; unsuccessful alleles are in **bold**.

Table 23.6 Colonization of mice by human disease isolates that vary in number of successful alleles

Strain	Inoculum[a]	Day 4 (faecal sample)[b]	Day 21 (caecal sample) [c]
TSB	0	0 (0/10 mice)	0 (0/10 mice)
11168	7.5×10^6	1.6×10^4 (6/10 mice)	1.1×10^6 (6/10 mice)
E9795412 (2)	2.3×10^7	0 (0/8 mice)	3 (1/8 mice)
E961009 (3)	3.1×10^6	2.5×10^6 (7/8 mice)	1.0×10^6 (7/8 mice)
3128 (1)	3.8×10^7	0 (0/8 mice)	0 (0/8 mice)
33291 (1)	3.4×10^7	0 (0/8 mice)	0 (0/8 mice)

[a]Inoculum cfu/ml.
[b]Average cfu/g faecal pellet for eight mice (E9795412, E961009, 3128, 33291) or 10 mice (11168).
[c]Average cfu/g caecal sample for eight mice (E9795412, E961009, 3128, 33291) or 10 mice (11168). The ratios in parentheses represent the number of mice colonized per group. The numbers highlighted in grey represent the number of successful alleles.

C. jejuni isolates from animals carry specific predominant alleles of Cj0045

To determine if phase variation in *Cj0045* to a successful allele is a common event in nature, *C. jejuni* strains were isolated from Canadian geese, turkeys, free range chickens, and swine. Fragment analysis was conducted on strains isolated from geese and turkeys before and after passage through the GI tract of commercial broiler chickens (Artymovich *et al.*, 2013). Strains isolated from swine and free range chickens were also subjected to fragment analysis but only after passage through the GI tract of commercial broilers (Table 23.7). The

Table 23.7 Fragment analysis of contingency gene *Cj0045* in *C. jejuni* isolated from wild birds, domesticated birds, swine, or after passage through commercial chickens

C. jejuni isolate/source	Frequency, successful allele	Frequency, unsuccessful allele
111957 Canadian goose	28% (10G HPT)[d]	67% (11G HPT)
111957 Commercial chickens[a]	49% (10G HPT)	44% (11G HPT)
6067 Turkey	6% (13G HPT)	94% (14G HPT)
6067 Commercial chickens	48% (13G HPT)	48% (14G HPT)
6461 Commercial chickens[b]	66% (10G HPT)	31% (11G HPT)
12394 Commercial chickens[c]	75% (10G HPT)	10% (11G HPT)

[a]111957 originally isolated from Canadian goose and then passaged through commercial chickens. Sample represents pooled colonies from caecal samples obtained at necropsy of five independent birds after 21 days' colonization.
[b]Originally isolated from swine and passaged through commercial chickens as in footnote (a) above.
[c]Originally isolated from free range chickens and passaged through commercial chickens as in footnote (a) above.
[d]HPT, homopolymeric tract.

data demonstrated that a successful allele (10G or 13G) of *Cj0045* is carried by *C. jejuni* populations isolated from geese (28%) and turkeys (6%) and that the frequency of the successful allele in the population increases during passage through commercial broiler chickens to allele frequencies of 49% and 48%, respectively. *C. jejuni* populations isolated from free range chickens and swine also carried a successful allele of *Cj0045* as the predominant allele (allele frequencies of 66% and 75%, respectively) after passage through the GI tract of broilers.

Discussion

The successful 10G allele of Cj0170 is a positive regulator of motility in strain KanR4

We previously observed that the 8G allele of *Cj0170* is not present at high frequency in *C. jejuni* KanR4 populations isolated from the GI tract of broiler chickens or mice. Furthermore, we observed that the successful 10G allele is the predominant allele in successful KanR4 populations (successfully colonize mice) and that the 9G allele is the predominant allele in unsuccessful populations (do not colonize mice) isolated from chickens (Kim *et al.*, 2012). Based on these data, it is reasonable to propose that the 10G successful allele of *Cj0170* encodes a protein with modified function that positively regulates motility of *C. jejuni* and its ability to colonize in mice. In contrast, we propose that the 9G allele encodes either a non-functional protein or a protein with modified function, but this protein does not positively regulate motility in *C. jejuni* strain KanR4. *C. jejuni* motility is required for colonization and disease in humans and animals (Dasti *et al.*, 2010; Guerry, 2007) supporting the idea that loss of motility in strains DM0170-2 and DM0170-6 directly results in loss of ability to colonize mice. Since colonization is an early and critical step in the disease process, we argue that the successful 10G allele of *Cj0170* is an important virulence factor for strain KanR4 in the mouse model. We do not know if the protein encoded by the 8G allele is functional and confers virulence. Future studies will focus on the differential roles of proteins encoded by these *Cj0170* alleles in motility and colonization.

Successful alleles of Cj0045 are directly involved in mouse colonization

We previously observed that phase variation from a predominant 11G unsuccessful allele of *Cj0045* to predominant 9G or 10G successful alleles occurred in strain 33292 after inoculation into mice but prior to necropsy at d21 and the occurrence of the 9G or 10G successful allele was strongly associated with mouse colonization (Kim *et al.*, 2012). Since stable colonization is established by day 4 post inoculation in most mice (Artymovich *et al.*, 2013), we theorized that phase variation to a successful allele prior to day 4 would support the idea that phase variation contributed directly to colonization. We recently demonstrated that phase variation from a predominant unsuccessful allele to predominant successful alleles in *Cj0045* occurred prior to day 4 post inoculation when *C. jejuni* establishes stable colonization of mice. We argue that this observation confirms a direct role for successful alleles of *Cj0045* in mouse colonization. In support of this argument, the unsuccessful 8G and 11G alleles of *Cj0045* were greatly underrepresented in *C. jejuni* isolated from faecal pellets at day 4 and from caecal tissue day 21 post inoculation suggesting a strong selection bias against these alleles. We hypothesize that the 8G and 11G alleles are unsuccessful because they encode proteins in which the *Cj0045* open reading frame (ORF) is extended by a single amino acid. This results in an overlap between the translation stop codon of *Cj0045* and the translation initiation codon of *Cj0044* immediately downstream and this, in theory, could reduce the ability to reinitiate translation of the Cj0044 protein (Kim *et al.*, 2012). Future studies will determine if *Cj0045*, *Cj0044* or both genes are associated with colonization and disease in mice.

Conservation of contingency loci in a large number and proportion of human disease isolates support their role in human colonization and disease

We demonstrated that *Cj0045*, *Cj0685*, *Cj0170*, and *Cj1420* are conserved among 16 human disease isolates including strains *C. jejuni* 11168, 33292, and 81-176 analysed as part of previous work (Kim *et al.*, 2012). Nucleotide sequence analysis of the complete ORF in *Cj0045*, *Cj0685*, *Cj0170*, and *Cj1420* in several randomly chosen human isolates among the 16 isolates analysed to date (Table 23.5) suggest that these genes are nearly 100% conserved. The high degree of sequence conservation in these contingency genes and their presence in the 16 isolates analysed to date strongly support the hypothesis that these five contingency genes may be important for colonization of humans.

Presence of successful alleles of Cj0045 and Cj1420 may predict colonization success in mice and humans

Designation as a successful allele using fragment analysis data is based on two criteria described previously (Kim *et al.*, 2012): (i) the allele is present in > 50% of the population (a predominant allele); and (ii) the allele strongly associates with ability to colonize C57BL/6 IL-10 deficient mice. In contrast, unsuccessful predominant alleles are not associated with colonization. Emergence of a successful *Cj0045* allele in several different strains of *C. jejuni* during passage through a variety of wild and domesticated animals demonstrates that phase variation at contingency locus *Cj0045* to a successful allele is not limited to *C. jejuni* strain 33292 and Ross 308 broilers; phase variation to a successful *Cj0045* allele appears to occur frequently in many different *C. jejuni* strains in wild and domesticated animals.

Successful alleles of *Cj0045* and *Cj1420* were observed in the majority of 16 human disease isolates analysed to date. Nucleotide sequence analysis of a 120 bp region surrounding the homopolymeric tract in *Cj0045* in six *C. jejuni* strains performed in our laboratory and complete sequence data for *Cj0045* in four additional *C. jejuni* strains, conducted by others, demonstrated that this region is identical in sequence to *Cj0045* in strain 33292 in 9 of the 10 *C. jejuni* isolates analysed (data not shown). These data strongly suggest that detection of 9G and/or 10G successful alleles of *Cj0045* in human isolates is significant because these alleles are directly involved in mouse colonization in strain 33292.

Allele diversity may provide a useful indicator of virulence

We hypothesized that *C. jejuni* strain 11168 was a successful colonizer of mice without passage through chickens because it carried a high level of genetic diversity at the 19 contingency loci under analysis in our laboratory. To test this hypothesis, we developed an allele distribution score (ADS) to measure the relative number and distribution of alleles at contingency loci. We utilized fragment analysis data to calculate allele distribution (AD) at 19 contingency loci (Table 23.8). We then generated an allele distribution score (ADS), which represents the sum of ADs for all 19 loci. We demonstrated that as ADS increases, one observes an increase in the number of alleles at contingency loci and a more even distribution of these alleles (Artymovich *et al.*, 2013). Based on this observation, one could reasonably argue that in a comparison of two *C. jejuni* strains, a strain with a higher ADS exhibits a higher level of genetic diversity than the strain with a lower ADS.

To illustrate this concept, we calculate ADS for a hypothetical strain carrying five contingency loci, all with one allele. AD for each gene would be calculated in the same way: $AD = 1/1+1 = 0.5$ [here the decimal fractions for high and low allele frequency would be 1.0 (100%) and 0.0 (0%)]. ADS is the sum of AD at all five genes, so $ADS = 0.5 + 0.5 + 0.5 + 0.5 + 0.5 = 2.5$. For a hypothetical strain carrying 4 loci with one allele and one locus with five alleles (frequencies of 50%, 30%, 8%,

Table 23.8 Calculation of allele distribution (AD) in *C. jejuni* strains grown under standard laboratory culture conditions in inoculum populations (CI)

Strain	11168			KanR4			CamR2		
Locus	Alleles X	Freq. Dif. Y	AD	Alleles X	Freq. Dif. Y	AD	Alleles X	Freq. Dif. Y	AD
0031–32	3	0.58	1.9	3	0.73	1.7	2	0.44	1.4
0045	6	0.39	4.3	4	0.83	2.2	4	0.75	2.3
170–171	3	0.51	2.0	3	0.84	1.6	3	0.84	1.6
0685	2	0.58	1.3	2	0.90	1.1	2	0.94	1.0
1139	3	0.59	1.9	2	0.86	1.1	3	0.92	1.6
1144–45	3	0.43	2.1	2	0.50	1.3	3	0.71	1.8
1305 [a]	4	0.70	2.4	3	0.53	2.0	3	0.53	2.0
1318 [b]	3	0.72	1.7	3	0.27	2.4	3	0.29	2.4
1321	3	0.61	1.9	3	0.78	1.7	3	0.80	1.7
1325–26	3	0.52	2.0	3	0.80	1.7	3	0.77	1.7
1342	3	0.62	1.9	3	0.86	1.6	3	0.66	1.8
1420	2	0.82	1.1	2	0.92	1.0	1	1.00	0.5
1421	4	0.60	2.5	2	0.92	1.0	3	0.90	1.6
1422	4	0.50	2.7	2	0.72	1.2	3	0.67	1.8
1426	3	0.57	1.9	3	0.75	1.7	1	1.0	0.5
1429	4	0.59	2.5	3	0.72	1.7	3	0.70	1.8
ADS			25.5[c]			34.1 [c] (11.1)[d]			25.0[c]

Strain	33292			81-176		
Locus	Alleles X	Freq. Dif. Y	AD	Alleles X	Freq. Dif. Y	AD
0045	4	0.75	2.3	4	0.63	2.5
0170–171	3	0.87	1.6	3	0.87	1.6
0685	2	0.92	1.0	2	0.92	1.0
1305	3	0.62	1.9	3	0.87	1.6
1420	2	0.96	1.0	2	0.90	1.1
ADS			7.8[d]			7.8[d]

[a] AD based on aggregate of fragment analysis of three nearly identical loci, *Cj1305, 1306, and 1310.*
[b] AD based on aggregate of fragment analysis of two nearly identical loci, *Cj1318, 1335–36.*
[c] Allele distribution score based on 15 individual of small groups of closely related loci.
[d] Allele distribution score (ADS) based on five conserved loci.
Numbers in column 1 represent locus designations for single genes or groups of closely related genes.

7% and 5%), AD = 5/1.45 = 3.45 (for the single locus with five alleles) and 2.0 for the other four loci with a single allele. ADS = 3.45 + 2.0 = 5.45. So for the same five loci, the ADS is more than 2-fold higher in the strain that carries five alleles at one locus. AD is based on two assumptions and places roughly equal weight on these factors. First, AD increases as the number of alleles at a particular locus increases (numerator). Second, AD decreases as the frequency of one allele becomes more predominant relative to other alleles in a population (denominator).

As ADS increases, one observes an increase in the number of alleles at a contingency locus and the frequency of these alleles becomes more evenly distributed. As ADS decreases, one observes fewer alleles at a contingency locus and specific alleles become more predominant. We propose that ADS is directly proportional to the level of genetic diversity in a bacterial population; that is the higher the score, the more even the distribution of contingency gene alleles, and the greater the allele diversity for the contingency loci under analysis.

We calculated ADS for real inoculum populations of *C. jejuni* 11168 and two genetically marked strains (CamR2 and KanR4) derived from 11168 grown under standard culture conditions (Table 23.8; Artymovich *et al.*, 2013). ADS for 11168 was 34.1 while ADS for CamR2 and KanR4 were 25 and 25.5, respectively (up to a 27% reduction in allele distribution score. Passage of 11168, CamR2 and KanR4 through the chicken reservoir generated no significant change in ADS for strain 11168 but ADS increased significantly in CamR2 (15%) and KanR4 (11%) for successful populations that subsequently colonized mice (Kim *et al.*, 2012).

We also calculated ADS for human disease isolates 81-176 and 33292, however, we restricted the analysis to only 5 contingency loci (*Cj0045*, *Cj0170*, *Cj0685*, *Cj1305/06/10*, and *Cj1420*) because they are the only genes sufficiently conserved in these 3 human disease isolates to allow use of primers designed for strain *C. jejuni* NCTC11168 (Table 23.8). The ADS for strain 11168 was 11.1 while ADS for both 81-176 and 33292 was 7.8 (ADS up to 30% lower than 11168) suggesting that strain 11168 exhibits a higher level of genetic diversity at the 5 conserved contingency loci under analysis. The observed differences in ADS associated closely with differences in ability of populations of 11168, 33292, and 81-176 to colonize C57BL/6 IL-10-deficient mice (Kim *et al.*, 2012). It is important to note that we developed ADS to measure allele distribution (genetic diversity) with the same limited number of contingency loci in *C. jejuni* populations. In this specific application, ADS performed well. ADS would not be appropriate to conduct in depth analysis of genetic diversity at all loci in *C. jejuni* – this type of analysis would require more sophisticated tools. Our data suggest that analysis of genetic diversity in conserved contingency loci (using ADS) may provide a second tool (in addition to genotype at highly conserved contingency loci) for predicting ability of *C. jejuni* strains to colonize mice and possibly humans.

How do successful *C. jejuni* populations emerge in the chicken GI tract?

Taken together, our data strongly suggest that the chicken reservoir plays an important role in human disease by promoting emergence of successful *C. jejuni* sub-populations in strains with low levels of genetic diversity. We present the following model to explain the emergence of successful *C. jejuni* subpopulations in the chicken reservoir. A high level of 'innate' genetic diversity at specific conserved contingency loci in *C. jejuni* 11168 assures that bacterial populations of this strain consist of many subpopulations of cells that carry a wide variety of surface structure phenotypes and consequently a wide range of antigenic types. These successful sub-populations for strain 11168 already exist in the inoculum (CI populations) at sufficient levels to colonize mice but they may also be enriched by passage through chickens. For strains with high innate genetic diversity like 11168, both CI and chicken passed (CP) populations are predicted to be successful colonizers of mice. In contrast, the level of genetic diversity at select contingency loci is significantly lower in CI populations of *C. jejuni* strains like CamR2, KanR4, 33292, and 81-176. Even though successful subpopulations may be present in CI populations, the number of individual cells in these subpopulations may be insufficient to assure high frequency colonization of mice. However, passage through the chicken reservoir can enrich these low abundance populations or enable the emergence of new successful sub-populations. These new successful subpopulations carry an appropriate array of alleles at contingency loci that alter surface structure or regulate the expression of several virulence related genes. For *C. jejuni* strains with lower initial genetic diversity (like 33292, CamR2, and KanR4), the model predicts low-frequency colonization by CI populations with low GDI. However, colonization can be enhanced by chicken passage due to selection for appropriate shifts in allele frequency (phase variation) at specific contingency loci and this enables the successful population to colonize mice at high frequency in agreement with our experimental data. Future work will focus on understanding the mechanisms by which the chicken and mouse GI tract select for specific subpopulations of *C. jejuni*.

Focus of future work

Mutation rate versus natural selection

Our current hypothesis is that mutation frequency within homopolymeric tracts in *C. jejuni* is independent of growth conditions and that selective pressure in the growth environment determines genotype at contingency loci. As cells that carry unsuccessful alleles at one or more contingency loci arise in a population due to random mutation, these cells do not compete effectively and are eliminated from the population while cells that carry successful alleles compete and thrive. This results in enrichment of successful alleles in the population. However, our experience in experiments CM1 and CM2 tells us that enrichment for successful alleles only occurred in about 50% of the *C. jejuni* populations passaged through the chicken GI tract and this was observed in three different *C. jejuni* strains. These data suggest that the GI tract in individual chickens may vary sufficiently so that different groups of birds may exert a differential influence on genotype at contingency loci.

Is a successful allele at a particular contingency locus strain specific?

We demonstrated that a successful allele at contingency locus *Cj0170* in *C. jejuni* strain KanR4 promotes colonization and disease in the mouse model. We must now determine if a successful allele at contingency locus *Cj0170* also promotes colonization in other *C. jejuni* strains that do not colonize mice or if this is this a strain-specific event.

Identification of targets to block human illness

We believe that study of conserved contingency loci is a logical approach to identify effective cellular targets to prevent human disease. We believe this approach is important for at least two reasons. First, since these conserved loci are present in most *C. jejuni* strains, treatments based on these conserved targets should be effective against most strains of *C. jejuni* thereby increasing the treatment efficacy. Second, because specific alleles of these conserved contingency loci are enriched in the disease reservoir, proteins encoded by specific alleles should be present on most cells in the population. This also should increase treatment efficacy. Using our knowledge of conserved contingency loci in *C. jejuni* and their role in disease, our long term goal is to develop vaccines targeted to proteins encoded by the conserved contingency loci that would be effective in chickens. In support of this approach, many of the proteins encoded by contingency loci are expected to localize at or near the cell surface providing better access to antibodies raised to a vaccine. We also wish to understand the specific mechanisms by which these proteins encoded by contingency loci contribute to disease so that pharmaceuticals can be developed specifically to block their activity in the human host.

Conclusions

We utilize a fundamentally new approach to identify key virulence factors in *C. jejuni* and to understand how they contribute to disease. We monitor the adaptation of *C. jejuni* populations within the chicken reservoir by tracking rapid shifts in allele frequency at contingency loci. Our data make several significant contributions to our understanding of the role of contingency loci in virulence (Artymovich *et al.*, 2013; Kim *et al.*, 2012; Wilson *et al.*, 2010). First, passage of *C. jejuni* through the chicken GI tract can promote emergence of successful populations with greatly enhanced virulence in mice. Second, mutations in homopolymeric tracts and natural selection in the chicken GI tract enrich successful alleles at contingency loci *Cj0045, Cj1420, and Cj0170*; a shift to a predominant (occurs in > 50% of the population) successful allele for a single contingency locus can generate a successful population. Third, natural selection during mouse colonization further enriches successful alleles in successful *C. jejuni* populations. Fourth, single colony isolates that carry successful alleles of *Cj0045* and *Cj1420* or *Cj0170* colonize and cause disease symptoms in mice; single colony isolates that carry unsuccessful alleles at these loci either do not colonize, or they experience a rapid shift to a successful allele and then colonize. Fifth, predominant successful alleles of *Cj1420* and *Cj0045/44* were detected in 14 of 16 and 12 of 16 randomly chosen *C. jejuni* human disease isolates, respectively, strongly suggesting that natural selection for successful alleles also occurs in the human GI tract. This series of discoveries advanced our understanding of how the chicken reservoir and specific *C. jejuni* contingency loci promote human illness. The chicken reservoir clearly provides a favourable growth environment for *C. jejuni* and a population increase in the reservoir increases the probability that humans will be exposed to this pathogen upon consumption of poultry products. Furthermore, *C. jejuni* populations adapt genetically in the GI tract of chickens. Based on our studies in the mouse model described here, we propose that chicken passage promotes emergence of *C. jejuni* populations that carry an appropriate mix of successful alleles at 4 highly conserved contingency loci and are, therefore, more likely to colonize and cause disease in humans. Also, 4 of the 19 contingency loci we analyse routinely are highly conserved in 16 *C. jejuni* human disease isolates analysed to date suggesting that they are among the specific loci involved in successful colonization and disease in humans. Our recent data demonstrate that *Cj0170* regulates motility and is directly associated with virulence in mice while *Cj0045* and *Cj1420* are strongly associated with colonization and disease in the mouse model. As the next logical step in this line of research, we will identify the function of *Cj0170, Cj0045,* and *Cj0044* proteins, design specific inhibitors of their activity (e.g. polyclonal antibodies if they are cell surface proteins), and then test the efficacy of these inhibitors in preventing disease in mice. We will use a similar process to identify key virulence factors in humans and to design inhibitors that block their activity and disease. Based on this work, we expect to generate effective pharmaceuticals or vaccines that block activity of proteins encoded by successful alleles of highly conserved contingency genes and prevent disease in humans.

References

Artymovich, K.A., Kim, J.S., Linz, J.E., Hall, D.F., Kelley, L.E., Kalbach, H.L., Kathariou, S., Gaymer, J., and Paschke, B. (2013). A successful allele of *Campylobacter jejuni* contingency gene *Cj0170* regulates motility: successful alleles of *Cj0045* occur at high frequency in human and animal isolates. Food Microbiol. *34*, 425–430.

Ashgar,S.S., Oldfield,N.J., Wooldridge, K.G., Jones, M.A., Irving, G.J., Turner, D.P., and a'Aldeen, D.A. (2007). CapA, an autotransporter protein of *Campylobacter jejuni*, mediates association with human epithelial cells and colonization of the chicken gut. J. Bacteriol. *189*, 1856–1865.

Balbontin, R., Rowley, G., Pucciarelli, M.G., Lopez-Garrido, J., Wormstone, Y., Lucchini, S., Garcia del Portillo, F., Hinton, J.C.D., and Casadesus, J. (2006). DNA adenine methylation regulates virulence gene expression in *Salmonella enterica* Serovar Typhimurium. J. Bacteriol. *188*, 8160–8168.

Batz, M.B., Hoffmann, D., and Morris, J.G. (2011). Report. Ranking the risks: the 10 pathogen–food combinations with the greatest burden on Public Health. University of Florida, Emerging Pathogens Institute.

Chollangi, S., Thompson, J.W., Ruiz, J.C., Gardner, K.H., And Bruick, R.K. (2012). Hemerythrin-like domain with F-box and leucine-rich repeat protein 5 (FBXL5) communicates cellular iron and oxygen availability by distinct mechanisms. J. Biol. Chem. *28*, 23710–23717.

Dasti, J.A., Tareen, A.M., Lugert, R., Zautner, A.E., and Gros, U. (2010). *Campylobacter jejuni*: A brief overview on pathogenicity-associated factors and disease-mediating mechanisms. Int. J. Med. Microbiol. *300*, 205–211.

Ewing, C.P., Andreishcheva, E., and Guerry, P. (2009). Functional characterization of flagellin glycosylation in *Campylobacter jejuni* 81-176. J. Bacteriol. *191*, 7086–7093.

Falker, S., Schmidt, M.A., and Heusipp, G. (2005). DNA methylation in *Yersinia enterocolitica*: role of the DNA adenine methyltransferase in mismatch repair and regulation of virulence factors. Microbiology *151*, 2291–2299.

French, C.E., Bell, J.M.L., and Ward, F.B. (2007). Diversity and distribution of hemerythrin-like proteins in prokaryotes. FEMS Microbiol. Lett. *279*, 131–145.

Guerry, P. (2007). *Campylobacter* flagella: not just for motility. Trends Microbiol. *15*, 456–461.

Guerry, P., Ewing, C.P., Schirm, M., Lorenzo, M., Kelly, J., Pattarini, D., Majam, G., Thibault, P., and Logan, S. (2006). Changes in flagellin glycosylation affect *Campylobacter* autoagglutination and virulence. Mol. Microbiol. *60*, 299–311.

Guerry, P., and Szymanski, C.M. (2008). *Campylobacter* sugars sticking out. Trends Microbiol. *16*, 428–435.

Guerry, P., Szymanski, C.M., Prendergast, M.M., Hickey, T.E., Ewing, C.P., Pattarini, D.L., and Moran, A.P. (2002). Phase variation of *Campylobacter jejuni* 81-176 lipooligosaccharide affects ganglioside mimicry and invasiveness *in vitro*. Infect. Immun. *70*, 787–793.

Gundogdu, O., Bentley, S.B., Holden, M.T., Parkhill, J., Dorrell, N., and Wren, B.W. (2007). Re-annotation of the *Campyloabacter jejuni* 11168 genome sequence. BMC Genomics *12*, 162.

Hendrixson, D.R. (2006). A phase-variable mechanism controlling the *Campylobacter jejuni* FlgR response regulator influences commensalism. Mol. Microbiol. *61*, 1646–1659.

Howard, S.L., Jagannathan, A., Soo, E.C., Hui, J.P., Aubry, A.J., Ahmed, I., Karlyshev, A.V., Kelly, J.F., Jones, M.A., Stevens, M.P., Logan, S., and Wren, B.W. (2009). *Campylobacter jejuni* glycosylation island important in cell charge, legionimic acid biosynthesis, and colonization of chickens. Infect. Immun. *25*, 2544–2556.

Humphrey, T., O'Brien, S., and Madsen, M. (2007). Campylobacters as zoonotic pathogens: a food production perspective. Int. J. Food Microbiol. *117*, 237–257.

Kakuda, T., and DiRita, V.J. (2006). Cj1496c encodes a *Campylobacter jejuni* glycoprotein that influences invasion of human epithelial cells and colonization of the chick gastrointestinal tract. Infect. Immun. *74*, 4715–4723.

Kim, J.S., Artymovich, K.A., Hakk, D.F., Smith, E.J., Fulton, R., Bell, J., Dybas, L., Mansfield, L.S., Tempelman, R., and Linz, J.E. (2012). Passage of *Campylobacter jejuni* though the chicken reservoir or mice promotes phase variation in contingency genes Cj0045 and Cj0170 that strongly associates with colonization and disease in a mouse model. Microbiology *158*, 1304–1316.

Linton, D., Gilbert, M., Hitchen, P.G., Dell, A., Morris, H.R., Wakarchuk, W.W., Gregson, N.A., and Wren, B.W. (2000b). Phase variation of a beta-1,3 galactosyltransferase involved in generation of the ganglioside GM1-like lipo-oligosaccharide of *Campylobacter jejuni*. Mol. Microbiol. *37*, 501–514.

McNally, D.J., Lamoureux, M.P., Karlyshev, A.V., Fiori, L.M., Li, J., Thacker, G., Coleman, R.A., Khieu, N.H., Wren, B.W., Brisson, J.R., Jarrell, H.C., and Szymanski, C.M. (2007). Commonality and biosynthesis of the O-methyl phosphoramidate capsule modification in *Campylobacter jejuni*. J. Biol. Chem. *282*, 28566–28576.

Mansfield, L.S., Bell, J.A., Wilson, D.L., Murphy, A.J., Elsheikha, H.M., Rathinam, V.A., Fierro, B.R., Linz, J.E., and Young, V.B. (2007). C57BL/6 and congenic interleukin-10-deficient mice can serve as models of *Campylobacter jejuni* colonization and enteritis. Infect. Immun. *75*, 1099–1115.

Oshima, T., Wada, C., Kawagoe, Y., Ara, T., Maeda, M., Masuda, Y., Hiraga, S., and Mori, H.. (2002). Genome-wide analysis of deoxyadenosine methyltransferase mediated control of gene expression in *Escherichia coli*. Mol. Microbiol. *45*, 673–695.

Parkhill, J., Wren, B.W., Mungall, K., Ketley, J.M., Churcher, C., Basham, D., Chillingworth, T., Davies, R.M., Feltwell, T., Holroyd, S., *et al.* (2000). The genome sequence of the food-borne pathogen *Campylobacter jejuni* reveals hypervariable sequences. Nature *403*, 665–668.

Pei, Z., and Blaser, M.J. (1993). PEB1, the major cell-binding factor of *Campylobacter jejuni*, is a homolog of the binding component in Gram-negative nutrient transport systems. J. Biol. Chem. *268*, 18717–18725.

Poli, V.F.S., Thorsen, L., Olesen, I., Wik, M.T., and Jespersen, L. (2012). Differentiation of the virulence potential of *Campylobacter jejuni* strains by use of gene transcription analysis and a Caco-2 assay. Intl. J. Food Microbiol. *155*, 60–68.

Prendergast, M.M., Tribble, D.R., Baqar, S., Scott, D.A., Ferris, J.A., Walker, R.I., and Moran, A.P. (2004). *In vivo* phase variation and serologic response to lipooligosaccharide of *Campylobacter jejuni* in experimental human infection. Infect. Immun. *72*, 916–922.

Snelling, W.J., Matsuda, M., Moore, J.E., and Dooley, J.S. (2005). *Campylobacter jejuni*. Lett. Appl. Microbiol. *41*, 297–302.

Srikhanta, Y.N., Fox, K.L., and Jennings, M.P. (2010). The phase varion: phase variation of type II DNA methyltransferases controls coordinated switching in multiple genes. Nature Rev. Microbiol. *8*, 196–206.

Wassenaar, T.M., Wagenaar, J.A., Rigter, A., Fearnley, C., Newell, D.G., and Duim, B. (2002). Homonucleotide stretches in chromosomal DNA of *Campylobacter jejuni* display high frequency polymorphism as detected by direct PCR analysis. FEMS Microbiol. Lett. *212*, 77–85.

Wilson, D., Rathinam, V.A.K., Qi, W., Wick, L.M., Landgraf, J., Bell, J., Plavonich-Jones, A., Parrish, J., Finley, R.L., Mansfield, L.S., and Linz, J. (2010). Genetic diversity in *Campylobacter jejuni* is associated with differential colonization of broiler chickens and C57BL/6J IL10- deficient mice. Microbiology *156*, 2046–2057.

Xiang, S.L., Zhong, M., Cai, F.C., Deng, B., and Zhang, X.P. (2006). The sialic acid residue is a crucial component of *C. jejuni* lipooligosaccharide ganglioside mimicry in the induction Guillain-Barré syndrome. J. Neuroimmunol. *174*, 126–132.

Yao, R., Burr, D.H., and Guerry, P. (1997). CheY-mediated modulation of *Campylobacter jejuni* virulence. Mol. Microbiol. *23*, 1021–1031.

Young, K.T., Davis, L.M., and Dirita, V.J. (2007). *Campylobacter jejuni*: molecular biology and pathogenesis. Nat. Rev. Microbiol. *5*, 665–679.

The Long Road Ahead: Unravelling the Immune Response to *Campylobacter* in the Chicken 24

Paul Wigley and Suzanne Humphrey

Abstract

Chicken is the most important reservoir for *Campylobacter*, yet relatively little is known about its interaction with the chicken immune system. *Campylobacter* infection of the chicken gastro-intestinal tract elicits an initial inflammatory response, but this differs in nature and is lower in magnitude than the well-understood response to *Salmonella*. *Campylobacter* is recognized by the host via a panel of Toll-like receptors (TLRs) notably TLR-4 and TLR-21. However, the response result is little pathology leading several observers to consider there is a tolerogenic response in the chicken. *Campylobacter* may also be invasive leading to liver infection and there is clear evidence of activation of adaptive immunity from recent transcriptional studies. The nature and role of adaptive responses is poorly understood, though maternally derived antibody is considered to protect against early colonization. This chapter reviews what is known of the immune response to *Campylobacter* in the chicken and the prospects of control through vaccination or breeding for genetic resistance. We will also focus on areas where future research can begin to address the considerable gaps in our knowledge.

Introduction

Chicken is a major vehicle for foodborne *Campylobacter* infection of humans. Therefore, control – for example through vaccination within the avian host – is seen as a potentially useful way of limiting human campylobacteriosis (de Zoete *et al.*, 2007). However, our understanding of the biology of *Campylobacter* in chickens is surprisingly limited. Many questions remain about the immune response to *Campylobacter* in the chicken, and the role this plays in colonization, pathogenesis and ultimately clearance of infection. Furthermore, the majority of studies use laboratory specified pathogen free chickens with high challenge doses that have only limited relevance to infection and immunity in commercial broiler chicken flocks where *Campylobacter* needs to be controlled. In this chapter we review current knowledge of the immune response to *Campylobacter* and the gaps in the knowledge that need to be addressed by future research.

The chicken immune system

In broad terms the avian and mammalian immune systems have much in common, though there are key differences. The immune system of birds and mammals perform the same function, but the avian system is smaller and simpler. The major receptors such as MHC and T cell receptors show less polymorphism in birds including chicken, there are only three immunoglobulin types and no IgG subclasses. However, the more compact immune system of birds causes less genetic stress and has a lower physiological burden to maintain compared to the larger and more complex mammalian system. Nevertheless, despite 200 million years of evolutionary divergence both systems ultimately function in similar ways.

Like mammals, pattern recognition of pathogens through receptors such as Toll-like receptors (TLRs) are important in innate immune activation. As in humans birds have 10 known TLRs, though only five are common to both birds and mammals (Temperley *et al.*, 2008). Key to recognition of Gram-negative bacterial pathogens in the chicken are TLR-2 that recognize peptidoglycan, TLR-4 that binds lipolysaccharide, TLR-5 that recognizes flagellin and TLR-21 which recognizes unmethylated CpG DNA commonly found in bacteria. In mammals CpG motifs are recognized by TLR-9 that is absent in the chicken (Keestra *et al.*, 2010). As in mammals biding of ligand to receptor leads to signalling via Myd88-dependent pathways that leads to expression of a range of cytokines and chemokines that attract and activate leucocytes. Perhaps the key leucocytes in the avian innate system are the heterophil, a polymorphonuclear cell that is broadly equivalent to the mammalian neutrophil (Farnell *et al.*, 2003), and the macrophage that shows similar biology to its mammalian equivalents (He *et al.*, 2011). Chicken thrombocytes also show phagocytic and antibacterial activity in addition to their main role in haemostasis (St Paul *et al.*, 2012; Wigley *et al.*, 1999). Experimental infection with *Salmonella* Typhimurium or Enteritidis illustrates the innate response to enteric infection in the chicken. Invasion of the intestinal epithelium leads to recruitment of heterophils and macrophages to the gut following signalling via pro-inflammatory cytokines such as interleukin 6 (IL-6) and chemokines such as CXCL1, -2 and -4 (Withanage *et al.*, 2004, 2005a). Furthermore LPS, flagellin and CpG oligonucleotides have been shown to stimulate both macrophages and heterophils *in vitro* (Iqbal *et al.*, 2005; Temperley *et al.*, 2008; Xie *et al.*, 2003). The result of this is an inflammatory response that largely limits infection to the gastrointestinal tract, though host adapted serovars of *Salmonella* such as *S. Pullorum* that lack flagella, are able to invade without inflammation (Chappell *et al.*, 2009). The chicken also has a larger population of γδ T lymphocytes than mammals (Berndt and Methner, 2001). These cells are important effectors of innate immunity at mucosal surfaces. Unlike conventional T lymphocytes they do not require prior antigenic exposure and are often regarded as a bridge between innate and adaptive systems. γδ T lymphocytes play a vital role in the initiation of adaptive responses to *Salmonella* infection (Berndt *et al.*, 2006; Pieper *et al.*, 2011).

As in mammals, chickens have two main branches of adaptive immunity namely humoral or antibody responses, initiated through T helper 2 (T_H2) cells, and cellular responses initiated through T helper 1 (T_H1) cells. The role of T_H17 lymphocytes have not been explored in any detail in the chicken but they are important in the response to bacteria in the mammalian intestinal tract. As in mammals, antigens are presented via Class I MHC on a range of cells or via Class II MHC on specialized antigen-presenting cells including dendritic cells and macrophages (Wu and Kaiser, 2011). As discussed previously, there is less polymorphism in both MHC and avian T cell receptors meaning that their antigenic repertoire is considerably smaller than in mammals (Kaufman, 2000). An example of the effect of this can be seen in the close association between MHC haplotypes and resistance to the herpes virus that causes Marek's Disease. Susceptibility is considered to be the result of the lack of polymorphism within the MHC and the fact that birds have only two MHC Class I alleles in each individual compared to six alleles in mammals. Despite this more compact system, chickens are capable of mounting effective antibody and cellular responses to a wide range of pathogens. The best-studied enteric pathogens of the chicken in terms of adaptive immunity are *Salmonella enterica* and the protozoan parasite *Eimeria*. In both cases, infection elicits humoral (antibody) responses and ultimately clearance. Subsequent protection is associated with cellular or T_H1 responses (Babu *et al.*, 2004; Withanage *et al.*, 2005b). Functional studies of the role of antibody in chicken salmonellosis have been possible through use of bursectomy-removal of the bursa of Fabricius, the primary lymphoid organ associated with the development and maturation of B lymphocytes. The most elegant of these studies utilized *in ovo* surgical bursectomy (Beal *et al.*, 2006). This is technically demanding but has no impact on T lymphocyte function, as is the case in chemical or hormonal bursectomy, and does not allow migration of B cells from the bursa when bursectomy is performed post-hatch. The use of bursectomy showed that the absence of antibody had no impact on either clearance or subsequent re-challenge with *S. Typhimurium* indicating that antibody was not required for immune clearance or protection. Furthermore, unpublished studies show that adoptive transfer of splenic lymphocytes partially protects against systemic *Salmonella* infection (Wigley, Powers and Smith, unpublished) and transfer of gut-associated intraepithelial lymphocytes protects against intestinal infection (Powers and Smith, unpublished). Indeed given the long-association between *Salmonella* and chickens, as well as the long history of vaccination in its control, it is the best understood bacterial system in terms of both pathogenesis and immunity. Although this cannot directly inform our understanding of immunity to *Campylobacter*, a very different pathogen, the techniques and approaches may be applied in understanding *Campylobacter* infection and immunity.

The innate immune response to *Campylobacter*

As with any bacterial enteric pathogen the first interaction between *Campylobacter* and the host requires a response to innate challenges. Even before the interaction with cells and receptors associated with innate immunity, *Campylobacter* has to overcome the physical obstacles to infection associated with the chicken gastrointestinal tract. These include the acid environment of the crop, gizzard and proventriculus, intestinal flow and secreted antimicrobial peptides prior to reaching the main sites of colonization in the caeca and distal ileum. Cell-based models show that *Campylobacter* interaction with epithelial cells elicits expression of pro-inflammatory chemokines CXCLi1 and CXCLi2 considered the orthologues of human interleukin 8 (IL-8), though levels of expression are considerably lower than seen in *Salmonella* and the cell lines used were not chicken intestinal cells (Larson *et al.*, 2008; Smith *et al.*, 2005).

In vivo studies also indicate that there is the initiation of an inflammatory response accompanied by a modest influx of heterophils to the intestine (Smith *et al.*, 2008). It is unclear to what extent *Campylobacter* is invasive or acts as an intracellular pathogen, though some invasion has been described in chicken primary epithelial cell models (Byrne *et al.*, 2007), *in vivo* it is considered that *Campylobacter* colonizes the mucin layer of the epithelial surface with little pathology of the intestinal epithelium (Hermans *et al.*, 2012). Chicken mucin prevents invasion of human epithelial cells whereas human mucin does not and it has been postulated that in the chicken mucin prevents invasion (Alemka *et al.*, 2010). The interaction between *Campylobacter* and host pattern recognition receptors is also distinct from that seen with *Salmonella*. Recognition of flagellin via TLR5 is a key process in *Salmonella* infection whereas *Campylobacter* evades recognition through TLR5 by alteration of flagella structure, particularly through glycosylation of flagellin subunits (de Zoete *et al.*, 2010b). However *Campylobacter* is recognized by other chicken TLRs *in vitro* including TLR1, 2, 4 and 21 leading to activation of the NF-κβ pathway and potential initiation of the immune response (de Zoete *et al.*, 2010a; Howard *et al.*, 2009). Further *in vivo* studies also show increased expression of TLR-21 along with TLR-4 early in infection suggesting these perhaps play the key roles in recognition of *Campylobacter*.

Infection also leads to increased CXC chemokine expression but not the pro-inflammatory cytokine IL-6 (Meade *et al.*, 2009; Shaughnessy *et al.*, 2009). The response contrasts with that of *S.* Typhimurium both temporally and in magnitude – in essence the host response is slower and lower. One key feature of *Campylobacter* infection is that the expression of antimicrobial peptides is described including a number of gallinacins, chicken β-defensins, and cathelicidin-2 (Meade *et al.*, 2009; Shaughnessy *et al.*, 2009; van Dijk *et al.*, 2012). This is in contrast to *Salmonella* and is likely to play a key role in evading clearance.

In summary the innate and inflammatory response to *Campylobacter* is considerably lower than that to *Salmonella* in the chicken. This to an extent could be considered tolerogenic allowing asymptomatic colonization of the caeca (Hermans *et al.*, 2012). However, it is wrong to consider this as anergic as *C. jejuni* is clearly recognized by TLR-4 and TLR-21, there is a degree of immune activation that subsequently leads to antibody responses and there are local and systemic changes to phagocyte populations. Furthermore, invasive campylobacteriosis of the chicken is an increasing occurrence leading to chicken liver becoming a major source of human infection. Liver infection in chickens may be asymptomatic (Cox *et al.*, 2005, 2009) or associated with vibrionic hepatitis (Jennings *et al.*, 2011). Clearly, invasive infection is an indication that *C. jejuni* is not merely a commensal under certain conditions, though as yet it is unclear whether this is related to the bacterial strain, changes in host immunity or indeed a combination of these.

Adaptive immunity to *Campylobacter*

Whilst we have at least a basic understanding of interaction between *Campylobacter* and the innate immune system, our understanding of the adaptive and cellular responses is rudimentary. It is clear that specific systemic IgG (or IgY) and mucosal IgA antibody responses are elicited by infection, or immunization (Cawthraw *et al.*, 1994; Widders *et al.*, 1996), and that these responses are directed to major antigens – particularly flagellin (de Zoete *et al.*, 2007; Widders *et al.*, 1998) and to other surface-expressed antigens such as LPS or LOS. In the absence of functional studies it is not clear to what extent antibody is protective, though some indication may come from the 'lag phase' – the period when *C. jejuni* rarely colonizes birds in production prior to 21 days of age. The apparent protection in chicks is considered to be driven by the presence of specific maternally derived antibodies that decline over the first three weeks of life (Cawthraw and Newell, 2010; Sahin *et al.*, 2003). This indicates that antibody is likely to play a protective role to *C. jejuni* infection in the chicken. However it should be considered that in the absence of truly functional studies these observations should be considered as an *association* between maternal antibody and protection and that other changes such as those to the microbiota occur in the first three weeks of life that may impact on susceptibility.

Even less is known about cellular responses. A recent study of invasive campylobacteriosis showed that liver infection elicits a T cell response (Jennings *et al.*, 2011). Perhaps the best indication we have of the nature of the adaptive immune response is coming from systems-based and transcriptional studies of the host response based on microarrays or RNA Sequencing. The most comprehensive of these studies has utilized mRNA sequencing in the intestine up to 48 h post challenge (Connell *et al.*, 2012). Perhaps unsurprisingly, there is increased expression of genes associated with innate immunity including TLRs, but there is also a considerable increase in expression of immunoglobulin and B lymphocyte-associated genes, genes of the T-cell receptor complex and key cytokines and their receptors associated with activation of adaptive immunity. The power of such approaches may improve our understanding and have potential to direct future immunological studies.

Regulation of the immune response

An area of huge importance is the interaction between the immune system, the commensal microbiota and pathogens in gut homeostasis. As *Campylobacter* persists for a considerable time in the gut, it is likely to be constantly re-stimulating the immune system. Without regulation of inflammatory innate responses, damage to the gut and enteritis will result. The same can be said to occur for other components of the microbiota. Therefore, such regulation by the immune system, and particularly T regulatory cells (T_H3 or T Regs), is crucial in controlling inflammation in the gut. In broad terms T Regs produce regulatory cytokines such as interleukin 10 (IL-10), that regulate pro-inflammatory innate and T_H1 responses. Little is known about immune regulation in the chicken, though there is evidence from *Salmonella* infection models that they may play a key role in preventing enteritis. *Salmonella* elicits a strong inflammatory response in the intestine. In young chicks this is largely uncontrolled leading to considerable inflammation of the ileum and caeca accompanied by high levels of bacterial invasion (Withanage *et al.*, 2004). In older chicks, however, inflammatory responses are more transient and infection is accompanied by expression of the regulatory cytokines TGF-β4 and IL-10 (Setta *et al.*, 2012; Withanage *et al.*, 2005b). The result would appear to be regulation of inflammation leading to bacterial colonization without enteritis, with gradual clearance by the adaptive immune response. Understanding immune regulation in the *Campylobacter*-infected gut may reveal more information about the factors that influence colonization success and the innate response to infection.

Vaccination to control *Campylobacter*

Vaccination is perhaps the most obvious approach to *Campylobacter* control (de Zoete *et al.*, 2007). However, producing an effective and safe vaccine for use in broiler chickens is more complex than simply producing an antibody response that may or may not be protective. Indeed without understanding the protective response, attempts to produce vaccines are largely empirical. Though such approaches have been successful in control of *Salmonella* in laying hens, the challenge of protecting broilers is much greater so that only broiler breeder flocks are *Salmonella*-vaccinated. The reasons for this include: (i) safety considerations – as there is a risk of the more effective live attenuated vaccines entering the food chain; (ii) practical and economic considerations – as vaccines may be too costly for a production system; (iii) immunological reasons – as it is hard to elicit a protective response to *Salmonella* in young chickens (Beal *et al.*, 2004). Vaccination is more economically viable in laying hens or breeders that may have a production life of up to two years. The short life of the broiler, typically between 5 and 6 weeks, presents particular challenges. Nevertheless if an effective vaccine can be produced for broilers, or indeed breeder birds, it would be valuable to the poultry industry.

A number of approaches using inactivated, subunit and peptide vaccines have been considered and shown to be immunogenic, at least in respect of antibody production (de Zoete *et al.*, 2007; Fernando *et al.*, 2008; Widders *et al.*, 1998). However, none have led to a commercial vaccine. The problem associated with such approaches is clearly illustrated by the use of formalin-killed bacteria as an experimental vaccine delivered subcutaneously with adjuvant. This leads to high levels of IgG antibody but no protection to *Campylobacter* challenge (Okamura *et al.*, 2012). An alternative approach is to use vectored vaccines in systems known to elicit good enteric immunity; *Salmonella Typhimurium* and *Eimeria tenella* (Buckley *et al.*, 2010; Clark *et al.*, 2012). Both approaches use the *C. jejuni* amino acid binding protein CjaA as antigen expressed either within *S. Typhimurium* AroA attenuated vaccine carrier or in an attenuated *Eimeria tenella* laboratory isolate. Both achieved modest success, conferring significant levels of protection in reducing colonization levels in experimental challenge by approximately 1 $\log_{10}$ cfu per gram of caecal content. Clearly such an approach has potential but there is some distance to go before a licensed commercial vaccine is produced and given the use of live genetically modified vectors in the food-chain, considerable regulatory hurdles to face. Nevertheless it illustrates that successful *C. jejuni* vaccines are unlikely to be simple produce and are likely to require novel approaches to developing vaccines, delivery systems and adjuvants to achieve success. It may be that the cost of vaccination is beyond the means of the industry for routine use in broilers and as such may be used in broiler breeders to reduce risk of transmission in hatcheries and to increase levels of potentially protective maternally derived antibody.

Genetic resistance – selective breeding towards *Campylobacter* control

An alternative approach to vaccination or feed modification or probiotics is breeding to increase innate resistance to *Campylobacter* colonization (Kaiser *et al.*, 2009). Such an approach is particularly desirable to the poultry industry as it requires minimum intervention and is cost effective once resistant birds have been bred. Approaches to understand the underlying mechanisms have been made in inbred White Leghorn lines where differences in both the level and length of colonization were found between resistant Line 6 and susceptible Line N chickens (Boyd *et al.*, 2005). Approaches including mapping quantitative trait loci (QTL) and single nucleotide polymorphism (SNP) typing have been applied to identify the loci associated with resistance (Kaiser *et al.*, 2009). In the USA studies have focused on broiler lines showing differences in colonization with *C. jejuni*. Microarray-based transcriptional approaches of host response in spleen and caeca show marked differences in expression of immune-associated genes (Li *et al.*, 2011, 2012). In the caeca there is increased expression in resistant birds of genes associated with activation of innate immunity including the *MAPK* pathway and tumour necrosis factor superfamily receptors (Li *et al.*, 2011), whilst in the spleen there is increased expression of genes associated with lymphocyte activation and induction of antibody production (Li *et al.*, 2012). These findings suggest that resistance is likely to have a strong immunological basis.

Conclusions

Campylobacter infection of the chicken clearly leads to an immune response though beyond descriptions of interaction with the innate immune system it is barely characterized. Infection produces antibody responses, but the only association with protection is from maternally derived antibodies during the 'lag phase' in young chicks. Many studies compare the response between *Campylobacter* and *Salmonella* which tend to indicate that the innate and inflammatory response to *Campylobacter* is of lower magnitude and different to that of *Salmonella*. Whilst this is a logical approach it should be considered that inflammatory invasion is central to *Salmonella* pathogenesis and is not seen with all avian pathogens. Perhaps as a consequence it is considered that the response to *Campylobacter* is tolerogenic to some extent it is when compared to *Salmonella*. Nevertheless there is evidence that *Campylobacter* infection activates both innate and adaptive immune responses.

Understanding immunity is central to developing vaccination as a control strategy in the chicken. Simple vaccines that produce antibody responses alone are not likely to be sufficient and understanding what elements of the immune response are protective is important for rational vaccine design. Novel approaches to vaccine design, adjuvants and delivery systems may all be required. The use of live *Salmonella* or *Eimeria* as a vaccine vectors offer some possibility of protection through there would be considerable legal and logistical issues to overcome. Perhaps an immunological approach to control may not be sufficient and will need to be part of a multi-component approach including selective breeding, improved bird welfare and biosecurity. What is clear is that we have a long way to go in understanding the immunobiology of *Campylobacter* in the chicken.

Future trends – the long road ahead

Although it is frequently carried in the gastrointestinal tract without disease, from a purely immunological standpoint it is difficult to consider *Campylobacter* as a commensal as has been recently suggested (Hermans *et al.*, 2012). Whilst it is true that experimentally *C. jejuni* does not frequently cause enteritis, it can be argued that this is the same for *Salmonella*. In fact it is likely that regulation of the inflammatory response in the chicken gut is key to the absence of enteritis in salmonellosis. However, chickens can and do get diarrhoea following infection in the field and the risk of liver infection and the public health consequences of this are clearly an increasing problem. To what extent this is related to a changing pathogen or to the host is unclear, but given the ever-increasing growth and feed conversion rate (the rate feed is converted to muscle/protein) of the broiler chicken it is likely that this will be offset by changes in immune function. It is clear that poor welfare has an impact on bird health and carriage of foodborne pathogens (Humphrey, 2006). To what extent this is due to immunological changes is a particular challenge to address.

Future studies could aim to address fundamental immunological questions such as: What are the immune mechanisms that underlie *Campylobacter* clearance and protection to subsequent re-infection? How does immune regulation in the gut influence colonization by *C. jejuni* and prevent pathology? Can immunity be harnessed through effective vaccination? To what extent does the modern broiler chicken and intensive production impact on immune function and infection including *Campylobacter*? The approaches used to identify innate responses are by their nature rapid and longer and more complex experiments are required to address these questions. Advances in technology such as DNA and RNA sequencing will allow a better descriptive understanding of host responses and can inform functional studies. Perhaps the first major obstacle is understanding the relative role of humoral and cellular immunity in protection. Well designed experiments in broiler chickens may identify associations between T_H1 and T_H2 responses in clearance and better inform vaccine development. On-going work continues to develop vaccines, breed resistant chickens, investigate the use of probiotics, and develop functional studies using techniques such as adoptive transfer of lymphocytes or bursectomy.

Understanding the role of immune regulation and the influence of factors including chicken stress and welfare on *Campylobacter* colonization remain a considerable challenge. Technologies and reagents to effectively dissect responses at a cellular level in the chicken gut will enhance understanding of regulation but other factors are more complex. There is still much to learn about the immunobiology of avian campylobacteriosis, especially the adaptive immune response, and applying knowledge towards the control of *Campylobacter* in the chicken represents a long road ahead.

Acknowledgements

We wish to thank all members of the Zoonotic Infections in Pigs and Poultry Research Group, University of Liverpool. We acknowledge the support of the Biotechnology and Biological Sciences Research Council, European Union and the poultry industry for their support of *Campylobacter* research in poultry at Liverpool.

References

Alemka, A., Clyne, M., Shanahan, F., Tompkins, T., Corcionivoschi, N., and Bourke, B. (2010). Probiotic colonization of the adherent mucus layer of HT29MTXE12 cells attenuates *Campylobacter jejuni* virulence properties. Infect. Immun. 78, 2812–2822.

Babu, U., Dalloul, R.A., Okamura, M., Lillehoj, H.S., Xie, H., Raybourne, R.B., Gaines, D., and Heckert, R.A. (2004). *Salmonella enteritidis* clearance and immune responses in chickens following *Salmonella* vaccination and challenge. Vet. Immunol. Immunopathol. *101*, 251–257.

Beal, R.K., Wigley, P., Powers, C., Hulme, S.D., Barrow, P.A., and Smith, A.L. (2004). Age at primary infection with *Salmonella enterica* serovar *Typhimurium* in the chicken influences persistence of infection and subsequent immunity to re-challenge. Vet. Immunol. Immunop. *100*, 151–164.

Beal, R.K., Powers, C., Davison, T.F., Barrow, P.A., and Smith, A.L. (2006). Clearance of enteric *Salmonella enterica* serovar *Typhimurium* in chickens is independent of B-cell function. Infect. Immun. *74*, 1442–1444.

Berndt, A., and Methner, U. (2001). Gamma/delta T cell response of chickens after oral administration of attenuated and non-attenuated *Salmonella typhimurium* strains. Vet. Immunol. Immunopathol. *78*, 143–161.

Berndt, A., Pieper, J., and Methner, U. (2006). Circulating gamma delta T cells in response to *Salmonella enterica* serovar *enteritidis* exposure in chickens. Infect. Immun. *74*, 3967–3978.

Boyd, Y., Herbert, E.G., Marston, K.L., Jones, M.A., and Barrow, P.A. (2005). Host genes affect intestinal colonisation of newly hatched chickens by *Campylobacter jejuni*. Immunogenetics *57*, 248–253.

Buckley, A.M., Wang, J., Hudson, D.L., Grant, A.J., Jones, M.A., Maskell, D.J., and Stevens, M.P. (2010). Evaluation of live-attenuated *Salmonella* vaccines expressing *Campylobacter* antigens for control of *C. jejuni* in poultry. Vaccine *28*, 1094–1105.

Byrne, C.M., Clyne, M., and Bourke, B (2007). *Campylobacter jejuni* adhere to and invade chicken intestinal epithelial cells *in vitro*. Microbiology *153*, 561–569.

Cawthraw, S.A., and Newell, D.G. (2010). Investigation of the presence and protective effects of maternal antibodies against *Campylobacter jejuni* in chickens. Avian Dis. *54*, 86–93.

Cawthraw, S., Ayling, R., Nuijten, P., Wassenaar, T., and Newell, D.G. (1994). Isotype, specificity, and kinetics of systemic and mucosal antibodies to *Campylobacter jejuni* antigens, including flagellin, during experimental oral infections of chickens. Avian Dis. *38*, 341–349.

Chappell, L., Kaiser, P., Barrow, P., Jones, M.A., Johnston, C., and Wigley, P. (2009). The immunobiology of avian systemic salmonellosis. Vet. Immunol. Immunopathol. *128*, 53–59.

Clark, J.D., Oakes, R.D., Redhead, K., Crouch, C.F., Francis, M.J., Tomley, F.M., and Blake, D.P. (2012). Eimeria species parasites as novel vaccine delivery vectors: anti-*Campylobacter jejuni* protective immunity induced by Eimeria tenella-delivered CjaA. Vaccine *30*, 2683–2688.

Connell, S., Meade, K.G., Allan, B., Lloyd, A.T., Kenny, E., Cormican, P., Morris, D.W., Bradley, D.G., and O'Farrelly, C. (2012). Avian resistance to *Campylobacter jejuni* colonization is associated with an intestinal immunogene expression signature identified by mRNA sequencing. PLoS One 7, e40409.

Cox, N.A., Hofacre, C.L., Bailey, J.S., Buhr, R.J., Wilson, J.L., Hiett, K.L., Richardson, L.J., Musgrove, M.T., Cosby, D.E., Tankson, J.D., *et al.* (2005). Presence of *Campylobacter jejuni* in various organs one hour, one day, and one week following oral or intracloacal inoculations of broiler chicks. Avian Dis. *49*, 155–158.

Cox, N.A., Richardson, L.J., Buhr, R.J., and Fedorka-Cray, P.J. (2009). *Campylobacter* species occurrence within internal organs and tissues of commercial caged Leghorn laying hens. Poult. Sci. *88*, 2449–2456.

van Dijk, A., Herrebout, M., Tersteeg-Zijderveld, M.H., Tjeerdsma-van Bokhoven, J.L., Bleumink-Pluym, N., Jansman, A.J., Veldhuizen, E.J., and Haagsman, H.P. (2012). *Campylobacter jejuni* is highly susceptible to killing by chicken host defense peptide cathelicidin-2 and suppresses intestinal cathelicidin-2 expression in young broilers. Vet. Microbiol. *160*, 347–354.

Farnell, M.B., He, H., and Kogut, M.H. (2003). Differential activation of signal transduction pathways mediating oxidative burst by chicken heterophils in response to stimulation with lipopolysaccharide and lipoteichoic acid. Inflammation *27*, 225–231.

Fernando, U., Biswas, D., Allan, B., Attah-Poku, S., Willson, P., Valdivieso-Garcia, A., and Potter, A.A. (2008). Serological assessment of synthetic peptides of *Campylobacter jejuni* NCTC11168 FlaA protein using antibodies against multiple serotypes. Med. Microbiol. Immunol. *197*, 45–53.

He, H., Genovese, K.J., and Kogut, M.H. (2011). Modulation of chicken macrophage effector function by T(H)1/T(H)2 cytokines. Cytokine *53*, 363–369.

Hermans, D., Pasmans, F., Heyndrickx, M., Van Immerseel, F., Martel, A., Van Deun, K., and Haesebrouck, F. (2012). A tolerogenic mucosal immune response leads to persistent *Campylobacter jejuni* colonization in the chicken gut. Crit. Rev. Microbiol. *38*, 17–29.

Howard, S.L., Jagannathan, A., Soo, E.C., Hui, J.P., Aubry, A.J., Ahmed, I., Karlyshev, A., Kelly, J.F., Jones, M.A., Stevens, M.P., *et al.* (2009). *Campylobacter jejuni* glycosylation island important in cell charge, legionaminic acid biosynthesis, and colonization of chickens. Infect. Immun. *77*, 2544–2556.

Humphrey, T. (2006). Are happy chickens safer chickens? Poultry welfare and disease susceptibility. Br. Poult. Sci. *47*, 379–391.

Iqbal, M., Philbin, V.J., Withanage, G.S., Wigley, P., Beal, R.K., Goodchild, M.J., Barrow, P., McConnell, I., Maskell, D.J., Young, J., *et al.* (2005). Identification and functional characterization of chicken toll-like receptor 5 reveals a fundamental role in the biology of infection with *Salmonella enterica* serovar *typhimurium*. Infect. Immun. *73*, 2344–2350.

Jennings, J.L., Sait, L.C., Perrett, C.A., Foster, C., Williams, L.K., Humphrey, T.J., and Cogan, T.A. (2011). *Campylobacter jejuni* is associated with, but not sufficient to cause vibrionic hepatitis in chickens. Vet. Microbiol. *149*, 193–199.

Kaiser, P., Howell, M.M., Fife, M., Sadeyen, J.R., Salmon, N., Rothwell, L., Young, J., Poh, T.Y., Stevens, M., Smith, J., *et al.* (2009). Towards the selection of chickens resistant to *Salmonella* and *Campylobacter* infections. Bull. Mem. Acad. R. Med. Belg. *164*, 17–25; discussion 25–16.

Kaufman, J. (2000). The simple chicken major histocompatibility complex: life and death in the face of pathogens and vaccines. Philos. Trans. R. Soc. Lond. B. Biol. Sci. *355*, 1077–1084.

Keestra, A.M., de Zoete, M.R., Bouwman, L.I., and van Putten, J.P. (2010). Chicken TLR21 is an innate CpG DNA receptor distinct from mammalian TLR9. J. Immunol. *185*, 460–467.

Larson, C.L., Shah, D.H., Dhillon, A.S., Call, D.R., Ahn, S., Haldorson, G.J., Davitt, C., and Konkel, M.E. (2008). *Campylobacter jejuni* invade chicken LMH cells inefficiently and stimulate differential expression of the chicken CXCLi1 and CXCLi2 cytokines. Microbiology *154*, 3835–3847.

Li, X.Y., Swaggerty, C.L., Kogut, M.H., Chiang, H.I., Wang, Y., Genovese, K.J., He, H., Pevzner, I.Y., and Zhou, H.J. (2011). Caecal transcriptome analysis of colonized and non-colonized chickens within two genetic lines that differ in caecal colonization by *Campylobacter jejuni*. Anim. Genet. *42*, 491–500.

Li, X., Swaggerty, C.L., Kogut, M.H., Chiang, H.I., Wang, Y., Genovese, K.J., He, H., McCarthy, F.M., Burgess, S.C., Pevzner, I.Y., *et al.* (2012). Systemic response to *Campylobacter jejuni* infection by profiling gene transcription in the spleens of two genetic lines of chickens. Immunogenetics *64*, 59–69.

Meade, K.G., Narciandi, F., Cahalane, S., Reiman, C., Allan, B., and O'Farrelly, C. (2009). Comparative *in vivo* infection models yield insights on early host immune response to *Campylobacter* in chickens. Immunogenetics *61*, 101–110.

Okamura, M., Tominaga, A., Ueda, M., Ohshima, R., Kobayashi, M., Tsukada, M., Yokoyama, E., Takehara, K., Deguchi, K., Honda, T., *et al.* (2012). Irrelevance between the induction of anti-*Campylobacter* humoral response by a bacterin and the lack of protection against homologous challenge in Japanese Jidori chickens. J. Vet. Med. Sci. *74*, 75–78.

Pieper, J., Methner, U., and Berndt, A. (2011). Characterization of avian gammadelta T-cell subsets after *Salmonella enterica* serovar *Typhimurium* infection of chicks. Infect. Immun. *79*, 822–829.

Sahin, O., Luo, N., Huang, S., and Zhang, Q. (2003). Effect of *Campylobacter*-specific maternal antibodies on *Campylobacter jejuni* colonization in young chickens. Appl. Environ. Microbiol. *69*, 5372–5379.

Setta, A.M., Barrow, P.A., Kaiser, P., and Jones, M.A. (2012). Early immune dynamics following infection with *Salmonella enterica* serovars *Enteritidis*, *Infantis*, *Pullorum* and *Gallinarum*: cytokine and chemokine gene expression profile and cellular changes of chicken cecal tonsils. Comp. Immunol. Microbiol. Infect. Dis. *35*, 397–410.

Shaughnessy, R.G., Meade, K.G., Cahalane, S., Allan, B., Reiman, C., Callanan, J.J., and O'Farrelly, C. (2009). Innate immune gene expression differentiates the early avian intestinal response between *Salmonella* and *Campylobacter*. Vet. Immunol. Immunopathol. *132*, 191–198.

Smith, C.K., Kaiser, P., Rothwell, L., Humphrey, T., Barrow, P.A., and Jones, M.A. (2005). *Campylobacter jejuni*-induced cytokine responses in avian cells. Infect. Immun. *73*, 2094–2100.

Smith, C.K., Abuoun, M., Cawthraw, S.A., Humphrey, T.J., Rothwell, L., Kaiser, P., Barrow, P.A., and Jones, M.A. (2008). *Campylobacter* colonization of the chicken induces a proinflammatory response in mucosal tissues. FEMS Immunol. Med. Microbiol *54*, 114–121.

St Paul, M., Paolucci, S., Barjesteh, N., Wood, R.D., Schat, K.A., and Sharif, S. (2012). Characterization of chicken thrombocyte responses to Toll-like receptor ligands. PLoS One *7*, e43381.

Temperley, N.D., Berlin, S., Paton, I.R., Griffin, D.K., and Burt, D.W. (2008). Evolution of the chicken Toll-like receptor gene family: a story of gene gain and gene loss. BMC Genomics *9*, 62.

Widders, P.R., Perry, R., Muir, W.I., Husband, A.J., and Long, K.A. (1996). Immunisation of chickens to reduce intestinal colonisation with *Campylobacter jejuni*. Br. Poult. Sci. *37*, 765–778.

Widders, P.R., Thomas, L.M., Long, K.A., Tokhi, M.A., Panaccio, M., and Apos, E. (1998). The specificity of antibody in chickens immunised to reduce intestinal colonisation with *Campylobacter jejuni*. Vet. Microbiol. *64*, 39–50.

Wigley, P., Hulme, S.D., and Barrow, P.A. (1999). Phagocytic and oxidative burst activity of chicken thrombocytes to *Salmonella*, *Escherichia coli* and other bacteria. Avian Pathol. *28*, 567–572.

Withanage, G.S.K., Kaiser, P., Wigley, P., Powers, C., Mastroeni, P., Brooks, H., Barrow, P., Smith, A., Maskell, D., and McConnell, I. (2004). Rapid expression of chemokines and proinflammatory cytokines in newly hatched chickens infected with *Salmonella enterica* serovar *typhimurium*. Infect. Immun. *72*, 2152–2159.

Withanage, G.S., Wigley, P., Kaiser, P., Mastroeni, P., Brooks, H., Powers, C., Beal, R., Barrow, P., Maskell, D., and McConnell, I. (2005a). Cytokine and chemokine responses associated with clearance of a primary *Salmonella enterica* serovar *Typhimurium* infection in the chicken and in protective immunity to rechallenge. Infect. Immun. *73*, 5173–5182.

Withanage, G.S.K., Wigley, P., Kaiser, P., Mastroeni, P., Brooks, H., Powers, C., Beal, R., Barrow, P., Maskell, D., and McConnell, I. (2005b). Cytokine and chemokine responses associated with clearance of a primary *Salmonella enterica* serovar *Typhimurium* infection in the chicken and in protective immunity to rechallenge. Infect. Immun. *73*, 5173–5182.

Wu, Z., and Kaiser, P. (2011). Antigen presenting cells in a non-mammalian model system, the chicken. Immunobiology *216*, 1177–1183.

Xie, H., Raybourne, R.B., Babu, U.S., Lillehoj, H.S., and Heckert, R.A. (2003). CpG-induced immunomodulation and intracellular bacterial killing in a chicken macrophage cell line. Dev. Comp. Immunol. *27*, 823–834.

de Zoete, M.R., van Putten, J.P., and Wagenaar, J.A. (2007). Vaccination of chickens against *Campylobacter*. Vaccine *25*, 5548–5557.

de Zoete, M.R., Keestra, A.M., Roszczenko, P., and van Putten, J.P. (2010a). Activation of human and chicken toll-like receptors by *Campylobacter* spp. Infect. Immun. *78*, 1229–1238.

de Zoete, M.R., Keestra, A.M., Wagenaar, J.A., and van Putten, J.P. (2010b). Reconstitution of a functional Toll-like receptor 5 binding site in *Campylobacter jejuni* flagellin. J. Biol. Chem. *285*, 12149–12158.

Index

A

Adaptive immunity *see* Host immunity
Adhesins *see* Adhesion
Adhesion
- adhesins 63, 78, 90, 101–102, 107–109, 162–163, 177–181, 191, 196, 314
- biofilm formation 101, 136, 150–151, 159, 170, 177–183
- cell adhesion 63, 101, 163

Aerotolerance 143, 145, 147, 150–152
Antigenic variation 25, 161

B

Bacterial cell envelope (capsule)
- capsular polysaccharide 59, 61–62, 94, 102, 125, 180, 324
- capsule biosynthesis 45, 322, 325
- extracellular matrix 160, 177, 180
- glycocalyx 75, 83
- lipooligosaccharide (LOS) 25, 45, 59, 61, 77, 94, 125, 180, 198–199, 324–325

Bacterial conjugation 4, 9
Biofilm formation *see* Adhesion
Biosecurity 213, 288, 293–295, 297, 299, 301, 352
Birth-and-death model of evolution 34
Broiler chicken 94, 213–214, 216, 279, 287–289, 316, 323, 338–339, 347, 351–353

C

Campylobacter coli 3–6, 13, 16, 33, 35–36, 38, 44–47, 50, 90, 100–101, 112, 116–118, 121, 126–127, 143, 147–148, 152, 196, 213–217, 223–227, 229, 256, 258–259, 261–262, 266, 279, 289, 294, 303, 313
- *C. coli clade 1* 5

C. curvus 44–47, 49, 50, 225
C. fetus 43–47, 100, 223, 225–228
C. helveticus 223, 225–228
C. hominis 44–48
C. insulaenigrae 223, 225–226, 228
C. lanienae 223, 225–226, 228
C. lari 44–47, 50, 105, 223, 225–228, 266–267
C. rectus 44–45, 180
C. showae 44–45
C. sputorum 225
C. upsaliensis 43–46, 50, 223, 225–228
Caecum *see* Gastrointestinal tract colonization
Capsule *see* Bacterial cell envelope
Catalase 126–128, 130, 133, 146–149, 153, 182, 197
Cattle colonization 17, 35, 43, 112, 114, 152, 222, 228, 231–234, 243–244, 246–247, 250, 280–281, 292, 313
Cell invasion 84–85, 99, 101, 149, 153, 163, 183, 193
Chemotaxis 50, 90–92, 95, 102, 104, 179, 336
Chicken immune system *see* Host immunity
Clonal frame 10–13, 16, 20
Colon *see* Gastrointestinal tract colonization
Commensalism 83–85, 94–95, 144, 313, 315
Concerted evolution 33–38, 48–49
Contingency loci *see* Hypervariable regions
Core genome 10, 46–47
CRISPR-Cas system 46, 48, 56
Cytochrome C peroxidases 128, 137
Cytolethal distending toxin 77, 93, 102, 162, 194, 196, 199, 316

D

Disease in chickens 311, 315
DNA repair 26, 38, 43, 49, 58
DNA uptake 4, 9
Duodenum *see* Gastrointestinal tract colonization
Duplicated genes 33–35, 38

E

Effector delivery *see* Protein secretion systems
Eggs 265, 291
Enterobactin *see* Iron acquisition
Epidemiological 24, 218, 223, 241–242, 248, 253, 265, 278, 280–281, 287–288, 302, 304, 313
Epistasis 28
Epithelium *see* Gastrointestinal tract colonization
Estimates of r/m 9, 12, 14, 17, 20
Extracellular matrix *see* Bacterial cell envelope

F

Ferredoxin *see* Iron acquisition
Ferrichrome *see* Iron acquisition
Flagellum
- motility 23–24, 26, 60, 90, 103, 160, 179
- role 24, 91, 159, 324
- *flaA* 36–37, 60, 91, 161, 178
- genes 60, 160–161

glycosylation 27, 178
hypermotility 25

G

G. mellonella infection model 62, 134, 136, 191, 197–200
Gastrointestinal tract colonization 43, 75–76, 84–85, 91–92, 99, 153, 189, 347–349, 352
caecum 64, 66–67, 75, 80, 83–85, 92, 94, 112, 136, 143, 152, 196, 202, 213, 216–218, 301–302, 314, 323–324, 327–330, 334–336, 338–340, 351
colon 75, 81, 89, 93–94, 112, 170, 193, 328, 332, 335
duodenum 75, 94
ileum 93–94, 112, 349–350
epithelium 75–76, 83–84, 89–91, 95, 99–101, 107, 153, 160, 163, 192, 196, 314, 348–349
stomach 75, 67, 99, 149, 152–153, 192
Gene conversion 34–38
Gene flow 3–5, 13, 43, 49, 232, 234, 281
Genealogical trees 10–11, 13
Genetic hitchhiking 28, 30, 64–65, 67
Genome synteny 49–50
Glycans 76, 81, 83, 89, 92–95, 167, 170, 179
Glycosylation 37, 45, 76, 89, 92–93, 95, 111, 163, 325
Goblet cells 75–76, 89
Guillain–Barré syndrome 25, 43, 61, 69, 143, 199, 242, 321

H

Haem 111–114, 116, 127–128, 148
Helianthus 5–6
Homologous recombination 4, 9, 20–22, 26–27, 34, 39
Homopolymeric DNA tract *see* Hypervariable regions
Host adaptation 29–30, 189, 192, 201, 234, 281
Host generalism 4, 23, 202
Host immunity
adaptive immunity 200, 347–348, 350
chicken immune system 347
NF-κB 76
Host passage experiments 23–26, 28, 30
Host transmission 43, 66–67, 69, 94, 143–146, 148, 150–152, 183, 214, 216, 218, 241, 243, 246, 254, 261, 266, 278, 280–281, 289–291, 296–297, 299, 302, 304, 313, 351
Housekeeping genes 9–10, 14, 36, 39, 55, 224, 254
Human niche 99
Hybrids, hybridization 3–6, 215–217
Hyper-invasiveness 99, 105–106
Hypermotility *see* Flagellum
Hypervariable regions
contingency loci 27, 45, 55, 57, 322, 325–326, 329, 333
homopolymeric DNA tract 23–25, 27–30, 68, 321–322, 325–327, 333–334, 338–340, 343–344
polyA/T, polyC/G tracts 27–28, 55–58, 60–61, 63–64

I

Ileum *see* Gastrointestinal tract colonization
Incubation period (infection) 265
Infection dose 25, 152, 159, 265, 290, 299
Introgression 3–5
Iron acquisition
enterobactin 111–112, 114–115, 118
ferredoxin 147
ferrichrome 111, 114–115, 120
iron homeostasis 145, 149–150
rhodotorulic acid 114–115, 120
salmochelin 111, 115, 118–120

L

Linkage disequilibrium (LD) 10–11, 13, 20
Lipooligosaccharide *see* Bacterial cell envelope
LOS *see* Bacterial cell envelope

M

Metabolism
asparagine 80
glutamate and aspartate 77–80, 85, 101, 202
l-fucose 6, 45, 77, 81–82, 92–93, 202
serine 77–78
proline 79
Miller–Fisher syndrome 43
MLST 4, 9–10, 14, 16, 215, 217, 241, 245
Motility *see* Flagellum
Mouse model of infection 24–25, 27, 79, 104, 199, 202, 321, 323, 327, 329, 339, 343–344
Mucus, gastric mucin 75–76, 81, 83, 85, 89, 152
Multilocus sequence typing *see* MLST
Multiple carriage 214–216, 218
Mutability 23, 27, 29–30, 57–59, 69
mutS 34, 58–59

N

Natural transformation 27, 43, 45, 47, 322
Noradrenaline 116, 316–317

O

Outbreak 144, 218, 234, 241, 247–248, 253–257, 260–261, 266, 278

P

Paneth cells 75–76
PFGE 24, 26, 234, 241, 243
Phase variation 24, 28, 55, 58, 67–68, 181, 246–248, 279–280, 331–332
Phylogenetic split times 227–228
Pig colonization 24, 77, 82–83, 85, 93, 112–114, 120, 152, 163, 165, 189, 191, 195–196, 221, 254, 260, 279, 292, 313, 338–339
Plasmids 4–5, 43–44
pTet 44
pVir 44
Plasticity 45–46, 61
PolyA/T, polyC/G tracts, *see* Hypervariable regions
Protein secretion systems 47, 77, 103, 159, 164, 167, 180
effector delivery 103, 107, 159–160, 162, 164–166, 168–169, 171, 348

Q

Quorum sensing 135, 180–182

R

Reactive oxygen species 144–146, 125–126, 147
Regulatory RNAs 48–49
Rhodotorulic acid *see* Iron acquisition

S

Salmochelin *see* Iron acquisition
Seasonality 218, 241, 243, 245, 247, 256–257, 261, 278, 294–296
Sheep colonization 43, 144, 152, 195, 221–222, 231, 233–234, 280–281, 313
Sigma factor 48, 102, 161–162
Simple sequence repeats 24, 55–60, 68–69
Single locus variants 14–16, 20–21, 228–229
SLVs *see* Single locus variants
Speciation 3–4
SSRs *see* Simple sequence repeats
Stomach *see* Gastrointestinal tract colonization
Superoxide dismutase 127, 130, 146, 148, 151, 182

T

TLR *see* Toll-like receptors
Toll-like receptors 76, 347
Transcriptome 43, 47, 50, 81, 135

V

Vaccines, vaccination 116–117, 120, 170–171, 200–201, 288, 299, 343, 347, 349, 351–353
Variable number tandem repeats 34, 56, 89
VNTR *see* Variable number tandem repeats

W

Wild birds colonization 5, 144, 213, 221–225, 227, 229, 231, 234, 238, 247, 260, 265–266, 276–281, 292, 339

Z

Zebra fish embryo 200

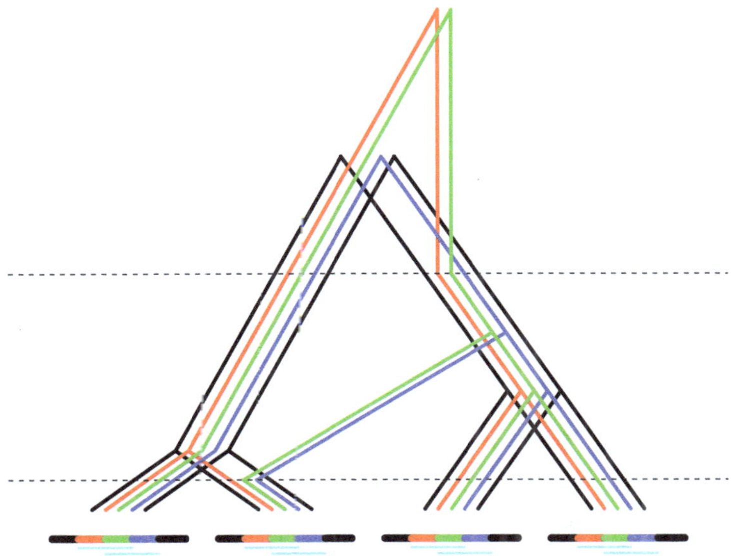

Figure 2.1 Example of the genealogical trees for four isolates in one gene. The effect of recombination is that different sites in the gene have different genealogical trees. Here we have two recombination events that have affected the genealogical relationship of the isolates (the position back in time of these are denoted by the dashed lines, the regions they affect by the light blue lines under the gene fragments). The black tree corresponds the clonal frame of the sample: the genealogy at regions unaffected by recombination. The other trees represent the genealogies of regions affected by the either one or both recombination events.

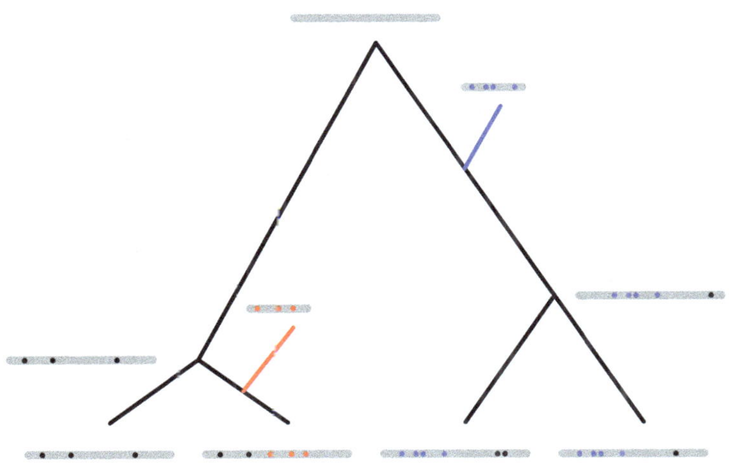

Figure 2.2 Simplified clonal frame model for genealogies of Fig. 2.1. Origin of recombinant regions is ignored, and the two sequences imported at recombination events are just modelled as introducing a number of new nucleotide differences. In this schematic the grey bars represent the ancestral sequence, black dots indicate nucleotide changes introduced by mutation on the clonal frame, and red and blue dots those introduced by the two respective recombination events.

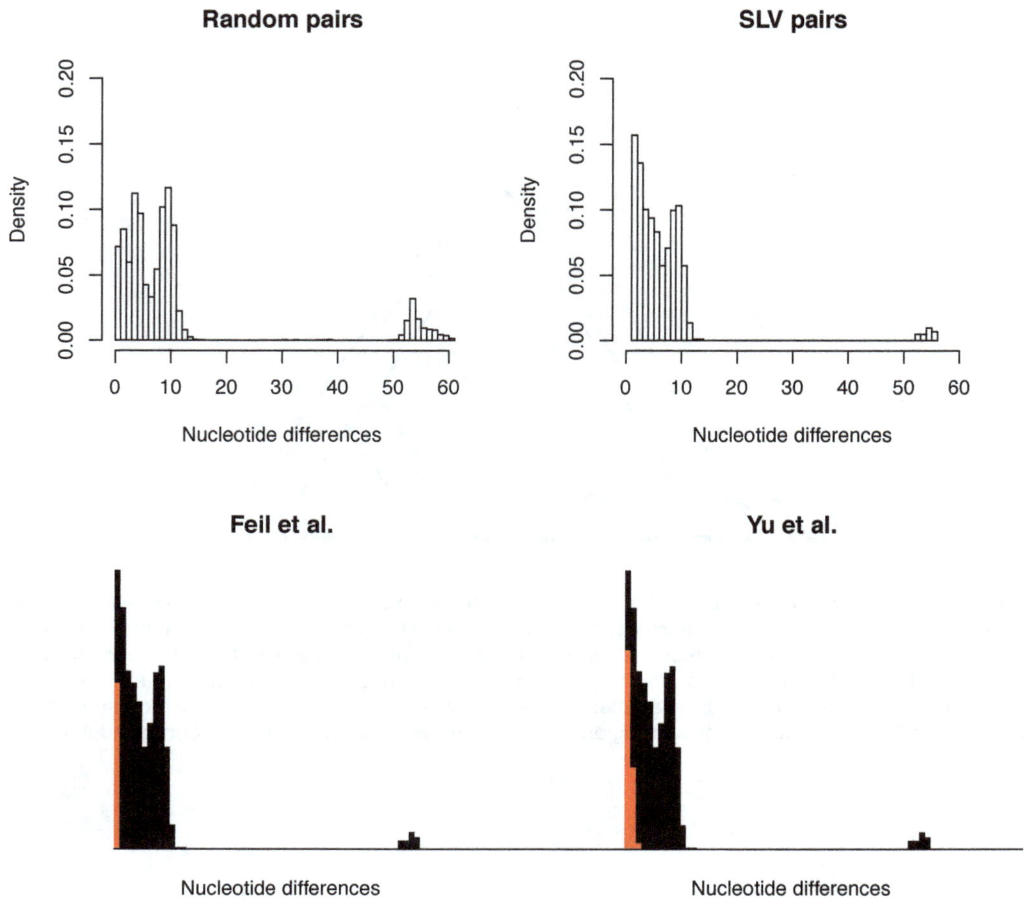

Figure 2.3 Nucleotide differences at one of the MLST alleles for a random sample of pairs of *C. jejuni* isolates (top left), and for SLV pairs (top right). The method of Feil *et al.* (2000), which was used to analyse SLVs from *C. jejuni*, is to assign a proportion of the SLVs with one nucleotide difference to be due to mutation (bottom left plot; red), and all others to be due to recombination (bottom left plot; black). The method of Yu *et al.* (2012) uses a simple model that assumes the data from SLV pairs is a mixture of pairs where the changes are due solely to mutation, and those that are due to recombination. Simplistically this models the data for SLV pairs to be a mixture of a geometric distribution for the number of differences (mutation only case) and the nucleotide differences observed from random pairs (recombination case). It then calculates the expected proportion of SLV that are due to mutation only (bottom right plot; red) or recombination (bottom right plot; black) for each value of the number of nucleotide difference. Note the two methods have different ways of collecting the SLV pairs to analyse (see text for details).

REF1 ACTAGCCAAAGGC

REF2 ATGAGGCCGACCC

ISO2 ACGTGGCCTATGC

ISO1 ACGTGGCCGACTC

Figure 2.4 Example of model in Biggs *et al.* (2011), for a pair of isolates and two reference genomes. The model gives a distribution for the DNA sequence one isolate (ISO1) conditional on the sequence of a closely related isolate (ISO2), and two more distantly related reference genomes (REF1 and REF2). The DNA sequence of ISO1 is a mosaic of the DNA sequences of ISO2, REF1 and REF2 with mutation. In this example the nucleotide at the first eight sites are copied from ISO2, and for the remaining five sites from REF2. Additionally there is a mutation at site 12. The last five sites would be classified as a recombinant region.

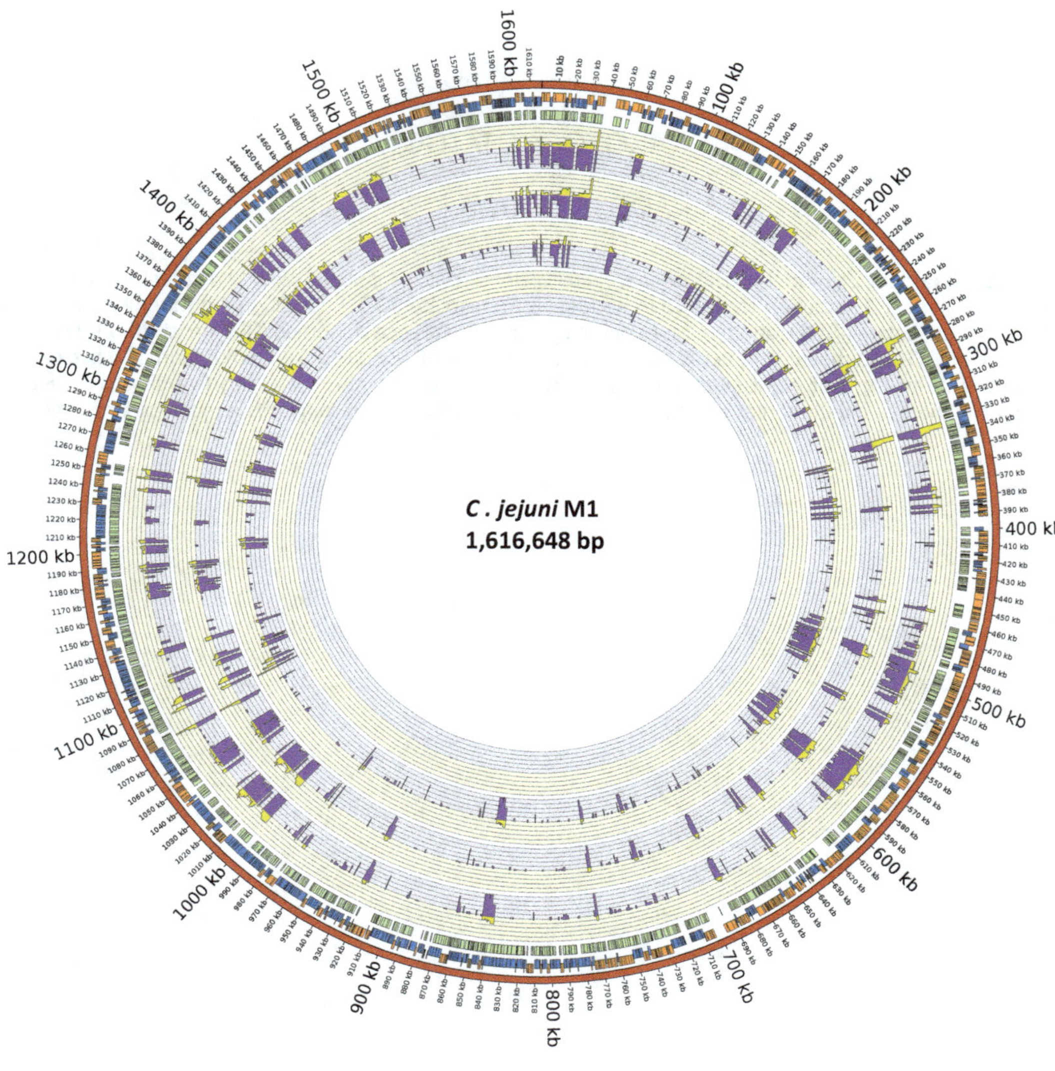

Figure 2.6 Genome pairs mapped against the M1 *C. jejuni* genome. The outer three rings are: M1 ideogram (red); coding genes (orange: forward; blue: reverse); genes under analysis (green). For the remainder of the plot, results for the 4 pairs of *C. jejuni* isolates, from outside to inside: P722b and S150a; P104a and S150a; P722b and P104a; P544b and P104a. The yellow histogram represents the percentage nucleotide difference between the orthologous genes (with a maximum of 10%), and the purple represents the estimated proportion of the gene that is in a recombinant region.

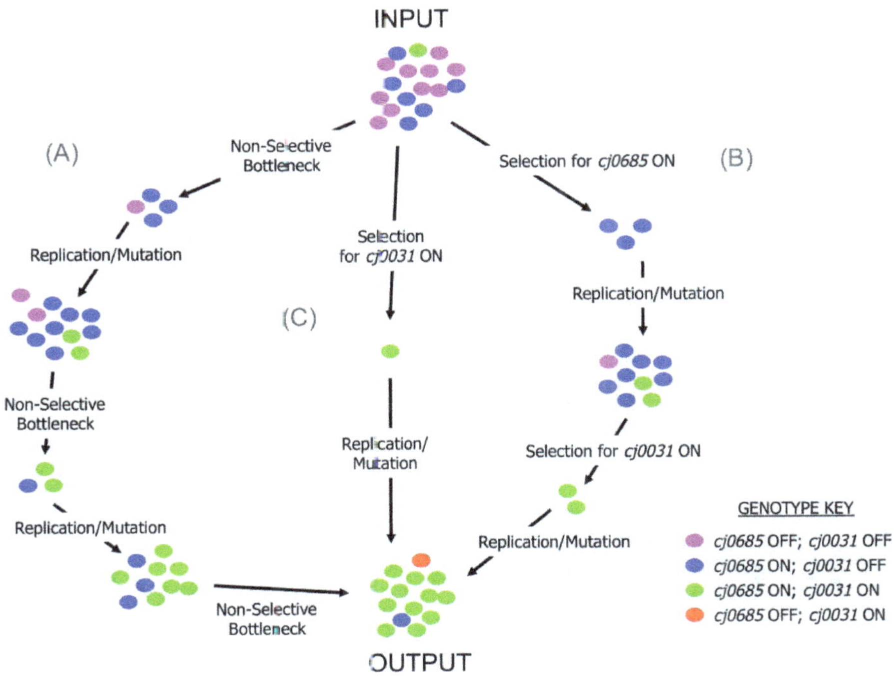

Figure 6.3 Potential pathways for altering the proportions of phase variants within a population. A bacteria containing two phase-variable genes is shown switching from one distribution of expression states to another (the *INPUT* and *OUTPUT* populations). Three pathways to change in the population are shown and a fourth, mutational drift, is outlined in Fig. 6.4. (A) Non-selective bottlenecks. The population is reduced to a small number of cells randomly selected from the initial population. A series of bottlenecks leads to a gradual change in the dominant phase variants. (B) Selection. Sequential changes in the environment select first for the ON state of one gene and then for the ON state of the second gene. This leads to a major change in population structure. (C) Hitch-hiking. Selection is imposed for the ON state of *cj0031*, which exists as a rare variant in the initial population. This variant is in the ON state for the second gene (*cj0685*), which 'hitch-hikes' to prominence in the population due to linkage with the *cj0031* ON state. Note that this population has an identical phenotype to population (B).

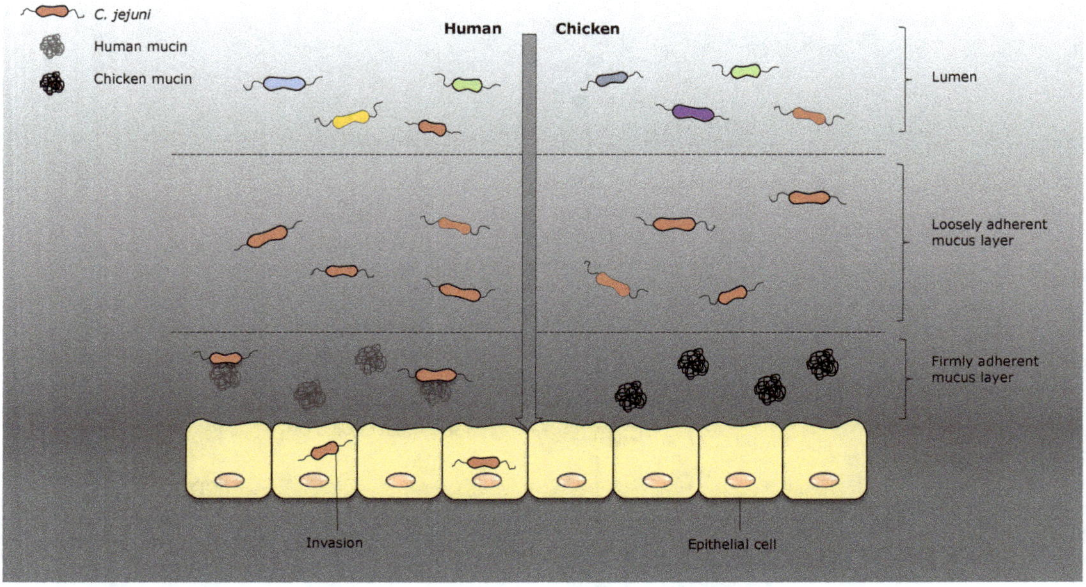

Figure 8.1 A hypothetical model of *C. jejuni* infection in the human versus chicken gut. The spiral shape and bipolar flagella of *Campylobacter* make it ideally suited to life in the mucosal layer. Penetration and colonization of the dense, inner mucus layer in humans results in a pathogenic effect on intestinal epithelium (Left panel), whereas failure to transit the dense mucus layer in the chicken intestine (Right panel) results in a commensal lifestyle. This may arise from a species-specific mucin interaction, as illustrated.

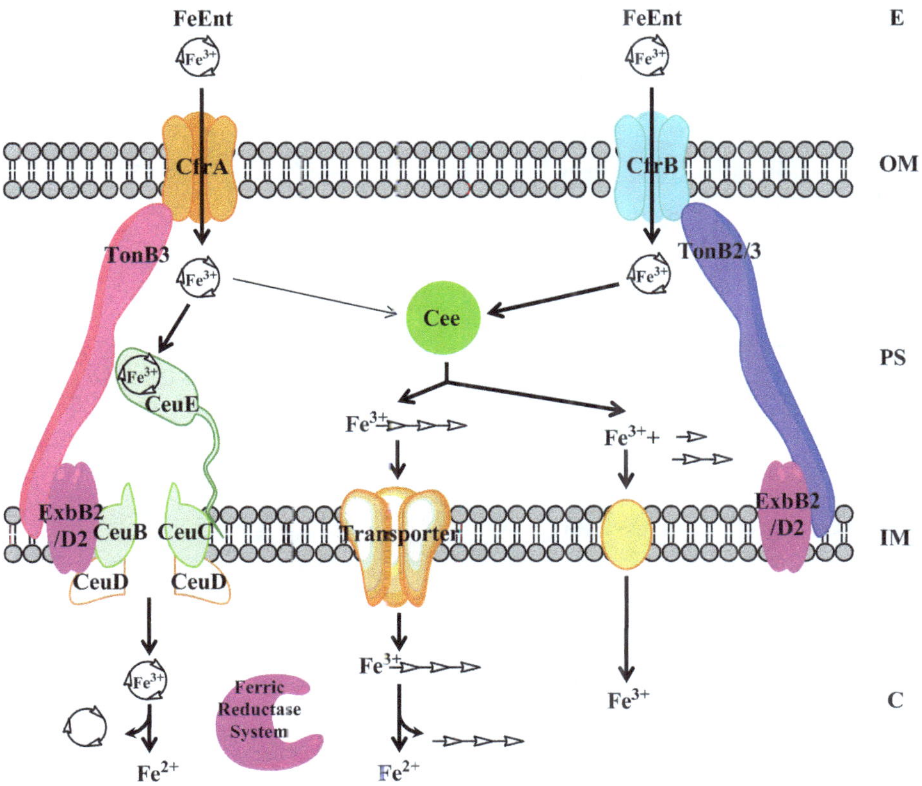

Figure 10.2 The proposed new model of FeEnt acquisition system in *Campylobacter.* ◯, Ent, or cyclic trimer of DHBS; ▻▻▻, linear trimer of DHBS; ▻▻, dimer of DHBS; ▻, DHBS monomer. The thinner arrow line indicates an alternative pathway for utilizing CfrA-transported FeEnt when CeuBCDE system is not functional. E, extracellular environment; OM, outer membrane; PS, periplasmic space; IM, inner membrane; C, cytoplasm (this figure was cited from Zeng *et al.* (2013) with slight modification).

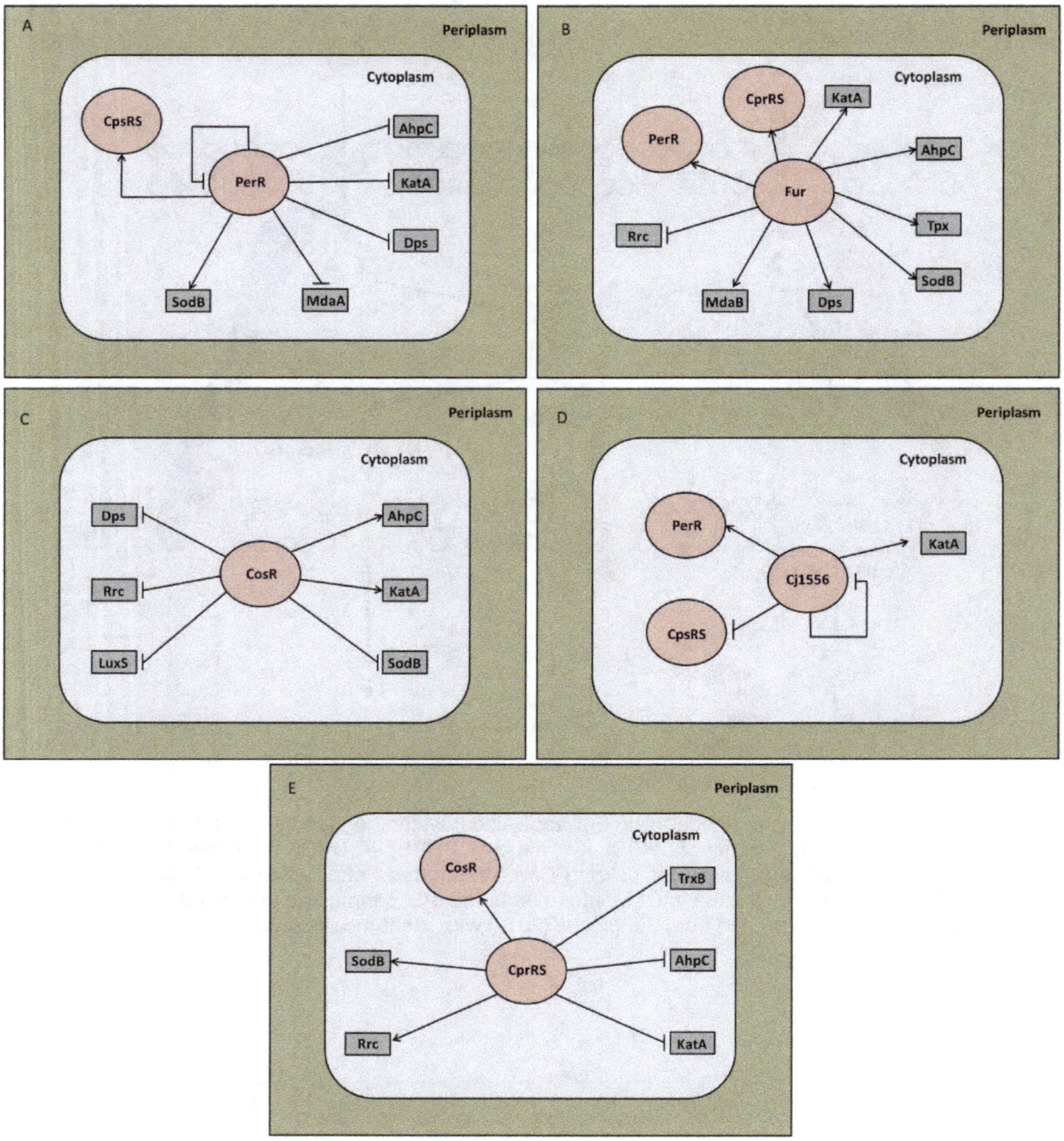

Figure 11.1 *C. jejuni* oxidative stress regulators. (A) PerR; (B) Fur; (C) CosR; (D) Cj1556; (E) CprRS. Proteins involved in countering ROS within *C. jejuni* are displayed in grey boxes. Regulatory proteins are displayed in pink boxes. CC indicates cytochrome C peroxidase. Arrows indicate activation of gene expression whilst a halt indicates suppression of gene expression. This is determined from published microarray expression studies. If mutation of a gene encoding a regulator results in up-regulation of gene expression, it is assumed that the regulator normally suppresses expression of that gene in the wild-type strain. The opposite is true for genes down-regulated in microarray expression studies.

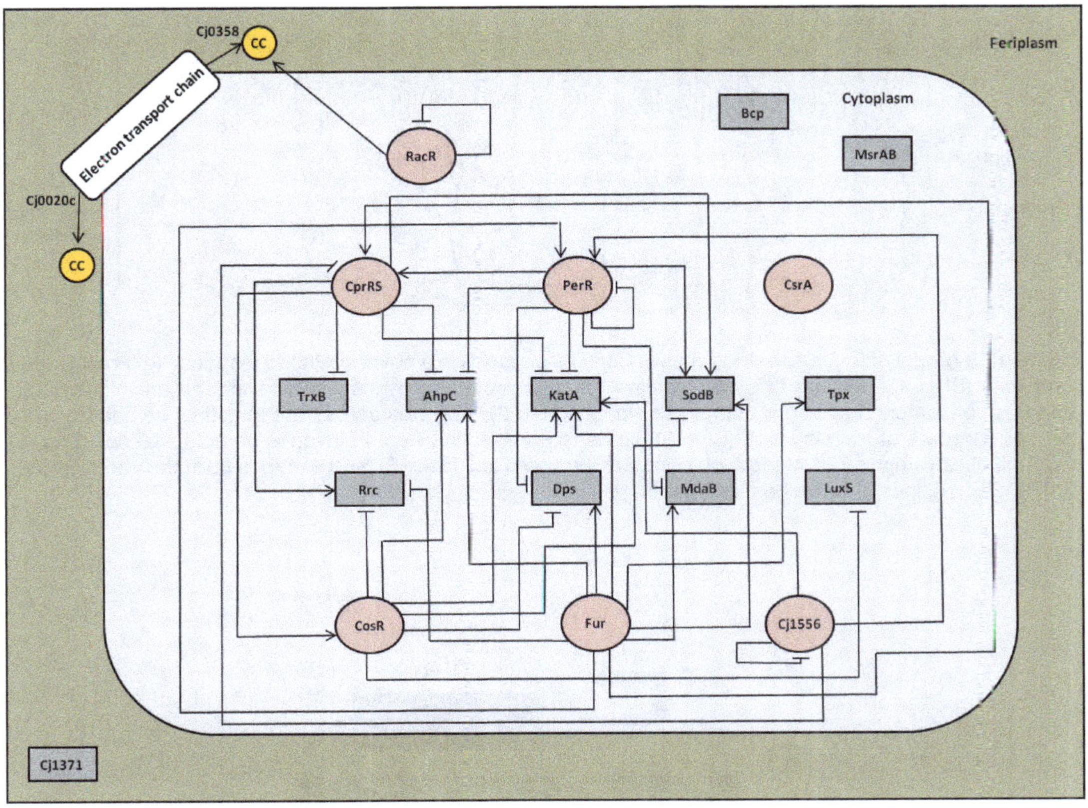

Figure 11.2 The regulation of the *C. jejuni* oxidative stress response. Proteins involved in countering ROS within *C. jejuni* are displayed in grey boxes. Regulatory proteins are displayed in pink boxes. CC indicates cytochrome C peroxidase. Arrows indicate activation of gene expression whilst a halt indicates suppression of gene expression. This is determined from published microarray expression studies. If mutation of a gene encoding a regulator results in up-regulation of gene expression, it is assumed that the regulator normally suppresses expression of that gene in the wild-type strain. The opposite is true for genes down-regulated in microarray expression studies.

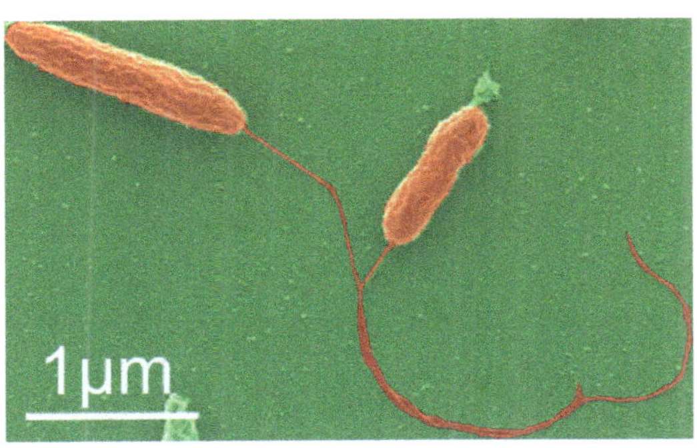

Figure 13.1 Scanning electron micrograph of *C. jejuni* with flagellum. Image courtesy of D.R. Samuelson (unpublished).

Figure 13.2 Model of *C. jejuni* pathogenesis. *C. jejuni* engages in a novel synthetic response (effector protein synthesis) (Step 1), delivers effector proteins (i.e. Cia proteins) to host cells to induce an inflammatory response, ultimately resulting in barrier disruption (Step 2), migrates across the intestinal cell barrier via a paracellular route of translocation (Step 3), binds to fibronectin (Step 4), invades an epithelial cell (Step 5), and multiplies within a vacuole to high numbers intracellularly (Step 6), and escapes from the vacuole (Step 7). The dashed lines indicate host cell death.

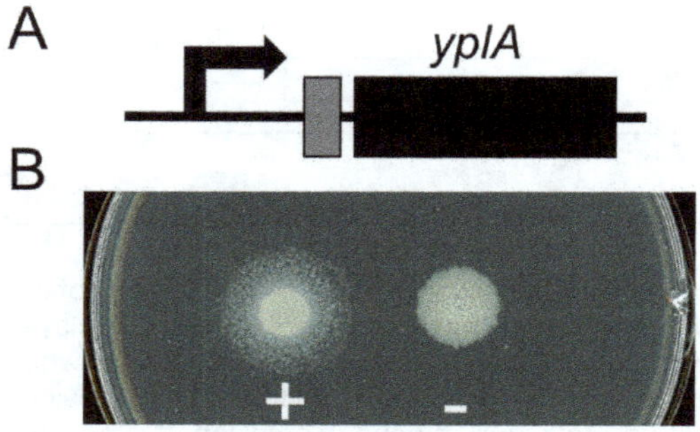

Figure 13.3 Identification of proteins secreted by the flagellum. (A) Schematic of the secretion reporter construct with the constitutive chloramphenicol acetyltransferase (*cat*) promoter (black arrow), a restriction site used for cloning *C. jejuni* fragments (or other pathogens) with potential flagellar secretion signals (grey), and the *yplA* reporter gene. (B) *Y. enterocolitica* harbouring a plasmid with the reporter construct plated on phospholipase indicator plates. A zone of haze indicates secretion of the YplA phospholipase enzyme.

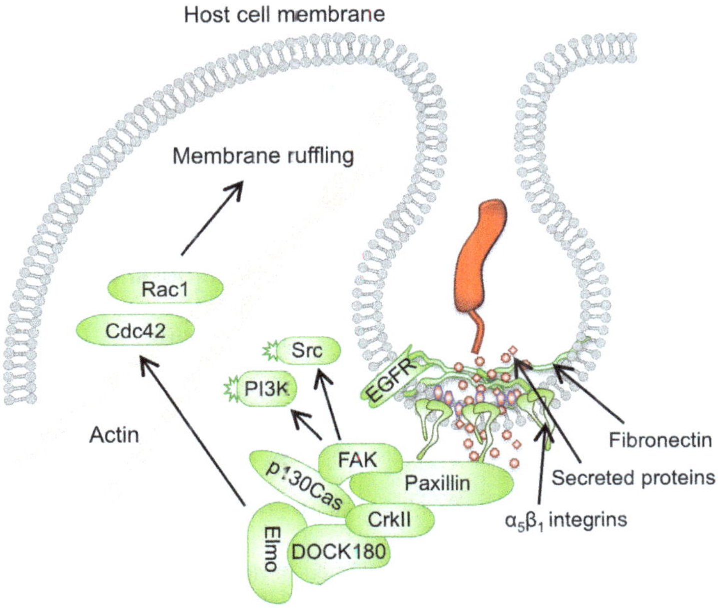

Figure 13.5 Model of host cell signalling. *C. jejuni* secretes proteins that function to modify host cell processes. Initial binding of the bacterium to fibronectin induces $\alpha_5\beta_1$ integrin occupancy and clustering, ultimately resulting in activation. The integrins activate the EGF receptor and FAK. Activated FAK associates with c-Src and PI3K, which contributes to the activation of the focal complex (i.e. FAK, paxillin, p130Cas, CrkII, DOCK180, ELMO). The activated focal complex results in the activation of Rho G proteins including Rac1 and Cdc42. The ultimate result is increased actin cytoskeletal rearrangement leading to membrane ruffling, which is necessary for maximal invasion of host cells by *C. jejuni*.

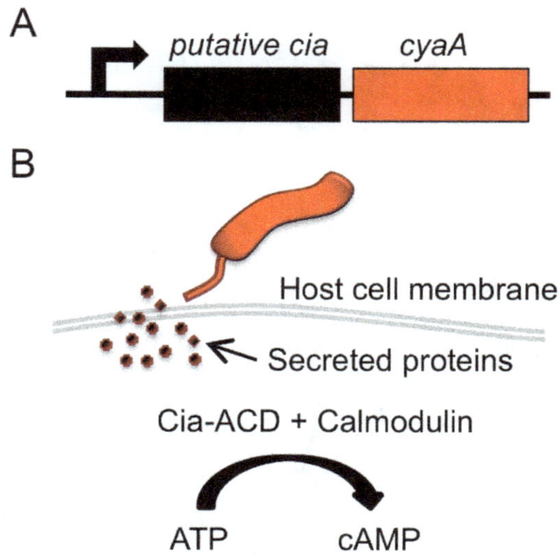

Figure 13.7 The adenylate cyclase delivery assay is used to determine the proteins that are delivered to epithelial cells. (A) Schematic of delivery constructs, including the native promoter (black arrow) with gene of interest (i.e. *cia*) in black and *cyaA* (adenylate cyclase domain, ACD) in red. (B) *C. jejuni* harbouring plasmid encoding reporter construct secretes and delivers the protein to cells (red and black dots). The ACD of the fusion protein is activated in the presence of calmodulin in the cytosol of host cells and catalyses cAMP from ATP, which is measured by ELISA.

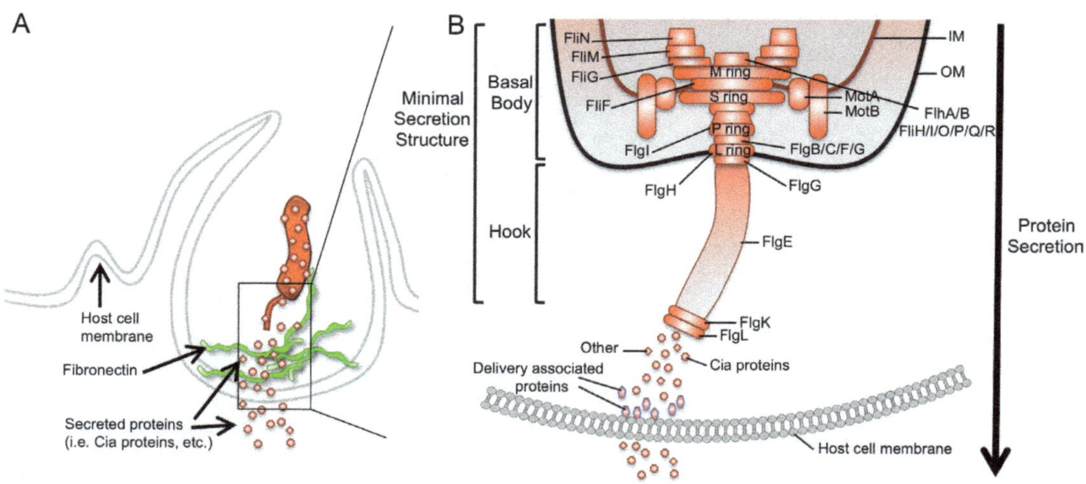

Figure 13.8 Model of *Campylobacter* protein secretion. (A) The proteins secreted by the flagellum are surface-exposed and are delivered to the host cytosol in a mechanism that is not fully defined. (B) Boxed inlet demonstrating the minimal secretion structure of *C. jejuni*. The secreted proteins include the Cia proteins and putative delivery associated proteins. The flagellar filament (not shown) may still be involved during the secretion of effector proteins, although it is not essential. IM, inner membrane; OM, outer membrane.

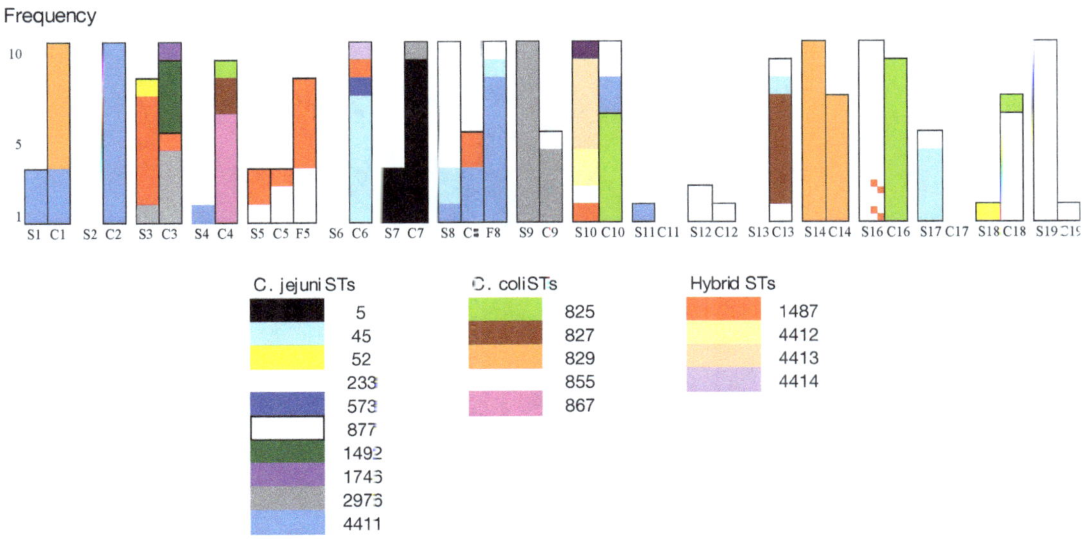

Figure 16.2 The *Campylobacter* MLST types isolated from swab of the cloacal opening (S), caecal contents (C) and faecal samples (F), grouped by individual bird are shown. (No *Campylobacter* was recovered from Bird 15). The birds were from the same commercial free-range broiler breeder flock aged 57 weeks, detailed in Table 16.1. Following direct culture onto mCCDA and depending on growth, up to ten *Campylobacter* colonies were sub-cultured and typed. Different STs are represented by different colours.

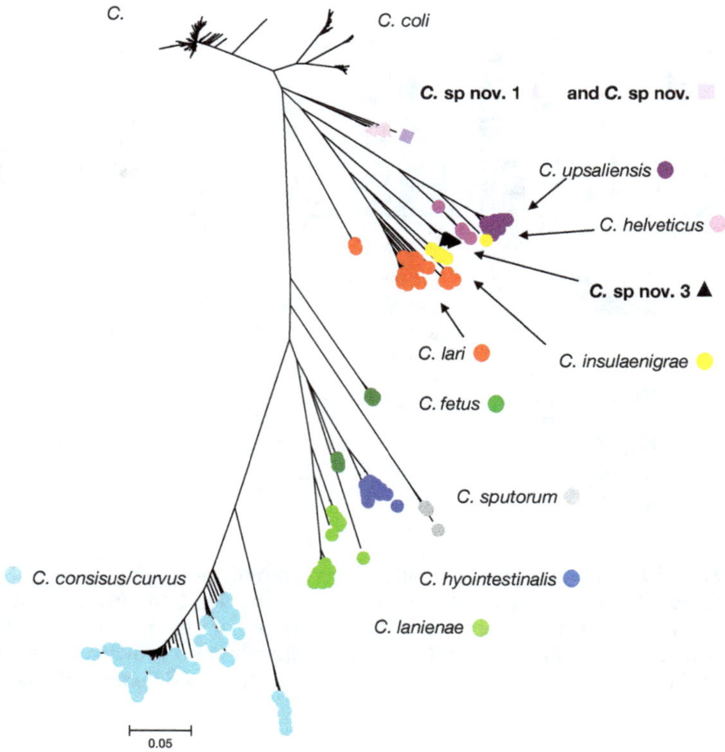

Figure 17.2 Neighbour-Joining phylogeny of the *glnA* locus including isolates from water and wild birds in New Zealand and the proposed new species: *C.* sp nov. 1, 2 and 3.

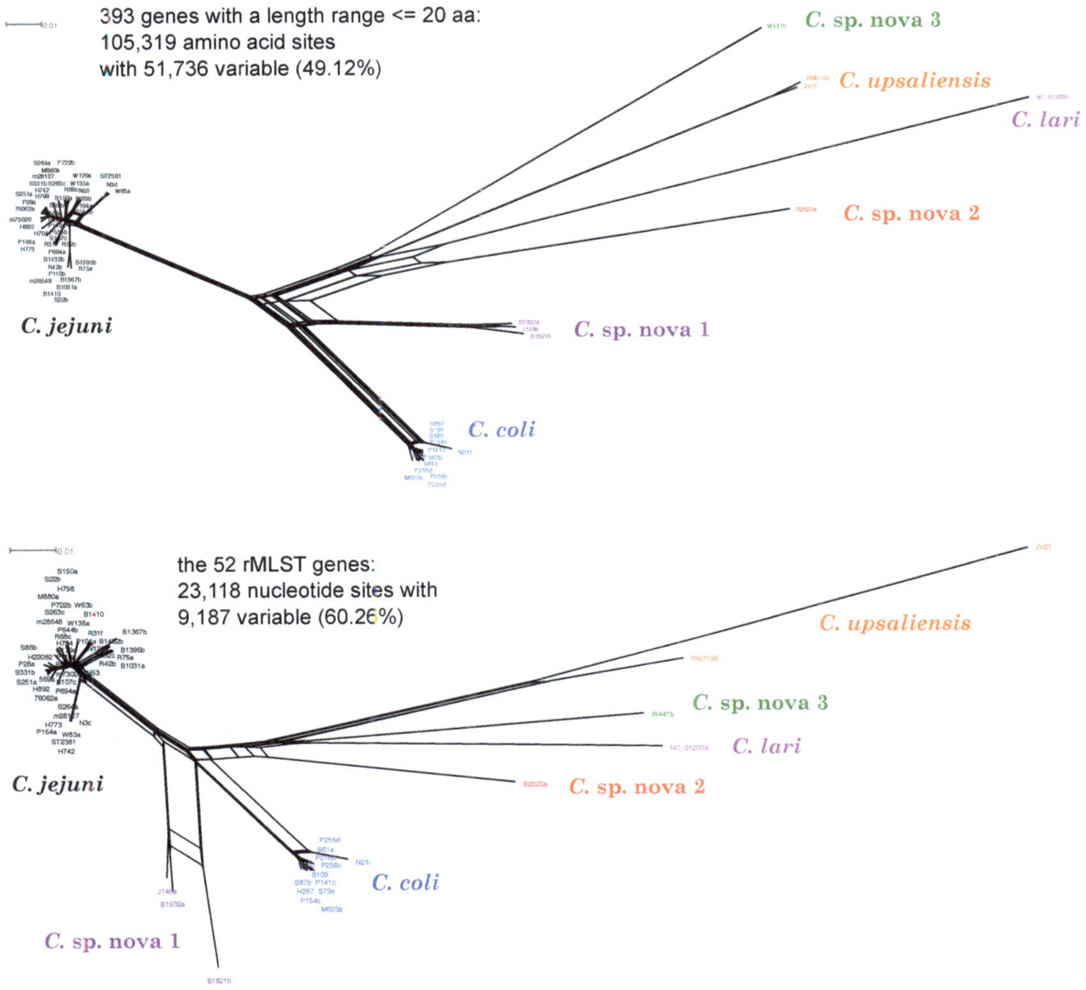

Figure 17.3 NeighbourNet plots showing the relatedness of isolates from New Zealand and *C. lari* and *C. upsaliensis* reference strains, based on full genome analysis. The top plot shows the phylogeny based on 393 concatenated genes, the bottom plot is based on a concatenation of the 52 ribosomal genes.

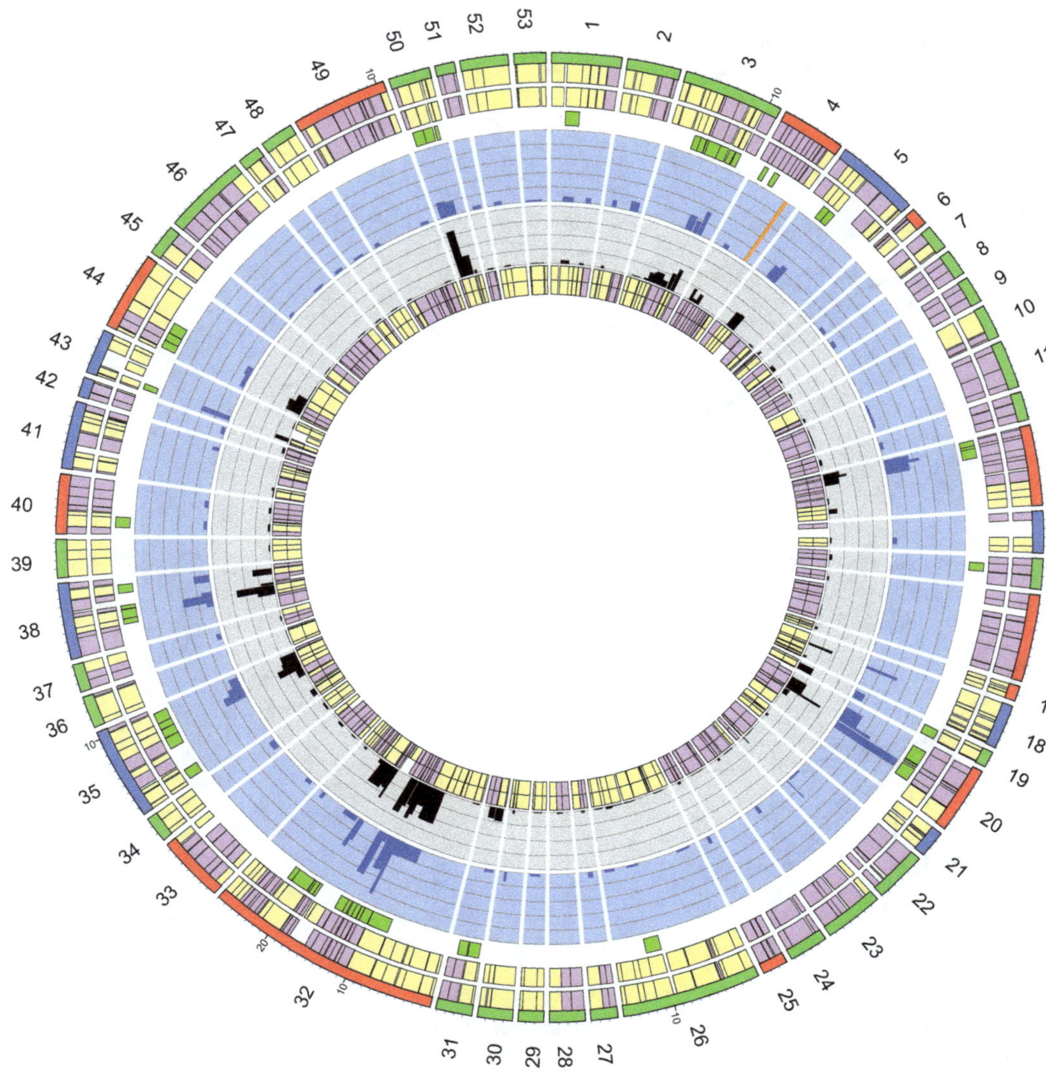

Figure 17.9 Circos plot of orthologous genomic regions for two *C. jejuni* ST-474 genomes isolated in New Zealand where sequence differences were found (taken from Biggs *et al.*, 2011). The figure shows 37 detectable regions. Orthologous gene pairs in P110b and H22082 were calculated using OrthoMCL with a 'pi_cutoff' value of 0.95. The tracks from outside to inside are; genome region name and size (green, regions of same length in P110b and H22082; blue, the region from P110b is longer; red, the region from H22082 is longer); genes and sizes for P110b; genes and sizes for H22082; genes coloured cyan showing evidence of recombination; histograms for P110b (orange) and H22082 (red) showing the nucleotide coverage from the short reads plotted as an average over the length of the gene, along with the standard deviations for the coverage as round circles; histograms showing the number of sequence differences between the genes at the protein level (dark blue on a light blue background) and DNA level (black on a grey background) as a fraction of protein or gene length respectively; a repeat of the gene sizes and locations in P110b and H22082. Gene orientations are shown in purple and yellow for the forward and reverse strand respectively. The nucleotide coverage histograms have the same scale but a different magnitude, 180 and 120 for P110b and H22082 respectively. Similarly the sequence difference histograms have the same scale but a different magnitude, 0.05 and 0.04 for the protein and the DNA histograms respectively.

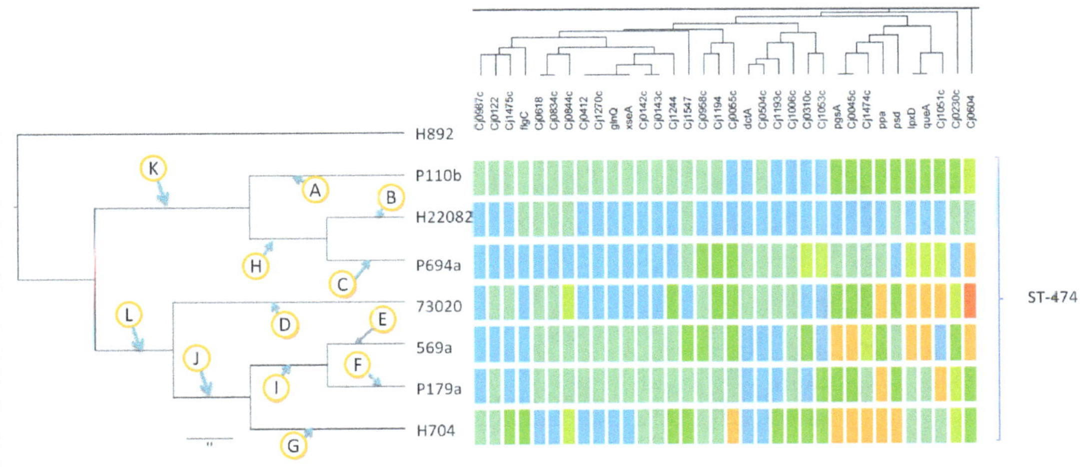

Figure 17.10 Maximum parsimony (MP) cladogram based on seven full-genome sequences of *C. jejuni* ST-474 isolated from humans, poultry and ruminants in New Zealand, and an isolate of ST-48 (horizontal dendrogram), and clustering of loci with similar character states across isolates (vertical dendrogram). The MP tree was drawn based on 1667 loci, of which 274 showed allelic variation between the isolates. 83 of these were parsimony informative. The 34 loci shown in the matrix did not fit perfectly on the tree, requiring more than one change, and are coloured according to their allelic designation (i.e. different colours represent different alleles). The loci are clustered to identify similar patterns of allelic variation across the ST-474 genomes. The circles labelling each branch (A–L) are referred to in Table 17.4.

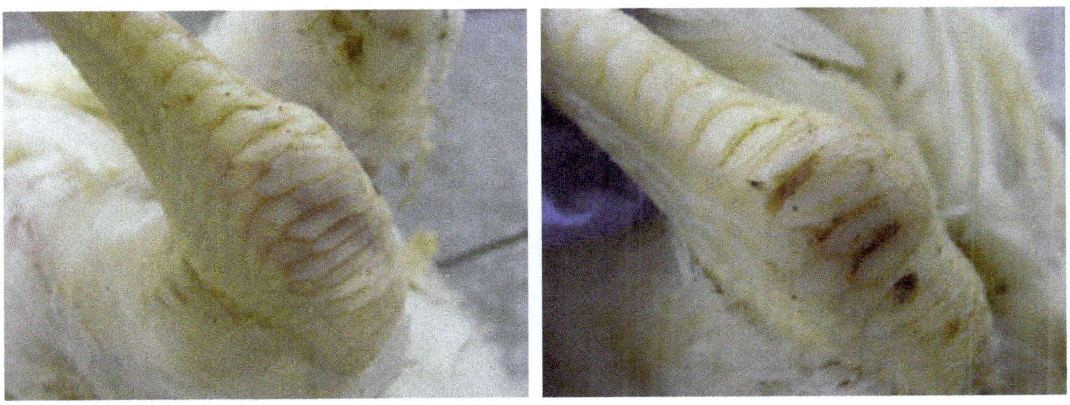

Figure 22.1 Hock marks on broiler chickens. Normal (left) and inflamed (right) hocks. Superficial marks shown here can progress to erosion of the skin as a result of exposure to wet litter.

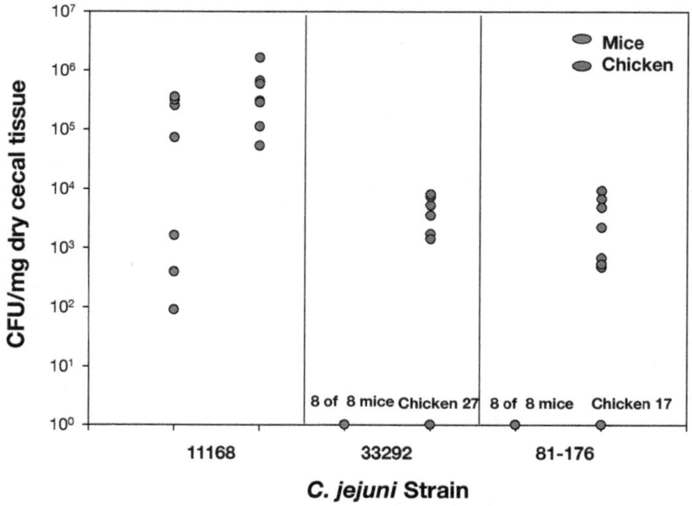

Figure 23.1 *C. jejuni* colonization of C57BL/6J IL10 deficient mice and Ross 308 broiler caecal tissue at 14 day post-inoculation. *C. jejuni* strains 11168, 33292 and 81-176 are represented on the x-axis. Eight mice and eight chickens were sacrificed on day 14 (d14) and caecal tissue homogenized, serially diluted, and plated onto a selective growth medium. Mice inoculated with strains 33292 and 81-176 were culture negative at d14 post-inoculation, with the limit of detection in these experiments ranging from 150 to 300CFUgram^{-1} of dry caecal tissue. Chicken 17 produced a relatively low *C. jejuni* concentration of 5×10^{-1}CFUmg^{-1} dry caecal tissue while chicken 27 did not produce a detectable level of *C. jejuni*. Each animal was housed separately in an individual cage.

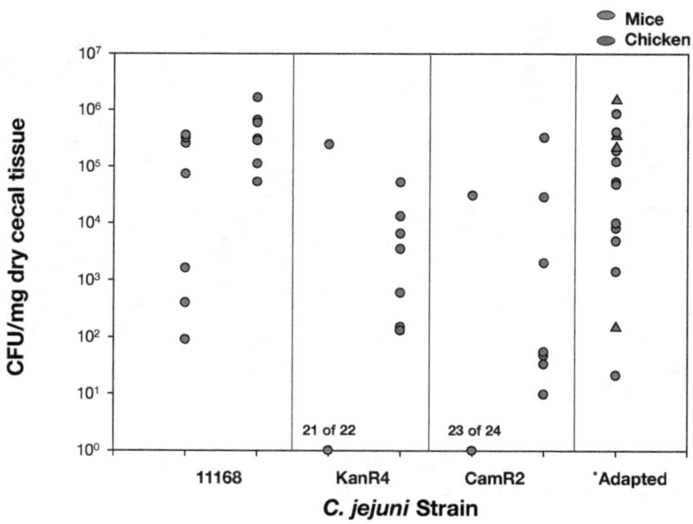

Figure 23.2 Colonization of C57BL/6J IL10 deficient mice and Ross 308 broiler caecal tissue at d14 or d16 post-inoculation by genetically marked *C. jejuni* 11168. Colonization data for wild-type *C. jejuni* 11168 in mouse and chicken caecal tissue are presented for comparison with antibiotic resistant (genetically marked) 11168 variants. 43 of 46 mice inoculated with a *C. jejuni* population prepared under standard laboratory culture conditions (inoculum population) were culture negative for *C. jejuni* at d14 post-inoculation. Mouse-adapted cultures of 11168-Tn5CamR6 (chloramphenicol resistant variant indicated by a red circle in the Adapted column) and 11168–*23S*KanR4 (kanamycin resistant variant indicated by a red triangle in the Adapted column) were recovered from caecal tissue on d 14 post-inoculation from mouse 712 and 682 respectively. The limit of detection in these experiments ranged from 150 to 900CFUgram^{-1} of dry caecal tissue. Each animal was housed separately in an individual cage.

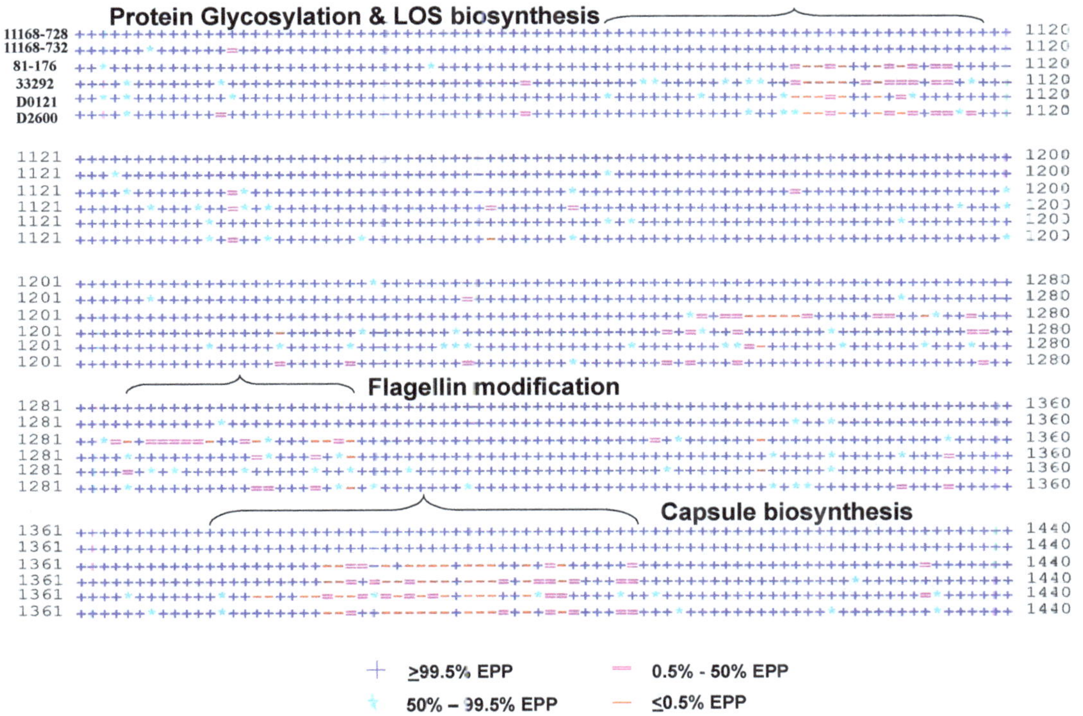

Figure 23.3 Competitive microarray DNA–DNA hybridization of *C. jejuni* chromosomal DNAs. Glass slides carrying several copies of a complete *C. jejuni* 11168 cDNA microarray were hybridized to a chromosomal DNA probe prepared from *C. jejuni* 11168 (reference probe) plus fluorescently labelled chromosomal DNA from either *C. jejuni* strain 33292 or strain 81-176. *C. jejuni* strains D0121 and D2600 are distantly related human isolates and are included here to provide more robust analysis between chromosomes. A GACK value of –0.5 represents a 100% chance of sequence divergence and an Estimated Probability of Presence (EPP) ≤0.5%. A GACK score of 0.5 represents a 100% chance of identical sequence and an EPP ≥99.5%.

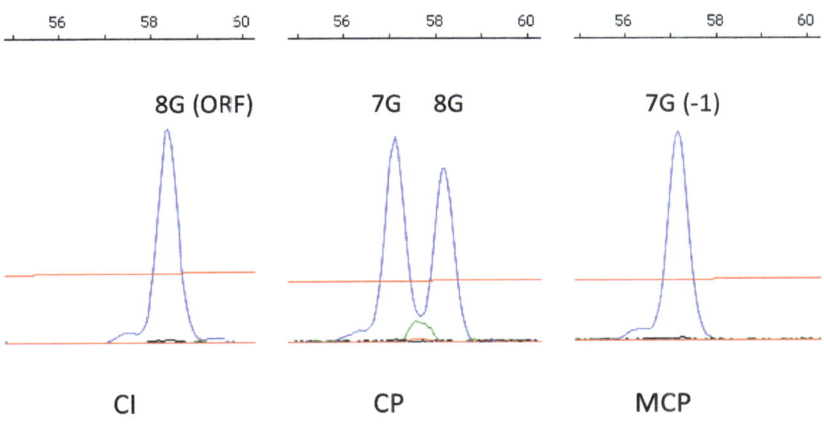

Figure 23.4 Fragment analysis of the homopolymeric tract (HT) at contingency locus *Cj1139 (wlaN)*. DNA samples from an inoculum population prepared under standard laboratory culture conditions (CI), from chicken-passaged populations (CP), and from CP populations passaged through mice (MCP) of strain CamR2 were analysed by fragment analysis. PCR products labelled with 6-FAM fluorescent dye were separated by electrophoresis on capillary columns to resolve alleles. The number of nucleotides in the HT was revealed by parallel sequence analysis of each sample.

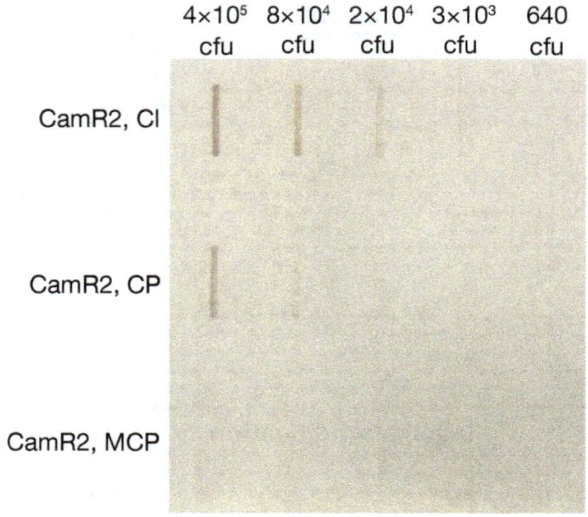

Figure 23.6 Cholera toxin binding assay. A cholera toxin binding assay was used to detect a terminal galactose residue (GM-1-like structure) on LOS of CamR2 CI, CP, and MCP populations. An allele of contingency locus *Cj1139* that carries a homopolymeric tract with 8G encodes an enzyme that produces a GM-1-like LOS structure. The 8G allele of Cj1139 occurred at 94% frequency in the CamR2 CI population. The 8G and 7G allele frequencies in the CP population were 45% and 55%, respectively. The 7G allele frequency was 100% in the MCP population. The designated concentrations of *C. jejuni* cells were boiled for 5 min and cell lysates were loaded onto a PVDF membrane and dried. Then, the terminal galactose residue was detected by peroxidase conjugated to cholera toxin B subunit and a DAB staining kit.

www.ingramcontent.com/pod-product-compliance
Lightning Source LLC
LaVergne TN
LVHW062216200926
848484LV00006B/1408